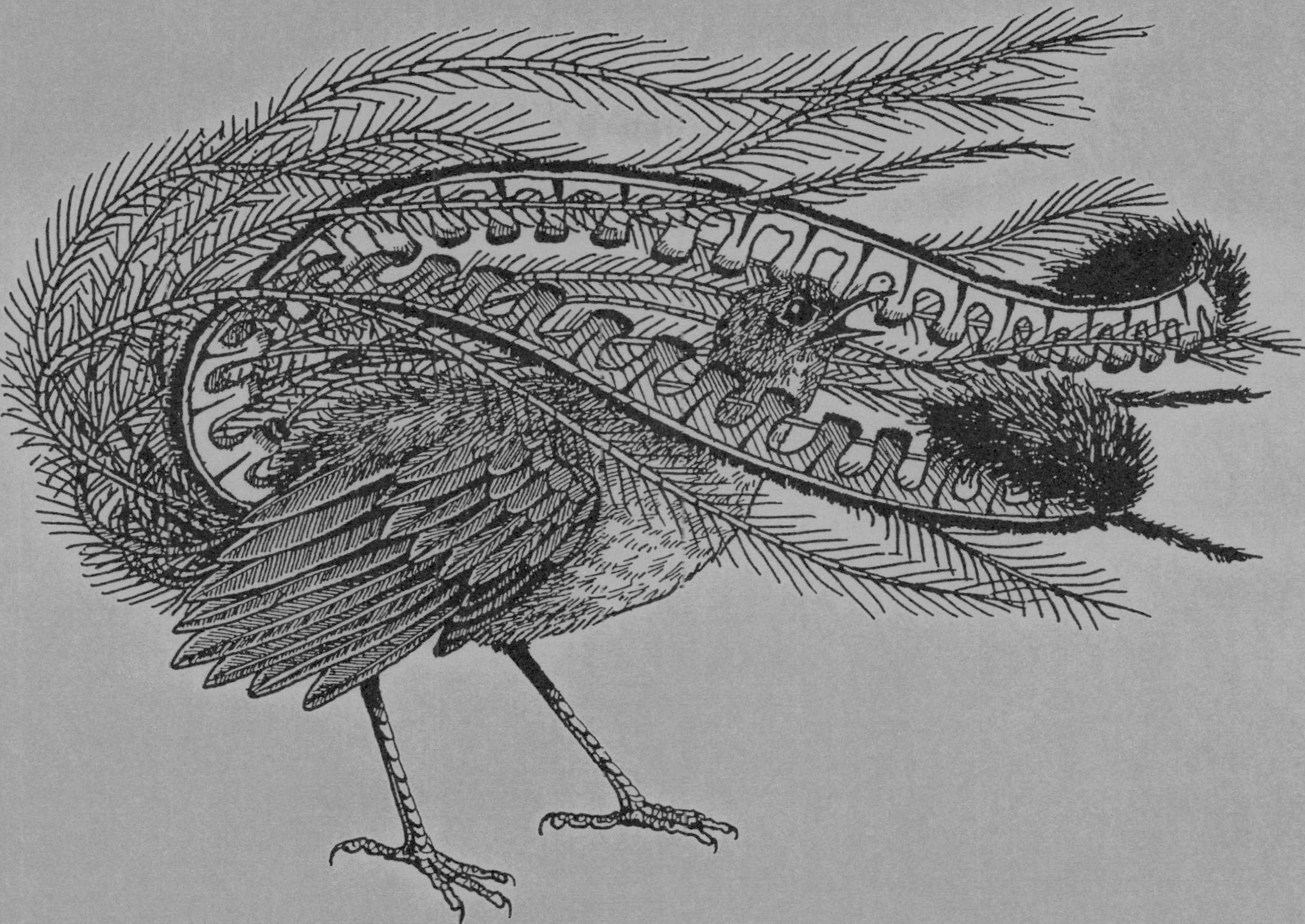

BIRDS OF EUCALYPT FORESTS AND WOODLANDS:

ECOLOGY, CONSERVATION, MANAGEMENT

BIRDS OF EUCALYPT FORESTS AND WOODLANDS:

ECOLOGY, CONSERVATION, MANAGEMENT

Edited by

A. Keast, H. F. Recher, H. Ford and D. Saunders

Illustrated by

D. McFarland, D. Milledge, J. Trompp

Published by

Surrey Beatty & Sons Pty Limited

In association with

The Royal Australasian Ornithologists Union

DEDICATION

We dedicate this book to the memory of Harry Bell whose studies of forest and woodland birds in Australia and New Guinea have done much to extend our understanding of the ecology of the Australasian avifauna and who was an inspiration to all of us.

ISBN—0 949324 06 X
National Library Canberra

Published December, 1985

PRINTED IN AUSTRALIA BY
SURREY BEATTY & SONS PTY LIMITED
43 Rickard Road, Chipping Norton, NSW 2170

Rainbow Lorikeet, *Trichoglossus haematodus*

FOREWORD

The Royal Australasian Ornithologists Union (RAOU) was established in 1901 and is the senior society in Australia for those interested in birds, both professional and amateur ornithologists. It has published throughout this century "The Emu", which is recognised worldwide as a leading scientific journal in this field. Other publications have been produced from time to time, for instance there are regular newsletters for members of the RAOU, conservation statements of ornithological importance, and reproductions of theses not otherwise available to the interested public. The most notable publication was "The Atlas of Australian Birds" in 1984. This was the result of a study for five years by thousands of individuals co-ordinated from RAOU Headquarters, which defined, for the first time, the distribution of all bird species found on this continent. It has been widely acclaimed both in Australia and overseas.

Another RAOU activity has been the holding annually of a congress, followed by a campout in the field. These occur in a different State each year, and in November 1982 the congress was in Armidale, New South Wales, the subject being "Birds of Eucalypt Forests and Woodlands". The participants, both speakers and audience, were stimulated by the facts presented and the views expresed and, as so often happens on such occasions, many said "these proceedings ought to be published". Professor Allen Keast was especially enthusiastic and persuaded me, as President of the RAOU at the time, to do something to implement this proposal. The RAOU has limited resources of its own and relies greatly on the help of volunteers, given willingly and continuously, who believe that "action speaks louder than words", so I asked Allen Keast if he himself would take over the promotion of this project. To his great credit and my gratitude he agreed to do so, despite the fact that he lives in Canada, provided that Hugh Ford in Armidale and Harry Recher in Sydney would join him on the editorial committee. This trio, with the addition of Denis Saunders from Western Australia, have worked tirelessly since then to produce the present book, and the RAOU is proud to be associated with them in presenting a very important series of papers on the subject of birds in forests and woodlands.

The book aims to provide the facts obtained by studies of the birds themselves, their relations with their surroundings, their adaptation to changing conditions, and the effects, both intentional and unintentional produced by people. Not all the papers were read at the Congress and the editors invited other participants to fill gaps in the presentation. The time lapse since the Congress has also allowed some more recent research to be included. All in all, it is an exciting, stimulating, and sobering experience to read this book. Many answers to the problems raised are still unknown and, if we are to continue enjoying our Australian forests and woodlands, much effort is still needed. This means trained personnel and intelligent amateurs willing to study a particular area, and also the support of governments, universities, companies, foundations and private individuals to provide the funds.

On behalf of the RAOU I wish to thank all those who have contributed to the success of this book, especially the editors and authors, and trust that it will be a stimulus to further research and a better understanding of the need to conserve and manage Australia's forests and woodlands and their associated plants and animals.

Dr. Norman Wettenhall,
June 1985

Clinking Currawong, *Strepera versicolor arquta,* in Highland White Gum

CONTENTS

Front Cover: Varied Lorikeet *Psitteuteles versicolor*. Photographed by Hans and Judy Beste at Mt. Isa, Queensland.

Back Cover: Forest is a tall open-forest dominated by Narrow-leaved Peppermint *Eucalyptus radiata* on Coast Range Road, Victoria. Photographed by Harry Recher. Inset: Golden Whistler *Pachycephala pectoralis* at nest. Photographed by Norman Chaffer.

Photographs courtesy of the National Photographic Index of Birds and the Australian Museum, Sydney.

Beautiful Firetail, *Emblema bella*

Yellow Wattlebird, *Anthochaera paradoxa*

LIST OF ILLUSTRATIONS

HABITAT (After Appendix) Continued

BIRDS (after Appendix)

Forty-spotted Pardalote, *Pardalotus quadragintus*

Spotted Pardalote, *Pardalotus punctatus*

CHAPTER 1

Eucalypt Forests, Woodlands and Birds: An Introduction

Harry F. Recher[1]

AUSTRALIA is a dry continent. Over most of Australia, drought of varying duration is the normal pattern. For this reason, at the time of European colonization in 1788 forests covered less than 15 percent of the continent and were restricted to the high rainfall areas along the northern and eastern coasts and in the southwest (Leeper 1970; Specht *et al.* 1974). Away from the coasts the forests graded into much more extensive woodlands. These areas are the most important part of the Australian continent. Here, where the rainfall exceeds 500 mm per annum, all agriculture and 99% of human settlement is located. This book is about the forest and woodland zones of Australia, as a habitat, and protector of the bulk of Australia's natural history legacy. It focuses on the birds, the most obvious and spectacular animal group. It brings together the results of ten years concentrated work on the interrelationships of eucalypt forest and woodland birds and, applies this new knowledge on their ecology to the conservation and management of Australia's avifauna.

Apart from the tropical north, the settlement of Australia by Europeans proceeded rapidly and by the 1880's farming and the harvest of timber affected most of the continent's forests and woodlands. By the 1980's nearly half of Australia's forests had been cleared. Large areas of woodlands were also cleared or ring-barked to provide land for farming or were grazed by a host of introduced herbivores ranging in size from rabbits and hares, to cattle and feral camels. Against this background, Australians discovered their environment and their forests.

The wave of environmental consciousness that spread over Australia during the 1970's focused on a wide variety of issues, but none was grasped more fervently than the conservation of forests and their preservation as national parks. For environmentalists the establishment of woodchip industries in Tasmania, Western Australia and New South Wales, the clearing of forests to establish plantations of exotic pines and to provide more land for farming, the death of trees on lands used for grazing, and the logging of rainforests were tangible and highly visible threats to what forests remained. The controversies that ensued were often heated and pitted environmentalists on one side against foresters and the timber industry on the other. A consequence of this new concern for Australia's forests was the proliferation of research related to forests, forestry and forest management. Much of the research focused on birds; their ecology, conservation status and how they were affected by forest management.

Prior to 1970 little had been published about the biology or ecology of forest and woodland birds in Australia. A few workers had conducted censuses of avian communities (Recher 1984) and there were accounts of the biology of a small number of species. However, most of the information available was anecdotal and ill-suited for use in resolving problems related to the management of forests and forest wildlife. The situation now is much better. The research initiated during the 1970's provides a base on which to assess the impact of forest utilization and to develop plans of management.

In this book we have brought together the current state of knowledge about the ecology of Australia's eucalypt forest and woodland birds. Rainforest avifaunas have been excluded, not because they are unimportant or unknown, but because they are an entity unto themselves and would require another book for proper consideration. Much of the material presented here is new and relates to forestry practices. Indeed, the last section of the book is principally concerned with the effects of forest utilization and managment on birds. The first two sections of the book provide the biological and ecological information needed for the formulation of plans of management for the conservation of forest birds. The first section deals with studies of avian communities. The second treats the biology of individual species or groups of closely related species. Not all the research on forest and woodland birds could be covered in the individual chapters.

[1]The Australian Museum, P.O. Box A285, Sydney South, New South Wales 2000.
Pages 1-10 *in* BIRDS OF EUCALYPT FORESTS AND WOODLANDS: ECOLOGY, CONSERVATION, MANAGEMENT ed. by A. Keast, H. F. Recher, H. Ford and D. Saunders. Royal Australasian Ornithologists Union and Surrey Beatty & Sons.

The sytheses at the end of each section attempt to broaden the coverage by integrating the research published here with previously published studies and with other work in progress.

The remainder of this introductory chapter provides background information needed for individual chapters and the syntheses. Terms are defined and information is presented about Australia's eucalypt forests and woodlands; their vegetation and structure, the birds that live in them, the extent of forests and woodlands, and the trees that dominate them. The introduction concludes with some brief comments about the direction of future research and the responsibilities of scientist and bird-watcher alike in ensuring that research on the ecology of forest and woodland birds is continued.

EUCALYPT: THE DOMINANT TREE

Apart from the closed-forests (rainforests), Australian forests and woodlands are dominated by a single group of trees — the eucalypts. Approximately 500 species of eucalypts have been described with only three species occurring naturally outside of Australia and the dry lowlands of New Guinea (Pryor and Johnson 1981). The systematics of eucalypts *Eucalyptus* and the closely related angophoras *Angophora* is a subject of considerable debate among botanists. Nine subgenera (including *Angophora*) have been proposed (Pryor and Johnson 1971, 1981) and it is possible that these will be elevated to generic rank in the near future. The subgenera of eucalypts differ in a variety of morphological and physiological characters. Many of these differences are relevant to an understanding of the ecology of forest and woodland birds. For example, Lambert and Turner (1983) found that nutrient levels in the foliage of eucalypts within the subgenus *Symphyomyrtus* were significantly higher than found in eucalypts within the subgenus *Monocalyptus*. There was little variation within a species even when trees growing on nutrient-rich soils were compared with those of the same species on poorer sites. However, species of *Symphyomyrtus* were associated primarily with nutrient-rich sites while *Monocalyptus* dominated the poorer soils. At Eden in southeastern New South Wales where this work was done, forests dominated by *Symphyomyrtus* species support the greatest numbers and kinds of birds (Recher *et al.* 1980). Whether there is a relation between foliage nutrient levels, insect abundance and the richness of avian communities remains to be seen, but it presents a plausible hypothesis.

Eucalypts probably evolved as the Australian continent became increasingly arid during the early tertiary (Beadle 1981). Except as emergents eucalypts are absent from closed-forests and are one of several groups of Australian plants that have adapted to infertile soils, aridity and fire (Beadle 1981; Pryor and Johnson 1981).

The foliage of eucalypts is sclerophyllous — that is, the leaves are leathery, the number of stomates is reduced, and the leaves have a high oil content. In addition the leaves of most eucalypts hang vertically on slender flexible petioles which can twist to reduce the area exposed to the sun. Foliage is relatively sparse and often clustered near the ends of branches which with the vertically hanging leaves creates an open canopy. These are adaptations which reduce the heating and drying effects of the hot Australian sun (Pryor and Johnson 1981). They also affect the kinds and abundance of insects on the foliage and influence the behaviour of birds which forage for insects in the canopy. For example, the similarity between species of eucalypts in the kinds of leaves they possess may limit the number of co-existing foliage gleaners.

The bark of eucalypts is another of their distinctive features. The bark of eucalypts can be fibrous (stringybarks), flakey or tessellated (boxes, bloodwoods), or hard and deeply furrowed (ironbarks). Many species shed all or part of their bark annually from branches and trunk (gums and peppermints). The uniqueness of these bark characteristics and the difficulty of separating related species in the field commonly results in eucalypts being grouped by the type of bark — that is, as gums, stringybarks, boxes etc. Birds respond to the types of bark as different foraging substrates and some co-existing congeners (e.g. treecreepers, *Climacteris*), differ in their preferences for bark types (Noske 1985).

The shedding of bark, branches, twigs and leaves leads to the annual accumulation in eucalypt forests of litter at rates of 4-10 tonnes/ha (O'Connell and Menage 1982; Walker 1979). The greatest accumulations occur in moist forests on relatively fertile soils (e.g. Mountain Ash, *E. regnans*, in Victoria and Karri, *E. diversicolor,* forests in West Australia). The accumulation of litter coupled with the hot, dry conditions which prevail over most of Australia during the summer or in the northern dry season, has meant that fire is a frequent event in forests and woodlands. Fires can therefore be extremely intense with the hottest fires occurring at infrequent intervals in the moister types of forest.

Most of the eucalypts resist fire. Thick bark (or a stocking of thick bark on the trunks of gums and peppermints), proventitious buds on the trunk under the bark, and lignotubers are means by which eucalypts resist the heat of fire and which permits them to grow new foliage immediately after being burnt. Seeds are held (often for some years) in hard woody capsules and released when heated or on the death of the tree. Thus eucalypts which are fire sensitive (e.g. Mountain Ash, Karri) and which are killed by intense burns, as well as species which survive burning, seed in heavily after a fire.

These responses to fire also allow eucalypts to resist dry conditions or drought (Gill 1975, 1981). The rapid growth of foliage enables eucalypts to use nutrients released from the litter by fire. The young foliage is therefore succulent, high in nutrients and densely packed along the trunk or in sprouts from the lignotuber. Such foliage commonly supports high population densities of insects and is therefore attractive to birds.

Another important feature of eucalypts is the large amount of energy-rich carbohydrates produced. Many species produce large quantities of nectar and rely on birds and/or mammals as well as insects for pollination (Recher 1981). Sap and manna produced in response to injury or insect attack, and lerp and honeydew secreted by sap-sucking insects are other important sources of carbohydrates used by birds (Paton 1980). Manna and lerp are found on the foliage while honeydew is more often associated with insects on branches and twigs or under loose and peeling bark.

Although abundant, carbohydrates are not uniformly available through the year or in all types of forest and woodland. Carbohydrates associated with loose bark are found only on trees which shed bark while some species are more likely to support large populations of insects producing manna, honeydew and lerp than others. Possibly manna, honeydew and lerp are most often associated with trees in the subgenus *Symphyomyrtus* with their higher foliage nutrient levels. The variation in carbohydrate abundance extends to the pattern of flowering. Not all eucalypts bloom each year and there can be considerable local and regional differences between populations and species in the times of flowering. These variations in the abundance and availability of carbohydrates affect the abundance of nectar-feeders and other birds which use these energy-rich foods (Recher and Holmes 1985; Pyke 1985). Loyn (1985) has also shown how some honeyeaters (Meliphagidae) exclude other forest birds from habitats supporting dense populations of lerp producing insects.

FORESTS AND WOODLANDS

In 1970 a system of classiying plant communities based on the height of the dominant life form (i.e. tree, shrub or herb) and the projected foliage cover of the tallest layer of vegetation was introduced (Specht 1970). This system avoided the ambiguities and semantic problems associated with descriptive terms such as rainforest, dry sclerophyll and wet sclerophyll, and at the same time conveyed information about the structure of the community being discussed. Specht's terminology, as it is known, is followed throughout this book. Table 1 sets out the terminology necessary for eucalypt forests and woodlands. For reference the equivalent older descriptions are as follows. Rainforests are now described as 'closed-forest' where the projected cover of the canopy foliage covers 70-100 percent of the ground. Wet and dry sclerophyll are embodied in the terms 'open-forest' with projected foliage cover of 30-70 percent. Woodlands remain known as woodlands, but are defined as plant communities where trees are the tallest lifeform and the foliage cover is sparse (10-30 percent).

Sixty percent of the Australian continent receives less than 500 mm of rain annually (Fig. 1) and is dominated by shrub- and grasslands (Fig. 2) (Leeper 1970). Forests and woodlands cover the remaining third of the land with forests restricted to areas receiving more than 800 mm of rain annually (Fig. 2) (Specht 1981a, b). Closed forests dominate the highest rainfall areas (> 1000 mm annually) and are confined to the east coast.

Climate and the pattern of plant growth vary across the continent (Figs 1 and 2) (see Fitzpatrick and Nix 1970; Nix 1976 for details). Rain falls evenly throughout the year in southeastern Australia and Tasmania. In southwestern Australia most rain falls during winter while in northern Australia the main rainfall is monsoonal coming during the summer. Maximum growth indices occur when moisture is adequate and temperatures are moderate (Nix 1976). Thus, peak plant growth in southeastern Australia occurs from August to January, in southwestern Australia from May to October, and northern Australia from December to March (Nix 1976). Within each region there is local variation with changes in elevation, topography and soils.

The tallest and most complex open-forests (tall open-forest or wet sclerophyll) develop on moist, rich soils. Annual rainfall may exceed 1000-1500 mm with that of the driest month exceeding 50 mm (Ashton 1981). Depending on the frequency of fire, such sites are potentially capable of supporting closed-forest (rainforest) and in many places a rainforest understorey develops with eucalypts remaining only as emergents. Regeneration of the eucalypts occurs when, and if, the forest is burnt. For this reason it is common forestry practice to burn wet sclerophyll habitats after logging to ensure the regrowth of eucalypts. On sites where the rainfall in the driest month is less than 25-30 mm, the understorey of tall open-forest is dominated by a mixture of wattles and eucalypts (Ashton 1981). Shrub and ground vegetation is often well-developed and rich in species creating a multi-layered forest of three or four layers (canopy, sub-canopy, shrub and ground vegetation).

With increasing dryness and poorer soils, tall open-forest grades into open-forest. Open-forests were previously described as dry sclerophyll forest

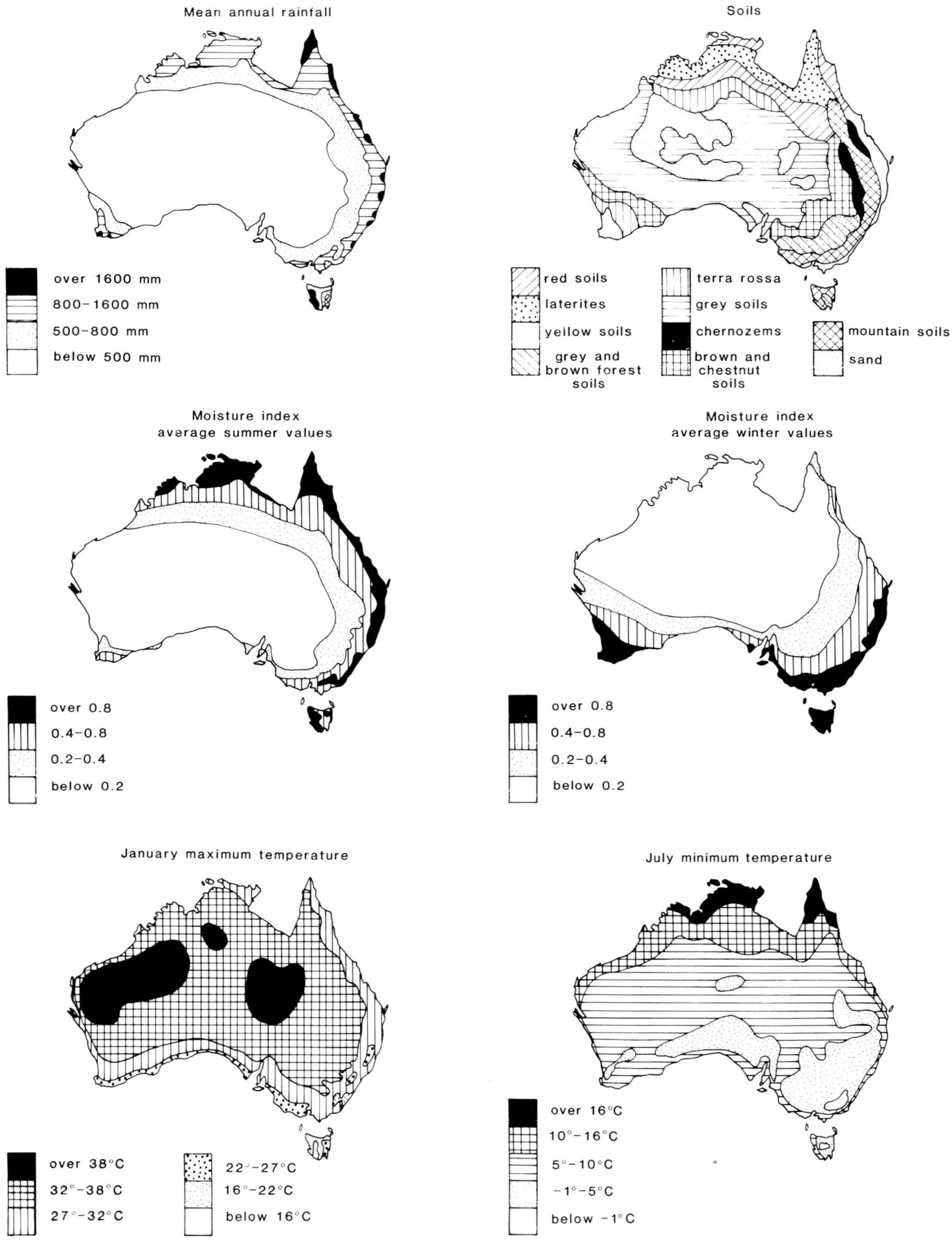

Fig. 1. Australian Climate. Australia is a flat continent with little topographic relief. Most of Australia is dry and receives less than 500 mm of rain annually. Summer temperatures are hot and although cooler during the winter, there are few places where temperatures fall below freezing (0°C). Eastern and southeastern Australia receive rain throughout the year, but in northern Australia rainfall is monsoonal and falls during the summer. Southwestern Australia has a winter rainfall pattern. The patterns of rainfall and temperatures are combined to produce moisture indices which show the times of the year when rainfall exceeds evaporation. (From Recher, Lunney and Dunn 1979).

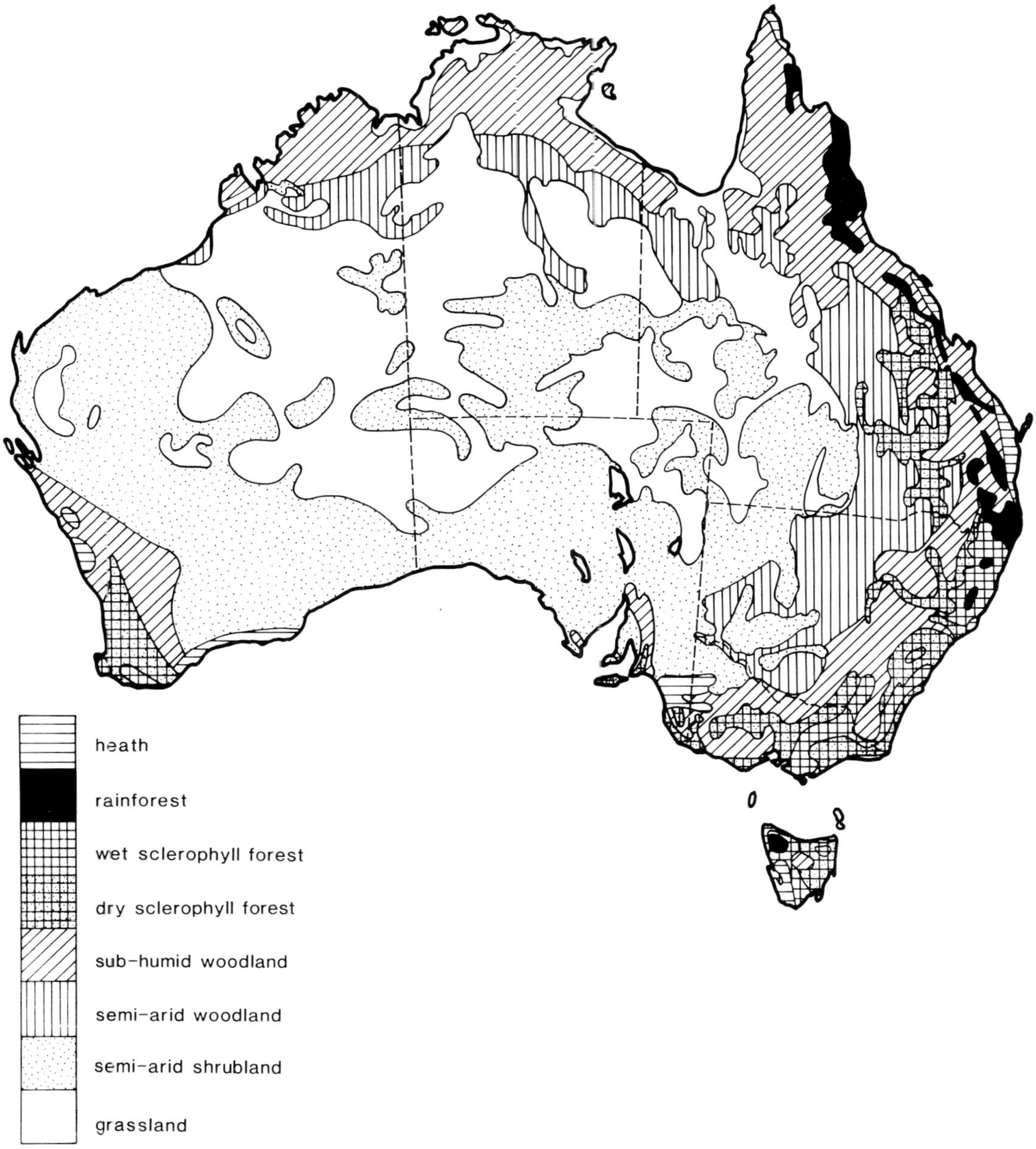

Fig. 2. Distribution of Vegetation in Australia. Forests and woodlands are restricted to the areas of higher rainfall along the eastern, northern and southwestern coasts. Most of the continent receives less than 500 mm of rain annually and is dominated by shrub- and grasslands. (From Recher, Lunney and Dunn 1979).

and generally lack the structural and floristic complexity of tall open-forests. An exception to this occurs on some soils particularly poor in nutrients where a floristically rich and heath-like shrub layer develops with banksias *Banksia* and she-oaks *Casuarina* forming an understorey with the dominant eucalypts. Such forests are typical of sandstones and coastal sandy soils in eastern and southwestern Australia. With increasing dryness open-forest grades into woodland.

Woodlands occupy 20-25 percent of the Australian continent and 80 percent of eucalypts are woodland species (Gillison and Walker 1981). The subgenus *Monocalyptus* is poorly represented in woodlands and most woodland species belong to the subgenera *Symphyomyrtus, Blakella, Corymbia, Eudesmia,* and *Idiogenes. Monocalyptus* is therefore confined almost exclusively to coastal forests.

Woodlands are open habitats (10-30 percent foliage projective cover) with few shrubs and a ground vegetation dominated by grasses. The term woodland and savannah are often interchangeable with savannah referring to the grass component (Gillison and Walker 1981).

While it is easy to describe the major structural features of forests and woodlands, the plant species

Table 1. Structural forms of vegetation in Australia (from Specht *et al.* 1974).

Life Form and Height of Tallest Stratum*	Projective Foliage Cover of Tallest Stratum*			
	Dense (70-100%)	Mid Dense (30-70%)	Sparse (10-30%)	Very Sparse† (<10%)
Trees ‡ >30 m	Tall closed-forest*	Tall open-forest	Tall woodland §	Tall open-woodland§
Trees‡ 10-30 m	Closed-forest*	Open-forest	Woodland	Open-woodland§
Trees‡ 5-10 m	Low closed-forest*	Low open-forest	Low woodland	Low open-woodland
Shrubs‡ 2-8 m	Closed-scrub	Open-scrub	Tall shrubland	Tall open-shrubland
Shrubs‡ 0-2 m	Closed-heath	Open-heath	Low shrubland	Low open-shrubland§
Hummock grasses 0-2 m	—	—	Hummock grassland	Open hummock grassland§
Herbs (incl. moss, ferns, hemicryptophytes, geophytes, therophytes, hydrophytes, helophytes)	Closed-herbland††	Herbland††	Open-herbland††	—
	Closed-tussock grassland	Tussock grassland	Open-tussock grassland	—
	Closed-grassland	Grassland	Open-grassland	—
	Closed-herbfield	Herbfield	Open-herbfield	—
	Closed-sedgeland	Sedgeland	Open-sedgeland	—
	Closed-fernland	Fernland	Open-fernland§	—
	Closed-mossland	Mossland	Open-mossland§	—

* Isolated trees (emergents) may project from the canopy of some communities (Richards, Tansley and Watt, *Imp. For. Inst. Pap.* No. 19, 1939, 6). In some closed-forests, emergent *Araucaria, Acacia,* or *Eucalyptus* species may be so frequent that the resultant structural form may be classified better as an open forest.

† Some ecologists prefer to ignore scattered trees and shrubs, equivalent to emergents in a predominantly grassland, heath or shrubland formation.

‡ A tree is defined as a woody plant more than 5 m tall, usually with a single stem. A shrub is a woody plant less than 8 m tall, frequently with many stems arising at or near the base (slightly modified from Beadle and Costin, *Proc. Linn. Soc. NSW.* 77: 1952, 61).

§ These formations are rare in Australia.

†† Appropriate names for the community will depend on the nature of the dominant herb.

composition of canopy, subcanopy, shrub and ground vegetation is far more complex and variable. Woodlands, for example, vary in plant species richness from 3 to 80 species/ha (Gillison and Walker 1981). Moreover not all open-forests and woodlands are dominated by eucalypts. In the north, deciduous woodlands are dominated by the non-eucalypt genera *Bauhinia* (Papilionaceae) and *Terminalia* (Combretaceae) (Gillison and Walker 1981). *Angophora* is often a co-dominant with eucalypts in open-forests of eastern Australia and in some habitats *Melaleuca, Casuarina, Callitris* and *Acacia* are the dominant trees.

If eucalypts appear to dominate the Australian landscape, it is by virtue of their size as trees. Wattles *Acacia* are even more diverse and widespread than eucalypts. More than 600 species of *Acacia* have been described in Australia (Burbidge 1963). *Acacia* shrublands and low woodlands replace eucalypt communities in the drier regions of the continent (Gillison and Walker 1981; Johnson and Burrows 1981). In the winter rainfall zone of southern Australia wattles predominate where the annual rainfall is less than 250 mm. In northern Australia wattles replace eucalypts where the rainfall drops below 350 mm annually. Low open-forests of Brigalow, *A. harpophylla,* extend to near the coast in central Queensland where rainfall exceeds 900 mm annually (Johnson and Burrows 1981). Wattles are also prominent in the understorey of open- and tall-open forests (Gill 1981; Ashton 1981).

Wattles and she-oaks may be particularly important for their role in nitrogen fixation. Wattles, which regenerate rapidly, are instrumental in restoring the nitrogen balance within forest ecosystems following fires (Ashton 1981). Wattles, she-oaks, and other understorey plants also add an important element of diversity to eucalypt communities. The different kinds of foliage, bark and seeds associated with these plants offer different kinds of food and foraging opportunities for birds. In their chapter Milledge and Recher (1985) point out how the addition of plant genera to the forest understorey appears to increase the number of co-existing bird species. In a similar manner the patchy distribution of plant communities which develop in response to changes in elevation, topography and soil conditions is exploited by forest and woodland birds. Many species shift seasonally between habitats or congregate in habitats where food (e.g. nectar, seeds) may become locally abundant. These responses of the avifauna form the basis of discussion in the following sections of this volume which deals with bird communities and populations.

FOREST AND WOODLAND BIRDS

About 720 species of birds have been recorded in Australia (Pizzey 1980). Of these, 530 are terrestrial and fresh-water species (Keast 1961). Keast (1961, 1981) assigned species to the habitat (e.g. rainforest, woodland, marsh) in which each was most common. Birds of open-forests and woodlands accounted for 234 species with 149 species occuring predominantly in woodland and 85 species in open-forest. Another 80-85 species were most abundant in closed-forest (Keast 1981, Brereton and Kikkawa 1963). Pizzey (1980) described the habitat of 315 species as including open-forest or woodland for at least part of the year. Of these, there were 109 species of non-passerine and 206 species of passerine birds. Among the non-passerines were 46 species of parrots (Cacatuidae, Psittacidae and Loriidae), 16 species of hawks (Pandionidae, Accipitridae and Falconidae), and 13 species of pigeons (Columbidae). The most diverse groups of passerines were the honeyeaters (Meliphagidae) with 61 species, the old-world thrushes and flycatchers (Muscicapidae) with 34 species, and the thornbills, warblers and scrubwrens (Acanthizidae) with 24 species.

In broad terms the number of species of birds in Australia increases from south to north (Kikkawa and Pearse 1969). However, the latitudinal trend in species richness is not linear (Pianka and Schall 1981) and species number varies in several ways. For example, there are fewer species in island forest habitats (e.g. on Tasmania and Kangaroo Island) than on the nearest mainland than can be accounted for by latitudinal differences (Abbott 1975). Similarly forests in southwestern Australia and the Mount Lofty Ranges in South Australia which are isolated from the main areas of forest and woodland in eastern and northern Australia also have fewer species of birds (Keast 1976). Although the total number of woodland species is greater than that of forests, for equal sized areas forest avifaunas are richer in species than woodlands (Keast 1961). Recher (1969, 1971) demonstrated that bird species diversity in southeastern Australia was correlated with the complexity of the vegetation as measured by the vertical distribution of foliage (foliage height diversity). Comparison of census results from a variety of forest and woodland habitats in southeastern Australia shows that on average species richness and population densities of birds are greatest in tall open-forest and least in woodland (Recher 1984, 1985).

Spencer (1896) recognized three zoogeographic regions in Australia; Bassian, Torresian and Eyrean. The Bassian corresponds closely with the forested areas of southeastern and southwestern Australia (Keast 1961). The Torresian region encompasses the tropical forests and woodlands of coastal northern Australia and has affinities with New Guinea. The remainder of the continent including most of the woodlands falls within the Eyrean region. The numerical analysis of the geographical distribution of land birds in Australia by Kikkawa and Pearse (1969) confirmed the faunal regions proposed by Spencer and other early workers. The faunal areas identified by Kikkawa and Pearse (1969) can be separated at both the species and generic level. For species a primary division separates the species rich forests and woodlands of eastern Australia from the western parts of the continent. At the generic level there is a primary separation of the northern and eastern regions from the central and southwest zones. The divisions are roughly concentric and conform to continental patterns of rainfall, temperature and vegetation. In discussing the pattern of species richness of woodland and eucalypt forest avifaunas, Keast (1961) noted that there was a relatively greater replacement of species from north to south in woodlands than in forests. As several of the papers in this volume point out, a number of forest birds have very wide distributions and a relatively small number of species dominate the forest aviauna.

AVIAN ADAPTATIONS TO A DRY CONTINENT

Adaptations of Australian birds to forest and woodland habitats are the subject of papers in the second section of the book. The ways in which birds forage, their social and reproductive behaviour, and their habitat requirements are related to the kinds of food and spatial resources available to them.

The papers in the second section deal with individual species such as the Rufous Scrub-bird, *Atrichornis rufescens,* and groups of closely related species such as the thornbills *Acanthiza*, but Australian birds as a whole are distinguished by a suite of adaptations to the continent's prevailing patterns of seasonal rainfall, aridity, and poor soils. Yom-Tov (ms) found that Australian passerines have significantly smaller sized clutches than passerines in the Northern Hemisphere (northern Africa, North America). Yom-Tov suggests that the difference may be related to the unreliability of rainfall and poor soils in Australia. If food resources are neither predictable due to unreliable rainfall nor abundant as a result of low productivity on nutrient poor soils, than there should be strong selection for small clutches, rapid maturation of young, frequent nesting and long breeding seasons which would permit birds to respond rapidly to favourable conditions. Yom-Tov found that the incubation and fledging periods of Australian passerines did not differ from those of passerines on other continents, and that among Australian birds clutch size was positively correlated with the frequency of nesting and the length of the breeding season. Comparison of the last two variables with birds on other continents was not possible, but it appears that if small clutches have evolved in response to environmental conditions in Australia, then other factors besides those mentioned must intervene to ensure reproductive success.

Birds of 'wet' forests (rainforest, wet sclerophyll) have significantly smaller clutches than birds nesting in 'semi-arid' formations (open-forest and woodlands) (Kikkawa 1974). A small clutch is typical of tropical forest birds (Kikkawa 1974) and may be an adaptation to reduce the incidence of nest predation by reducing the frequency with which parents must visit the nest (Skutch 1949, 1967). Cody (1966) argued differently and suggested that small clutch sizes indicated low mortality rates and long life allowing energy and time to be diverted from reproduction to other activities such as defence of territories and predator avoidance. The arguments concerning small clutch sizes among wet forest or tropical forest birds are not easily resolved, but they do present hypotheses for testing in Australia. Information on predation, mortality, and longevity among forest and woodland birds is practically non-existent and yet such data are needed to fully understand the ways birds have adapted to the Australian environment.

Between 10 and 15 percent of the species of forest and woodland birds in Australia are communal or co-operative breeders (Dow 1980). Co-operative nesting is not restricted to Australia and approximately equivalent numbers of species in Africa (Grimes 1976), North America (Woolfenden 1976), and Eurasia (Zahavi 1976) nest co-operatively. However, co-operative nesting has figured prominently in efforts to link the behaviour and ecology of birds to the vagaries of the Australian climate and through that to limits in the amount of available food. Rowley (1976), for example, contended that co-operative breeding was an adaptation to an erratic climate. Thomas (1974) argued that it was 'an adaptation to a basically sparse food supply', while Dow (1980) suggested that the Australian environment is conducive to the persistence of birds in groups and that co-operative breeding should occur more frequently in predictable than less stable environments. In the absence of data on the long-term availability of food and the differential survival or reproductive success of birds nesting co-operatively versus those forming simple pair bonds, it is again impossible to resolve these different views. As Dow (1980) concluded, 'it is most unlikely that a single ecological variable, or even a few, can be found to account for the widespread incidence of communal breeding by birds in Australia'.

In the second section, papers such as Bell's on the social system of thornbills and Noske's on treecreepers make a start in resolving issues by linking the behaviour and ecology of birds to the structure of the habitat they occupy and the distribution and abundance of available food. When such research can be combined with long-term studies such as Marchant's work on Yellow Robins, *Eopsaltria australis*, we will be in a better position to understand why birds respond to the Australian environment in the ways they do.

Although evergreen, eucalypt forests and woodlands are highly seasonal environments and seasonal changes determine the timing of reproductive activities, post-breeding dispersal and migration. Nix (1976) showed that the annual cycle of birds and timing of reproduction could be predicted from seasonal patterns of plant growth. Thus the peak nesting of birds in northern Australia is timed to occur with the onset of the wet season and in southern Australia it commences as temperatures increase during the spring. It is at these times that food resources are most likely to be abundant.

Australia is not noted for its spectacular migrations. Nonetheless, movements — migration, dispersal and nomadic wanderings — are common among forest and woodland birds throughout Australia. About half the birds on the forest plots in southeastern New South Wales studied by Recher and his colleagues migrated or dispersed into other habitats during the winter (Recher *et al.* 1983). Most movements of forest and woodland birds are 'diffuse' with individuals and small groups moving slowly over short distances to lower elevations or north along the coast in winter and south in spring. Unlike migrants in the Northern Hemisphere such birds seldom accumulate fat reserves, but feed as they travel. Apart from the highest elevations, insects, nectar and other carbohydrates are available throughout the year in most habitats. Many birds, honeyeaters for example, congregate wherever food is particularly abundant with the result that there can be considerable local and year to year variation in the abundance and distribution of birds during the non-breeding season. Such variation creates special difficulties when considering the conservation and management of forest and woodland avifaunas. In particular, conservation can not be guaranteed simply by establishing a system of reserves. Reserves or national parks may be the keystones of a conservation network, but to allow for the movements of birds and their requirements during the non-breeding season, it will be necessary to ensure a continuity of forest and woodland habitats between parks.

CONSERVATION AND MANAGEMENT

No continent has been more seriously affected by European civilization than Australia. The absence of mountain refuges, a generally flat topography and benign (if erratic) climate enabled Europeans to colonize the entire continent and convert most of it into farms and grazing lands within 100 years of the first settlement. Coupled with the clearing of forests, logging and changed fire regimes, the impact has been devastating and for quite considerable areas we will never know what the original landscape or vegetation was like. Yet the Australian avifauna has mostly managed to survive. No species is extinct and although a number of birds, including several woodland species, are endangered, the prospects for their conservation and survival must be rated fairly high given current social attitudes towards conservation and extinction.

Conservation and management are the themes of the final section. The papers in that section deal specifically with the problems created by forestry practices, fire, agriculture and mining. In addition many of the papers in the first two sections are based on research initiated in response to the effects of forest management on birds.

One can not be happy about all the changes that have taken place in Australia during the past 200 years. However, it is encouraging that our reaction has produced a scientific framework on which to proceed with the conservation and management of the forest and woodland avifauna of Australia. The papers presented in this book represent a benchmark in ornithological research in Australia. However, the research presented did not proceed in isolation nor could it have been done without using the work of earlier ornithologists as stepping-stones. The uniqueness of the work presented here is in integrating ecological and behavioural studies with an objective of applying the results to the practical problems of wildlife management.

ACKNOWLEDGEMENTS

Many people have contributed to the success of the book, but we would especially like to thank Lynne Albertson, Greg Gowing and Judy Turley-Recher for handling an editor's work during long periods when he was otherwise engaged. David McFarland and David Milledge contributed their skills as artists with David McFarland providing the pen and ink sketches of birds and David Milledge the frontispiece. Photographs were contributed by H. and J. Beste, N. Chaffer, H. Cogger, S. Craig, H. Ford, C. and D. Frith, S. Gilmore, D. Lunney, D. Milledge, R. Noske, P. O'Connor, H. Recher, D. Saunders, M. Seyfort, G. Smith, A. G. Wells and B. Wykes. The maps used in the introduction were prepared by David Milledge for *A Natural Legacy* published by Pergamon Press (Aust.) and we appreciate being able to use them here.

REFERENCES

ABBOTT, I., 1975. Density and species diversity of bird populations in *Eucalyptus* forests in Victoria, Bass Strait Islands, and Tasmania. *Proc. Roy. Soc. Vict.* **87**: 187-96.

ASHTON, D. H., 1981. Tall open-forests. Pp. 121-151 *in* Australian Vegetation ed. by R. H. Groves. Cambridge University Press, Sydney.

BEADLE, N. C. W., 1981. Origins of the Australian angiosperm flora. Pp. 409-426. *in* Ecological Biogeography of Australia ed. by A. Keast. Dr. W. Junk, The Hague.

BELL, H. L., 1985. The social organization and foraging behaviour of three syntopic thornbills *Acanthiza* spp. Pp. 151-63 *in* Birds of Eucalypt Forests and Woodlands: Ecology, Conservation, Management ed. by A. Keast, H. F. Recher, H. Ford and D. Saunders. Royal Australasian Ornithologists Union.

BRERETON, J. LE G. AND KIKKAWA, J., 1963. Diversity of avian species. *Aust. J. Sci.* **26**: 12-14.

BURBIDGE, N. T., 1963. Dictionary of Australian plant genera. Angus and Robertson, Sydney.

CODY, M. L., 1966. A general theory of clutch size. *Evolution* **20**: 174-84.

DOW, D. D., 1980. Communally breeding Australian birds with an analysis of distributional and environmental factors. *Emu* **80**: 121-40.

FITZPATRICK, E. A. AND NIX, H. A., 1970. The climatic factor in Australian grassland ecology. Pp. 3-26 *in* Australian Grasslands ed. by R. M. Moore. Aust. Nat. Univ. Press, Canberra.

GILL, A. M., 1975. Fire and the Australian flora: a review. *Aust. For.* **38**: 1-25.

GILL, A. M., 1981. Patterns and processes in open-forests of Eucalyptus in southern Australia. Pp. 152-176 *in* Australian Vegetation ed. by R. H. Groves. Cambridge Univ. Press, Sydney.

GILLISON, A. N. AND WALKER, J., 1981. Woodlands. Pp. 177-197 *in* Australian Vegetation ed. by R. H. Groves. Cambridge Univ. Press, Sydney.

GRIMES, L. G., 1976. Co-operative breeding in African birds. *Proc. 15th Int. Ornith. Congress.* Pp. 667-73.

JOHNSON, R. W. AND BURROWS, W. H., 1981. *Acacia* open-forests, woodlands and shrublands. Pp. 198-226. *in* Australian Vegetation ed. by R. H. Groves. Cambridge Univ. Press, Sydney.

KEAST, A., 1961. Bird speciation on the Australian continent. *Bull. Mus. Comp. Zool.*, Harvard **123**: 303-495.

KEAST, A., 1976. Ecological opportunities and adaptive evolution on islands, with special reference to evolution on the isolated forest outliers of Southern Australia. *Proc. 16th Int. Ornith. Congress.* Pp. 573-84.

KEAST, A., 1981. The evolutionary biogeography of Australian birds. Pp. 1581-1636. *in* Ecological Biogeography of Australia ed. by A. Keast. Dr. W. Junk, The Hague.

KIKKAWA, J., 1974. Comparison of avian communities between wet and semi-arid habitats of eastern Australia. *Aust. Wildl. Res.* **1**: 107-16.

KIKKAWA, J. AND PEARSE, K., 1969. Geographical distribution of land birds in Australia — a numerical analysis. *Aust. J. Zool.* **17**: 821-40.

LAMBERT, M. AND TURNER, J., 1983. Soil nutrient-vegetation relationships in the Eden area, NSW III. Foliage nutrient relationships with particular reference to *Eucalyptus* subgenera. *Aust. For.* **46**: 200-09.

LEEPER, G. W. (ed.), 1970. The Australian Environment. 4th edn. (rev.) CSIRO Australian and Melbourne University Press, Melbourne.

LOYN, R., 1985. Ecology, distribution and density of birds in Victorian forests. Pp. 33-46 *in* Birds of Eucalypt Forests and Woodlands: Ecology, Conservation, Management ed. by A. Keast, H. F. Recher, H. Ford and D. Saunders. Royal Australasian Ornithologists Union.

MILLEDGE, D. AND RECHER, H. F., 1985. A comparison of forest bird communities of the New South Wales south and mid-north coasts. Pp. 47-52 *in* Birds of Eucalypt Forests and Woodlands: Ecology, Conservation, Management ed. by A. Keast, H. F. Recher, H. Ford and D. Saunders. Royal Australasian Ornithologists Union.

NIX, H. A., 1976. Environmental control of breeding, post-breeding dispersal and migration of birds in the Australian region. *Proc. 16th Int. Ornith. Congress.* Pp. 273-305.

NOSKE, R., 1985. Adaptations of trunk-foraging birds: ecology and food. Pp. 193-204 *in* Birds of Eucalypt Forests and Woodlands: Ecology, Conservation, Management ed. by A. Keast, H. F. Recher, H. Ford, and D. Saunders. Royal Australasian Ornithologists Union.

O'CONNELL, A. M. AND MENAGE, P. M. A., 1982. Litter fall and nutrient cycling in Karri (*Eucalyptus diversicolor* F. Muell.) Forest in relation to stand age. *Aust. J. Ecol.* 7: 49-62.

PATON, D. C., 1980. The importance of manna, honeydew and lerp in the diet of honeyeaters. *Emu* 213-26.

PIANKA, E. R. AND SCHALL, J. J., 1981. Species densities of Australian vertebrates. Pp. 1677-1694 *in* Ecological Biogeography of Australia ed. by A. Keast. Dr. W. Junk, The Hague.

PIZZEY, G., 1980. A Field Guide to the Birds to Australia. Collins, Sydney.

PRYOR, L. D. AND JOHNSON, L. A., 1971. A Classification of the Eucalypts. Aust. Nat. University, Canberra.

PRYOR, L. D. AND JOHNSON, L. A. S., 1981. Eucalyptus, the universal Australian. Pp. 499-536. *in* Ecological Biogeography of Australia ed. by A. Keast. Dr. W. Junk, The Hague.

PYKE, G., 1985. The relationships between the abundances of honeyeaters and their food resources in open-forests near Sydney. Pp. 65-77 *in* Birds of Eucalypt Forests and Woodlands: Ecology, Conservation, Management ed. by A. Keast, H. F. Recher, H. Ford, and D. Saunders. Royal Australasian Ornithologists Union.

RECHER, H. F., 1969. Bird species diversity and habitat diversity in Australia and North America. *Am. Nat.* **103**: 75-80.

RECHER, H. F., 1971. Bird Species Diversity: a review of the relationship between species number and environment. *Proc. Ecol. Soc. Aust.* **6**: 135-52.

RECHER, H. F., 1981. Nectar-feeding and its evolution among Australian vertebrates. Pp. 1637-1648 *in* Ecological Biogeography of Australia ed. by A. Keast. Dr. W. Junk, The Hague.

RECHER, H. F., 1984. Use of bird census procedures in Australia: a review. Pp. 3-14 *in* Methods of Censusing Birds in Australia ed. by S. J. J. F. Davies. R.A.O.U. Report No. 7.

RECHER, H. F., 1985. Synthesis: A model of forest and woodland bird communities. Pp. 129-35 *in* Birds of Eucalypt Forests and Woodlands: Ecology, Conservation, Management ed. by A. Keast, H. F. Recher, H. Ford, and D. Saunders. Royal Australasian Ornithologists Union.

RECHER, H. F., ROHAN-JONES, W. AND SMITH, P., 1980. Effects of the Eden Woodchip industry on terrestrial vertebrates with recommendations for management. Forestry Commission of New South Wales Research Note 42.

RECHER, H. F., GOWING, G., KAVANAGH, R., SHIELDS, J. AND ROHAN-JONES, W., 1983. Birds, resources and time in a tablelands forest. *Proc. Ecol. Soc. Aust.* **12**: 101-124.

RECHER, H. F., AND HOLMES, R. T., 1985. Foraging ecology and seasonal patterns of abundance in a forest avifauna. Pp. 79-96 *in* Birds of Eucalypt Forests and Woodlands: Ecology, Conservation, Management ed. by A. Keast, H. F. Recher, H. Ford, D. Saunders. Royal Australasian Ornithologists Union.

RECHER, H. F., LUNNEY, D. AND DUNN, I., 1979. A Natural Legacy: Ecology in Australia. Pergamon Press, Sydney.

ROWLEY, I., 1976. Co-operative breeding in Australian birds. *Proc. 16th Int. Ornith. Congress.* Pp. 657-666.

SKUTCH, A. F., 1949. Do tropical birds rear as many young as they can nourish? *Ibis* **91**: 430-55.

SKUTCH, A. F., 1967. Adaptive limitation of the reproductive rate of birds. *Ibis* **109**: 579-99.

SPECHT, R. L., 1970. Vegetation. Pp. 44-67 *in* The Australian Environment ed. by G. W. Leeper, 4th edn. (rev.), CSIRO Australian and Melbourne University Press, Melbourne.

SPECHT, R. L., 1981a. Major vegetation formations in Australia. Pp. 163-298 *in* Ecological Biogeography of Australia ed. by A. Keast. Dr W. Junk, The Hague.

SPECHT, R. L., 1981b. Ecophysiological principles determining the biogeography of major vegetation formations in Australia. Pp. 299-334 *in* Ecological Biogeography of Australia ed. by A. Keast. Dr W. Junk, The Hague.

SPECHT, R. L., ROE, E. M., AND BOUGHTON, V. H. (eds.), 1974. Conservation of major plant communities in Australia and Papua New Guinea. *Aust. J. Bot. Suppl.* No. 7, 667 pp.

SPENCER, B., 1896 (?1898). Summary of results of the Horn Expedition to Central Australia. Dulav and Co., London.

THOMAS, D. G., 1974. Some problems associated with the avifauna. Pp. 339-65 *in* Biogeography and Ecology in Tasmania ed. by W. D. Williams. Dr W. Junk, The Hague.

WALKER, J., 1979. Aspects of fuel dynamics in Australia, CSIRO *Aust. Div. Land Usi. Res. Tech.* Memo 79/7.

WOOLFENDEN, G. E., 1976. Co-operative breeding in American birds. *Proc. 16th Int. Ornith. Congress.* Pp. 674-684.

YOM-TOV, Y. (ms) The reproductive rates of Australian passerines.

ZAHAVI, A., 1976. Co-operative breeding in Eurasian birds. *Proc. 16th Int. Ornith. Congr.* Pp. 685-694.

CHAPTER 2

Australia's Forest and Woodland Avifauna: An Ecological and Evolutionary Perspective

Allen Keast[1]

The contemporary avifauna of Australia is diverse with 535 species of terrestrial and fresh water birds. Of these, 85 species are confined to eucalypt forests and 149 species to woodlands. Many of the more important ecological roles are filled by members of endemic groups which evolved in isolation on the Australian continent since its separation from Antarctica nearly 50 million years ago. The avifauna and the eucalypts which dominate Australia's forests and woodlands are intimately interdependent and have evolved together over the last 10 million years as Australia became progressively warmer and drier. Changes in historic times following European settlement, most of them inadequately documented, have been deleterious and many birds are threatened or have become very rare. There is a critical need for research to document these changes and to determine the ecology of forest and woodland bird species, as a background to more effective management of Australia's forests and woodlands and their bird life.

INTRODUCTION

OF THE 535 species of terrestrial and fresh water birds in Australia about 85 are confined to, or have their centre of abundance in open eucalypt forest, and 149 species are predominantly eucalypt woodland dwellers. Together they account for nearly one half of the Australian land birds. Most of the endemic families of Australian birds are centered in the forests and woodlands (Table 1). These include the honeyeaters, butcherbirds and magpies, woodswallows, megapodes, treecreepers and sittellas, thornbills and wrens. Parrots, pigeons and kingfishers, and honeyeaters have been strikingly successful in Australia's forest and woodlands, and illustrate the unique character of this avifauna. The parrots, for example, are far more diversified in Australia than anywhere else in the world and a high proportion of parrot species are forest and woodland inhabitants.

What is the nature of the Australian forest and woodland avifauna? What kinds of birds compose it? What is the relationship of the Australian avifauna to that of the rest of the world? What are its origins? What is the future of the Australian forest and woodland avifauna and, indeed of the forests and woodlands with which they are interdependent?

Birds of the Eucalypt Forests and Woodlands, Endemism and Diversity

The Australian land and fresh water avifauna is made up of 75 families, 240 genera, and about 535 species of birds (Keast 1961, 1981; Schodde 1982). Australia has about 14 endemic families of birds, amounting to 25-30% of the total avifauna; 90% of its genera and probably 95% of its species are endemic. This is because of the continent's long isolation. Australia has been an island since it separated from Antarctica 50 million years ago. Only South America, which was similarly isolated until the formation of the Panamanian land-bridge two to three million years ago, has as many endemic bird families.

Sharing the Australian forests and woodlands with these endemics are Australian representatives of cosmopolitan families, groups that are world-wide in occurrence, but that have had a long history in Australia. These groups include the hawks and eagles (*Accipitridae*), cuckoos (*Cuculidae*), pigeons and doves (*Columbidae*), owls (*Strigidae*), and kingfishers (*Alcedinidae*). Some of these have been exceedingly successful in Australia. Australia has 12 genera and 52 species of parrots, Australia and New Guinea combined have 36 genera and 88 species. Thus half of the world's parrot species are endemic

[1]Biology Department, Queen's University, Kingston, Canada

Pages 11-15 *in* BIRDS OF EUCALYPT FORESTS AND WOODLANDS: ECOLOGY, CONSERVATION, MANAGEMENT ed. by A. Keast, H. F. Recher, H. Ford and D. Saunders. Royal Australasian Ornithologists Union and Surrey Beatty & Sons.

to the Australo-Papuan region (Table 1). Likewise half of the world's species of kingfishers and one-third of the world's species of pigeons are unique to this region. As with the parrots, the Australo-Papuan assemblages of kingfishers and pigeons include some of the world's most striking and beautiful forms.

A third division is recognisable in the forest and woodland birds, members of groups that are highly diversified elsewhere in the world and that have only one or a few species in Australia. These include the swallows, swifts, orioles and white-eyes. The Silvereye, *Zosterops lateralis*, perhaps the most abundant native bird of the major cities, is a member of this group.

The Origin of Australia's Avifauna

What is the history of the avifauna of Australia? When did birds first occur in Australia? If birds colonized from outside where did they come from?

Too little is known to adequately answer these questions, but there are some clues. The discovery of a fossil feather in rocks of Early Cretaceous age in Victoria (Talent *et al.* 1976) showed that birds have lived in Australia for at least 120 million years. the first fossil bird, *Archeopteryx,* a toothed creature with feathers but a reptile-like tail, lived in Bavaria 180,000,000 years ago. Thus, Australia had acquired birds quite early in the history of the group. Unfortunately, from the Cretacous through until the late Miocene, 10,000,000 years ago, there is a gap in the Australian bird fossil record.

There is every reason to believe Australia must have had a diversified avifauna throughout this time. A hundred million years ago Australia, South America and Antarctica were joined as part of a great southern continent, Gondwana. Australia remained adjacent to Antarctica until the Eocene some 50,000,000 years ago when Australia separated and started to drift north.

At that time Antarctica was not the frigid continent that it is today. It supported conifers and the fore-runners of the Proteaceae, to which banksias, waratahs, and grevilleas belong. We know nothing of what land birds occurred in Australia at this time, but elsewhere modern bird types were already extant. These included ratites (the giant flightless birds that include Emu, Cassowary, Rhea and Ostrich), cormorants, rails, pelicans, sandpipers, flamingoes, grebes, ibises, ducks, petrels, cranes, gulls and hawks (Brodkorb 1971; Rich and van Tets 1982). Australia most assuredly shared many of these at this time.

About 20 million years ago Australia reached its present proximity to Asia. From this time onwards Australia must have received an increasing number of colonizers from the northwest. Probably some of our most distinctive genera were among these colonizers. Unfortunately, there is no way we can say which groups came from the south and which entered from the north, but colonization of birds from Asia into Australia certainly occurred and continues to occur. Mayr (1944) postulated that the Australian avifauna had been built up by a series or waves of colonization from Asia over a prolonged period. Endemic families would thus represent the oldest stocks, endemic genera subsequent colonizers, while cosmopolitan species such as the Peregrine Falcon, *Falco peregrinus*, and Barn Owl, *Tyto alba*, would be recent colonizers. Mayr's theory was, however, developed before it had been demonstrated that Australia had formerly been in the south. Accordingly, he did not allow for the possibility of an original southern home for Australia's distinctive endemic families.

There is now much speculation as to which of our birds are descendents of an old southern element and which evolved from colonisers reaching Australia from the north. (Serventy 1972; Rich 1975, 1976; Cracraft 1976; Keast 1981; Schodde 1982). However, with little knowledge of evolutionary rates and hampered by the poor fossil record, we still do not know which, or how many of Australia's endemic families had a southern origin. The flightless birds almost certainly are of southern origin. Pelicans, cormorants, many waterfowl, and hawks, are other possibilities since they had already evolved when Australia was joined to Antarctica and South America.

What does the fossil record tell us about Australia's birds of ten million years ago? The Miocene avifauna was highly diversified. A group of large flightless birds, the dromornithids, now extinct, were widespread. Hawks, ducks, nightjars, pelicans, rails, grebes, plovers, pigeons, and flamingoes were present (Rich and van Tets 1982). Flamingoes are in many ways the most surprising fossil discovery; today they occur almost everywhere in the world except Australia. They became extinct before the late Pleistocene, apparently with the drying up of the giant lakes in south and central Australia. More recent fossil deposits, aged a million years ago and less, include many modern forms. Examples are button-quails, owls, frogmouths, parrots, and kingfishers.

A better understanding of the history of Australia's birds is awaited. But in the absence of fossils there are some interesting data from biochemical studies. Techniques such as DNA hybridization are providing a much better appreciation of the relationships of living forms (Sibley 1976; Sibley and Ahlquist 1985). On this basis it has been shown that Australia's robins, flycatchers, and sittellas, that had formerly been assumed to be Australian representatives of overseas familes have their relationships with the endemic butcherbirds and magpies, woodswallows, whistlers, fantails, and others. The truly endemic

Table 1. Relative representation of some cosmopolitan and endemic families of birds that are especially important in forest and woodland, numbers of genera and species. Modified from Keast (1981).

Family	Australia proper	Australia and New Guinea	Oriental Region	Africa	South America	North America	Palearctic
			COSMOPOLITAN GROUPS				
Hawks and Eagles (Accipitridae)	13-18	18-31	17-31	26-52	25-54	11-25	19-45
Falcons (Falconidae)	1-6	1-7	3-10	2-9	8-22	1-6	1-15
Quail (Phasianidae)	1-4	3-5	34-46	10-44	—	5-6	25-53
Cuckoos (Cuculidae)	5-11	8-20	9-27	7-20	7-21	4-7	3-9
Pigeons and Doves (Columbidae)	18-21	19-58	10-35	5-22	9-47	8-13	3-22
Parrots (Psittacidae)	12-52	36-88	4-10	3-16	25-111	2-2	1-2
Owls (Strigidae and Tytonidae)	2-8	4-11	7-26	11-21	12-29	11-17	13-24
Kingfishers (Alcedinidae)	4-10	8-24	6-17	4-14	2-5	2-3	3-7
Swifts (Apodidae)	1-2	1-4	5-10	3-17	7-21	4-5	3-8
Swallows (Hirudinidae)	3-4	3-5	4-7	6-32	11-20	9-12	3-10
Crows (Corvidae)	1-5	2-7	11-21	4-8	1-14	10-16	10-27
			ENDEMIC GROUPS				
Mound-builders (Megapodidae)	3-3	5-8	1-1	—	—	—	—
Lyrebirds (Menuridae)	1-2	1-2	—	—	—	—	—
Fantails (Rhipidurinae)	1-4	1-13	1-5	—	—	—	1-1
Whistlers (Pachycephalinae)	3-13	6-30	1-1	—	—	—	—
Australian Warblers (Acanthizinae)	13-40	13-52	1-1	—	—	—	—
Fairy-wrens (Malurinae)	3-18	6-23	—	—	—	—	—
Pardalotes (Pardalotidae)	1-5	1-5	—	—	—	—	—
Honeyeaters (Meliphagidae)	23-71 (including 'chats')	32-102	—	—	—	—	—
Woodswallows (Artamidae)	1-6	1-7	1-1	—	—	—	—
Australian Magpies (Cracticidae)	3-8	3-11	1-1	—	—	—	—
Bowerbirds (Ptilorhynchidae)	5-9	9-18	—	—	—	—	—

element, in other words, is much greater than was formerly believed.

The History of Australia's Forest and Woodland Avifauna

What about the history, specifically, of Australia's forest and woodland avifauna? This segment of the avifauna presumably began to evolve its present structure with the drying up of the continent and the ascendency of *Eucalyptus, Acacia* and scleropyllous vegetation at the end of the Miocene. Prior to this the southern half of the continent had been covered by temperate rainforest in which the broad-leafed Antarctic Beech, *Nothofagus*, was prominent. This change is well documented in the fossil pollen record. Knowing how dependent many honeyeaters are on the nectar of *Eucalyptus*, pardalotes on the lerps and scales of its leaves, and parrots for nest sites in its hollow trunks and limbs, these eucalypt dependent groups probably began major radiations after the Miocene no earlier than 10,000,000 years ago.

Although frequently connected as a single land mass, New Guinea has a closed-forest avifauna and Australia an open country one. The two avifaunas largely replace each other. One hundred and three genera of birds are confined to Australia, and 117 to New Guinea. For species the figure is 381 and 397.

Does the New Guinea avifauna provide any clues as to what that of mainland Australia was like ten million years ago? In its composition the answer is probably no because New Guinea is rich and tropical, old Australia would have been mostly cool temperate. The contrast between the rainforest adapted avifauna of contemporary New Guinea and that of the open eucalypt forests and woodlands of Australia can be seen when the former colonize northern Australia, as they have in many cases (Kikkawa *et al.* 1981; Kikkawa 1982). The former keep strictly to rainforests in their new home. By contrast where southern Australian forest genera occur in New Guinea they are confined to the cool highlands (Schodde and Calaby 1972). The tree-creeper, *Climacteris (leucophaea) placens*, and the thornbill, *Acanthiza murina*, are examples. Thus in colonizing New Guinea, Australian birds carry their habitat preferences with them, just as do the New Guinea rainforest species.

The Division of Ecological Roles in Australian Eucalypt Forest and Woodland Birds

The Australian eucalypt forest and woodland avifauna is highly diversified. All the major ecological groups of birds are present. The relative representation in Australia of 11 major cosmopolitan groups of birds is shown in Table 1. Australia has 13 genera and 18 species of hawks and eagles, one genus and six species of falcons, 18 genera and 21 species of pigeons and doves and, as noted, 12 genera and 52 species of parrots. There are four species of quail, 11 cuckoos, 10 kingfishers, 4 swallows and 5 crows. These groups fill the same ecological roles that they do elsewhere. The hawks are the major diurnal predators and the owls the nocturnal ones. Swallows and swifts are the insect feeders of the air. The parrots and pigeons are the major seed-eaters and arboreal frugivores. Finches and parrots take small seeds from the ground. Crows fill the role of large-bodied scavengers in Australia as elsewhere.

The ecological roles of the endemic Australian bird groups can in part be determined on body morphology and, simply, by watching their habits. Thus, the honeyeaters are the major nectar-feeders, butcherbirds and magpies the major terrestrial predators, thornbills and warblers the foliage feeding insectivores, treecreepers and sittellas the major bark-foraging insectivores, while the woodswallows are, like the swallows, aerial insectivores.

The papers in this volume take our understanding of the ecology of Australian birds much further. For the first time we have the insights necessary to understand how ecological roles are divided, and the needs and contributions of different groups to the Australian forest and woodland ecosystem. The ecology of the treecreepers, thornbills, scrub-birds, honeyeaters, and parrots now emerges as more complex than formerly thought. The relations within the forest and woodland ecosystem are much more intimate and the components of the ecosystem, the forests and birds, are much more interdependent than was formerly suspected.

The Conservation of Australian Forest and Woodland Birds

Australia's forest and woodland avifauna is striking and unique; it is a national asset. Today as never before this heritage is getting the appreciation in Australia that it deserves. Yet, like the forests and woodlands that is its home, it has been, and is, under considerable stress. The final section of this book, details some of the forces working to the disadvantage of Australia's forests and their wildlife. That there has been a slippage of our national ornithological resources is very evident. Naturalists of the last generation can point to dozens of examples of species being absent from areas where once they were common.

A year ago this writer was horrified, on flying over western New South Wales to find how little of western New South Wales still carries native woodland and mallee vegetation. Mallee is almost gone from New South Wales, the victim of clearing for wheat. The little reserve at Rankins Springs where, twenty years ago, Dr. H. J. Frith (Frith 1959) carried out his classic studies on the Mallee Fowl, *Leipoa ocellata*, stood out clearly from the air. Yet the Mallee Fowl is now gone: the area set aside was too small. Twenty years ago striking honeyeaters like the Regent Honeyeater, *Xanthomyza phrygia*, and the unique mistletoe-berry feeding Painted Honeyeater, *Grantiella picta*, were regular spring and summer visitors to the western suburbs of Sydney, visiting the flowering and fruiting mistletoes. In recent years the Regent Honeyeater has only been seen a couple of times. The Painted Honeyeater has not been seen for twenty years. Are these species, which wintered in Queensland, victims of the wholesale clearing of the Brigalow, *Acacia harpophylla*, forests in central Queensland?

The papers in these volumes show that an active group of biologists are now working on Australia's forest and woodland birds. A start has been made in developing a forest policy that will take into account all the demands and needs of economics, preservation and conservation. For the first time the effects of fires, logging, forest die-back, and drought, are being studied in an integrated manner. Land managers are starting to think of management of the plants, mammals, and birds within the forests and not just of timber production or grazing.

The papers in these volumes represent a new generation of investigations. The community studies provide parallel data on eucalypt forests and woodlands equivalent to that obtained for rainforests by Jiro Kikkawa in his pioneer studies of that vegetation formation. Studies on groups and individual species give new insights into habits, needs, and contributions to forest welfare. They provide an understanding of wider problems like the role of habitat shrinkage and change in the near-extinction of the Noisy Scrub-bird, *Atrichornis clamosus*, and the perils to parrot populations of the rapid reduction in mature eucalypts that, though senescent in the sense of timber production, provide the nesting holes for parrots. Without research we would not have recognised these problems much less have been able to focus on answers. If this volume achieves nothing more than to alert Australians to the problems of the present and the perils ahead, it will have achieved a major objective.

REFERENCES

BRODKORB, P., 1971. Origin and evolution of birds. Pp. 19-55 *in* Avian Biology 1 ed. by D. S. Farner and I. R. King. Academic Press, New York.

CRACRAFT, J., 1976. Avian evolution on southern continents: influences of palaeogeography and palaeoclimatology. Proc. 16 Inter. Ornith. Congr., *Canberra*. Pp. 40-52.

FRITH, H. J., 1959. Breeding of the mallee fowl, *Leipoa ocellata* Gould (Megapodiidae). *Aust. Wildl. Res.* **4**: 31-60.

KEAST, A., 1961. Bird speciation on the Australian continent. *Bull. Mus. Comp. Zool. Harvard* **123**: 303-495.

KEAST, A., 1981. The evolutionary biogeography of Australian birds. Pp. 1586-1635 *in* Ecological Biogeography of Australia ed. by A. Keast. Junk, The Hague.

KIKKAWA, J., 1982. Ecological association of birds and vegetation structure in wet tropical forest of Australia. *Aust. J. Ecol.* 7: 325-45.

KIKKAWA, J., MONTEITH, G. B. AND INGRAM, G., 1981. Cape York Peninsula: major region of faunal interchange. Pp. 1695-1742 *in* Ecological Biogeography of Australia ed. by A. Keast. Junk, The Hague.

MAYR, E. 1944. Timor and the colonization of Australia by birds. *Emu* **44**: 113-30.

RICH P. V., 1975. Changing continental arrangements and the origin of Australia's non-passeriform avifauna. *Emu* **75**: 97-112.

RICH, P. V., 1976. The history of birds on the island continent of Australia. Proc. 16 Intern. Ornith. Congr., Canberra: Pp. 53-65.

RICH, P. V. AND VAN TETS, G. F., 1982. Fossil birds of Australia and New Guinea, phylogenetic, and biostratigraphic input. Pp. 236-384 *in* The Fossil Vertebrate Record of Australasia ed. by R. P. Rich and E. M. Thompson. Monash University, Melbourne.

SCHODDE, R., 1982. Origin, adaptation and evolution of birds in arid Australia, pp. 191-225 *in* Evolution of the Flora and Fauna of Arid Australia ed. by W. R. Barker and P. J. M. Greenslade. Peacock, Frewville, South Australia.

SCHODDE, R. AND CALABY, J. H., 1972. The biogeography of the Australo-Papuan bird and mammal faunas in relation to Torres Strait. Pp. 257-300 *in* Bridge and Barrier; The Natural and Cultural history of Torres Strait ed. by D. Walker. Aust. Nat. U., Dept. Biogeog. Geomorphol. Pub. BG/3.

SERVENTY, D. L., 1972. Causal ornithogeography of Australia. Proc. 15 Int. Ornith. Congress, The Hague. Pp. 574-84.

SIBLEY, C. G., 1976. Protein evidence of the relationships of some Australian passerine birds. Proc. 16 Int. Ornith. Cong., Canberra: 557-70.

SIBLEY, C. G. AND AHLQUIST, J. E., 1985. Relationships of Australian birds. Proc. 18. Int. Ornith. Cong., Moscow, in press.

TALENT, J. A., DUNCAN, P. M. AND HANDLEY, P. L., 1976. Early Cretaceous feathers from Victoria. *Emu* **76**: 81-6.

Eastern Rosella, *Platycercus eximius*

Bird Communities in Forests and Woodlands

Co-ordinator — H. F. Recher

Eastern Whipbird, *Psophodes olivaceus*

INTRODUCTION

A COMMUNITY is an assemblage of species which share a common environment. The community can be the populations of all the organisms occuring together or, for the sake of simplicity, the community can be restricted to a single group or taxon of plants or animals. The papers in this section deal with communities of eucalypt forest and woodland birds.

The study of community ecology owes a great deal to the study of birds. Birds are large, conspicuous and mostly diurnal. They also rely on their visual and auditory senses to find food, locate mates and move about. These are attributes which also describe most ecologists and help explain why birds have figured importantly in ecological research. The ways in which birds use and share resources, their morphological differences and similarities, and the features of the environment which limit the number of co-existing species provided the empirical information used to formulate and test most of the theory of community ecology developed during the 1960's and 1970's.

Australia's contribution to community ecology has been limited. This is because field ornithology and community studies were late in starting in Australia and the data needed to formulate and test hypotheses has been lacking. This is changing and as illustrated by the papers in this section, Australian studies of avian communities will contribute importantly during the 1980's to advances in our understanding of animal communities.

The first two papers of the section describe forest and woodland bird communities in Victoria and discuss reasons for differences in the kinds of birds and numbers of species found in different types of forest. The next two papers compare forest communities along the coast of New South Wales. In each of the four papers the work discussed was initiated in response to a conservation or forest management issue. Sandy Gilmore surveyed woodland habitats in central and western Victoria to determine whether the different kinds of bird communities were represented in nature conservation areas or if there was a need to expand the reserve system to include particular kinds of communities. The work reported by Richard Loyn in eastern Victoria and by Harry Recher and David Milledge for coastal New South Wales was initiated in response to the concern expressed by environmentalists about the effects of logging on forest wildlife. The report by Jim Shields and his colleagues was also part of a study of logging effects and formed part of an Environmental Impact Statement prepared by the Forestry Commission of New South Wales for its proposal to log closed- and tall open-forests on the catchment of the Hastings River in the northern part of the state. Each of the papers describes the avifauna, the species present and their abundance and relates these to the kinds of vegetation in which the community is found. The work reported provides a basis on which to discuss the relationship between species richness and habitat structure or diversity. This is done in the synthesis concluding the section.

The papers by Richard Holmes and Harry Recher, Allen Keast, and Graham Pyke as well as those by Loyn and Gilmore tackle another aspect of community ecology — what are the resources used by birds?, and how do birds apportion these resources in space and time?. Holmes and Recher describe the annual cycle of a forest and woodland avifauna on the Southern Tablelands of New South Wales and investigate the pattern of use of food resources among the most common species. Graham Pyke describes how the numbers of honeyeaters in woodland near Sydney vary through the year according to the amount of nectar available. Their studies and those of Keast who compares the species composition and foraging habits of birds in eastern Australia with those in the far north and southwest of the continent are an extension of the studies presented in the first four papers.

The work of Keast, who also describes how birds share time during the day for vocalizing, and that of Recher, Holmes and Pyke rely heavily on basic information about the biology of birds and their distribution and abundance. The papers presented in this section show that a start has been made in obtaining the data needed for an understanding of community ecological problems in Australia. Moreover these studies and others like them provide the basis on which to evaluate the conservation requirements of Australian birds and to formulate plans of management for natural areas.

Harry F. Recher

Fan-tailed Cuckoo, *Cuculus pyrrhophanus*

CHAPTER 3

The Influence of Vegetation Structure on the Density of Insectivorous Birds

A. M. Gilmore[1]

Breeding bird populations were censused on eleven woodland and one heathland plots, and estimates made of the proportional use of substrates by insectivores. Four guilds of insect eating birds were identified and their densities were correlated with the quantities of foraging substrates. The numbers of species of insectivorous birds on a plot were correlated with vegetation biomass and canopy height, but not to the number of plant species. The results suggest that structure and composition of insectivorous bird communities is largely determined by structure of the vegetation. Many birds in different guilds appear to co-exist without obvious interaction.

INTRODUCTION

MOST insectivorous birds forage on a narrow range of substrates. The availability of particular foraging substrates may therefore determine the presence and abundance of particular species and foraging guilds, but quantitative analyses of the relationships between the abundance of bird species and the abundance of their foraging substrates have seldom been attempted in Australia. Most often the relationship has been inferred, not proven. Pattemore and Kikkawa (1974) concluded that insectivorous birds in the lower strata of rain-forest increased after logging because there was an enormous increase in the amount of understorey foliage. This presumably led to an improvement in the availability of food and/or shelter. Gepp (1976) also found that bird species occurred only in habitats that contained suitable foraging substrates. Ford and Bell (1981) compared bird communities in a series of woodland plots subject to varying degrees of crown dieback. They concluded that foliage density at several levels influenced the densities of birds by affecting the abundance of food, nesting sites and refuges. Cheal *et al.* (1979) related abundance of various bird species to the changing quantities of their foraging substrates during post fire succession, as did Recher *et al.* (1980) for post-logging regeneration. Loyn *et al.* (1980) pointed out that the abundances of bird populations in their study appeared to be inversely related to insect populations. They also implied that after logging, the presence of particular bird species and guilds was influenced by the change in the availability of foraging substrates. Group size in Grey-crowned Babblers, *Pomatostomus temporalis* (King 1980), Halls Babblers, *P. halli* (Brown and Balda 1977) and Superb Fairy-wrens, *Malurus cyaneus* (Nias 1984) is more closely related to vegetation structure than to the size of the home range. Woinarski (1978) showed the abundance of some species was proportional to the quantity of their foraging substrate(s).

The hypothesis investigated in this study was that the abundance of co-existing insectivorous birds which utilize a particular foraging substrate is proportional to the quantities of that substrate. In this paper a preliminary analysis is presented of data which concentrates on vegetation features as they relate to the availability and use of foraging substrates by insectivorous birds, the structure and composition of bird communities and various community attributes such as biomass and species richness. For the purpose of this study, honeyeaters were included as insectivores. Although most honeyeaters utilize nectar for energy all require insects as a source of protein.

STUDY AREAS

Four study areas were located in central Victoria, four in western Victoria, two in the Victorian Alps, and two in central Queensland (Table 1). The

[1]Arthur Rylah Institute for Environmental Research, Fisheries and Wildlife Division, Ministry for Conservation, 123 Brown Street, Heidelberg, Victoria 3084. Present address: "Morinda", Mafeking Road, Federal, New South Wales 2480.

Pages 21-31 *in* BIRDS OF EUCALYPT FORESTS AND WOODLANDS: ECOLOGY, CONSERVATION, MANAGEMENT ed. by A. Keast, H. F. Recher, H. Ford and D. Saunders. Royal Australasian Ornithologists Union and Surrey Beatty & Sons.

Table 1. Vegetation Structural Parameters of the Census Plots.

Vegetation Parameter	Census Plot											
	Central Victoria				Western Victoria				Alpine		Queensland	
	Red gum	Grey box	Mixed age iron-bark	Pole stand iron-bark	Black box	Heathy open w/land	Heath	Mallee broom-bush	Snow gum shrubby	Snow gum grassy	Gidgee	Poplar box
Number of Trees	268	2370	3015	1894	5680	3880	0	est. 10000	5000	6000	5825	2150
Basal Area (m^2)	118.4	96.7	140.6	139.8	165.9	81.4	0	0	125	126	122.8	83.4
Canopy Height (m)	13	17	15	17	12	6.5	1.5	4.5	12	11	11	15
Foliage cover of Trees (%)	10	18	26	26	25	7	0	8	30	24	27	8
Canopy Volume of Trees (m^3/ha)	8.23	8.39	3.46	4.86	4.89	2.37	0	0.200	4.16	3.64	6.81	1.68
Canopy cover of Shrubs (%)	0	15.5	15	22	15	42	70	n.a.	n.a.	n.a.	4.5	1
No. of Plant Species	7	12	14	14	68	95	48	55	70	40	20	20
Annual Litter Fall (grams/m^2)	470[1]	140[2]	202[2]	202[2]	194[2]	105[2&3]	90[3]	114[4]	267[5]	214[5]	111[6]	47[7]

1. Data from Briggs and Maher (1983); 2. Hutchings and Oswald (1975); 3. Specht and Rayson (1957); 4. Specht (1966); 5. Park (1977); 6. Tunstall (1973); 7. Walker (1979).

woodland classification of Gillison and Walker (1981) has been used for plot descriptions although detailed structural measurements can be found in Table 1.

Central Victoria

In central Victoria one plot was located in Reedy Lake State Forest and three in nearby Rushworth State Forest. The Reedy Lake plot was a tall open woodland of River Red Gum, *Eucalyptus camaldulensis*. The plots in Rushworth State Forest were tall woodlands one of which was dominated by Grey Box, *E. microcarpa,* and the others by Red Ironbark, *E. sideroxylon*.

Within Rushworth State Forest, logging had changed the structure of the forest from a tall woodland of thirty trees per hectare with a mean diameter of 1 m (Newman 1961) to one more densely stocked with younger and smaller diameter trees (Table 1). The mean annual rainfall for Rushworth is 525 mm with most falling in winter and spring.

The Grey Box woodland (GB) had a few Yellow Gum, *E. leucoxylon,* which were flowering at the time of the census. There were many dead branches on the ground following removal of the largest trees, several years before. There was a medium density shrub stratum consisting of Varnish Wattle, *Acacia verniciflua,* 1.5 m in height and Drooping Cassinia, *Cassinia arcuata,* and Fringe Myrtle, *Calytrix tetragona,* to 0.75 m in height. Ground cover was sparse.

The second plot was a mixed Red Ironbark woodland (MIB) located on a low hill. Grey Box were scattered throughout and a few Red Box, *E. polyanthemos,* and Red Stringybark, *E. macrorhyncha,* occurred near the hill top. The plot had a patchy understorey with sparse shrubs to 0.7 m including Spreading Wattle, *A. diffusa,* and Golden Wattle, *A. pycnantha,* to 3 m, Ground cover was principally lichens with scattered grass tussocks and prostrate shrubs of *Astroloma humifusum*. Bedrock outcrops occurred through the skeletal soils. The third plot was a pole stand of Red Ironbark (PIB) which was floristically similar to the previous plot. It had been managed to reduce the number of coppice shoots and the tree size was relatively uniform. There was a stratum of xeromorphic shrubs 0.7 m high including Golddust Wattle, *A. acinacea*, Gorse Bitter Pea, *Daviesia ulicifolia,* and Fringed Heath Myrtle, *Micromyrtus ciliatus*. A prominant but sparse ground cover of Wallaby Grass, *Danthonia* sp. and prostrate shrubs was present.

The River Red Gum plot (RG) was dominated by mature trees with a sparse ground cover of sedges and grasses with occasional forbs. A few fallen trees and Red Gum saplings to 1 m high were the only other features. The Red Gum plot had not been heavily logged but the plot was grazed by sheep and cattle.

Western Victoria

The plots in western Victoria consisted of tall woodland, medium open woodland, low woodland and heathland in the Little Desert. Average rainfall is 423 mm falling principally in winter and spring. Information on the physiography and soils of this area is in Blackburn *et al.* (1967) and vegetation in McMahon (1977), Specht (1966), Cheal *et al.* (1979) and Specht *et al.* (1958).

The tall woodland plot (BB) was dominated by Black Box, *E. largiflorens,* with some Yellow Gum and Green Mallee, *E. viridis*. There was a sparse shrubby understorey with occasional Violet Honey-myrtle, *Melaleuca wilsonii,* to 6 m. Stumps and many coppice shoots indicated past removal of posts from this area.

The medium open woodland (H/W) was dominated by multistemmed Brown Stringybark, *E. baxterii,* with some Yellow Mallee, *E. incrassata,* and Yellow Gum. There was a medium dense stratum of xeromorphic shrubs 0.5 m to 1.0 m high.

The low woodland (MB) was a mallee-broombush association with three distinct strata, Multi-stemmed eucalypts including Red Mallee, *E. foecunda*, Yellow Mallee and Green Mallee dominated the plot. The middle stratum was dominated by *Melaleuca uncinata* and *Baeckia behrii* (collectively broombush). There was a diverse ground cover of ephemeral herbs and small shrubs.

The heathland plot (H) was a typical *Banksia ornata-Casuarina pusilla* sand heath. It had a diverse collection of xeromorphic shrubs, with an average height of 1 to 1.5 m.

Victorian Alps

The two plots in the Victorian Alps were situated on the Dargo High Plains, a plateau 1620 m above sea level. Mean annual rainfall was estimated at 1600 mm, falling mainly in winter and spring (L.C.C. 1977). Information on the physical environment and vegetation of the region is available in Rowe (1967) and Land Conservation Council report (1977). The area was grazed by cattle during summer.

One plot was Snow Gum, *E. pauciflora,* tall woodland (SG-S) with a diverse mixture of shrubs, ferns and herbs to 0.5 m high including Spiny-headed Mat Rush, *Lomandra longifolia*, Alpine Hovea, *Hovea longifolia,* and Mother Shield-fern, *Polystichum proliferum*. There were occasional taller shrubs to 1 m including Dogwood, *Cassinia aculeata*, Dusty Daisy Bush, *Olearia phlogopappa*, Rough Coprosma, *Coprosma hirtella,* and Mountain Pepper, *Drimys xerophila*. The trees had many dead branches and fallen trees were common.

The second plot was also a Snow Gum tall woodland (SG-G) but differed from the preceding plot in that the trees had a lower mean diameter, the shrub stratum was sparser and there was a herbaceous ground cover of Snow Grass, *Poa australis*, Bidgee Widgee, *Acaena anserinifolia,* and Pennywort, *Hydrocotyle* sp.

Central Queensland

The two plots in central Queensland were a Gidgee, *Acacia cambagei,* medium woodland 20 km northwest of Tambo and a tall open woodland dominated by Poplar Box, *E. populnea,* 6 km north of Tambo. Annual average rainfall is 529 mm which falls mainly in summer and autumn.

The Poplar Box plot (PB) had a few other tree species including Broad-leaved Ironbark, *E. melanophloia,* and White Cypress Pine, *Callitris collumellaris*. Scattered shrubs and small trees included Wilga, *Geigera parviflora, Eremophila* sp. and *Cassia* spp. A medium dense groundcover was dominated by grasses. The Gidgee woodland (Gid) had a few emergent Bottle Trees, *Brachychiton rupestre,* with other vine thicket species including Leopardwood, *Flindersia maculosa*. There was a high proportion of standing dead stems and logs. Some trees had been removed for poles. There was a sparse cover of succulent shrubs including Climbing Saltbush, *Rhagodia nutans,* and Ruby Saltbush, *Enchylaena tomentosa,* and much bare or leaf strewn ground.

METHODS

Each census plot was a square of ten hectares. Plant species were listed from throughout the plot. Estimates of the abundance of foraging substrates (1. canopy foliage, 2. bark, 3. air space within and between tree canopies, and 4. litter) on each plot were determined from vertical profiles drawn to scale, made above random line transects, combined with information on tree density and diameters collected by measuring the circumference of all trees on an area equal to one-fifth of the plot. On the plots dominated by Snow Gums and mallee eucalypts where the number of stems was exceptionally high, a smaller area was sampled. Foliage cover was determined using photographs to determine crown type (Walker and Tunstall 1981). Canopy height and crown depth were measured with an Abney level. The area of canopy intersected was expressed as a proportion of the total area between canopy and ground. This was used as an estimate of the volume of space occupied by tree canopies.

Annual litter fall for each plot was estimated from data in Walker (1979) and Briggs and Maher (1983). As there had been no recent fires or logging activities, litter fall was assumed to be in equilibrium with litter decomposition. Annual litter fall was used as a measure of the amount of organic matter available for detritivores and therefore as an index of the amount of substrate available to ground foragers. The Red Gum plot was excluded from the analysis as the decomposition rates of litter in this periodically inundated area were erratic.

An index for estimating the area of bark was calculated, based on the geometry of the branches and boles of trees. Details can be obtained from the author.

Aerial foragers require foliage as a source of invertebrate production and they use the space within and between individual tree canopies for foraging. Total canopy volume should therefore reflect within tree foraging substrate plus a 'sphere of influence' around each tree .The cube root of the average canopy volume per tree squared was used as an index of surface area of tree crown and adjacent 'air space' available as foraging substrate for this guild.

Each plot was censused by three observers once a day for ten successive days. Counts alternated between morning (two to four hours after sunrise)

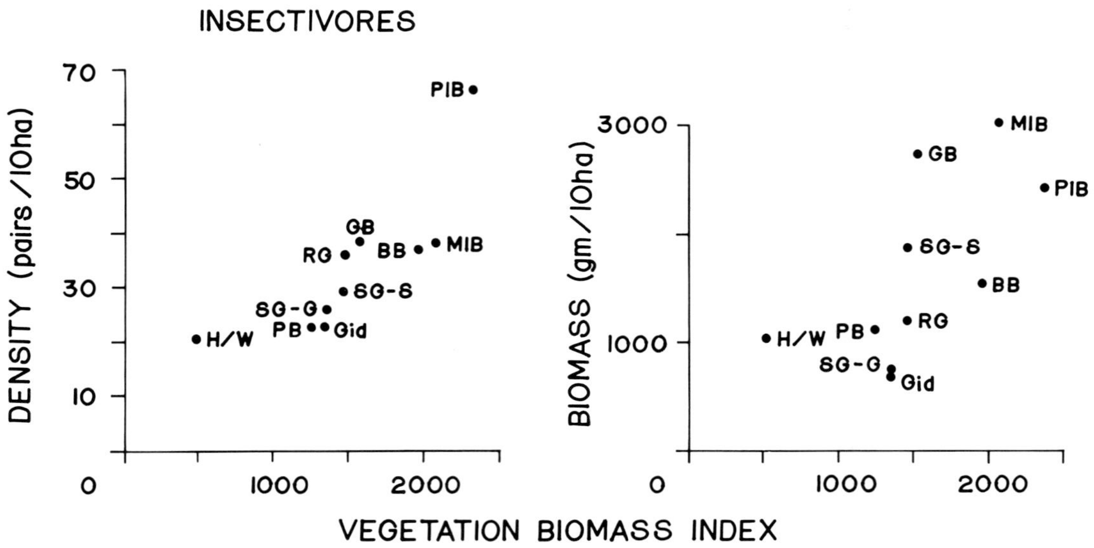

Figs 1a and 1b. Community attributes of insectivorous birds as a function of vegetation biomass index (a) density of insectivores on each plot ($r = 0.83$, $p < 0.01$) (b) total biomass of insectivores on each plot ($r = 0.67$, $p < 0.05$).

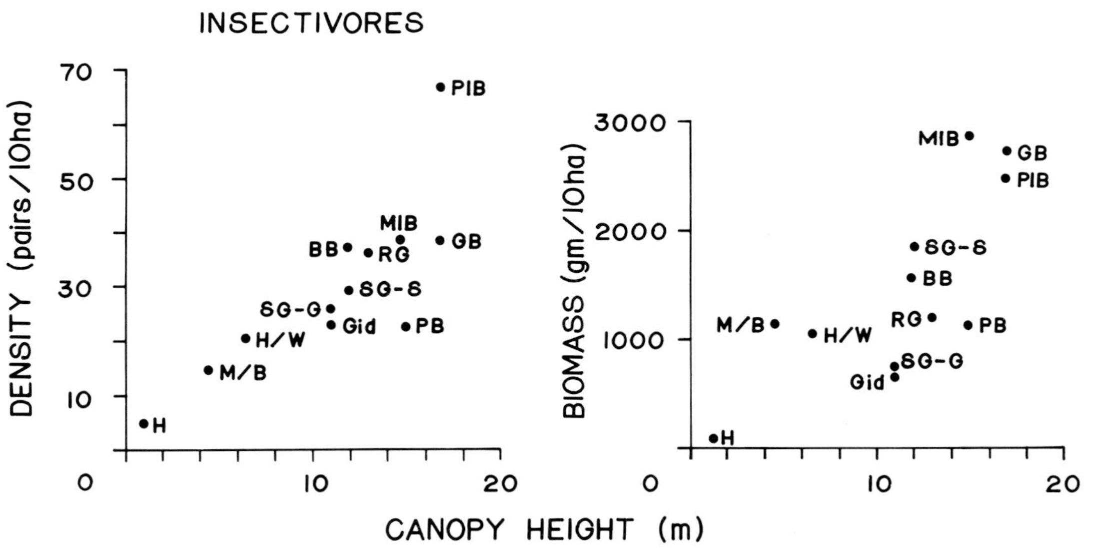

Figs 1c and 1d. Community attributes of insectivorous birds as a function of canopy height (c) density of insectivores on each plot ($r = 0.81$, $p < 0.01$) (d) total biomass of insectivores on each plot ($r = 0.75$, $p < 0.01$).

and evening (two and five hours before sunset). Observers walked approximately 20 m apart and the central observer plotted the position of all birds seen or heard. At the end of each row the observers moved across and returned along the adjacent parallel strip. Counts began at different corners each day so that over the course of the ten days each plot was traversed from four different directions. Breeding bird territories were plotted using the procedures of the International Bird Census Committee (1969) and the number of territories used to estimate the abundance of species.

The central Victorian plots were censused in October and November 1976. The plots in western Victoria were censused in October and November 1978. The plots in the Victorian Alps were censused during December 1977 and the central Queensland plots during August 1978. In addition, following the breeding census, the central Victorian plots were censused at approximately 35 day intervals throughout the year.

Densities and biomass of each species of bird on each plot were divided between foraging substrates

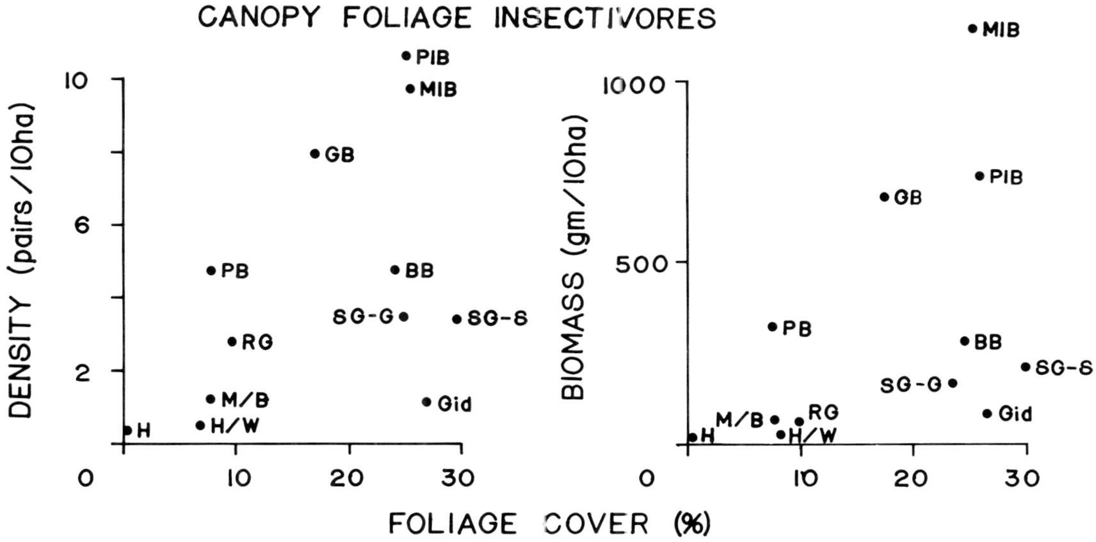

Figs 2a and 2b. Attributes of the canopy foliage feeding guild as a function of foraging substrate (a) density of canopy foliage insectivores ($r = 0.65$, $p < 0.05$) (b) biomass of canopy foliage insectivores ($r = 0.54$, $p < 0.10$).

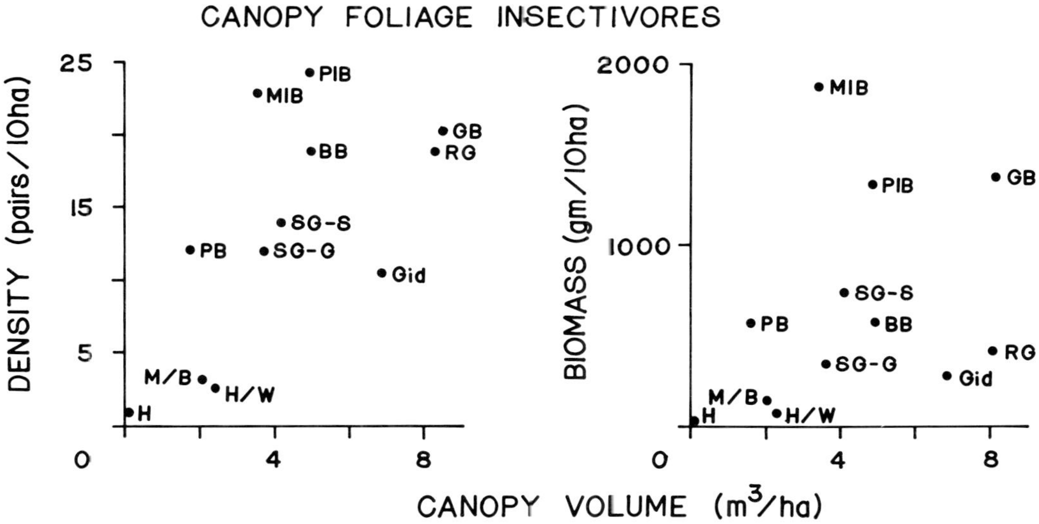

Figs 2c and 2d. Attributes of canopy foliage feeding guild as a function of an alternative index of their foraging substrate (c) density of canopy foliage insectivores ($r = 0.65$, $p < 0.05$) (d) biomass of canopy foliage insectivores ($r = 0.37$, n.s.).

in the proportion that each species utilized that substrate as estimated from the literature (Wykes 1983; Ford and Paton 1976; Abbott 1975; Keast 1970; Hunt 1979; Simpson 1973; and J. Woinarski pers. comm.) Bird species were grouped into guilds according to the substrates they used when foraging. Comparisons of similarity in bird species composition between pairs of plots were made using presence or absence of species and the Jaccard coefficient

$$\frac{a}{a + b + c}$$

where a is the number of species shared by both sites, b and c are the number of species restricted to each of the pair of sites (Clifford and Stephenson 1975).

RESULTS

The densities of the most abundant insectivorous birds on each plot are given in Table 2. Thirty-six resident species with less than one pair per plot have been excluded, as have all species recorded on fewer than four visits or where species were obviously not resident. A complete list of the birds on each plot is available from the author.

Table. 2. Densities of birds on census plots.

Scientific name	Census Plot Common name	Red gum	Grey box	Mixed age ironbark	Pole stand ironbark	Black box	Heathy woodland	Heathland	Mallee broom-bush	Snow gum shrubby	Snow gum grassy	Gidgee	Poplar box	Average weight	Insectivore principal foraging substrates
Cuculus pallidus	Pallid Cuckoo			<1	1							+	+	85	GFB30-40-30
Cuculus pyrrhophanus	Fan-tailed Cuckoo		+	+	1	<1	+			<1	+			42	FGB30-40-30
Chrysococcyx lucidus	Shining Bronze-Cuckoo			+	1	+								21	F100
Halcyon sancta	Sacred Kingfisher	1	<1	<1										43	GV50-50
Cecropis nigricans	Tree Martin	14	+											15.7	A100
Coracina novaehollandiae	Black-faced Cuckoo-shrike	+	+	1.5	1.5	<1							+	140	F100
Drymodes brunneopygia	Southern Scrub-robin								3					35.1	G100
Petroica phoenicia	Flame Robin									8	7			13.6	G100
Petroica multicolor	Scarlet Robin					3	<1							12	GF66-33
Petroica goodenovii	Red-capped Robin			<1	1.5							2	+	10	GA70-30
Melanodryas cucullata	Hooded Robin							1						21.3	GA70-30
Eopsaltria australis	Eastern Yellow Robin		3	3										20.1	GB25-75
Microeca leucophaea	Jacky Winter					<1						+	2	11.9	AG50-50
Falcunculus frontatus	Crested Shrike-tit	+	2	<1										27	B100
Pachycephala inornata	Gilbert's Whistler		1.5	+					2					29.5	FG50-50
Pachycephala pectoralis	Golden Whistler				+	1	<1		2					26	FB20-80
Pachycephala rufiventris	Rufous Whistler		+	2	1.5	1.5	+	<1			1	3	1.5	26	FB 20-80
Colluricincla harmonica	Grey Shrike-thrush	+	1.5	1	+	1.5	<1		+	+	<1			70	GB10-90
Rhipidura fuliginosa	Grey Fantail			1.5	1.5	1	+			0.5	1	0.5	<1	9.5	AFB50-20-30
Rhipidura leucophrys	Willie Wagtail	+	+			2							+	21	GA50-50
Pomatostomus superciliosus	White-browed Babbler		<8inds				<7inds		5inds					41	GB70-30
Malurus cyaneus	Superb Fairy-wren			<1	6	2	2	<1					<4inds	10	GBF50-10-40
Malurus lamberti	Variegated Fairy-wren						1	10inds	1					8	GBF40-10-50
Sericornis frontalis	White-browed Scrubwren									1.5	1			13.8	FGB50-30-20
Sericornis sagittatus	Speckled Warbler				1							1		13.5	G100
Smicrornis brevirostris	Weebill				5	6	<1						5	6	ABF45-35-20
Gerygone fusca	Western Gerygone			<1	+							+	1	7	AB70-30
Acanthiza pusilla	Brown Thornbill			<1	6.5	1	3	2	<1	2	2			7.5	FB20-80
Acanthiza uropygialis	Chestnut-rumped Thornbill											1	2	6.3	GBF10-80-10
Acanthiza reguloides	Buff-rumped Thornbill			3.5	14	3	4							7.7	BG70-30
Acanthiza iredalei	Slender-billed Thornbill							7						7.5	FG50-50
Acanthiza chrysorrhoa	Yellow-rumped Thornbill											3	0.5	8.5	GBF40-30-30
Acanthiza nana	Yellow Thornbill				+							7		5.5	FBA10-70-20
Acanthiza lineata	Striated Thornbill			+	2	5	+			2	1			7.5	BFA50-30-20
Daphoenositta chrysoptera	Varied Sittella			+	<7inds		+					+	<3inds	14	B100
Climacteris leucophaea	White-throated Treecreeper			1	2	2	<1			0.5	<1			23	B B70-30 (low)(high)
Climacteris affinis	White-browed Treecreeper											2	1	20	B100
Climacteris picumnus	Brown Treecreeper	2.5	1.5	1		<1							0.5	34	BG50-50
Anthochaera carunculata	Red Wattlebird	+	2	2	1.5									130	FB40-60
Philemon corniculatus	Noisy Friarbird	+	+									+	1	92	F100
Philemon citreogularis	Little Friarbird	+	+									+	1	92	F100
Lichenostomus chrysops	Yellow-faced Honeyeater									3	3			17.4	FBA50-30-20
Lichenostomus virescens	Singing Honeyeater											2		22.7	AFGB30-20-20-10
Lichenostomus leucotis	White-eared Honeyeater				4	1	+		2	1	1			19	BF70-30
Lichenostomus melanops	Yellow-tufted Honeyeater		4	2										21	FBN33-7-50
Lichenostomus fuscus	Fuscous Honeyeater		14	12										18	FBGAN30-10-5-8-37
Lichenostomus penicillatus	White-plumed Honeyeater	4											+	19	FBA33-33-33
Melithreptus brevirostris	Brown-headed Honeyeater			+	3	+	+		+	1	1		<1	15	BF70-30

Table 2. Continued.

Scientific name	Census Plot Common name	Red gum	Grey box	Mixed age ironbark	Pole stand ironbark	Black box	Heathy woodland	Heathland	Mallee broom-bush	Snow gum shrubby	Snow gum grassy	Gidgee	Poplar box	Average weight	Insectivore principal foraging substrates
Lichmera indistincta	Brown Honeyeater											3		18?	A100
Phylidonyris novaehollandiae	New Holland Honeyeater					+	3	1	+					22	A100
Phylidonyris melanops	Tawny-crowned Honeyeater							2						18	GA70-30
Dicaeum hirundinaceum	Mistletoe bird				3							4		9.5	FRUIT
Pardalotus punctatus	Spotted Pardalote			+	1	+								9	BFA73-15-12
Pardalotus striatus	Striated Pardalote	15		1.5		4	<1			8	6		1.5	12	BFA82-7-6
Zosterops lateralis	Silvereye			+	10	+	<1		<1	+	1	+		11	B F F20-33-42 (high)(low)
Oriolus sagittatus	Olive-backed Oriole	+	1.5	1.5	+									97	F100
Artamus superciliosus	White-browed Woodswallow		+	+				11nests						41	AG90-10
Artamus cyanopterus	Dusky Woodswallow		4	<1				2nests						37	AG90-10
Total species consistently present		25	35	46	33	29	24	15	17	17	18	32	36		
Total number of breeding species		11	19	28	22	23	16	12	11	16	14	13	24		
Total number of breeding pairs		71	44.5	47	72	40.5	20.5	20(+13)	14.5	32	28	29.5	24.5		
Number of breeding insectivore species		5	15	23	20	21	18	12	11	14	13	11	21		
Number of insectivore breeding pairs		36.5	39	38.5	67	37.5	20.5	5(+3)	15	29.5	26	25.5	23		
Insectivore Biomass (gm)		1208	2736	2886	2427	1573	1049	29.5	1148	1868+	765	677	1128		

*F — foliage; B — bark; A — air; G — ground; V — vertebrates; N — nectar; C — carrion.

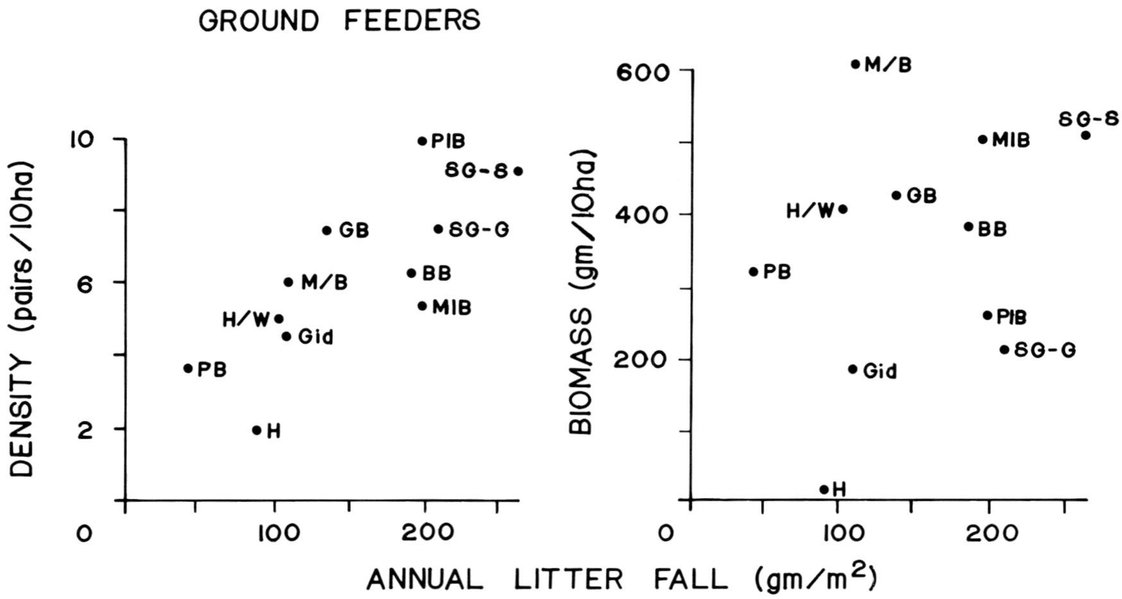

Figs 3a and 3b. Attributes of the ground feeding guild as a function of foraging substrate (a) density of ground feeders ($r = 0.78$, $p < 0.01$) (b) biomass of ground feeders ($r = 0.30$, n.s.).

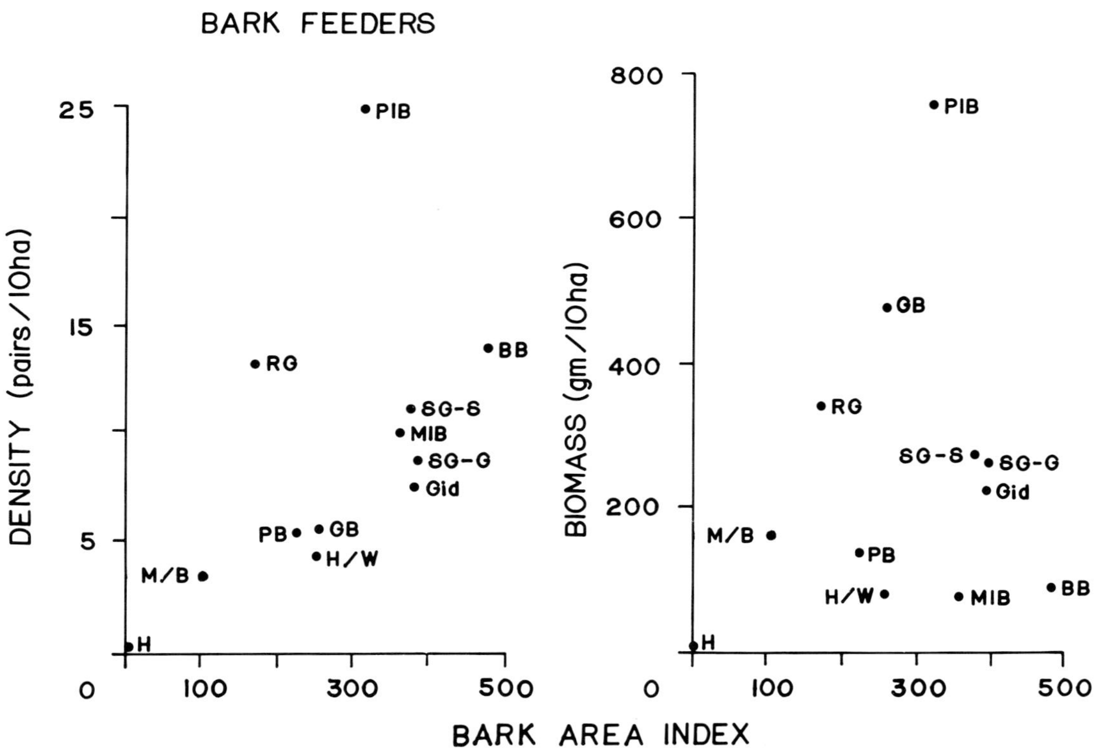

Figs 3c and 3d. Attributes of the bark feeding guild as a function of foraging substrate (c) density of bark feeders ($r = 0.77$, $p < 0.01$) (d) biomass of bark feeders ($r = 0.56$, $p < 0.10$).

Various vegetational parameters were used as independent variables to investigate the relationship between vegetative structure and insectivore density. An index of plant biomass, determined by multiplying the canopy height by basal area, gave the best predictions (Fig. 1a, $r = 0.83$, $p < 0.01$). Canopy height itself was almost as good as biomass at predicting total insectivore densities (Fig. 1c, $r = 0.81$, $p < 0.01$). It has the advantage of being easy to measure. The regression equation relating insectivore density to canopy height is $y = 2.6x + 0.77$ (where y = No. of pairs/10 ha and x = canopy height in metres). The results of Loyn *et al.* (1980) were used to test the generality of the regression. Predictions from this equation with respect to insectivore densities tended to be underestimated in taller forests.

Estimates of the abundance of foraging substrates on each plot were used as independent variables to investigate the relationship between the amount of substrate available and the abundance of birds in each foraging guild. The abundance of foliage feeders in the canopy was not significantly correlated to canopy cover ($r = 0.49$, n.s.). Taking account of the gaps between the branches and leaves to give a measure of foliage cover improved the relationship (Fig. 2a, $r = 0.65$, $p < 0.05$). The measure of the volume of space bounded within tree crowns (Fig. 2c) had similar predictive power. When the abundance of ground feeders on each plot was compared with annual litter fall, a significant correlation was found (Fig. 3a, $r = 0.78$, $p < 0.01$). A significant correlation was also found between the abundance of bark foragers and an index of bark area (Fig. 3c, $r = 0.77$, $p < 0.01$).

For insectivorous birds and all guilds except "canopy air" foragers, densities were better predicted by foraging substrate quantity than was consumer biomass (Figs 1a-d, 2a-d, 3a-d, 4a, b).

There was no apparent relationship between similarity in composition of plant species and insectivorous bird species for pairs of plots in central and western Victoria. However, when pair wise comparisons are made between all eight plots and plant genera are used for determining floristic similarity a significant correlation is found ($r = 0.48$, $p < 0.01$) (Table 3).

Some plots shared many of the same bird species, although they did not share any plant species. For example most insectivorous species in the Red Gum woodland were also present in the Poplar Box woodland. The Gidgee woodland shared almost 50% of its resident insectivore species with the pole stand Red Ironbark woodland. Allowing for replacement by congeneric species within the same guild, over 90% of the ecological equivalents are present.

DISCUSSION

Many other structural parameters vary in a way that is related to increasing canopy height (O'Neill and De Angelis 1981). It is unlikely that any species of bird responds solely to the height of the canopy. However, as an index of structural complexity, canopy height appears to be a simple and reliable estimator of insectivore densities for woodlands and low forests. Predictions of insectivore densities from forests of greater height tend to be underestimated.

In all guilds, except aerial foragers, abundances were more accurately predicted by the amount of foraging substrate than was consumer biomass. These results suggest that birds apportion habitat according to the availability of foraging substrate rather than prey availability within that substrate.

Although pairs of plots often shared bird and plant species, this is mostly because both are related to plant structure. The genus *Eucalyptus* frequently occurred in the canopy, *Acacia* and *Melaleuca* in the middle understorey and *Leucopogon, Calytrix, Hibbertia, Dillwynia, Lomandra* and *Danthonia* in the shrub layer. The negative correlation between the number of resident bird species and the number of plant species may be related to the difference between the diverse sclerophyll flora of nutritionally poor soils and the lower number of plant species growing in adjacent savannah communities on soils of higher nutrient status. The results suggest that most of the plant species are relevant to insectivores only through their proportional contribution to the total biomass of the vegetation; a large number of rare plant species contribute little to the habitat of insectivorous birds. Support for this interpretation is provided by Gullan and Norris (1981) who found it necessary to invoke structural criteria in explaining the presence of particular foraging groups within floristically determined vegetation types.

Specht (1981) pointed out that while foliage projective cover in regenerating vegetation tends to remain constant with age, many slender stems are replaced by fewer stems of increasing girth. These changes are borne out by the results of Anderson (1981) on leaf area index. Changes in the absolute and relative values of the various foraging substrates are presumably an important factor causing turnover in species composition of co-existing birds through time. A corollary of these and other studies of stand dynamics is that one might expect less change in the densities of foliage feeders and ground feeders than bark or aerial feeders throughout a progressive secondary succession.

Co-existing species in a community frequently interact through competition, predation or mutualism. Such interactions have been regarded as important in determining the structure of the community. Trees provide foliage, bark and 'canopy air' as foraging substrates. This enables complementary foraging guilds to co-exist and enables us to classify bird communities. Kikkawa (1982) from a study of birds in Australian wet tropical forests, postulated that . . . "most species follow independent distributions to one another . . ."

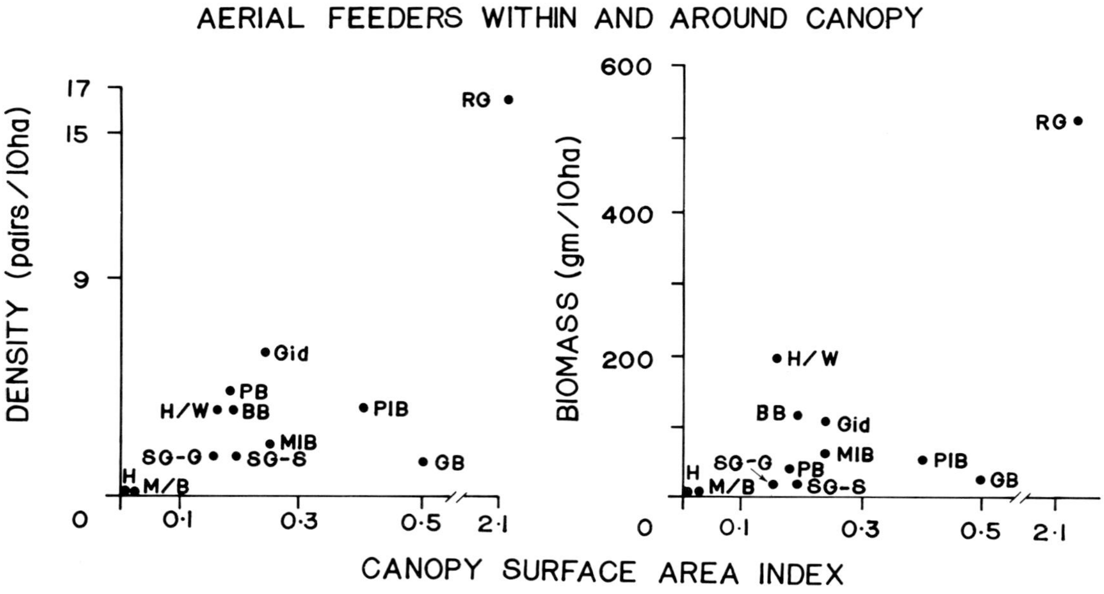

Figs 4a and 4b. Attributes of the guild of aerial insectivores which feed within and around the canopy (a) density of aerial feeders ($r = 0.90$, $p < 0.01$) (b) biomass of aerial feeders ($r = 0.91$, $p < 0.01$).

He explained patterns of co-existence as reflecting a ". . . common response to similar features of the habitat." The results from woodlands reported in this study support his hypothesis. The replacement of species by congeners which used similar foraging substrates in different plots indicates that competitive interactions may be important in restricting members of a foraging guild. However, bird communities are chiefly artifacts of the architecture of the vegetation.

Table 3. Insectivore similarity and floristic similarity between pairs of the eight Central and Western Victorian plots.

		Insectivore Similarity							
		BB	H/W	H	M/B	RG	GB	MIB	PIB
Floristic Similarity	BB		.53	.11	.24	.13	.15	.33	.46
	H/W	.13		.15	.38	.05	.13	.29	.41
	H	.04	.23		.12	.03	.05	.15	.10
	M/B	.12	.13	.07		.03	.10	.10	.15
	RG	0	0	0	0		.38	.26	.19
	GB	.10	.04	.05	.05	.06		.35	.16
	MIB	.08	.10	.09	.10	.04	.21		.48
	PIB	.08	.06	.06	.12	.05	.21	.58	

ACKNOWLEDGEMENTS

Bill Bren, Keith Cherry, Steve Craig, Bill Emison, Ann Gedye, David Hespe, Greg Horrocks, Jane Marcius, Rosy Jordon, Alicia McShane, Judy McVicar. Peter Menkhorst, Ken Norris and Robin Wilson helped with the census and vegetation measurements in Victoria. Cecily Allen, Helen Aston, Peter Disher, Ann Forbes, Kath Fritz, Bill Horton, Kath Hough, Judy McVicar, Dorothy Moroney, Chris Morris, Penny Paton, Pauline Reilly, Terry Robinson, Jill Rossiter, Ann Sheppard, Ian Stringer and Elizabeth Turnbull helped with the census and vegetation measurements in Queensland. John Woinarski provided unpublished foraging data. Paul Gullan, Andrew McMahon and Neville Walsh kindly assisted with plant identification.

REFERENCES

Abbott, I., 1975. Density and species diversity of bird populations in *Eucalyptus* forests in Victoria, Bass Strait Islands and Tasmania. *Proc. R. Soc. Vict.* **87**: 187-96.

Anderson, M. C., 1981. The geometry of leaf distribution in some southeastern Australian forests. *Agric. Meteor.* **25**: 195-205.

Blackburn, G., Bond, R. D. and Clarke, A. R. P., 1967. Soil development in relation to stranded beach ridges of county Lowan, Victoria. *CSIRO Soil Publ.* No. 24.

Briggs, S. V. and Maher, M. T., 1983. Litter fall and leaf decomposition in a River Red Gum, *Eucalyptus camaldulensis,* swamp. *Aust. J. Bot.* **31**: 307-16.

Brown, J. L. and Balda, R. P., 1977. The relationship of habitat quality to group size in Hall's Babblers *(Pomatostomus halli). Condor* **79**: 312-20.

Cheal, P. D., Day, J. C. and Meredith, C. W., 1979. Fire in National Parks of north west Victoria. Nat. Parks Service, Victoria.

Clifford, H. T. and Stephenson, W., 1975. An Introduction to Numerical Classification. Academic Press, London.

Ford, H. A. and Bell, H., 1981. Density of birds in eucalypt woodland affected to varying degrees by dieback. *Emu* **81**: 202-8.

Ford, H. A. and Paton, D. C., 1976. Resource partitioning and competition in honeyeaters of the genus Meliphaga. *Aust. J. Ecol.* **1**: 281-87.

Gepp, B., 1976. Bird species distribution and habitat diversity in an exotic forest in South Australia. *Aust. For.* **39**: 269-87.

Gillison, A. N. and Walker, J., 1981. Woodlands *in* Australian Vegetation ed. by R. H. Groves. Cambridge University Press.

GULLAN, P. K. AND NORRIS, K. C., 1981. Floristic classifications, small mammals and birds. Pp. 181-191 *in* Vegetation Classification in Australia ed. by A. N. Gillison, and D. J. Anderson. CSIRO in association with ANU Press, Canberra.

HUNT, T., 1979. Habitat preferences of birds in the Big Desert. *VORG Notes* **15**(2): 43-53.

INTERNATIONAL BIRD CENSUS COMMITTEE, 1969. Recommendations for an international standard for a mapping method in bird census work. *Bird Study* **16**: 248-55.

KEAST, J. A., 1970. Adaptive evolution and shifts in niche occupation in island birds. *Biotropica* **2**: 61-75.

KIKKAWA, J., 1982. Ecological association of birds and vegetation structure in wet tropical forests in Australia. *Aust. J. Ecol.* **7**: 325-45.

KING, B. R., 1980. Social organisation and behaviour of the Grey-Crowned Babbler, *Pomatostomus temporalis*. *Emu* **80**: 59-76.

LAND CONSERVATION COUNCIL, 1977. Report on the Alpine Study Area, Melbourne.

LOYN, R. H., MACFARLANE, M. A., CHESTERFIELD, E. A. AND HARRIS, J. A., 1980. Forest utilisation and the flora and fauna in Boola Boola State Forest in southeastern Victoria. *Bull. No. 28.* Forests Commission Victoria, Melbourne.

MCMAHON, A., 1977. The effect of fire on heath floristics in the Little Desert National Park, Victoria. B.Sc. (Hons.) Thesis, Latrobe University, Bundoora, Victoria.

NEWMAN, L. A., 1961. The Box-Ironbark forests of Victoria, Australia. *Bull. For. Comm. of Victoria.* 14.

NIAS, R., 1984. Territory quality and group size in the Superb Fairy-wren. *Emu* **84**: 178-81.

O'NEILL, R. V. AND DE ANGELIS, D. L., 1981. Comparative productivity and biomass relations of forest ecosystems. Pp 411-450 *in* Dynamic Properties of Forest Ecosystems ed. by D. E. Reichle.

PATTEMORE, V. AND KIKKAWA, J., 1974. Comparison of Bird Populations in logged and unlogged Rainforest Wiangarie State Forest, NSW. *Aust. For.* **37**: 188-98.

RECHER, H. F., ROHAN-JONES, W. AND SMITH, P., 1980. Effects of the Eden Woodchip Industry on terrestrial vertebrates with recommendations for management. *Forestry Commission of NSW Research Note* No. 42: 1-83.

ROWE R. K., 1967. A study of the Land in the Victorian Catchment of Lake Hume. Soil Conservation Authority, Victoria.

SIMPSON, K. N. G., 1973. Birds of the Murray River Region between Mildura and Renmark, Australia — breeding, behavioural and feeding records. *Mem. Nat. Mus. Vict.* **3**: 253-74.

SPECHT, R. L., 1966. The growth and distribution of mallee-broombush (*Eucalyptus incrassata-Melaleuca uncinata* association) and heath vegetation near Dark Island Soak, Ninety-Mile Plain, South Australia. *Aust. J. Bot.* **14**: 361-71.

SPECHT, R. L., 1981. Foliage projective cover and standing biomass. *In* Vegetation Classification in Australia ed. by A. N. Gillison and D. J. Anderson. CSIRO in association with ANU Press, Canberra.

SPECHT, R. L., RAYSON, P. AND JACKMAN, M., 1958. Dark Island Heath VI Pyric succession changes in the composition, coverage, dry weight and mineral nutrient status. *Aust. J. Bot.* **6**: 58-88.

WALKER, J., 1979. Aspects of fuel dynamics in Australia. *CSIRO Divn. of Land Use Res. Tech. Memo* 79/7.

WALKER, J. AND TUNSTALL, B. R., 1981. Field estimation of foliage cover in Australian woody vegetation. *CSIRO Divn. Land Use Res. Tech. Memo* 81/19.

WOINARSKI, J. C. Z., 1978. The influence of vegetation heterogeneity on bird distribution. B.Sc. Hons. Thesis, Monash University.

WYKES B., 1983. Resource partitioning and the role of competition in structuring *Lichenostomus* honeyeater communities in southern Victoria. Ph.D. Thesis, Monash University.

Bell Miner, *Manorina Melanophrys*

CHAPTER 4

Ecology, Distribution and Density of Birds in Victorian Forests

Richard H. Loyn[1]

Bird communities and populations were studied in a range of Victorian forests, from the coast across the mountains to the flood-plain of the Murray River. They are discussed in relation to the foraging ecology of common species with comments on management and the roles of birds in controlling insects.

The main divisions in bird communities are between those of wet gullies or mountain forests, those of drier foothill forests and those of the drier forest types on the north side of the Great Dividing Range. Important features of the environment which affect the bird community include shrub structure, successional stage of the forest, presence of nectar-rich flowering shrubs, presence of arborescent wattles, abundance of tree hollows, presence of smooth-barked or rough-barked eucalypts, infestation by canopy insects and whether eucalypts are producing flows of nectar or other carbohydrate exudates. In foothills bird populations are higher in gullies than on ridges, and generally high where shrub layers are well developed or where there are abundant food sources in the canopy.

Insectivorous birds play active roles in determining the health of the forest. This has been illustrated by experiments where we translocated Bell Miners, *Manorina melanophrys,* from unhealthy forest infested with psyllids, *Glycaspis* spp. Other common birds moved in and eradicated the infestations. Various implications of this are discussed.

INTRODUCTION

THIS paper describes the bird communities of ten major vegetation types in the forested parts of Victoria, and discusses ways in which populations of birds with different requirements vary according to features of vegetation and management history.

The information comes from several years of fieldwork with what is now the Ecological Studies Section of the State Forests and Lands Service, Victoria. This research is part of a series of studies to determine effects of forest operations on the ecology of the forests and their wildlife and to provide information to assist in managing forests. The most detailed of these studies have been in the mountains and foothills of eastern Victoria including Gippsland, where the aim was to assess the impact of harvesting eucalypts for sawlogs and pulpwood, and to understand the factors determining distribution of species so that effective measures could be developed for their conservation in productive forest. In many parts of Victoria surveys were made to assist in local management decisions, and other studies investigated effects of operations such as grazing, flood regulation, burning and firewood cutting as well as the effects of some wildlife species on the habitats they occupy.

This paper presents an overall view, without going into detail on individual studies. Finally some work is described which shows the ability of common forest birds to control insects and determine tree health. A hypothesis is presented to relate the distribution of canopy insectivores to productivity of wood or resources eaten by birds in different forest types.

METHODS

To give a broad picture of the variation in Victorian forests, data are drawn from a number of studies in different parts of the State. The aims of the studies varied and hence different methods were used these are summarised in Table 1. Quantitative data tabulated refer to the spring and summer seasons which embrace the main breeding season for most species. The following descriptions are based on observational and quantitative data from other seasons or places, in addition to the tabulated

[1]Forests Commission Victoria, (now Department of Conservation, Forests and Lands), Mountain Research Station, Sherbrooke, Victoria 3789.

Pages 33-46 *in* BIRDS OF EUCALYPT FORESTS AND WOODLANDS: ECOLOGY, CONSERVATION, MANAGEMENT ed. by A. Keast, H. F. Recher, H. Ford and D. Saunders. Royal Australasian Ornithologists Union and Surrey Beatty & Sons.

Table 1. Methods used to estimate bird populations.

Vegetation type	Methods used	Place	Region	Date	Reference
Coastal Forests	5 Area Searches (3 ha, 20 mins) in heathy forest with gully	East Wingan	East Gippsland	Jan 83	unpublished
Rainforest Gullies	Area Search (intensively, of one 6 ha patch)	Reedy Creek	East Gippsland	Dec 76	Loyn and Macfarlane (in prep.)
Foothill Gullies (Mountain Grey Gum/ Messmate)	Territory Mapping (3 areas)	Boola Boola	Central Gippsland	75/76	Loyn (1980)
Foothill Gullies (Manna Gum/Peppermint)	25 Area Searches (3 ha, 20 mins)	Cooaggalah	East Gippsland	Jan 83	unpublished
Foothill Ridges (regrowth)	Territory Mapping (7 areas aged 6-40 years), silvertop/ stringybark	Boola Boola	Central Gippsland	75/76	Loyn (1980)
Foothill Ridges (mature)	Territory Mapping (3 areas), silvertop/stringybark	Boola Boola and Black Range	Central Gippsland	75/76	Loyn (1980)
Mountain Ash	Territory Mapping (3 areas aged 39-70 years)	Toolangi	Central Highlands	77/78	Loyn (in press)
Alpine Ash	27 Area Searches (3 ha, 20 mins)	East Kiewa	N.E. Vic.	Dec 82	Loyn and Woinarski (unpublished)
Snow Gum	20 Area Searches (3 ha, 20 mins)	Mt Baw Baw and Mt Anderson	Central Gippsland	Dec 82 Oct 83	unpublished
Alpine Herbfield	General observations only	Various	Eastern Victoria	Various	unpublished
Red Ironbark	30 Area Searches (3 ha, 20 mins)	Rushworth	N-Central Vic.	Nov 81	unpublished
Grey Box/Red Box	19 Simple Transects (20 mins)	Reef Hills	Benalla, N.E. Vic.	79/80	unpublished
River Red Gum	10 Area Searches (4 ha, 1 hour)	Barmah	mid Murray	Oct/Nov 77 & 78	unpublished

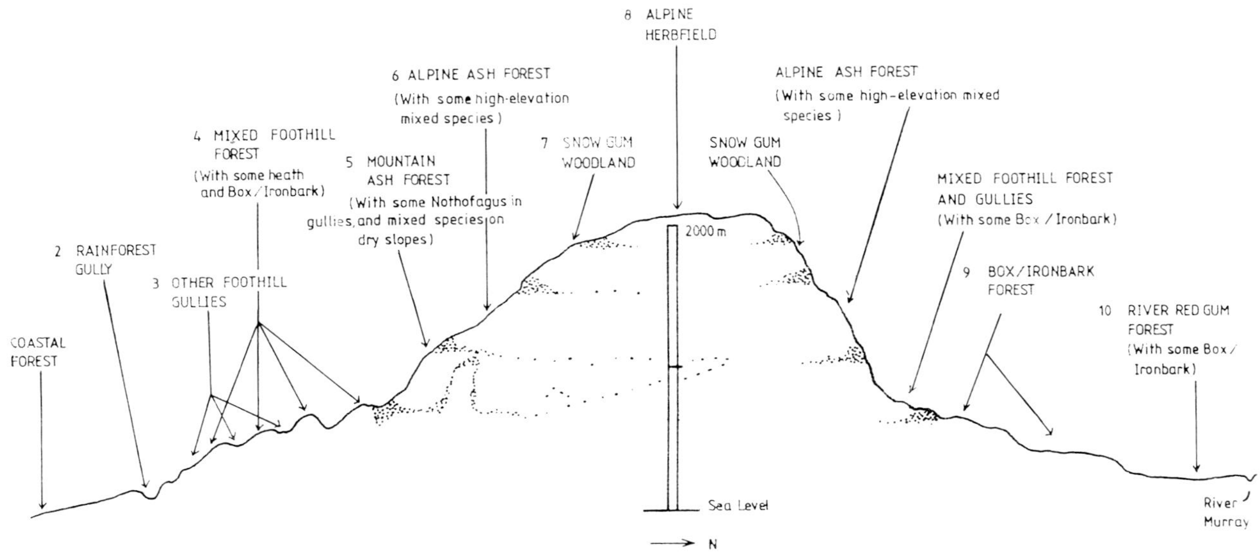

Fig. 1. Simplified cross-section of eastern Victoria showing main vegetation communities.

studies. Data on food and feeding ecology are based on direct observations.

VEGETATION COMMUNITIES

The ten communities discussed embrace a large proportion of the eucalypt forests or woodland in Victoria except mallee (a distinctive community of multi-trunked eucalypts which grows on sandy soils in the arid northwest). These communities can be arranged on a transect northwards from the coast across the Great Dividing Range to the floodplain of the River Murray (Fig. 1). They occupy environments with widely differing characters such as rainfall and drainage, and vary greatly in structure, height and understorey vegetation (Tables 2 and 3). The tallest, straightest trees grow in the mountain forests of Mountain Ash, *Eucalyptus regnans,* and Alpine Ash, *E. delegatensis*, and the former can be the tallest flowering plants in the world. The wettest of these forests are the rainforest gullies and the driest are the Box/Ironbark forests on the rainshadow of the

Table 2. Floristic characteristics of vegetation communities[1].

Vegetation type	Dominant tree species	Dominant understorey or middle storey plants
Coastal Forests	Various stringybarks (e.g. *E. globoidea*), silvertop *E. sieberi*, manna gum *E. viminalis*, southern mahogany *E. botryoides* and other species.	heathy shrubs, with many nectar-rich plants such as *Banksia*.
Rainforest Gullies	Lilly-pilly *Eugenia smithii*, with some emergent gums e.g. mountain grey gum *E. cypellocarpa*.	gully ferns and shrubs, vines and epiphytes; many fruit-bearing species.
Foothill Gullies (a)	*E. cypellocarpa*; messmate *E. obliqua*; some narrow-leafed peppermint *E. radiata* (latter mainly on slopes or broad flat gullies).	gully ferns and tall shrubs, and wattle trees *Acacia*.
Foothill Gullies (b)	*E. viminalis* and *E. radiata*.	gully ferns and tall shrubs, and wattle trees.
Foothill Ridges	*E. sieberi*, *E. obliqua*, *E. globoidea* and other stringybarks.	sclerophyllous shrubs, sometimes heathy.
Mountain Ash	*E. regnans* (with some pure or mixed stands of shining gum (*E. nitens*) and rainforests of myrtle beech (*Nothofogus cunninghamii*)).	gully ferns and tall shrubs, and wattle trees.
Alpine Ash	*E. delegatensis* (with some others as above)	gully ferns and tall shrubs, and wattle trees.
Snow Gum	*E. pauciflora*	sclerophyllous shrubs or grasses.
Alpine Herbfield	none	alpine plants
Red Ironbark	*E. sideroxylon*, often in mixture with box species	grasses and sclerophyllous shrubs
Grey Box/Red Box	*E. microcarpa*, *E. polyanthemos* and other box and stringybark species	grasses and sclerophyllous shrubs
River Red Gum	*E. camaldulensis*	grasses and sedges.

1 See Frankenberg (1971) for a more detailed classification.

Divide, above the influence of floodwaters from the River Murray. A dense middle storey of tall shrubs and arborescent wattles (e.g. Silver Wattle and Blackwood, *A. melanoxylon*) is usually found in wet gullies and Mountain Ash forests. The understorey of foothill forests is shorter and usually dense except on the driest slopes, or in the first year after a fire. The most widespread and varied of these forest types are the mixed eucalypt foothill forests, and these are the forests which have been studied extensively in Victoria and New South Wales.

BIRD POPULATIONS

Estimates of absolute bird density in foothill forests and Mountain Ash forests were obtained by territory mapping (Table 4). Intensive area searches at Reedy Creek in East Gippsland gave estimates of density which were probably lower than true values; they are also shown in Table 4. The value for territories in mature forest on foothill ridges was inflated by one area with unusually high bird populations; evidence from other counts suggests that such forests generally support lower bird populations than the wetter Mountain Ash forests or gullies. The rainforest gully at Reedy Creek supported higher bird populations than mature forest on ridges, and mature forest supported higher populations than regrowth. Counts in mature Box/Ironbark and River Red Gum, *E. camaldulensis,* forests showed that total bird populations were slightly higher than in foothill forests but lower than in foothill gullies. In the following section relative populations of birds and species composition are emphasised as data are more easily compared from different study methods.

BIRD COMMUNITIES

Relative abundances of 56 common forest birds in different forest types are given in Table 5, as percentages of total individual birds observed in each community. It can be seen that some species are confined to either the wetter or the drier forest types whereas many occur with varying abundance in a wide range of forests. Some genera include species in each category, e.g. the whistlers (Olive Whistler, *Pachycephala olivacea,* in wet forest, Rufous Whistler, *P. rufiventris,* mainly in dry forest and Golden Whistler, *P. pectoralis,* in a wide range of forest types). In some cases this reflects differences in foraging behaviour, and the species co-exist where suitable habitats occur together; for example Rufous Whistlers feed in the eucalypt canopy, Golden Whistlers mainly in the tall shrub layer and Olive Whistlers in the understorey and damp ground below; all three can co-exist in broad foothill gullies. In winter drier habitats are chosen; Rufous Whistlers migrate from southern Victoria and Golden Whistlers feed more extensively in the eucalypt canopy, vacating their wettest habitats and extending into drier forest.

Some other species forage in similar ways and usually occupy different forest types (e.g. Striated Thornbill, *Acanthiza lineata,* and Weebill, *Smicrornis brevirostris,* among eucalypt foliage in the canopy, the latter in drier forest). Brown and Yellow Thornbills, *A. pusilla* and *A. nana,* replace each other similarly (brown among understorey and Silver Wattles, yellow in drier forest among Black Wattles, *Acacia mearnsii,* understorey and eucalypts), though all four co-exist locally along

Table 3. Structural and environmental characters of vegetation communities.

Vegetation type	Precipitation (mm per year)	Fire history	Tree height at maturity (m)†	Usual Structure contributed: (a) by trees	(b) by tall shrubs and middle storey	(c) by understorey
Coastal Forests	600-800	2	10-40	MUF or O	0-1	1-2
Warm Temperate Rainforest Gullies	800	0	10-30	MUC	1-2	1
Foothill Gullies (Mountain Grey Gum/ Messmate)	800-1200	1	40-60	MUF	2	1-2
Foothill Gullies (Manna Gum/Peppermint)	800-1200	1	40-50	MUF	1	1-2
Foothill Ridges (regrowth)	600-1200	2	-	MEF	0-1	1-2
Foothill Ridges (mature)	600-1200	2	15-60	MUF	0-1	0-2
Mountain Ash	1200+	R	40-100	PEF	1-2	1-2
Alpine Ash	1000+	R	25-80	PEF	1	0-2
Snow Gum	1200+	1	1-20	PUW-PEF	0-1	0-2
Alpine Herbfield	1200+	0	-	0	0	0-2
Red Ironbark	400-800	2	10-30	) P or M,	0-1	0-1
Grey Box/Red Box	400-800	2	10-25	) E or U,) F or W	0-1	0-1
River Red Gum	400-800	1	12-45	PUF	0	0

† partly from Costermans (1981)

Fire History: 0 = very rare; 1 = rare understorey or crown fires; 2 = more frequent understorey fires (intervals of 2-30 years); R = rare crown fires, which are the natural means of regeneration.

Tree Structure: P = pure stand; M = mixture of species; E = even-aged stand; U = uneven-aged stand; C = closed forest; F = open forest; W = open woodland; 0 = no trees.

Tall Shrub Structure : 0 = very sparse tall shrubs; 1 = scattered tall shrubs and wattle trees *Acacia*; 2 = dense tall shrubs and wattle trees *Acacia* (closed canopy)

Understorey Structure : 0 = grassy or bare ground; 1 = scattered low shrubs or ferns; 2 = dense low shrubs or ferns

Table 4. Total bird populations in gullies and on ridges of foothill forests and in Mountain Ash forests.

Vegetation type	Territories km^{-2} (mean of 3 areas)	Birds observed km^{-2} (intensive summer search at Reedy Creek)	Birds observed km^{-2} (mean of 4 intensive May searches at Reedy Creek)
Rainforest Gullies	-	2133	2850
Foothill Gullies (Mountain Grey Gum/ Messmate)	2081	947	1450
Foothill Ridges (regrowth 6-14 years)	307	-	-
Foothill Ridges (regrowth 15-40 years)	719	-	-
Foothill Ridges (mature)	1303	691	664
Mountain Ash (39-70 years)	1177	-	-

with one or two primarily ground-feeding thornbills (Buff-rumped, *A. reguloides* or its open-country counterpart, Yellow-rumped, *A. chrysorrhoa).* The robins, *Petroica* spp., provide a complex example. One species breeds only in the driest communities (Red-capped Robin, *P. goodenovii*); two breed only in the wettest (Pink Robin, *P. rodinogaster,* feeding mainly from the ground, and Rose Robin, *P. rosea,* from the foliage of Silver Wattle) and two breed in a wide range of intermediate forest types (Scarlet

Robin, *P. multicolor* and Flame Robin, *P. phoenicea).* The last two defend territories against each other where they occur together (Loyn 1980), but the Scarlet Robin is primarily a resident foothill species whereas Flame Robins extend in summer into wetter and higher habitats in the mountains, migrating to open country in the winter and rarely remaining over winter in foothill or mountain forests.

The main foraging behaviour of each species is shown in Table 5. Fifteen groups of species can be recognised on this basis, and combined populations of all species in each group are shown in Table 6. Some of the groups are better defined than others; one well-defined group consists of birds that feed mainly on invertebrates from damp ground or understorey (e.g. Olive Whistler, Pilotbird, *Pycnoptilus floccosus* and Eastern Whipbird, *Psophodes olivaceus),* and their abundances vary in similar ways with changing habitat. Honeyeaters are difficult to classify as they feed in different ways in different forest types, and are best discussed separately for each community. Before discussing the communities separately, some general comments should be made about introduced birds, hole-nesting birds, and the importance of other habitat features including successional stage of the forest.

Introduced birds are common in suburban and agricultural habitats and on forest edges but not in extensive forests where they rarely comprise more than 0.2% of the population. Blackbirds, *Turdus merula,* are widespread in wet gullies and in Mountain Ash forests at low density (0.2% of the population in Mountain Ash, less in foothill gullies, more in gullies choked with introduced Blackberry, *Rubus fruticosus).* European goldfinches, *Carduelis carduelis,* visit recently harvested forests to feed on seeds of herbs and grasses, and occur more rarely in older forests where they feed at flowers of eucalypts or wattles. Common Starlings, *Sturnus vulgaris,* nest in hollow trees on forest edges and in some open woodland, but usually fly to permanent clearings to feed.

Birds which nest in tree hollows are often considered to be among the most sensitive to harvesting operations as hollows do not form in young vigorous trees; in intensively managed forests these birds rely on retained trees or systems of retained vegetation for nest-sites. Tree hollows are in short supply in some forests and experimental provision of nest-boxes has encouraged high populations of some hole-nesting birds (Calder *et al.* 1983). Relative populations of birds which use tree hollows are shown in Table 7, classified into those that use small hollows (treecreepers *Climacteris* spp., Striated Pardalote, *Pardalotus striatus*, Tree Martin, *Cecropis nigricans* and Sacred Kingfisher, *Halcyon sancta)* and large hollows (parrots, owls, etc.). The highest relative population of hole-nesting birds is in mature forests of River Red Gum and the lowest is in regrowth on foothill ridges (aged 6-40 years) and Mountain Ash forest (aged 39-70 years). Populations of birds which use small hollows are specially low in Mountain Ash forest and this may be related to the generation gap which exists naturally in these forests between even-aged regrowth and large fire-killed stags. It may also be related to food supplies as the psyllids eaten by pardalotes are not common in Mountain Ash forests (F. G. Neumann, pers. comm.), and the rough bark where treecreepers forage persists only on the base of the trees and as hanging ribbons.

The abundance of hole-nesting birds in forests of River Red Gum reflects both the age of these forests and the rate at which they form hollows. Formation and use of hollows in Victoria has been studied by Ambrose (1982), Calder *et al.* (1983) and Manderson (1979), and they have shown that hollows form most readily in trees of the smooth-bark gum subgenus *Symphyomyrtus.* Striated Pardalotes are common in stands of this subgenus, and nest in tree hollows, whereas Spotted Pardalotes, *Pardalotus punctatus,* are common in rough barked eucalypts (subgenus *Monocalyptus*) and nest in holes in the ground (Table 5).

Some of the differences between forests dominated by *Symphyomyrtus* and *Monocalyptus* may be related to food supplies rather than availability of hollows. Different subgenera of psyllids occur in the foliage of the two eucalypt subgenera (Ward and Neumann 1983); presumably this affects the distribution of the pardalotes, and their nesting behaviour has evolved to take advantage of the most abundant nest sites in the different forest types. The amount of persisting rough bark is usually less in *Symphyomyrtus* than *Monocalyptus*; hence populations of treecreepers are low in pure stands of the former except in River Red Gum where Brown Treecreepers, *Climacteris picumnis,* feed from the forest floor and fallen branches as well as from trunks (Table 6). Crested Shrike-tits, *Falcunculus frontatus,* often feed at separating bark and are fairly common in stands of this subgenus (Table 5).

Old trees can provide different food sources from young trees. Infestations of Mistletoe (Loranthaceae) are common in old stands and they attract honeyeaters and Mistletoebirds, *Dicaeum hirundinaceum.* Infestations of canopy-feeding insects may also be more common and they are discussed in the final section of this chapter. Birds which feed from eucalypt flowers and from bark are more abundant in mature than regrowth foothill forests (Table 6), and flocks of honeyeaters sometimes

Table 5. Relative populations of birds in different vegetation communities (as % of total populations in each community).

Species	Feeding category	Tree hollow nesting	coastal	rainforest	Foothill gullies		Foothill ridges		Mt. Ash	Alpine Ash	Snow Gum	Red Iron-bark	Box	River Red Gum
					Mt. Grey Gum/ Mess-mate	Manna Gum/ Pepper-mint	regrowth	mature						
Gang-gang Cockatoo	ST	H	1.6	+	0.1	1.6	+	0.6	0.5	0.2	+	W	W	0
Sulphur-crested Cockatoo	SG	H	+	0	+	+	0	+	+	0	0	0.2	0.2	3.7
Crimson & Yellow (=Y) Rosellas	ST	H	C	3.8	0.8	1.9	0.2	1.8	2.4	2.8	0.7	1.8	W	2.3(Y)
Fan-tailed Cuckoo	I		C	0.6	2.0	0.1	0.5	1.2	0.4	0.3	1.0	+	+	+
Horsfield's Bronze-Cuckoo	IC		+	0	0	0	0.7	0.1	0	0	+	0.5	0.2	+
Shining Bronze-Cuckoo	IC		C	0.6	+	C	0.3	2.4	1.0	0.1	+	0.2	0.2	+
Laughing Kookaburra	V	H	3.2	+	0.6	0.5	1.6	1.0	0.6	1.2	0.7	1.3	0.4	0.2
Sacred Kingfisher	V	h	+	0	0	0.6	0	+	0	0.2	0	2.3	+	3.2
Superb Lyrebird	ID		+	1.9	0.8	0.3	C	0.4	0.7	0.4	0.3	0	0	0
Tree Martin	IA	h	C	0	0.3	1.1	+	+	+	+	0	+	0.2	2.5
Black-faced Cuckoo-shrike	IC		0.5	+	1.0	0.1	0.8	1.5	0.1	+	+	1.4	0.6	0.9
White's Thrush	ID		W	+	0.6	+	+	0	1.2	1.0	0	0	0	0
Rose Robin	IS		+	1.3	0.8	0.2	0	0	3.6	+	0	0	0	0
Pink Robin	ID		W	W	W	W	W	W	0.5	+	+	W	W	0
Flame Robin	IG		+	0	0	0.5	0.8	1.6	1.0	2.6	7.6	0	W	W
Scarlet Robin	IG		C	0	0	+	2.4	0.6	0	0	0	0.2	+	+
Eastern Yellow Robin	ID		0.5	3.2	5.2	4.0	4.8	3.3	5.6	0.4	+	1.3	2.0	0
Crested Shrike-tit	IB		+	0.6	0.2	0.6	+	+	1.3	+	0	0.5	+	0.7
Olive Whistler	ID, F		+	+	C	0.1	+	0	1.4	C	10.4	0	0	0
Golden Whistler	IS		0.5	4.5	8.6	2.8	4.1	5.3	7.0	0.2	0.3	W	W	W
Rufous Whistler	IC		1.6	0	0.7	2.6	1.0	2.4	0	0.6	+	13.0	3.0	3.0
Grey Shrike-thrush	I		C	1.9	3.5	2.2	4.7	4.4	3.7	4.0	1.4	10.6	1.4	3.5
Rufous Fantail	IS		0.5	5.1	1.1	2.6	+	0	6.4	0.1	0	0	0	0
Grey Fantail	I		5.3	1.9	7.3	7.0	6.8	6.0	8.5	7.8	1.0	0.4	1.2	1.6
Willie Wagtail	IG		0	0	0	0	0	0	0	0	0	1.1	1.2	1.6
Eastern Whipbird	ID		2.6	2.6	0.9	0.4	+	0	1.8	0.1	0	0	0	0
Superb Fairy-wren	IG		4.7	+	1.8	1.6	6.3	4.3	+	+	0	1.1	3.4	4.4
Pilotbird	ID		C	3.2	C	+	+	+	1.3	0.4	+	0	0	0
White-browed Scrubwren	ID		1.0	6.4	8.1	2.2	6.8	1.3	11.8	7.3	6.9	+	+	3.0
Weebill	IC		0	0	0	0	0	0	0	0	0	2.0	7.6	3.9
Brown Gerygone	IS		0	12.2	W	0.2	0	0	0	0	0	0	0	0
Brown Thornbill	IS		2.6	7.7	15.4	6.7	23.7	12.3	9.4	6.1	6.2	2.2	0.8	+
Buff-rumped Thornbill	IG, IB		C	0	0	0	+	+	0	0	+	1.8	2.6	3.0
Striated Thornbill	IC		3.2	1.3	7.6	7.6	16.9	14.8	4.2	11.3	2.4	1.0	2.0	3.7
Varied Sittella	IB		C	+	+	1.4	+	1.4	+	+	0	+	2.4	0.9
White-throated Treecreeper	IB	h	8.9	1.3	7.6	5.2	6.0	5.6	2.1	1.4	+	4.5	4.2	1.6
Red-browed Treecreeper	IB	h	+	2.0	1.4	3.0	0.4	1.8	0.7	0.8	0	0	0	0
Brown Treecreeper	IBG	h	0	0	0	0	0	0	0	0	0	3.8	0.8	8.1
Red Wattlebird	N		4.7	3.8	+	+	+	0.6	+	3.6	9.7	2.0	+	0
Little Friarbird	N		0	0	0	0	0	0	0	0	0	0	0.2	0.9
Lewin's Honeyeater	F		0	3.2	0.8	+	0	0	+	0	0	0	0	0
Yellow-faced Honeyeater	N		5.3	2.6	3.8	5.6	2.1	4.7	+	16.5	0.7	W	0	0
White-eared Honeyeater	X		3.2	0	0.3	C	1.9	0	2.6	+	17.6	0.2	+	0
Fuscous Honeyeater	N		0	0	0	0	0	0	0	0	0	19.2	13.6	0
White-plumed Honeyeater	N		0	0	0	0	0	0	0	0	0	0	6.6	24.5
Brown-headed Honeyeater	N		1.0	+	0.7	1.3	+	1.0	4.2	0.5	0.7	0.7	2.0	0.5
White-naped Honeyeater	N		3.2	+	2.2	3.0	+	1.6	+	2.8	0.7	W	0	0
Crescent Honeyeater	X		+	0.6	2.0	3.1	0.4	2.7	2.1	0.3	6.6	0	0	0
New Holland Honeyeater	N		20.5	+	0	3.8	+	+	+	0	0	0	0	0
Eastern Spinebill	N		2.6	3.2	1.6	0.2	2.2	4.7	2.7	+	0	0	0	0
Spotted Pardalote	IC		3.2	0.6	0.9	4.7	0.4	2.8	3.6	11.0	2.1	0.9	5.4	W
Striated Pardalote	IC	h	0.5	W	3.9	2.6	0.9	5.9	1.2	9.9	10.4	5.2	9.4	10.2
Silvereye	F		C	C	3.2	2.0	1.0	1.0	3.3	3.3	1.0	0.4	0.4	0.7
Dusky Woodswallow	IA		+	0	0	3.8	+	+	0	0	0	4.3	2.4	2.8
Pied Currawong	V		2.1	2.6	0.3	0.7	+	0.1	0.3	2.0	2.1	W	W	W
Grey Currawong	V		C	0	+	0.2	0.1	0.1	0.6	0.6	3.1	0.2	0	0

Species marked : C are common in the habitat, but were not recorded on these counts;
+ occur locally in the habitat;
W occur as winter visitors in the habitat;
0 occur in the habitat only as vagrants or not at all.

Feeding categories refer to those listed in Table 6, viz. (in order), IG, ID, IS, I, IC, IB, IBG, N, X, F, SG, ST, V, IA.

Birds that nest in large tree hollows are marked H and those that nest in small tree hollows are marked h.

Table 6. Relative populations of groups of birds with similar foraging patterns in different vegetation communities (as % of total population in each community).

Percentage of population feeding on:	Coastal forest	Rainforest gullies	Foothill Gullies: Mountain Grey Gum /Messmate /Peppermint	Foothill Gullies: Manna Gum/ Peppermint	Foothill Ridges: regrowth	Foothill Ridges: mature	Mountain Ash	Alpine Ash	Snow Gum	Alpine Herb-field	Red Iron-bark	Box	River Red Gum
Invertebrates from bare ground (e.g. Superb Fairy-wren, Scarlet Robin) IG	4.7	+	1.8	1.9	11	7.5	1.0	2.6	8.5	+	7.1	14	11
Invertebrates from dense understorey or ground below (e.g. Superb Lyrebird, White-browed Scrubwren, Olive Whistler) ID	4.2	17	15	7.0	12	5.0	24	8.8	18	+	1.2	2.4	3.0
Invertebrates in shrub and middle layers (e.g. Brown Thornbill, Golden Whistler) IS	3.7	31	26	13	28	18	26	6.4	6.5	0	3.6	0.8	+
Invertebrates at all levels (e.g. Grey Fantail, Grey Shrike-thrush) I	6.3	4.5	13	9.5	12	9.0	13	12	3.4	+	13	6.4	5.3
Invertebrates in canopy (e.g. Striated Thornbill, pardalotes) IC	8.9	5.8	15	19	21	30	11	34	15	0	24	29	22
Invertebrates from trunks and branches (e.g. tree-creepers, Varied Sittella) IB	8.9	3.8	9.8	11	6.4	8.8	5.1	2.3	+	0	4.9	6.6	3.2
Invertebrates from trunks, branches and bare open ground (Brown Treecreeper) IBG	0	0	0	0	0	0	0	0	0	0	3.8	0.8	8.1
Nectar, invertebrates and plant or insect exudates (e.g. honeyeaters) N	48	9.6	8.3	14	4.3	15	6.9	23	12	0	27	25	26
Plant or insect exudates mainly from bark with some nectar and invertebrates (e.g. White-eared and Crescent Honeyeaters) X	3.2	0.6	2.4	3.1	2.3	+	4.8	0.3	24	0	0.2	+	0
Fruit, nectar and invertebrates (e.g. Silvereye) F	+	20	4.7	1.8	1.0	1.9	3.3	3.3	1.0	+	0.4	0.8	0.9
Seeds close to ground and some invertebrates (e.g. firetails) SG	1.6	1.3	+	6.7	+	+	+	+	0	0	0.4	2.8	5.3
Seeds, fruits, fern spore and invertebrates at all levels (e.g. rosellas) ST	1.6	3.8	0.9	3.5	0.2	2.4	2.9	3.0	1.4	+	3.9	0.8	2.3
Vertebrates and large invertebrates (e.g. raptors, currawongs) V	6.3	2.6	1.9	2.4	2.7	2.0	2.1	4.1	7.6	+	6.5	1.2	5.6
Flying insects above canopy or in large open spaces below (e.g. hirundines, woodswallows) IA	2.6	+	0.4	7.3	+	+	+	+	1.4	+	4.6	9.0	6.5

+ = regularly present in community but locally or in very low numbers (less than 0.2% of population).

gather to feed at the bark of trees that appear unhealthy, either in dense young stands which are self-thinning, or in old stands containing senescent trees.

Other differences between young and old forest are associated mainly with changes in the shrub layer. In foothill forests, floristic changes after harvesting eucalypts are small (Loyn *et al.* 1980, 1983a) but the shrub layer is initially reduced for a few years after which regrowth becomes shrubbier than mature forest. These changes are reflected in bird populations in the different age-classes (Loyn 1980).

From Table 6 it is clear that birds which feed from bare ground are most common in dry forest types with little understorey. Birds which feed from shrub or understorey layers or the damp ground below are most common in wet forest types with well-developed shrub layers, specially in cold, wet forests of Mountain Ash and gullies of Mountain Grey Gum/ Messmate, *E. cypellocarpa* and *E. obliqua*, and in warmer rainforest gullies in East Gippsland. Some bird species which feed on insects in the tall shrub layer (e.g. Golden Whistler, Grey Fantail, *Rhipidura fuliginosa*) are abundant in these habitats in summer but vacate the first two in winter, although they remain at similar altitudes in drier forest. Altitude plays an indirect role in determining bird communities, and only one species (Pink Robin) is confined to mountain forests, and it only in the breeding season (see under Mountain Ash forests below).

Table 7. Relative populations of hole-nesting birds in different vegetation communities (as % of total population in each community).

Percentage of population in:	Coastal forest	Rainforest gullies	Foothill gullies: Mountain Grey Gum /Messmate	Foothill gullies: Manna Gum/ Peppermint	Foothill ridges: Regrowth	Foothill ridges: Mature	Mountain Ash	Alpine Ash	Snow Gum	Red Ironbark	Box	River Red Gum
Birds which nest in small tree hollows	9.5	3.2	13.5	11.4	7.5	13.3	4.1	12.2	10.4	15.5	14.7	25.7
Birds which nest in large tree hollows	4.7	12.2	1.7	5.2	1.8	3.6	3.5	4.5	2.4	6.9	3.6	7.9
Total:	14.2	15.4	15.2	16.6	9.3	16.9	7.6	16.7	12.8	22.4	18.3	33.6

In the following account the birds of each vegetation community are discussed in turn, emphasising the commonest species present, the main food sources and any scarce species for which a community is important.

Coastal Forests

These varied forests on sandy soils often have a heathy understorey and abundant nectar-rich flowering plants such as *Banksia*. Nectivorous birds (e.g. New Holland Honeyeater, *Phylidonyris novaehollandiae*) are usually the most abundant species (Tables 5 and 6). Some honeyeaters are found primarily in these habitats (Tawny-crowned Honeyeater, *Phylidonyris melanops* in open heath, Little Wattlebird, *Anthochaera chrysoptera* among tall banksias). Lorikeets *(Psittacidae)* are common, specially in East Gippsland when Red Bloodwood, *E. gumifera,* and Rough-barked Apple, *Angophora floribunda,* flower. In most respects the bird fauna resembles that of dry mixed forests in the foothills. Birds that feed from bare open ground (e.g. Scarlet Robin and Buff-rumped Thornbill, *Acanthiza reguloides)* are common in sandy parts with sparse understorey. Tree martins often nest in holes in gnarled coastal Manna Gums, *E. viminalis*. Some scarcer species such as Emu, *Dromaius novaehollandiae,* and Chestnut-rumped Hylacola, *Sericornis pyrrhopygius,* occur locally. It should be noted that the data for the coastal forests in Tables 5, 6 and 7 refer to just one relatively wet site, and do not reflect the full range of coastal forests.

Swampy treeless heaths often occur among these forests and support low populations of birds of few species, primarily small insectivores with Southern Emu-wren, *Stipiturus malachurus,* often being the commonest. They are the main habitat for seed-eating Ground Parrots, *Pezoporus wallicus*. Taller thickets of Paperbark, *Melaleuca* spp. and Teatree, *Leptospermum* spp. attract some common forest insectivores as well as honeyeaters and some scarcer species (e.g. Eastern Bristlebird, *Dasyornis brachypterus* and seed-eating Beautiful Firetail, *Emblema bella).*

Substantial areas of coastal forest in Victoria are no longer subject to harvesting. Management of fire may be critical to the maintenance of heathland for Ground Parrots and other species (Meredith and Isles 1981).

Warm Temperate Rainforest Gullies

Closed thickets of warm temperate rainforest have developed in some coastal and foothill gullies of East Gippsland. Plant species diversity is high but much less than in the more extensive rainforests further north in New South Wales or Queensland (D. Cameron, pers. comm.).

In sheltered rainforest gullies (e.g. Reedy Creek), bird populations are about twice as dense as in ordinary gullies (Table 4), though Male (1979) found fewer birds in coastal rainforest than in adjacent eucalypt forest. Rainforest gullies are the main Victorian habitat for two bird species, the Brown Gerygone, *Gerygone mouki* (a common small resident insectivore) and the Black-faced Monarch, *Monarcha melanopsis* (a summer migrant which takes insects from the canopy). Black-faced Monarchs also breed in some other wet communities (e.g. in forests of Cut-tail, *E. fastigata*) and Brown Gerygones extend in winter into other foothill gullies but appear to breed exclusively in rainforest. Characteristic gully birds which feed from damp shady ground (e.g. Superb Lyrebird, *Menura novaehollandiae*, Pilotbird), from branches of tall shrubs and vines (Large-billed Scrubwren, *Sericornis magnirostris*), or from fruits (e.g. Australian King-parrot, *Alisterus scapularis,* Lewin's Honeyeater, *Meliphaga lewinii,* Satin Bowerbird, *Ptilonorhynchus violaceus*), are generally commoner than in other gullies (Tables 5 and 6). Sooty Owls, *Tyto tenebricosa,* inhabit some rainforest gullies (feeding on small mammals) and otherwise are confined to other wet gullies and to wet forests of Mountain Ash.

Rainforest gullies are protected from harvesting operations by prescription in Victoria and provide a valuable refuge for many birds. They are usually too wet to be affected by fire but many were burnt by the severe fires of 1983.

Other Foothill Gullies

Typically these contain a luxurious middle storey of tree ferns, shrubs and wattles below a fairly sparse overstorey of tall eucalypts mostly of the subgenus *Symphyomyrtus*, with rough-barked Messmate and Narrow-leaf Peppermint, *E. radiata*, on adjacent slopes or the flat bottoms of broader gullies. Manna Gum and Narrow-leaf Peppermint are dominant on sunnier sites.

Bird populations are usually twice as dense as in adjacent mixed eucalypt forests on ridges (Table 4). Species which feed from damp shady ground or in tall shrubs are abundant, especially in cold wet gullies (Table 6). In foothills many of these birds are virtually confined to gullies. Insectivores which feed in the middle storey are more abundant in summer than in winter Silver Wattle trees are attractive to various insectivores, notably Rose Robin and Fantails *Rhipidura* spp. Exudates from Silver Wattle are sometimes taken by honeyeaters.

The foliage of Narrow-leaf Peppermint and Manna Gum is susceptible to attack by insects such as psyllids and populations of canopy-feeding insectivores are high (e.g. psyllid-eating White-naped Honeyeater, *Melithreptus lunatus* and Spotted Pardalote). Bark-gleaning Red-browed Treecreeper, *Climacteris erythrops* and Crested Shrike-tit are also more common than in other forest types. Birds which feed in the middle layers (e.g. Golden Whistler) are less abundant than in cold wet gullies which have a more luxuriant middle layer (Table 6). Birds which feed from open ground (e.g. Superb Fairy-wren, *Malurus cyaneus*), on grass seeds (e.g. Red-browed Firetail, *Emblema temporalis*) or from open spaces between trees (e.g. Tree Martin, Dusky Woodswallow, *Artamus cyanopterus*) are common in these and other broad, sunny gullies but only occur rarely in cold wet gullies. The broader gullies are the most diverse forest habitats and often support populations of localised species such as Powerful Owl, *Ninox strenua*, Yellow-tufted Honeyeater, *Lichenostomus melanops,* Beautiful Firetail and Satin Bowerbird.

Harvesting in gullies has usually been selective and current policy is not to log within 20 m or more of permanent streams. Retained gully strips provide a valuable refuge for many bird species.

Mixed Foothill Forest on Ridges

These include a wide range of forests covering a large area of the State. The dominant trees are rough-barked species which survive most fires, resulting in uneven-aged regeneration. Common tree species include Silvertop, *E. sieberi*, Messmate, Yertchuk, *E. consideniana,* White Stringybark, *E. globoidea,* Brown Stringybark, *E. baxteri* and Yellow Stringybark, *E. muellerana*, usually growing as various mixtures. Trees characteristic of foothill gullies or Box/Ironbark forest may also occur on wet or dry slopes respectively.

The commonest birds are two small insectivores, Brown and Striated Thornbills, from the coast to the mountains (Table 5). Brown Thornbills feed primarily from foliage of narrow-leaved plants in the understorey (extending into the canopy where this includes trees such as Silver Wattle) while Striated Thornbills feed primarily from foliage of broad-leaved plants, especially eucalypts in the canopy. Both co-exist with overlapping territories but Brown Thornbills are more numerous in regrowth and Striated Thornbills are more numerous in mature forest. In winter Striated Thornbills form small mobile flocks, often joined by other species. Both are scarce in tall even-aged regrowth with sparse vegetation between ground and canopy. Where grazing or fire has removed the understorey, their dominant position is taken by pardalotes which glean insects from canopy foliage and nest in holes in bare ground (Spotted Pardalote) or trees (Striated Pardalote). Pardalotes are most common where trees are mature.

The commonest honeyeaters are Yellow-faced Honeyeaters, *Lichenostomus chrysops* (feeding on canopy insects and nectar) and Eastern Spinebills, *Acanthorhynchus tenuirostris* (feeding on nectar of understorey plants and mistletoe, and insects). Other honeyeaters are generally scarce except where particular eucalypts or mistletoes flower heavily, or at other temporary rich sources of food. The most attractive nectar-bearing trees include Red Ironbark, *E. sideroxylon* and Yellow Stringybark.

Birds that feed from bare open ground are far more numerous on ridges than gullies (Table 6) and many species are absent from gullies (e.g. Scarlet Robin, Buff-rumped Thornbill, Spotted Quail-thrush, *Cinclosoma punctatum* and some scarcer birds such as White-throated Nightjar, *Caprimulgus mysticalis* and Painted Button-quail, *Turnix varia*). All occur in recently burnt or harvested areas. Some others are common on ridges but occur also in the broader gullies; these include ground-feeding Common Bronzewing, *Phaps chalcoptera,* Flame Robin and Superb Fairy-wren as well as canopy-feeding Gang-gang Cockatoo, *Callocephalon fimbriatum* and Rufous Whistler.

Bird populations are low in the first few years after harvesting or wildfire and include a small number of open-country birds which are absent at later stages (Loyn 1980). However, many common farmland birds do not colonise such areas. Forest birds return as suitable habitat develops. Regrowth aged 5 to 30 years tends to be more shrubby than mature forest (Loyn *et al.* 1980) with higher populations of understorey birds but lower populations of canopy birds (Table 6). In dense regrowth aged 5 to 10 years there is a continual process of thinning in which suppressed stems die, and branches abscise from remaining trees. Insect infestations are common at this stage and help produce sweet exudates attractive to honeyeaters (e.g. White-eared Honeyeaters, *Lichenostomus leucotis*). The dense low vegetation attracts some birds otherwise found mainly in gullies (e.g. Eastern Whipbird). No species is confined to this intermediate stage of succession, though some are more numerous than at other stages (e.g. Brush Bronzewing, *Phaps elegans,* below stands of wattles when they have reached maturity).

In our Gippsland studies we identified a group of birds found mainly in mature forest. Similar observations were made in southeastern New South Wales by Recher *et al.* (1980). The group included some which nest in tree hollows (e.g. Gang-gang Cockatoo, Red-browed Treecreeper), some which feed on fruit or nectar of mistletoe, *Amyema pendulum* (Mistletoebird and honeyeaters) and others which feed on insects in the eucalypt canopy (e.g. Satin Flycatcher, *Myiagra cyanoleuca,* Varied Sittella, *Daphoenositta varia*, White-naped Honeyeater and Spotted Pardalote). Most of these species occur locally in regrowth, especially in patches of retained trees or outside the breeding season. From the study at Boola Boola, it was concluded that most species had survived in fair numbers despite systematic harvesting for more than 30 years; about 25% of the forest remained unharvested. Lorikeets and large owls were rare in that forest and appeared to be the two groups most sensitive to forest operations. Most birds which inhabit these forests are widespread in Victoria, but two cockatoos (Glossy Black Cockatoo, *Calyptorhynchus lathami* and Red-tailed Black Cockatoo, *C. magnificus,* are confined respectively to stands containing Black She-oak, *Casuarina littoralis,* in East Gippsland, and to forests of Brown Stringybark in the southwest of the State.

Mountain Ash Forest

Many of Victoria's Mountain Ash forests have regenerated from extensive fires in 1939 and contain a scattering of large dead trees (stags) from the previous generation, and a dense middle storey of tall shrubs and wattle trees. Cool temperate rainforest (often dominated by Myrtle Beech, *Nothofagus cunninghamii*) grows in wet sheltered gullies and slopes but generally there is little floristic difference between gullies and slopes or ridges, all supporting a mesic type of vegetation.

Populations of breeding birds are high though not as high as in foothill gullies (Table 4). Species composition is similar to the wetter foothill gullies and birds that feed from bare open ground (e.g. Superb Fairy-wren) are absent except along roads and in stands burnt or cleared within three years. Species that feed among or below tall shrubs make up a high proportion of the bird fauna and species that feed in the eucalypt canopy are scarce (Table 6). For example, White-naped and Yellow-faced Honeyeaters occur mainly when Mountain Ash flowers or during outbreaks of phasmatids. Only about 65 bird species occur regularly in Mountain Ash forests (compared with about 100 in mixed foothill forests or other varied forest types); notable absentees include Scarlet Robin and Buff-rumped Thornbill (which feed largely from bare ground) and Horsfield's Bronze-cuckoo, *Chrysococcyx basalis* and Rufous Whistler (which feed mainly from the canopy in drier forest types).

Mountain Ash forests contain the main breeding habitats for Pink Robins which breed at maximum density (about 30 territories km^{-2}) in sheltered mountain gullies containing cool temperate rainforest (e.g. of Myrtle Beech). Such gullies occur mainly in Mountain Ash forests, and to a lesser extent in Alpine Ash or Snow Gum, *E. pauciflora*. Similar dense stands of Southern Sassafras, *Atherosperma moschatum* and other species occur as a middle storey in high elevation forests of Shining Gum, *E. nitens,* and are another important habitat for Pink Robins. Pink Robins also breed at lower densities elsewhere in Mountain Ash forests, e.g. in the damp shady environment below stands of Silver Wattle. They winter in many types of forest but rarely breed below the altitudinal limit of Mountain Ash; they are the only Victorian bird which is virtually confined to mountains in the breeding season.

In winter many of the abundant middle storey insectivores (e.g. Grey Fantail, Golden Whistler) leave these forests, though they remain in drier forest at similar altitude. The main winter food sources in Mountain Ash are terrestrial invertebrates, various exudates and invertebrates behind hanging bark, and nectar of Mountain Correa, *Correa lawrenciana,* in the understorey. The latter is taken extensively by resident honeyeaters and by Striated Thornbills. Birds which use these food sources remain common over winter. (In many respects high elevation forests of Shining Gum are similar but their more diverse rainforest middle storey is a richer winter habitat attracting flocks of fruit-eating birds such as Satin Bowerbird, Pied Currawong, *Strepera graculina*, and Silvereye, *Zosterops lateralis*).

Mountain Ash trees are valuable as timber, and regeneration is achieved by clear-felling and burning. A few open-country birds occur for about three years after wildfire or clear-felling (e.g. Richard's Pipit, *Anthus novaehollandiae,* Australian Kestrel, *Falco cenchroides,* Blue-winged Parrot, *Neophema chrysostoma*) but regeneration is rapid and stands soon become populated by the same understorey and shrub bird species that are common in older forest. Stands of 1939 regrowth (aged about 40 years) support a similar bird fauna to much older stands, and only Sooty Owls occur primarily in old Mountain Ash (> 200 years). Canopy-feeding birds associated with mature forest in the foothills are generally uncommon in Mountain Ash of any age. Mountain Ash trees grow tall and straight, forming few hollows in their life-span (which rarely exceeds 400 years) and suffering little from ground fires or insect damage. Hence the main source of hollows is in the dead stags, which can remain standing for decades. The needs of hollow-nesting species must be considered in long-term management, and systems of retained vegetation (e.g. streamside strips) may need to be managed on long rotations for this purpose.

Alpine Ash Forest

Many of Victoria's mature Alpine Ash forests have regenerated from extensive wildfires in 1851; such fires play an essential role in the regeneration of the species which grows at altitudes subject to cold wet winters with regular snow cover (Grose 1960).

Although regeneration follows a similar pattern to Mountain Ash, the bird community is quite different. The middle storey is usually sparse and birds inhabiting this layer and the ground below are scarce (Table 6), though some (e.g. Pilotbird) occur at low density among suitable understorey. A more luxuriant middle storey develops in gullies and occasionally elsewhere, and this supports similar birds to Mountain Ash forests.

Birds that feed in the canopy are much more common than in Mountain Ash forests (Table 6). They are a similar selection to those in foothill forests but with more honeyeaters (especially Yellow-faced) and fewer thornbills. Most birds that feed from bare open ground are absent though Flame Robins are common in summer. Alpine Ash forests are not the main habitat for any bird species. Alpine Ash trees are valuable as timber and regeneration is achieved by clear-felling and burning as with Mountain Ash and Shining Gum.

Snow Gum Woodland or Forest

The highest mountain forests are dominated by open stands of low smooth-barked multi-stemmed Snow Gum.

Summer bird populations are less dense than in Mountain Ash or gully forests, and are further reduced in winter. An important food source is provided by exudates on the smooth and separating bark, and this is used extensively by three of the commonest birds (White-eared Honeyeater, Crescent Honeyeater, *Phylidonyris pyrrhoptera,* and Red Wattlebird, *Anthochaera carunculata*) (Tables 5 and 6). As in most forests of smooth-barked eucalypts, the Striated Pardalote is a common foliage-gleaning insectivore. Canopy-feeding birds make up a lower proportion than in Mountain Ash forests (Table 6). They include some summer migrants which are absent from Mountain Ash forests (e.g. Rufous Whistler). Treecreepers are rare (Table 6) as the bark is usually smooth from a low height.

Snow Gum woodland appears too wet or cold for ground-feeding Scarlet Robin or Superb Fairy-wren, which are usually absent even from stands with an open grassy understorey. Buff-rumped Thornbills occur locally. Flame Robins and Spotted Quail-thrushes can tolerate these conditions and Flame Robins are commoner than in other forest types. Where the understorey is dense and shrubby many species typical of gully forests occur (e.g. Pilotbird, Olive Whistler and Superb Lyrebird); Olive Whistlers are especially common and widespread (Table 5). One scarce species which occurs locally in grassy Snow Gum woodland at all times of year is the Emu Snow Gum trees are rarely used for timber and the main management problems concern control of grazing, fire and recreation.

Alpine Herbfield

Few birds inhabit alpine herbfield above the treeline, and none are confined to it. The commonest insectivores are ground-feeding Flame Robin (mainly near trees) and Richard's Pipit (more widely). Other birds include Brown Falcon, *Falco berigora,* Australian Kestrel, nomadic flocks of Little and Australian Ravens, *Corvus mellori* and *C. coronoides,* and various species which fly there to feed from adjacent forest. The Pipit, raptors and Little Raven are open-country birds which avoid uncleared forest. Migrant flocks of White-throated Needletails, *Hirundapus caudacutus*, appear in late summer, feeding conspicuously over mountain ridges as well as all types of forest. Other birds occur in very low numbers, including Peregrine Falcons, *Falco peregrinus,* migrant Latham's Snipe, *Gallinago hardwickii* and occasionally other raptors and water birds.

Box/Ironbark Forest

These rough-barked eucalypts grow in dry areas mainly on the north slope of the Great Dividing Range. Common species include Grey Box, *E. microcarpa,* Red Box, *E. polyanthemos,* Red Ironbark,

Long-leaf Box, *E. goniocalyx,* But-But, *E. bridgesiana,* Yellow Box, *E. melliodora* and Red Stringybark, *E. macrorhyncha,* growing as pure stands or as various mixtures. They are prolific producers of nectar, and Red Ironbark in particular attracts large flocks of honeyeaters. White Ironbark, *E. leucoxylon,* (a related but smooth-barked species), is also attractive to honeyeaters and lorikeets, but grows in western Victoria and has not yet been included in our studies.

The bird fauna differs markedly from that in the forest types above, with many common species replaced by dry-country counterparts. Fuscous Honeyeaters, *Lichenostomus fuscus,* are the commonest birds. They take nectar but spend most of the time gleaning insects from the canopy. On the northern slope of the Great Dividing Range, Fuscous Honeyeaters extend in summer into mixed species foothill forests but on the southern slope they are local even in the limited areas of Box/Ironbark forests. Small stands of Box/Ironbark within River Red Gum forests or farmland tend to be populated by White-plumed Honeyeaters, *L. penicillatus,* instead of Fuscous Honeyeaters.

The common thornbills are Weebill (replacing Striated Thornbill in the canopy, especially among Red Box), and Yellow Thornbill (replacing Brown Thornbill in shrubs, regenerating eucalypts and Black Wattle trees, *Acacia mearnsii*). Replacement is not complete and Brown and Striated Thornbills occur at low density. In open stands of Yellow Box, Yellow-rumped Thornbills, *Acanthiza chrysorrhoa,* replace Buff-rumped Thornbills as the common ground-feeders; otherwise Buff-rumped Thornbills are widespread (feeding from branches as well as the ground) while Yellow-rumped Thornbills are scarce in natural forest and common in farmland.

A number of bird species are commoner in Box/Ironbark forest than in drier or wetter vegetation communities. They include Fuscous Honeyeater, Yellow-tufted Honeyeater, Speckled Warbler, *Sericornis sagittatus,* White-browed Babbler, *Pomatostomus superciliosus,* Diamond Firetail, *Emblema guttata,* and lorikeets. Species that feed from bare open ground (e.g. Red-rumped Parrot, *Psephotus haematonotus,* White-winged Chough, *Corcorax melanorhamphos*), or on nectar are abundant (Table 6). Common forest species occur locally (e.g. Eastern Yellow Robin, *Eopsaltria australis*) or as winter visitors (e.g. Golden Whistler, White-naped Honeyeater). Birds characteristic of drier vegetation occur as summer migrants (e.g. Rainbow Bee-eater, *Merops ornatus*), irregular visitors (e.g. White-browed and Masked Wood-swallows, *Artamus superciliosus* and *A. personatus*), or resident species (e.g. Red-capped Robin). A few scarce species have Box/Ironbark forests as an important habitat, notably Regent Honeyeater, *Xanthomiza phrygia,* Painted Honeyeater, *Grantiella picta,* Barking Owl, *Ninox connivens* and Turquoise Parrot, *Neophema pulchella.*

Some Box/Ironbark forest has regenerated after extraction of mining timber last century. Recent harvesting operations are less intense but terrain is generally flat and does not provide the protection that it has done in foothills and mountains; often the oldest trees are now in roadside strips outside reserved forest. Artificial fire-dams are often the only source of standing water and are used extensively for drinking by honeyeaters and seed-eating birds.

River Red Gum Forest

River Red Gum is the most widespread eucalypt in Australia but the largest forests occur in the flood-plain of the Murray and Murrumbidgee Rivers where they depend on periodic flooding for growth and regeneration (Dexter 1967). Trees can live for over 800 years; old trees are often highly branched with many hollows.

The commonest bird is the White-plumed Honeyeater which occupies the niche filled by Fuscous Honeyeater in Box/Ironbark forest, as a gleaner from foliage. White-plumed Honeyeaters have become common in farmland trees and town parks but are rare in most other Victorian forest types. The nectar of River Red Gum is not attractive to birds. Little Friarbirds, *Philemon citreogularis,* are common in summer but the few other species of honeyeaters are scarce and all feed mainly on insects or plant and insect exudates.

The next commonest birds are Striated Pardalote and Brown Treecreeper (Table 5). Birds that feed from bare ground (including Brown Treecreeper) make up a higher proportion of the community than in other forest types. Many ground-feeding birds which are common in grassy River Red Gum forest have adapted well to farmland. These include three abundant black-and-white species (Willie Wagtail, *Rhipidura leucophrys,* Australian Magpie, *Gymnorhina tibicen* and Australian Magpie-lark, *Grallina cyanoleuca*), none of which is common in other forest types (Loyn and Middleton 1981). Many of the ground-feeding birds are capable of feeding in other situations during a flood (e.g. Brown Tree-creeper and Buff-rumped Thornbill on trunks and branches).

River Red Gum forests support a mixture of species typical of dry woodland and foothill forest (Table 5). Birds which feed in the sparse shrub and understorey layers are scarce (Table 6). Hole-nesting birds are common (Table 7), and include conspicuous species such as Sulphur-crested Cockatoo, *Cacatua galerita,* Yellow Rosella, *Platycercus flaveolus* and Sacred Kingfisher. As in Box/Ironbark, some forest birds appear as winter visitors (Flame Robin, Golden Whistler, Pied Currawong). Many water-birds enter during a flood.

Scarce landbirds include Emu, Superb Parrot, *Polytelis swainsonii,* and Barking Owl, all at low densities.

Scattered stands of River Red Gum and Forest Red Gum, *E. tereticornis,* occur in Gippsland (S.E. Victoria) but many of the birds mentioned are absent, including White-plumed Honeyeater and Brown Treecreeper.

River Red Gum provides valuable durable timber but harvesting operations are not intense and more difficult management problems concern the regulation of floodwaters and grazing by cattle.

INSECTS, BIRDS AND FOREST MANAGEMENT

The ability of birds to control insects has been investigated in many parts of the world (Otvos 1979), and in Australia, Readshaw (1968) concluded that Pied Currawongs could control local outbreaks of stick insects (Phasmatidae) as long as such outbreaks were not too numerous. Generally it is not easy to obtain definitive results without expensive exclusion experiments. In southeastern Australia a situation exists where insectivorous birds are constantly excluded from sections of forest by Bell Miners. Heavy infestations of psyllid insects occur in these sections of forest and the Bell Miners feed largely on the psyllid larvae and their protective lerps (e.g. Paton 1980). In the first of several experiments we removed Bell Miners (Loyn *et al.* 1983b), and observed an immediate influx of common forest birds which reduced the psyllid population to low levels within four months. The main bird species involved were Striated Thornbill, White-naped Honeyeater, Spotted Pardalote and Crimson and Eastern Rosella, *Platycercus elegans* and *P. eximius,* the rosellas scraping infested leaves through their beaks to remove insects. Numbers observed are shown in Table 8. Treecreepers and sittellas appeared simultaneously, searching for food along branches. When psyllid populations crashed, the numbers of these birds dropped to low levels and honeyeaters, pardalotes and sittellas virtually disappeared.

The main interest of this and subsequent experiments was to show the ability of common birds to control insects, and the value of interspecific territoriality to Bell Miners. But another intriguing aspect was that the invading forest birds were species that had been found to be more abundant in mature forest than in regrowth (Loyn 1980). This suggests that the mature forest sustained a higher level of insect productivity than either the regrowth or the forest from which we removed Bell Miners. At Boola Boola the activity of flying insects was found to be low in mature forest (Loyn *et al.* 1980), but the canopy was not sampled.

Table 8. Numbers of birds and psyllids observed on experimental site before and after removal of Bell Miners (birds as mean per ten 20-minute searches).

	April-July 1981	Aug & Sep 1981	Oct-Dec 1981	Jan 1982-Mar 1983
	(Bell miners removed July/August)			
(A) Bell Miners	188	7	0	0
(B)				
Crimson Rosella	8	26	9	6
Eastern Rosella	0	9	10	0
Striated Thornbill	0	78	24	14
White-naped Honeyeater	5	90	6	5
Spotted Pardalote	3	17	4	5
Striated Pardalote	0	13	33	1
(C)				
White-throated Treecreeper	0	10	3	5
Red-browed Treecreeper	0	4	4	0
Varied Sittella	0	24	2	0
Psyllids on foliage	abundant	declining rapidly	very few	hardly any
Psyllid lerps collected in 24 trays (mean per month)	1920	211	6	3
Epicormic crown cover (%)	32.5		33.3	37.3 (Feb 1982)

Group A (Bell Miners) were feeding mainly on psyllids and their associated lerps and honeydew, often taking the psyllid nymph but not its protective lerp.

Group B fed mainly on psyllids and lerps while they were in the study area.

Group C were bark-gleaners.

Birds that fed in the understorey or on other foods in the canopy, and uncommon species are not shown. Data from control plots and other years showed seasonal variation but no peak in bird numbers in Aug / Sept.

As a hypothesis, I suggest that populations of insectivorous birds in the forest canopy are an indicator of insect productivity. In the removal experiment insect numbers were high when Bell Miners were defending their food source, but dropped to much lower levels after predation by other species. At these levels honeyeaters and sittellas could no longer be supported except as occasional visitors. However, some forest types support high populations of these species, including some of the old forest at Boola Boola. Forest types with high populations of these birds (Table 6) are those of relatively low timber productivity (Snow Gum, Box/Ironbark and River Red Gum), and this may indicate that much of the trees' resources are going into production of edible resources (bark exudates, nectar and insects respectively), rather than wood. In Mountain Ash forests and regrowth in the foothills, populations of insectivorous birds in the eucalypt canopy are low (Table 6) and timber productivity is high.

Yet insects do inhabit regrowth and the potential for outbreaks exists. Hence, it is valuable to have a reservoir of birds capable of controlling such outbreaks. A strategic network of mature forest offers the best chance of conserving all species, specially those such as honeyeaters, pardalotes and sittellas which specialise at feeding in areas of high food productivity and have shown themselves capable of controlling insect outbreaks.

ACKNOWLEDGEMENTS

I thank the many people who have helped at various times with fieldwork, or have provided helpful comments, and Ruth Blackbourn for typing the final draft.

REFERENCES

AMBROSE, G. J., 1982. An ecological and behavioural study of vertebrates using hollows in eucalypt branches. Ph.D. Thesis, La Trobe University, Bundoora, Vic. (unpubl.), 419 pp.

CALDER, T. G., GOLDING, B. G. AND ANDERSON, A. D., 1983. Management for arboreal species in the Wombat State Forest. Environmental Report 16, Monash University, Graduate School of Environmental Science, Clayton, Vic. 108 pp.

COSTERMANS, L. F., 1981. Native trees and shrubs of southeastern Australia, Rigby, Melbourne, 422 pp.

DEXTER, B. D., 1967. Flooding and regeneration of River Red Gum, *Eucalyptus camaldulensis*, Dehn. FCV Bulletin 20, 35 pp.

FRANKENBERG, J., 1971. Nature conservation in Victoria. A survey. VNPA. Sydney, 145 pp.

GROSE, R. J., 1960. The silviculture of *Eucalyptus delegatensis* R. T. Baker. Ph.D. Thesis, School of For., Univ. of Melbourne (unpubl.).

LOYN, R. H., 1980. Bird populations in a mixed eucalypt forest used for production of wood. *Emu* **80**: 145-56.

LOYN, R. H. (in press). Bird populations in successional forest of Mountain Ash, *Eucalyptus regnans. Emu* **85**.

LOYN, R. H. AND MACFARLANE, M. A., in prep. Changes in bird and mammal populations after harvesting eucalypts for sawlogs and pulpwood in East Gippsland.

LOYN, R. H. AND MIDDLETON, W. G. D., 1981. Eucalypt dieback and wildlife in rural areas. Pp. 95-111 *in* "Eucalypt Dieback in Forests and Woodlands" ed. by K. M. Old, G. A. Kile and C. P. Ohmart. CSIRO, Australia.

LOYN, R. H., MACFARLANE, M. A., CHESTERFIELD, E. A. AND HARRIS, J. A., 1980. Forest utilisation and the flora and fauna of Boola Boola State Forest in southeastern Australia. FCV Bulletin **28**: 80 pp.

LOYN, R. H., FAGG, P. C., PIGGIN, J. E., MORTON, A. G. AND TOLHURST, K., 1983a. Changes in the composition of understorey vegetation after harvesting eucalypts for sawlogs and pulpwood in East Gippsland. *Aust. J. Ecol.* **8**: 43-53.

LOYN, R. H., RUNNALLS, R. G., FORWARD, G. Y. AND TYERS, J., 1983b. Territorial Bell Miners and other birds affecting populations of insect prey. *Science* **221**: 1411-13.

MALE, E. B., 1979. Habitat preferences: the importance of lowland rainforest pockets for some species of birds in the Wingan region of East Gippsland. Hons. Thesis, Monash Univ., Clayton, Vic., 47 pp.

MANDERSON, A. D., 1979. Aspects of forest management for natural hollows in the Wombat State Forest. M. Env. Sc. Thesis, Monash Univ., 79 pp.

MEREDITH, C. W. AND ISLES, A., 1981. A study of the Ground Parrot, *(Pezoporus wallicus)* in Victoria. Min. Conserv., Melbourne (unpubl.).

OTVOS, I. S., 1979. The effects of insectivorous bird activities in forest ecosystems: an evaluation. Pp. 341-74 *in* "The Role of Insectivorous Birds in Forest Ecosystems" ed. by J. G. Dickson, R. N. Conner, R. R. Fleet, J. A. Jackson and J. C. Kroll. Academic Press, New York.

PATON, D. C., 1980. The importance of manna, honeydew and lerps in the diet of honeyeaters. *Emu* **80**: 213-26.

READSHAW, J. L., 1965. A theory of phasmatid outbreak release. *Aust. J. Zool.* **13**: 475-90.

RECHER, H. F., ROHAN-JONES, W. AND SMITH, P., 1980. The effects of woodchipping at Eden on wildlife with recommendations for management. Forests Commission of NSW, Research Note **42**: 83 pp.

WARD, B. K. AND NEUMANN, F. G., 1983. Eucalypt dieback in foothill forests of the Dandenong Ranges. *For. Tech. Papers* **29**: 10-14.

CHAPTER 5

A Comparison of Forest Bird Communities on the New South Wales South and Mid-north Coasts

D. R. Milledge[1] and H. F. Recher[2]

Birds were counted in eucalypt forests near Eden on the south coast and at Kendall on the mid-north coast of New South Wales. At Kendall, a higher rainfall, richer soils and a warmer, more equitable climate has resulted in generally taller forests with more complex shrub and understorey layers than at Eden.

Bird communities at Kendall were richer than at Eden. About half of the species were common to both localities, but Kendall sites were distinguished by species with tropical affinities and Eden sites by species of temperate origin. Some closely related and ecologically similar species pairs had one member replaced by the other between localities. Other pairs had both members present at Kendall and only one at Eden. This is seen as a response to the greater structural and floristic diversity of the forests at Kendall.

Neither species richness nor abundance was correlated with differences in the structural diversity of forest types. In both localities, some sites with low structural diversity had particularly rich avifaunas. An abundance of nectar-feeders in low, dry open forest and fruit-eaters in tall, wet open forest at Kendall accounted for most of the differences in avifaunal richness. This suggests that floristic diversity is an important factor affecting bird species diversity in eucalypt forests.

INTRODUCTION

FOR birdwatchers the forest avifauna of eastern Australia can appear monotonously uniform. Kikkawa (1968) commented that 80 percent of the sclerophyll or eucalypt forest birds which characterised sites in northeastern Queensland (16°-19°S) also occurred in similar habitats in northeastern New South Wales (28°-32°S). Of the 90 species that Kikkawa registered in his censuses of forest and woodland birds in northeastern New South Wales, 71 (80%) occur in similar habitats in southeastern New South Wales (37°S) (Recher *et al.* 1980). The degree of similarity is not surprising; eucalypt forest and woodland extends in a band from Cape York (11°S) to Tasmania (43°S) and apart from Bass Strait there are no physical barriers to the movement of birds nor significant changes in habitat.

Kikkawa (1968) based his analysis on the presence and absence of species. However, sclerophyll forest bird communities not only have the same species, but the relative abundance of species also appears to be constant throughout. In this paper we compare the avifaunas of forests on the south and mid-north coast of New South Wales. Generally, in Australia the number of bird species increases from southern (temperate) to northern (tropical) latitudes (Kikkawa and Pearse 1969). The avifauna also changes along an east-west moisture gradient (Keast 1959 Kikkawa 1968) with species number decreasing with increasing aridity. The comparison we make is between coastal forests with similar rainfall. However, there is a smaller variation in seasonal temperatures on the north than the south coast Our objective is to describe how forest avifaunas differ within a region and how they change along a latitudinal gradient. These changes are then related to broad vegetational parameters.

Study Sites

We used vegetation structure and topography to group mature forests at Eden and Kendall into three broad divisions; 1) low, dry open-forest on the

[1]Upper Coopers Creek Road, Rosebank via Lismore, New South Wales 2480.
[2]The Australian Museum, P.O. Box A285, Sydney South, New South Wales 2000.

Pages 47-52 *in* BIRDS OF EUCALYPT FORESTS AND WOODLANDS: ECOLOGY, CONSERVATION, MANAGEMENT ed. by A. Keast, H. F. Recher, H. Ford and D. Saunders. Royal Australasian Ornithologists Union and Surrey Beatty & Sons.

Table 1. Dominant tree and understorey species at Eden and Kendall.

	Eden		Kendall	
	canopy	understorey	canopy	understorey
low, dry open-forest (coastal plain)	Yertchuk *Eucalyptus consideniana* White Stringybark *E. globoidea* Woolybutt, *E. longifolia* Silvertop Ash, *E. sieberi*	Black Oak *Casuarina littoralis* Saw Banksia *Banksia serrata* wattles, *Acacia* spp. small eucalypts	White Stringybark Grey Ironbark *E. paniculata* Red Bloodwood *E. gummifera* North Coast Scribbly Gum *E. signata* Narrow-leafed Red Gum *E. seeana* Grey Gum, *E. propinqua* White Mahogany *E. acmenioides*	Black Oak Broad-leaved Paperbark *Melaleuca quinquenervia* Prickly Paperbark *M. stypheloides* teatrees *Leptospermum* spp. Pink-tipped Bottlebrush *Callistemon salignis* Saw Banksia small eucalypts
tall, dry open-forest (slopes and ridges)	Silvertop Ash White Stringybark Yertchuk Blue-leaved Stringybark *E. agglomerata*	small eucalypts Black Oak wattles persoonias, *Persoonia* spp. hakeas, *Hakea* spp.	Blackbutt, *E. pilularis* Tallowwood, *E. microcorys* White Mahogany Red Mahogany, *E. resinifera* Pink Bloodwood *E. intermedia*	small eucalypts Forest Oak, *C. torulosa* wattles Blackwood *Acacia melanoxylon* Scentless Rosewood *Synoum glandulosum* Blueberry Ash *Elaeocarpus reticulatus*
tall, wet open-forest (gullies and creek flats)	Monkey Gum, *E. cypellocarpa* Yellow Stringybark *E. muellerana* Messmate, *E. obliqua*	Small-leafed Lilly Pilly *Acmena smithii* Dogwood, *Pomaderris apetala* Honeywood, *Bedfordia salicina* Blackwood tree ferns, *Dicksonia* sp., *Cyathea* spp.	Tallowwood Blue Gum, *E. saligna* Flooded Gum, *E. grandis* Brushbox *Lophostemon confertus* Turpentine *Syncarpia glomulifera* Blackbutt	small eucalypts Forest Oak Crab Apple *Schizomeria ovata* Corkwood *Caldcluvia paniculata* Coachwood *Ceratopetalum apetalum* Black Booyong *Heritiera actinophyllum* Yellow Carabeen *Sloanea woolsii* Maidens Blush *S. australia* White Bollywood *Neolitsea dealbata* Red Ash *Alphitonia excelsa* walnuts etc. *Cryptocarya* spp. *Endiandra* spp. Bangalow Palm *Archontophoenix cunninghamiana* Socketwood *Daphnandra micrantha* Sassafras *Doryphora sassafras*

coastal plain, 2) tall, dry open-forest on ridges and slopes, and 3) tall, wet open-forest in gullies or along creeks (Table 1). The canopy of low, dry open-forest was less than 15 m in height and had a projected foliage cover of less than 50 percent. In places, this type of forest tended towards woodland (canopy cover < 30%). There was a well developed shrub layer with an abundance of nectar-rich plants (e.g. *Callistemon* spp., *Melaleuca* spp.). The plant species composition of the shrub layer contained a number of species common on local heathlands. Tall, dry open-forest had a canopy 25 to 30 m in height at Eden and up to 50 m at Kendall. Projected foliage cover was between 60 and 80 percent. Understorey and shrub vegetation was variable and relatively poor in species. Tall, wet open-forest had a canopy exceeding 30 m with a projected foliage cover between 60 and 80 percent. There were four distinct and continuous layers of vegetation; ground, shrub, understorey and canopy.

Most sites had been affected by logging, but mature and senescent trees were abundant. The Kendall sites had a greater number of plant species than the sites at Eden. For example, at Eden the canopy was mostly dominated by one or two species of eucalypt, but at Kendall four or five species were usually co-dominant (Table 1). All of the sites were below 200 m in elevation. Characteristic tree and understorey species for the study sites are given in

Table 1. The forests at Eden are described in more detail by Recher *et al.* (1980) and those at Kendall by Milledge (1979).

METHODS

Three sites in each of the three types of forest at Eden and Kendall were selected for comparison. Allowing for differences in vegetation, the sites selected were as similar as possible between the two localities in topography, aspect and history of disturbance. The Eden sites were censused in November 1975 and the Kendall sites in November 1978. A fixed width transect 420 m long by 120 m wide was used to count birds. The census procedure used is described by Recher *et al.* (1983). In this study, each site was counted three times for two hours during the early morning. The data from the three censuses were combined to provide an estimate of the relative abundances of the bird species at each site.

RESULTS

A total of 85 species were recorded; 60 species occurred on the Eden sites and 66 species at Kendall; 41 species occurred at both Eden and Kendall; 19 species were registered only at Eden and 25 species occurred only at Kendall (Table 2). Many of the species recorded at one locality and not the other were uncommon, wide ranging birds (e.g. Whistling Kite, *Haliastur sphenurus*, Little Falcon, *Falco longipennis*) which, although present in both regions, had a low probability of occurrence during a census. When such species are excluded from consideration, only seven species present in the Eden forests were absent from Kendall while twelve species at Kendall did not occur at Eden.

If we consider only those species confined to one region or the other, the avifauna at Eden was characterised by birds of Bassian (southeastern) origin and that at Kendall by birds of Torresian (northern) origin. At Eden, birds of Bassian origin included the Gang-gang Cockatoo, *Callocephalon fimbriatum*, Pilotbird, *Pycnoptilus floccosus*, Crescent Honeyeater, *Phylidonyris pyrrhoptera* and Beautiful Firetail, *Emblema bella*. Birds of Torresian origin at Kendall included the Brush-turkey, *Alectura lathami*, Brown Pigeon, *Macropygia amboinensis*, Pale-yellow Robin, *Tregellasia capito*, Spectacled Flycatcher, *Monarcha trivirgatus*, Log-runner, *Orthonyx temminckii*, Yellow-throated Scrubwren, *Sericornis citreogularis*, Scarlet Honey-eater, *Myzomela sanguinolenta*, Regent Bowerbird, *Sericulus chrysocephalus* and Green Catbird, *Ailuroedus crassirostris*.

Although the regional avifaunas were distinguished by birds of Bassian or Torresian origins, these species were restricted to a narrow range of forest habitats. With the exception of the Scarlet Honeyeater, all of the birds listed above at Kendall were mainly confined to tall, wet open-forest in gullies. To a lesser extent this was also the situation at Eden. Species such as the Pilotbird and Crescent Honeyeater were more typical of wet forest than of drier more open forests on ridges and slopes. Generally, regional differences were not strongly pronounced, with a relatively small number of species common to both regions accounting for the majority of individuals.

The twelve most abundant species at Eden accounted for 70 percent of individuals recorded during censuses. At Kendall the twelve most abundant species accounted for 61 percent of individuals. Eight of the twelve were the same for both regions and accounted for nearly half the birds recorded (Table 3). With the exception of the Brown Warbler, *Gerygone mouki*, which was restricted to wet forest in gullies or on adjacent slopes, the most abundant birds occurred in all types of forest.

Twenty-three species at Eden and twenty-four at Kendall were recorded in the three forest types sampled (Table 2). Fourteen of these occurred in all forest types in both regions. These abundant and widely occurring species are common in forest habitats throughout southeastern Australia including forest at higher elevations. Eight of the most abundant species in tall, open-forest on the Southern Tablelands of New South Wales (800 m a.s.l., Recher unpubl. data) and nine of the most abundant species from a variety of forest habitats in central Gippsland (680 m a.s.l., Loyn 1980) are included among the twelve most abundant species at Eden and Kendall.

In a few instances, a species present in one region was replaced by a closely related species in the other. The Musk Lorikeet, *Glossopsitta concinna*, and the Superb Blue Wren, *Malurus cyaneus*, at Eden were replaced by the Scaly-breasted Lorikeet, *Trichoglossus chlorolepidotus*, and the Variegated Wren, *M. lamberti*. More often, closely related or ecologically similar species were added at Kendall. These added species then shared habitats at Kendall with a species which occurred alone at Eden. This happened most often in the tall, wet open-forest of gullies, but was not restricted to this habitat. For example, Yellow and Striated Thornbills, *Acanthiza nana*, and *A. lineata*, shared dry forest habitats at Kendall, but only the Striated Thornbill occurred at Eden. Species pairs represented by one species at Eden and by both in tall, wet open-forest at Kendall included the Eastern Yellow Robin/Pale-yellow Robin, *Eopsaltria australis/Tregellasia capito*, Black-faced Flycatcher/Spectacled Flycatcher, *Monarcha melanopsis/M. trivirgatus*, Eastern Whipbird/Log-runner, *Psophodes olivaceus/Orthonyx temminckii*, White-browed Scrubwren/Yellow-throated Scrub-

Table 2. The total numbers of individuals of species counted during censuses in low, dry open-forest (p) on coastal plains, tall, dry open-forest (s) on slopes, and tall, wet open-forest (g) in gullies at Eden and Kendall. Three sites were censused in each type of forest at each locality. Data were collected during November 1975 at Eden and September-November 1978 at Kendall. Species recorded only at Eden are notated E, only at Kendall notated K.

		Eden			Kendall		
	site	p	s	g	p	s	g
K	Whistling Kite				2		
E	Brown Goshawk	1					
K	Collared Sparrowhawk						4
K	Grey Goshawk						1
K	White-bellied Sea-eagle				1		
K	Little Eagle				2		1
E	Little Falcon	1	1				
K	Brush-turkey						5
K	Brown Pigeon						4
K	Wonga Pigeon				1		
K	Glossy Black Cockatoo				3		10
E	Gang-gang Cockatoo	4	4	4			
K	Rainbow Lorikeet				2		
K	Scaly-breasted Lorikeet				4	2	
E	Musk Lorikeet	4	1				
E	Little Lorikeet	4					
K	Australian King Parrot				12	6	
	Crimson Rosella		5	2	17	10	
K	Eastern Rosella				1		
E	Pallid Cuckoo	2					
E	Brush Cuckoo	1					
	Fan-tailed Cuckoo	7	7	9	3	3	
E	Horsfield's Bronze-cuckoo	2	6	3			
	Golden Bronze-cuckoo	3	2	5		2	5
	Laughing Kookaburra	4	4	6	2		
	Sacred Kingfisher	1			5		5
	Superb Lyrebird		4	5			3
	Black-faced Cuckoo-shrike	5	5	1	8	7	
K	Little Cuckoo-shrike				6		
	Cicadabird	3			1		3
	Ground Thrush			3			3
	Rose Robin			11			1
E	Scarlet Robin	2					
	Eastern Yellow Robin	15	10	13	13	58	31
K	Pale-yellow Robin						22
	Shrike-tit			2		3	5
	Golden Whistler	7	13	44	2	39	54
	Rufous Whistler	38	25	2	48	32	1
	Grey Shrike-thrush	17	14	2	7	16	13
	Black-faced Flycatcher	1	2	4	8	16	37
K	Spectacled Flycatcher						16
	Leaden Flycatcher	16	6		25	27	3
E	Satin Flycatcher			9			
	Rufous Fantail		1	18			8
	Grey Fantail	38	19	57	25	75	34
K	Logrunner						8
	Eastern Whipbird	1	2	7			13
E	Superb Blue Wren	28	32	4			
K	Variegated Wren				62	20	2
	Southern Emu-wren	16			8		
E	Pilotbird			1			
	Large-billed Scrubwren			13			30
K	Yellow-throated Scrubwren						13
	White-browed Scrubwren	1	5	54	2	4	7
E	Chestnut-rumped Heathwren	1					
	Brown Warbler			67		19	138
K	White-throated Warbler				6	8	1
	Brown Thornbill	23	12	82	10	27	31
K	Yellow Thornbill				35	20	
	Striated Thornbill	50	108	19	18	76	54
	Orange-winged Sittella		4		11	11	
	White-throated Treecreeper	29	23	10	19	23	12
	Red-browed Treecreeper	2	1		1	6	1
E	Red Wattlebird	7		1			
	Little Wattlebird	6			4		
K	Noisy Friarbird				6		
	Lewin's Honeyeater			3	1	16	61
	Yellow-faced Honeyeater	53	14	26	151	81	20
E	Brown-headed Honeyeater		3				
	White-naped Honeyeater	12			29	13	29
E	Crescent Honeyeater	9		15			
E	New Holland Honeyeater	12		2			
	Eastern Spinebill	36	5	14	7	16	16
K	Scarlet Honeyeater				45	16	7
	Mistletoebird	1			98	2	6
	Spotted Pardalote	7	8	8	16	15	10
	Silvereye			21	20	13	7
	Red-browed Finch	3		1	13	1	25
E	Beautiful Firetail	2					
	Olive-backed Oriole	1			8	1	
	Satin Bowerbird	1	1	1		1	12
	Regent Bowerbird						
K	Green Catbird						2
	Pied Currawong	3	6	2	3	4	6
E	Australian Raven	6		1			

wren, *Sericornis frontalis/S. citreogularis,* and Satin Bowerbird/Regent Bowerbird-Green Catbird, *Ptilinorhynchus violaceus/Sericulus chrysocephalus-Ailuroedus crassirostris*. The first species of each pair was the one present at Eden.

The forests at Kendall had more species of birds and higher population densities than forests at Eden (Table 4). The difference in species number was greatest between tall, wet open-forest and least between low, dry open-forest sites. Tall, wet open-forest also had higher population densities at Kendall, but the greatest differences in abundance were between tall, dry open-forests on ridges and slopes. In this type of forest the numbers of birds at Kendall were nearly twice those at Eden. Low, dry open-forest sites had similar numbers of species at Eden and Kendall. However, because of the abundance of the Yellow-faced Honeyeater, *Meliphaga chrysops* and Mistletoebird, *Dicaeum hirundinaceum,* the numbers of birds were greater on the Kendall sites than at Eden.

A larger number of raptors, ground foraging species and fruit-eating species accounted for the greater numbers of birds in tall, wet open-forest at Kendall (Table 4). The number of understorey and canopy foraging insect-eating species in low, dry open-forest was also greater at Kendall. Apart from this difference, the number of insectivores in different vegetation layers was about the same at all sites. Low, dry open-forest and tall, dry open-forest sites at both localities had about the same diversity of nectar-feeders, fruit-eaters, seed-eaters and raptors, but abundances were greater at Kendall.

Low, dry open-forest is structurally less diverse than tall, dry open-forest or tall, wet open-forest. Generally at Eden there were only two layers of vegetation in low, dry open-forest compared with

Table 3. The twelve most abundant species at Eden and Kendall.

Eden	Total	No. sites present/ total sites	Kendall	Total	No. sites present/ total sites
Striated Thornbill *Acanthiza lineata*	177	8/9	Yellow-faced Honeyeater *Meliphaga chrysops*	252	8/9
Brown Thornbill *Acanthiza pusilla*	117	8/9	Brown Warbler *Gerygone mouki*	157	5/9
Grey Fantail *Rhipidura fuliginosa*	114	9/9	Striated Thornbill *Acanthiza lineata*	148	6/9
Yellow-faced Honeyeater *Meipbaga chrysops*	93	9/9	Grey Fantail *Rhipidura fuliginosa*	134	9/9
Brown Warbler *Gerygone mouki*	67	3/9	Mistletoebird *Dicaeum hirundinaceum*	106	5/9
Rufous Whistler *Pachycephala rufiventris*	65	6/9	Eastern Yellow Robin *Eopsaltria australis*	102	9/9
Golden Whistler *Pachycephala pectoralis*	64	8/9	Golden Whistler *Pachycephala pectoralis*	95	7/9
Superb Blue Wren *Malurus cyaneus*	64	6/9	Variegated Wren *Malurus lamberti*	84	6/9
White-throated Treecreeper *Climacteris leucophaea*	62	9/9	Rufous Whistler *Pachycephala rufiventris*	81	7/9
White-browed Scrubwren *Sericornis frontalis*	60	5/9	Lewin's Honeyeater *Meliphaga lewinii*	78	7/9
Eastern Spinebill *Acanthorhynchus tenuirostis*	55	8/9	White-naped Honeyeater *Melithreptus lunatus*	71	7/9
Eastern Yellow Robin *Eopsaltria australis*	38	9/9	Brown Thornbill *Acanthiza pusilla*	68	8/9
12 most abundant species	976 individuals % of total individuals 70		12 most abundant species	1376 individuals % of total individuals 61	
8 most abundant species common to both localities	735 individuals % of total individuals 53		8 most abundant species common to both localities	1037 individuals % of total individuals 46	

Table 4. A breakdown of species richness and numbers of individuals to trophic levels for low, dry open-forest (p), tall, dry open-forest (s) and tall, wet open-forest (g) sites at Eden and Kendall.

	Species						Individuals					
	Eden			Kendall			Eden			Kendall		
	p	s	g	p	s	g	p	s	g	p	s	g
raptors	2	1	1	3	+	3	2	1	1	5	+	6
ground foragers	5	4	7	3	1	10	22	20	46	20	58	104
low understorey foragers	11	8	10	9	8	10	127	95	197	113	115	181
tall understorey plus canopy foragers	15	13	16	16	18	16	192	207	260	258	377	398
nectarivores	8	3	3	6	4	3	131	20	41	219	115	43
frugivores	2	2	2	2	4	7	2	6	3	99	32	45
graminivores	3	1	2	3	1	2	9	4	5	17	1	35
totals	46	32	41	42	36	51	485	353	553	731	698	812

three in tall, dry open-forest and four in tall, wet open-forest. The same relationships hold at Kendall, but tall, dry open-forest may have had a better developed understorey than at Eden. Therefore, differences in species number cannot be explained solely by differences in the structural complexity of the vegetation.

If structural complexity was the principal determinant of bird species diversity in forest habitats we would expect low, dry open-forest to have had the fewest species and tall, wet open-forest the greatest number. Instead, species diversity was least in tall, dry open-forest. Other factors must be considered to explain the differences in bird species number between forest habitats within each region and within forest types between Eden and Kendall.

Low, dry open-forest was floristically more diverse than tall, dry open-forest. Mistletoes infected a large proportion of trees and with other nectar-rich plants provided food for large numbers of nectar-feeders (honeyeaters, lorikeets and the Silvereye, *Zosterops lateralis*). Mistletoe also provided fruit on which the Mistletoebird is dependent. These kinds of birds accounted for the greater species richness and abundance of individuals in low, dry open-forest when compared to tall, dry open-forest (Table 3).

In a similar way, the greater floristic diversity of tall, wet open-forest provided food resources not available in other forest habitats. Probably, gully and creek-flat habitats also had richer soils and more equitable moisture regimes than forests on the slopes and ridges. Total biomass was greater and food was probably more abundant, which could support greater numbers of birds and enable more species to co-exist. An increase in bird species diversity may have come about through a finer division of resources or through the availability of additional kinds of resources.

Fruit was a particularly important resource in tall, wet open-forest at Kendall and fruit-eating birds accounted for part of the difference in species number within this forest type between regions. There were also more ground foraging birds in tall, wet open-forest at Kendall which also may have been due in part to differences in floristic diversity. Fruit, for example, may be used by ground-dwelling species such as the Brush-turkey. Site productivity may also have been important in increasing the numbers of ground-dwelling birds. The milder climate at Kendall probably enhanced year round productivity and led to increased litter production. Turnover rates among ground invertebrates were likely to have been higher and the total amount of food available to ground-dwelling birds increased relative to the more pronounced seasonal environment at Eden.

Tall, dry open-forest had a poorer avifauna than low, dry open- or tall, wet open-forest (Table 4). The difference was greatest at Eden. Ridge and slope forests at Eden tended to be very open with poorly developed ground, shrub and understorey vegetation. Most often the understorey was an extension of the canopy and consisted of the same species of trees. The number of plant species per site was low relative to the other type of forest. Poor soils and relatively frequent fires probably contributed to the sparse growth of ground and shrub vegetation. Productivity on these sites was probably low. Reduced floristic diversity, sparse vegetation and low productivity would account for the generally smaller number of birds and fewer species found in these forests at Eden. At Kendall, tall, dry open-forests were more diverse than at Eden. In particular, the understorey and shrub vegetation was better developed (denser) and richer in plant species. These differences help explain the richer avifauna in the tall understorey and canopy of tall, dry open-forest at Kendall.

Evidence for the greater productivity of forests at Kendall can be seen in the increased abundance of many species which occurred at both places (Table 2). Numbers of the Eastern Yellow Robin, Golden and Rufous Whistlers, *Pachycephala pectoralis* and *P. rufiventris*, Black-faced Flycatcher, Leaden Flycatcher, *Myiagra rubecula*, Brown Warbler, Yellow-faced Honeyeater, White-naped Honeyeater, *Melithreptus lunatus* and Spotted Pardalote, *Pardalotus punctatus,* were noticeably greater at Kendall than at Eden. Only a few species were more abundant at Eden. Most notable among these were the White-browed Scrubwren, Brown Thornbill, *Acanthiza pusilla*, and Striated Thornbill. Possibly their numbers at Kendall were affected by the presence or greater abundance of ecologically similar species such as the Large-billed Scrubwren, *S. magnirostris* and Yellow-throated Scrubwren in the case of the White-browed Scrubwren; Large-billed Scrubwren and Brown Warbler in the case of the Brown Thornbill, and Yellow Thornbill in the case of the Striated Thornbill.

ACKNOWLEDGEMENTS

The work of the Australian Museum at Eden was supported by a grant from Harris-Daishowa Pty Ltd. Dick Smith of Dick Smith Electronics Pty Ltd funded the research at Kendall with a grant to the Australian Museum. The Forestry Commission of New South Wales gave permission for work to be carried out in State Forests and staff at both Eden and Kendall gave generously of their time to assist in the projects.

REFERENCES

Keast, A., 1959. Australian birds: their zoogeography and adaptations to an arid continent. *In* Biogeography and Ecology in Australia ed. by A. Keast, R. L. Crocker and C. S. Christian. Junk, The Hague. pp. 89-114.

Kikkawa, J., 1968. Ecological association of bird species and habitats in eastern Australia; similarity analysis. *J. Anim. Ecol.* 37: 143-65.

Kikkawa, J. and Pearse, K., 1969. Geographical distribution of land birds in Australia — a numerical analysis. *Aust. J. Zool.* **17**: 821-40.

Loyn, R. H., 1980. Bird populations in a mixed eucalypt forest used for production of wood in Gippsland, Victoria. *Emu.* **80**: 145-56.

Milledge, D., 1979. The Camden Haven Wildlife Refuge Study. The Australian Museum, Sydney.

Recher, H. F., Milledge, D. R., Smith P. and Rohan-Jones, W., 1983. A transect method to count birds in eucalypt forest. *Corella.* 7: 49-54.

Recher, H. F., Rohan-Jones, W. and Smith, P., 1980. Effects of the Eden woodchip industry on terrestrial vertebrates with recommendations for management. *For. Comm. of NSW Research Note* No. 42.

Brush-turkey, *Alectura lathami*

Sooty Owl, *Tyto tenebricosa*

CHAPTER 6

Forest Avifauna of the Upper Hastings River

J. M. SHIELDS[1], R. P. KAVANAGH[1] and W. G. ROHAN-JONES[1,2]

THE abundance of birds and the richness of bird species occurring in terrestrial habitats is directly related to the dominant vegetation regime of any geographic region (Udvardy 1959). In northern New South Wales, the original vegetation pattern was dominated by a continuous mosaic of rainforest, tall open-forest and dry open-forest (Baur 1983). These forests support a rich and varied avifauna (Frith 1975; Morris *et al.* 1981).

The pattern of vegetation has been altered since European settlement of the area in the early 19th century (R. Aust. Hist. J. 1940) and it can be assumed that the distribution and abundance of terrestrial birds has changed accordingly. In coastal regions and accessible inland areas, large tracts of land have been cleared for agriculture and most forests have been intensively logged (Pople and Cowley 1981). Relatively untouched forest remains in the rugged and inaccessible ranges that rise 50 to 80 km from the coast. In 1980, when logging was proposed for the mountainous region comprising the upper catchment of the Hastings River, an integrated investigation was initiated to determine basic factors of the environment and the effect of timber harvest on the forest ecosystem. A study of forest birds was part of this investigation.

The purpose of the bird study was to determine species composition of the avian communities and the relative abundance of birds in the major forest types of the area. The immediate and long term effects on avian communities due to structural and floristic changes in the forest environment caused by logging were also investigated. Specific questions asked were: "Is there a difference in the diversity or abundance of birds in different types of forest?" and "Within these forest types, is there a difference in the diversity or abundance of birds when forest structure is manipulated?"

STUDY SITES

Bird counts were done at 11 study sites in Mt. Boss State Forest. These study sites were located in the upper catchment of the Hastings River, near the headwaters of its major tributary, the Forbes River (Map 1), 80 km inland from Port Macquarie, northern New South Wales. Altitude above sea level ranged from 650 m to 1220 m (Table 1).

The topography of this region is varied, typical of the catchments of coastal rivers. A flat coastal plain occurs to the east, giving way to steeply sloping valleys and peaks at the head-waters of the drainage systems. The New England tablelands border the area on the west. The climate of the upper reaches of the catchments is generally cool and moist. Average annual rainfall at a location central to the study sites (Cameron's Camp) was 2053 mm from 1969 to 1979 (For Comm. 1981). Rainfall is higher in the summer than winter. Summer temperatures at Cameron's Camp are 22.7°C (max.), 13.9°C (min.). while average winter temperatures were 13.3°C (max.), 5.3°C (min.) (For. Comm. 1981).

The upper reaches of the Hastings and Forbes catchments have a continuous cover of moist forest, giving way to dry sclerophyll forest towards the tablelands in the west. Forest types vary with moisture, soil type and altitude. Study sites were located in four major vegetation associations (Tables 1 and 2).

Three study plots were located in cool temperate rainforest (sites 1-3, Tables 1 and 2), which is characterized by the dominance of Antarctic Beech, *Nothofagus moorei*. Site 1 was unlogged, and dominated by a tall closed canopy of Antarctic Beech with a Coachwood, *Ceratopetalum apetalum,* understorey. Study site 2 was logged 1961/3, at which time removal of merchantable stems of Antarctic Beech was standard practice. This left an open forest with few mature stems. Remnant canopy trees were severely affected by dieback, caused in part by a wild fire before logging, but otherwise due to increased exposure following logging. Site 3 was logged in 1977/78. No Beech was harvested in this operation, which followed current logging

[1]Forestry Commission of New South Wales, Forest Research Division, P.O. Box 100, Beecroft 2119.
[2]National Parks and Wildlife Service, New South Wales, P.O. Box 22, Parramatta 2150.
Pages 55-64 *in* BIRDS OF EUCALYPT FORESTS AND WOODLANDS: ECOLOGY, CONSERVATION, MANAGEMENT ed. by A. Keast, H. F. Recher, H. Ford and D. Saunders. Royal Australasian Ornithologists Union and Surrey Beatty & Sons.

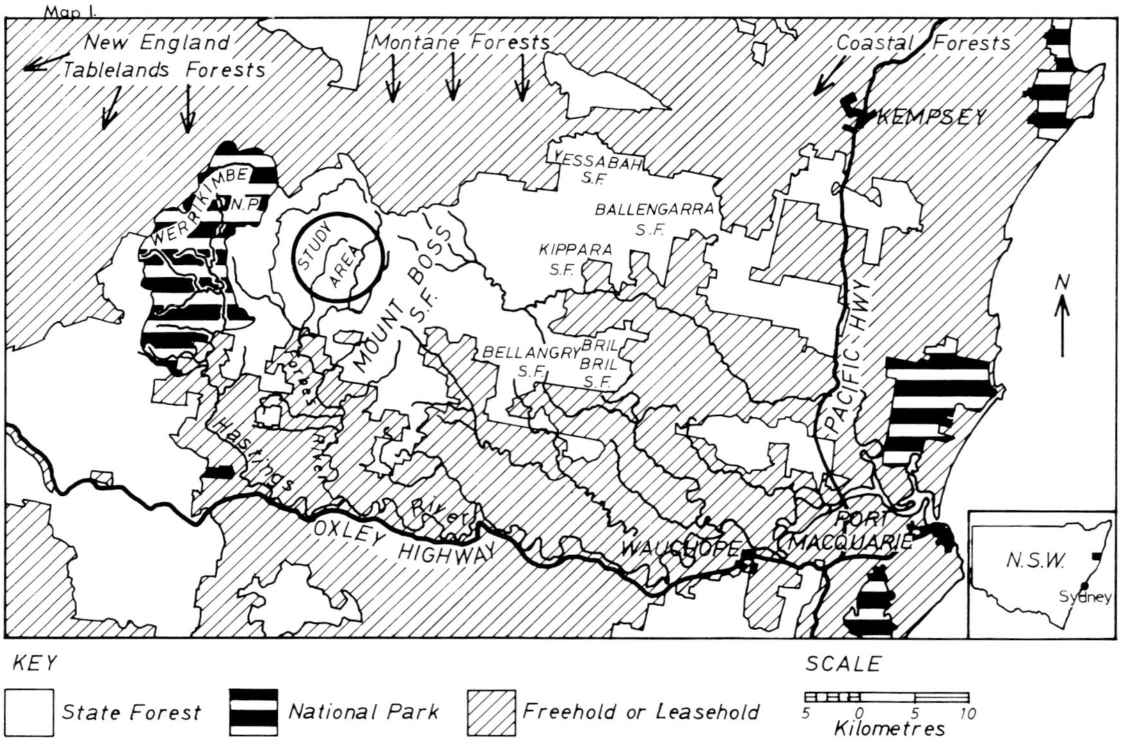

Map 1. Location of study area in Mt Boss State Forest in the Upper Hastings River region of New South Wales.

prescriptions by removing only understorey stems of Coachwood and Sassafras, *Doryphora sassafras*. There was a closed canopy of healthy Antarctic Beech at this site.

There were three study sites (sites 4-6, Tables 1 and 2) in mixed warm temperate/subtropical rainforest. Site 4 was unlogged, and carried tall closed-forest dominated by Yellow Carabeen, *Sloanea woolsii*, Coachwood, Sassafras and Red Carabeen, *Geissois benthamiana*. Study site 5 carried similar species to Site 4 as well as Black Booyong, *Heritiera actinophylla,* and Socketwood, *Daphnandra micrantha*. Study site 5 was heavily logged 1956/57. At the time of this study, the canopy was closed and 20-30 m in height. Study site 6 was logged to maximum utilization standards in 1979/80. Isolated stems of Sassafras remained at this site, although there was no continuous canopy above eight metres.

There were three study sites in tall open-forest (sites 7-9, Tables 1 and 2). The unlogged forest on site 7 was dominated by Tallowwood, *Eucalyptus microcorys*, Sydney Blue Gum, *E. saligna,* and Brush Box, *Tristania conferta,* and Forest Oak, *Casuarina torulosa*. At the time of our study, remnant stems and regrowth had formed a closed canopy of Tallowwood, Sydney Blue Gum and Brush Box after logging (1963) at study site 8. Recent logging (1978) had taken place on study site 9, where few trees remained of the original stand of Tallowwood, Sydney Blue Gum, Brush Box and Forest Oak.

There were two study sites in dry open-forest (sites 10-11, Tables 1 and 2). Study sites with different forest structure were selected on the basis of moisture regime rather than logging intensity, as this forest type is not logged intensively in the region. The dry open-forest at site 10 carried Messmate, *E. obliqua*, New England Blackbutt, *E. andrewsii*, Diehard Stringybark, *E. cameronii,* and Narrow-leaf Peppermint, *E. radiata*, together with a xeric understorey. Some stems were removed in a selective logging operation 1968/69. An open forest at site 11 was dominated by Manna Gum, *E. viminalis,* Snow Gum, *E. pauciflora*, Narrow-leaved Peppermint and New England Blackbutt, and had a mesic understorey. Both sites are referred to as dry open-forest because of similarity in forest structure and species composition of the overstorey.

METHODS

Censuses were conducted in late October in 1980, 1981, and 1982. This time of year was chosen to coincide with maximum breeding activity of forest birds and the presence of migrant species that leave the area in winter. The study was continued through three years to reduce the variability inherent in single year, single season studies (Wiens 1981).

Birds were counted using the point count method (Ralph and Scott 1980). Ten points (four in 1980) were located at 100 m intervals along a transect through each study site. A point consisted

Table 1. Physical site data and floristics.

Study Site No.	Forest Type and Treatment	Major Tree Species	Major Understorey Species	Altitude Above Sea Level
1	Cool temp. rainforest Unlogged	*Nothofagus moorei* *Ceratopetalum apetalum* *Sloanea woolsii*	*Tasmannia insipida* *Ripogonum discolor* *Cyathea leichhardtiana*	1000
2	Cool temp. rainforest Old logging	*Nothofagus moorei* *Ceratopetalum apetalum* *Doryphora sassafras*	*Callicoma serratifolia* *Tasmannia insipida* *Ripogonum discolor*	1000
3	Cool temp. rainforest Recent logging	*Nothofagus moorei* *Ceratopetalum apetalum* *Doryphora sassafras*	*Cryptocarya meisnerana* *Tasmannia insipida*	1100
4	Warm temp. sub-tropical rainforest Unlogged	*Doryphora sassafras* *Sloanea woolsii* *Geissois benthamiana*	*Ripogonum discolor* *Tasmannia insipida* *Cyathea leichhardtiana*	910
5	Warm temp. sub-tropical rainforest Old logging	*Doryphora sassafras* *Sloanea woolsii* *Geissois benthamiana* *Heritiera actinophylla* *Daphnandra micrantha*	*Dendrocnide excelsa* *Tasmannia insipida* *Cryptocarya meisnerana* *Cissus spp.* *Rubus spp.*	890
6	Warm temp. sub-tropical rainforest Recent logging	*Ceratopetalum apetalum* *Sloanea woolsii* *Doryphora sassafras*	*Prostanthera spp.* *Cissus spp.* *Rubus spp.*	640
7	Tall open forest Unlogged	*Eucalyptus saligna* *Eucalyptus microcorys* *Lophostemon confertus* *Casuarina torulosa*	*Schizomeria ovata* *Ackama paniculata* *Elaeocarpus reticulatus*	910
8	Tall open forest Old logging	*Eucalyptus saligna* *Eucalyptus microcorys* *Lophostemon confertus*	*Callicoma serratifolia* *Acacia binervata*	670
9	Tall open forest Recent logging	*Eucalyptus saligna* *Eucalyptus microcorys* *Lophostemon confertus* *Casuarina torulosa*	*Cryptocarya rigida* *Elaeocarpus reticulatus* *Eupomatia rigida*	670
10	Dry open forest Logged dry	*Eucalyptus obliqua* *Eucalyptus andrewsii* *Eucalyptus cameronii* *Eucalyptus radiata*	*Proteaceae* *Rutaceae* *Xanthorrhoea spp.* *Epacridaceae* legumes	1140
11	Dry Open forest Unlogged moist	*Eucalyptus andrewsii* *Eucalyptus viminalis* *Eucalyptus pauciflora* *Eucalyptus radiata*	*Proteaceae* *Rutaceae* *Xanthorrhoea* spp. *Epacridaceae* legumes *Cyathea* spp.	1000

of a circle with a radius of 50 m. All birds heard or seen within this circle were counted for ten minutes at each point. This procedure was repeated four times on each transect, yielding a total of 40 point counts (16 in 1980) in each habitat per year. Distance from the observer to the bird was recorded to allow calculation of the number of birds per hectare. Only birds recorded within 30 m of the observer were used in the calculation of density to reduce bias due to decreased detectability at greater distances.

These data were analysed in terms of the mean number of species per point (expressed as species per hectare), mean number of species per transect and the mean number of individuals. The mean number of species per point is a measure of the diversity of species occurring within a forest type or treatment. The mean number of species per transect is a measure of bird species richness across the total of ten hectares of habitat sampled over a 1000 m transect. The mean number of birds per hectare is a measure of the population density of birds. In

Table 2. Forest Structure and Habitat Diversity.

Study Site No.	Stand Ht. (m)	Canopy Cover %	Shrub Cover %	Layers of Vegetation	No. Plant Species	No. Eucalypt Species	Open Habitat %
1	30-40	90	90	3	79	0	1
2	10-15	50	80	2	53	0	3
3	25-35	80	80	4	70	1	10
4	25-35	90	50	3	98	0	1
5	20-30	90	70	4	104	0	5
6	10+	8	70	2+	110	0	85
7	40-50	90	60	4	112	3	2
8	30-40	90	60	4	79	3	5
9	8+	8	70	2+	125	3	80
10	20	40	15	3	105	4	40
11	20	60	10	3	102	4	25

+ Plus emergent to 30 m in height.

Table 3. Frequency of occurrence of 32 bird species in 11 habitats 1981-1982. *See Appendix 1 for Scientific names.

Study Site No.	1 #CTRF UL	2 CTRF OL	3 CTRF RL	4 WTST UL	5 WTST OL	6 WTST RL	7 TO UL	8 TO OL	9 TO RL	10 DO M	11 DO D
Brown Pigeon*				2.5+	15	20					
King Parrot*	20	73	50	4	23	13	5	23	16	4	1.25
Crimson Rosella*	75	73	68	21	28	35	49	35	53	66	41
Pallid Cuckoo*										4	12
Fan-tailed Cuckoo*	11	76	68	6	13	24	24	13	33	50	51
Welcome Swallow*							5			1.25	6
Scaly Thrush*	6	13	6	6		1.25	2.5	1.25			
Rose Robin*	26	25	26	21	15	21	30	15	16	5	
Scarlet Robin*			1.25							34	26
Yellow Robin*	38	28	32	36	35	41	39	40	17	17	6
Olive Whistler*	21	23	12.5	12.5	23	2.5	1.25				
Rufous Whistler*							1.25	5	6	56	77
Golden Whistler*	73	70	45	78	55	75	80	56	78	35	13
Grey Shrike-thrush*	27		24	7		17	30	43	27	40	47
Spectacled Flycatcher*				6							
Leaden Flycatcher*										15	13
Southern Logrunner*	6		6	4	23	11	2.5	13			
Whipbird*	41	18	40	23	30	48	56	58	44	24	17
Large-billed Scrubwren*	10	5	5	21	8	5	5	20	6		
Yellow-throated Scrubwren*	22	25	27	16	38	21		15	8		
Brown Thornbill*	67	63	72	30	63	51	45	47	45	63	34
Striated Thornbill*	8	3	10	5	8	6	32		17	49	61
Brown Warbler*	76	76	81	93	83	77	76	93	50	2.5	1.25
Varied Sittella*							9		1.25	5	2.5
Lewin Honeyeater*	26	18	45	58	80	90	65	90	70	6	6
Yellow-faced Honeyeater*									29	31	32
White-naped Honeyeater*										18	20
Red-browed Firetail*			2.5	1.25		4			5	8	18
Spotted Pardalote*					3	6	25	3	23	34	21
Striated Pardalote*									7	2.5	9
Paradise Riflebird*	8			6		14	15		5		
Dusky Woodswallow*									5	2.5	1.25
N =	80	40	80	80	40	80	80	40	80	40	40

+ Values are the percentage of points at which a bird species was recorded (N = number of points sampled in each habitat).

# CTRF = Cool Temperate Rainforest;
WTST = Warm Temperate/Subtropical Rainforest;
TO= Tall Open-Forest;
DO= Dry Open-Forest;
UL= Unlogged;
OL= Old Logging;
RL= Recent Logging;
M= Moist understorey;
D= Dry understorey.

Table 4. Census Results.

A. Mean Bird Species per hectare 1980-1982.

	Cool Temperate RF*			Warm Temp./Sub RF*			Tall Open			Dry Open	
#	UL	OL	RL	UL	OL	RL	UL	OL	RL	M	D
Mean	7.5	7.3	7.8	6.3	7.2	7.5	7.5	7.4	6.1	7.8	6.6
S.E.	2.9	2.6	1.8	2.7	2.6	1.2	2.9	2.1	1.3	1.8	1.4
N=	96	56	96	96	56	96	96	56	96	80	80

(N is the number of spot census points)

B. Mean Bird Species per Transect (10 hectares) 1980-1982.

	Cool Temperate RF			Warm Temp./Sub RF			Tall Open			Dry Open	
#	UL	OL	RL	UL	OL	RL	UL	OL	RL	M	D
Mean	33.6	28.5	34	32	33.5	36.7	34.6	32.5	36	41.5	41.7
S.E.	4.4	4.5	7.0	4.7	0.5	1.6	4.6	0.5	6.5	4.5	5.0
N=	12	8	12	12	8	12	12	8	12	8	8

(N is the number of transects)

C. Mean Number Birds per hectare 1980-1982.

	Cool Temperate RF			Warm Temp./Sub RF			Tall Open			Dry Open	
#	UL	OL	RL	UL	OL	RL	UL	OL	RL	M	D
Mean	32.4	31.4	33.7	25.9	33	38.4	31.5	35.5	25.9	31.5	31.5
S.E.	3.9	5.1	4.6	4.0	5.9	5.3	4.1	5.8	3.6	1.9	1.9
N=	96	56	96	96	56	96	96	56	96	80	80

(N is the number of spot census points)

*Cool temperate RF = Cool temperate rainforest;
Warm temp/sub RF = Warm temperate/subtropical rainforest.
UL = Unlogged; OL = Old logging; RL = Recent logging; M = Moist understorey; D = Dry understorey.

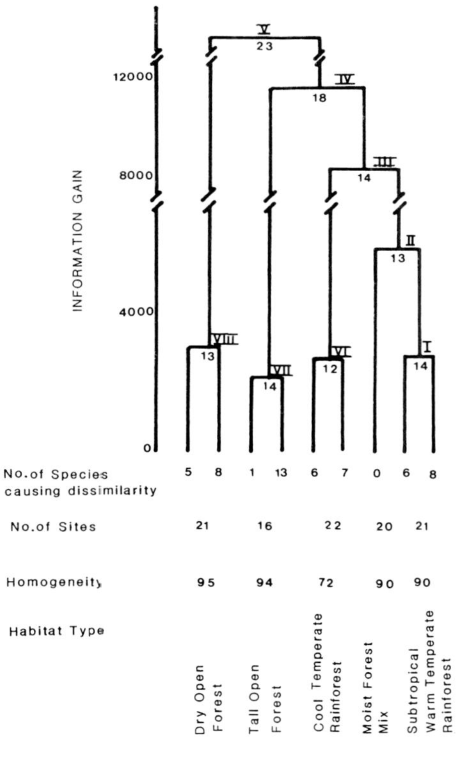

Fig. 1. Site groupings of forest types.

addition, the ratio between the number of ten minute counts on which a bird species was recorded to the total number of ten minute counts done at each study site was calculated (Table 3). This ratio provides a basis for comparing the frequency of occurrence of a bird species between habitats.

The Mann-Whitney U-test was used to test the significance of the difference in the yearly values obtained at each study site for mean number of bird species per hectare, mean number of bird species per transect and the mean number of individual birds per hectare (Table 4). Dissimilarity analysis (NIASM from TAXON, CSIRONET) was used to group habitat types according to species richness and to the abundance of birds at each census point (Fig. 1).

RESULTS

In this study, 117 bird species were recorded, and 133 bird species have been confirmed from the Hastings region (this study, For. Comm. 1981; R.A.O.U., 1982). There were 53 species of insect-eating birds. Many of these were ubiquitous. A common suite of forest insectivores includes Fan-tailed Cuckoo, *Cuculus pyrrhophanus*, Rose Robin, *Petroica rosea*, Yellow Robin, *Eopsaltria australis*, Golden Whistler, *Pachycephala pectoralis*, Southern Logrunner, *Orthonyx temminkii*, Brown Warbler, *Gerygone mouki,* and Brown Thornbill, *Acanthiza pusilla* (Table 3).

Other major groups include the Pigeons, (Columbidae) and the Honeyeaters, (Meliphagidae), Honeyeaters were found primarily in the eucalypt forests although two species, Scarlet Honeyeater, *Myzomela sanguinolenta,* and Lewin Honeyeater, *Meliphaga lewinii*, occurred in rainforest. The pigeons were primarily fruit eaters (five species) and confined to rainforest, although two species, Wonga Pigeon, *Leucosarcia melanoleuca,* and Brush Bronzewing, *Phaps elegans,* were also found in other forest types and habitats. None of the Pigeons were present in large numbers, although Brown Pigeon, *Macropygia amboinensis,* was common in moist forests.

A group of large, omnivorous species were distributed through the moist elements of the Hastings forest (rainforest and moist eucalypt forests). Canopy dwellers in this group include Satin Bowerbird, *Ptilonorhynchus violaceus*, Green Catbird, *Ailuroedus crassirostris,* and Paradise Riflebird, *Ptiloris paradiseus*. Ground feeders include Brush Turkey, *Alectura lathami*, Noisy Pitta, *Pitta versicolor,* and Superb Lyrebird, *Menura novaehollandiae*. Satin Bowerbird was found commonly in moist forests throughout the region, while the other species were irregularly distributed.

Three species which have limited and disjunct distributions in New South Wales (Morris *et al.* 1981) were recorded in this study: Rufous Scrub Bird, *Atricornis rufescens*, Olive Whistler, *Pachycephala olivacea,* and Forest Raven, *Corvus tasmanicus*. Rufous Scrub Bird was recorded in all habitats (Table 3), showing a preference for habitat with a thick understorey (Table 2). Olive Whistler was common in the moist forest types, showing preference for those containing Antarctic Beech (Table 3). Forest Raven was recorded infrequently throughout the region.

Four species of owls were recorded in the Hastings region. The Boobook Owl, *Ninox novaeseelandiae,* was recorded in all forest types. Powerful Owl, *N. strenua,* and Sooty Owl, *Tyto tenebrosica,* were recorded infrequently in the moist forest types. Masked Owl, *T. novaehollandiae,* was recorded once in dry open-forest.

Frequency of Species Occurrence

The frequency of occurrence of bird species over the four unlogged treatments show a sharp difference between the moist forest types (cool temperate rainforest, warm temperate/subtropical rainforest and wet sclerophyll forest) and dry open-forest (Table 3). Using the frequency of occurrence data, it is possible to define one suite of bird species associated with the moist forests and one suite associated with dry open-forest. The moist forest suite included King Parrot, *Alistersus scapularis*, Scaly Thrush, *Zoothera dauma*, Rose Robin, Southern Logrunner, Eastern Whipbird, *Psophodes olivaceus*, Yellow-throated Scrubwren, *Sericornis citreogularis,* and Brown Warbler, among others. The mean frequency of occurrence of this species group was 24% on moist forest plots and 1.9% on open plots. Another suite of species was common in dry open-forest and absent or uncommon in the moist forest types (Table 3). Pallid Cuckoo, *Cuculus pallidus*, Scarlet Robin, *Petroica multicolor*, Rufous Whistler, *Pachycephala rufiventris*, Yellow-faced Honeyeater, *Lichenostomus chrysops,* and Striated Pardalote, *Pardalotus striatus,* represent a suite of birds with a mean frequency of occurrence of 30% in dry, open plots and 1.8% in the moist forest types.

The comparative frequency of occurrence of three pairs of congeneric species illustrates the difference between moist, dense forest and open, dry forest. Rose Robin and Scarlet Robin are common in the Hastings region but differ in frequency of occurrence across forest types (Table 3). Rose Robin was most abundant in moist habitats, showing a slight preference for tall open-forest (30% occurrence) over cool temperate rainforest (26% occurrence) or warm temperate/subtropical rainforest (21% occurrence). Rose Robin was uncommon on the open plots (5% occurrence). Scarlet Robin was common in dry open-forest (34% occurrence) and absent from the moist forest types. Rufous Whistler virtually replaces Olive Whistler at the transition between moist and dry habitat (Table 3). Rufous Whistler was common in open habitat (occurring on 56% of the counts), uncommon in tall open-forest (1.25% occurrence) and absent from rainforest. Olive Whistler occurred in all moist habitats (cool temperate rainforest: 27%; warm temperate/subtropical rainforest: 12.5%; tall open-forest: 1.25%) and was absent from dry open-forest (Table 3). Lewin Honeyeater was ubiquitous in tall open-forest (occurring in 90% of the counts), abundant in warm temperature/subtropical forest (58% occurrence), common in cool temperate rainforest (26% occurrence) and rare in dry open-forest (6% occurrence). Yellow-faced Honeyeater was common in dry open-forest (31% occurrence) and absent from the moist forest types.

Of the 32 species in Table 3 (which presents data from 1981 and 1982), nine bird species were restricted to eucalypt forest and did not occur at all in the rainforest sites. Two species from Table 3, Spectacled Flycatcher, *Monarcha trivergatis,* and Brown Pigeon, were restricted to rainforest and did not occur in eucalypt sites. Of the 117 species recorded over the entire census period (1980-1982), only two were restricted to rainforest (Spectacled Flycatcher and Pale Yellow Robin, *Tregallasia capito*). Both of these species were found only in unlogged warm temperate/subtropical rainforest. Sixteen species from the total of 117 were confined to eucalypt forest.

Site Classification by Distribution and Abundance of Birds

One hundred forest census sites were subjected to NIASM dissimiliarity analysis to give grouping according to the population density of birds and the species richness of each site. The results show that the dry open-forest sites separate from all other sites at high level of information gain (Fig. 1). The moist forests (tall open-forest, cool temperate rainforest, warm temperate/subtropical rainforest) were grouped together, separating into discernible groups at a lower level of information gain. The three moist forest types did appear to have distinct avian communities, as indicated by the homogeneity of the groupings. That is, a high percentage of the sites from each of these forest types were placed in the same group by the analysis procedure (94%, tall open-forest; 72%, cool temperate rainforest; 90%, warm temperate/subtropical rainforest). However, considerable overlap among the avian communities was indicated by the site grouping comprised of mixed moist forest census sites.

The Number of Bird Species

There was no significant difference in the mean number of bird species per hectare between unlogged forest types ($P > 0.05$). There was no significant difference in the mean number of bird species per transect at the 0.05 level, although there was a difference between dry open-forest and all other forest types at the 0.2 level of significance. Basically, forest types were similar in terms of species richness, although dry open-forest showed a difference that was almost significant when transects were compared. The two unlogged rainforest types supported fewer bird species than the two unlogged sclerophyll forest types (Table 4). Dry open-forest had the highest mean number of species per hectare (7.8) and per transect (41.5). Tall open-forest had the second highest mean number of species per transect (34.6) and was similar to cool temperate rainforest in mean number of species per hectare (both 7.5). Cool temperate rainforest had a lower mean number of bird species per transect (33.6). Warm temperate/subtropical rainforest had the lowest mean value for number of bird species per hectare (6.3) and number of bird species per transect (32).

The plots with a eucalypt component had higher bird species counts than the plots dominated by rainforest trees (Table 4). The plot with the highest bird species counts (dry open-forest) had a large component of open habitat (40% cover) which was essentially absent from the other habitats (Table 3).

Population Density of Birds

There was no significant difference in the number of birds per hectare between unlogged forest types ($P > 0.05$). The mean number of birds per hectare varied by less than one individual bird across three forest types (cool temperate rainforest, tall open-forest and dry open-forest). The only habitat which varied was warm temperate/subtropical rainforest. This forest type carried an average of 5.3 fewer birds per hectare than the other unlogged forest types. Over the three year study period, cool temperate rainforest supported the largest mean number of birds per hectare (32.4) of any unlogged forest (Table 4). Tall open- and dry open-forest both carried 31.5 birds per hectare. Warm temperate/subtropical rainforest supported the lowest number of birds per hectare (25.9).

Avifauna of Logged Forests

Census results from different forest treatments (logging histories) within forest types are presented to illustrate the effects of manipulation of forest structure and floristics on the diversity and abundance of birds (Table 4). Logged sites had the highest measures of open habitat and plant species (Tables 1 and 2). There were no significant differences in diversity or abundance of birds ($P > 0.05$) between logged and unlogged sites. The recently logged sites in cool temperate rainforest and warm temperate/subtropical rainforest had the highest measures of avian diversity and abundance (Table 4). Measures of avian diversity and abundance in tall open-forest and dry open-forest were variable across the forest treatments sampled, and revealed no repeated pattern during this study.

DISCUSSION

Interaction between avian communities and habitat have been studied over a broad spectrum of vegetation and forest types in Australia by Kikkawa 1968, Holmes 1973 and Recher 1969. The effects of alteration to forest environments have been examined by Disney and Stokes 1976; Driscoll 1977; Recher *et al.* 1980 and Loyn 1980, among others. Studies of forest alteration have concentrated on the effects of establishing plantations of exotic species or major logging operations in the sclerophyll forest of southern Australia (see also Smith 1985 and Kavanagh *et al.* 1985). Results have indicated a significant reduction in the number of birds and the number of bird species in treated forest, followed by varying degrees of recovery. Limited studies of the effects of logging in the moist forests of northern New South Wales have been carried out by Pattemore and Kikkawa 1975 and Binns 1983. Neither of these studies investigated the relationship between avian communities and a range of forest types and treatments, which was our intention. The discussion that follows relates the observed distribution and abundance of birds to the habitat parameters of the forests sampled.

The mean numbers of bird species censused in unlogged forest were not significantly different across the four forest types sampled, indicating that the bird species are evenly distributed throughout the forest environment. Despite the overall uniformity of the avian community, two major species groups were revealed by the frequency with which they occurred across forest types (Table 3) and by the grouping of sites from dissimilarity analysis of species occurrence (Fig. 1). One suite of species was associated with open forests containing a eucalypt element, while the other was typical of moist forests with a closed canopy and a dense rainforest understorey.

Dry open-forest had the most diverse and distinctive avian community in terms of bird species occurrence (Table 4). This forest type harboured the suite of bird species associated with open eucalypt habitat. The number of eucalypt tree species present (Table 3) is one measure of the availability of foraging resources (nectar, carbohydrate exudates) required by some bird species that were restricted to sclerophyll habitat (Rainbow Lorikeet, *Trichoglossus haematodus,* and White-naped Honeyeater, *Meliphaga lunatus*). Nesting hollows, required by many species of Australian birds (Disney and Stokes 1976), are more common in eucalypt forest than rainforest (Mackowski, pers. comm.). Birds that forage in open habitat (Leaden Flycatcher, *Myiagra rubecula*, Welcome Swallow, *Hirundo neoxena*) or that forage on open ground (Scarlet Robin, Fantailed Cuckoo) were common in dry open-forest and absent or less common in other habitats.

Tall open-forest shared some of the above habitat features with dry open-forest (e.g. nectar, hollows, carbohydrates). Open habitat, however, was a unique feature of dry open-forest. Structurally, tall open-forests resembled the rainforest types. Subsequently, the two sclerophyll forests had bird species in common that were dependent on eucalypt resources, but open country birds were restricted to the drier forest.

Tall open-forest also had many bird species in common with the two rainforest types (Table 3). This uniformity within the avian community reflects an overlap of species that forage in the complex association of understorey plants that is a common feature of rainforest and moist eucalypt forest (Webb 1968). Golden Whistler, Brown Warbler and Large-billed Scrub Wren, *Sericornis magnirostris,* all feed on insects taken from the sub-canopy or upper shrub layer. Rufous Fantail, *Rhipidura rufifrons,* and Black-faced Flycatcher, *Monarcha melanopsis,* forage in lower shrubs, while Yellow-throated Scrubwren, *S. frontalis*, Scaly Thrush and Southern Logrunner are primarily ground feeders. All of these birds were common in tall open-forest as well as rainforest (Table 3).

Patches of moist understorey can be found along stream courses even in dry open-forest, resulting in the sporadic intrusion of "moist" bird species into this habitat. Patches of dry habitat did not intrude into rainforest. Thus, bird species associated with dry open-forest seldom occur in rainforest.

The uniform nature of the avian community across all forest types was accentuated by the lack of birds specifically associated with rainforest. Birds that were largely confined to rainforest habitat (e.g. Top-knot Pigeon, *Lopholaimus antarcticus*, Emerald Dove, *Chalcophaps indica*, Pale Yellow Robin, Spectacled Flycatcher) were found to be uncommon in the Hastings region in terms of occurrence and abundance. Many rainforest species which could be expected in the region (Morris *et al.* 1981) were not recorded at all. This group includes Red-crowned Pigeon, *Ptilopus regina*, Yellow-eyed Cuckoo-shrike, *Coracina lineata,* and Rufous Shrike Thrush, *Colluricincla megarhyncha.* The absence of these species from the Hastings study sites contributed to the comparatively low number of species recorded in warm temperate/subtropical rainforest (Table 4). About two hundred bird species could be expected to occur in the Hastings region, but only 133 have been confirmed (this study, For. Comm. 1981). Rainforest birds comprised a suite of birds conspicuous by their absence.

The lack of significant difference in mean numbers of birds per hectare indicate that the forest types censused supported similar numbers of birds. Although differences were small, warm temperate/subtropical rainforest supported the lowest number of birds per hectare (Table 4). There was little difference in the mean number of birds in the other three forest types, where the populations of birds varied by less than one individual per hectare. Lower bird numbers in the warm temperate/subtropical forest can be related to the lack of a eucalypt component and lack of rainforest fruits. Birds that require resources supplied by eucalypts were largely excluded from this habitat. Many birds associated with rainforest are dependent on fruits, which appear at irregular intervals (Crome 1975; Frith 1975). None of the rainforest trees were fruiting during the census periods.

Avifauna of Forest Treatments

No significant differences in bird diversity or abundance related to forest treatment resulted from our census work. This is in agreement with the results of Pattemore and Kikkawa 1975 and Binns 1983, who worked in subtropical rainforest in northern New South Wales, and with some studies from North America (e.g. Webb *et al.* 1977) and Europe (Slagsvold 1977). However, studies of eucalypt logging in southeastern Australia have revealed significant differences between logged and unlogged forest (Loyn 1980; Pattemore 1980; Smith 1985; Kavanagh *et al.* 1985). In these areas, the removal of trees for woodpulp as well as for sawn timber resulted in an intensity of logging greater than that generally practised in the Hastings region, while forest growth potential is generally lower (For. Comm. 1981; Bridges 1983). The lower intensity of logging and higher growth potential of the Hastings forests were factors involved in reducing the difference among the avian communities of the logged and unlogged forests sampled in this study.

Open habitat was increased in logged areas. Bird species typical of open habitat were present in logged treatments of the closed forest types, and absent or uncommon on the unlogged sites in these forest types (Table 3).

CONCLUSIONS

Early research into the relationship between birds and a wide range of habitats indicated that avian diversity and abundance were predicted by the structure of vegetation (MacArthur and MacArthur 1961), with structurally complex habitats supporting more birds than simple habitats. Later studies indicated that floristic characteristics of habitat were the dominant factors in determining the structure of the avian community within a single form of vegetation (Williamson 1971). Current theory, which is supported by this study, recognises that both structural and floristic parameters influence the relationship between birds and habitat (Cody 1974).

Our study found that bird species were uniformly distributed across forest types and treatments in the Hastings region of northern New South Wales. The most important structural characteristic determining the observed differences in bird species diversity was the amount of open habitat, whether occurring naturally or as a result of logging. The number of bird species in logged forests was also increased by remnant canopy trees, dense regeneration at ground level and the presence of nearby unlogged reserves. In relation to bird species number, the most important floristic character was the component of eucalypts present or absent in each habitat or treatment. Bird species dependent on rainforest were uncommon or not recorded in the region, and the two rainforest types sampled were comparatively depauperate in terms of bird species occurrence. The population density of birds was uniform across forest types and treatments.

These data and conclusions are the result of census studies, and not a direct measure of reproductive or foraging use of habitat by the avian community. To develop detailed land management principles, it will be necessary to evaluate the functional strategy that birds employ to exploit forest resources in northern New South Wales. Current options indicated by this study for the management of bird communities in a multiple use forest would include retention of canopy elements within logged areas and maintaining a system of unlogged forest reserves within the region of forest operations. The extent and distribution of retained vegetation necessary for the conservation of birds depends on the intensity of logging, site characteristics and the reproductive and foraging requirements of the resident bird community.

ACKNOWLEDGEMENTS

The Forestry Commission of New South Wales provided facilities and support for this programme of research. The authors would like to acknowledge the field assistance provided by Steve Wallace, Walter Boles and Glen Holmes. Ian Bevege, Hugh Dowden, Gary King, R. A. Curtin and John Mills provided invaluable comments on the manuscript. Bill Chapman and Gary King supplied detailed descriptions of the vegetation occurring on the study sites.

REFERENCES

BAUR G. N., 1983. Notes on the silviculture of major New South Wales forest types. 1. Moist Coastal Hardwood Types. For. Comm. New South Wales Internal Report.

BINNS, D., 1983. Birds of a logged and unlogged rainforest. Forest and Timber. Vol. 19(2).

BRIDGES, R. G., 1983. Integrated logging and regeneration in silvertop ash-stringybark forests of the Eden region. For. Comm. NSW Res. Paper Number 2.

CODY, M. L., 1974. Competition and the structure of bird communities. Princeton Univ. Press. Princeton, N.J. U.S.A.

CROME, F. H. J., 1975. The ecology of fruit pigeons in tropical northern Queensland. *Aust. Wildlife Res.* 2: 155-85.

DISNEY, H. J. DE S. AND STOKES, A., 1976. Birds in pine and native forests. *Emu.* 76: 133-38.

DRISCOLL, P. V., 1977. Comparison of bird counts from pine and native forests. *Aust. Wildlife Res.* 4: 281-88.

FORESTRY COMMISSION OF NEW SOUTH WALES, 1981. Proposed rainforest logging operations in the Hastings Catchment: Environmental Impact Statement. For. Comm. New South Wales, Sydney.

FRITH. H. J., 1975. Wildlife Conservation. Angus and Robertson, Sydney.

HOLMES, G., 1973. The bird species diversity of some subtropical Australian forests. Unpubl. Thesis. B.Sc. (Hons.) U.N.E.

KAVANAGH, R. P., SHIELDS, J. M., RECHER, H. F. AND ROHAN-JONES, W. G., 1985. Birds in a logged-unlogged forest mosaic in the Eden woodchip region. Pp. 273-81 *in* Birds of Eucalypt Forests and Woodlands: Ecology, Conservation, Management ed. by A. Keast, H. F. Recher, H. Ford and D. Saunders. Royal Australasian Ornithologists Union.

KIKKAWA, J., 1968. Ecological association of bird species and habitats in eastern Australia; similarity analysis. *J. Anim. Ecol.* **36**: 143-66.

LOYN, R. H., 1980. Bird populations in a mixed eucalypt forest used for production of wood in Gippsland, Victoria. *Emu.* **80**: 145-56.

MACARTHUR, R. H. AND MACARTHUR, J. W., 1961. On bird species diversity. *Ecology.* **42**: 594-98.

MORRIS, A. K., MCGILL, A. R. AND HOLMES, G., 1981. Hand list of Birds in New South Wales. New South Wales Field Onithologists Club. Sydney.

PATTEMORE, V., 1980. Effects of the pulpwood industry on wildlife in Tasmania. Part 3: Succession in bird communities and their habitat. Tasmanian N.P.W.S. Tech. Rep. 80/1.

PATTEMORE, V. AND KIKKAWA, J., 1975. Comparisons of bird populations in logged and unlogged rainforest at Wiangarie State Forest. *Aust. For.* **37**: 189-98.

POPLE, G. D. AND COWLEY, N. B., 1981. The rainforest inventory. For. Comm. New South Wales, Sydney.

RALPH, C. J. AND SCOTT, J. M. (eds), 1980. Estimating Numbers of Terrestrial Birds. Proceedings of an International Symposium held at Asilomar, California, October 26-31, 1980. Studies in Avian Biology No. 6.

ROYAL AUSTRALASIAN ORNITHOLOGISTS UNION, 1982. Records of Congress field survey, Mt. Boss State Forest. RAOU Field Atlas Records, Dec., 1982.

ROYAL AUSTRALIAN HISTORICAL JOURNAL, 1940. Cedar and the cedar getters. *Royal Aust. Historical Journal and Proceedings* **25**(2): 131-56.

RECHER, H. F., 1969. On the relationship between bird species diversity and habitat diversity in Australia and North America. *Amer. Nat.* **103**: 75-9.

RECHER, H. F., ROHAN-JONES, W. G. AND SMITH, P., 1980. Effects of the Eden woodchip industry on terrestrial vertebrates with recommendations for management. For. Comm. of New South Wales, Res. Note 42.

SLAGSVOLD, T., 1977. Bird population changes after clearance of deciduous scrub. *Biol. Cons.* **12**: 229-44.

SMITH, P., 1985. Woodchip logging and birds near Bega, New South Wales. Pp. 259-71 *in* Birds of Eucalypt Forests and Woodlands: Ecology, Conservation, Management ed. by A. Keast, H. F. Recher, H. Ford and D. Saunders. Royal Australasian Ornithologists Union.

UDVARDY, M. D. F., 1959. Notes on the ecological concepts of habitat, biotope and niche. *Ecology.* **40**: 725-28.

WEBB, L. J., 1968. Environmental relationships of the structual types of Australian rainforest vegetation. *Ecology* **49**: 296-311.

WEBB, W. L., BEHREND, D. T. AND LAISORN, B., 1977. Effects of logging on songbird populations in a northern hardwood forest. *Wildl. Monogr.* **55**: 35p.

WIENS, J. A., 1981. Single-sample surveys of communities: are the revealed patterns real? *Amer. Nat.* **117**: 90-8.

WILLIAMSON, P., 1971. Feeding ecology of the Red-eyed Vireo (*Vireo olivaceus*) and associated foliage gleaning birds. *Ecological Monographs* **41**: 129-52.

CHAPTER 7

The Relationships Between Abundances of Honeyeaters and Their Food Resources in Open Forest Areas Near Sydney

Graham H. Pyke[1]

In open forest near Sydney honeyeater density (or honeyeater energy requirement) was positively correlated, both seasonally and spatially, with the rate of production of nectar energy per unit area. However, these correlations were weak. There were no significant seasonal or spatial correlations between honeyeater density (or honeyeater energy requirement) and average biomass of flying insects caught by sticky traps.

INTRODUCTION

HONEYEATERS (Meliphagidae) are ideal subjects for determining relationships between their abundances and the levels of productivity of their food resources. Many species appear to obtain virtually all their energy from nectar or non-nectar carbohydrate sources such as manna or lerp and their protein from flying insects (Recher and Abbott 1970; Paton 1980; Pyke 1980). The productivities of nectar and non-nectar energy and the availability of flying insects are easily estimated (Gill and Wolf 1975; Hainsworth and Wolf 1972; Paton 1980; Pyke 1980, 1983a; Wolf 1975).

Very few studies of these relationships have, however, been carried out. Collins and Briffa (1982) found that, for each of three areas within a 259 hectare reserve near Perth, Western Australia, honeyeaters were present only when there were appropriate nectar-producing flowers present. They also detected movement from one area to another by a few birds (Collins and Briffa 1982). However, they were unable to examine the relationship between honeyeater abundance and nectar productivity in a more quantitative fashion (see Collins and Briffa 1982). In a companion study to the present, I compared seasonal patterns of honeyeater abundance within circles of radius 20 m located in heathland habitat near Sydney with seasonal patterns of productivity of nectar energy (no non-nectar carbohydrate energy sources detected) and of abundance of flying insects (honeyeaters caught almost all their insects by hawking them from the air) (Pyke 1983a). I found no apparent relationship between these patterns (Pyke 1983a). I also found that differences in honeyeater density between nearby circles within a study site were unrelated to differences in nectar energy productivity (Pyke 1983a). A number of authors have suggested that movement patterns of honeyeaters are closely related to seasonal patterns of flowering in different places (Collins 1980; Collins and Briffa 1982; Ford and Pursey 1983; Keast 1968). However, in view of the largely anecdotal nature of supporting evidence and the results of my study of heathland honeyeaters (Pyke 1983a), this claim should be further evaluated.

I report here the results of a study, identical in design to my study of heathland honeyeaters (Pyke 1983a), carried out in open forest areas adjacent to the heathland areas in the other study. The aim of both studies was to determine, as precisely as possible, the relationships between honeyeater density, energy productivity and insect abundance.

METHODS

Only those methods unique to the present study are described in detail here. Other methods are reported elsewhere (Pyke 1983a, b).

The present study extended from December, 1980 to March, 1983. It was carried out in the following areas: (a) the West Head region of Ku-ring-gai Chase National Park near Post 28, 30 km north of

[1]Department of Vertebrate Ecology, The Australian Museum, 6-8 College Street, Sydney, New South Wales, Australia.
Pages 65-77 *in* BIRDS OF EUCALYPT FORESTS AND WOODLANDS: ECOLOGY, CONSERVATION, MANAGEMENT ed. by A. Keast, H. F. Recher, H. Ford and D. Saunders. Royal Australasian Ornithologists Union and Surrey Beatty & Sons.

Sydney); and (b) the Brisbane Water National Park (on west side of road from Pearl Beach to Patonga, just north of road to Warrah Trig, 35 km north of Sydney). In each of these areas study sites were established in open-forest habitat (Benson and Fallding 1981) adjacent to the heathland areas considered in the companion study (see Pyke, 1983a). The dominant tree species in these open-forest areas are *Eucalyptus haemastoma, E. gummifera* and *E. oblonga*. Beneath these trees is a shrubby understorey composed mostly of *Lambertia formosa, Petrophile pulchella, Isopogon anemonefolius, Grevillea buxifolia, Acacia myrtifolia, Gompholobium grandiflorum* and *Phylloda phylicoides*. The trees are mostly 10 to 15 m tall, while the shrubs are about 1 to 1.5 m tall. At each study site circles of radius 20 metres were permanently located and marked with coloured surveyors' tape at 10 and 20 metre distances from the centres. At West Head there are two woodland circles, about 20 m apart, established on February 25th, 1981. At Brisbane Water there are six open-forest circles. Two of these were established on December, 15th, 1980 in a portion of this study area that last burnt in December 1976 (R. Bradstock, pers. comm.). Two more were established on March, 8th, 1981 in the same portion of this study area, but in an area where *Telopea speciosissima* (Waratah) is relatively abundant. Flowering in the four circles just mentioned was very similar, however, and they are combined in the analysis below. The last two woodland circles were established on September, 8th and 17th, 1981 in a portion of this study area that last burnt in December, 1979 (pers. obs.). Little or no flowering occurred in these last two circles prior to September, 1981 (see discussion in Pyke, 1983c). All woodland circles at Brisbane Water are located within a 150 m x 50 m area. Every three weeks for all circles estimates were made of the density of honeyeaters, daily energy production and insect abundance. These estimates were obtained in the following ways:

(i) *Honeyeater Density*

An observer, sitting on top of a 2.5 m ladder located at the centre of a circle, would scan one half of the circle every minute and record the following: species identity of birds sighted (plus age and sex if known), distance of each bird from observation point, behaviour of each bird (e.g. taking nectar, sitting). For heathland birds, I found that data obtained by the two of the observers employed in the present study (author and Helen Ponton) could be combined and used to compare *relative* abundances of different species in the various circles within a study site at different times of the year (Pyke 1983b). The same applies to the method when applied to open-forest birds by the three observers employed in the present study (author, Helen Ponton and David Allen) (Pyke, unpublished data). In the present study, presumably because the vegetation is extremely similar in both study sites, it is also possible to use the census method to compare the two study sites (Pyke, unpublished data).

On each census day up to December 17th, 1981, an early morning and a late morning census were carried out (see Pyke 1983a). After December 17th, 1981, only early morning censuses were carried out. The number of observers varied from one to two (author, G.H.P., alone; Helen Ponton alone; G.H.P. and H.P.; G.H.P. and David Allen). An observer spent approximately 16 minutes at a circle before moving to another (see Pyke 1983a). At West Head each observer visited each circle twice during a census period. At Brisbane Water each observer usually visited each circle once, but on a few occasions when bird densities were very low, one observer visited half the circles once each while the other observer did the same for the other circles.

(ii) *Energy Production*

With the possible exception of White-eared Honeyeaters, *Meliphaga leucotis*, which occurred in low numbers in both study areas, no honeyeaters were observed feeding on non-nectar carbohydrate sources and no such energy sources were detected in the foraging locations of other honeyeaters. Consequently nectar is assumed to be the energy source for all other honeyeaters observed in the present study.

Flower densities were estimated by counting the flowers within the circles. This was difficult for the flowers of the *Eucalyptus* spp. and consequently the estimates of flower density for these species are at best approximations and at worst relative indices. This problem does not, however, affect the conclusion of the study (see below).

In the present study, as in the companion heathland study (Pyke 1983a), rates of nectar production had to be estimated differently for each plant species. For the *Eucalyptus* spp. visited by the honeyeaters *(E. oblonga* and *E. gummifera)* almost all nectar-feeding by the birds occurred early in the morning and high levels of honeybee, *Apis melifera*, feeding kept nectar at very low volumes for the rest of the day. Consequently the daily production of *Eucalyptus* nectar available to the honeyeaters was estimated by the average dawn standing crop of nectar. Nectar production for *Lambertia formosa* was estimated in the manner described in Pyke (1983a). Nectar production in *Telopea speciosissima* and *Banksia serrata* was estimated as the average nectar volume present in flowers (inflorescences) from which nectar had been removed and that had been bagged with mosquito netting for 24 hours.

For all plant species the sugar concentrations of nectar samples were measured using a pocket refractometer compensated to 20°C (see

Table 1. Body weights and daily energy requirements of honeyeaters.

Species	Mean Body weight (g)	n	s.e.	Daily energy requirements (kJ)
White-cheeked Honeyeater*	17.99	403	0.09	59.14
New Holland Honeyeater*	19.55	182	0.13	62.44
Little Wattlebird*	69.45	12	4.79	142.86
Eastern Spinebill*	11.08	184	0.08	43.09
White-eared Honeyeater*	21.63	54	3.3	66.70
Yellow-faced Honeyeater*	17.78	247	0.27	58.68
White-naped Honeyeater*	14.50	50	2.0	51.37
Tawny-crowned Honeyeater*	19.00	23	3.3	61.28
Noisy Friarbird*	122.00	1	—	206.40
Red Wattlebird*	110.00[1]	—	—	192.90
Lewin Honeyeater*	30.60	2	—	83.65

[1]From Ford and Paton (1982). *See Appendix 1 for Scientific names.

Table 2. Summary of open-forest observations of nectar-feeding by various honeyeater species.

Honeyeater Species	*Lambertia formosa*	*Banksia serrata*	*Eucalyptus oblonga*	*Eucalyptus gummifera*	*Telopea speciosissima*	Totals
White-cheeked Honeyeater *(Phylidonyris nigra)*	164	53	35	30	4	286
Little Wattlebird *(Anthochaera chrysoptera)*	72	17	8	7	5	109
New Holland Honeyeater *(Philidonyris novaehollandiae)*	89	18	6	11		124
Eastern Spinebill *(Acanthorhynchus tenuirostris)*	50	3	3	2	3	61
White-eared Honeyeater *(Meliphaga leucotis)*	6	10	1	1		18
Tawny-crowned Honeyeater *(Phylidonyris melanops)*			2			2
Noisy Friarbird *(Philemon corniculatus)*		4				4
ALL HONEYEATERS	381	105	55	51	12	604

Hainsworth and Wolf 1972). These sugar concentrations were transformed into mg sucrose equivalents per 100µl of nectar and multiplied by the daily volumes of nectar produced to yield estimates of daily energy production per flower. With further multiplication by the numbers or density of flowers, the rates of energy production per circle or per unit area are obtained.

(iii) *Insect Abundance*

The abundance and sizes of flying insects were estimated using the "sticky trap" method described in Pyke (1983a). At West Head sticky trap samples were obtained every three weeks from March 25th, 1981, until February 16th, 1982. Sticky trap samples were obtained in the Brisbane Water 1976-burn area between March 26th, 1981, and March 10th, 1982, and in the 1979-burn area between October 20th, 1981 and February 17th, 1982.

(iv) *Energetic Requirements of Honeyeaters*

Daily requirements for each honeyeater species were estimated from their average body weights (Table 1) and the equation of Walsberg (1980) (log of Daily Energy Requirements in kJ = log 8.96 + 0.653 log W where W is body weight in g) for birds that do not feed on the wing (see Table 1).

RESULTS

(a) *Seasonal Patterns of Flowering*

Only five forest plant species were utilized as nectar sources by honeyeaters. These were *Lambertia formosa, Telopea speciosissima, Banksia serrata, E. gummifera* and *E. oblonga* (Table 2). *Banksia spinulosa* and *B. marginata* (West Head only) flowered in low numbers and were not seen to be visited by the birds.

Most forest plant species have a single annual flowering peak which occurs at approximately the same time in the two study sites (Figs 1, 2, 3). *Lambertia formosa*, however, has two annual flowering peaks, one in November, the other around March (Figs 1, 2).

Flowering levels vary markedly with location and year. For *Lambertia formosa* and *Telopea speciosissima* flower densities are greatest in the Brisbane Water 1979-burn area, lower in the 1976-burn area and lowest at West Head which last burnt in 1965 (Figs 1, 2, 3). This pattern and other evidence indicates that flowering in these species is a function of the time since the last fire (Pyke 1983b).

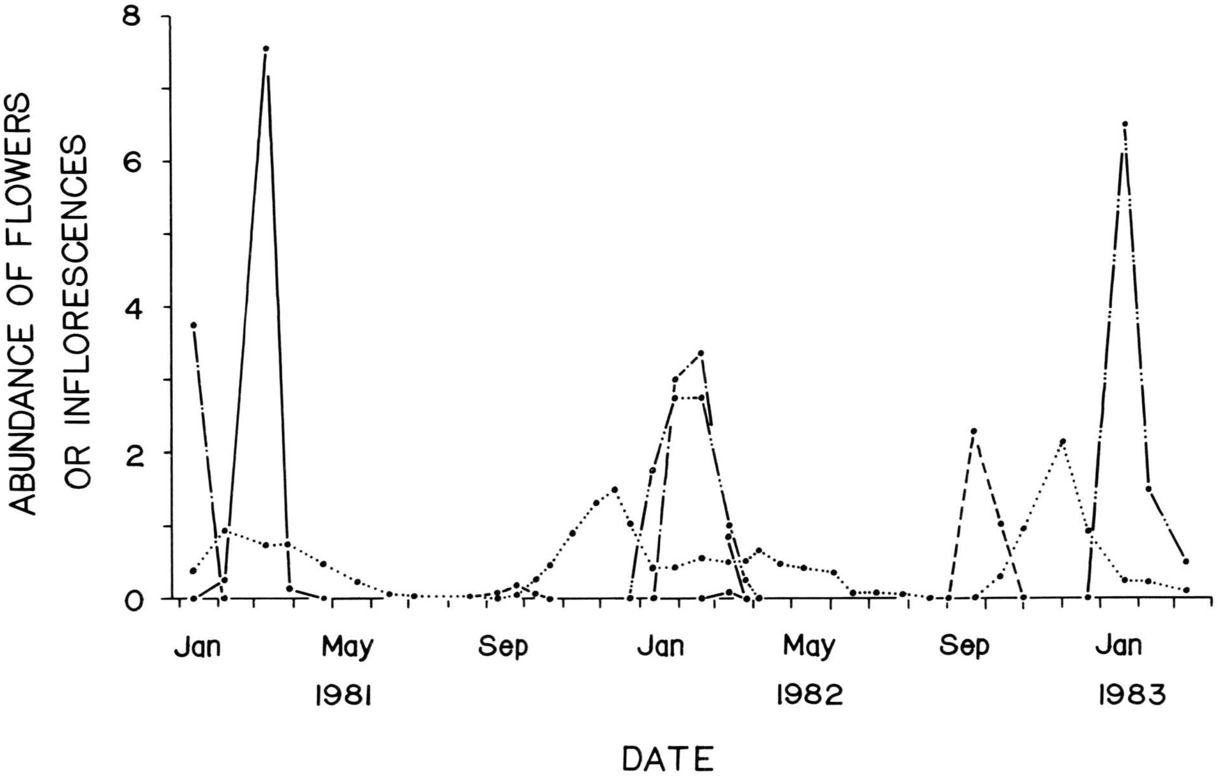

Fig. 1. Numbers of flowers or inflorescences per circle in the Brisbane Water study area, burnt in 1976. ——— *Eucalyptus gummifera* (No. flowers/400), ——— . ——— *E. oblonga* (No. flowers/400), •.........• *Lambertia formosa* (No. inflorescences/60), ——— .. ——— *Banksia serrata* (No. inflorescences), and — — — — — — *Telopea speciosissima* (No. flowers/200).

Table 3. Summary of daily nectar productions and nectar concentrations for various plant species.

	Daily nectar production (μl)			Nectar concentration (% wt/vol)			Daily energy production (J)	
Plant Species	Mean	s.e.	n	Mean	s.e.	n	Mean	s.e.
Lambertia formosa								
(per flower) —a	21.8	1.2	575	12.8	0.3	112	43.1	2.6
—b	75.0	2.0	329	15.3	0.2	125	188.8	5.4
—c							116.0	3.0
—d	22.8	2.2	183	14.7	0.2	119	51.8	
—e	45.9	2.0	198	15.9	0.1	204	113.4	
—f							82.6	
Eucalyptus oblonga (per flower)	0.19	0.05	39	15	(estimated[1])		0.44	
E. gummifera (per flower)	12.9	1.5	29	10.5	0.2	10	21.1	2.4
Banksia serrata								
(per inflorescence) —g	1928	597	10	18.6	0.6	20	6005	1999
—h	1952	527	10	26.0	0.7	16	7724	2012
—i	2469	557	10	32.0	2.8	9	10903	2192
Telopea speciosissima (per flower)	9.7	1.5	66	24.4	0.9	32	36.7	6.0

a, d — Nectar produced between late afternoon (mid morning) of first day of flower's opening and late afternoon (mid morning) on next (i.e. last) day of flower's life.

b, e — Nectar produced during bud stage and up to late afternoon (mid morning) on first day of flower's life.

c, f — Combined average energy production per flower, independent of flower age.

d, e, f — January 11th, 28th; February 17th, 1982.

[1] — Estimated from literature values for *Eucalyptus* spp.

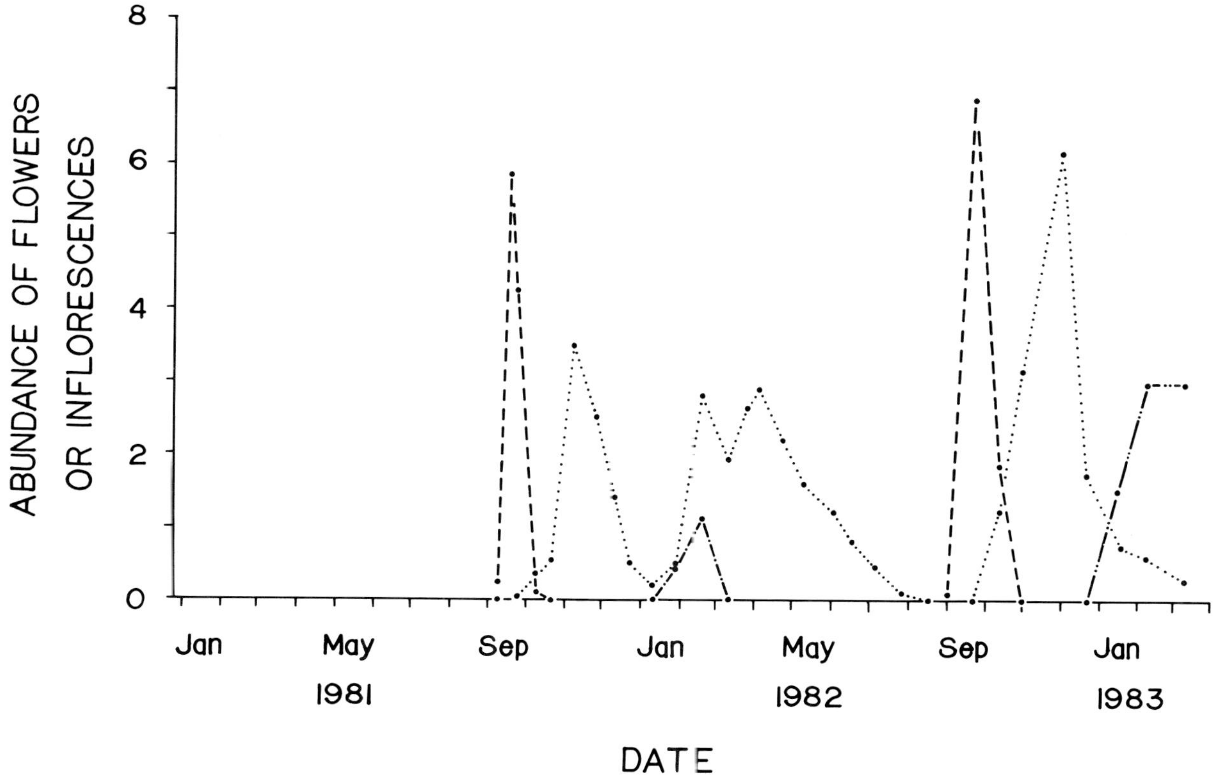

Fig. 2. Numbers of flowers or inflorescences per circle in the Brisbane Water study area, burnt in 1979. ________ *Eucalyptus gummifera* (No. flowers/400), ______ . ______ *E. oblonga* (No. flowers/400), ●.....................● *Lambertia formosa* (No. inflorescences/120), ______ .. ______ *Banksia serrata* (No. inflorescences), and _ _ _ _ _ _ *Telopea speciosissima* (No. flowers/200).

Table 4. Summary of data on abundance of flying insects.

	Number of flying insects per sticky trap						
	January 1st, to June 30th			July 1st, to December 31st			
Location	Mean	s.e.	n	Mean	s.e.	n	Probability[2]
West Head	3.64	0.34	102	7.40	0.49	129	$<10^{-5}$
Brisbane Water (1976 burn)	2.86	0.21	109	6.17	0.55	148	$<10^{-5}$
Brisbane Water (1979 burn)	3.11	0.38	56	6.81	1.14	95	$<10^{-5}$
Probability[1]	0.56			0.60			
Brisbane Water (combined)	2.95	0.19	165	6.42	0.56	243	$<10^{-5}$
Probability[3]	0.08			0.19			

[1] Comparing 1976 and 1979 burn areas (Student t-test).
[2] Comparing West Head and combined Brisbane Water areas (Student t-test).
[3] Comparing two time periods (Student t-test).

Eucalyptus gummifera flowered extensively in both study sites in February and March of 1981 but produced few or no flowers in 1982 (Figs 1, 2, 3). *Banksia serrata* showed the reverse pattern (Figs 1, 2, 3).

(b) *Seasonal Patterns of Nectar Production and Concentration*

Most plant species showed no significant seasonal variation in nectar production and concentration. Data for these species are summarized in Table 3. For *Banksia serrata* nectar concentration increased significantly during its flowering season (Table 3, P's < 0.0s, Student t-test).

Estimates of daily energy production per flower (or per inflorescence) are obtained for each plant species by multiplying together the nectar production (μl), the nectar concentration (mg sucrose equivalents per 100 μl of nectar) and the energy value of sucrose (15.5 J/mg — Heinrich 1975).

(c) *Seasonal Patterns of Energy Production*

Daily energy production per circle shows two distinct annual peaks in both study sites, one in February-March and another in October-November (Figs 4, 5, 6). The first peak results from flowering by *Lambertia formosa* and either *Eucalyptus*

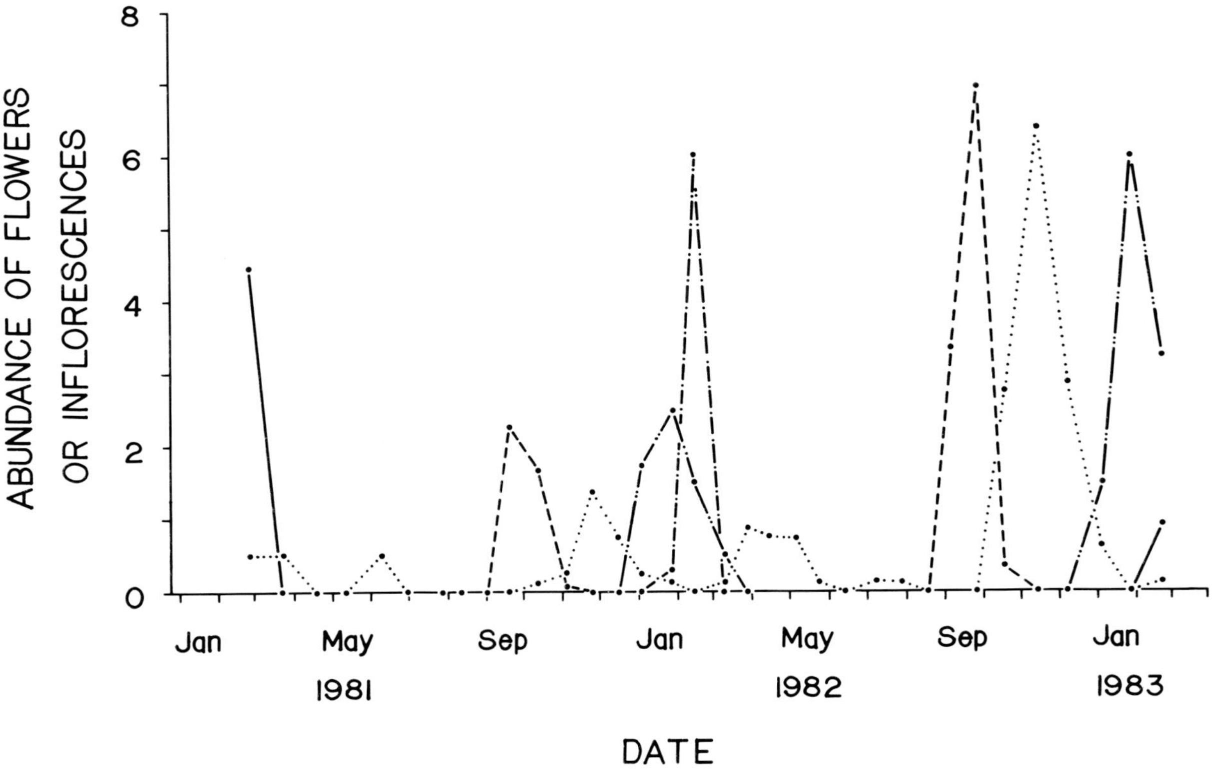

Fig. 3. Numbers of flowers or inflorescences per circle at West Head. ——— *Eucalyptus gummifera* (No. flowers/200), ——·—— *E. oblonga* (No. flowers/200), •·········• *Lambertia formosa* (No. inflorescences/4), ——··—— *Banksia serrata* (No. inflorescences/2), and – – – – – – *Telopea speciosissima* (No. flowers/20).

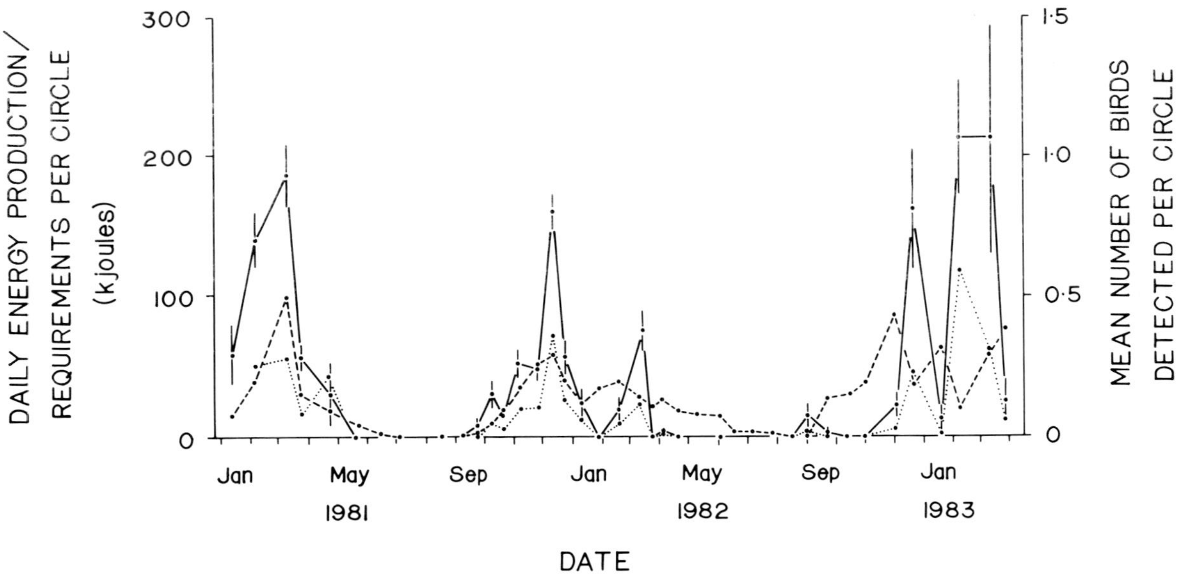

Fig. 4. Daily energy production, mean numbers of honeyeaters and daily energy requirements of honeyeaters per circle for Brisbane Water study area, burnt 1976. ——— mean number of birds, – – – – – daily energy production, ············ requirements per circle.

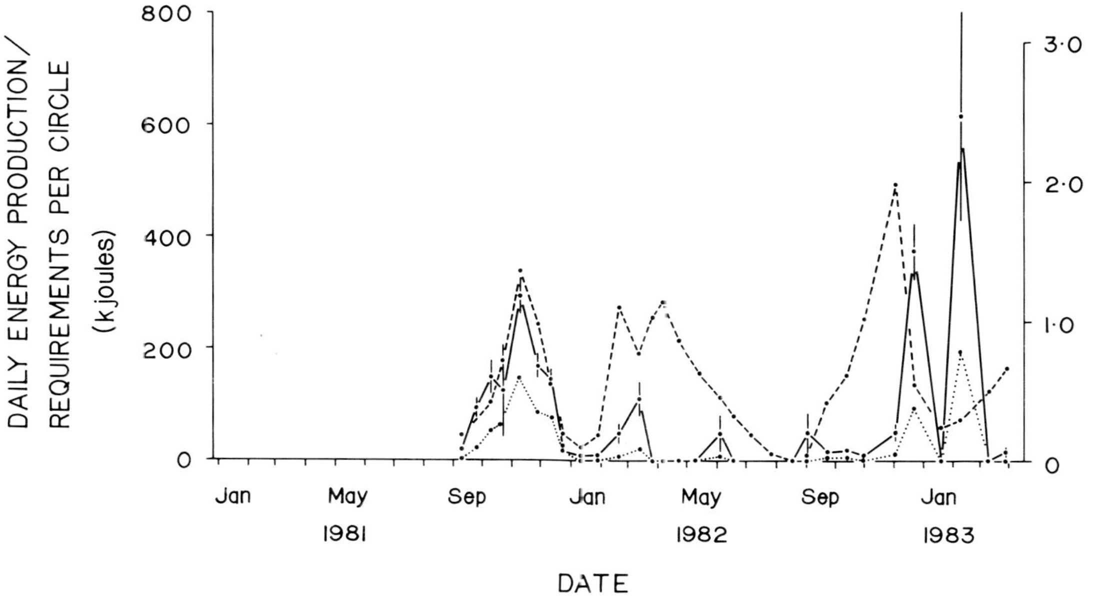

Fig. 5. Daily energy production, mean numbers of honeyeaters and daily energy requirements of honeyeaters per circle for Brisbane Water study area, burnt in 1979. ———— mean number of birds, – – – – – daily energy production, ... requirements per circle.

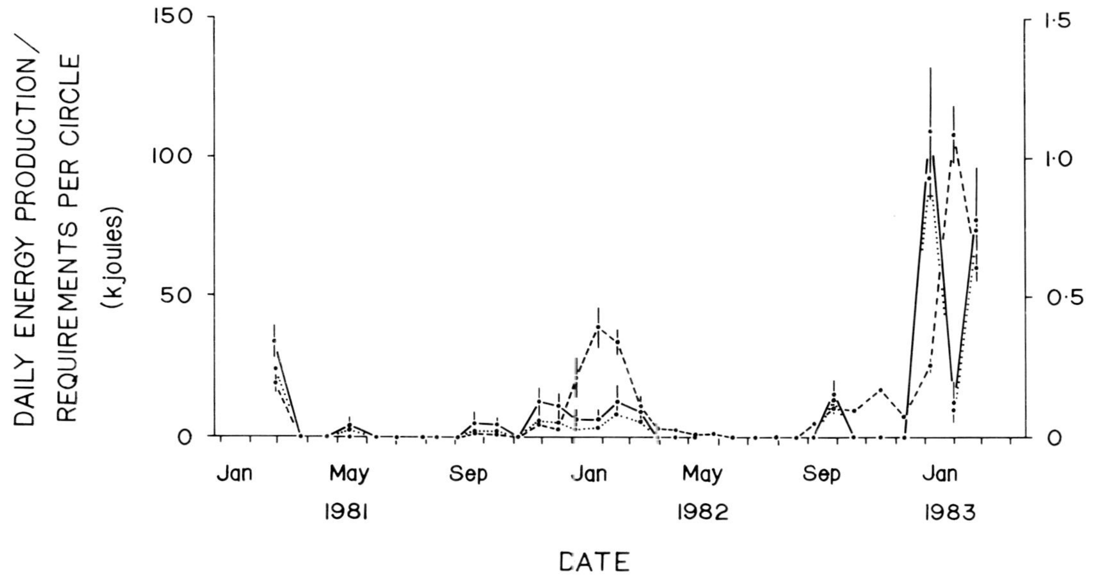

Fig. 6. Daily energy production, mean numbers of honeyeaters and daily energy requirements of honeyeaters per circle at West Head. ———— mean number of birds, – – – – – daily energy production, ... requirements per circle.

gummifera or *Banksia serrata*. The second peak is almost entirely due to *Lambertia formosa*.

(d) *Seasonal Patterns of Abundance, Size and Composition of Flying Insects*

Almost all the animals caught by the sticky traps were flying insects. A few (1.4%, n = 3394) spiders and ants were also caught, presumably after walking onto the traps, and will not be considered further.

Diptera made up most (90.4%) of the 3,345 insects caught in both study sites with Hymenoptera accounting for 2.9%, Lepidoptera 0.6%, Coleoptera 5.8%, and Hemiptera 0.6%. Over the whole year there was no difference between West Head and the Brisbane Water 1976-burn area in the composition of insects caught. Diptera comprised 92.9% (n = 1302) of the insects caught at West Head and 91.9% (n = 1228) in the Brisbane Water 1976-burn area (P = 0.32; Student t-test based on normal approximation

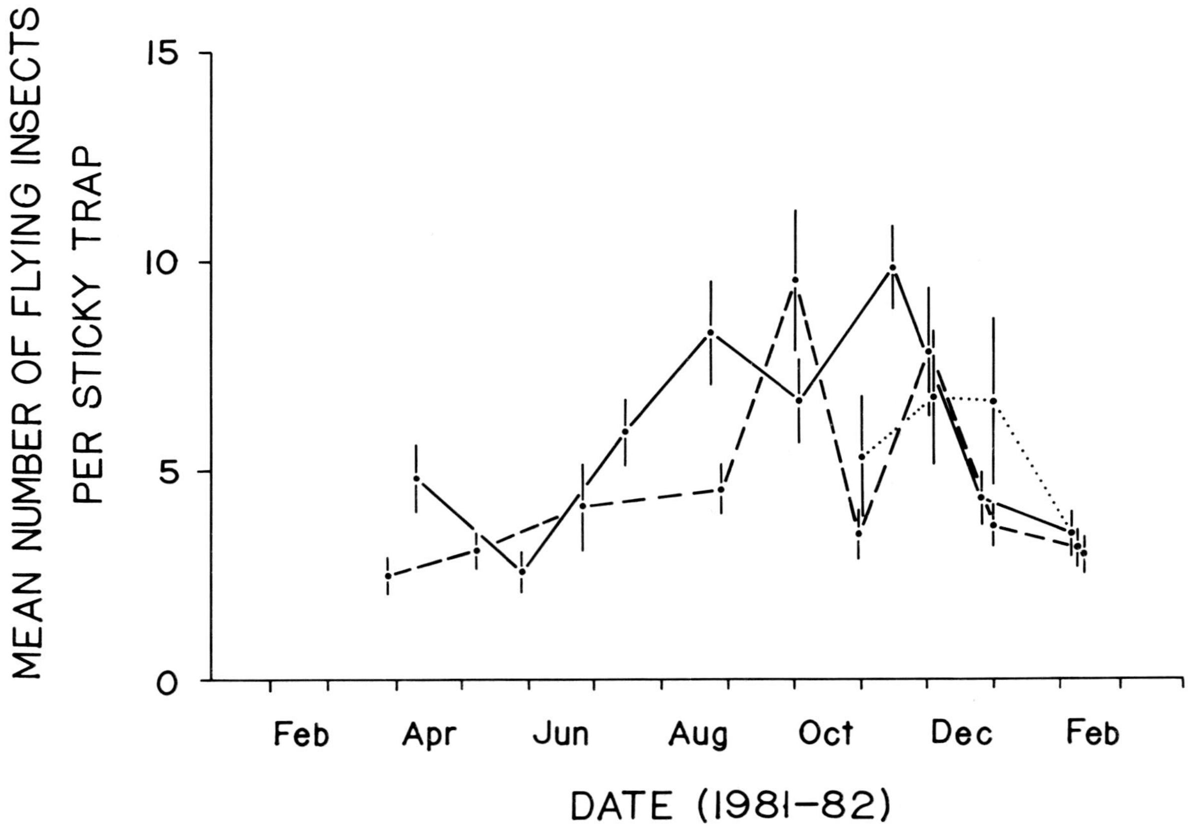

Fig. 7. Mean numbers of flying insects per sticky trap at different times in three study areas. _ _ _ _ _ Brisbane Water area burnt in 1976, ________ West Head, ... Brisbane Water area burnt in 1979.

to binomial). During the period when sticky trap samples were obtained in the Brisbane Water 1979-burn area (i.e. October 21st, 1981 to February 17th, 1982) the percentage of Diptera in the samples was slightly lower but there were no significant differences between areas (e.g. Brisbane Water 1976-burn area — 86.5%, n = 541; 1979-burn area — 84.2%, n = 815; P = 0.28, Student t-test based on normal approximation to binomial).

The abundance of flying insects tended to be higher between July 1st, and December 31st, than in the first half of the calendar year in both study sites (Fig. 7, Table 4). No significant differences were found between the 1979-burn and the 1976-burn portions of the Brisbane Water study site (Table 4). For both halves of the year the average number of insects caught per sticky trap was higher at West Head than for the combined Brisbane Water samples; however, the differences are not significant (Table 4).

For most of the year mean size of flying insects follows the same seasonal pattern in all areas, being low in the winter months of May through August and relatively high the rest of the year (Fig. 8, Table 5). At West Head, however, mean insect size dropped almost to winter levels in December before increasing again (Fig. 8). No significant differences occurred between the 1979- and 1976-burn portions of the Brisbane Water study site (Table 5). The average insect size was significantly lower at West Head than for the combined Brisbane Water samples during both time periods (Table 5). Average insect biomass per trap exhibited peaks during September to December and March to April and troughs at other times, especially during May to August (Pyke 1983c).

(e) *Honeyeater Behaviour*

The behaviour of honeyeaters recorded during scans is summarized in Table 6. Birds that were observed hawking are assumed to have been catching flying insects (Recher and Abbott 1970). Birds that were gleaning are assumed to have been harvesting insects from the vegetation as no other food resources were apparent in these locations. Probing leaf and stem exils by White-eared Honeyeaters may have yielded carbohydrate-rich sap. It is not known what birds obtained while feeding on bark. No seasonal behaviour patterns were apparent.

The top four species in Table 6 all rely heavily on nectar rather than insects for food. White-eared Honeyeaters, on the other hand, utilized nectar for

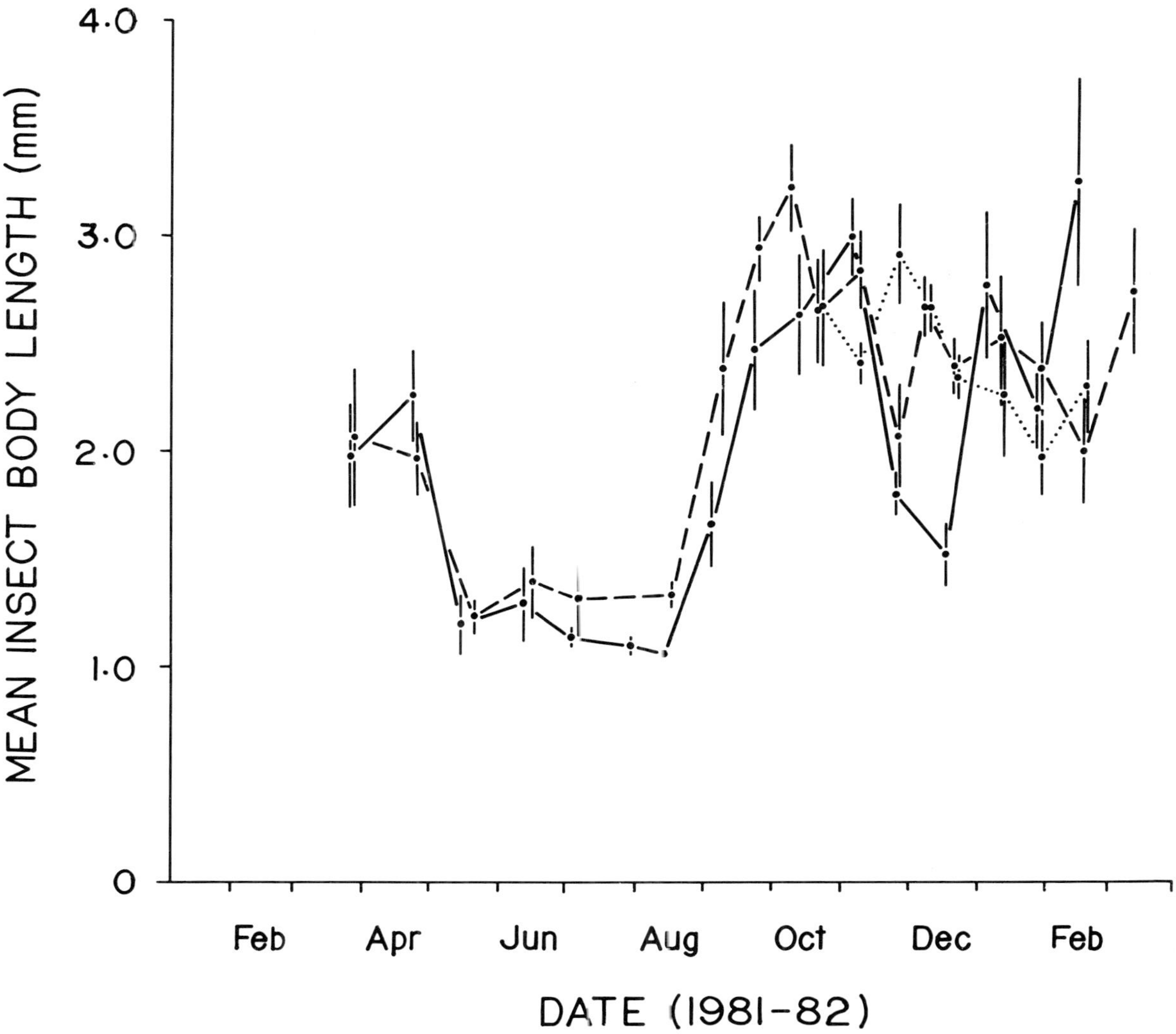

Fig. 8. Mean insect body lengths for flying insects caught in three study areas at different times. _ _ _ _ _ Brisbane Water area burnt in 1976, ________ West Head, ... Brisbane Water area burnt in 1979.

Table 5. Summary of data on size of flying insects

	Insect Body Length (mm)						
	May-August			September-April			
Location	Mean	s.e.	n	Mean	s.e.	n	Probability[1]
West Head	1.12	0.03	421	2.29	0.07	896	$< 10^{-5}$
Brisbane Water (1976 burn)	1.32	0.08	239	2.62	0.06	990	$< 10^{-5}$
Brisbane Water (1979 burn)	—	—	—	2.46	0.06	823	
Probability[2]	—			0.06[4]			
Brisbane Water (combined)	1.32	0.08	239	2.55	0.04	1813	
Probability[3]	0.003			0.007			

[1] Comparing two time periods (Student t-test).
[2] Comparing 1976 and 1979 burn areas (Student t-test).
[3] Comparing West Head and combined Brisbane Water (Student t-test).
[4] If observations are further restricted to the period (October 21st to February 17th) when insect samples were taken in the 1979-burn area, the difference between the two areas is slightly smaller, in the opposite direction and still non-significant

less than half of the feeding observations and may have obtained much of their energy from plant sap (Table 6). No nectar feeding was observed for Lewin Honeyeater, *M. lewinii*, and Yellow-faced Honeyeater, *M. chrysops*, (Table 6), two species that were seldom present in the study sites (Tables 6 to 8). The following comparisons of honeyeater density with nectar-energy productivity will therefore focus on the top four species of Table 6.

Table 6. Summary of observations of honeyeater behaviour. *See Appendix 1 for Scientific names.

Species	Perching	Flying	Nectar	Hawking	Gleaning	Probing axibs	Bark	n
White-cheeked Honeyeater*	31.2%	13.2%	54.3%	0.7%	0.6%			536
New Holland Honeyeater*	36.3%	9.3%	51.8%	1.3%	1.3%			226
Little Wattlebird*	66.8%	7.2%	23.2%		2.5%		0.2%	488
Eastern Spinebill*	44.6%	8.0%	38.4%	5.4%	3.6%			112
Noisy Friarbird*	72.7%	9.1%	18.2%					22
White-eared Honeyeater*	31.7%	6.7%	26.7%		15.0%	18.3%	1.7%	60
Lewin Honeyeater*	83.3%				16.7%			6
Yellow-faced Honeyeater*		100.0%						5

Table 7. Honeyeater densities at West Head (Mean number per circle ± 1 s.e.

Date	White-cheeked Honeyeater	New Holland Honeyeater	Little Wattlebird	Eastern Spinebill	White-eared Honeyeater
26/ 2/81	0.083 ± 0.083	0.208 ± 0.045	0.042 ± 0.042	—	0.042 ± 0.042
25/ 3/81	—	—	—	—	—
21/ 4/81	—				
13/ 5/81	0.021 ± 0.021			0.021 ± 0.021	
11/ 6/81					
2/ 7/81					
30/ 7/81					
13/ 8/81					
3/ 9/81					
22/ 9/81				0.045 ± 0.045	
14/10/81				0.047 ± 0.027	
5/11/81					0.031 ± 0.022
25/11/81				0.125 ± 0.048	0.031 ± 0.022
17/12/81		0.031 ± 0.022		0.078 ± 0.034	0.141 ± 0.045
5/ 1/82				0.063 ± 0.044	
27/ 1/82		0.031 ± 0.031		0.031 ± 0.031	
16/ 2/82		0.125 ± 0.061			
12/ 3/82	0.094 ± 0.053				0.063 ± 0.044
29/ 3/82	—				
16/ 4/82					
7/ 5/82					0.094 ± 0.053
7/ 9/82					
29/ 9/82	0.109 ± 0.040		0.047 ± 0.027		
19/10/82					
16/11/82					
10/12/82					
4/ 1/83	0.125 ± 0.061	0.656 ± 0.173	0.313 ± 0.128		
28/ 1/83	0.031 ± 0.031	0.063 ± 0.044	0.031 ± 0.031		
23/ 2/83	0.031 ± 0.031	0.438 ± 0.144	0.313 ± 0.128		

(f) *Seasonal Patterns of Honeyeater Abundance*

The average honeyeater density (for the top four species of Table 6) appears to consistently show three peaks — in about September, November and February (Figs 4 to 6). There are no obvious differences among the species in seasonal patterns of abundance (Tables 7 to 9).

Honeyeaters tended to be absent from the study areas when little or no nectar was being produced (Figs 4 to 6). Overall, however, the seasonal correlations between honeyeater density (top five species in Table 6) and nectar-energy productivity are low (Table 10). These correlations are little changed when total daily energy requirements are substituted for density (Table 10). The seasonal correlations between honeyeater density or daily energy requirements and average insect biomass per trap are all non-significant (Table 10).

(g) *Spatial Patterns of Honeyeater Abundance*

Nectar energy productivity and honeyeater density (or honeyeater energy requirements) were correlated spatially. During the study period nectar energy produced per circle per day was always greater for the Brisbane Water 1979-burn area than for 1976-burn area and, with the exception of one occasion, was greater for the latter area than for West Head. On 18 of 24 occasions the honeyeater density (or energy requirements) was greater for the Brisbane Water 1979-burn area than for the 1976-burn area. On 25 of 29 occasions the latter area had a higher honeyeater density (or energy requirements) than the West Head area.

Differences in honeyeater density between the three study areas did not correspond to similar differences in insect biomass per sticky trap. For

Table 8. Honeyeater densities in Brisbane Water 1976-burn area (Mean number per circle ± 1 s.e.).

Date	White-cheeked Honeyeater	New Holland Honeyeater	Little Wattlebird	Eastern Spinebill	White-eared Honeyeater	Noisy Friarbird	Red Wattlebird	Tawny-crowned Honeyeater	Yellow-faced Honeyeater	Lewin Honeyeater
11/ 1/81	0.273 ± 0.108	0.018 ± 0.018								
5/ 2/81	0.485 ± 0.104	0.091 ± 0.042	0.106 ± 0.058	0.015 ± 0.015	0.030 ± 0.021					
10/ 3/81	0.659 ± 0.087	0.114 ± 0.045	0.023 ± 0.017	0.136 ± 0.039						
27/ 3/81	0.258 ± 0.044			0.023 ± 0.013						
23/ 4/81			0.008 ± 0.008	0.008 ± 0.008						
19/ 5/81										
15/ 6/81										
6/ 7/81										
18/ 8/81										
8/ 9/81										
24/ 8/81				0.042 ± 0.029						
9/10/81	0.069 ± 0.031	0.056 ± 0.034		0.028 ± 0.020						
21/10/81	0.017 ± 0.017	0.067 ± 0.033								
10/11/81	0.141 ± 0.035	0.021 ± 0.010	0.068 ± 0.022	0.031 ± 0.015						
26/11/81	0.132 ± 0.030		0.088 ± 0.029	0.022 ± 0.013						
10/12/81	0.301 ± 0.046	0.110 ± 0.038	0.301 ± 0.046	0.088 ± 0.027	0.015 ± 0.010					
21/12/81	0.103 ± 0.038	0.044 ± 0.025	0.118 ± 0.041	0.015 ± 0.015						
11/ 1/82			0.063 ± 0.031	0.063 ± 0.038						
28/ 1/82										
17/ 2/82	0.031 ± 0.022		0.047 ± 0.035	0.016 ± 0.016	0.156 ± 0.076					
10/ 3/82	0.139 ± 0.043	0.014 ± 0.042	0.042 ± 0.037	0.153 ± 0.049				0.028 ± 0.020		
26/ 3/82										
5/ 4/82	0.016 ± 0.016									
22/ 4/82									0.031 ± 0.022	
2/ 6/82										
30/ 8/82				0.075 ± 0.043						
21/ 9/82				0.016 ± 0.016						
11/10/82										
1/11/82										
30/11/82	0.047 ± 0.027			0.063 ± 0.031						
22/12/82	0.500 ± 0.161	0.125 ± 0.061	0.063 ± 0.044		0.031 ± 0.031					0.156 ± 0.081
18/ 1/83				0.031 ± 0.031			0.031 ± 0.031			
7/ 2/83	0.281 ± 0.098	0.094 ± 0.069	0.313 ± 0.111	0.031 ± 0.031	0.156 ± 0.081	0.344 ± 0.105				
8/ 3/83	1.031 ± 0.410			0.031 ± 0.031						
27/ 3/83	0.063 ± 0.044		0.063 ± 0.063							

Table 9. Honeyeater densities in Brisbane Water 1976-burn area (Mean number per circle ± 1 s.e.).

Date	White-cheeked Honeyeater	New Holland Honeyeater	Little Wattlebird	Eastern Spinebill	White-eared Honeyeater	Noisy Friarbird	Red Wattlebird	White-naped Honeyeater	Yellow-faced Honeyeater	Lewin Honeyeater
17/ 9/81		0.019 ± 0.019		0.058 ± 0.033						
24/ 9/81	0.055 ± 0.023		0.110 ± 0.029	0.086 ± 0.027						
9/10/81		0.344 ± 0.109	0.219 ± 0.083	0.031 ± 0.031						
21/10/81		0.072 ± 0.051	0.407 ± 0.088							
10/11/81	0.198 ± 0.056	0.084 ± 0.036	0.906 ± 0.110							
26/11/81	0.088 ± 0.033	0.038 ± 0.022	0.550 ± 0.077							
10/12/81	0.075 ± 0.030		0.513 ± 0.076							
21/12/81			0.025 ± 0.025	0.500 ± 0.036	0.025 ± 0.025					
11/ 1/82				0.032 ± 0.032						
28/ 1/82	0.032 ± 0.032				0.068 ± 0.044					
17/ 2/82				0.157 ± 0.068				0.032 ± 0.032		
10/ 3/82	0.125 ± 0.062			0.285 ± 0.115						
26/ 3/82										
5/ 4/82										
22/ 4/82									0.032 ± 0.032	
2/ 6/82				0.188 ± 0.136						
30/ 8/82				0.200 ± 0.138						
21/ 9/82	0.063 ± 0.044									
11/10/82	0.052 ± 0.037			0.021 ± 0.015						
1/11/82				0.031 ± 0.031						
30/11/82	0.016 ± 0.016	0.172 ± 0.050		0.016 ± 0.016						
22/12/82	0.365 ± 0.127	1.000 ± 0.134	0.058 ± 0.033	0.077 ± 0.038	0.019 ± 0.019					0.019 ± 0.019
18/ 1/83					0.063 ± 0.044					
7/ 2/83	1.750 ± 0.749		0.281 ± 0.088	0.094 ± 0.069		0.344 ± 0.105				
8/ 3/83										
27/ 3/83	0.063 ± 0.044									

Table 10. Summary of seasonal correlations between honeyeater density or biomass per unit area and nectar energy productivity or insect biomass per trap.

	Nectar energy productivity			Insect biomass per trap		
	West Head	Brisbane Water 1976-burn	Brisbane Water 1979-burn	West Head	Brisbane Water 1976-burn	Brisbane Water 1979-burn
Honeyeater density	0.44 (n=29, P=0.02)	0.48 (n=34, P=0.01)	0.00 (n=26, P=1.00)	0.07 (n=16, P>0.70)	0.43 (n=17, P=0.09)	0.36 (n=8, P=0.36)
Honeyeater daily energy requirements	0.44 (n=29, P=0.02)	0.35 (n=34, P=0.04)	0.10 (n=26, P>0.70)	0.05 (n=16, P>0.70)	0.34 (n=17, P=0.20)	0.46 (n=8, P=0.26)

example, marked differences in honeyeater density occurred between the two Brisbane Water areas (Tables 8 and 9) although there was no significant difference in insect biomass per trap.

DISCUSSION

In the present open-forest study sites honeyeater density (or energy requirements) was correlated, both seasonally and spatially, with the rate of production of nectar energy per unit area. These correlations were not, however, very strong. A possible explanation for these results is as follows. Movement of honeyeaters may tend to occur through open-forest areas similar to those in the present study during times when nectar production is greatest. Such a situation could have arisen as the result of co-evolution of bird behaviour and plant flowering. The timings and magnitudes of honeyeater movement could also, however, be partly determined by other factors such as levels of nectar production in other areas. If some birds tend to be resident, at least temporarily, and if site quality increases with increasing nectar production, then a spatial correlation between honeyeater density and nectar-energy productivity would be expected. Movements of transients and lack of omniscience on the part of residents could easily, however, prevent this correlation from being perfect.

In heathland areas adjacent to the present open-forest study there was *no* apparent relationship either seasonally or spatially, between honeyeater density/or daily energy requirements) and nectar energy productivity (Pyke 1983a). This result contrasts somewhat with that of the present study and it is not clear how to reconcile the results from the two habitats. In both habitats New Holland, *Phylidonyris novaehollandiae*, and White-cheeked, *P. nigra* Honeyeaters account for most of the honeyeaters observed (Table 2 and Pyke 1983a). Both species are commonly observed nesting and in nesting-related activities in the heathland, but such activities are seldom observed for these species in the open-forest (personal observations), despite relatively lengthy periods of high nectar production. The abundance, movements and behaviour of these honeyeaters are apparently affected by both habitat and nectar production in a given area and by these and other factors in other areas frequented by the birds.

ACKNOWLEDGEMENTS

The present study was supported by a grant from the Australian Research Grants Committee. Helen Ponton and David Allen helped collect the field data. The NSW National Parks and Wildlife Service permitted the research to be carried out in Ku-ring-gai and Brisbane Water National Parks.

REFERENCES

BENSON, J. S. AND FALLDING, H., 1981. Vegetation survey of Brisbane Water National Park and environs. *Cunninghamia* **1**: 79-113.

COLLINS, B. G., 1980. Seasonal variations in the abundance and food preferences of honeyeaters (Meliphagidae) at Wongamine, Western Australia. *West. Aust. Nat.* **14**: 207-12.

COLLINS, B. G. AND BRIFFA, P., 1982. Seasonal variation of abundance and foraging of three species of Australian honeyeaters. *Aust. Wildl. Res.* **9**: 557-69.

FORD, H. A. AND PATON, D. C., 1982. Partitioning of nectar sources in an Australian honeyeater community. *Aust. J. Ecol.* **7**: 149-59.

FORD, H. A. AND PURSEY, J. F., 1983. Status and feeding of the Eastern Spinebill, *Acanthorhynchus tenuirostris,* at New England National Park, northeastern New South Wales **82**: 203-11.

GILL, F. B. AND WOLF, L. L., 1975. Economics of feeding territoriality in the golden-winged sunbird. *Ecology* **56**: 333-45.

HAINSWORTH, F. R. AND WOLF, L. L., 1972. Crop volume, nectar concentration and hummingbird energetics. *Comp. Biochem. Physiol.* **42A**: 359-66.

HEINRICH, B., 1975. Energetics of pollination. *Ann. Rev. Ecol. Syst.* **6** 139-70.

KEAST, A., 1968. Seasonal movements in the Australian honeyeaters (Meliphagidae) and their ecological significance. *Emu* **67**: 159-209.

PATON, D. C., 1980. The importance of manna, honeydew and lerp in the diets of honeyeaters. *Emu* **80**: 213-26.

PYKE, G. H., 1980. The foraging behaviour of Australian honeyeaters: a review and some comparisons with hummingbirds. *Aust. J. Ecol.* **5**: 343-69.

PYKE, G. H., 1983a. Seasonal pattern of abundance of honeyeaters and their resources in heathland areas near Sydney. *Aust. J. Ecol.* **8**: 217-33.

PYKE, G. H., 1983b. Analysis of an instantaneous census method for heathland birds. *Aust. Wildl. Res.* **10**: 521-6.

PYKE, G. H., 1983c. Relationship betweeen time since the last fire and flowering in *Telopea speciosissima* R. Br. and *Lambertia formosa* Sm. *Aust. J. Bot.* **31**: 293-6.

RECHER, H. F. AND ABBOTT, I. J., 1970. The possible ecological significance of hawking by honeyeaters and its relation to nectar feeding. *Emu* **70**: 90.

WALSBERG, G. E., 1980. Energy expenditure in free-living birds: Patterns and diversity. Pp. 300-305 *in* The XVII Int. Ornithol. Congr. Berlin 1980.

WOLF, L. L., 1975. Energy intake and expenditures in a nectar feeding sunbird. *Ecology* **56**: 92-104.

White-cheeked Honeyeater, *Phylidonyris nigra*

CHAPTER 8

Foraging Ecology and Seasonal Patterns of Abundance in a Forest Avifauna

H. F. Recher[1] and R. T. Holmes[2]

In forests on the Southern Tablelands of New South Wales birds were most abundant and species richness greatest during spring and summer. This coincided with the peak abundance of insects and other foods used by birds. Most species also nested during these months. The majority of the avifauna was insectivorous or nectarivorous. A few species ate seeds or fruits or preyed on other vertebrates. Species often differed in size, habitat, foraging behaviour, the substrates on which prey were located, and foraging heights. Using these differences it is possible to assign species to groups which used the same resources and foraging behaviours. The community organisation developed in this way is useful for understanding how birds use their environment, for determining which resources are important for their survival and for developing plans of management that take wildlife into account.

INTRODUCTION

IN A FOREST we are soon aware of the sounds and movements of birds. With little effort we can identify species and recognize individuals. As we watch, it becomes obvious that species differ in their habits, where they feed and nest, and what foods they use. Should we stay through the day, a routine emerges as birds awaken, feed, interact with each other and return to their roosts for the night. Night birds replace those of the day and the pattern is repeated. Were we possessed with remarkable endurance and remained through the year, the birds about us would change daily with births and deaths, migrations and wanderings. Between years events would be affected by weather, fires or human activities, but the pattern of events would be little changed. The recognition, description and explanation of such patterns is central to the study of community ecology and necessary for the development of sound, scientifically based plans of management for forest wildlife.

In this paper we describe the basic patterns of a eucalypt forest avifauna; where and how birds feed, when they nest, changes in numbers through time, and how they use space. Our objective is to explain how birds respond to temporal and spatial changes in their environment and to identify the resources needed by birds for their survival. We then use this information to analyse the structure of a eucalypt forest avifauna and show briefly how such information can be applied to problems of conservation and forest management.

THE STUDY AREA

Three study plots were located adjacent to the Bondi State Forest near the town of Bombala (36° 54′S, 149°14′E) on the Southern Tablelands of New South Wales at an elevation of 800-860 m. Each plot was 10 ha in area and divided into two census transects (420 m × 120 m) positioned side by side. Two plots (Woodlots 1 and 2) were censused each December from 1976 through 1980. The third plot (Woodlot 3) was censused only in December 1980. Censuses were also conducted on Woodlots 1 and 2 during most months in 1978 and from October 1979 to January 1981.

Woodlots 1 and 2 were mostly regrowth forest and woodland 20 to 80 years old. Mature and senescent trees with numerous hollows for hole-nesting birds were abundant on Woodlot 2, but not on Woodlot 1. Both plots were grazed by sheep and cattle, and timber removed for fencing. Woodlot 3 was located in Victoria in an area of mature forest for which there was no evidence of recent logging and which was not grazed by domestic stock. The three plots were within 5 km of each other.

About half of Woodlot 1 was woodland dominated by Snow Gum, *Eucalyptus pauciflora* and Black Sally,

---

[1]The Australian Museum, P.O. Box A285, Sydney South 2000.
[2]Department of Biological Sciences, Dartmouth College, Hanover, N.H., 03755, U.S.A.
Pages 79-96 *in* BIRDS OF EUCALYPT FORESTS AND WOODLANDS: ECOLOGY, CONSERVATION, MANAGEMENT ed. by A. Keast, H. F. Recher, H. Ford and D. Saunders. Royal Australasian Ornithologists Union and Surrey Beatty & Sons.

E. stellulata. The woodland graded into pasture and was open with a grass *Poa* ground cover and small patches of Matt Rush, *Lomandra longifolia,* and Bracken Fern, *Pteridium*. This half had few shrubs. The forested half of Woodlot 1 was dominated by Narrow-leaved Peppermint, *E. radiata,* and Manna Gum, *E. viminalis,* on well-drained soils and Swamp Gum, *E. ovata,* where drainage was impeded. Canopy heights in the woodland varied from 10 to 20 m and in the forest from 15 to 30 m. The woodland lacked a subcanopy, but in the forest the subcanopy was dominated by young eucalypts with a few Blackwood, *Acacia melanoxylon,* and Silver Wattle, *A. dealbata.* The forest also had extensive areas of shrubs dominated by Blackthorn, *Bursaria spinosa*, Silver Wattle and Holly-leaved Lomatia, *Lomatia ilicifolia.* The ground cover in the forest was sparse and consisted mostly of patches of Bracken and Matt Rush with some grass. There was a deep layer of litter.

Woodlot 2 was more uniform. Except for a woodland fringe of Snow Gum and Black Sally along one side, it was mostly an open dry forest dominated by Narrow-leaved Peppermint and Mountain Gum, *E. dalrympleana.* The canopy averaged 16-17 m with individual trees to 25 m. The subcanopy consisted entirely of small eucalypts. Woodlot 2 lacked ground vegetation and there were few shrubs. The litter layer tended to be sparse.

On Woodlot 3, Narrow-leaved Peppermint, Manna Gum and Mountain Gum dominated the canopy with small numbers of Brown Barrel, *E. fastigata,* and Monkey Gum, *E. cypellocarpa,* also present. The canopy averaged 22 m in height with individual trees to 40 m. Subcanopy, shrub and ground vegetation was well-developed. Young eucalypts and Silver Wattle formed a subcanopy to 13 m in height. The shrub layer was diverse, but Gippsland Waratah, *Teleopea oreades*, Holly-leaved Lomatia, Blackthorn, and wattles (*A. melanoxylon, A. dealbata,* and *A. mucronata*) contributed most of the foliage. The shrub layer varied from 1 to 4 m in height. Ground vegetation consisted of ferns, sedges and various forbs.

The climate at Bombala is cool temperate. Monthly rainfall is variable and there is no seasonal peak. Between 1970 and 1982 rainfall on the Bondi State Forest averaged about 1000 mm per year (Forestry Commission, NSW weather records). However, during this study conditions were drier. Between 1976 and 1981 rainfall averaged 820 mm annually and only 1978 had more than 1000 mm of rain. The Southern Tablelands was declared a "drought area" during 1981 and 1982. Characteristic of the tablelands, summers on the study plots tended to be warm during the day, but cool at night. Winters are cold, but temperatures during the day are seldom below freezing. Snow is infrequent.

METHODS

We used a fixed width transect procedure for counting birds. The method is described by Recher *et al.* (1983b) and compared with other census procedures on the same plots by Shields and Recher (1984). Briefly, birds were counted separately along the transects using detections based on sight and sound. Counts were of two hours duration and were repeated on four consecutive days, weather permitting. Counts commenced within 30 minutes of dawn during the summer and somewhat later in winter. Each transect was censused twice early in the morning and twice late in the morning. Our concern was with the relative number of birds and not with absolute densities. Therefore in most instances we have used an index of avian abundance based on the total number of individuals detected during the four counts. Data from the two transects have been combined to provide an index for the woodlot. We also attempted to locate all nests on the woodlots during 1979/80 and 1980/81. Each nest was marked and the progress of the nest monitored throughout the time it was active.

Bird foraging behaviour was studied from October 1980 to January 1981. Approximately equal amounts of time were spent on each woodlot with observers visiting a different plot each day. At the start of each two-week period we recorded data for each foraging bird as it was encountered. However, once we had recorded 100 or more foraging movements for a species on a woodlot, we ceased making observations on that species for that woodlot and concentrated on other birds.

For each individual we recorded a maximum of five attempts at prey capture (prey attack). Since we could not always determine success, a prey attack was recorded whether or not it was successful. For each attempt at prey capture we recorded the behaviour of the bird, and the height and substrate of the prey. Height was estimated to the nearest metre, but in the analysis heights have been grouped according to vegetation layers. The layers used were ground and debris (0-0.2 m), shrubs (0.2-4.0 m), subcanopy (4.0-10.0 m), and canopy (> 10 m).

Major substrates exploited by birds were ground, air and vegetation. If the prey was on a plant, the plant was identified and the position of the prey described according to whether it was on the trunk or stem, a large or small branch, a twig or foliage. If the bark on a trunk or branch was loose (flaking) or hanging, this was noted. Dead substrates were distinguished from live. In the analysis substrates have been grouped into six categories; ground (including debris and litter), tree trunks, loose bark (on trunks or branches, or hanging), branches (large and small), foliage (leaves and twigs), and air. Additionally two kinds of food were equated with substrates; nectar (flowers) and seeds (e.g. grass seed heads, eucalypt capsules, wattle seed pods).

Nine types of foraging or prey attack behaviour were recognized. Definitions follow Recher *et al.* (1985):

1. *Pounce* — the bird drops or flies down from a perch to take a prey item from the ground or low vegetation. The bird lands, simultaneously taking the prey and then returns to a perch.

2. *Probe* — in probing the bird extracts prey from within or under a substrate. This includes taking prey from soil and litter, and from crevices, loose bark and soft wood.

3. *Glean* — in gleaning the bird walks or hops along, searching for and taking prey from the substrate surface at relatively close distances.

4. *Hang-Glean* — the same as glean, but the bird hangs upside down in a relatively stationary posture while taking prey.

5. *Hover* — a bird in flight remains stationary for brief periods while taking prey from a substrate.

6. *Snatch* — in snatching, a flying bird grabs or snatches a prey item from a substrate other than in the air. In most cases, the bird spots the prey when perched. Then, depending on the distance and position of the prey, the bird takes flight or simply jumps up without pausing as when "hovering" to grab the item.

7. *Hawk* — the bird flies from a perch to take a flying insect.

8. *Nectar* — visiting flowers to feed on nectar.

9. *Seeds* — visiting seed heads or capsules to feed on seeds.

Numbers of arthropods on the major substrates used by birds were assessed as part of the study of foraging ecology. The substrates sampled were ground, bark of trees, foliage of trees, and air (flying insects). In addition the abundance of nectar and eucalypt seed was scored by counts of flowers and seed capsules, but other carbohydrates (e.g. manna, honeydew, grass seed) were not sampled. All substrates and flowers were sampled monthly from July 1980 through June 1981. Estimates of food abundance are therefore slightly out of sequence with the bird census data. The procedures used in sampling food resources available to forest and woodland birds are described by Recher *et al.* (1983a) and by Gowing and Recher (1984).

The Forest Avifauna

Recher *et al.* (1980) recorded 143 species of terrestrial birds during their survey of forest habitats in southeastern New South Wales. One hundred and fifteen species were found in the Bombala district and of these, 81 were considered to be forest or woodland species. The remainder were introduced species (4), vagrants (6), and open country birds (24).

The avifauna of the woodlots was typical of forests and woodlands in southeastern New South Wales. Between 1976 and 1982, we recorded 90 species of birds on the three plots. Fifty-six of these are common and widely distributed in forests and woodlands. Fifty species of birds were recorded nesting on the woodlots between 1976 and 1981. Of these, 38 species nested on one or more of the plots each year.

Woodlot 3 had the richest avifauna and the greatest number of individuals. Averaging the results from the two transects, in December 1980 we recorded 32 species and 168 individuals on Woodlot 3 for a two hour count. Woodlot 1 averaged 26 species and 88 individuals while Woodlot 2 averaged 22 species and 89 individuals. The woodland half (C transect) of Woodlot 1 was less rich than the forested portion (G transect). In December 1980, the woodland census averaged 24 species and 82 individuals compared with 28 species and 94 individuals in the forest census for a two hour count. The difference between the two transects was consistent between seasons and from year to year (Recher *et al.* 1983a).

Although most of the common birds were abundant on all woodlots, there were differences in relative abundances and in the overall composition of the avifaunas on each plot (Table 1). On Woodlot 3 the ground, shrub and subcanopy vegetation was denser and more diverse than on Woodlots 1 and 2. There was also more ground litter and debris. Probably because of these differences, birds which foraged among litter and in the lower layers of vegetation were more abundant on Woodlot 3 than on the other plots. Eastern Yellow Robin, Eastern Whipbird, Superb Lyrebird, Brown Thornbill, Golden Whistler and White-browed Scrubwren were typical of this group. White-naped Honeyeater and Striated Thornbill which foraged mainly in the subcanopy and canopy were also more abundant on Woodlot 3.

In contrast, woodland birds and species which frequented open areas in the forest were more abundant on Woodlots 1 and 2. White-winged Chough and Australian Magpie were large ground foragers absent from Woodlot 3, but common in woodland areas. The small ground-foragers Buff-rumped Thornbill, Yellow-rumped Thornbill and Superb Blue Wren were restricted to woodland habitats and were common in the Snow Gum-Black Sally woodland on Woodlot 1 and along the edge of Woodlot 2. Flame and Scarlet Robins, Rufous Whistler and Yellow-faced Honeyeater are birds of

Table 1. Size and relative abundance of the most common bird species using the woodlots during summer 1980/81*.

Species and Main Food		Average[c] Weight (gms)	Abundance[a] Woodlot I	II	III	Number of Foraging[d] Acts Recorded
SEED-EATERS						
Gang-gang Cockatoo	*Callocephalon fimbriatum*	219	0.5[b]	0.3	4.3[b]	280
Crimson Rosella	*Platycercus elegans*	116	5.5	8.5[b]	14.5[b]	1309
INSECT-EATERS						
Fan-tailed Cuckoo	*Cuculus pyrrhophanus*	46	4.3[b]	0.6[b]	0.6[b]	100
Superb Lyrebird	*Menura novaehollandiae*	746	1.0[b]	+	2.6	62
Black-faced Cuckoo-shrike	*Coracina novaehollandiae*	110	1.5[b]	3.3[b]	1.3	75
Ground Thrush	*Zoothera dauma*	106	+[b]	0.3	4.3[b]	225
Rose Robin	*Petroica rosea*	9	0.3[b]	0	6.0[b]	195
Flame Robin	*P. phoenicea*	13	7.6[b]	13.1[b]	3.6[b]	830
Scarlet Robin	*P. multicolor*	13	4.3[b]	1.3[b]	0	465
Eastern Yellow Robin	*Eopsaltria australis*	20	5.5[b]	4.0[b]	16.6[b]	408
Shrike-tit	*Falcunculus frontatus*	29	1.8[b]	0.5	5.6	518
Golden Whistler	*Pachycephala pectoralis*	25	3.3	1.8	9.5[b]	551
Rufous Whistler	*P. rufiventris*	26	15.5[b]	15.8	6.3[b]	983
Grey Shrike-thrush	*Colluricincla harmonica*	76	5.6[b]	3.0[b]	5.3[b]	258
Black-faced Flycatcher	*Monarcha melanopsis*	24	2.0[b]	1.0	1.3	283
Satin Flycatcher	*Myiagra cyanoleuca*	18	6.6[b]	4.3[b]	4.8[b]	429
Rufous Fantail	*Rhipidura rufifrons*	10	0.5[b]	0.3	3.3[b]	303
Grey Fantail	*R. fuliginosa*	9	19.3[b]	19.0[b]	20.3[b]	2354
Eastern Whipbird	*Psophodes olivaceus*	62	0	0	4.5[b]	133
Superb Blue Wren	*Malurus cyaneus*	10	5.0[b]	4.8[b]	0	803
White-browed Scrubwren	*Sericornis frontalis*	13	0.5[b]	0.3	30.3	711
Brown Thornbill	*Acanthiza pusilla*	7	6.8[b]	7.1[b]	26.6[b]	1832
Buff-rumped Thornbill	*A. reguloides*	8	3.8[b]	0	0	559
Yellow-rumped Thornbill	*A. chrysorrhoa*	9	+[b]	0	0	378
Striated Thornbill	*A. lineata*	7	8.8[b]	22.8[b]	55.1[b]	2331
Orange-winged Sittella	*Neositta chrysoptera*	12	+[b]	+[b]	1.3	454
White-throated Treecreeper	*Climacteris leucophaea*	22	6.3[b]	6.6[b]	8.3	1224
Red-browed Treecreeper	*C. erythrops*	23	1.5[b]	2.5[b]	5.0	1238
Spotted Pardalote	*Pardalotus punctatus*	8	1.3	0.6	3.3	317
Striated Pardalote	*P. striatus*	12	4.6[b]	11.5[b]	10.5[b]	810
White-winged Chough	*Corcorax melanorhamphos*	379	0	2.8[b]	0	250
Dusky Woodswallow	*Artamus cyanopterus*	36	0	0.8[b]	0	145
Australian Magpie	*Gymnorhina tibicen*	314	0.3	2.0	0	622
NECTAR-FEEDER						
Red Wattlebird	*Anthochaera carunculata*	98	4.3	+[b]	4.0	254
Yellow-faced Honeyeater	*Meliphaga chrysops*	17	14.1[b]	4.5[b]	7.6[b]	909
White-eared Honeyeater	*M. leucotis*	25	2.8	0.3	0	614
Brown-headed Honeyeater	*Melithreptus brevirostris*	15	2.5	3.1	9.1	235
White-naped Honeyeater	*M. lunatus*	15	17.8[b]	23.3[b]	52.3[b]	2805
Crescent Honeyeater	*Phylidonyris pyrrhoptera*	16	0.3	0	1.8	177
Eastern Spinebill	*Acanthorhynchus tenuirostris*	11	1.0[b]	0.8	2.5	157
Silvereye	*Zosterops lateralis*	10	0.8	0	7.5	530
					Total:	27,116

[a] Abundance is based on the average number of individuals recorded during counts of four hours each on four mornings during December 1980. Each count covered the 10ha plot.
[b] Breeding on the woodlot was confirmed on at least one occasion between 1976 and 1981.
[c] Size measurements obtained from mist-netted birds on Woodlots 1 and 2.
[d] Only species where 50 or more prey attacks were recorded are considered.
\+ present on plot, but not recorded during censuses.
* From Recher *et al.* (1985).

woodland and open forest and were also more abundant on Woodlots 1 and 2 than on Woodlot 3.

Variation in the Composition of the Avifauna

The number of individuals and species of birds recorded during December censuses on Woodlots 1 and 2 are compared in Table 2. In most instances there was relatively little difference in either species richness or the number of birds between years. However, on three of the four transects there was an increase in the number of species during 1978 and on all transects there was a decrease in the number of individuals during 1980 (Recher *et al.* 1983a).

In contrast to the other years of the study, rainfall in 1978 was above average. The increased number of species during 1978 may therefore be related to more favourable conditions for birds. If conditions were better, it did not result in increased estimates of abundance. The increased number of species was due to the presence of a greater number of

Table 2. Number of species and individuals ($\bar{x} \pm$ S.E.) during December of birds on each transect (5ha) of woodlots 1 and 2[1].

Transect/Year	Habitat	1976	1977	1978	1979	1980
A. Species						
Woodlot 1C	Woodland	21 ± 2	23 ± 1	27 ± 2	23 ± 2	24 ± 0
1G	Forest	n.d.	25 ± 1	27 ± 1	27 ± 1	28 ± 2
Woodlot 2C	Forest	23 ± 1	22 ± 2	28 ± 2	20 ± 1	23 ± 1
2G	Forest/ Woodland Fringe	n.d.	20 ± 1	31 ± 2	21 ± 2	21 ± 1
B. Individuals						
Woodlot 1C	Woodland	97 ± 6	96 ± 6	100 ± 6	99 ± 17	82 ± 4
1G	Forest	n.d.	104 ± 11	102 ± 8	138 ± 8	94 ± 10
Woodlot 2C	Forest	110 ± 12	113 ± 9	113 ± 3	116 ± 11	103 ± 5
2G	Forest/ Woodland Fringe	n.d.	97 ± 6	108 ± 10	118 ± 20	75 ± 7

[1] adapted from Recher *et al.* (1983a).

uncommon birds (e.g. Cicadabird, *Coracina tenuirostris*, Rufous Fantail) and the more even distribution of common birds (e.g. thornbills, Grey Fantail) which increased the frequency with which these species were detected during counts.

In comparison with other years, conditions in 1980 during the drought were poor. There was little soil moisture and most annual plants and ferns failed to grow. On Woodlots 1 and 2 the average number of individuals detected during counts declined from September through December. In the same period in 1978, numbers increased (Fig. 1). In addition to the smaller number of birds recorded during censuses, fewer were caught for banding (Shields and Recher 1984).

There are several reasons which might explain the smaller number of birds in 1980. Probably due to the drought, reproductive success on the woodlots was poor. The species affected included Grey Fantail, Eastern Yellow Robin, Rufous Whistler, Flame Robin and Satin Flycatcher each of which was normally abundant and contributed a large number of young birds to the December counts. Apart from the scarcity of juveniles, there was also a decrease in the numbers of resident adults during the summer. In addition the numbers of juvenile Golden Whistler and Ground Thrush which usually dispersed from adjacent moist forests through the plots in December were less than in other years.

Another effect of drought is illustrated by the invasion of the Southern Tablelands during 1980 by White-browed Woodswallows, *Artamus superciliosus* (Recher and Schulz 1983). Two colonies totalling 60 pairs were established on Woodlot 2 where they nested and fledged young. We had not previously recorded this bird in the area and their presence was almost certainly the result of extreme dryness further west in their normal nesting habitats.

In contrast to the differences between years in the composition of the summer avifauna, there were few differences between winter bird communities (Fig. 1). Transects differed in species number and total number of birds between years in both the May and July counts, but the differences were not consistent. Most of the birds which forage on the ground (e.g. Flame Robin) or take flying insects (e.g. Grey Fantail) are absent during the winter. As these species were the ones which appeared most affected by the drought, fewer differences are to be expected between winter bird communities in dry and wet years.

Seasonal events

The greatest variation in the numbers of individuals or species between years is small when compared with the changes that occur between summer and winter (Recher *et al.* 1983a). Reproduction, the dispersal of young and migration are seasonal events which result in substantial changes in the abundance and species composition of the avifauna. These events can be related to changes in the distribution and abundance of food and the energy requirements of birds at different seasons.

Species richness and the number of individuals on the woodlots is greatest during spring and summer and lowest in winter (Fig. 1). The abundance of invertebrates and the number of invertebrate taxa which can be sampled is also highest in summer and least in winter (Recher *et al.* 1983a). Our sample of the kinds of food which birds feed on is probably a reasonable index of the abundance of prey available to birds. Thus, in winter, when the energy requirements of birds are high because of low temperatures and short days, the invertebrate prey of birds was less abundant than in summer. Food resources should also decline in line with Nix's (1976) growth indices for plants as conditions

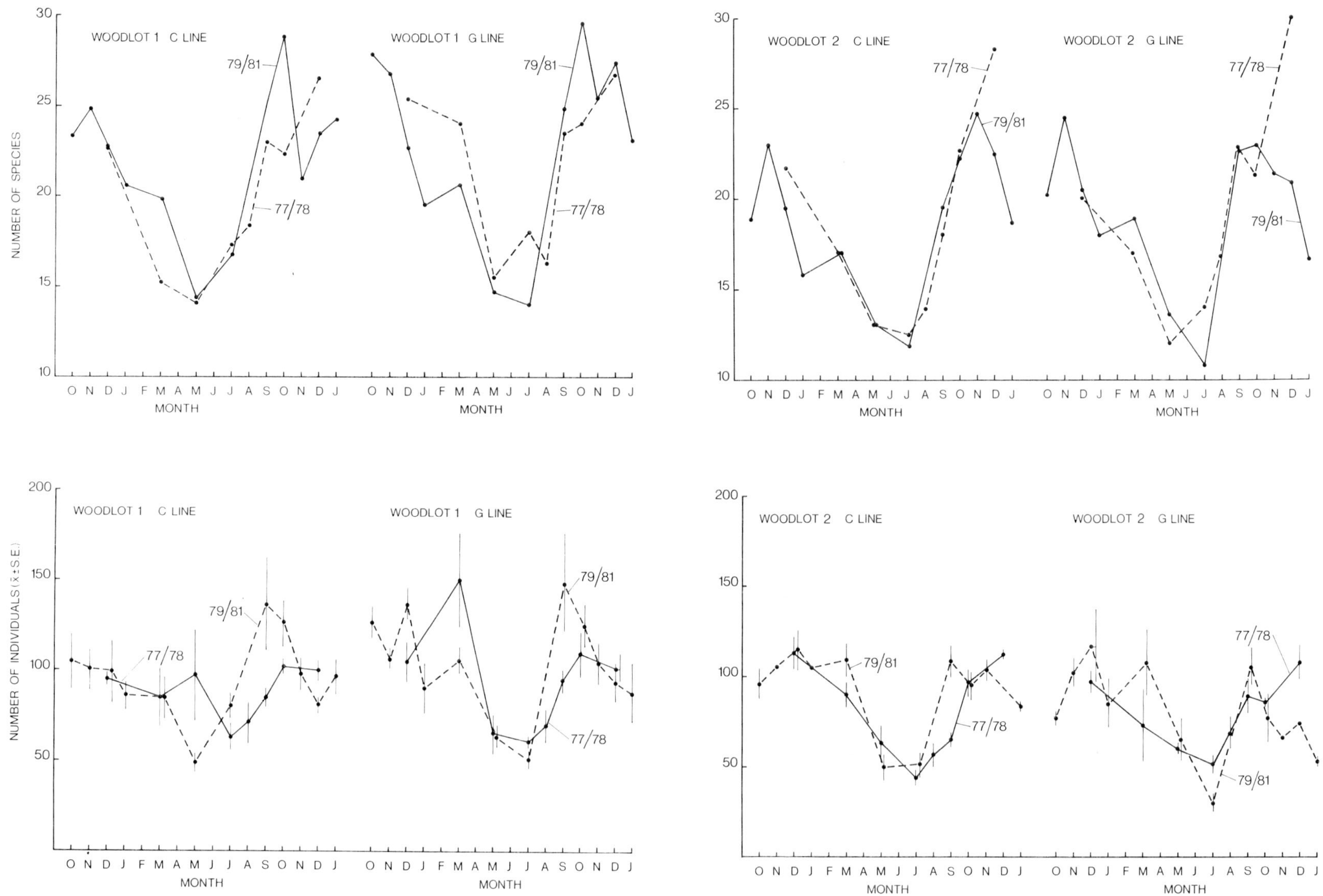

Fig. 1. Average number of species and individuals detected in a 2 h count on each of the transects on Woodlots 1 and 2 during censuses in 1977/78 and 1979/81. On Woodlot 1 the habitat along the 'c' transect was a Snow Gum-Black Sally woodland and along the 'g' transect it was a Peppermint-Manna Gum forest. On Woodlot 2 the 'g' transect was bordered by woodland. (Adapted from Recher *et al.* 1983a).

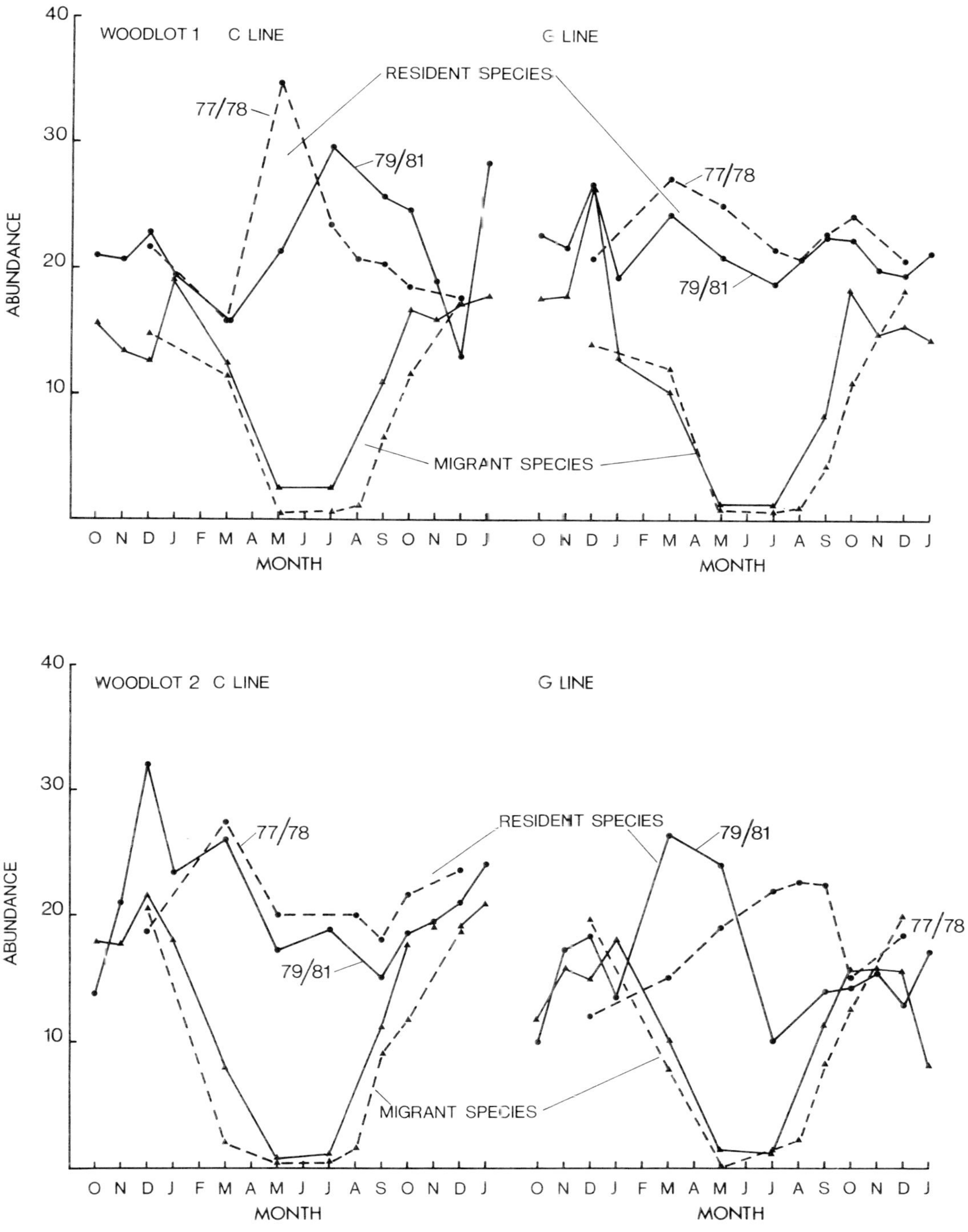

Fig. 2. Relative abundance of migrant and resident insectivorous birds on Woodlots 1 and 2, 1977/78 and 1979/81. Abundance is based on the average number of detections of all species during each 2 h count. (Adapted from Recher *et al.* 1983a).

become hotter and drier in late summer and during drought.

Movement of Birds

About half the species which nested on the woodlots were absent during the winter or present in very low numbers (Fig. 2). Nine species including Golden-bronze Cuckoo, *Chrysococcyx plagosus*, Rufous Whistler, Satin Flycatcher, and Rufous Fantail migrate north of New South Wales for the winter. All are insectivorous. Three honeyeaters (Yellow-faced Honeyeater, Red Wattlebird and White-naped Honeyeater) and the Silvereye also migrated. Many of the other birds which nested on the woodlots departed during the winter (Fig. 2), but their movements are harder to describe. Perhaps they could be called 'diffuse migrants'. Some probably

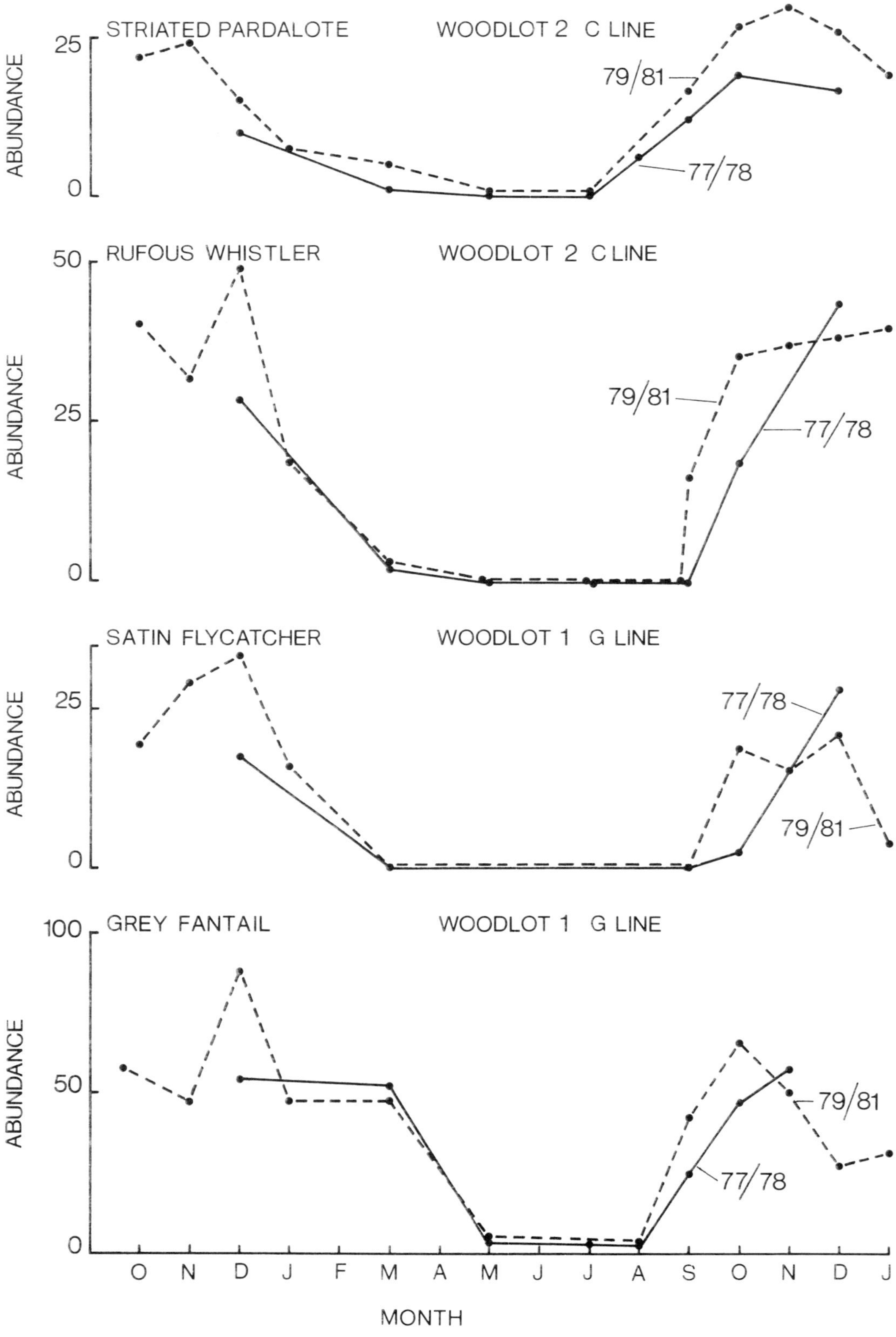

Fig. 3. Relative abundance of Striated Pardalote, Rufous Whistler, Satin Flycatcher and Grey Fantail, 1977/78 and 1979/81. The species are migrants absent during winter, but migrating at different times. Abundance is based on total detections during census, but only the transect on which the species was most abundant is illustrated. Patterns of change were the same for all transects. (Adapted from Recher *et al.* 1983a).

'drift' northerly, but large numbers also shift to lower elevations and 'winter' in coastal forests at the same latitude as the study plots. Such altitudinal migration is probably an important part of the post-breeding dispersal of such species as Grey Fantail, Striated Pardalote and Golden Whistler.

Regardless of the nature of the movements of these birds, they are not synchronous; species departed and returned at different times (Fig. 3) (Recher *et al.* 1983a). Altitudinal migrants began to return in August, but some latitudinal migrants did not return until late October or November (Fig. 3). Species, such as Satin Flycatcher and Black-faced Flycatcher, which returned late were among the first to depart.

At times large numbers of migrating or dispersing birds passed through the woodlots. Honeyeaters, Silvereye, Flame Robin, Ground Thrush and Striated Pardalote were the most conspicuous. The most abundant of these were the Silvereye, Yellow-faced Honeyeater, Red Wattlebird, and White-naped Honeyeater (Fig. 4). Generally we observed Red Wattlebird and White-naped Honeyeater moving south in September and Yellow-faced Honeyeater and Silvereye in October.

Apart from migrants we did not record large numbers of nectar-feeders on the woodlots and the aggregations that occurred did not always coincide with censuses. What appeared to happen was a shift of small numbers of Yellow-faced Honeyeater, Brown-headed Honeyeater, Crescent Honeyeater and White-eared Honeyeater to the woodlots from nearby pine plantations and eucalypt forest whenever nectar-rich trees or shrubs on the plots were in bloom. White-eared and Crescent Honey-eaters were also more abundant on the woodlots during autumn and winter when eucalypts (especially Swamp Gums) shed bark exposing insects, and carbohydrates such as honeydew and sap (Recher *et al.* 1983a).

Reproduction and Nest Site Selection

Nix (1976) demonstrated that the peak nesting period at Canberra in the Australian Capital Territory for terrestrial birds occurred during late spring and early summer. This is when conditions are most favourable for plant growth. Marchant (1981) also found a 'short and sharply defined' breeding season from early August to late December in coastal forest at Moruya (35°52′S, 150°04′E). Although Canberra (35°17′S, 149°08′E) is north of Bombala and at a lower elevation (565 m), it is further inland and the two areas should have similar plant growth indices.

Resident species began nesting on the woodlots in winter (Recher *et al.* 1983a). Lyrebirds had eggs by July and Striated Thornbills began nest building in late July and August. By October all resident species had eggs or young. Migrants began to nest as soon as they returned to the plots.

Nesting was initiated at the same time that invertebrate food resources were increasing (Fig. 5) and young were fledged when food was most abundant (Recher *et al.* 1983a). Thus, the greatest number of species and individuals nested from late September to early December or about the same period as at Canberra. The number of active nests on the woodlots declined rapidly during December and by January few birds were nesting. However, Red-browed Finch, *Emblema temporalis*, in response to the ripening of grass and weed seeds, initiated nesting at this time.

A necessary resource for successful nesting is a suitable nest site. Although birds may adapt to circumstances and nest in "atypical" locations, the species which nested on the woodlots were highly selective in the sites they chose for nests. No two species selected the same kind of nest site and the presence or absence of suitable nest sites may have contributed to some of the differences between the woodlots in the composition of their avifaunas.

A total of 789 nests of 45 species were found on the woodlots between 1976 and 1981. Data are most complete for Woodlots 1 and 2 and differences between these plots illustrate the importance of nest sites for birds. Ten species (22%), including the Starling, *Sturnus vulgaris*, nested exclusively in tree hollows or cavities. Of the four most abundant hole nesters, Crimson Rosella, White-throated Tree-creeper, Red-browed Treecreeper and Striated Pardalote, 66 nests were found on Woodlot 2 where tree hollows were abundant compared with 14 nests on Woodlot 1 where hollows were uncommon.

Species which did not nest in hollows placed nests in a variety of positions and heights above the ground. Apart from the Spotted Pardalote which nested in a burrow excavated into the ground, no birds on the woodlots nested on the ground. However, two species, the Superb Blue Wren and White-browed Scrubwren, placed their nest in thick ground vegetation usually within 50 cm of the ground. Four species, Brown Thornbill, Rufous Fantail, Golden Whistler and Red-browed Finch, nested in shrubs and thick vegetation between one and three metres above ground. The abundance of these species on Woodlot 3 and their scarcity on Woodlot 2 and the woodland area of Woodlot 1 may in part be related to the absence of suitable nest sites in the latter areas. The Grey Fantail also nested in shrubs, but of 123 nests, 89 (72%) were positioned between 2 and 13 m above ground. The Grey Fantail was abundant on all three woodlots. Similarly the four species, Striated Thornbill, Shrike-tit, Yellow-faced Honeyeater and White-naped Honeyeater, which placed their nests in the outer foliage of trees

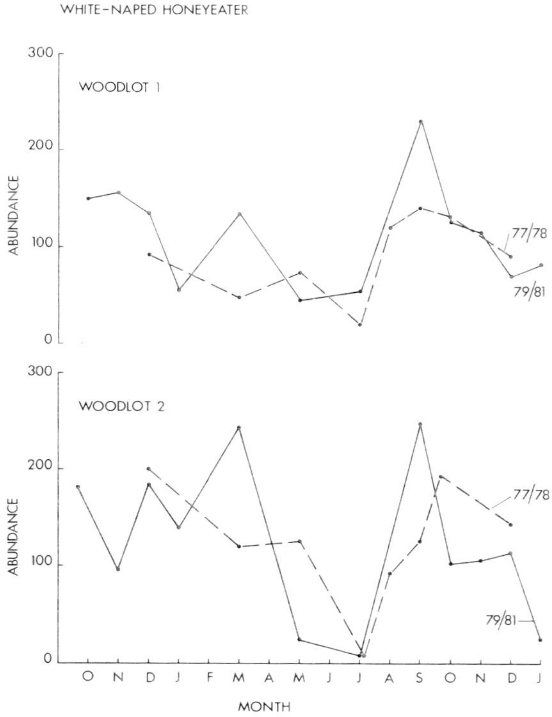

Fig. 4. Abundance of White-naped Honeyeaters based on total detections on woodlots during censuses in 1977/78 and 1979/81. The White-naped Honeyeater was the most abundant nectar-feeding bird on the woodlots and its movements greatly affected the total numbers of nectarivorous birds detected in a census. (Adapted from Recher *et al.* 1983a).

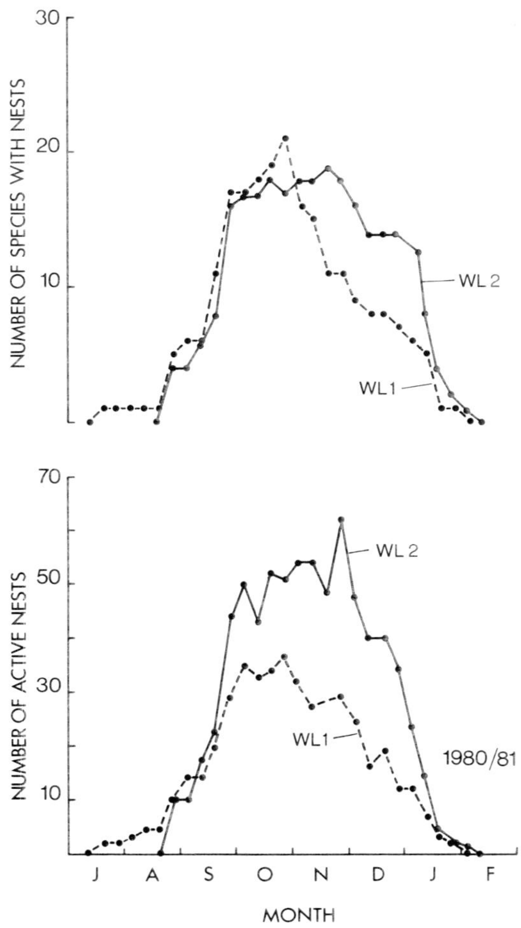

Fig. 5. Numbers of species of birds and active nests on Woodlots 1 and 2 each fortnight, 1980/81 breeding season. (Adapted from Recher *et al.* 1983a).

nested on all three plots. The remaining 23 species nested mainly on the branches or the forks of trees and large shrubs, but differed in the height of their nests, the size of the branch or tree selected for nesting and nest materials. These differences will be discussed in another paper.

Foraging Patterns

The ways in which foraging birds differed in their use of the woodlots can be described by the heights at which they foraged, the kinds of food taken, the substrates of their prey (food), and the feeding or prey attack behaviour used.

Sufficient data were obtained to analyse the foraging patterns of 41 species (Recher *et al.* 1985). Twenty-nine of these species fed primarily on insects (Table 1). There were two species of parrots, Gang-gang Cockatoo and Crimson Rosella, which fed mostly on eucalypt seeds. The Silvereye and seven species of honeyeaters fed on a variety of carbohydrates (i.e. nectar, manna, honeydew and sap), but also took insects. The two pardalotes, Striated and Spotted, probably fed mainly on lerps gleaned from the foliage of eucalypts (see Woinarski, this Volume) but are considered insectivorous in this analysis. In addition to the birds for which we have sufficient data for detailed analysis, the Red-browed Finch and European Goldfinch, *Carduelis carduelis,* fed mainly on grass and weed seeds. The Brown Goshawk, *Accipiter fasciatus*, which was the only common raptor, took birds, but mainly fed on rabbits. Some other large birds, such as Laughing Kookaburra, *Dacelo novaeguineae* and Pied Currawong, *Strepera graculina*, commonly fed on lizards and large insects.

In the following the woodlots are treated as a single plot grading from a Snow Gum-Black Sally woodland (Woodlot 1) through open, dry forest (Woodlots 1 and 2) to tall, moist forest (Woodlot 3). Although the plots differed in vegetation and degree of disturbance, they had similar avifaunas and shared the most abundant species in common (Table 1). There were differences among the plots in the ways or places that species foraged which were related to differences in the kinds and density of vegetation, but these do not affect the general patterns described here.

Table 3. Height distribution of birds in eucalypt forest and woodland at Bondi (% of prey attacks)*.

Vegetation Layer	Ground 0 m ≤ 0.2 m	Shrub 0.2 m ≤ 4.0 m	Subcanopy 4.1 m — 10.0 m	Canopy > 10.0 m	x̄ foraging Height (± SD)	Foraging Height Diversity (H′)
CANOPY						
Gang-gang Cockatoo		7	21	72	13.1 ± 5.3	0.75
Black-faced Cuckoo-shrike	7	4	38	51	11.4 ± 6.0	1.03
Brown-headed Honeyeater		6	46	48	11.4 ± 5.3	0.88
Crimson Rosella	1	9	44	46	10.5 ± 5.2	0.94
Striated Pardalote		21	39	40	9.3 ± 5.3	1.06
						mean H′ = 0.93
SUBCANOPY						
Satin flycatcher	1	13	49	37	9.5 ± 4.5	1.03
Shrike-tit		17	47	36	9.3 ± 5.0	1.02
Spotted Pardalote		21	43	36	9.9 ± 5.4	1.06
Orange-winged Sittella		5	63	32	8.3 ± 3.7	0.80
White-naped Honeyeater		21	47	32	9.6 ± 7.0	1.05
Rufous Whistler	3	20	47	30	8.4 ± 5.1	1.14
Striated Thornbill		27	45	28	8.5 ± 5.7	1.07
Yellow-faced Honeyeater	2	25	51	23	7.4 ± 4.4	1.10
White-eared Honeyeater	1	33	48	18	6.8 ± 4.6	1.07
White-throated Treecreeper	3	44	40	13	5.6 ± 4.2	1.10
Grey Fantail	9	43	36	12	5.8 ± 5.5	1.20
Red-browed Treecreeper	1	43	44	12	6.3 ± 4.7	1.03
Black-faced Flycatcher	1	40	48	11	6.6 ± 5.0	1.01
Dusky Woodswallow	18	35	47		4.6 ± 2.8	1.03
						mean H′ = 1.05
SHRUB						
Silvereye	5	43	28	24	6.8 ± 5.7	1.21
Red Wattlebird	2	65	16	17	5.9 ± 5.8	0.95
Golden Whistler	3	52	30	15	5.9 ± 5.5	1.09
Rose Robin	10	55	23	12	4.2 ± 4.5	1.15
Crescent Honeyeater		74	20	6	4.7 ± 3.8	0.71
Brown Thornbill	9	69	18	4	3.1 ± 3.4	0.91
Eastern Spinebill		99	1		2.2 ± 0.8	0.06
Rufous Fantail	21	74	5		1.3 ± 1.5	0.70
Eastern Whipbird	38	58	4		1.1 ± 1.6	0.81
						mean H′ = 0.84
GROUND						
Grey Shrike-thrush	38	23	28	11	4.6 ± 5.7	1.31
Buff-rumped Thornbill	58	29	12	1	1.5 ± 2.6	0.98
Flame Robin	64	22	13	1	1.6 ± 2.9	0.91
Scarlet Robin	68	25	7		1.1 ± 2.3	0.80
Eastern Yellow Robin	77	17	5	1	1.0 ± 2.4	0.75
White-browed Scrubwren	75	25			0.3 ± 0.8	0.57
Superb Blue Wren	80	20			0.2 ± 0.8	0.50
Fan-tailed Cuckoo	86	12	1	1	0.5 ± 1.8	0.48
Yellow-rumped Thornbill	93	4	3		0.3 ± 1.4	0.30
Ground Thrush	97	3			0.2 ± 0.3	0.14
Australian Magpie	98	2			0.0 ± 0.3	0.10
White-winged Chough	100				0.0 ± 0.0	0.00
Superb Lyrebird	100				0.0 ± 0.0	0.00
						mean H′ = 0.53
% of all prey attacks	18	32	33	17		*Total mean H′ = 0.83*

*From Recher *et al.* (1985).

Foraging Height Distribution

The height distributions of foraging birds are presented as corresponding to the four layers of vegetation; ground, shrub, subcanopy and canopy (Table 3). However, these divisions are arbitrary and, apart from some ground-dwellers, most birds foraged over a wide range of heights.

Thirteen species of birds were ground foragers. Five of these, Superb Lyrebird, Ground Thrush, Yellow-rumped Thornbill, White-winged Chough and Australian Magpie, were ground dwellers. Four species, Eastern Whipbird, Superb Blue Wren, White-browed Scrubwren, and Buff-rumped Thornbill, foraged on the ground and in low vegetation and debris. Fan-tailed Cuckoo, Eastern Yellow Robin, Scarlet Robin and Flame Robin fed mainly on ground-dwelling invertebrates, but hunted from low perches (< 1.5 m) from which they pounced on prey. Laughing Kookaburra and Sacred Kingfisher, *Halcyon sancta*, for which there are only limited data, were also "ground pouncers", but used higher perches (> 2 m). The Grey Shrike Thrush foraged extensively on the ground (38% of foraging manoeuvres), but also hunted high in the vegetation wherever there were large branches.

Eight species foraged principally in the shrub layer. The most conspicuous and abundant of these were the Brown Thornbill and Golden Whistler, but in tall open-forest Rose Robin and Rufous Fantail were also typical of this layer. Most of the foraging observations of Red Wattlebird, Crescent Honeyeater and Eastern Spinebill were of birds feeding on the nectar of Gippsland Waratah, a tall shrub. However, these three honeyeaters visit a wide range of flowers including those of eucalypts and mistletoe and are not strictly birds of the shrub layer.

The remaining 19 species foraged widely through the shrub, subcanopy and canopy layers. Only two of these, the Gang-gang Cockatoo and the Black-faced Cuckoo-shrike, foraged predominantly in the canopy. The others are harder to characterise. A number of species, including Striated Thornbill, Spotted and Striated Pardalotes, White-naped and Yellow-faced Honeyeaters foraged almost exclusively on prey taken from eucalypt foliage and were as likely to forage in saplings or the subcanopy as in the crowns of tall trees. This was also the case with bark foragers such as White-throated and Red-browed Treecreepers and White-eared Honeyeaters. Grey Fantails and Dusky Woodswallows exploited the air space below the canopy to ground level, while the Satin Flycatcher foraged mainly in openings throughout the subcanopy and canopy.

Substrates of Prey

Ground, bark (trunks and branches), foliage (including petioles and twigs), and air were the major substrates exploited by birds on the woodlots. Foliage was the most important substrate used by birds foraging above the ground and accounted for 35% of the foraging manoeuvres recorded and was the principal foraging substrate used by 13 species (Table 4). Pardalotes, White-naped Honeyeater and Striated Thornbill foraged almost exclusively on the leaves and small outer branches of eucalypts. Other species tended to use a wider range of vegetation. For example, 65% of the foraging of Brown Thornbills was in wattles, blackthorn, and other shrubs.

Bark was the next most important substrate exploited by insect-eating birds. Twenty-five percent of foraging manoeuvres involved food items on or under bark surfaces on the trunks and branches of trees. Five species, White-throated and Red-browed Treecreepers, Orange-winged Sittella, and White-eared Honeyeater, were bark specialists (Table 4). Red-browed Treecreepers foraged on gums and probed under loose bark significantly more often than White-throated Treecreepers (χ^2; $p < 0.01$). The White-throated Treecreeper foraged mainly on peppermints and other rough-barked species. Such differences serve to highlight the importance of bark as a foraging substrate in forests and the diversity of foraging opportunities it provides birds. The Orange-winged Sittella, for example, foraged mainly on branches and selected dead substrates whereas the treecreepers preferred live substrates.

Loose and hanging bark is a distinctive and important substrate in eucalypt forests and accounted for a third of all bark foraging. All the eucalypts on the woodlots shed the bark from the extremities of their branches and the gums shed bark from along the length of their branches and most of the length of the trunk. Loose bark sheltered a rich invertebrate fauna and was the principal foraging substrate for Shrike-tits and White-eared Honeyeaters. For five species, including Brown-headed Honeyeater, Red-browed Treecreeper, and Grey Shrike-thrush, loose bark was the second most important substrate on which they foraged (Table 4). Shrike-tits took mainly insects and spiders from under bark or from within coiled, hanging bark strips. Honeyeaters and other birds probably took sap and honeydew, in addition to insects. Visual inspection showed that carbohydrates, such as sap and honeydew, were abundant under bark as it first loosened and was shed. Newly shed bark was therefore attractive to birds and attracted them in the same way that birds were attracted to nectar-rich flowers. For example, White-eared and Crescent Honeyeaters aggregated on Woodlot 1 during the winter when Swamp Gum shed its bark. In winter these birds foraged almost exclusively on Swamp Gums, probing under the bark at the point it loosened from the tree.

Ground foraging accounted for 18% of foraging manoeuvres while air as a substrate for prey accounted for 11% of prey attacks by insect-eating birds (Table 4). Flying insects were the most important category of prey for four species, Grey and Rufous Fantails, Satin Flycatcher and Dusky Woodswallow. Apart from the Satin Flycatcher and Black-faced Cuckoo-shrike which commonly fed in openings in the canopy and purely aerial feeders such as Spine-tailed Swift, *Hirundapus caudacutus,* and Tree Martin, *Petrochelidon nigricans,* which foraged above the canopy, most aerial foraging occurred in the open spaces below the canopy. Aerial foraging by Rufous Fantail, and Eastern Yellow, Scarlet and Flame Robins were mostly of insects flying near the ground. In contrast, Rose Robin and Grey Fantail foraged for flying insects from ground level into the canopy.

Prey Attack Behaviour

Most insect-eaters searched for prey by moving steadily through the vegetation or across the ground

Table 4. Substrate of prey taken by bird species (percent of prey attacks)*.

Substrate	Ground	Bark: Tree trunks	Bark: Branches	Bark: Loose bark	Foliage	Air	Flowers (nectar)	Eucalypt capsules (seed)
GROUND FORAGERS								
Superb Lyrebird	100							
White-winged Chough	100							
Australian Magpie	98				1	1		
Ground Thrush	98		2					
Yellow-rumped Thornbill	93		1	2	3	1		
Superb Blue Wren	85		2		9	4		
Fan-tailed Cuckoo	84	10	4		1	1		
White-browed Scrubwren	82	1	5	1	11			
Eastern Yellow Robin	77	8	2	1	5	7		
Scarlet Robin	68	9	3	1	3	16		
Flame Robin	64	9	2		6	19		
Buff-rumped Thornbill	58	3	16	6	16	1		
Eastern Whipbird	53	6	32	6	3			
Grey Shrike-thrush	41	8	17	22	12			
BARK FORAGERS								
Tree Trunks								
White-throated Treecreeper	2	80	11	6	1			
Red-browed Treecreeper		56	14	28	2			
Branches								
Orange-winged Sittella		4	80	16				
Loose Bark								
Shrike-tit		2	4	90	4			
White-eared Honeyeater		6	13	58	19	3	1	
FOLIAGE FORAGERS								
Striated Pardalote			5	1	92	2		
Spotted Pardalote			10		90			
Striated Thornbill		1	12	5	81	1		
White-naped Honeyeater		1	7	11	77		4	
Black-faced Flycatcher	1	2	10	1	65	21		
Yellow-faced Honeyeater	1	1	9	10	66	2	11	
Golden Whistler	2	5	13	3	67	10		
Brown Thornbill	7	4	14	5	65	5		
Rufous Whistler	2	6	16	3	59	14		
Black-faced Cuckoo-shrike	8	1	25		59	7		
Brown-headed Honeyeater	2		16	33	48		1	
Silvereye	5		27	2	41		25	
Rose Robin	11	14	2	9	38	26		
AERIAL FORAGERS								
Grey Fantail	4	2	2		14	78		
Dusky Woodswallow	11	13	4		4	68		
Satin Flycatcher	1	3	3		32	61		
Rufous Fantail	15	3			28	54		
NECTAR FEEDERS								
Crescent Honeyeater		2	2	6	5	1	84	
Eastern Spinebill	1	1	1	9	3	2	83	
Red Wattlebird	2		7		24	2	65	
SEED EATERS								
Gang-gang Cockatoo								100
Crimson Rosella	1		2	1	6			90
% of total foraging acts	18	8	8	9	37	11	3	6

*From Recher *et al.* (1985).

and gleaned prey from the substrate. Gleaning accounted for 54% of all prey attacks and was the principal foraging behaviour of 14 species (Table 5). Gleaners included the thornbills, treecreepers, pardalotes, Orange-winged Sittella, and wrens. The behaviour of the Striated Thornbill was distinguished by the frequency with which it gleaned insects from foliage while hanging upside down ("hang-glean"). On the ground, Eastern Whipbird, Ground Thrush, Grey Shrike-thrush, White-winged Chough and Australian Magpie, gleaned insects and other invertebrates from the litter and the uppermost layer of soil. The litter was disturbed and moved aside mostly with the beak, but the Ground Thrush and White-winged Chough did some raking or scratching with their feet. The Superb Lyrebird foraged principally by raking the litter with its feet and commonly scratched aside the soil to depths of

Table 5. Percent of prey attack manoeuvers by insectivorous birds on the woodlots during the summer of 1980/81*.

Species/Manoeuvre	Pounce	Probe/Prise	Glean	Hang-glean	Hover	Snatch	Hawk
POUNCERS							
Fan-tailed Cuckoo	86		2		1	10	1
Eastern Yellow Robin	75	2	3		13	7	
Scarlet Robin	68		3		1	12	16
Flame Robin	52	1	13			15	19
PROBERS/PRISERS							
Superb Lyrebird		100					
Shrike-tit		90	8	1		1	
Ground Thrush	1	76	23				
White-winged Chough		70	30				
GLEANERS							
Orange-winged Sittella			99	1			
Yellow-rumped Thornbill			98			1	1
White-browed Scrubwren		1	97			2	
Superb Blue Wren			92			4	4
Buff-rumped Thornbill		3	91	1	1	3	1
White-throated Treecreeper		11	88	1			
Striated Pardalote			86	3	3	7	1
Spotted Pardalote			85	3	4	7	1
Red-browed Treecreeper		19	80	1			
Brown Thornbill		1	75	5	4	11	4
Eastern Whipbird		34	64			2	
Striated Thornbill		1	62	23	7	6	1
Grey Shrike-thrush	1	18	61			20	
Australian Magpie		43	56				1
SNATCHERS							
Black-faced Cuckoo-shrike	7	1	12			74	6
Golden Whistler	1	1	16		1	71	10
Black-faced Flycatcher			8		3	68	21
Rufous Whistler	2	1	19		1	61	16
Rose Robin	10		5			59	26
HAWKERS							
Grey Fantail			8		1	13	78
Dusky Woodswallow	15		4			12	69
Satin Flycatcher	1		1		2	36	60
Rufous Fantail	3		12		1	30	54
% of all foraging acts	5	8	54	4	1	14	14

*From Recher *et al.* (1985).

5 or 6 cm. Snatching was the second most common prey attack behaviour used by gleaners. The Shrike-thrush and Brown Thornbill, in particular, took prey in this manner (Table 5). Interestingly, gleaners did not pounce and rarely hawked. Snatching accounted for 14% of recorded foraging manoeuvres and was the principal prey attack behaviour of whistlers, Black-faced Cuckoo-shrike, Black-faced Flycatcher and Rose Robin. These birds also took prey by hawking, gleaning and pouncing.

Snatchers moved continuously through the vegetation, searching for and locating prey from a distance. For example, the average distance that Rufous Whistler moved in snatching prey was 70 cm and for Golden Whistlers the distance was 60 cm. Most prey taken by this manoeuvre was large (> 10 mm) and located on substrates above the bird.

Hawking accounted for 14% of all recorded prey attacks and was the most common foraging behaviour of fantails, Satin Flycatcher and Dusky Woodswallow (Table 5). Three types of behaviour distinguished birds which took flying insects. Martins and swifts behaved as aerial hawkers. Grey and Rufous Fantails moved rapidly through and over vegetation, rapidly changing perches and foraging heights. Fantails fed mainly on small insects (< 2 mm) and sometimes hovered in swarms of insects taking prey in quick succession. Often the insects taken by Grey Fantails were ones disturbed as the birds moved through the foliage or moved rapidly up and down tree trunks. Dusky Woodswallows and Satin Flycatchers used perches from which they scanned the surrounding air space for insects. Both species frequently changed perches.

Of the remaining foraging behaviours, probing/prising accounted for 8%, pouncing for 5%, and hovering for 1% of all recorded manoeuvres (Table 5). The importance of probing and pouncing to particular species of forest birds has been discussed in the preceeding sections. Although hovering was an uncommon behaviour, it accounted for 13% of the prey attacks of Eastern Yellow Robins and 7% of the manoeuvres of Striated Thornbills.

Table 6. Foraging behaviour of Honeyeaters and Silvereyes on the woodlots during summer 1980/81. Numbers are percent of "prey attacks" for each species*.

Behaviour	Glean/Hang-glean		Probe	Snatch	Hawk	Visit flowers
Food	Manna, lerp and honeydew, insects		Sap, manna and honeydew, insects	insects	insects	nectar
Substrate	Foliage	branches loose bark	loose and hanging bark, trunks	foliage, twigs	air	flowers
White-eared Honeyeater	17	19	58	2	3	1
Brown-headed Honeyeater	48	34	17			1
Silvereye	46	29				25
Red Wattlebird	18	7		8	2	65
Crescent Honeyeater	1	2	8	4	1	84
Eastern Spinebill	2	11		2	2	83
Yellow-faced Honeyeater	63	9	11	4	2	11
White-naped Honeyeater	76	14	5	1		4

*From Recher *et al.* (1985).

Nectar-feeders

Apart from taking nectar, honeyeaters and the Silvereye used the same array of foraging behaviours as insect-eaters (Table 6). Probing loose bark was the most common foraging behaviour of White-eared Honeyeater and also distinguished the foraging of the Brown-headed Honeyeater from that of the congeneric White-naped Honeyeater. White-naped Honeyeaters were the most abundant honeyeaters on the woodlots and foraged almost exclusively by gleaning food, either insects or carbohydrates, from the foliage of eucalypts. Hawking and snatching insects were uncommon behaviours of nectar-feeders and were used mainly by Red Wattlebirds which foraged for large insects in the canopy.

Community Organisation

Although species broadly overlapped, those which foraged in similar ways or exploited the same substrates can be grouped together. By so organising the species within a community into groups a picture of community organisation or structure emerges which can be compared with communities in other types of forest or on other continents. It is possible in this way to identify what resources are important for birds, to detect communities where the basic structure may have been disturbed by human activities, and to compare the patterns of evolution among communities with very different histories or occupying very different environments.

One possible array of groups for the avifauna on the woodlots is illustrated in Figure 6. The groupings are hierarchial and schematic, but are useful in summarising the ways in which eucalypt forest and woodland birds forage for food and which resources are important for the survival of particular species.

Nectar-feeders and parrots are separated from insect-eaters on the basis of diet. The two parrots differ in size (Table 1) and use different substrate. The Crimson Rosella forages more on the ground and in the winter exploits loose bark more often than the Gang-gang Cockatoo.

Among the insect-eaters, there is a primary separation between birds which obtained their food from the ground and those which fed above ground. Above ground, bark foragers are separated from aerial feeders and birds which took their prey from foliage. Beyond these major divisions there are other, more subtle differences between species in the substrates used or in prey attack behaviour. For example, the preferences of Orange-winged Sittellas for dead surfaces, the use of fibrous bark by White-throated Treecreepers, and the preference of Red-browed Treecreepers and Shrike-tits for loose bark separates these four species of bark foragers. The Shrike-tit is distinguished from the Red-browed Treecreeper by its use of hanging bark strips which the treecreeper avoids. Among the foliage gleaners, Brown and Striated Thornbills are separated by foraging height and the use of hang-gleaning by Striated Thornbills. Hang-gleaning also separates the Striated Thornbill from the pardalotes. Size and foraging behaviour separate the group of aerial foragers with the two fantails separated by foraging height and habitat.

Size is also important in separating ground foragers. The largest insect-eaters (including the Laughing Kookaburra) are ground foragers. Birds the size of a Lyrebird or Magpie are clearly different from small ground foragers such as thornbills and wrens Among the small ground foragers there is a separation by habitat with the White-browed Scrubwren most abundant in forest and the Superb Blue Wren and thornbills most common in woodland. The robins and Fan-tailed Cuckoo which are separated from other ground foragers by their prey attack behaviour can also be separated into mainly forest and woodland species. The Eastern Yellow Robin is primarily a forest bird, while the Scarlet and Flame Robins are woodland species. Scarlet and

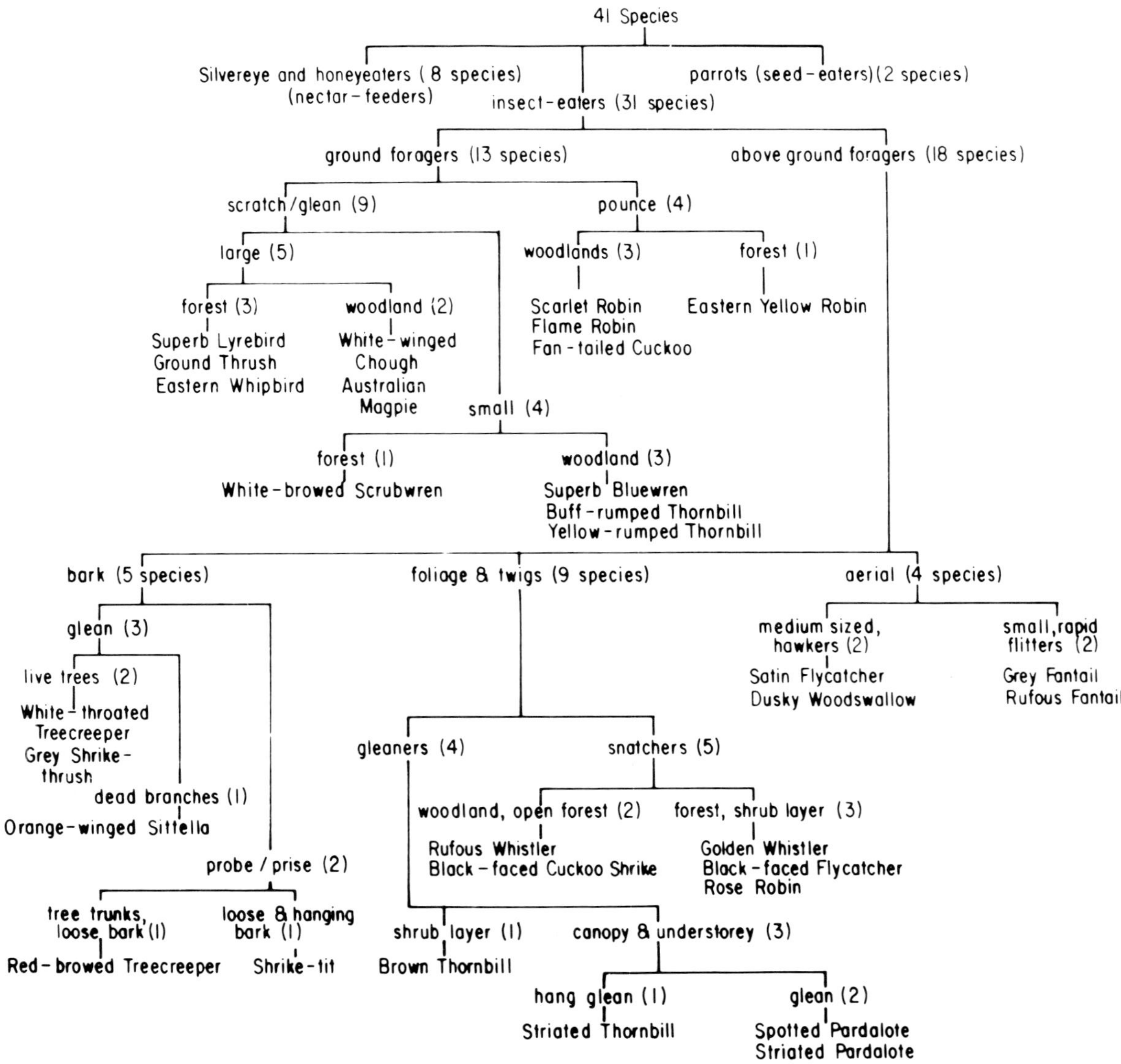

Fig. 6. Community organisation of insectivorous birds on the Bondi woodlots. Number of species in parenthesis.

Flame Robins defended territories against each other (Recher unpubl. data; Loyn 1980). The Fan-tailed Cuckoo occurred in both habitats, but is a much larger bird and on the woodlots fed exclusively on lepidoptera larvae.

DISCUSSION

The climate of southeastern Australia is mild, the vegetation is evergreen, and invertebrate prey are active throughout the year at all except the highest altitudes. Seeds, nectar and other carbohydrates are also available summer and winter. Although the breeding season and movements of long-distance migrants may be, as described by Marchant (1981), short and sharp, almost all changes in the avifauna are protracted and sometimes difficult to document. Non-migrants in particular may have long breeding seasons with multiple nestings. Movements of these species appear to be local or diffuse with different individuals moving into and out of the same habitats. Nonetheless there are patterns and our knowledge of the ways in which birds use forests and woodlands can be used to develop plans of management which will ensure the survival of wildlife.

Breeding, dispersal of young, migration and shifts in the use of habitats occur in response to changes in the distribution, abundance and availability of food (Recher *et al.* 1983a). The magnitude of changes in food resources and the composition of the avifauna are affected by climatic conditions or regional events such as the blossoming of nectar-rich flowers in other areas. However, the pattern of change remains essentially the same with most breeding occurring during spring and summer and most movements of birds occurring from late summer to spring.

Each of the woodlots had a distinctive avifauna. Although we treated them as a unit here, the spatial variation in the composition of the avifauna highlights the similarities in the use of foraging resources on all plots. The presence or absence of species and changes in the relative abundances of others is related to the spatial and temporal differences in the kinds of food and foraging substrates available to birds on each woodlot.

Regardless of differences in the composition of the avifauna, the same foraging groups (Fig. 6) occurred on all plots. Lyrebirds, for example, were abundant on Woodlot 3 and absent from Woodlot 2, but "large ground foragers" were represented on Woodlot 2 by the White-winged Chough. Similarly bark foragers, foliage gleaners and nectar-feeders occurred on all plots, but Red-browed Treecreepers occurred only where there were trees with loose bark, Brown Thornbills were absent from woodlands which lacked a shrub layer, and nectar-feeders were abundant only when flowers were in bloom. Species richness increased as the number of foraging opportunities increased (e.g. by adding loose bark or providing flowers), but species were added within the existing broad pattern of community organisation. That is, species were added to groups rather than groups being added to the community

The pattern of seasonal events and the effect of variation in prey or foraging substrates can also be used to develop or refine plans of management for the conservation of forest and woodland birds. Previous recommendations for the management of forest avifaunas in southeastern Australia depended on a knowledge of which types of forest species used and which forests had the richest avifaunas. Recommendations for wildlife management sought to maximise breeding species diversity by conserving particular types of forest or maintaining a diverse age class structure of trees (e.g. Recher *et al.* 1980; Loyn 1980). Except in a general way, information which explained why particular species occurred in one type of forest and not another was not available. Nor was consideration given to the requirements of birds outside the breeding season or during migration. Thus management of wildlife could only be approached in the most general way and without detailed consideration of individual species or even groups of species.

There is a basic conservation requirement of providing birds with alternative feeding areas as different plants blossom or trees shed bark, as migratory stopovers, and as refuges from places affected by drought or fire. The long distances that birds move requires regional management. It achieves little to reserve breeding habitat, if the avifauna is deprived of a place to winter. Nix (1976) observed that the wintering areas for southeastern bird populations have been extensively disturbed by agricultural development and clearing. The intensification of forest management since the 1960's has expanded the nature of this disturbance with large areas of forest cleared for pine plantations or clearfelled for the production of woodchips.

Forest and woodland birds in Australia rely on a regular supply of food along whatever paths are followed in their migrations. Land managers therefore need to ensure continuity of habitats. It is especially important that major reserves be linked by corridors of relatively undisturbed vegetation.

Management policies which change the kinds of food or foraging substrates available to birds will affect the composition of the avifauna. Some species may benefit, but only at the expense of others. Reduction of fuel loads by regular prescribed burning, for example, will reduce the suitability of forests for birds such as the Brown Thornbill and Superb Lyrebird by reducing or removing the shrub and litter layers on which these species depend. They may be replaced by birds such as the Buff-rumped Thornbill and White-winged Chough which require drier, more open habitats. Species richness may be unchanged, but there will be a different avifauna as seen by comparing the birds on Woodlot 2 with those on Woodlot 3. If management practices reduce the abundance of trees which have loose or peeling bark or prevent these trees from maturing and thereby reduce the total amount of loose bark substrates, species such as the Red-browed Treecreeper and Shrike-tit would be disadvantaged. Knowing how birds forage and the temporal and spatial patterns of forest and woodland avifaunas provides the means to predict the kinds of changes which will follow particular management practices. The application of this knowledge to management should ensure that adverse effects are minimized.

ACKNOWLEDGEMENTS

R. Ferster, G. Gowing, R. Kavanagh, W. Rohan-Jones, M. Shulz, J. Shields and P. Smith assisted with the fieldwork. J. Caldwell allowed us to work on his woodlots at Bondi and the Forests Commission of Victoria permitted us to work in Victoria. The Forestry Commission of NSW encouraged the work and provided accommodation at its Bondi camp. A. Keast commented on earlier drafts of the manuscript. The work on foraging ecology was conducted while R. T. Holmes was a Visiting Fellow at the Australian Museum and supported by grants from the National Science Foundation (U.S./Australian Co-operative Science Programme), the National Parks and Wildlife Service (NSW), the Forestry Commission of NSW and Dartmouth College, USA. The participation of the Australian Museum was supported by a grant from Harris-Daishowa Pty Ltd.

REFERENCES

GOWING, G. AND RECHER, H. F., 1984. "Length versus weight relationships for invertebrates from forests of southern NSW". *Aust. J. Ecol.* **8**: 5-8.

LOYN, R. H., 1980. "Bird populations in a mixed eucalypt forest used for production of wood in Gippsland, Victoria". *Emu* **80**: 145-56.

MARCHANT, S., 1981. "The breeding season at Moruya, New South Wales". *Corella* **5**: 19-25.

NIX, H. A., 1976. "Environmental control of breeding, post-breeding dispersal and migration of birds in the Australian region". *Proc. 16th Int. Ornith. Congress:* 273-305.

RECHER, H. F., ROHAN-JONES, W. AND SMITH, P., 1980. "Effects of the Eden woodchip industry on terrestrial vertebrates with recommendations for management". *Forestry Commission of NSW. Research Note* No. 42, Sydney, 1980.

RECHER, H. F. AND SCHULZ, M., 1983. "Observations on the breeding of White-browed Woodswallows". *Corella* **7**: 1-6.

RECHER, H. F., GOWING, G., KAVANAGH, R., SHIELDS, J. AND ROHAN-JONES, W., 1983a. "Birds, resources and time in a tablelands forest". *Proc. Ecol. Soc. Aust.* **12**: 101-23.

RECHER, H. F., SMITH, P., MILLEDGE, D. AND ROHAN-JONES, W., 1983b. 'A transect method for sclerophyll forest'. *Corella* **7**: 49-54.

RECHER, H. F., HOLMES, R. T., SCHULZ, M., SHIELDS, J. AND KAVANAGH, R., 1985. Foraging patterns of breeding birds in eucalypt forest and woodland of southeastern Australia. *Aust. J. Ecol.* **10**: (in press).

SHIELDS, J. M. AND RECHER, H. F., 1984. 'Breeding birds censuses: an evaluation of four methods for use in sclerophyll forest'. *Corella* **8**: 29-41.

WOINARSKI, J., 1985. Foliage gleaners of the treetop, the pardalotes. Pp. 165-75 *in* Birds of Eucalypt Forest and Woodlands: Ecology, Conservation, Management ed. by A. Keast, H. F. Recher, H. Ford and D. Saunders. Royal Australasian Ornithologists Union.

CHAPTER 9

Bird Community Structure in Southern Forests and Northern Woodlands: A Comparison

Allen Keast[1]

Comparison of eucalypt forest and woodland avian communities at Hawkesbury River, NSW, Margaret River, southwestern Australia, South Alligator River, Northern Territory, and Lockhardt River, Cape York, in October-November, showed differences in species richness and avian biomass. The major ecological roles were utilized by the same taxonomic groups, and commonly the same species or allopatric congeners. Ecological counterparts were hence similar in body form. Marked similarities in community structure suggests minimal isolation between the north and south. Some differences between the east and southwest, and the numerically impoverished nature of the latter suggest that the west has had a greater degree of isolation and less capacity to support a diversity of species.

Data were analysed on use of vertical living space, biomass of birds supported by different vertical levels, and degree of ecological overlap between species. Widespread species had the same niches in different communities, and there was much overlap in feeding zone suggesting that, within the communities the species were behaving semi-independently and not significantly influencing each other's ecology.

INTRODUCTION

ALTHOUGH extensive studies have been carried out on the distribution of bird species throughout the Australian continent, on geographic variation in morphology, and role of isolation in speciation (Keast 1961; Ford 1986), no comparative geographic study has been made on geographic variation in the bird communities and their structure. The subject is of particular interest in that the north and south, and east and west, differ in climate as environments (Nix 1982, 1983), belong to distinct biogeographic or avifaunal regions (Spencer 1900; Kikkawa and Pearse 1969; Serventy and Whittell 1976), and differ in vegetation (Specht *et al.* 1974). Moreover, they have had different climatic histories, and histories of isolation. The continent had a Pleistocene/post-Pleistocene history of alternating dry/wet phases (Bowler 1976, 1982) which resulted in the forest and woodland zones expanding, contracting, and being disrupted. The forests of southwestern Australia and Tasmania, and the woodlands of the Northern Territory and Cape York continue to be isolated (Keast 1961; Macdonald 1966; Ford 1986). Climatic changes, and changes in the areas and distribution of the forest/woodland habitat must surely also have had a differential effect on bird community structure in different places.

The present paper has as its objective a comparative study of bird communities located in four segments of the Australian continent, southeast, southwest, northwest, and northeast. It focuses on species composition, origins and relationships, numbers of species, species richness and biomass, foraging habits, numbers and biomass of birds feeding in each vertical zone, and what factors may influence these. The degree of ecological (feeding zone) and morphological overlap between the component species in each region is quantified.

METHODS

Four sites were chosen as representative of the major areas: Hawkesbury River near Ebenezer in central New South Wales; Margaret River, southwestern Western Australia; South Alligator River in Arnhemland, Northern Territory; and Lockhart River, Cape York, Queensland (Fig. 1). In each area studies were carried out in October-November, early in the breeding season. Periods of 2-3

[1]Biology Department, Queen's University, Kingston, Canada.

Pages 97-116 *in* BIRDS OF EUCALYPT FORESTS AND WOODLANDS: ECOLOGY, CONSERVATION, MANAGEMENT ed. by A. Keast, H. F. Recher, H. Ford and D. Saunders. Royal Australasian Ornithologists Union and Surrey Beatty & Sons.

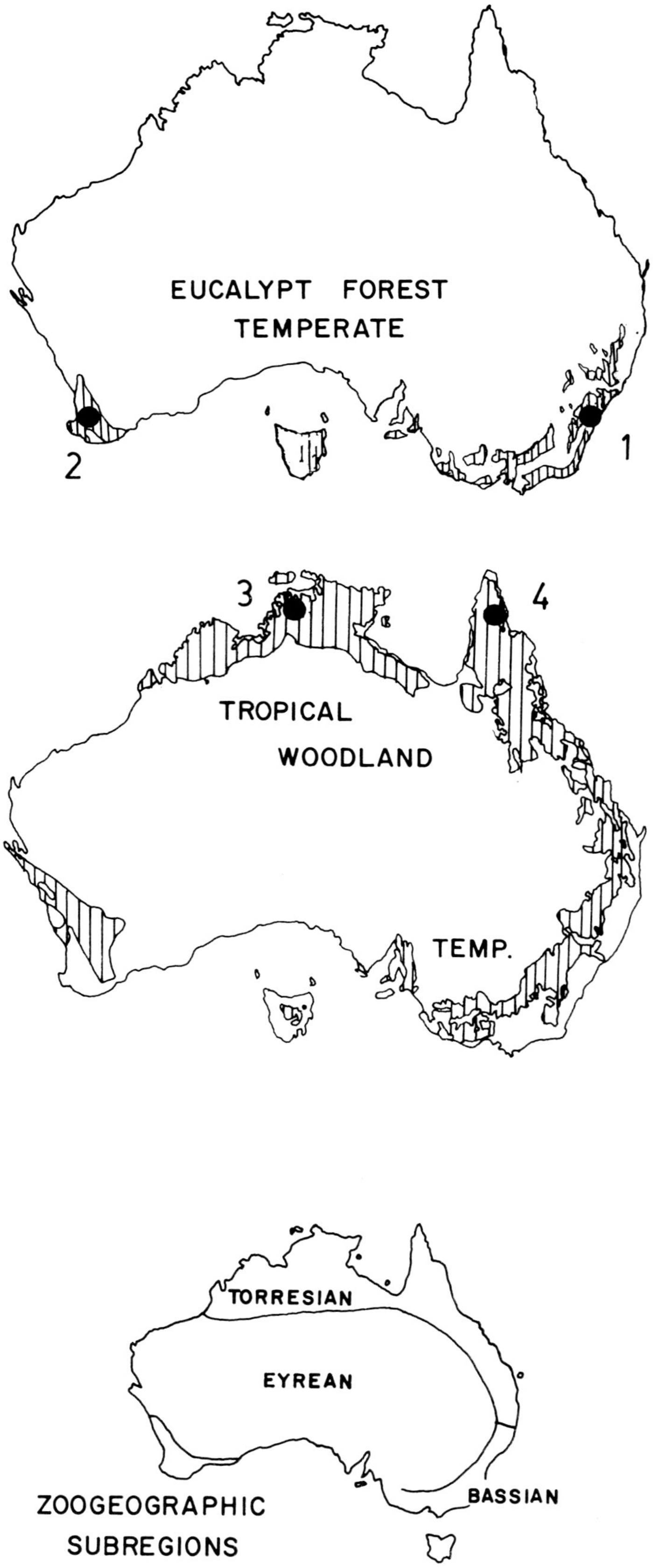

Fig. 1. The four study sites, geographic locations. 1. Hawkesbury R., NSW. 2. Margaret R., southwest Australia. 3. South Alligator R., Northern Territory. 4. Lockhart R., Cape York, Queensland. *Bottom figure:* The zoogeographic subregions of Australia, as per Spencer (1900) and Serventy and Whittell (1976).

Table 1. Climatic characteristics of southeast, southwest, northwest and northeast of continent, and vegetation data from the Hawkesbury River, Margaret River, South Alligator River and Lockhart River study sites. Climatic data is from Nix (1982, 1983). Vegetation classification is that of Specht *et al.* (1974) and Specht (1981).

	Hawkesbury River, Ebenezer, New South Wales	Margaret River, Western Australia	South Alligator River Darwin, Northern Territory	Lockhart River, Cape York, Queensland
Latitude, longitude	33.85 S 150.4 E	33.65 S 116.1 E	12.75 S 132.8 E	13.35 S 143.4 E
Vegetation category (former)	Sclerophyll forest	Sclerophyll forest	Sub-tropical woodland	Sub-tropical woodland
Specht's vegetation designation	Open-forest	Open-forest	Woodland	Woodland — Open woodland
Rainfall, mean annual, mm	1100	900	1200	1350
Rainfall, seasons	May-October and November-April	May-October	November-April	November-April
Temperature, mean daily max., hottest week °C	25-30	<25	35-40	30-35
Temperature, mean daily min., coldest week °C	<5	5-10	15-20	15-20
Tree height max. (m)	30	35	25	27
mean (m)	20(15-30)	25(20-35)	20(15-25)	17(15-27)
density (10,000 m^2)	700(450-900)	650(400-800)	520(410-650)	400(180-520)
canopy %	20-40	20-40	20-35	10-40
Tall shrubs (> 2 m)				
mean height	4.2(2.5-6.0)	3.9(2.5-6.0)	3.5(3.0-5.0)	3.6(3.0-5.0)
density (1000 m^2)	1200-2900(1900)	900-1400(1200)	800-1300(1000)	200-500(420)
Low shrubs (< 2 m) % cover	45-50	20	5-10	5
Grass and Cover	10-25	10-15	25-40	35-60
Ground, bare %	25-45	40-45	40-45	35-60
Dominant trees, No. species > 3% of total	12	9	8	5
Tall shrubs, No. species > 3% of total	22	15	10	7
Trunk diameter range (cm) trees	30-55	30-60	30-40	20-35

weeks were spent in each region, working in areas of about 20 km^2. The Hawkesbury data were gathered in 1980 and 1981, the Margaret in 1977 and 1980, the South Alligator in 1976 and 1982, and the Lockhart in 1978. Other than in the case of the Lockhart the work was preceded by an earlier trip when initial feeding zone data were gathered.

Data on bird species composition and populations were obtained by the strip count method (Emlen 1971, 1977; Burnham *et al.* 1980). Strips of 30 m width (Ford and Bell 1981) and 1 km long were used. Such strips gave areas of 3 ha. The counts were made during the morning feeding period, from half to 3½ hours after sunrise. Pre-planned routes were followed, paths and roads being avoided. Only birds seen were counted though many were initially located from calls. A target of 20-25 counts was set for each site.

The following habitat determinations were made: foliage height profiles (MacArthur and MacArthur 1961; MacArthur 1964; Recher 1969); degree of canopy cover; tree and tall shrub (height 2-6 m) densities; trunk diameters; percent low shrub (height < 2 m) and grass cover; and percentage of bare ground (Table 1). Foliage height profiles and degree of canopy cover were determined at 20 points chosen at random through the study area. Tree and shrub densities were determined from a series of twenty 200 m transects, with strip widths standardized at 2 m.

Counts of numbers of dominant trees and tall shrubs were obtained similarly. Flowering plants on which honeyeaters were feeding were collected for identification.

Feeding heights were determined on the basis of 15 vertical divisions or strata (Figs 2-4, Table 2). Feeding individuals were watched for periods of up to three minutes and individual feeding actions (pecks, pursuit flights) to a limit of 20 for each individual were recorded. Emphasis was placed on the more abundant species. Data were gathered throughout the feeding period and throughout the respective areas. The initial goal was to obtain 500 observations per species. Swallows and the White-rumped Swiftlet, *Collocali spodiopygia,* that foraged above the canopy were assumed to be feeding exclusively there.

Objectives were (a) to determine for each species the distribution of feeding effort (i.e. percentage of feeding actions relative to the vertical division of

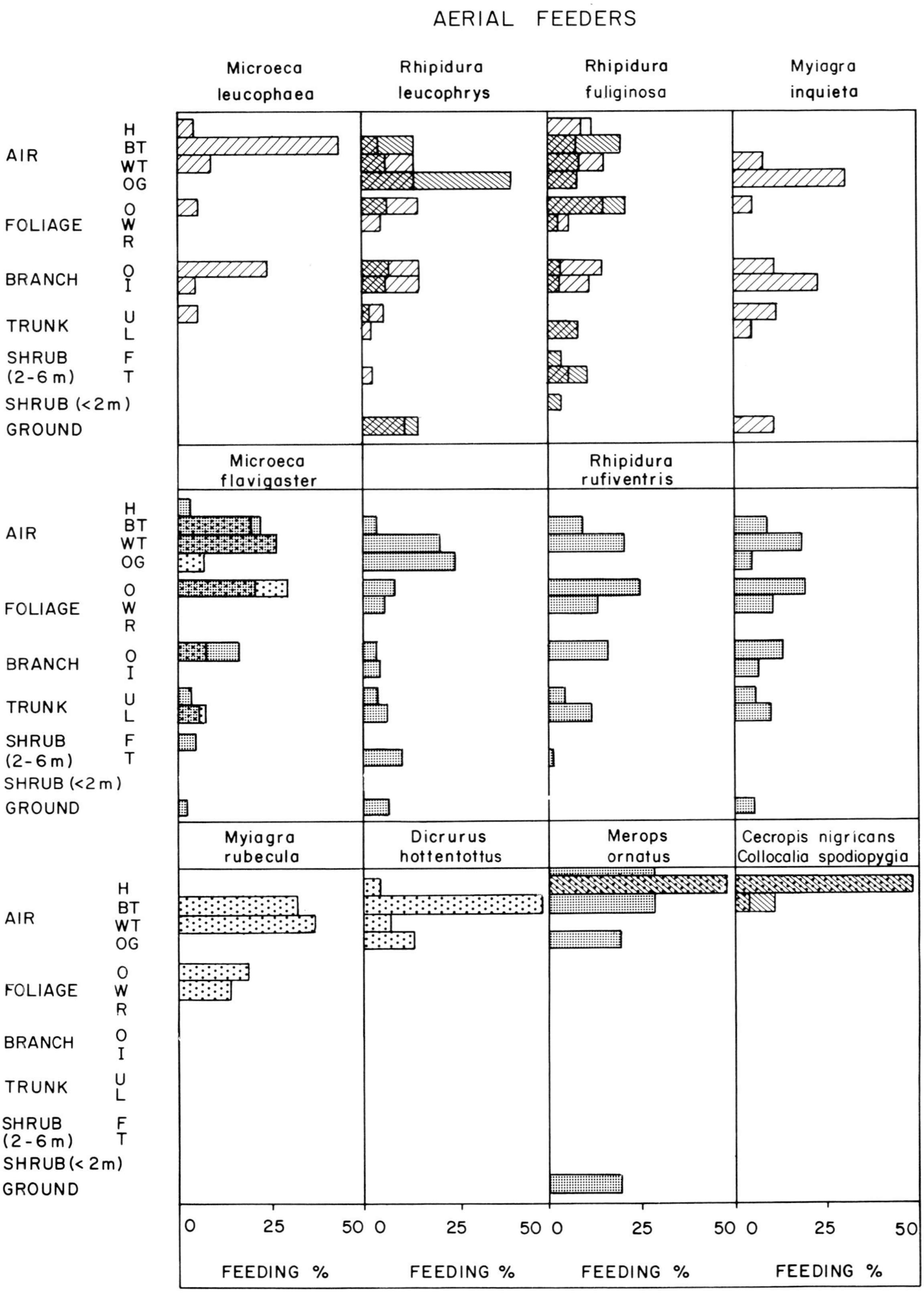

Fig. 2. The aerial feeders of the four communities, division of feeding effort. Cross-hatching (broad) — Hawkesbury River, cross-hatching (narrow) — Margaret River, spotting (fine) South Alligator River, spotting (coarse) — Lockhart River.

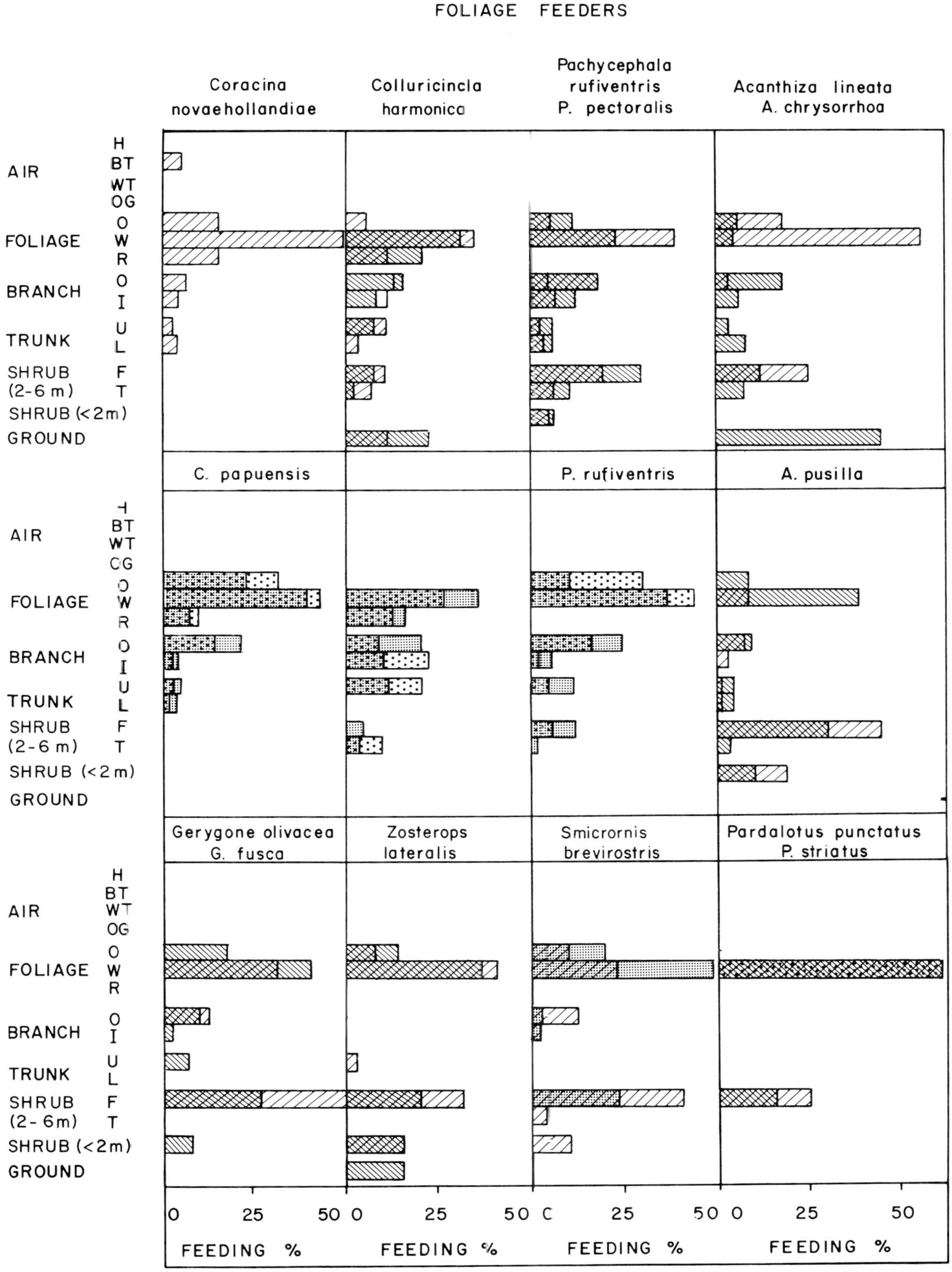

Fig. 3. The foliage feeders of the four communities, division of feeding effort. Cross-hatching (broad) — Hawkesbury River, cross-hatching (narrow) — Margaret River, spotting (fine) South Alligator River, spotting (coarse) — Lockhart River.

Table 2. The distribution of feeding effort relative to vertical feeding zone, insect eaters, in terms of biomass (g/ha) and % representation October-November, the four study sites compared. Nectar-feeding is excluded.

	Hawkesbury River		Margaret River		South Alligator River		Lockhart River	
	g	%	g	%	g	%	g	%
Air								
above trees	2.1	0.5	14.2	6.7	17.9	6.2	12.4	7.2
between trees	16.6	4.0	4.8	2.3	24.4	8.4	12.6	7.4
within trees	3.7	0.9	1.4	0.7	21.0	9.9	6.3	3.7
over ground	2.0	0.5	3.5	1.6	17.1	5.9	3.4	2.0
TOTAL	24.4	5.9	22.9	11.3	80.4	30.4	30.7	20.3
Foliage								
outer	25.8	6.2	27.7	13.0	23.1	8.0	16.0	9.3
within	94.4	22.8	51.4	24.2	77.7	26.8	58.2	34.0
reaching into	21.9	5.3	8.1	3.8	20.9	7.2	4.7	2.7
Shrubs > 2 m								
foliage	104.8	25.3	35.8	16.9	13.8	4.8	10.5	6.1
TOTAL	246.9	59.6	123.0	57.9	135.5	46.8	89.4	52.1
Tree Branches								
outer	21.4	5.2	15.0	7.1	21.2	7.3	18.7	10.9
inner	17.8	4.3	7.4	3.5	12.6	4.3	13.1	7.7
TOTAL	39.2	9.5	22.4	10.6	33.8	11.6	31.8	18.6
Trunks								
upper	12.9	3.1	5.2	2.4	14.0	4.8	6.2	3.6
lower	19.1	4.6	3.1	1.5	13.1	4.5	1.5	0.9
TOTAL	32.0	7.7	8.3	3.9	27.1	9.3	7.7	4.5
Shrubs > 2 m								
trunks and branches	16.5	4.0	4.1	1.9	6.9	2.4	4.2	2.5
TOTAL	87.7	21.2	34.8	16.4	67.8	23.3	43.7	25.6
shrubs, tall grass < 2 m	24.2	5.8	7.7	3.6	1.8	0.6	3.4	2.0
Ground	31.1	7.5	22.9	10.8	4.4	1.5	—	—
TOTAL	414.3		212.3		289.9		171.2	

living space, (b) quantify the amount of overlap in use of the vertical zones by different guild members, (c) determine if the zonal feeding of species varied geographically and (d) calculate the total avian biomass supported at each vertical level.

Biomass supported per feeding zone for each species was calculated as the product of the proportion of its feeding in that zone and the biomass/ha for that bird species. Thus, if a 20 gm bird with a population of 10 birds/ha did 50% of its feeding among the outer leaves, then the amount of biomass supported by that zone was 100 gm/ha. Data from the various species were added to give the total.

Overlaps in space use were calculated for pairs of species in two major feeding guilds; aerial-feeders and small foliage-gleaners. For this the formula of Schoener (1970) was used;

$$\alpha = 1 - 0.5\,(\sum_{i=1}^{n} |P_{xi} - P_{yi}|)$$

where n is the number of vertical feeding zones and P_{xi} is the proportion of feeding zone (i) used by species x and P_{yi} is the proportion of its use by species y.

For the morphological studies wing and bill lengths of seven adult males of each species from each population were measured. Bills were measured from the base of the skull to the tip and wing lengths from angle of the folded wing to the tip. The species weights used in the biomass calculations are from Ford and Bell (1981) and Serventy and Whittell (1976). Where weights were unavailable, estimates based on body size, have been allocated. These are shown in brackets in Table 3.

Characteristics of the Study Sites

The geographic location and major characteristics of the four sites are given in Table 1. The Hawkesbury and Margaret sites were vegetated by temperate open-forest, South Alligator by woodland, and Lockhart by woodland-open woodland (classification of Specht *et al.* 1974). While the mean annual rainfall is similar, the seasonal distribution differs. The Hawkesbury has two rainy periods, in late winter-early spring and summer-autumn, the Margaret only one period, in late winter-early spring. The two northern sites were tropical woodlands having a monsoonal climate with summer rains and dry winters. The two northern localities also have warmer winters and hotter summers than the southern ones.

All sites were characterized by good tree development and mature trees. Canopies were open. Canopy cover ranged from 20% in open sections to 40% in denser ones. Tree density was lowest at Lockhart. Foliage profiles were generally

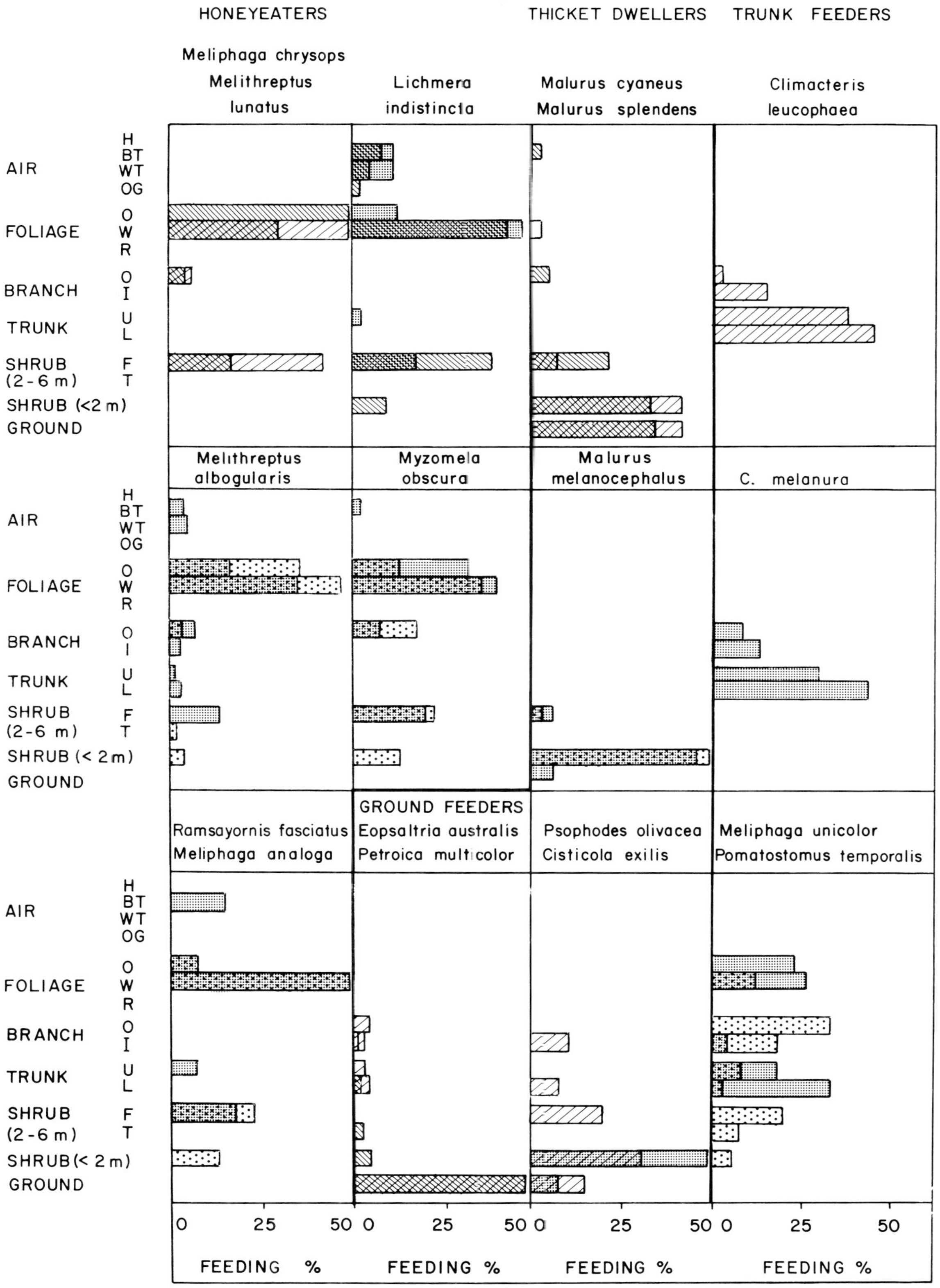

Fig. 4. The honeyeaters, thicket dwellers, trunk feeders and ground feeders at the four communities, division of feeding effort. Cross-hatching (broad) — Hawkesbury River, cross-hatching (narrow) — Margaret River, spotting (fine) South Alligator River, spotting (coarse) — Lockhart River.

Table 3. Avian communities at four study sites, southeastern eucalypt forest (Hawkesbury River), southwestern forest (Margaret River), northwestern woodland (South Alligator River), and northeastern woodland (Lockhart River). Figures are numbers of individuals per hectare, based on series of 30 individual strip counts of 3 ha. The species are allocated to primary diet categories on the basis of published data on food Lea Gray (1935, 1936) and Rose (1973) feeding habits at the time of the surveys.

Symbols: IFS, IFM, IFL — insects, foliage, small medium, and large, respectively; IT — insects, trunks and branches; IS — insects, low shrubs and tall grass; IG — insects, ground; AIH — aerial insects, above trees; AIL — aerial insects taken at tree level or below; N — nectarivore; NV — nectar and vegetable matter; NI — nectar and insects; GG — granivore, ground; GT — granivore, trees; P — predator on large insects and small vertebrates.

Body weights are from Ford and Bell (1981), and Serventy and Whittell (1976). Where no weight is available an estimated one has been allocated — see brackets.

Marginal symbols — X — precise feeding data given in Figs 2-4; O — wing and bill lengths plotted in Figure 5.

Species		Diet Category	Weight	Hawkesbury River		Margaret River		South Alligator River		Lockhart River	
			g	Nos.	Wt (g)	Nos.	Wt (g)	Nos.	Wt (g)	Nos.	Wt (g)
Columbidae		GG	56					0.8	44.8	0.45	25.2
Geopelia striata											
G. humeralis		GG	147					0.02	2.9	0.03	4.4
Phaps chalecoptera		GG	395	0.2	79						
Petrophassa smithii		GG	(300)					0.2	60		
Psittacidae											
Trichoglossus haematodus		NV	(90)							0.4	36
T. rubritorquatus		NV	(85)					0.2	17		
Psitteuteles versicolor		N	(45)					0.5	22.5		
Glossopsitta porphyrocephala		N	70			0.2	14.0				
G. pusilla		N	(70)	0.1	7.0						
Aprosmictus erythropterus		ST	130					0.02	2.6		
Platycerus eximius		GTG	110	0.1	11.0						
P. isterotis		GTG	(90)			0.2	18.0				
Barnardius zonarius		GTG	222			0.04	9.0				
Cuculidae											
Cuculus pallidus		IFM	(69)	0.08	4.8						
C. pyrrhophanus		IFM	50			0.1	5.0	0.04	2.0		
Chrysococcyx lucidus		IFM	23	0.1	2.3	0.4	9.2				
C. malayanus		IFM	(20)					0.1	2.0		
Centropus phasianinus		IGL	670					0.02	14.0	0.1	6.7
Apodidae											
Collocalia spodiopygia	X	AIH	(16)							0.2	3.2
Alcedinidae											
Dacelo gigas (novaeguineae)		P	305	0.2	61	0.2	61				
D. leachii		P	300					0.03	90	0.14	42
Halcyon macleayii		P	(45)					0.07	3.2	0.34	15.3
H. sancta		P	45	0.8	36			0.4	18		
Meropidae											
Merops ornatus	X	AIHLG	70			0.01	0.4	1.2	48	0.2	8.0
Coraciidae											
Eurystomus orientalis		AIHL	140	0.02	2.8			0.1	14.0	0.1	14.0
Hirundinidae											
Hirundo neoxena		AIH	(20)	0.1	2.0						
Cecropis nigricans	X	AIH	(20)			0.7	14.0	0.2	4.0		
Campehagidae											
Coracina novahollandiae	X	IFL	130	0.6	78			0.2	26.0		
C. papuensis		IFL	(80)	0.01	0.8			0.7	5.6	0.5	4.0
Lalage sueurii		IFM	26					0.02	0.5		
Muscicapidae											
Petroica multicolor	X	IG	13.2			0.4	5.3				

Table 3. Continued.

Species		Diet Category	Weight	Hawkesbury River		Margaret River		South Alligator River		Lockhart River	
			g	Nos.	Wt (g)	Nos.	Wt (g)	Nos.	Wt (g)	Nos.	Wt (g)
Eopsaltria georgiana		IG	(19)			0.02	0.4				
E. australis	X	IG	19.6	0.4	7.8	0.2	3.9				
Microeca flavigaster	XO	AIL	(14)					0.6	8.4	0.4	5.6
M. leucophaea	X	AIL	(15)	0.5	7.5						
Falcunculus frontalis		IT	(25)					0.01	0.25		
Pachycephala pectoralis	XO	IFM	24	0.2	4.8	1.3	31.0				
P. rufiventris	XO	IFM	25.5	1.4	35.7	0.06	1.5	0.3	7.7	0.5	11.5
Colluricincia harmonica	X	IFL	64	0.9	57.6	0.4	25.6	0.4	25.6	0.3	19.2
Myiagra rubecula	XO	AIL	14							0.25	3.5
Myiagra inquieta	XO	AIL	24	0.1	2.4	0.2	4.8	0.5	12.0		
Rhipidura fuliginosa	XO	AIL	8.2	1.3	10.7	1.2	9.8				
R. rufiventris	XO	AIL	(14)					0.15	2.0		
R. leucophrys	XO	AIL	22.3	0.2	4.5	0.2	4.5	0.9	20.1		
Orthonychidae											
Psophodes olivaceus	X	IS	(70)	0.4	28.0						
Timaliidae											
Pomatostomus temporalis	X	IT	(70)							0.3	21.0
Sylviidae											
Cisticola exilis	X	IS	(8)					0.1	0.8		
Maluridae											
Malurus cyaneus	XO	ISG	9	1.1	9.9						
M. splendens	XO	ISG	(9)			0.6	5.4				
M. lamberti		IS	(9)	0.4	3.6						
M. elegans		IS	(10)			0.02	0.2				
M. melanocephalus	XO	IS	6.5					0.2	1.3	0.1	0.9
Acanthizidae											
Origma solitaria		IG	(20)	0.3	6.0						
Smicrornis brevirostris	XO	IFS	(7)	0.6	4.2			1.2	8.4		
Gerygone fusca	XO	IFS	(10)			0.9	9.0				
G. olivacea	XO	IFS	10	0.7	7.0					0.06	0.6
Acanthiza pusilla	XO	IFS	7	0.3	2.1	0.7	4.9				
A. inornata		IFS	(8.3)			0.04	0.3				
A. reguloides		ITG	8.3	0.2	1.7						
A. chrysorrhoa	XO	IFG	(8.5)			1.3	11.0				
A. nana	O	IFS	(7.3)	1.0	7.3						
A. lineata	XO	IFS	7.3	1.3	9.5						
Climacteridae											
Climacteris leucophaea	XO	IT	23.6	0.9	21.2						
C. rufa	O	IT	(34)			0.03	1.0				
C. melanura	XO	IT	(34)					0.4	13.6		
Meliphagidae											
Anthochaera carunculata		IFL	125			0.2	25.0				
A. chrysoptera		IFL	80	0.2	16.0						
Philemon buteroides		NI	(90)					0.1	9.0	0.3	27
P. argenticeps		IFL	(90)					0.09	8.1		
P. corniculatus		NI	107	0.1	10.7						
P. citreogularis		NI	(70)					0.07	4.9		
Entomyzon cyanotis		IFL	(110)					0.02	2.2		
Meliphaga lewinii	O	IFM-Fruit	(30)	0.2	6.0						
M. analoga	XO	IFMN	(15)							0.4	6.0

Table 3. continued.

Species		Diet Category	Weight	Hawkesbury River		Margaret River		South Alligator River		Lockhart River	
			g	Nos.	Wt (g)	Nos.	Wt (g)	Nos.	Wt (g)	Nos.	Wt (g)
M. gracilis	O	NI	(12)							0.16	1.92
M. (Lichenostomus) chrysops	XO	IFM	17	2.2	37.4						
M. unicolor	XO	ITF	(30)					0.3	9.0		
M. flava		IFM	(30)							0.02	0.6
Melithreptus gularis	O	IFM	(19)					0.05	0.95		
M. albogularis	XO	IFM	(14)					2.1	29.4	1.4	19.6
M. lunatus	XO	IFM	14.7	0.2	2.9	1.5	22.0				
Lichmera indistincta	XO	NI	(13)			0.6	7.8	0.7	9.1		
Phylidonyris novaehollandiae		NI	(25)			0.4	10.0				
Ramsayornis fasciatus	XO	NI	(10)					0.5	5.0		
Conopophila albogularis	O	IFM	(10)					0.3	3.0		
Acanthorhynchus tenuirotris		NI	12	0.4	4.8						
A. superciliosus		NI	(11)			0.04	0.4				
Myzomela obscura	XO	IFMN	(10)					0.2	2.0	0.3	3.0
M. sanguinolenta		IFS	(9)	0.6	5.4						
Dicaeidae											
Dicaeum hirundinaceum		FI-fruit	8	0.2	1.6	0.4	3.2	0.3	2.4	0.3	2.7
Pardalotidae											
Pardalotus punctatus	XO	IFS	8	1.6	12.8	0.4	3.2				
P. striatus	XO	IFS	12			0.6	7.2			0.2	2.4
Zosteropidae											
Zosterops lateralis	XO	IFS	10.3	1.1	11.3	1.6	16.5				
Ploceidae											
Emblema temporalis		GG	10.9	0.4	4.4						
Poephila bichenovii		GG	10.5	0.2	2.1			0.3	3.2		
P. personata		GG	(12)					0.2	2.4		
P. acuticauda		GG	(13)					0.05	6.5		
Oriolidae											
Oriolus sagittatus		IFL	96	0.2	19.2			0.3	28.8		
O. flavocinctus		IFL	(90)					0.04	3.6	0.1	9
Dicruridae											
Dicrurus hottentottus	X	AIL	(90)					0.03	2.7	0.2	18
Grallinidae											
Grallina cyanoleuca		IGL	89	0.5	44.5	0.02	1.8	0.6	53.4		
Cracticidae											
Cracticus quoyi		P	(95)							0.1	9.5
C. torquatus		P	84	0.3	25.2	0.1	8.4	0.06	5.0		
Gymnorhina tibicen		P	307	0.2	61.4	0.3	92.1				
Corvidae											
Corvus coronoides		P	600	0.2	120	0.1	60.0				

similar except at the shrub level. A tall shrub layer (height 2-5 m) was marked at Hawkesbury, moderately developed at Margaret and South Alligator, and sparce at Lockhart. There was a well developed low shrub layer (< 2 m) at Hawkesbury. This was absent in the north, but was partly compensated for by grass cover that ranged from 25-60%. At all sites there were extensive areas of open ground.

Tree species richness, and tall shrub species richness was highest at Hawkesbury, intermediate at Margaret and South Alligator and lowest at Lockhart. The high vegetation diversity at Hawkesbury resulted from several factors including its varied topography. The Margaret, South Alligator, and Lockhart regions were topographically flat and the vegetation was uniform throughout. The Hawkesbury and Margaret sites had creeks with running water and the South Alligator had water-filled lagoons nearby. The Lockhart site was the driest with little standing water.

The major trees at Hawkesbury were *Eucalyptus gummifera, E. exima, E. paniculata, E. piperata,* and *Casuarina torulosa*. The tall shrub layer included *Ceratopetalum gummiferum* and *Pomaderris lanigera* in the gullies, and stands of *Acacia decurrens* and *A. longifolia, Banksia serrata, Melaleuca* sp. and *Exocarpus cuppressiformis* elsewhere. The low shrub layer was diverse and dominated by sandstone elements including, *Banksia spinulosa*, four species of *Acacia, Persoonia, Isopogon, Lambertia, Grevillea, Comesperma*, a dozen legumes, and at least ten major species of grasses. The major nectar producers at the time of the study were *E. paniculata* and some introduced *Erythrina*.

The Margaret River supported *Eucalyptus marginata, E. calophylla, E. diversicolor* and other eucalypts. The tall shrub layer was dominated by *Melaleuca*, large *Acacia, Banksia, Exocarpus*, and others. The low shrub layer included a wide variety of genera including *Banksia, Acacia, Bossoiea* and *Pimelia*. Wetter areas in low-lying places supported extensive stands of *Melaleuca, Acacia* and *Lepidosperma*. *Eucalyptus diversicolor* and *Melaleuca* were the main nectar-producers.

Major trees of the South Alligator were *Eucalyptus tetradonata, E. miniata* and *E. dichromophoea*. The tall shrub layer included *Melaleuca*, especially *M. symphyocarpa* and *M. viridiflora, Tristania lactiflora, Erythyophloem chlorestachys*, and *Pandanus* near the lagoons. Other major elements were *Terminalia, Petalostigma, Calytrix, Xanthostemon*, and some *Barrington*. *Sorghum* and other Poaceae covered the woodland floor. Several trees and shrubs were flowering. Of these, *Planchonia careya, Melaleuca* and *Eucalyptus tetradonata* were the most important sources of nectar.

The Lockhart had the fewest species of *Eucalyptus* and tall shrubs. Major trees were *Eucalyptus tetradonata, E. polycarpa, E. dichromphoea*, and *E. papuana*. Major shrubs included *Grevillea glauca, Melaleuca, Acacia rothii* and *Alphitonia excelsa*. *Sorghum plumosa, Schizochyrum* and *Thaumastochloa* were the main grasses. *Grevillea* and *Eucalyptus tetradonata* were the main nectar-producing species.

RESULTS

Comparison of the Four Sites in Terms of Bird Species Richness

The bird species present in each community, their abundances and biomass are given in Table 3. Table 4 integrates these data at the family level and Table 5 expresses them in terms of representation of the major 'ecological types' of birds, based on way of life and diet. Hawks, other large birds of prey and the larger parrots, which commonly flew over the study sites but which could not be regarded as members of the immediate community are not included.

The sites differed in avifaunal richness. Numbers of species occurring were: Hawkesbury, 47; Margaret, 37; South Alligator, 49; Lockhart, 29. Biomass figures (g/ha) were, respectively, 902, 507, 668 and 321. Numbers of individual birds/ha were: Hawkesbury, 23; Margaret, 15; South Alligator 16; Lockhart, 8. These figures are generally in line with those obtained for Australian eucalypt forests and woodlands by other workers, vide: woodlands near Armidale, NSW, 22.5 birds/ha (Ford and Bell 1981); 7.0 birds/ha for Canberra (Bell 1980); 15.2-47.2 birds/ha for Victorian sclerophyll forests (Loyn 1980); and 16 birds/ha for a dry sclerophyll forest in Tasmania (Recher *et al.* 1971), and 15-20 and 14-23 birds/ha for tall open-forest and open-forest, respectively, at Eden, NSW (Recher 1984).

Why do species numbers and biomass differ so strikingly between the four sites? High bird species numbers at Hawkesbury partly resulted from its two types of habitat: richer Wianamatta shale/open-forest and more open sandstone vegetation. This also explained the high floristic diversity. The area supported some bird species characteristic of each vegetation type. For example, three small foliage-gleaning warblers co-occurred, the Striated Thornbill, *Acanthiza lineata*, a typical open-forest species, occurred in the gullies; the Yellow Thornbill, *A. nana*, in the stands of *Acacia decurrens*, and the Weebill, *Smicrornis brevirostris*, a predominantly dry-country species, on the top of the escarpment. It also possessed two other *Acanthiza*, the Brown and Buff-tailed Thornbills, *A. pusilla* and *A. reguloides*. The Margaret site had only two, the Brown Thornbill and Yellow-tailed Thornbill, *A. chrysorrhoa*, which was mainly a ground feeder. The only member of this group in the north was the Weebill. Two habitat specialists, the sandstone-dwelling Rock Warbler or Origma, *Origma solitaria*, and thicket-dwelling Eastern Whipbird, *Psophodes olivaceus*, were also unique to the Hawkesbury. The next richest site, the South Alligator, partly owed its avifaunal richness to the presence of 11 species of honeyeaters (Meliphagidae). Some of these were associated with lagoon-edge thickets, others more with the open woodland, and several concentrated at flowering trees. By contrast only seven meliphagids were present at the Hawkesbury, five at Margaret, and six at Lockhart. The South Alligator also had three pigeons, three kingfishers, three cuckoo-shrikes, three finches and two orioles, compared to one or two members of these groups elsewhere.

The lower figures for bird species numbers, and numbers of individuals at Lockhart and Margaret

Table 4. Representation of different taxonomic categories (families) of birds, number of individuals and biomass/(g/ha) and % representation (g) October-November, the four study sites compared.

	Hawkesbury River			Margaret River			South Alligator River			Lockhart River		
	No.	Wt (g/ha)	%	No.	Wt (g/ha)	%	No.	Wt (g/ha)	%	No.	Wt(g/ha)	%
Columbidae	0.2	79	8.8	—	—	—	1.02	107.7	16.1	0.5	29.6	9.2
Psitacidae	0.2	18.1	2.0	0.41	41.0	8.1	0.72	42.1	6.3	0.4	36	11.2
Cuculidae	0.18	7.1	0.8	0.5	14.2	2.8	0.2	18.0	2.7	0.1	6.7	2.1
Apodidae	—	—	—	—	—	—	—	—	—	0.2	3.2	1.0
Alcedinidae	1.0	97	10.8	0.2	61	12.0	0.5	111.2	16.6	0.5	57.3	17.9
Meropidae	—	—	—	0.1	0.04	0.01	1.2	48.0	7.2	0.2	8.0	2.5
Coraciidae	0.02	2.8	0.3	—	—	—	0.1	14.0	2.1	0.1	14.0	4.4
Hirundinidae	0.1	2.0	0.2	0.7	14.0	2.8	0.2	4.0	0.6	—	—	—
Campephagidae	0.6	78.8	8.7	—	—	—	0.9	32.1	4.8	0.5	4.0	1.2
Muscicapidae	5.1	13.1	14.5	3.8	86.9	17.1	2.9	76.1	11.4	1.4	39.8	12.4
Orthonychidae	0.4	28.1	3.1	—	—	—	—	—	—	—	—	—
Timaliidae	—	—	—	—	—	—	—	—	—	0.3	21.0	6.5
Sylviidae	—	—	—	—	—	—	0.1	0.8	0.1	—	—	—
Maluridae	1.4	13.5	1.5	0.6	5.6	1.1	0.2	1.3	0.2	0.13	0.9	0.3
Acanthizidae	4.4	37.8	4.2	2.8	25.2	5.0	1.2	8.4	1.3	0.06	0.6	0.2
Climacteridae	0.9	21.2	2.4	0.03	1.0	0.2	0.4	13.6	2.0	—	—	—
Meliphagidae	3.9	83.2	9.2	2.6	65.3	12.9	4.5	82.7	12.4	2.6	58.1	18.1
Dicaeidae	0.2	1.6	0.2	0.4	3.2	0.6	0.3	2.4	0.4	0.34	2.7	0.8
Pardalotidae	1.6	12.8	1.4	1.0	10.4	2.1	—	—	—	0.2	2.4	0.7
Zosteropidae	1.1	11.3	1.3	1.6	16.5	3.3	—	—	—	—	—	—
Ploceidae	0.6	6.3	0.7	—	—	—	0.4	12.1	1.8	—	—	—
Oriolidae	0.2	19.2	2.1	—	—	—	0.34	32.4	4.9	0.1	9.0	2.8
Dicruridae	—	—	—	—	—	—	0.03	2.7	0.4	0.2	18.0	5.6
Grallinidae	0.5	44.5	4.9	0.02	1.8	0.4	0.6	53.4	8.0	—	—	—
Cracticidae	0.5	86.6	9.6	0.4	100.6	19.9	0.06	5.0	0.7	0.1	9.5	3.0
Corvidae	0.2	120.0	13.3	0.1	60.1	11.9	—	—	—	—	—	—
TOTAL	23.3	901.9		15.3	506.8		16.0	668		7.9	320.8	

were due to different factors. The Lockhart lacked many species that, from their broad geographic ranges, might have been expected to occur. There was only one cuckoo, no *Rhipidura* flycatchers, and no finches. This can be ascribed to the lower density of trees and tall shrubs and low tree and shrub species richness. Low species numbers at Margaret, by contrast, reflected the avifaunal impoverishment of the southwestern forest block. It is isolated by 2000 km of arid terrain and, though floristically rich, it has little more than half the passerine birds of the eastern forest block (Keast 1975, 1976a and b).

There were major differences in the taxonomic composition of the four avifaunas. A few genera were restricted, alternatively, to the north (*Ramsayornis, Conopophila, Collocalis, Dicrurus*) or the south (*Origma, Psophodes, Eopsaltria, Petroica, Acanthorhynchus*). Species were restricted to one or other area, e.g. the Masked and Long-tailed Finches, *Peophila personata* and *P. acuticauda,* to the north, and Red-browed Finch (Firetail), *Aegintha temporalis*, to the south. In several cases allopatric species replaced each other (e.g. in *Microeca, Dacelo, Meliphaga*). This species replacement also occurred in the southeast relative to southwest (different species of *Acanthorhynchus*). Such differences have been used as the basis for defining biogeographic sub-regions (Fig. 1) (Spencer 1900; Serventy and Whittell 1976).

Virtually all major taxonomic groups were present throughout, i.e. flycatchers, honeyeaters, warblers, cuckoos, pardalotes, cuckoo-shrikes, parrots, treecreepers, and swallows. A large percentage of the total species occurred both in the southern forests and northern woodlands. If allopatric congeneric species, which are ecological equivalents (see later) are added to this the degree of similarity is marked.

The structure of the four communities must also be looked at in another framework, the relative abundances, and relative biomass of the major taxonomic and ecological groups of birds. This will reflect the capacity of the region to support such taxonomic or ecological types (Tables 3 and 4).

The four families with the greatest numbers of individuals at Hawkesbury were, in order of abundance; muscicapids, acanthizids, honeyeaters, and pardalotids. At Margaret it was muscicapids, acanthizids, honeyeaters, and white-eyes. At South Alligator it was honeyeaters, muscicapids, bee-eaters, and pigeons, and at Lockhart, where overall numbers were much lower, honeyeaters and muscicapids, followed by pigeons, kingfishers and cuckoo-shrikes were most abundant.

Table 5. Number of individuals, biomass (g/ha), and % representation of species in major dietary categories, October-November, the four study sites compared.

	Hawkesbury River			Margaret River			South Alligator River			Lockhart River		
	No.	Wt (g/ha)	%	No.	Wt (g/ha)	%	No.	Wt (g/ha)	%	No.	Wt (g/ha)	%
Insect-eaters, foliage, small	7.2	59.6	6.6	4.2	41.1	8.1	1.2	8.4	1.3	0.1	3.0	0.9
medium	4.1	87.9	9.7	3.4	68.7	13.5	2.9	45.6	6.8	0.6	31.7	9.9
large	2.0	171.6	19.0	0.6	50.5	9.9	1.8	97.9	14.7	0.2	32.2	10.0
Insect-eaters, trunks and foliage	—	—	—	—	—	—	0.3	9.0	1.3	—	—	—
Insect-eaters, trunks and branches	0.9	21.2	2.4	0.03	1.0	0.2	0.4	13.9	2.1	0.1	21.0	6.5
Insect-eaters, trunks and ground	0.2	1.7	0.2	—	—	—	—	—	—	—	—	—
Insect-eaters, low shrubs	0.8	31.6	3.5	0.02	2.9	0.6	0.3	2.0	0.3	0.03	0.9	0.3
Insect-eaters, ground	1.2	58.3	6.5	1.9	23.3	4.6	0.6	67.4	10.1	0.03	6.7	2.1
Insect-eaters, low shrubs and ground	1.1	9.9	1.1	0.6	5.4	1.1	—	—	—	—	—	—
Aerial insectivores feeding above trees	0.1	4.8	0.5	0.7	14.0	2.7	0.3	18.0	2.7	0.3	17.2	5.4
Aerial insectivores feeding at tree level or below	2.1	25.1	2.8	1.6	19.1	3.7	2.2	45.2	6.8	1.0	22.1	8.4
Aerial insectivores also feeding on ground	—	—	—	0.01	0.4	0.1	1.2	48.0	7.2	0.2	8.0	2.5
Nectarivores	0.1	7.0	0.8	0.2	14.0	2.7	0.5	25.5	3.8	—	—	—
Nectar and vegetable matter feeders	—	—	—	—	—	—	0.2	17.0	2.5	0.4	36.0	11.2
Nectarivore-insectivores	0.5	15.5	1.7	1.00	18.2	3.6	1.6	30.0	4.5	1.2	37.9	11.8
Granivore, ground	0.8	85.5	9.5	—	—	—	1.6	118.8	17.8	0.5	29.6	9.2
Seed-eaters, trees	—	—	—	—	—	—	0.02	2.6	0.4	—	—	—
Seed-eaters, trees and ground	0.1	11.0	1.2	0.2	27.0	5.3	—	—	—	—	—	—
Frugivore-insectivores	0.4	7.6	0.8	0.4	3.2	0.6	0.3	2.4	0.4	0.3	2.7	0.8
Predators on large insects and small vertebrates	1.7	303.6	33.7	0.7	221.5	43.4	0.6	117.2	17.5	0.5	66.8	20.8
TOTAL	23.3	901.9		15.3	510.4		16.0	667.9		7.9	320.8	

Families with the greatest biomass at Hawkesbury were corvids, kingfishers, cracticids, honeyeaters, columbids and cuckoo-shrikes; at Margaret, cracticids, muscicapids, honeyeaters, kingfishers and psittacids; at South Alligator, kingfishers, columbids, honeyeaters and muscicapids and, at Lockhart, honeyeaters, kingfishers, muscicapids and columbids.

Reasons for some of these differences between numbers and biomass are readily apparent. The warbler and flycatcher adaptive zones can only be utilized by birds of small body size, but they are capable of supporting large numbers of individuals. Large body size, however, is a pre-requistite to feeding on small terrestrial reptiles and large terrestrial insects. The corvids, cracticids and kingfishers, that constituted this guild were few in numbers but their biomass was high. The open forest floor also favours ground-feeding granivores like pigeons.

Honeyeaters had a high biomass at all sites. The occurrence of some of these was linked to the presence of flowering trees and shrubs. Nectar-feeding lorikeets were common at Margaret River. Honeyeaters are, of course, also insectivorous. The success of the meliphagids in Australia, as judged by number of species and high biomass, attests to the success of the mixed insectivore/nectarivore feeding strategy.

Feeding Guilds and the Subdivision of Ecological Opportunities

Since the sites apparently varied in their capacities to support different ecological types of birds, it is appropriate to investigate further how ecological opportunities were divided up in each area. This can best be done by relating individual species to a series of dietary guilds, e.g. ground feeding granivores, frugivores, foliage-gleaning insectivores. Enough data are now available on diets (e.g. Lea and Gray 1935, 1936; Rose 1973 — see Table 3), and the utilization of vertical feeding space (Figs 2-4), to enable this to be done. On this basis the various species were grouped into 20 dietary and space-use categories (Table 5), and the numbers, percentage numbers, and biomass of individuals determined. The following conclusions emerged:

(1) Small insectivores that obtain all or much of their food from the foliage (thornbills, the weebill, warblers, silvereye, pardalotes) were abundant at Hawkesbury, moderately so at Margaret, but virtually absent from the north. Medium-sized insectivores of the outer branches and leaves (*Pachycephala*) and large-bodied ones (orioles, cuckoo-shrikes, and *Colluricincla*) were evenly distributed. They were uncommon at Lockhart. Only a few large forms occurred at Margaret.

(2) Trunk and branch feeding insect-eaters (e.g. treecreepers, shrike-tit, White-gaped Honeyeater, *Meliphaga unicolor*, occurred at all sites but in much smaller numbers. They were least common at Margaret and Lockhart.

(3) Insect-eaters of low shrubs (wrens, the Brown Thornbill, *Cisticola*) were most abundant at Hawkesbury and South Alligator, uncommon at the other sites. Those that fed partially from the shrubs and ground (wrens) and ground (robins, Rock Warbler, Yellow-tailed Thornbill) were absent from, or uncommon at, the northern sites.

(4) Aerial insect-eaters that fed above the trees (swallows, swiftlet), or at tree level or below (flycatchers), were equally abundant at all sites.

(5) Nectarivores and those that fed jointly on nectar and insects were equally abundant throughout.

(6) Ground-feeding granivores (pigeons, finches), arboreal seed-eaters, and those that fed both in the trees and from the ground (e.g. some parrots), were most abundant at Hawkesbury and South Alligator, least so at Margaret.

(7) Frugivores were represented only by the Mistletoe-bird, *Dicaeum hirundinaceum*. It occurred in small numbers everywhere.

(8) Predators that took insects and small invertebrates from the ground (corvids, cracticids, kingfishers), were equally abundant in all regions.

Differences in abundance and biomass of the different feeding guilds are unlikely to be fully explained until data on productivity, availability of different food types, and insect abundance cycles, are documented.

The Feeding Niches of the Insect-eaters, Vertical Feeding Zones, and Geographic Variation in these

The most important ecological group in Australian forests and woodlands are the insect-eaters. Many species occurred in each area raising questions of how insect resources were divided up between species, and how species arranged themselves relative to a resource that is distributed from ground to canopy, and in the air. Two other questions presented themselves. Is the pattern of division of insect resources the same throughout the continent and do widespread insectivorous species occupy the same feeding zone throughout their range?

The data on vertical feeding zones in the small-bodied insectivores at the four communities is given as follows: aerial feeders (Fig. 2), medium-sized and large insectivores of the branches and leaves (Fig. 3), small-bodied foliage-gleaners, clingers, and hoverers, and mixed insect- and nectar-feeders (Fig. 3); other honeyeaters, thicket dwellers, trunk feeders, and ground feeders (Fig. 4), and trunk feeders (Fig. 4).

The following role-utilization patterns emerged:

(a) *Aerial feeders that hawked above the trees.* Swallows and, at Lockhart, swiftlets filled this role. Artamids were absent from all sites.

(b) *Aerial insectivores, feeding at tree level and below.* At Hawkesbury the following species filled this role: the Jacky Winter, *Microeca leucophaea*, Willie Wagtail, *Rhipidura leucophrys*, and Grey Fantail, *R. fuliginosa*. At Margaret it was the two *Rhipidura*. At South Alligator River the Lemon-breasted Flycatcher, *Microeca flavigaster*, Northern Fantail, *Rhipidura rutiventris*, Willie Wagtail, Drongo, *Dicrurus hottentottus*, and the Dollar-bird, *Eurystomus orientalis*, were the major aerial feeders. At Lockhart it was the Lemon-breasted Flycatcher, Leaden Flycatcher, *Myiagra rubecula*, and the Drongo.

Wherever it occurred, *Microeca* concentrated its feeding mostly at the level of foliage and branches, as did the Leaden Flycatcher at Lockhart, and the Drongo at South Alligator and Lockhart. The two fantails fed at various levels but the Willie Wagtail did much feeding immediately over, or on, the ground.

The Jacky Winter occurred at one site and its northern replacement, the Lemon-breasted Flycatcher at two; the Willie Wagtail at three; the Northern Fantail at two; and the Restless Flycatcher, *Myiagra inquieta*, at two. In none of these species did the vertical distribution of feeding effort vary geographically.

(c) *Insect-eaters of the foliage.* The large-bodied insect-eaters here were the cuckoo-shrikes (that fed variously by picking insects off the terminal foliage and fluttering at it), orioles and, to some extent, friarbirds (Fig. 3). Whistlers fed at various levels but mainly within the foliage of trees and tall shrubs. The Grey Shrike-thrush, *Colluricincla harmonica,* fed at all levels (including on the ground at Margaret). Up to 50% of its time was spent at the ends of branches and within the leaves.

The Striated and Yellow Thornbills, White-throated Warbler, *Gerygone olivacea*, Silvereye, *Zosterops lateralis*, Weebill and Spotted Pardalote, *Pardalotus punctatus*, were the major small foliage insectivores at Hawkesbury; the White-tailed Warbler, *Gerygone fusca*, Silvereye, Brown Thornbill and two pardalotes at Margaret, the Weebill at South Alligator, and Striated Pardalote, *P. striatus*, at Lockhart. the members of this guild varied in their feeding habits, the thornbills feeding mainly within the leaves and sometimes hanging down from them, while the weebill did much hovering and the *Gerygone* snatching of prey from the leaves.

(d) *The Nectarivore/Insectivores*. Most of the honeyeaters spent time feeding both from the foliage and at flowers. Probably much of the former was directed at insect-eating. Honeyeaters, however, also consume manna and lerps (Paton 1980). At Hawkesbury the Little Wattlebird, *Anthochaera chrysoptera*, at Margaret the Red Wattlebird, *A. carunculata*, and in the north Friarbirds, *Philemon* spp., fed in the foliage. The Yellow-faced and White-naped Honeyeaters, *Meliphaga chrysops* and *Melithreptus lunatus,* at Hawkesbury, the White-naped Honeyeater at Margaret, and White-throated Honeyeater, *M. albogularis*, at South Alligator and Lockhart were never, or only rarely seen at flowers.

(e) *Insect-eaters of the low shrubs, tall grass, and ground*. The Whipbird, Brown Thornbill and two wren species occupied this zone at Hawkesbury, with the wrens also feeding on the open ground. Major occupants of this zone at Margaret were wrens with the Silvereye spending part of its time in low shrubs and, in the north, the Red-backed Wren, *Malurus melanocephalus*, and Fantail-warbler, *Cisticola exilis*, (Fig. 3). At Hawkesbury the Yellow Robin, *Eopsaltria australis*, Rock Warbler and, to some extent, the Buff-tailed Thornbill; at Margaret the Scarlet Robin, *Petroica multicolor*, and Yellow-tailed Thornbill were ground feeders.

(f) *Trunk and branch feeders*. Treecreepers were prominent at Hawkesbury and South Alligator, uncommon at Margaret, and absent at Lockhart. At South Alligator Crested Shrike-tits, *Falcunculus frontatus*, and at Lockhart Grey-crowed Babblers, *Pomatostomus temporalis*, fed here (Fig. 4). At South Alligator the White-gaped Honeyeater was partly a trunk feeder, and the Restless Flycatcher commonly picked up prey from the trunk and branches. At Margaret the Grey Shrike-thrush, Golden Whistler, *Pachycephala pectoralis*, Grey Fantail, and thornbills took prey from this zone to a greater extent than did their counterparts at Hawkesbury (Fig. 3).

In conclusion, the insectivores included both specialists, that fed entirely in one vertical zone (swallows, Striated and Yellow Thornbills), and generalists, that took their prey from a variety of substrates. Where such generalists occurred in more than one region the allocation of feeding effort relative to vertical levels sometimes varied slightly.

The Distribution of Insectivore Biomass Relative to Feeding Zone and Substrate

The total biomass of birds supported by aerial insects was similar at three sites but total and percent biomass were higher at South Alligator (Table 5). Aerial feeders here were most abundant near the lagoons and were partly supported by aquatic insects. The biomass of birds supported by foliage insects, including tall shrubs as well as trees, was highest at Hawkesbury, moderately so at Margaret, and low at the northern sites. The percentage of total avian biomass feeding in this zone was the same at all sites This suggests that the biomass of insects in the foliage was constant relative to other zones. Trunks and branches supported a similar biomass, and percent of total biomass at three sites, but the figure was slightly higher at Hawkesbury. The biomass, and percentage of total biomass, of birds feeding in low shrubs and from the ground was, as previously noted, high at the southern sites, low at the northern ones

These results confirm those in Table 5. Two additional features emerge. When a specialist feeding guild is under-represented, as with trunk and branch feeders at Margaret, other members of the community may compensate by using the available resources. This has previously been found for trunk-feeders in Tasmania (Keast 1968), and foliage-gleaners in southwestern Australia (Keast 1976a). Secondly the biomass of foliage-feeding honeyeaters in the north does indeed seem to compensate for the deficiency of small foliage-gleaning warblers there. That these may have been partially utilizing manna does not invalidate this finding.

Morphology, Way of Life, Morphological Distance Separating Species, and Ecological Overlaps between Species of Insectivores

Wing and bill lengths of the small and medium-sized insectivores in the four communities are plotted in Figure 5. Wing-length provides an indication of over-all body size in birds and bill length can be correlated with size and type of food eaten. How do the major morpho-ecological types of birds separate out relative to these dimensions, and how different are the co-occurring species in their measurements? Are measurements of body size and size of bill so different that co-occurring species are not in competition?

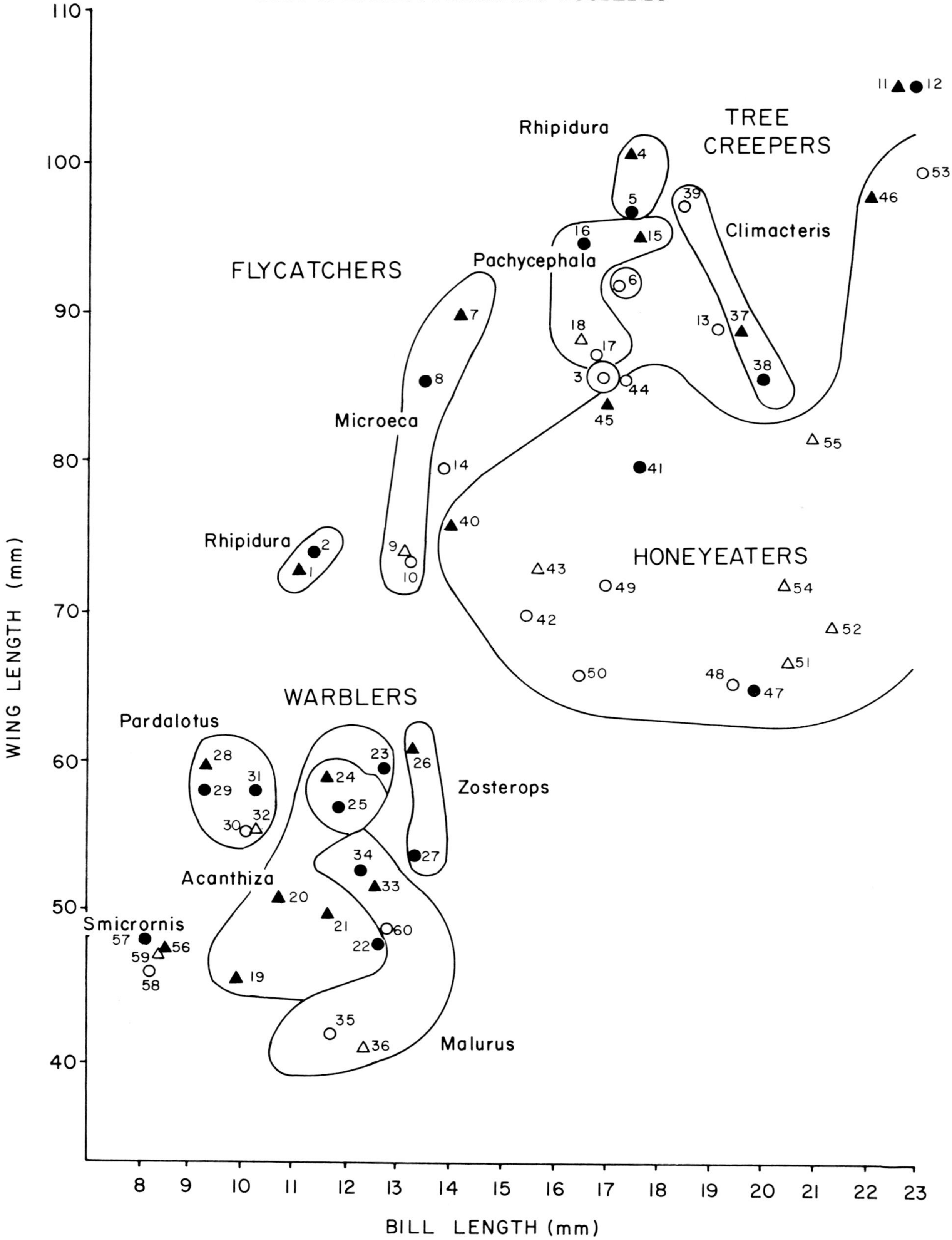

Fig. 5. Morphometric types of the four communities, small and medium sized insect-eaters and honeyeaters, wing lengths and bill lengths. Key: Symbols indicate localities as follows: Black Triangles — central New South Wales; black circle — southwestern Australia; white circle — coastal Northern Territory; white-triangle — Cape York. The species are: (1) and (2) *Rhipidura fuliginosa*; (3) *R. rufiventris*; (4)-(6) *R. leucophrys*; (7) and (8) *Microeca leucophaea*; (9) and (10) *M. flaviventer*; (11)-(13) *Myiagra inquieta*; (14) *M. rubecula*; (15), (17), (18) *Pachycephala rufiventris*; (16) *P. pectoralis*; (19) *Acanthiza nana*; (20) *A. lineata*; (21) and (22) *A. pusilla*; (23) *A. chrysorrhoa*; (24) *Gerygone olivacea*; (25) *G. fusca*; (26) and (27) *Zosterops lateralis*; (28) and (29) *Pardalotus punctatus*; (30)-(32) *P. striatus*; (33) *Malurus cyaneus*; (34) *M. splendens*; (35) and (36) *M. melanocephalus*; (37) *Climacteris leucophaea*; (38) *C. rufa*; (39) *C. melanura*; (40) and (41) *Melithreptus lunatus*; (42) and (43) *M. albogularis*; (44) *M. gularis*; (45) *Meliphaga* chrysops; (46) *M. lewini*; (47) and (48) *Lichmera indistincta*; (49) *Ramsayornis modestus*; (50) *Conopophila albigularis*; (51) and (52) *Myzomela obscura*; (53) *Meliphaga* unicolor; (54) *M. gracilis*; (55) *M. notatus*; (56)-(59) *Smicrornis brevirostris*.

Table 6. Proportionate differences in bill length, forest/woodland birds.

Species with smallest bill and length	Other members in size sequence, proportionate increases in length
	AERIAL INSECTIVORE GUILD
Hawkesbury River *Rhipidura fuliginosa* (11.6 mm)	1.22, *Microeca leucophaea*; 1.55, *Rhipidura leucophrys*; 1.28, *Myiagra inquieta*.
Margaret River *Rhipidura fuliginosa* (11.7 mm)	1.54, *R. leucophrys*; 1.15, *Lichmera indistincta*; 1.3, *Phylidonyris novaehollandiae*.
South Alligator River *Microeca flavigaster* (13.2 mm)	1/3, *Rhipidura ruviventris*; 1.0, *R. leucophrys*; 1.12, *Myiagra inquieta*.
Lockhart River *Myiagra rubecula* (13.8 mm)	1.27, *Rhipidura leucophrys*.
	FOLIAGE INSECTIVORE GUILD
Hawkesbury River *Smicrornis brevirostris* (8.5 mm)	1.08, *Pardalotus punctatus*; 1.16, *Acanthiza lineata*; 1.13, *Zosterops lateralis*; 1.16, *Myzomela sanguinolenta*; 1.13, *Meliphaga chrysops*; 1.04, *Pachycephala rufiventris*; 1.43, *Colluricincla harmonica*; 1.26, *Coracina novaehollandiae*; 1.16, *Philemon corniculatus*.
Margaret River *Pardalotus punctatus* (9.2 mm)	1.12, *P. striatus*; 1.07, *Acanthiza pusilla*; 1.02, *zosterops lateralis*; 1.29, *Pachycephala pectoralis*; 1.07, *Melithreptus lunatus*; 1.15, *Lichmera indistincta*; 1.32, *Colluricincla harmonica*.
South Alligator River *Smicrornis brevirostris* (8.1 mm)	1.25, *Pardalotus striatus* (not present on the site at the time); 1.52, *Melithreptus albogularis*; 1.07, *Conophila albogularis*; 1.02, *Ramsayornis fasciatus* and *Pachycephala rufiventris*; 1.16, *Lichmera indistincta*; 1.07, *Myzomela obscura*; 1.18, *Colluricincla harmonica*; 1.12, *Coracina papuensis*; 1.36, *Philemon argenticeps*.
Lockhart River *Pardalotus striatus* (10.3 mm)	1.52, *Melithreptus albogularis*; 1.04, *Pachycephala rufiventris*; 1.26, *Myzomela obscura*; 1.01, *Meliphaga analoga*; 1.10, *Colluricincla harmonica*; 1.20, *Coracina novaehollandiae*; 1.65, *Philemon buceroides*.

Two findings emerge. The major guilds and genera appear as series of discrete size clusters. The pardalotes range in wing length from 54-62 mm and bill length from 9.3-10.5 mm, acanthizid warblers from 46-52 mm in wing length and 10.0-12.8 mm in bill length. Corresponding figures for the honeyeaters are 66 mm to over 100 mm in wing length, and 14-23 mm in bill length. (Larger species like the *Philemon* and *Anthochaera* are not plotted. The treecreepers (*Climacteris*) range in wing length from 84-98 mm and in bill length from 18.0-20.0 mm. Secondly within these groups, allopatric representatives in the different areas were similar in measurements. Do these similarities suggest that an optimum body form for these species roles has been reached, and can their wide distribution partly be explained by their efficiency in respective roles? Honeyeaters differ in that they form a highly varied assemblage.

Do the co-occurring insectivores differ in measurements by a minimum amount? Note here that Hutchinson (1959) suggested that for species of animals to be able to co-occur they had to differ in the sizes of their trophic structures by an order of 1.3. This concept of 'minimum' morphological separation later found expression in the community assembly rules of Diamond (1975) who suggested that additional species could only be added to a community in accordance with competitively driven constraints on minimum size spacing.

To test this concept members of two feeding guilds, aerial insectivores and foliage-feeding insectivores, were compared in terms of bill lengths (Fig. 5). This was done by arranging the members of each in order of increasing bill length and determining successive increases as ratios. The aerial feeder with the smallest bill in the Hawkesbury assemblage, the Grey Fantail, has a mean bill-length of 11.6 mm. The next species, the Jacky Winter, has a bill 1.22 times longer, the Willie Wagtail 1.55 times longer still, and between it and the fourth member, the Restless Flycatcher, the ratio is 1.28. The mean ratio of increasing bill length was 1.25. Comparable figures for the Margaret series are 1.33, South Alligator 1.14; at Lockhart, where only the Grey Fantail and Willie Wagtail occurred, 1.27. For the members of the foliage insectivore guild, with many more species (Table 6) mean bill length increases were as follows: Hawkesbury, 1.16; Margaret 1.15; South Alligator, 1.16, and Lockhart 1.23. In both cases the degree of difference is similar in all four communities. In each case the mean ratios approached but did not meet the original Hutchison figures or a comparative figure given by Schoener (1971). So far as the present series are concerned Dr.

Table 7. Overlap in the use of vertical space, insect-eaters, the four communities compared, October-November. (a) Aerial feeders, (b) Small-bodied foliage gleaners. Overlap index of Schoener (1970).

	AERIAL FEEDERS			
HAWKESBURY RIVER	*Rhipidura fuliginosa*	*R. leucophrys*	*Myiagra inquieta*	
M. leucophaea	0.60	0.51	0.34	
R. fuliginosa	—	0.72	0.44	
R. leucophrys	—	—	0.62	
MARGARET RIVER		*R. leucophrys*		
R. fuliginosa		0.37		
SOUTH ALLIGATOR RIVER	*Rhipidura rufiventris*	*R. leucophrys*	*Myiagra inquieta*	
M. flavigaster	0.71	0.45	0.75	
M. rufiventris	—	0.48	0.81	
M. leucophrys	—	—	0.65	
LOCKHART RIVER	*Microeca flavigaster*			
Myiagra rubeula	0.64			
	SMALL FOLIAGE GLEANERS			
HAWKESBURY RIVER	*A. pusilla*	*Gerygone olivacea*	*Zosterops lateralis*	
A. lineata	0.46	0.61	0.77	
A. pusilla	—	0.64	0.64	
G. olivacea	—	—	0.65	
SOUTH ALLIGATOR RIVER	*Ramsavornis fasciatus*	*Myzomela obscura*	*Conopophila albogularis*	*Melithreptu albogularis*
L. indistincta	0.81	0.71	0.54	0.79
R. fasciatus	—	0.67	0.56	0.66
M. obscura	—	—	0.56	0.75
C. albogularis	—	—	—	0.54
LOCKHART RIVER	*Meliphaga analoga*	*M. gracilis*	*Myzomela obscura*	
M. albogularis	0.56	0.41	0.59	
M. analoga	—	0.61	0.79	
M. gracilis	—	—	0.59	
MARGARET RIVER		*Gerygone fusca*	*Zosterops lateralis*	
A. pusilla		0.85	0.75	
G. fusca		—	0.77	

H. Recher points out to me that the ratios are most similar between the species with the most similar diets and Dr. Hugh Ford notes that the ratios are usually less than 1.10 where pairs differ in feeding habits, e.g. one is a gleaner and the other a snatcher or hoverer. Weins (1982) has already noted that where two species have different feeding habits they are hardly likely to influence each others' occurrence or evolution. Allowing for this qualification the fact that the mean morphological distances between the bills of the insectivores is so similar in each community is further evidence that the four communities, despite their wide separation, are similarly structured.

Ecological interactions between species within communities have also been assessed by calculating ecological overlaps. Such ratios were calculated here for two sets of guilds: aerial feeding flycatchers, and small foliage gleaners. Comparisons were on the basis of the use of vertical living space, using pairs of species (Table 7). In all communities overlap values prove to be high, in only one case falling below 0.45, on a scale of 0-1 (Schoener index). This is evidence that these species pairs are not reducing interspecific competition for feeding zones or substrate by segregating vertically.

DISCUSSION

The Australian environment is highly variable seasonally; this argues that most species must remain somewhat generalized and capable of adjusting their foraging habits as circumstances dictate. Secondly since birds are mobile, species can adjust densities according to availability of resources. Hence a tight degree of 'species packing' should not be expected. Yet there must have been some selective force directing the evolution of species differences. Presumably this has not been interspecific competition but the types of resource available.

There is currently much discussion among ecologists as to whether bird communities are structured. Are they integrated units with the species interacting or largely random and independent assemblages of species made up of combinations that can co-occur? Members of the MacArthur school (MacArthur 1972; Diamond 1978; Landres and McMahon 1980) have supported the former view and Bock *et al.* (1978), Simberlof (1980), Wiens and Rotenberry (1981) the latter. Kikkawa (1982), finding an apparent overlap of niches and lack of clear-cut resource partitioning among co-occurring species in Australian rainforest communities concluded that these communities are not tightly structured. The present study suggests that this is also the case in Australian eucalypt forest and woodland bird communities.

The most interesting finding that emerges from the present study is how similar the forest and woodland communities are throughout Australia. The same taxonomic types are present. The major ways of life and adaptive zones are occupied by the same families and genera. Many species occur in all four regions and they, or congeneric counterparts, fill the same niches. The bird communities are structured the same way, not only in how feeding zones are divided but, commonly, in the proportions of avian biomass occupying each.

These similarities can only be ascribed to two factors. Firstly the forests and woodlands must be sufficiently similar ecologically, and in their structure to provide a common set of ecological opportunities. Secondly, it is obvious that continuous interchange of genera and species between the different forest and woodland areas must have been possible. Possibly climatic shifts have accelerated this interchange, just as they have split species distributions and led to speciation (Keast 1961) in others. Where this speciation has occurred the allopatric daughter species have remained more or less ecological counterparts.

REFERENCES

BELL, H. L., 1980. The effects of power-line clearing on birds of dry sclerophyll forest at Black Mountain Reserve, Australian Capital Territory. *Corella* **4**: 8-19.

BOCK, C. E., RAPHAEL, M. AND BOCK, J. H., 1978. Changing avian community structure during early post-fire succession in the Sierra Nevada. *Wilson Bull* **90**: 119-23.

BOWLER, J. M., 1976. Aridity in Australia: age, origins and expression in aeolian land forms and sediments. *Earth Sci. Rev.* **12**: 279-310.

BOWLER, J. M., 1982. Aridity in the late Teritary and Quaternary of Australia. Pp. 35-45 *in* Evolution of the flora and fauna of arid Australia ed. by W. R. Barker and P. J. M. Greenslade. Peacock Publications, Frewville, South Australia.

BURNHAM, K. P., ANDERSON, D. R. AND LAAKE, H. L., 1980. Estimation of density from line transect sampling of biological populations.*Wildlife Monogr.* **72**: 1-202.

DIAMOND, J. M., 1975. Assembly of species communities. Pp. 342-444 *in* Ecology and evolution of communities ed. by M. L. Cody and J. M. Diamond. Beldnap Press, Cambridge, Mass.

DIAMOND, J. M., 1978. Niche shifts and the rediscovery of interspecific competition. *Am. Sci.* **66**: 322-31.

EMLEN, J. T., 1971. Population densities of birds derived from transect counts. *Auk* **88**: 323-42.

EMLEN, J. T., 1977. Estimating breeding season bird densities from transect counts. *Auk* **94**: 455-68.

FORD, H. A. AND BELL, H., 1981. Density of birds in eucalypt woodland affected to varying degrees by dieback. *Emu* **81**: 202-08.

FORD, J., 1986. Avian hybridization and allopatry in the region of the Einasleigh Uplands and Burdekin-Lynd Divide, northeastern Queensland. *Emu* (in press).

HUTCHINSON, G. E., 1959. Homage to Santa Rosalia or why are there so many kinds of animals? *Amer. Natur.* **93**: 145-59.

KEAST, A., 1961. Bird speciation on the Australian continent. *Bull. Mus. Comp. Zool.* **123**: 305-495.

KEAST, A., 1968. Competitive interactions and the evolution of ecological niches as illustrated by the Australian honeyeater genus *Melithreptus* (Meliphagidae) *Evolution* **22**: 762-84.

KEAST A., 1975. Zonal feeding in the birds of Culeenup Island, Yandurup. *W.A. Naturalist* **13**: 25-9.

KEAST A., 1976a. Ecological opportunities and adaptive evolution on islands, with special reference to evolution in the isolated forest outliers of Southern Australia. Proc. 16 Intern. Ornith. Congr., Canberra: 573-83.

KEAST, A., 1976b. The origins of adaptive zone utilizations and adaptive radiations, as illustrated by the Australian Meliphagidae. Proc. 16 Intern. Ornith. Congr., Canberra: 71-82.

KIKKAWA, J., 1982. Ecological association of birds and vegetation structure in wet tropical forests of Australia. *Aust. J. Ecol.* **7**: 325-45.

KIKKAWA, J. AND PEARSE, K., 1969. Geographic distribution of land birds in Australia: a numerical analysis. *Aust. J. Zool.* **17**: 821-40.

LANDRES, P. N. AND MCMAHON, J. A., 1980. Guilds and community organization: analysis of an oak woodland avifauna in Sonora, Mexico. *Auk* **97**: 351-65.

LEA, A. M. AND GRAY, J. T., 1935 and 1936. The food of Australian birds. *Emu* **34**: 275-92; **35**: 63-98, 145-78; 251-80; 335-46.

LOYN, R. H., 1980. Bird populations in a mixed eucalypt forest used for production of wood in Gippsland, Victoria. *Emu* **80**: 145-56.

MACARTHUR, R. H. AND MACARTHUR, J. W., 1961. On bird species diversity. *Ecology* **42**: 594-98.

MACARTHUR, R. H., 1964. Environmental factors affecting birds species diversity. *Am. Nat.* **98**: 387-97.

MACARTHUR, R. H., 1972. Geographical ecology. Harper and Row, New York.

MACDONALD, J. D., 1966. The status of some Australian treecreepers. *Emu* **66**: 113-15.

NIX, H. A., 1982. Environmental determinants of biogeography and evolution in Terra Australia. Pp. 47-66 *in* Evolution of the flora and fauna of arid Australia ed. by W. R. Harker, and P. J. M. Greenslade. Peacock Pubs, Frewville, South Australia.

NIX, H. A., 1983. Climate of tropical savannas. Pp. 37-62 *in* Tropical Savannas ed. by F. Bourliere. Elsevier, Amsterdam.

PATON, D. C., 1980. The importance of manna, honeydew and lerp in the diets of honeyeaters. *Emu* **80**: 213-26.

RECHER, H. F., 1969. Bird species diversity and habitat diversity in Australia and North America. *Am. Nat.* **103**: 75-80.

RECHER, H. F., 1984. Use of bird census procedures in Australia: a review. Report No. 7, Royal Australasian Ornithologists Union.

RECHER, H. F., THOMAS, D. G. AND MILLEDGE, D. R., 1971. Census of dry sclerophyll. *Emu* **71**: 40-43.

ROSE, A. B., 1973. Food of some Australian birds. *Emu* **73**: 177-83.

SCHOENER, T. W., 1965. The evolution of bill size differences among sympatric congeneric species of birds. *Evolution* **19**: 189-213.

SHOENER, T. W., 1970. Non-synchronous spatial overlap of lizards in patchy habitats. *Ecology* **51**: 408-18.

SERVENTY, D. L. AND WHITTELL, H. M., 1976. Birds of Western Australia. Univ. West. Australia Press, Perth W.A.

SIMBERLOFF, D., 1980. A succession of paradigms in ecology: essentialism to materialism and probabalism and reply. *Synthese* **43**: 3-39, 79-93.

SPECHT, R. L., ROE, E. M. AND BOUGHTON, V. H., 1974. Conservation of major plant communities in Australia and Papua New Guinea. *Aust. J. Bot. Suppl.* No. 7, 667 pp.

SPECHT, R. L., 1981. Major vegetation formations in Australia. Pp. 163-298 *in* Ecological Biogeography of Australia ed. by A. Keast. Dr W. Junk, The Hague.

SPENCER, B., 1900. Report on the work of the Horn Scientific expedition to central Australia. Narrative. 197. Dulau and Co., London

WIENS, J. A., 1982. On size ratios and sequences in ecological communities: Are there no rules? *Ann. Zool. Fennici.* **19**: 297-308.

WIENS, J. A. AND ROTENBERRY, J. T., 1981. Habitat associations and community structure of birds in shrubstep environments. *Ecol. Monogr.* **51**: 21-41.

Golden Whistler, *Pachycephala pectoralis*

Laughing Kookaburra, *Dacelo novaeguineae*

CHAPTER 10

Springtime Song, Periodicity and Sequencing, A Comparison of a Southern Forest and Northern Woodland Bird Community

Allen Keast[1]

Singing times, and song periodicity patterns were compared in avian communities at Hawkesbury River, New South Wales, and South Alligator River, Northern Territory, at the beginning of the breeding season. Bird song began about half an hour before sunrise, there was a marked dawn chorus when members of most species sang. An early morning period of concentrated song extended for about 2½ hours beginning shortly after sunrise. The amount of song then dropped precipitously, singing during the day was limited to a few species and these sang only intermittently.

Most species in both communities fell into one of three categories: (a) pre-dawn singers, (b) post-dawn singers with most song being concentrated into the early morning, and (c) day-long singers.

Species common to both communities, and representative congeners, had the same singing times in the two communities. This argued for singing times being more in the nature of species characters rather than independent responses to different backgrounds of competitor species in the two communities.

INTRODUCTION

BIRD SONG is a major feature of forest and woodland bird communities. It plays a role in territorial advertisement, mate attraction, and maintaining cohesion between members of species. Song reaches a maximum at the beginning of the breeding season, when it represents a major component of the time and energy budget of an individual. Bird song is most intense at dawn. Various hypotheses have been advanced in an attempt to explain why dawn is the most active song period. Dawn, since it is still too dark to feed, is an effective time for asserting territorial rights (Kacelnik and Krebs 1982). Also, conditions for sound transmission are best then (Anstrom 1976; Waser and Waser 1977; Henwood and Fabrick 1979). Song, along with feeding and other activities, tapers off during the morning. Birds rest during the warmest and dryest period of the day (Palmgren 1935), which helps reduce metabolic expenditure (Kacelnik and Krebs 1982).

For all members of bird communities to sing simultaneously would seem anomalous in that the major objective of the singer is to be heard over a distance. Interference effects and the obliteration of individual calls during the peak song period are thus potentially a problem. A way to minimize the difficulties would be for species to vary in periods of maximum song. Cody and Brown (1969) have suggested that "interspecific competition for broadcasting space" may lead to temporal separation of singing within the morning chorus. The idea has not been properly tested. Individual bird species within communities begin their morning singing at different times (Stokes 1979) but whether or not this supports the Cody-Brown ideas is not clear.

The present study, which supplements work on feeding within bird communities (Keast 1985), investigated daily song patterns of co-occurring species. Two communities were examined, a eucalypt forest on the Hawkesbury River, New South Wales, and a eucalypt woodland on the South

[1]Biology Department, Queen's University, Kingston, Ontario, Canada.

Pages 119-28 *in* BIRDS OF EUCALYPT FORESTS AND WOODLANDS: ECOLOGY, CONSERVATION, MANAGEMENT ed. by A. Keast, H. F. Recher, H. Ford and D. Saunders. Royal Australasian Ornithologists Union and Surrey Beatty & Sons.

Alligator River, Northern Territory. The work was carried out at the beginning of the breeding season. Objectives were: (a) to document the singing times and song periodicity patterns of the major species; (b) compare song patterns in the two communities and, (c) compare the singing patterns of species occurring in both. The latter was possible because many species either occurred in both or were represented by allopatric congeners in the second community. If the individual species, when these occurred in both, and species counterparts had similar patterns of song in the two communities, it would infer that the patterns were old and had developed during the evolution of the species. In contrast, differing singing times in the two communities would imply independent evolution in the two areas and suggest interspecies competition could have been a moulding factor as suggested by Cody and Brown (1969).

METHODS

Descriptions of the two study sites are given elsewhere (Keast 1985). The Hawkesbury site was located on the edge of a small escarpment, and combined the features of forest with dense under-shrubbery and a more open woodland with some grass on the top of the escarpment. The South Alligator site was in open woodland adjacent to a lagoon. In both areas, the audible range exceeded 100 m in every direction. This was important since individual birds wander while foraging (Keast 1984), and the members of the community had to be equally audible throughout the day. The work was carried out as follows: Hawkesbury River, New South Wales, October 6-15, 1981; South Alligator River, Northern Territory, September 29-October 8, 1982. Detailed records of bird song were obtained for five days within each period. Only sunny days were used. Data in Figures 1-6 are averages. Focus was on two components; documenting the times of day when each species sang and, secondly, measuring how the frequency of singing varied in the course of the day.

Recording began when the first bird started to call, which was 30 to 40 minutes before sunrise. It was continued until the last species ceased calling approximately 15 minutes after dusk. During the morning period of peak song, pre-dawn to about three hours after sunrise, recording was carried out over two minute periods at five minute intervals. Thereafter it was determined for five minutes at 30 minute intervals. Towards dusk, when some species increased their singing, records were made over five-minute periods at 15 minute intervals.

Frequency of singing was measured in terms of the percentage of each time period that the species sang. The bird species fell into two categories, continuous singers that sang for long periods of time, and phrase utterers. The former were given a rating based on the percentage of the recording time that the individual sang. For the latter a rating scale was set up based on the maximum number of phrases uttered per minute by that species during its time of most concentrated and continuous song. This was 12-14 phrases per minute for the Grey Fantail, *Rhipidura fuliginosa,* and White-throated Warbler, *Gerygone olivacea,* 45 in the Bronzewing Pigeon, *Phaps chalcoptera,* and so on. In these species, if the maximum number of phrases per minute was uttered during the recording period, a 100% frequency rating was given. A calling rate of half the maximum (six phrases per minute in the Grey Fantail) was rated at 50%, and a quarter of the maximum at 25%. The method thus produced precise data on how, during the species singing period, the amount of song varied.

Both study sites had 12-15 dominant bird species. It was not possible however, to quantify the singing of so many species simultaneously. Accordingly, the species that were singing at the beginning and end of the period were noted, then singing frequency was determined for groups of three to four species over a period of 30 seconds each.

Most species were represented by single calling males. Where two males sang in adjacent territories, as for example with the Rufous Whistler, *Pachycephala rufiventris,* and Yellow-faced Honeyeater, *Meliphaga (Lichenostomus) chrysops,* data on the two males was recorded separately, and an average developed. Usually, in these cases there was interaction between the two males in that when one sang so did the other. Where only one individual called and a second male was known to be in the immediate vicinity the song frequency figure for that species was adjusted downwards to take into account the absence of singing in the latter.

RESULTS

The song patterns of 12 of the common Hawkesbury River and 16 of the South Alligator River species are shown in Figures 1-6.

Hawkesbury Community

Bird song began about one half hour before sunrise, reached a peak shortly after daybreak, remained at high frequency for two to three hours, then declined. Few individuals called through the middle of the day and those that did sang only occasionally. In a few species there was a minor increase in bird song towards dusk. The last individual bird called 15-20 minutes after sunset. This pattern was thus similar to that in Europe and North America (e.g. Palmgren 1935).

The individual species differ considerably in their singing times. Three broad categories were discernible:

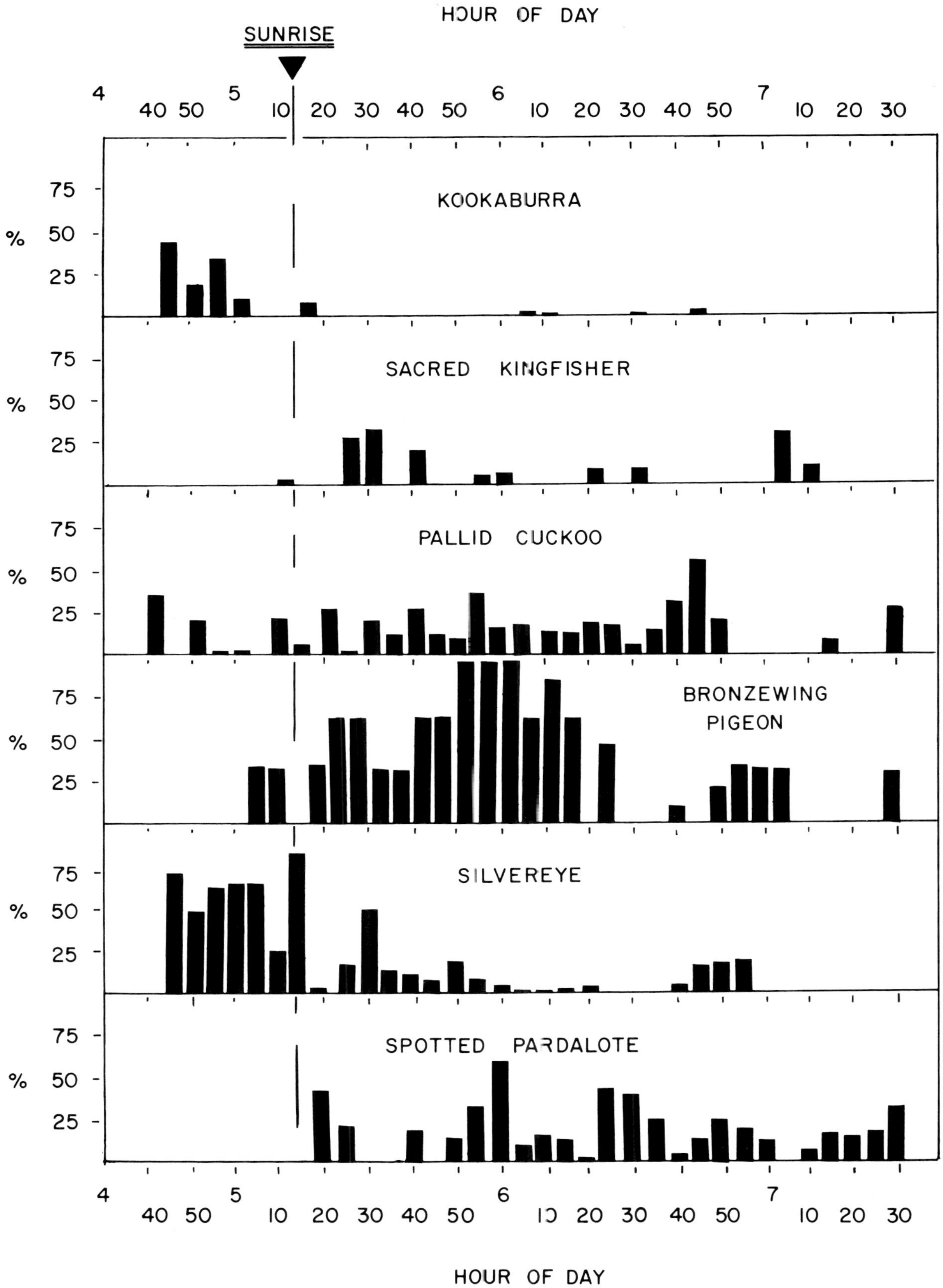

Fig. 1. Song periodicity at the Hawkesbury River site for some major bird species illustrating the pre-dawn and early morning song period (0430-0730). Song counts were made at five-minute intervals and averaged over five days. The vertical scales gives the amount or frequency of singing expressed as a percentage of the maximum amount possible.

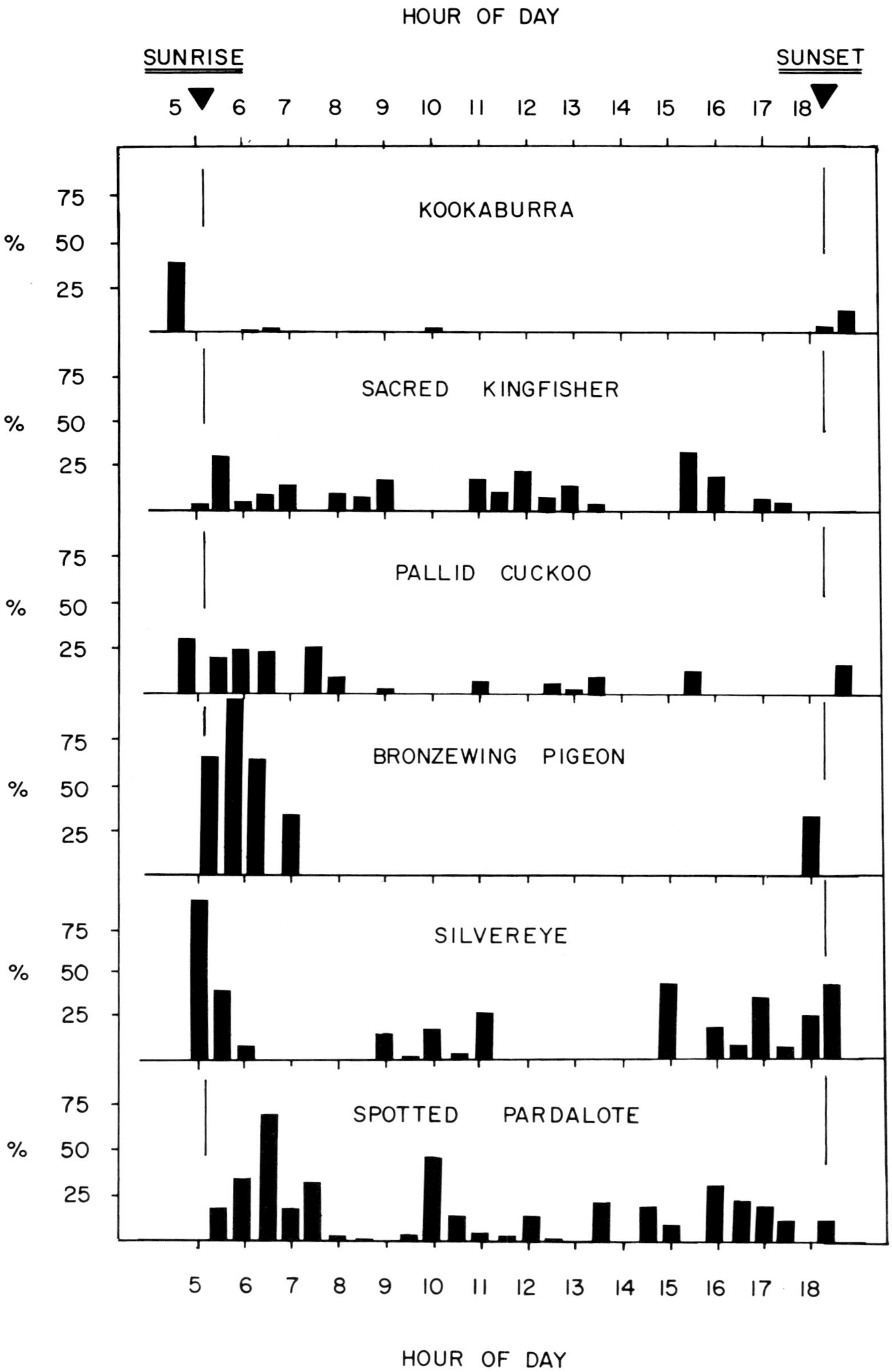

Fig. 2. Song periodicity at the Hawkesbury River site for the same bird species as in Figure 1, but illustrating the frequency of calling from sunrise to sunset determined at 30-minute intervals. Vertical scale is as in Figure 1.

(a) *Pre-dawn singers* that did most of their calling before daybreak. Examples were the Laughing Kookaburra, *Dacelo novaeguineae,* (Fig. 1) Little Wattlebird, *Anthochaera chrysoptera,* and Blue or Superb Fairy Wren, *Malurus cyaneus.* In the Laughing Kookaburra this took the form of a series of call-sequences with several individuals taking part. In the wattlebird it consisted of the loud, sustained, and continuous calling of many individuals. In the wren the twitter sequence instead of being uttered singly was repeated continuously for several minutes.

The pre-dawn singers ceased calling at dawn. They called only rarely, if at all, during the day.

(b) *Post Dawn-Early morning songsters,* began calling after dawn and called repeatedly for 1½-3 hours. Thereafter they ceased calling or, called only occasionally. Many species fell into this category. They included the Olive-backed Oriole, *Oriolus sagittatus,* and Bronzewing Pigeon, *Phaps chalcoptera,* (Figs 1, 2 and 3).

(c) *Day-long callers.* These had a concentrated period of song in the morning followed by intermittent calling throughout the day. They usually called once or twice every half-hour. Examples were the Sacred Kingfisher, *Halcyon sancta,* (Figs 1 and 2), Weebill, *Smicrornis brevirostris,* and Yellow-faced Honeyeater, (Fig. 3).

A number of species did not clearly belong to one or another of these categories. Thus the Silvereye, *Zosterops lateralis,* sang loudly and continuously before sunrise, and to a moderate degree post-dawn. At this time much of the calling became of a social, flock-contact kind. Individual birds occasionally sang for periods of many minutes during the day. There was a marked increase in song and social calling towards dusk. The Pallid Cuckoo, *Cuculus pallidus,* called occasionally during the night, as did the Willie Wagtail, *Rhipidura leucophrys.* It called continuously before dawn and in the early morning, and occasionally during the day. The Spotted Pardalote, *Pardalotus punctatus,* did not begin calling until shortly after sunrise. It called fairly frequently for the next couple of hours, and then throughout the day. It was one of the most frequent day-long callers.

The dawn chorus, when large numbers of birds sang at once, characterised the immediate pre-sunrise and post-sunrise period. It was then still too dark to feed. This accords with overseas observations. The dawn chorus was a blend of the calls of a great many species. It tapered off about daylight. At this time the observer could discern the print on his record sheet. Feeding of the birds did not begin actively until about half an hour after sunrise when the lower parts of the forest became well illuminated.

During the ensuing morning period of concentrated song not all species sang simultaneously. Individual species alternated between periods of high frequency calling and periods of silence. For example the Rufous Whistler, Grey Shrike-thrush, *Colluricincla harmonica*, Jacky Winter, *Microeca leucophaea*, and Olive-backed Oriole called continuously for five or fifteen minutes then became silent for 5-10 minutes. The birds were feeding actively at this time and called as they moved from tree to tree, or through the branches. Species with somewhat similar calls, e.g. Rufous Whistler, Golden Whistler, *Pachycephala pectoralis,* and Grey Shrike-thrush, often alternated in their bouts of song. The possibility that these species were listening to each other and restricting their major song periods to times when others were not calling, merits investigation.

Rufous Whistler and Grey Shrike-thrush had a repertoire of different note-phrases. They uttered one call-type for a period, then another. For example, the Rufous Whistler which uttered its loud ringing note sequence for some minutes, and then the "eachong" call. The significance of these different call-types is unknown.

Of interest more in that it indicates species differences than having any immediate territorial function was the time of first call of the different species. The Laughing Kookaburra and Pallid Cuckoo were regularly the first to call pre-dawn. The Little Wattlebird, White-throated Treecreeper, *Climacteris leucophaea*, Silvereye, Yellow-faced Honeyeater, and Jacky Winter, characteristically called within a few minutes of the Pallid Cuckoo and Kookaburra, with the order varying slightly from day to day. The last species to call at, or after, sunset were usually the Kookaburra, White-throated Treecreeper, Blue Wren, and Silvereye. The last calls were uttered 10-15 minutes after sunset.

South Alligator Community

Again the bird species could be grouped into pre-dawn, post-dawn-early morning, and day-long callers. The most important pre-dawn singers were the Blue-winged Kookaburra, *Dacelo leachii,* and the White-gaped Honeyeater, *Meliphaga unicolor,* (Figs 4 and 5). The former, however, called occasionally during the day. The major early morning songsters were the Lemon-breasted Flycatcher, *Microeca flavigaster,* and White-throated Honeyeater, *Melithreptus albogularis,* (Figs 4 and 5), Olive-backed Oriole, Brown Honeyeater, *Lichmera indistincta,* and Willie Wagtail (Fig. 6). Day-long callers were the Peaceful Dove, *Geopelia striata,* Weebill and, to a degree, the Grey Shrike-thrush. There was much social communication by Bee-eaters, *Merops ornatus,* in the afternoon. At South Alligator the Black-tailed

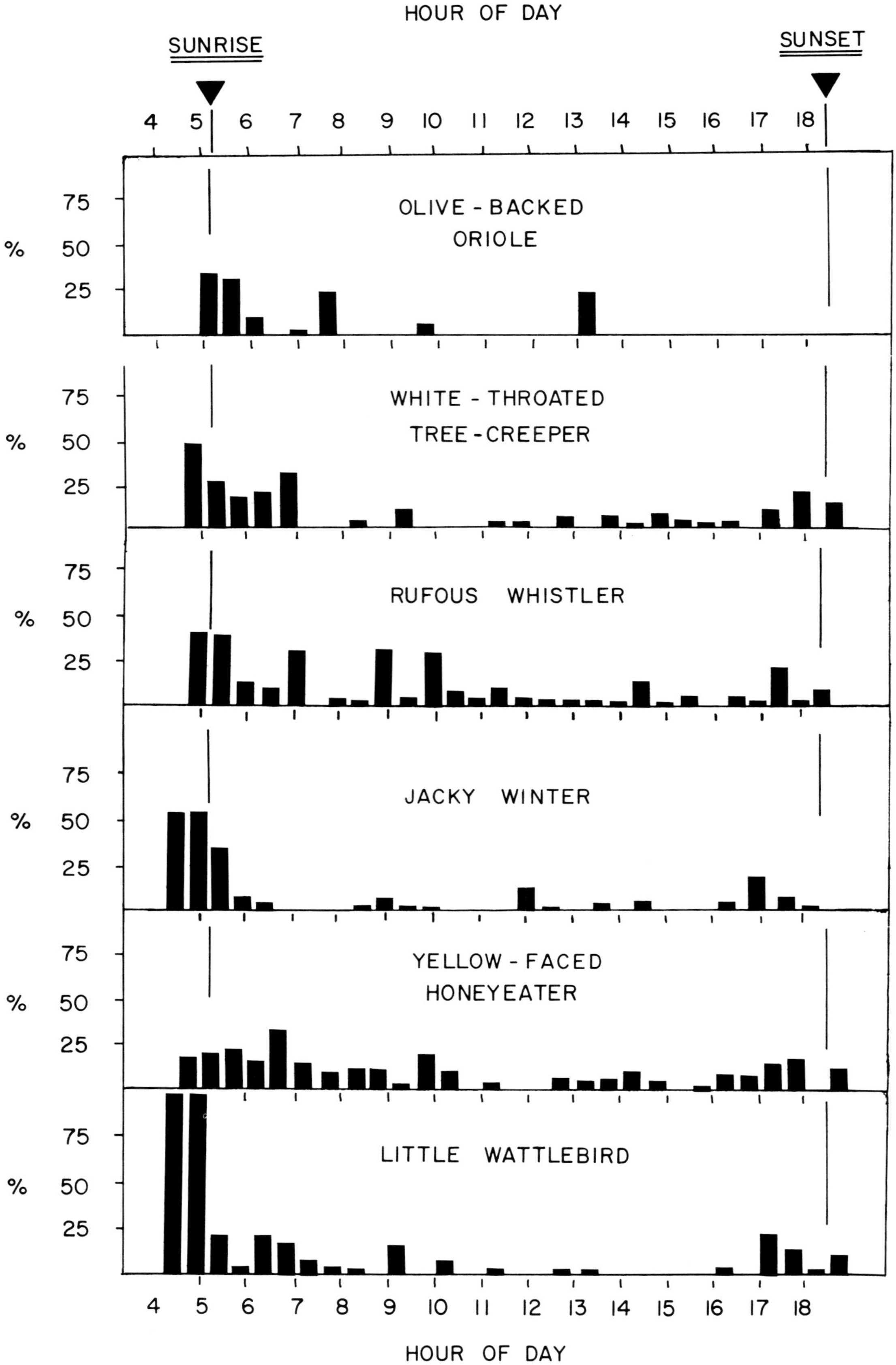

Fig. 3. Song periodicity at the Hawkesbury River site for additional species of birds and again illustrating the frequency of calling throughout the day as determined at 30-minute intervals. Vertical scale is as in Figure 1.

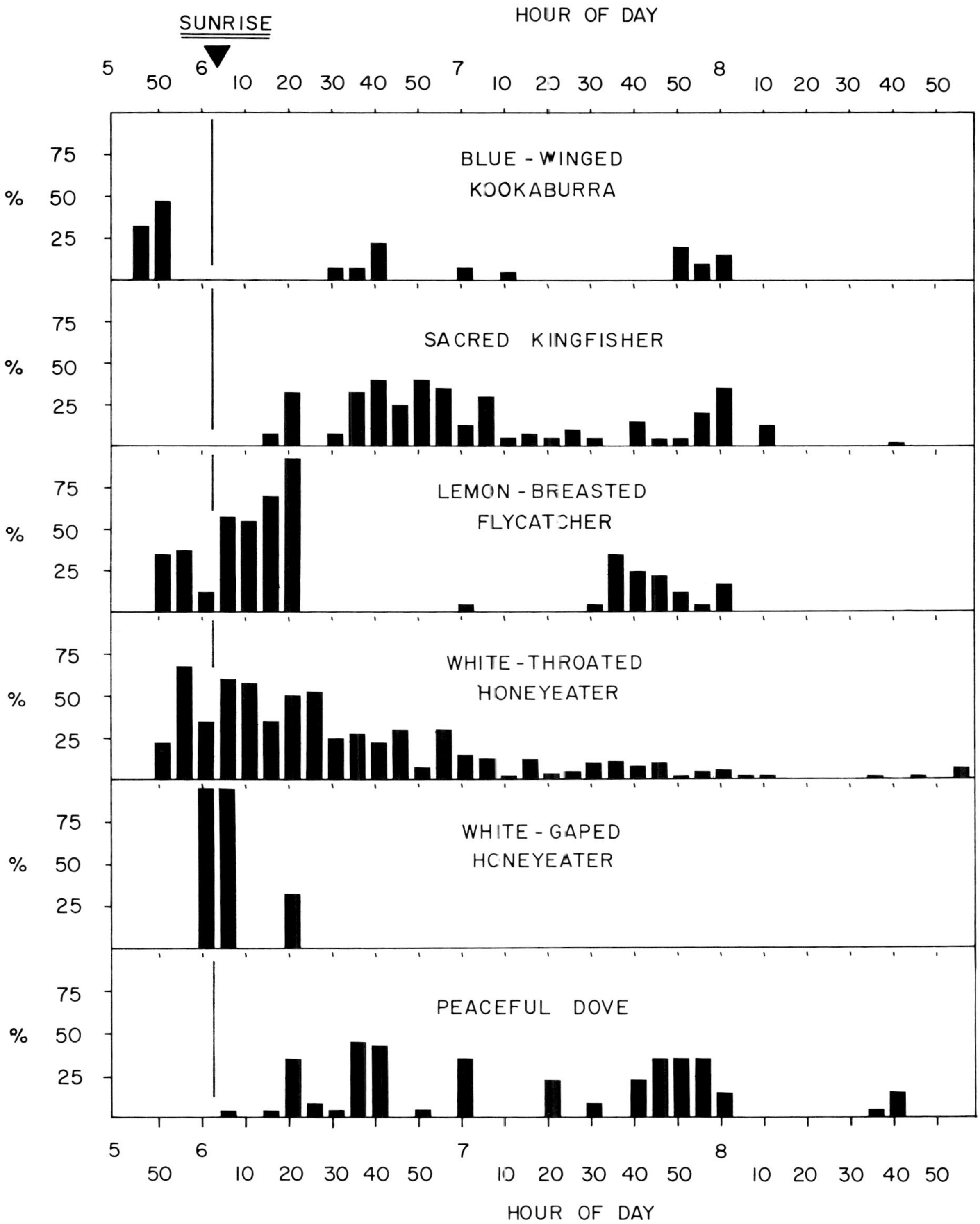

Fig. 4. Song periodicity at the South Alligator River site for some major bird species illustrating the pre-dawn and early morning song period (0540-0900) determined at five-minute intervals. The vertical scale is as in Figure 1.

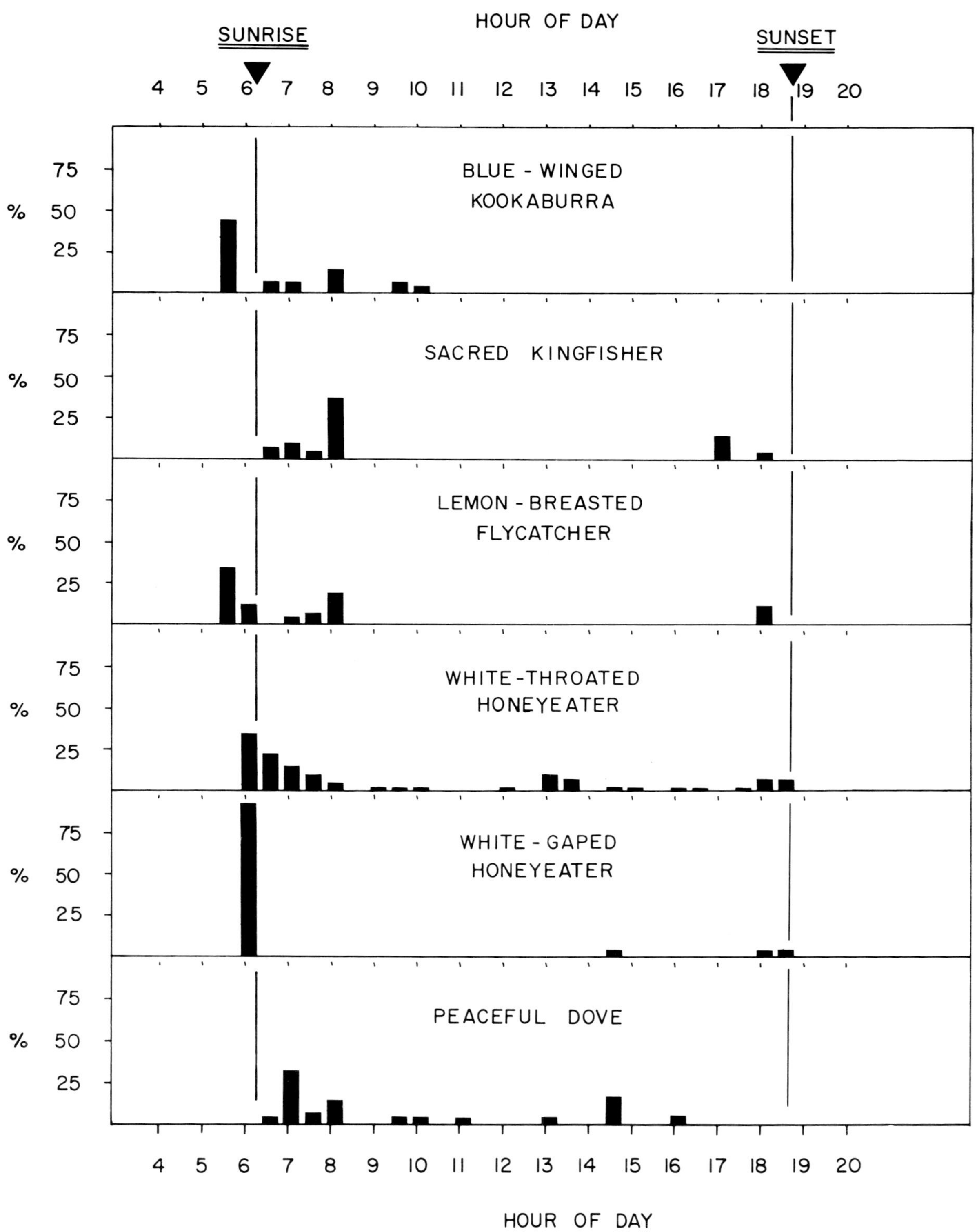

Fig. 5. Song periodicity at the South Alligator River site for the same species as in Figure 4, but illustrating the frequency of calling throughoutthe day as determined at 30-minute intervals. Vertical scale is as in Figure 1.

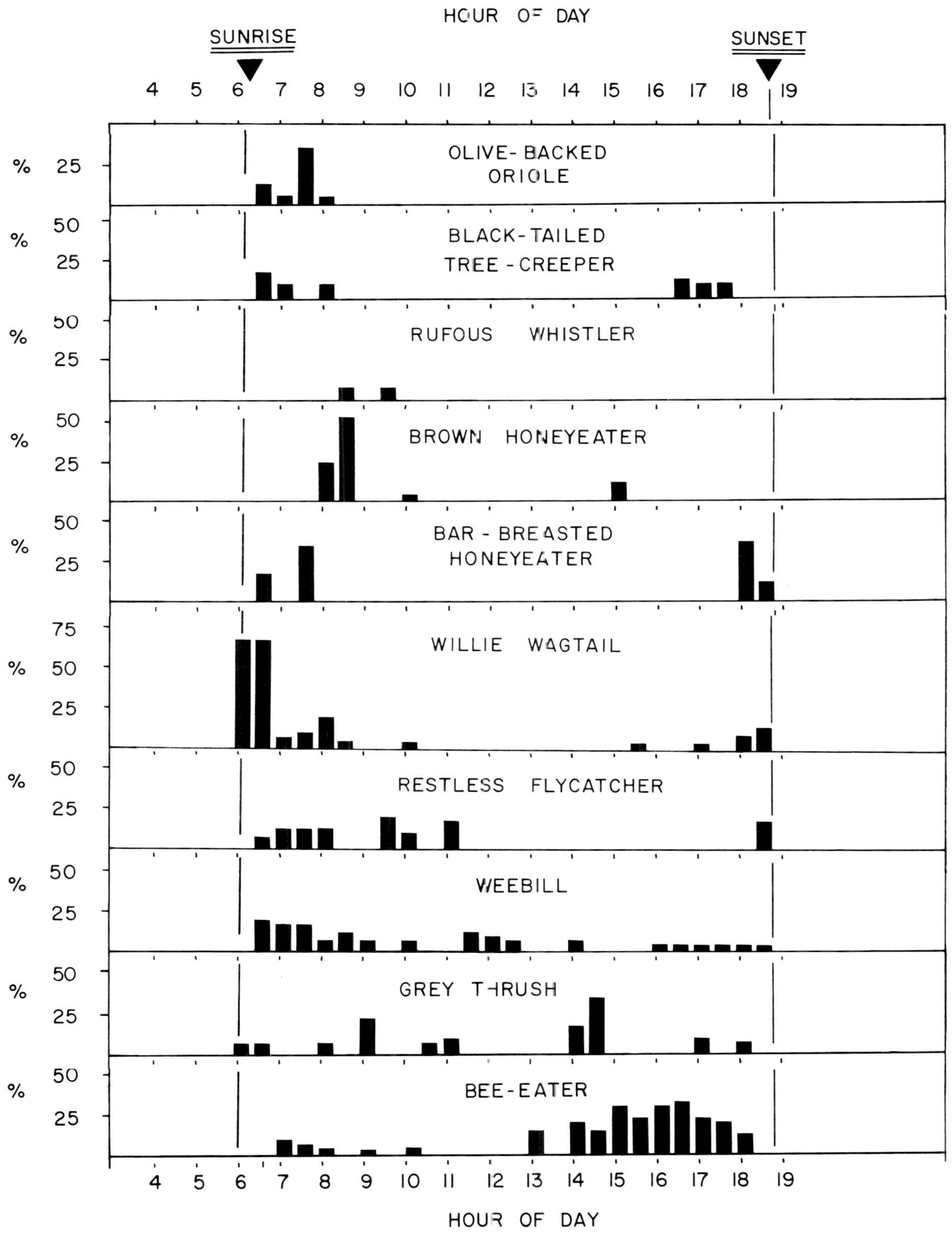

Fig. 6. Song periodicity at the South Alligator River site for additional species calling throughout the day as determined at 30-minute intervals. The vertical scale is as in Figure 1.

Treecreeper, *Climacteris melanotis,* called frequently in the early morning, was silent during the day, and called again towards sunset. This duplicated the pattern of the White-throated Treecreeper at Hawkesbury.

An interesting difference between the South Alligator and Hawkesbury areas was that while the early morning song period extended 2½-3 hours after dawn in both (compare Figs 1 and 4) it ceased more precipitously in the former area. Song during the day was much suppressed (compare Figs 2 and 3, with 5 and 6). This was probably a temperature effect as the heat became oppressive at the South Alligator site by mid-morning.

A second difference between the two areas was that the amount of singing was less in the South Alligator populations of some species than in the Hawkesbury ones, e.g. in the Olive-backed Oriole and Rufous Whistler (compare Figs 3 and 6). This partly reflected different population densities of these species in the two areas Keast (1985). It presumably also reflected the stage of breeding: in the south the breeding season extends from August to December (Robinson 1955; Marchant 1981) but in the Northern Territory most breeding is from November to March (Frith and Davies 1961).

DISCUSSION

Has competition for "broadcasting space", been a factor in the evolution of different signing times, as suggested by Cody and Brown (1969). This may well have been the case. However, it should be noted that the calls of co-occurring species vary greatly in wavelength, sequences and pitch, hence, few would be likely to drown each other out. The interference problem could also be minimized by members of different species simply listening to each other and calling when the other bird ceased. That birds do respond to each other's singing is illustrated by the prominence of counter-singing in some groups, i.e. the calling of one territorial male eliciting response from another.

The same basic song patterns (pre-dawn singers, early morning singers, and day-long singers) characterised both communities. When a species occurred in both areas it obviously belonged to the same song group. This also applied to allopatric counterparts in the two bird communities (e.g. Jacky Winter and Lemon-breasted Flycatcher). This finding indicates that singing times patterns were not evolved in each community independently and relative to competitors in that community. Rather, they are apparently characteristic of the particular taxonomic group. This would accord with the findings in Keast (1985) respecting feeding zones, that the individual members of the two bird communities had evolved their major adaptations elsewhere, rather than as a response to regional inter-species interactions.

ACKNOWLEDGEMENTS

This work was carried out while the writer held a research grant from the National Science and Engineering Council of Canada. I should like to thank this body for its support.

REFERENCES

Anstrom, G., 1976. Environmental influence on daily song activity of the reed bunting *(Amberiza schoeniclus* L.). *Zool. Suppl.* **2**: 1-82.

Cody, M. L. and Brown, J. H., 1969. Song asynchony in chapparal birds. *Nature* **272**: 778-80.

Frith, H. J. and Davies, S. J. J. F., 1961. Breeding seasons of birds in sub-coastal Northern Territory. *Emu* **61**: 97-111.

Henwood, K. and Fabrick, A., 1979. A quantitative analysis of the dawn chorus: temporal selection for communicatory optimation. *Amer. Natur.* **114**: 260-74.

Kacelnik, A. and Krebs, J. R., 1983. The dawn chorus of the Great Tit *(Parus major)* proximate and ultimate causes. *Behaviour* **83**: 287-310.

Keast, A., 1984. Assessment of community composition and species richness in contrasting habitats. Pp. 19-24 *in* Methods of censusing birds in Australia ed. by S. J. J. F. Davies. R.A.O.U., Report No. 7.

Keast, A., 1985. Bird community structure in southern forests and northern woodlands: a comparison. Pp. 97-116 *in* Birds of Eucalypt Forests and Woodlands: Ecology, Conservation, Management ed. by A. Keast, H. F. Recher, H. Ford and D. Saunders. Royal Australasian Ornithologists Union.

Marchant, S., 1981. The breeding season at Moruya, New South Wales. *Corella* **5**: 19-25.

Palmgren, P., 1935. Uber den Tagesrhythmus der Vogel in artkischen sommer. *Ornis. Fenn.* **12**: 107-21.

Robinson, A. H., 1955. Nesting seasons of Western Australian birds: a further contribution. *Western Australian Naturalist.* **4**: 187-96.

Stokes, T., 1979. Roosting and awakening in some inland Australian birds. *Corella* **3**: 77-81.

Waser, P. and Waser, M. S., 1977. Experimental studies of primate vocalisation specialisation for long distance propagation. *Tierpsychol.* **43**: 239-63.

CHAPTER 11

Synthesis: A Model of Forest and Woodland Bird Communities

Harry F. Recher[1]

THERE is a close relationship between the numbers and kinds of birds which occur together and the physical and biological attributes of the habitats in which they live. Communities vary with changes in latitude and elevation, the structure and plant species composition of the vegetation, and seasonal cycles of temperature and rainfall. Describing these patterns and the ways that birds respond to changes in their environment is the basis of all studies of plant and animal communities.

The papers presented in this section have described spatial and temporal patterns in the numbers and kinds of birds and their use of resources in eucalypt forests and woodlands. The patterns described have been interpreted in terms of changes in the structure and species composition of the vegetation, the seasonal and year to year variation in abundance or kinds of food available to birds, and the ways that birds differ among themselves in their foraging and breeding habits.

Ecologists have the ultimate goal of producing a general theory of ecology which will explain observed patterns and predict the consequences of changes to the environment. The goal is probably unattainable, but attempts to achieve it serve a useful purpose. By bringing disparate and, at times, contradictory observations together and attempting to mould these into a single theory or model, we learn where we are and can perhaps see more clearly where we should be going. In this synthesis I intend to bring together the ideas and information presented in the papers of the section and also drawing upon the literature produce a model of the community ecology of eucalypt forest and woodland birds. The model will describe the current state of our knowledge, provide a basis for formulating plans of management for the conservation of forest and woodland birds, and identify where research is needed.

DISTRIBUTION AND ABUNDANCE

Using presence/absence data, Kikkawa (1968) was able to show that structurally similar plant communities in northeastern New South Wales were alike in the species composition of their avifaunas. Habitats were ordered along a xeric-mesic gradient with a primary separation of closed-forest (rainforest) and tall open-forest (wet sclerophyll) from open-forest (dry sclerophyll) and woodlands. Among the dry formations, forest and tall woodlands differed significantly from other woodlands in the composition of their avifauna. Despite the separation of habitats, Kikkawa (1968) found that many species of birds occurred in a wide range of forests and woodlands. The species with greatest habitat fidelity were those associated with mesic habitats while birds which characterized open-forests and woodlands occurred more widely and were often present in moist forests. Recher (1975) obtained similar results at Myall Lakes on the New South Wales mid-north coast. Outside the breeding season the separation between habitats is not as pronounced (Kikkawa 1968). Kikkawa considered that the most important reason for the different results obtained between seasons was due to the large number of species which bred in moist forest, but migrated northwards (e.g. Black-faced Flycatcher, *Monarcha melanopsis,* Rufous Fantail, *Rhipidura rufifrons*) or wandered into dry forests and woodlands during the winter (e.g. Golden Whistler, *Pachycephala pectoralis*, Lewin's Honeyeater, *Meliphaga lewinii*).

Kikkawa's (1968) pioneering study identified a number of the major features which typify forest and woodland avifaunas in Australia. Two of the most important of these are the occurrence of many of the same species of birds in a wide range of habitats and the movement of birds from the habitats in which they breed into other types of forest and woodland without necessarily migrating.

[1]The Australian Museum, P.O. Box A285, Sydney South 2000.

Pages 129-35 *in* BIRDS OF EUCALYPT FORESTS AND WOODLANDS: ECOLOGY, CONSERVATION, MANAGEMENT ed. by A. Keast, H. F. Recher, H. Ford and D. Saunders. Royal Australasian Ornithologists Union and Surrey Beatty and Sons.

These two features of the Australian avifauna are illustrated in the chapters by Sandy Gilmore, Richard Loyn, and David Milledge and Harry Recher. Milledge and Recher (1985) compared census results from forests on the north and south coasts of New South Wales. Although the northern forests were richer in species, 62% of the birds recorded also occurred on the south coast. Of the twelve most abundant birds in each region, eight were the same for both localities. The wide distribution of species and the prevalence of a few 'most common' birds is typical of eucalypt forests and woodlands throughout Australia. In his paper, Loyn (1985) described the forest and woodland avifaunas of eastern Victoria along a transect from the coast, across the dividing range (elevation 2000 m a.s.1.), to the Murray River. Based on census data, Loyn considered 56 species of birds to be common in one or more vegetation associations along the transect. Of these, 58% occurred on nine or more of the 12 plots censused and 76% occurred on more than half the plots. Only a few species of moist forest and woodland birds had a restricted habitat distribution. Gilmore (1985) working in central and western Victoria and Shields *et al.* (1985) working in northern New South Wales, confirmed the wide distribution of species between habitats. Shields and his colleagues recorded 117 species of birds during their survey of the catchment of the Hastings River. Of these, two species (Spectacled Flycatcher, *M. trivirgatus* and Pale-yellow Robin, *Tregellasia capito*) were restricted to closed-forest (rainforest) and 16 species were recorded only in eucalypt forest.

Comparisons of avifaunas separated by greater distances still reveal remarkable similarities. Of 66 species of birds recorded by Keast (1985a) on two woodland plots — one near Sydney in southeastern Australia and the other on the South Alligator River in the Northern Territory — 13 occurred on both plots. There were 12 replacements among congeners so that despite the small number of species in common, the composition of the avifaunas is broadly similar. The most conspicuous differences between the two plots were at the generic level where of a total of 42 genera, 15 occurred only on the southern plot and four only on the northern one.

Although there may be similarities in the composition of bird communities and the same common species are shared, abundances vary considerably between different types of forest and woodland. The largest number of censuses which estimated species abundances have been conducted during the breeding season in southeastern Australia (Recher 1984). Most of these censuses were based on transect counts, but some data were obtained by mapping (e.g. Loyn 1980; Gilmore 1985) and nest searches (Bourke 1949). The plots censused span a wide latitudinal range (32°S to 41°S) and include a variety of forest types, but can be grouped into four structural formations: woodland, low open-forest, open-forest and tall open-forest. Within each group there is a range of estimated densities which should be used *only* as an index of real abundances (Pyke and Recher 1984).

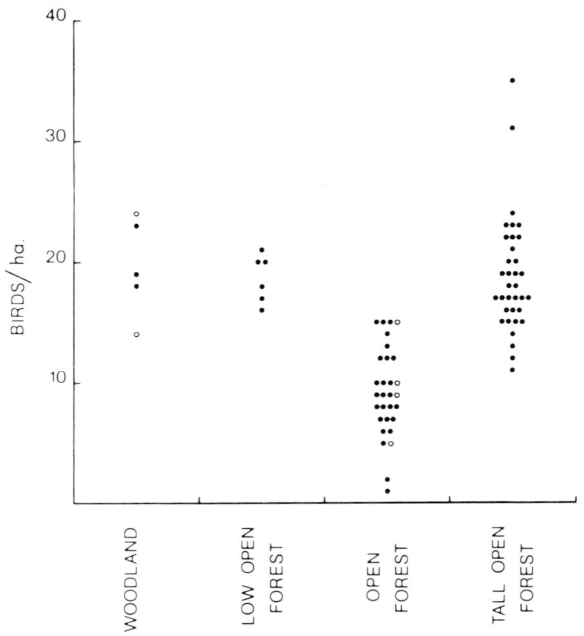

Fig. 1. The number of birds per hectare in a variety of forest and woodland plots from Tasmania, Victoria and New South Wales. Plots differed in size and different census procedures were used and numbers should be used only as a 'guide' to the relative abundances of birds in different habitats. The open circles counted breeding birds only, but in most instances observers attempted to count all the individuals on their plots. (From Recher 1984).

Open-forest has the smallest total densities (0.5-15 birds/ha) (Fig. 1). Most woodland, low open-forest and tall open-forest plots have broadly similar numbers (12-24 birds/ha) with the greatest densities (30-35 birds/ha) recorded in tall open-forest. Shields *et al.* (1985) found similar densities of birds in closed, tall open- and open-forests on the catchment of the Hastings River in northern New South Wales. The densities they estimated for unlogged forest (26-36 birds/ha) were greater than all except the richest plots in southeastern New South Wales and eastern Victoria.

PATTERNS OF DIVERSITY

Recher (1969) demonstrated that bird species diversity in Australian temperate habitats was correlated with the structural diversity of the vegetation (i.e. foliage height diversity). Bird species diversity increased as the number of layers of vegetation increased. However, the number of plant species may be important in explaining some of the variation in bird species number which occurs between structurally similar habitats (Recher and

Table 1. Coefficients of correlation between bird and vegetation data*.

Vegetation Data		Bird Data Species Diversity	Number of Species	Number of Individuals
Mature Eucalypt Forest and Regeneration Stages (69 plots)	Foliage height diversity**	0.48S	0.59S	0.48S
	Canopy height	0.50S	0.64S	0.55S
	Total foliage	0.41S	0.43S	0.31S
Mature Eucalypt Forest only (45 plots)	Foliage height diversity**	0.15N	0.28N	0.23N
	Canopy height	0.31N	0.43S	0.38S
	Total foliage	0.21N	0.20N	0.10N
	Canopy species diversity	0.12N	0.17N	0.08N
	Canopy species number	0.26N	0.29S	0.26N
	Understorey>3 m species diversity	0.26N	0.29S	0.15N
	Understorey>3 m species number	0.42S	0.46S	0.29S

*From Recher *et al.* 1980; **calculated from proportions of vegetation in 1.5 m layers; S, significant at 5% level; N, not significant at 5% level.

Abbott 1970; Milledge and Recher 1985). For example, the high species diversities recorded in some coastal New South Wales forests results from the large numbers of nectar-feeding birds attracted to these habitats by the abundance of nectar-rich flowers (Milledge and Recher 1985).

Gilmore (1985) compared woodlands which were similar in the structure of their vegetation, but differed in floristic composition. He found that the number of insectivorous bird species was negatively correlated with plant species diversity, but positively correlated with the number of plant genera. This differs from the results obtained in southeastern New South Wales by Recher *et al.* (1980). They reported significant correlations between the number of forest bird species and the number of plant species in the understorey (Table 1). A significant correlation also existed between the number of canopy tree species and the number of bird species. The different results reported by Gilmore (1985) and Recher *et al.* (1980) probably reflect the different intensities with which plant species were sampled. Recher *et al.* (1980) sampled only the canopy and understorey where the number of species is approximately equal to the number of genera. Gilmore (1985), on the other hand, was more thorough and surveyed the entire vegetation on his plots and thus recorded a much greater number of species including many in the same genus.

A number of workers have related bird species diversity to site conditions (e.g. Recher *et al.* 1980; Milledge and Recher 1985; Loyn 1980, 1985). They have suggested that bird species diversity is greatest in forests growing on nutrient rich soils with adequate moisture. Such plant communities are generally structurally complex and rich in plant genera and species (Recher 1985). Gilmore (1985) also found the most diverse avifaunas associated with plant communities growing on nutrient rich sites, but these woodlands were floristically poor compared with his plots on poorer soils. Gilmore noted that Specht (1981) had pointed out that the most diverse sclerophyll floras were associated with nutrient poor soils. However, Gilmore considered the correlation between bird species diversity and soil nutrients might be spurious and the results of inadequate sampling.

In most terrestrial habitats species richness and the abundance of birds are highly correlated (e.g. r=.9, for forests in southeastern New South Wales, Recher unpubl. data). The number of birds that can survive in a given area is ultimately determined by the amount of food available. Thus, it seems more reasonable to conclude that there is a relationship between abundance, species richness and productivity as a function of soil nutrients.

Gilmore (1985) found that the total abundance of insect-eating birds in woodlands of central Victoria was correlated with canopy height and an index of plant biomass. The abundance of species in different foraging guilds was correlated with the amount of available substrates. Gilmore also found that the abundance of canopy foliage feeders was correlated with foliage cover, ground feeders with annual litter fall, bark foragers with an index of bark area, and aerial foragers with an index of air space derived from canopy volumes. In contrast Pyke (1983, 1985a) obtained only weak correlation between the abundance of woodland honeyeaters and the amount of available nectar. There was no apparent relationship between food abundance and honeyeater numbers on nearby heaths. Pyke (1983, 1985a) concluded that the abundance of honeyeaters and their movements may be determined by, among other things, nectar production in other areas.

Given the apparent contradiction in results relating avian abundance and species diversity to food resources we need to consider carefully the relationships that exist between bird species diversity and habitat diversity. Specifically we need to consider whether the observations presented above are consistent with the apparent paradox that bird communities are more diverse on nutrient rich soils even though in some instances there are a smaller number of plant species creating the image of a less diverse environment. What habitat variables — food abundance, diversity of foraging opportunities, presence of other species, regional events — affect bird species diversity? Do all kinds of birds respond in the same ways to changes in their environment? An understanding of how birds forage, the kinds of foods and substrates they exploit, and how they differ among themselves in the use of these resources is necessary to answer such questions and resolve any conflicts in results.

FORAGING ECOLOGY

Recher and Holmes (1985) identified eight guilds of insect-eating birds plus three of nectar-feeders and two of seed-eaters in the forests and woodlands on the Southern Tablelands of New South Wales. They grouped birds into guilds on the basis of kinds of food taken, body size (weight), type of prey-attack behaviour, foraging height, and foraging substrate. Loyn (1985) used the types of food eaten, substrate and foraging height to group species of forest birds from eastern Victoria. Keast (1985a) and Gilmore (1985) also used the same criteria in assigning woodland birds from southern and northern Australia to foraging guilds. In all instances the categories are similar to those identified by Crome (1978) for avian communities inhabiting closed-forests in northern Queensland.

The categories of food used by eucalypt forest and woodland birds are invertebrates, carbohydrates (nectar, manna, honeydew, lerp and sap), seeds and other vertebrates. Some fruit is also taken, but compared with closed-forests most eucalypt habitats lack edible fruits for birds. The importance of carbohydrates and the large number of species which rely on them for all or part of their energy requirements distinguishes eucalypt communities from temperate zone forests on other continents (Recher and Holmes 1985). Seeds are less important although eucalypt seeds (taken while still in the capsule) are a major food item for parrots. In most eucalypt forests and woodlands insect-eating birds are the most abundant part of the avifauna while birds-of-prey and scavengers tend to be uncommon.

With the exception of loose, peeling or hanging bark of eucalypts, the substrates used (ground, tree trunks, branches, foliage, and air) and the foraging height categories (ground, shrub, subcanopy, and canopy layers) which characterize foraging guilds are the same in eucalypt habitats as in other forests (Recher and Holmes 1985). The shedding of bark by eucalypts creates a unique substrate exploited by a number of species, Crested Shrike-tit, *Falcunculus frontatus*, Red-browed Treecreeper, *Climacteris erythrops*, and White-eared Honeyeater, *Meliphaga leucotis,* are three species which specialize on this substrate and occur only where eucalypts which shed bark are abundant. Except as it is related to unique foods (e.g. manna) or special substrates (e.g. hanging bark strips), the prey-attack behaviour of eucalypt forest and woodland birds is similar to that elsewhere (Recher and Holmes 1985). Gleaning, probing, snatching, hawking, and pouncing are the most common behaviours.

The ways in which eucalypt forest and woodland birds differ among themselves in the ways they use (or apportion) resources are also similar to forest birds in general. Among insectivores, closely related species differ in foraging height (e.g. thornbills, *Acanthiza*), substrates (e.g. treecreepers, *Climacteris*), or occupy different habitats (e.g. whistlers, *Pachycephala*). Among co-existing congeners, differences in prey-attack behaviour appear to be less important than differences in the use of space and foraging substrates. Other important differences which are discussed in the second part of this Volume are differences in social behaviour, size, and movements between habitats.

TEMPORAL PATTERNS

Birds differ in the 'timing of their activities' and 'time' as such can be viewed as a resource analogous to food and space. Keast (1985b), for example, describes the pattern of song or calling activities of forest and woodland birds and shows how species differ in the times of day when they are most vocal. By calling at different times, interference between species is minimized. Other temporal patterns in the activities of forest birds are linked to seasonal events and climate.

Most birds in Australia nest over a period of five to six months with no significant differences from northern (e.g. Cape York) to southern (e.g. Tasmania) latitudes (Yom-Tov, ms). In southern Australia forest birds nest from late winter (July-August) to late summer (January-February). This coincides with maximum insect populations (Recher *et al.* 1983). In northern Australia nesting is initiated with the onset of the wet season (December) and continues into the beginning of the dry season (May-June). Nix (1976) has shown the timing of reproductive activities, post-breeding dispersal, and the migration of birds is linked to seasonal patterns of plant growth. Nesting occurs when conditions for the growth of plants are optimal. As conditions become colder at southern latitudes or drier in the north, many forest birds disperse or migrate. For example, on the Southern Tablelands of New South Wales, about half the species leave the forests or woodlands in which they nest (Recher and Holmes 1985).

DISCUSSION

The abundance of birds and the species richness of forest and woodland bird communities are related to the structure and productivity of the plant communities in which they reside. As habitats become more complex and the number of layers of vegetation increases, the kinds of foraging substrates also increases and a large number of bird species will occupy the habitat. As Gilmore (1985) demonstrated, it is the life forms (i.e. forb, shrub or tree) of the plants which determine the kinds of substrates available to birds and not plant species. An increase in plant species diversity will not, therefore, necessarily lead to an increase in bird species diversity. However, the addition of plant genera which provide different kinds of food (e.g. nectar, fruits, seeds) or have different kinds of foliage (e.g. the needle-like leaves of She-oaks *Casuarina*) increases foraging opportunities and leads to an increase in the number of co-existing species.

The relationship between the structure or complexity of the habitat and bird species diversity appears simple. The relationship between the abundance or production of food and species number is more complex. In its simplest form, the relation between the abundance of food and bird species diversity is seen in the seasonal cycles of reproduction and dispersal. Peak nesting periods and maximum community diversity occur when conditions for plant growth are optimal and food is most abundant. Birds depart as it becomes colder or drier, and return when conditions improve. More difficult to determine is any relationship between the numbers or diversity of forest and woodland birds and forest productivity. Can these be related in any way to soil nutrients or to the nutrient levels in the foliage of the dominant plant species?

Lambert and Turner (1984) showed that the leaves of eucalypts associated with nutrient rich soils are richer in nutrients than those associated with poorer soils. The difference is one between subgenera with eucalypts in the subgenus *Symphyomyrtus* having foliage high in nutrients even when growing in relatively poor sites and trees in the subgenus *Monocalyptus* having leaves low in nutrients. *Symphyomyrtus* dominates wherever the soils of forests and woodlands are high in nutrients. *Monocalyptus* is restricted to forests where it dominates the drier ridges and slopes with nutrient-poor soils. This pattern of eucalypt distribution and foliage nutrient levels, of course, parallels the most conspicuous pattern of bird species diversity and abundance. With some glaring exceptions, birds are most abundant and species diversity greatest in forests and woodlands dominated by eucalypts of the subgenus *Symphyomyrtus*. The relationship between abundance and species diversity and soil nutrient levels therefore seems fairly straight forward. Only information tying foliage nutrient levels to insect productivity is missing. This is an area where reseach would be particularly rewarding.

Both Loyn (1985) and Pyke (1985a, b) predicted that the abundance of birds would be correlated to the productivity of the foods they ate. Loyn *et al.* (1983) were able to show that the removal of Bell Miners *Manorina melanophrys* from forests where they excluded most insect-eating birds resulted in a rapid and transitory increase in the abundance and diversity of insectivorous birds. Numbers remained high until insect populations had been reduced to levels found in forests where Bell Miners were absent. If allowance is made for the spatial scale on which he worked, Pyke (1985) also demonstrated a relation between the numbers of birds and their food supply. The numbers of honeyeaters frequenting woodlands near Sydney was correlated with the production of nectar.

There are exceptions to the general correlation between soil nutrients and the abundance and diversity of forest and woodland birds. Pyke's honeyeaters are one of these exceptions. Another is the open forest and woodland habitats on the Southern Tablelands of New South Wales where Recher and Holmes (1985) conducted their studies. Their plots were located on nutrient-poor soils where many of the trees were stressed by site conditions and insect attack. Pyke's study areas are on Hawkesbury Sandstone where soil nutrient levels are exceptionally low. Loyn (1985) also noted that some forest and woodland habitats support permanent high populations of insect-eating birds, but have relatively low timber productivity. He suggested that this may indicate that a large proportion of the tree's resources are taken by the insects or lost as sap, manna and honeydew. Woodlands growing on Hawkesbury Sandstone are floristically diverse, have an abundance of plants which produce nectar-rich flowers, but support few insect-eating birds (Pyke 1983, 1985b). In this instance, it appears as if the vegetation was diverting its resources to the production of nectar.

I suggest that the general relationship between high nutrient levels, high insect production and an abundant avifauna is correct. This pattern will occur whenever site conditions are favourable, in which case it is permanent, or the pattern will be transitory as after fire when nutrient levels are temporarily increased as a result of nutrients being released from the litter when burnt. However, in some instances where the vegetation is stressed (e.g. shortage of water, over-grazing) or where nutrients are extremely limited as on the sandstone, insect and/or carbohydrate productivity will also be high. It is a matter of allocating resources. Where the vegetation is stressed, inadequate resources are available for the plants to fend off insect attack by, for example, producing toxins. On the sandstone the lack of nutrients in the foliage means low insect populations and reduced grazing pressure from herbivorous insects. Insects may also be unreliable as pollinators. My hypothesis is that the plant diverts large amounts of resources to nectar to attract bird (or mammal) pollinators (Carpenter 1978; Recher 1981) and can afford to do so in the absence of heavy grazing. Most of the nectar-feeding birds they attract will of necessity be produced in richer environments. Hence flowering and nectar production is concentrated during the winter when most birds are not breeding and can move freely between habitats.

The suggestions that I have made above, as well as those of Loyn (1985) and Pyke (1985) are speculative. The data needed to support or reject any of these ideas are lacking. However, the hypotheses are testable and in my view represent an important area of future research.

MANAGEMENT AND CONSERVATION

It would be simple to conclude that the best way to conserve forest and woodland birds is to reserve samples of their different kinds of habitats and ensure that the structural complexity of these habitats was maintained or even maximized. As a first approximation, this is probably adequate and is certainly an important goal in forests managed for timber production or in woodlands affected by agriculture. However, this approach is shallow and fails to recognize the needs of birds for alternative habitats during droughts nor does it allow for movements and migrations outside the breeding season. What is required are management and conservation strategies which manage forests and woodlands throughout Australia as a whole.

REFERENCES

BOURKE, P. A., 1949. The breeding population of a thirty-five acre 'timber paddock'. *Emu* **49**: 73-77.

CARPENTER, F. L., 1978. Hooks for mammal pollination? *Oecologia* **35**: 123-32.

CROME, F., 1978. Foraging ecology of an assemblage of birds in lowland rainforest in northern Queensland. *Aust. J. Ecol.* **3**: 195-212.

GILMORE, S., 1985. The influence of vegetation structure on the density of co-existing insectivorous birds. Pp. 21-31 *in* Birds of Eucalypt Forests and Woodlands: Ecology, Conservation, Management ed. by A. Keast, H. F. Recher, H. Ford and D. Saunders. Royal Australasian Ornithologists Union.

KEAST, A., 1985a. Bird community structure in southern forests and northern woodlands: a comparison. Pp. 97-116 *in* Birds of Eucalypt Forests and Woodlands: Ecology, Conservation, Management ed. by A. Keast, H. F. Recher, H. Ford and D. Saunders. Royal Australasian Ornithologists Union.

KEAST, A., 1985b. Springtime song, periodicity and sequencing, a comparison of a southern forest and northern woodland bird community Pp. 119-28 *in* Birds of Eucalypt Forests and Woodlands: Ecology, Conservation, Management ed. by A. Keast, H. F. Recher, H. Ford and D. Saunders. Royal Australasian Ornithologists Union.

KIKKAWA, J., 1968. Ecological association of bird species and habitats in eastern Australia; similarity analysis. *J. Anim. Ecol.* **37**: 143-65.

LAMBERT, M. AND TURNER, J., 1983. Soil nutrient-vegetation relationships in the Eden area, NSW III. Foliage nutrient relationships with particular reference to *Eucalyptus* subgenera. *Aust. For.* **46**: 200-09.

LOYN, R., 1980. Bird populations in a mixed eucalypt forest used for production of wood in Gippsland, Victoria. *Emu* **80**: 145-56.

LOYN, R., 1985. Ecology, distribution and density of birds in Victorian forests. Pp. 33-46 *in* Birds of Eucalypt Forests and Woodlands: Ecology, Conservation, Management ed. by A. Keast, H. F. Recher, H. Ford and D. Saunders. Royal Australasian Ornithologists Union.

LOYN, R., RUNNALLS, R., FORWARD, G. AND TYERS, J., 1983. Territorial Bell Miners and other birds affecting populations of insect prey. *Science* **221**: 1411-13.

MILLEDGE, D. AND RECHER, H. F., 1985. A comparison of forest bird communities of the New South Wales south and mid-north coasts. Pp. 47-52 *in* Birds of Eucalypt Forests and Woodlands: Ecology, Conservation, Management ed. by A. Keast, H. F. Recher, H. Ford and D. Saunders. Royal Australasian Ornithologists Union.

NIX, H., 1976. Environmental control of breeding, post-breeding dispersal and migration of birds in the Australian region. *Proc. 16th Int. Ornith Congr.* Pp. 273-305.

PYKE, G., 1983. Seasonal patterns of abundance of honeyeaters and their resources in heathland areas near Sydney. *Aust. J. Ecol.* **8**: 217-34.

PYKE, G., 1985a. The relationships between abundances of honeyeaters and their food resources in open forests near Sydney. Pp. 65-77 *in* Birds of Eucalypt Forests and Woodlands: Ecology, Conservation, Management ed. by A. Keast, H. F. Recher, H. Ford and D. Saunders. Royal Australasian Ornithologists Union.

PYKE, G., 1985b. Seasonal patterns of abundance of insectivorous birds and flying insects. *Emu* **85**: 34-39.

PYKE, G. AND RECHER, H., 1984. Censusing Australian birds: a summary of procedures and a scheme for standardization of data presentation and storage. Pp. 55-63 *in* "Methods of censusing birds in Australia" ed. by S. J. J. F. Davies. R.A.O.U. Report 7.

RECHER, H. F., 1969. Bird species diversity and habitat diversity in Australia and North America. *Am. Nat.* **103**: 75-80.

RECHER, H. F., 1975. Survey of the avifauna of Myall Lakes, NSW; report of the 1972 R.A.O.U. field-outing. *Emu* **75**: 213-25.

RECHER, H. F., 1981. Nectar-feeding and its evolution among Australian vertebrates. Pp. 1639-47 *in* "Ecological Biogeography of Australia" ed. by A. Keast. Junk, The Hague.

RECHER, H. F., 1984. Use of bird census procedures in Australia: a review. Pp. 3-14 *in* "Methods of censusing birds in Australia" ed. by S. J. J. F. Davies. R.A.O.U. Report 7.

RECHER, H. F., 1985. Forests, woodlands and birds. Pp. 1-10 *in* Birds of Eucalypt Forests and Woodlands: Ecology, Conservation, Management ed. by A. Keast, H. F. Recher, H. Ford and D. Saunders. Royal Australasian Ornithologists Union.

RECHER, H. F. AND ABBOTT, I., 1970. Some differences in the use of habitat by white-eared and white-cheeked honeyeaters. *Emu* **70**: 117-25.

RECHER, H. F., ROHAN-JONES, W. AND SMITH, P., 1980. Effects of the Eden woodchip industry on terrestrial vertebrates. *Research Note No. 42*, Forestry Commission of N.S.W.

RECHER, H. F. AND HOLMES, R. T., 1985. Foraging ecology and seasonal patterns of abundance in a forest avifauna. Pp. 79-96 *in* Birds of Eucalypt Forests and Woodlands: Ecology, Conservation, Management ed. by A. Keast, H. F. Recher, H. Ford and D. Saunders. Royal Australasian Ornithologists Union.

RECHER, H. F., GOWING, G., KAVANAGH, R., SHIELDS, J. AND ROHAN-JONES, W., 1983. Birds, resources and time in a tablelands forest. *Proc. Ecol. Soc. Aust.* **12**: 101-23.

SHIELDS, J., KAVANAGH, R. AND ROHAN-JONES, W., 1985. Forest avifauna of the Upper Hastings River. Pp. 55-64 *in* Birds of Eucalypt Forests and Woodlands: Ecology, Conservation, Management ed. by A. Keast, H. F. Recher, H. Ford and D. Saunders. Royal Australasian Ornithologists Union.

SPECHT, R., 1981. Ecophysiological principles determining the biogeography of major vegetation formations in Australia. Pp. 299-334 *in* "Ecological Biogeography of Australia" ed. by A. Keast. Junk, the Hague.

YOM-TOV, Y. (ms). The reproductive rates of Australian passerines.

Helmetted Honeyeater, *Lichenostomus cassidix*

Ecology and Adaptations of Forest and Woodland Birds

Co-ordinated by Hugh A. Ford

Buff-rumped Thornbill, *Acanthiza reguloides*

INTRODUCTION

THE PREVIOUS section, particularly Chapters 4, 8 and 9, has described bird communities in various eucalypt woodlands and forests. This has set the scene for the detailed description of the niches of some of the most important groups of forest birds, which is the aim of this section. The term niche usually includes the habitat, foraging sites and food of an animal. In Elton's (1927) original definition the enemies or predators of the animal were included. The importance of these is hard to determine except perhaps during the breeding season when predators influence breeding success.

All of the authors have something to say about these various components of the bird's niche, though the emphases are different. For such skulking birds as scrub-birds it is hard to record foraging behaviour and diet, but habitat can be measured. Simon Ferrier has done this painstakingly in Chapter 20. Sometimes observers have compared similar species in small areas, as Liz Cameron has for the fantails (Chapter 15), Richard Noske for treecreepers (Chapter 16) and Harry Bell for thornbills (Chapter 13). Other studies have covered a broader range of habitats as well as several species as in John Woinarski's study of the pardalotes and other leaf-foraging birds (Chapter 14) and Boyd Wykes' work on honeyeaters (Chapter 17). Although most of the species are primarily insectivorous, the herbivorous rosellas and nectarivorous lorikeets are reviewed by Ed Wyndham and Christine Cannon (Chapter 12). The highly nectarivorous New Holland Honeyeater, *Phylidonyris novaehollandiae* is David Paton's subject in Chapter 18. Breeding biology is the basis of the study by Stephen Marchant on Eastern Yellow Robins, *Eopsaltria australis* (Chapter 19) and has been compared in the thornbills, treecreepers and parrots.

Although many forest birds in Australia are quite generalized some show obvious morphological adaptations to their way of life. The pardalotes have short deep bills to prise off scale insects, fantails broad flat beaks to capture flying prey and shrike-tits powerful hooked beaks to tear bark apart. The thornbills and *Lichenostomus* honeyeaters show more subtle adaptations.

Many of the birds discussed in the following chapters are well known to ornithologists but only in the last decade have they received detailed quantitative study. The work presented here, much of it published for the first time, brings our knowledge of forest birds a giant step forward.

REFERENCES

ELTON, C. S., 1927. Animal Ecology. Sidgwick and Jackson; London.

Hugh A. Ford

Rainbow Lorikeet, *Trichoglossus haematodus*

CHAPTER 12

Parrots of Eastern Australian Forests and Woodlands: The Genera *Platycercus* and *Trichoglossus*

E. Wyndham[1] and C. E. Cannon[2]

THE ROSELLAS *Platycercus* spp. inhabit savannas and woodlands of peripheral parts of Australia (Fig. 1) and the trichoglossid lorikeets *Trichoglossus* spp. frequent woodlands and forests of eastern and northern Australia (Fig. 2). The current RAOU Checklist (Condon 1975) lists six species of rosellas and three species of trichoglossid lorikeets and several of these species have a number of sub-species (see Condon 1975).

The present Australian avifauna is derived from ancestral stock that inhabited the continent many millions of years ago (the autochthonous fauna, *sensu* Schodde 1982) and of more recent immigrants that have invaded it from the north (Keast 1981; Schodde 1982). Some 55 to 60 million years before present (m.y.b.p.) Australia separated from what remained of Gondwanaland and for an extended period, until the Miocene (26 m.y.b.p.) when it arrived in its present position, it drifted slowly north and was largely isolated by surrounding expanses of ocean (Crook 1981; Powell, Johnson and Veevers 1981). It is only in relatively recent times that terrestrial birds of south-east Asia have had easy access to Australia via New Guinea.

The rosellas are examples of the autochthonous stock, of southern origins, and the trichoglossid lorikeets are recent northern immigrants (Smith 1975). The present configuration of rosella species and sub-species evolved in Australia. Much of this evolution probably occurred during the Pleistocene when there were alternating phases of aridity and humidity (Bowler 1982). During periods of aridity the ancestral rosella stock, which presumeably inhabited sub-humid regions, was split into several isolated populations in the south-west, north and up the east coast, a situation propituous to speciation (Brereton, Richards and Williams 1969). Evolution of the trichoglossid lorikeets appears to have occurred outside Australia, possibly in New Guinea, and established species have colonized Australia in more recent times.

If this reconstruction of the phylogeny of the rosellas and trichoglossid lorikeets is correct, these genera are good groups in which to study adaptive and radiative processes in the Australian avifauna. The rosellas have representative species at southern, mid-latitude and northern extremities of the continent (see Fig. 1) and differences between species should reflect the selective forces acting in these parts. The trichoglossid lorikeets are examples of species that were pre-adapted to colonize the continent and have undergone little evolution *in situ*. In this chapter we review what is currently known of the ecology of the two genera, discuss how the genera utilize the forests and woodlands of eastern Australia and suggest worthwhile lines of future research.

Habitat useage by rosellas and lorikeets was studied in urban, agricultural and forested areas of southeastern Queensland and northeastern New South Wales. Cannon (unpubl.), while driving through this region, made periodic stops and recorded the numbers of individuals and groups of each species that were seen.

The diet and diurnal cycle of Eastern Rosellas, *P. eximius*, was studied near Warwick, Queensland (28°S, 152°E) and of Pale-headed Rosellas, *P. adscitus*, at the University of Queensland Veternary Farm, Moggil (27°S, 153°E) (Cannon 1981). These field studies of diet were supplemented by experiments on caged birds in which birds were offered a choice of feeding stations placed from 0 to 1.5 metres above the ground (Cannon unpubl.). Diets of Rainbow and Scaly-breasted Lorikeets, *T. haematodus*

[1]Department of Ecosystem Management, University of New England, Armidale, New South Wales 2351.
[2]C/o Queensland National Parks and Wildlife Service, P.O. Box 190, North Quay, Brisbane, Queensland 4000.
Pages 141-50 *in* BIRDS OF EUCALYPT FORESTS AND WOODLANDS: ECOLOGY, CONSERVATION, MANAGEMENT ed. by A. Keast, H. F. Recher, H. Ford and D. Saunders. Royal Australasian Ornithologists Union and Surrey Beatty & Sons.

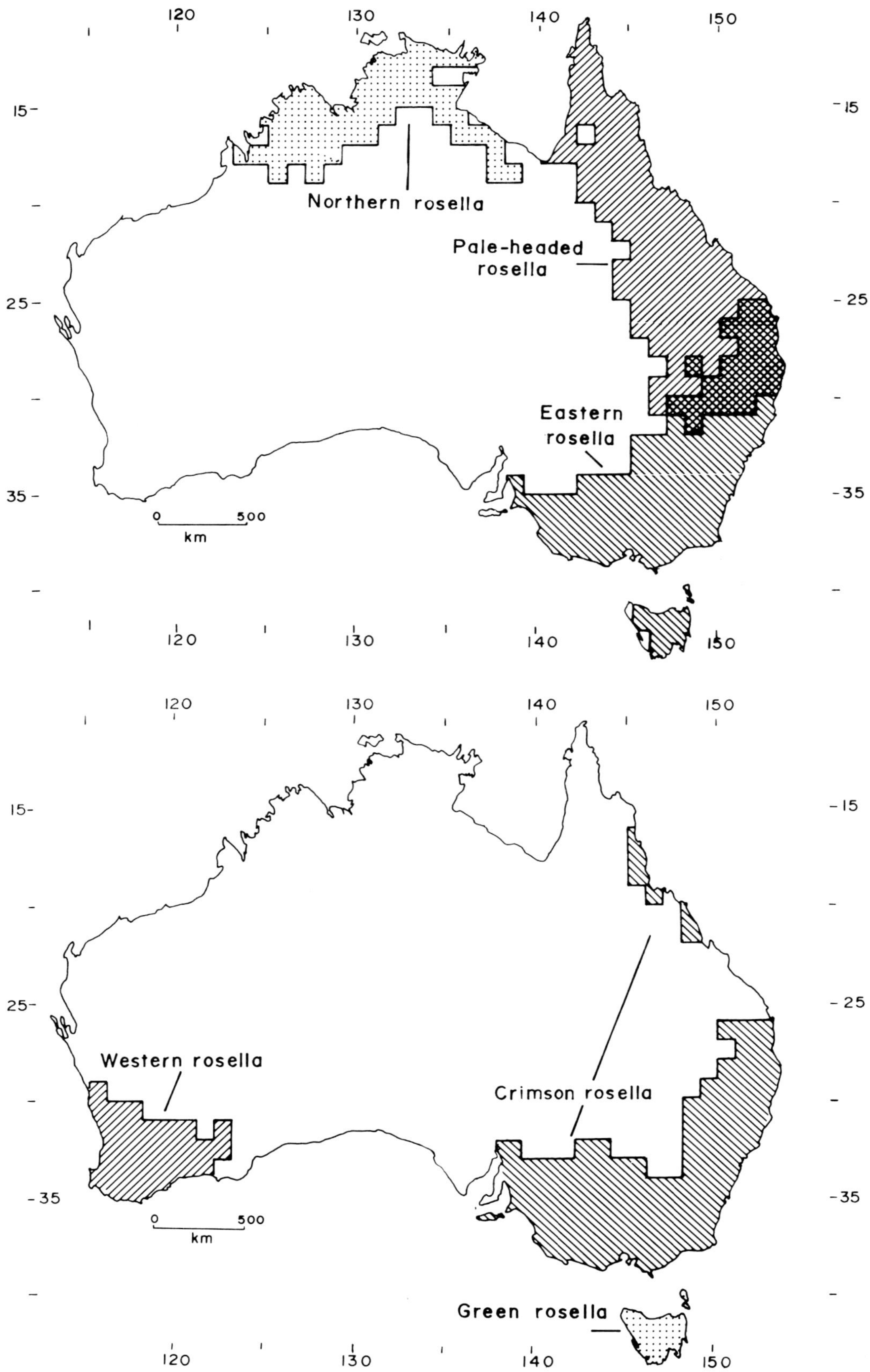

Fig. 1. Distribution of rosellas as ascertained by the RAOU Atlas Scheme. The Atlas Scheme collated historical records up to 1976 and between 1976 and 1981 recorded presence or absence of birds in 1° latitude and longitude grid blocks. The bottom map shows the distribution of the blue-cheeked (Green *Platycercus calidonicus* and Crimson *P. elegans*) and Western Rosella *P. icterotis* and the upper map shows the white-cheeked Rosellas (Eastern *P. eximius*, Pale-headed *P. adscitus* and Northern *P. venustus*).

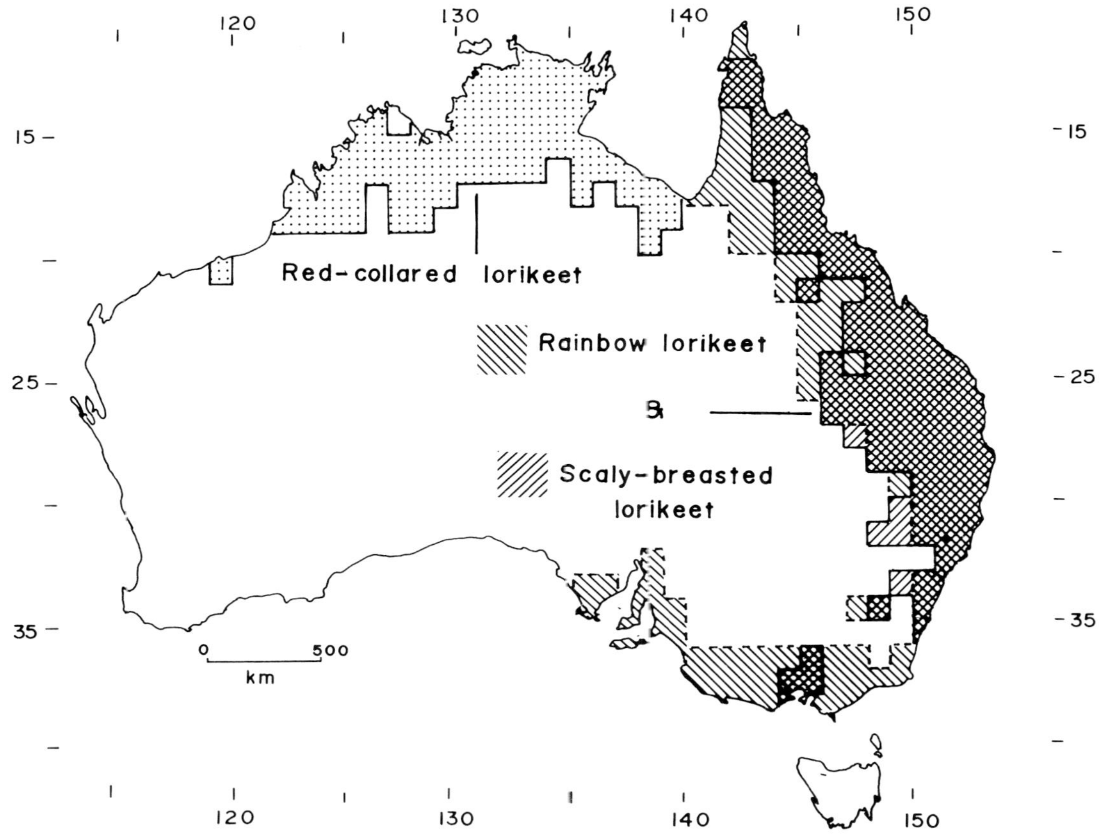

Fig. 2. Distribution of trichoglossid lorikeets; Red-collared, *Trichoglossus rubritorquis,* Rainbow, *T. haematodus,* and Scaly-breasted, *T. chlorolepidotus* Lorikeets. The Rainbow Lorikeet has been introduced into southwestern Western Australia.

and *T. chlorolepidotus* were studied in southeastern Queensland and northeastern New South Wales (Cannon 1979, 1984b). Diets of rosellas and lorikeets were studied by observations only. When a bird was seen feeding on a particular item this was scored as one observation, irrespective of the length of time spent feeding on that item.

Movements, social organisation and breeding of Eastern Rosellas was studied at Armidale, NSW (31°S, 152°E). Over 17 years rosellas were trapped and individually marked, firstly with coloured leg bands and later with patagial tags (Brereton and Pidgeon 1968). Artificial nest boxes were put up in the study area and most information on breeding of rosellas at Armidale comes from nests in these boxes. A few breeding pairs were kept in cages at the University of New England, Armidale.

Movements of Rainbow and Scaly-breasted Lorikeets were studied by Cannon (1984a) who banded and individually marked birds with paint on the bill at an artificial food source at the Currumbin Sanctuary, Currumbin, Queensland (28°S, 154°E).

Much information from these studies of rosellas and lorikeets is not published and in this account we draw on this unpublished material. We obtained additional information on breeding of rosellas from the RAOU Nest Record Scheme (NRS) and on movements of rosellas and lorikeets from the Australian Bird Banding Scheme. Aviculturists and field naturalists have built up background knowledge of the natural history of rosellas and lorikeets and this is ably summarized by Forshaw and Cooper (1981). We have supplemented our scientific studies with information from Forshaw and Cooper's accounts.

Sizes of rosellas and the trichoglossid lorikeets are given in Table 1. Both genera are brightly coloured and medium sized (*c.* 100 g) parrots. The Crimson, *P. elegans*, is the largest of the rosellas and the Western, *P. icterotis*, the smallest. Differences between plumages of the sexes range from almost none in the Crimson to considerable sexual dimorphism in the Western (see Forshaw and Cooper 1981 for details). Male Eastern Rosellas are larger than females (mean weights: male 117 g, female 102 g, difference statistically significant; Wyndham and Brereton 1982) and the weights given by Forshaw and Cooper (Table 1) suggest males are larger than females in all species. In all trichoglossid lorikeets males and females have similar plumage and the data in Table 1 suggest the sexes are the same size, or differ only slightly.

HABITATS AND FOOD

Rosellas

Rosellas inhabit savannas, woodlands and forests in high to medium rainfall regions, extending out

Table 1. Weights of rosellas and trichoglossid lorikeets. The data are from Forshaw and Cooper (1981) and show ranges of weight; Forshaw and Cooper did not give mean weight and sample sizes.

	Male	Female
Green Rosella	105-165	90-130
Crimson Rosella	123-169	112-146
Eastern Rosella	97-120	90
Pale-headed Rosella	131	112
Northern Rosella	92-112	88-92
Western Rosella	59-80	52-71
Rainbow Lorikeet	115-157	107-143
Red-collared Lorikeet	100-140	104-140
Scaly-breasted Lorikeet	82-89	70-83

into semi-arid parts where there are watercourses lined with trees. The genera does not inhabit semi-arid and arid grasslands or arid shrublands and is largely absent from rainforests. The Green Rosella, *P. calidonicus* does occur in rainforests (Ridpath and Moreau 1966) but favours sclerophyll forests and savannas (Green 1977). Crimson Rosellas may also be seen in rainforest (Forshaw and Cooper 1981): in the New England region Crimson Rosellas occur in rainforest that is surrounded by and contains a partial overstorey of wet sclerophyll forest (Wyndham pers. obs.).

In southeastern Queensland and northeastern New South Wales Crimson, Pale-headed and Eastern Rosellas are sympatric. Cannon (unpubl.) found that Crimson Rosellas occurred predominantly in forest areas while Pale-headed and Eastern Rosellas occurred predominantly in the more open agricultural areas (Table 2). Only 4% of 209 flocks of Eastern and Pale-headed Rosellas contained both species, despite their occurrence in the same type of habitat (Cannon unpubl.). The two species may remain separated because they use different feeding stations (see below). On the New England Tablelands, where Eastern and Crimson Rosellas are sympatric, mixed species flocks of these two species are rare (Brereton and Sourry 1959).

The diet of all rosellas consists of seeds of grasses, shrubs or trees; fruits, blossoms, buds, insects and insect larvae (Forshaw and Cooper 1981). The detailed study of the diet of Eastern and Pale-headed Rosellas at Warwick and Moggill showed that, although the diet of the two species consisted of similar types of food (seeds, fruits, flowers and insects), there were differences in the feeding stations (i.e. on the ground, perched above the ground on a log, stump or shrub, or feeding in a tree) and the type of plant (grass, herb, shrub or tree) from which each species took its food.

Fruits and seeds formed 81% of the Eastern Rosella's diet, flowers 7% and insects 12% (n = 1891), while fruits and seeds formed 83% of the

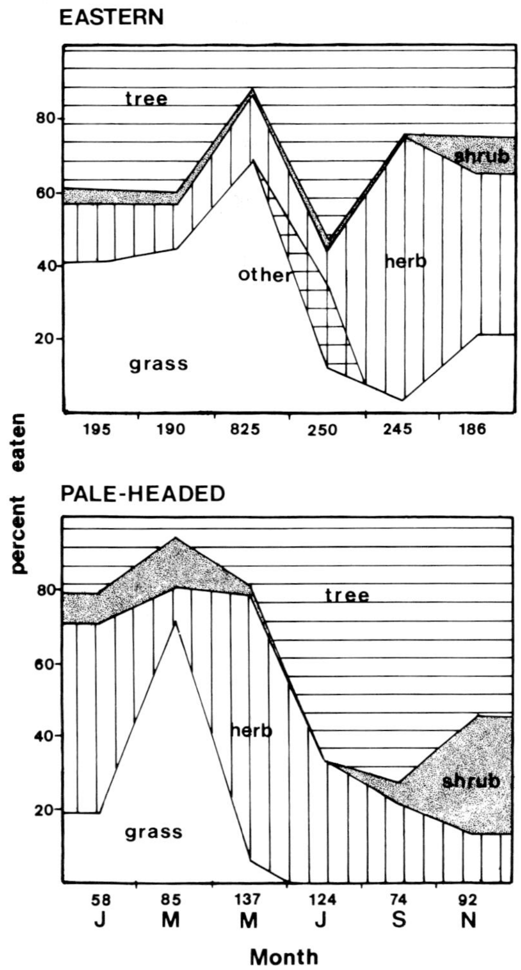

Fig. 3. The diet of Eastern and Pale-headed Rosellas (Cannon 1981, Fig. 1), showing the percentage of grasses, herbs, shrubs, trees and 'other' (mainly seeds from cattle dung) eaten each season. The number of observations per month are given on the abscissa.

Table 2. Habitat use by rosellas and lorikeets in southeastern Queensland and northeastern New South Wales. The area was surveyed by car between March 1978 and April 1979 and during stops in each habitat it was recorded whether each species was present. A total of 1517 stops were made, 10% in forest, 62% in agricultural and 28% in urban habitats.

Species	Percentage of stops in which species was seen: Forest	Agricultural	Urban	Total no. of stops in which species was seen
Crimson Rosella	69	31	0	26
Eastern Rosella	11	88	1	81
Pale-headed Rosella	12	75	12	122
Rainbow Lorikeet	10	31	59	229
Scaly-breasted Lorikeet	11	52	37	236

Pale-headed Rosella's diet, flowers 9% and insects 9% (n = 1531). The type of plant from which each species obtained its food is shown in Figure 3. Overall, Eastern Rosellas obtained 71% of their food (mainly fruit and seeds) from grasses and herbs, and in 69% of 1864 observations the birds were standing on the ground when they took items of food. In contrast, Pale-headed Rosellas obtained 50% of their food (seeds, fruits, flowers and insects) from trees and shrubs and in 82% of 559 observations the birds were perched above the ground when they took items of food. Eastern Rosellas when feeding on grasses and herbs usually stood on the ground, while Pale-headed Rosellas when feeding on grasses and herbs frequently perched above the ground, on logs, fences and shrubs. The mean height of feeding stations was 1.8 m in Eastern Rosellas (1864 observations) and 3.7 m in Pale-headed Rosellas (559 observations). In captive birds that were given a choice in height of feeding stations, Eastern Rosellas fed at a mean height of 0.8 m and Pale-headed Rosellas at 1.0 m (difference statistically significant).

In the field studies, Eastern Rosellas fed on 82 species of plants and Pale-headed Rosellas on 47 species. The greater number of species of plants in the diet of Eastern Rosellas may reflect a greater diversity of plants at Warwick, where Eastern Rosellas were studied, than at Moggill, where Pale-headed Rosellas were studied (unfortunately flora inventories were not made). Alternatively, or in addition, Eastern Rosellas fed predominantly on the ground and Pale-headed Rosellas fed predominately in shrubs and trees. Birds feeding on the ground move from plant to plant and thus they are likely to take food from a range of species; birds feeding in trees move within a tree or between a few trees and thus come into contact with fewer plants. Despite the more catholic diet of Eastern Rosellas, the bulk of the diet of both rosellas was obtained from a few species of plant. At any particular time of the year, Eastern Rosellas obtained 70% of its food items from an average of 5.5 species of plants and Pale-headed Rosellas obtained 80% from an average of 5.3 species.

Throughout the year at Warwick and Moggill the majority of food items eaten by both rosellas were fruits, seeds and flowers, but at times the birds took an amount of insect food. The insect food eaten was lerps (Psyllidae) and scale insects (Coccidae) attached to eucalypt leaves and was eaten in large amounts by Eastern Rosellas in March (22% of all items eaten) and July (50%) and by Pale-headed Rosellas in July (23%) and September (26%). Small amounts of insect food was taken by both species at other times of the year. Further work is needed to find why at times rosellas take appreciable amounts of insect food.

The diurnal cycles of the Eastern Rosella at Warwick and the Pale-headed Rosella at Moggill were similar (Fig. 4) and the differences that did occur were due to food type and the feeding stations used by each species. Although both rosellas spent between 80-85% of their time feeding and perching, the Eastern Rosella spent on average about 60% more time feeding than the Pale-headed Rosella. The annual time budget for each species was as follows, Eastern (N = 17 881 observations by instantaneous sampling method — see Altmann 1974) then Pale-headed Rosella N = 10 182 observations): feeding 49.7%, 29.7%; perching 35.1%, 50.7%; resting 5.0%,

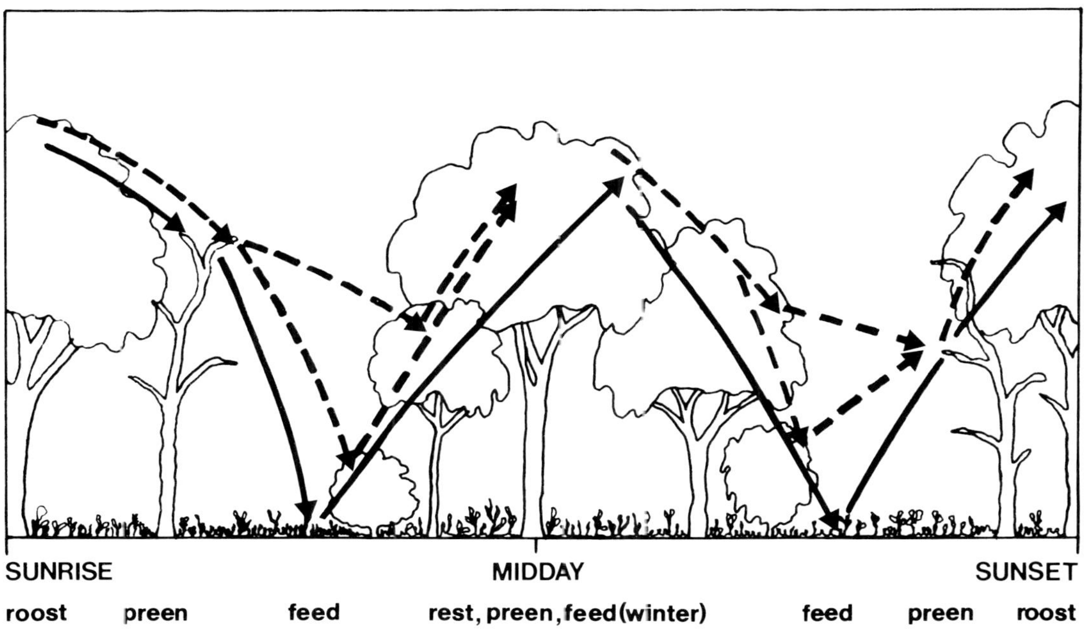

Fig. 4. Diurnal activity cycles of Eastern Rosellas (solid line) and Pale-headed Rosellas (dotted line) throughout a day. Observations were collected by instantaneous sampling (Altman 1974) throughout a day during a 12 month period.

3.7%; maintenance 7.1%, 8.8%; reproductive activity 1.2%, 4.6%; intra- and interspecific interactions 1.9%, 2.5%. Both species spent a greater proportion of the day feeding during the cooler months of the year (Eastern 50-60%, Pale-headed 45%) than during the summer (Eastern 30%, Pale-headed 20%).

For each species a 'typical day' followed a similar pattern. Around sunrise the birds moved out from roosting sites, perched on exposed branches in pairs or small groups and then moved to the feeding grounds. Eastern Rosellas in the morning fed on the ground in 80% of 943 observations while Pale-headed Rosellas fed on the ground in only 28% of 225 observations. Both species when not feeding on the ground fed in trees and shrubs. Intensive feeding, interspersed with short bouts of perching and preening, continued until late morning. Birds then moved to perch in trees and during the middle of the day maintenance activity (preening, calling softly, resting) predominated, although some feeding occurred, particularly in winter. In the afternoon there was a second main feeding session, again interspersed with bouts of preening and again Eastern Rosellas mostly fed on the ground (84.0% of 682 observations) while Pale-headed Rosellas fed mainly at higher levels (72.8% of 213 observations). Towards sunset both species moved towards roosting trees. They perched on exposed branches, bathed in the sun, preened, called and, as the light faded, moved into the foliage and became silent.

Lorikeets

Lorikeets of the genus *Trichoglossus* are typically a lowland species, found in areas where there are flowering trees (Forshaw and Cooper 1981). In southeastern Queensland and northeastern New South Wales groups of Rainbow Lorikeets were common in urban areas while Scaly-breasted Lorikeets frequented agricultural areas (Table 2). Mixed flocks of the two lorikeets did occur: at the Currumbin Sanctuary, where birds are fed each day, at feeding time a thousand or more birds regularly gathered that were made up of 90 to 95% Rainbow and the balance of Scaly-breasted Lorikeets. Elsewhere in the surveyed area only 5% of feeding groups contained both species.

Lorikeets feed mainly on flowers of trees and shrubs, taking nectar and, to a lesser extent, pollen (Churchill and Christensen 1970; Hopper and Burbidge 1979), and rarely come to the ground. In southeastern Queensland and northeastern New South Wales Rainbow Lorikeets took 87% of food items from flowers, 5% fruits, 4% leaf buds and 4% bark, insects, bread and honey (1207 observations) (Fig. 5; see also Cannon 1984b). Scaly-breasted Lorikeets took 99% of food items from flowers and the balance from fruits, leaf buds, bark, insects, bread and honey (1123 observations). Both lorikeets took food from a range of plant species (the Rainbow from 43 and the Scaly-breasted from 25); the Rainbow Lorikeet took 41% of items and the Scaly-breasted 71% of items from flowers of eucalypt trees. Other genera from which food was obtained, in order of importance, were *Melaleuca, Callistemon, Banksia* and *Tristania*.

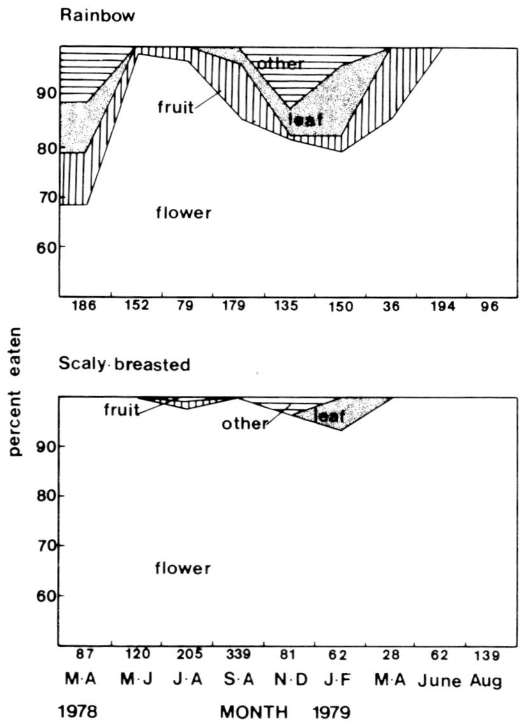

Fig. 5. The diet of Rainbow and Scaly-breasted Lorikeets (Cannon 1984b, Fig. 1), showing the percentages of flowers, fruits, leaf buds and other (including bark, insects and bread and honey) eaten each month. The number of observations per month are given on the abscissa.

MOVEMENTS AND SOCIAL ORGANISATION

Rosellas

Forshaw and Cooper (1981) describe all rosellas as sedentary. A total of 2288 Eastern Rosellas have been leg banded and, of 598 recaptures or recoveries, only one (banded as a juvenile) has been recovered away from the place of banding and this bird was recovered 3 km away (Purchase pers. comm.). A total of 1741 Crimson Rosellas have been banded and, of 146 recoveries, only six were recovered away from the place of banding. The most distant recovery was a bird banded as a juvenile and recovered 20 km from the place of banding (Purchase pers. comm.).

The study of Eastern Rosellas at Armidale established that adults are sedentary. Many marked adults were recaptured repeatedly over 12 months or longer: the longest record was of a female that was marked as an adult and recaptured 34 times over 9 years and 3 months. Surveys outside the areas where birds were trapped and individually marked, and casual observations while driving round the Armidale district, failed to detect marked birds. For four years trapping was conducted in three areas, each 8 km apart: there was no interchange of individuals between areas (Brereton 1963a).

One hundred and sixteen birds were tagged as nestlings and all disappeared from the trapped population within a few months of fledging. Two siblings disappeared after fledging but reappeared ten months later and were re-trapped repeatedly during the next two years. A third bird, a sibling of the previous two, disappeared after fledging, one and a half years later was re-trapped twice, and then disappeared again. Two explanations are viable: young birds disperse from their natal area within a few months of fledging or the patagial tags cause high mortality in recently fledged birds. Rowley and Saunders (1980) reported increased predation on cockatoos that had patagial tags compared to untagged birds.

Forshaw and Cooper (1981) state that all rosellas occur in pairs, small groups of about six birds and larger groups of up to a hundred birds. According to Forshaw and Cooper (1981), pairs and small groups predominate in spring and summer (the breeding season) and the small groups are family parties. The larger groups form in autumn and winter, particularly when the birds are feeding on the ground.

Brereton (1971) describes the social organisation of the Eastern Rosella. In spring a population consists of breeding pairs and small groups. Brereton (1963a, b) initially described the small groups as family parties comprised of a parental pair and its young from the previous year, but later he (Brereton 1971) states that the small groups are composed of young from the previous year, not necessarily siblings, and young adults (presumeably young birds in their second year) that are not the parents of the young birds. Chicks fledge in summer and remain dependent on their parents for about a month, then the young leave their parents and join up with other young birds. By autumn the population consists of the solitary breeding pairs and 'flocks' of up to a hundred birds and made up of young of the year and young adults. New groups of young adults and young of the year form within the flock in winter, and some birds form pairs. The flocks break up into the groups and pairs as spring approaches: the pairs become solitary and enter the breeding component of the population.

Lorikeets

Forshaw and Cooper (1981) state that the trichoglossid lorikeets are highly nomadic and claim that there are large regional fluctuations in numbers as the lorikeets move from area to area, following flowering of trees.

At Currumbin Sanctuary 63 Rainbow Lorikeets were individually marked and over 25 days they repeatedly visited the artificial food source, which indicates some short term stability in composition of flocks. However, over a 30 month period banded birds slowly disappeared from the flock feeding at the sanctuary. A hundred and eighty-four birds were banded and, of 153 recoveries, 12% were recovered away from the sanctuary, most within a radius of 5 to 10 kms, but the five most distant recoveries were 40 to 60 km away. Rainbow Lorikeets caught away from Currumbin and released into the flock at the Sanctuary returned several times during the next eighteen months to feed at the Sanctuary.

Fewer Scaly-breasted Lorikeets were banded (31 birds) or individually marked (11 birds). Individually marked birds over a short period repeatedly visited the artificial food source. No banded birds were recovered away from the Sanctuary but, because of the small number banded, the significance of this is not clear.

In the Currumbin Valley (southeastern Queensland) Cannon (unpubl.) found the mean flock size of Rainbow Lorikeets was 3.7 (1025 flocks) and of Scaly-breasted Lorikeets was 4.4 (427 flocks) birds. Both species usually occurred in groups of 5 or less birds. Larger feeding flocks, of up to 20 Rainbow or up to 50 Scaly-breasted Lorikeets, and flying flocks, of up to 37 Rainbow or 100 Scaly-breasted Lorikeets, sometimes occurred. Most lorikeet flocks were monospecific and, when mixed flocks did occur, the two species remained partially segregated (Cannon unpubl; see also Lavery and Blackman 1970). The Rainbow Lorikeet weighs about 40% more than the Scaly-breasted (Table 1) so it is likely that it excludes the Scaly-breasted when the two compete for food.

BREEDING BIOLOGY

Rosellas

At Armidale over eight years Eastern Rosellas started breeding (laid first eggs) between mid-November and mid-December and fledged young from early December to early February (40 nests). Nest records of the Eastern Rosella (126 nests) and Crimson Rosella (37 nests) submitted to the NRS over 26 years come from latitudes 25° to 40°S. Both species started nests between August and December. Eastern Rosellas started 80% of their nests between mid-October and mid-December and a breakdown of records into latitude bands showed there was no obvious difference in the time of breeding between latitudes 35°-40°S and 30°-35°S. Records from latitudes 25°-30°S were too few to compare with the two more southerly latitude bands.

Caged Eastern Rosellas will lay replacement clutches or two or three clutches in succession and within a few days of loss or fledging the earlier clutch (Brereton and Wyndham unpub.). Brereton (1963b) proposed that at Armidale wild Eastern Rosellas generally breed in spring and again in autumn, but subsequent field work produced no evidence of repeated clutches or autumnal breeding.

Rosellas nest in the hollows of trees or artificial structures (fence posts, stumps, nestboxes) that simulate tree hollows. Eggs are laid at one to two day intervals (Eastern Rosella, Wyndham pers. obs.; Crimson Rosella; Calder, Golding and Manderson 1979), incubation starts during the middle stages of laying (with the second egg in all parrots according to Forshaw and Cooper 1978) and, because of the protracted laying, in large clutches hatching is highly asynchronous (Eastern Rosella, Wyndham pers. obs.). Eastern Rosellas in some nests fledge all chicks during the same day, even though there is a few days difference in ages, while in other nests the younger chicks leave up to several days after the older ones (Wyndham pers. obs.).

Clutch size, from records submitted to the NRS, in the Eastern Rosella (51 clutches) ranged from three to nine with a modal (most common) size of six, and in the Crimson Rosella (17 clutches) ranged from two to seven with a modal size also of six. A breakdown of nests of the Eastern Rosella into the three 5° latitudinal bands (25°-30°, 30°-35°, and 35°-40°) showed no obvious association between clutch size and latitude. Forshaw and Cooper (1981) give a modal clutch size of four or five eggs in all rosellas except the Northern, *P. venustus,* for which they give a clutch size of two to four eggs.

Due to asynchronous hatching, the time from the laying of an egg to hatching will be longer in the earlier than the later eggs. In nests in which all chicks fledge on the same day, the older nestlings will have a longer nestling period than the younger ones. Thus the only periods that can be calculated from the NRS records, in which the eggs and young are not marked and the nests are mostly visited infrequently, is the time from laying to hatching the first egg, from hatching to fledging the first chick, and from laying the first egg to fledging the last chick. These periods are given for three rosellas in Table 3. There do not appear to be large differences in the length of the breeding cycle of these three species: the information is too sparce to pick up small differences. Forshaw and Cooper (1981) give the incubation period, presumeably the time from laying to hatching the later eggs, as 19 days in all rosellas. They (Forshaw and Cooper 1981) give a nestling period of five weeks for all rosellas except the Northern, for which they give a nestling period of seven weeks.

Two conventional measures of the breeding success of birds is nest success, the percentage of nests started that fledge one or more chicks; and egg success, the percentage of eggs laid that produce fledglings. From records submitted to the NRS the Eastern Rosella has a nest success of 44% (128 nests) and an egg success of 41% (532 eggs); the Crimson Rosella a nest success of 50% (38 eggs) and an egg success of 41% (106 eggs).

Golding (in Calder, Golding and Manderson 1979) reported that pairs of Crimson Rosellas in olive-green plumage, presumably birds approaching twelve months old, bred successfully. Wyndham and Brereton (unpubl.) observed an individually marked female Eastern Rosella breeding in the spring following its own birth, and Smith and Brereton (1976) found that birds in immature plumage (less than twelve months old) underwent the spring recrudescence of gonads. Thus it appears that in these two species birds can breed towards the end of their first year. Smith and Brereton (1976) propose that subordinate birds are inhibited from breeding by the dominant adult pairs. The extent to which the breeding of young birds is stopped in this way has yet to be ascertained.

Lorikeets

Breeding information on the trichoglossid lorikeets is restricted largely to avicultural records, which are summarized by Smith (1975) and Forshaw and Cooper (1981). The lorikeets nest in tree hollows, lay two, or rarely three eggs, incubation takes about 25 days and the nestling period is about eight weeks.

DISCUSSION

We have reported here on several independant studies, conducted in different areas and with differing objectives, but when considered together these studies provide some insights into how the rosellas and lorikeets utilize and co-exist in the forests and woodlands of eastern Australia. The species partition resources by utilizing different habitats, different strata of the vegetation, and having different dietary requirements.

Table 3. Duration of stages of breeding of rosellas from the RAOU Nest Record Scheme. There was little data on rosellas not shown and on any of the Trichoglossid Lorikeets.

Species	1st laid to 1st hatch n	$\bar{X} \pm$ s.e. (days)	1st hatch to 1st fledge n	$\bar{X} \pm$ s.e. (days)	1st laid to last fledge n	$\bar{X} \pm$ s.e. (days)
Eastern Rosella	7	24.2 ± 0.96	12	29.5 ± 0.72	8	53.4 ± 0.96
Crimson Rosella	0		3	32.3 ± 0.33	0	
Pale-headed Rosella	0		1	33	0	

Lorikeets feed in the upper foliage of trees and shrubs and eat nectar and pollen from flowers. The Rainbow Lorikeet frequents urban areas and the Scaly-breasted agricultural areas. The data presented above suggests that Rainbow Lorikeets take a greater amount of food from sources other than flowers than do Scaly-breasted Lorikeets; more detailed work may reveal other differences in the diet of these two species.

Crimson Rosellas frequent forested areas while Eastern and Pale-headed Rosellas frequent the more open, savanna-like agricultural areas. The Eastern and Pale-headed Rosella are partially segregated, since Eastern Rosellas feed predominantly on the ground while Pale-headed Rosellas feed predominantly in trees. These two rosellas co-exist with lorikeets in agricultural areas but feed on seeds, fruits and insects in addition to flowers and thus their diet differs substantially from that of the lorikeets.

Further insights into how lorikeets and rosellas utilize forests and woodlands come from studies of movements. Adult Eastern Rosellas at Armidale are sedentary and, while there is little information on other species, it is likely that all the rosellas make no more than local movements. Information on lorikeets is sparse: the common belief that they are nomadic establishes that flocks move from place to place but the distances moved by individuals and whether there is an underlying recurring annual pattern to movements have yet to be ascertained. Lorikeets fed at the Currumbin Sanctuary had a permanent year round supply of food at one place, which probably is unusual under natural conditions. Even so, some individuals did range 40 to 60 km away from the food source, which is further than any documented movements by banded rosellas.

The available data on rosellas with southern distributions show that these breed in spring and early summer. In the south in spring there is a flush of growth by grasses and legumes (Fitzpatrick and Nix 1970) and many of these plants flower and set seeds in late spring and summer. The breeding seasons of the rosellas probably coincide with the period of maximal abundance of food, although quantitive measurements of food abundance are needed to confirm this. It is unfortunate that there are few data on the breeding of rosellas in the northern half of Australia. In the north maximal growth of grasses and legumes occurs after the monsoonal rains, in late summer and autumn (Fitzpatrick and Nix 1970), and thus rosellas in these parts may breed at these times.

Rosellas, compared to the trichoglossid lorikeets, have large clutches (modal size of five versus two), short incubation periods (19 versus 25 days) and rapid development of nestlings (five versus eight weeks). Smith (1975) ascribes the lower clutch size and slower embryo and nestling development in lorikeets, compared to seed eating parrots, to the less nutritious "frugivorous" diet. In addition, the small clutches and slow development of lorikeets, a taxon of tropical origins, is consistant with the small clutches and slow development of tropical birds in general. The apparent reduction in clutch size (2-4 eggs) and slow nestling development (seven weeks) of the Northern Rosella is interesting. Field research on the breeding of this rosella, the tropical representative, is potentially rewarding.

None of the rosellas or trichoglossid lorikeets appears to be in immediate danger of extinction. All the rosellas have substantial parts of their range in areas that are used for agriculture or rangeland grazing. Generally, suitable food for rosellas has remained plentiful. There has been, and doubtless will continue to be, extensive clearing and mortality of trees over much of the range of the rosella complex. Trees, particularly eucalypts, are important to rosellas as sources of food, for nest sites and for shelter. In most places ample trees remain to ensure the continued survival of the rosellas. However, there could be some localized extinctions, possibly even complete loss of a sub-species, in areas that are extensively cleared or where disease causes widespread mortality of eucalypts, e.g. dieback on the New England Tablelands (see Chapter 29).

Appreciable areas of the lorikeets' ranges are being intensively developed, specially along the east coast, but to date this does not appear to be a serious threat to the birds. Natural vegetation in suburban areas is being partly retained or replaced by trees and shrubs that continue to provide food for lorikeets. The mobility of these birds should also ensure their survival, since they can move to the less developed offshore islands, hinterland or to northern parts of the continent. Free-flying lorikeets have become a part of the tourist industry in Queensland, as they are a feature of wildlife sancturies and parks in coastal resorts. Supplementary feeding provided by these sancturies and parks, and the increased public awareness of these birds should contribute to their conservation.

REFERENCES

ALTMANN, J., 1974. Observational study of behaviour: sampling methods. *Behaviour* 49: 227-67.

BOWLER, J. M., 1982. Aridity in the late Tertiary and Quarternary of Australia. Pp. 35-45 *in* "Evolution of the flora and fauna of arid Australia" ed. by W. R. Barker and P. J. M. Greenslade. Peacock: Frewville.

BRERETON, J. LE GAY, 1963a. Evolution within the Psittaciformes. *Proc. 13th Intern. Ornithol. Congr.* 499-517.

BRERETON, J. LE GAY, 1963b. The life cycles of three Australian parrots: some comparative and population aspects. *Living Bird* 2: 21-9.

BRERETON, J. LE GAY, 1971. Inter-animal control of space: the role of rank order, parent-offspring relations, and peer play, illustrated by comparative studies of Australian parrots. Pp. 69-91 *in* "Behaviour and environment" ed. by A. Esser. Plenum: New York.

BRERETON, J. LE GAY AND PIDGEON, R., 1968. Tagging methods for the Eastern Rosella. *Aust. Bird Bander,* June **1968:** 35-7.

BRERETON, J. LE GAY AND SOURRY, C., 1959. Some observations on the distribution and abundance of closely related parrots of the New England district of New South Wales. *Emu* **59:** 93-9.

BRERETON, J. LE GAY, RICHARDS, B. N. AND WILLIAMS, J. B., 1969. Australian ecosystems and their origins. Pp. 61-73 *in* "The last of lands" ed. by L. J. Webb, D. Whitelock and J. le Gay Brereton. Jacaranda: Brisbane.

CALDER, T. G., GOLDING, B. G. AND MANDERSON, A. D., 1979. Management for arboreal species in the Wombat State Forest. Unpubl. M.Env. Sc. Group Report: Monash University.

CANNON, C. E., 1979. Observations on the food and energy requirements of Rainbow Lorikeets, *Trichoglossus haematodus* (Aves: Psittacidae). *Aust. Wildl. Res.* **6:** 337-46.

CANNON, C. E., 1981. The diet of Eastern and Pale-headed Rosellas. *Emu* **81:** 101-10.

CANNON, C. E., 1982. Experimental aspects of food selection by Australian Rosellas. *Bird Behav.* **4:** 7-20.

CANNON, C. E., 1984a. Movements of lorikeets with an artificially supplemented diet. *Aust. Wildl. Res.* **11:** 173-79.

CANNON, C. E., 1984b. The diet of lorikeets *Trichoglossus* spp. in the Queensland-New South Wales border region. *Emu* **84:** 16-22.

CHURCHILL, D. M. AND CHRISTENSEN, P., 1970. Observations on pollen harvesting by Brush-tongued Lorikeets. *Aust. J. Zool.* **18:** 427-38.

CONDON, H. T., 1975. "Checklist of the birds of Australia, I. Non-passerines". RAOU: Melbourne.

CROOK, K. A. W., 1981. The break-up of the Australian-Antarctic segment of Gondwanaland. Pp. 2-14 *in* "Ecological biogeography of Australia" ed. by A. Keast. W. Junk: Hague.

FITZPATRICK, E. A. AND NIX, H. A., 1970. The climatic factor in Australian grassland ecology. Pp 3-26 *in* "Australian Grasslands" ed. by R. M. Moore. A.N.U. Press: Melbourne.

FORSHAW, J. M. AND COOPER, W. T., 1978. "Parrots of the World", 2nd edn. Landsdowne: Melbourne.

FORSHAW, J. M. AND COOPER, W. T., 1981. "Australian parrots", 2nd edn. Landsdowne: Melbourne.

GREEN, R. H., 1977. "The birds of Tasmania", 2nd edn. The author: Launceston.

HOPPER, S. O. AND BURBIDGE, A. A., 1979. Feeding behaviour of a Purple-crowned Lorikeet on flowers of *Eucalyptus buprostium. Emu* **79:** 40-42.

KEAST, A., 1981. The evolutionary biogeography of Australian birds. Pp. 1586-635 *in* "Ecological biogeography of Australia" ed. by A. Keast. W. Junk: The Hague.

LAVERY, H. J. AND BLACKMAN, J. G., 1970. Sorghum damage by lorikeets. *Qld. Agric. J.* **82:** 785-87.

POWELL, C. MCA., JOHNSON, B. D. AND VEEVERS, J. J., 1981. The early cretaceous break-up of eastern Gondwanaland, the separation of Australia and India, and their interaction with South-east Asia. Pp. 16-29 *in* "Ecological biogeography of Australia" ed. by A. Keast. W. Junk: The Hague.

RIDPATH, M. G. AND MOREAU, R. E., 1966. The birds of Tasmania: ecology and evolution. *Ibis* **108:** 348-93.

ROWLEY, I. AND SAUNDERS, D. A., 1980. Rigid wing-tags for Cockatoos. *Corella* **4:** 1-7.

SCHODDE, R., 1982. Origin, adaptation and evolution of birds in arid Australia. Pp. 191-24 *in* "Evolution of the flora and fauna of arid Australia" ed. by W. R. Barker and P. J. M. Greenslade. Peacock: Frewville.

SMITH, G. A., 1975. Systematics of parrots. *Ibis* **117:** 18-68.

SMITH, M. J. AND BRERETON, J. LE GAY, 1976. Annual gonadal and adrenal cycles in the Eastern Rosella, *Platycercus eximius* (Psittaciformes: Platyceroidae) *Aust. J. Zool.* **24:** 541-56.

WYNDHAM, E. AND BRERETON, J. LE GAY, 1982. Age classing and sexing Eastern Rosellas. *Corella* **6:** 89-91.

CHAPTER 13

The Social Organization and Foraging Behaviour of three Syntopic Thornbills *Acanthiza* spp.

H. L. Bell[1]

The Brown *Acanthiza pusilla,* Striated *A. lineata* and Buff-rumped Thornbills, *A. reguloides* were studied near Armidale, northeastern New South Wales.

Brown Thornbills lived as territorial pairs, whereas the other two species spent most of the year in clans of 10-20 birds. In the breeding seasons these separated into pairs or groups of a single female and two or three males. All members of a group helped to feed fledglings and nestlings of their group, and if their breeding attempts failed they assisted other groups of the same clan. Thornbills breed between July and December and success is low due to predators. Juvenile Brown Thornbills dispersed shortly after fledging, while in the two communal species dispersal occurred in spring, just before breeding.

Brown Thornbills gleaned from the foliage of shrubs, and Striated Thornbills from eucalypt foliage. Buff-rumped Thornbills fed on the ground and gleaned from bark, and to a lesser extent the foliage of shrubs and trees. Striated Thornbills often hung upside down while foraging, and occasionally hovered, while Brown Thornbills occasionally snatched insects from foliage. Buff-rumped Thornbills fed far more on the ground in autumn and winter, when they were in clans. The breeding pairs and groups fed mostly in trees, on leaves in spring and later on bark mostly. This seasonal change in foraging may reflect changes in abundance of insect prey on the different substrates, or alternatively larger groups may be less vulnerable to ground predators than are small groups.

Each species tended to retreat into its specialized niche when food became scarce. Consequently, overlap between any pair of species in foraging was inversely correlated with the abundance of arthropods. All species also fed more rapidly when food was scarce.

Striated and especially Buff-rumped Thornbills were nuclear species in mixed-species flocks outside the breeding season. Brown Thornbills sometimes joined these flocks. Each species altered its foraging when in a mixed flock and outnumbered by another thornbill species.

All three species show some morphological adaptations to their foraging specializations. Striated Thornbills have short beaks to pick insects, especially scales, off eucalypt leaves. Brown Thornbills have fine beaks to probe into the densely-packed foliage of shrubs, and Buff-rumped Thornbills have slightly decurved beaks to probe into fissures in bark.

INTRODUCTION

THE THORNBILLS *Acanthiza* spp, tiny 7-10 cm insect-eaters, are among the most prominent birds of the southern forests. They are particularly interesting because of the number of species, and their ecological diversity: while most take insects from the foliage others feed on the ground and at least one species (the Buff-rumped Thornbill, *A. reguloides*) forages off branches and bark. Although disagreement exists on the subject, there are probably twelve species of thornbills. Six are found in humid forests and woodlands of southeastern and far southwestern Australia, five are species of the arid interior and only two are completely tropical, being found in the mountains of north Queensland and New Guinea.

The present chapter is directed at an analysis of the basic biology, social organization and, particularly, feeding habits and breeding of three species, the Brown, *A. pusilla*, Buff-rumped and Striated Thornbills, *A. lineata*. In all these features they prove to be highly diversified. A final

[1]Department of Zoology, University of New England, Armidale, New South Wales 2351.
Pages 151-63 *in* BIRDS OF EUCALYPT FORESTS AND WOODLANDS: ECOLOGY, CONSERVATION, MANAGEMENT ed. by A. Keast, H. F. Recher, H. Ford and D. Saunders. Royal Australasian Ornithologists Union and Surrey Beatty & Sons.

discussion is directed at their role in the ecology of the forests.

STUDY AREA, MATERIALS AND METHODS

The study was from September 1978 to August 1981 (Bell 1983). The site was ca. 15 ha of fairly dense dry sclerophyll woodland (Beadle and Costin 1952) at Wollomombi Falls Reserve (30°32′S, 152°02′E; altitude 920 m), 42 km east of Armidale, NSW. The site was sharply demarcated on almost all sides by cleared lands and the cliffs and escarpments of the Macleay River gorge system. The woodland consisted mainly of *Eucalyptus caliginosa, E. laevopinea, E. melliodora, E. conica, E. viminalis* and *E. amplifolia* with a sparse but clumped understorey, mainly of *Acacia irrorata, Olearia viscidula, Cassinia* spp, *Jacksonia scoparia* and *Exocarpos cupressiformis*. Mean annual rainfall is 822 mm but rainfall was seriously deficient in the second and third years of study, which included the worst drought since 1902. Mean monthly maximum temperatures vary between 11 and 30°C, with mean minima dropping below freezing during June and July. At the height of the drought, in the winter of 1980, 37% of trees in the canopy, and 52% of shrubs in the understorey were either defoliated or dead, presumably from water-stress, and the foliage of the remainder was greatly thinned.

For foraging data the target each month was 750 foraging observations on 150 individuals (450 on 90 birds in the third year) for each species. Where possible observations were evenly divided between morning, mid-day and afternoon. Data for each observation included: individual (if identified), location, time, number of own or other species with which associated; and the main foraging data of: height, substrate, plant species foraged upon, foraging method used, speed of foraging and distance moved. Data were taken for five moves by each individual, a move being from one perch or position on a perch to another, or a sallying flight from and back to a perch. Heights recorded were: ground, 0-1 m, 1-2 m, 2-5 m, 5-10 m, 10-15 m and > 15 m. Substrates were: ground, trunks, limbs (i.e. branches thicker than the length of a thornbill), branches, leaves, air, and a number of minor categories included for analysis in those preceding. Foraging methods were: sallying (both prey and bird in flight), hovering, snatching (prey perched, bird flying), gleaning (both prey and bird perched), and hanging (bird upside down).

Colour-banding was aimed at establishing the social organization of each species and also any possible behavioural differences between individuals within a species. Continuous netting from June to August 1978 enabled almost all thornbills present to be colour-banded. From then on netting was intermittent, being specifically targetted at un-banded individuals observed in the area, or at young of the year, particularly those still with their parental groups. All observations of colour-banded birds were marked on a roll and in the event of an individual being missed, I made specific searches for it both in and adjacent to the study area, in order to establish whether it had in fact disappeared. Thus, the fortunes of individuals were accurately followed except for Striated Thornbills, whose colour-bands were often very difficult to identify accurately. Associations between colour-banded individuals were recorded. Sex of individuals was determined by a combination of methods, including wing measurements, behaviour and role in copulation.

Morphological data were taken in order to see if differences between species related to differences in their foraging ecologies. Data were taken from 86 live individuals, 17 skeletons and 58 museum specimens in the National Collection, CSIRO, Division of Wildlife Research, Canberra. Data taken were live weight; wing length; wing area; tail length; bill length, width, depth and curvature; foot span; front and rear toe length; front and rear claw length and curvature; length of femur, tibia, tarsus and total leg; and length of cranium.

Stomach samples of specimens taken under State permits were taken of approximately eighteen birds of each species. Samples were fairly evenly divided between 'warm' (October-March) and 'cool' (April-September) months, that is, periods of abundance and scarcity of insects. Direct observations were recorded, of food seen taken. Faecal samples, from netted birds, were also analyzed.

Abundance of arthropods was measured in order to establish relationships, if any, between behaviour and the availability of food. Abundance was measured for both the foliage and ground by monthly sampling. Each month 100 × 2 m sweeps were made along the ground with a 40 cm insect net, at ten sites randomly selected. Samples were taken from foliage of six plants: *Eucalyptus* spp (weighted according to frequency of main species), bipinnate *Acacia* spp, *Cassinia* spp, *Olearia viscidula, Jacksonia scoparia* and *Exocarpos cupressiformis*. These species represented 98% of all foliage present. About 2000 g of foliage was collected each month from randomly selected plants of each species, placed into plastic bags, sprayed with insecticide and weighed. Arthropods captured were later identified to order (family in some cases) and measured for length. Biomass was calculated using the length/weight correlations of Zug and Zug (1979) and Paton (1979). The final data were wet weight of arthropods/1000 g of foliage for each plant species and values for all plants combined were weighted in the proportions of canopy space occupied by each plant species present. Data were also available from Wollomombi for the abundance

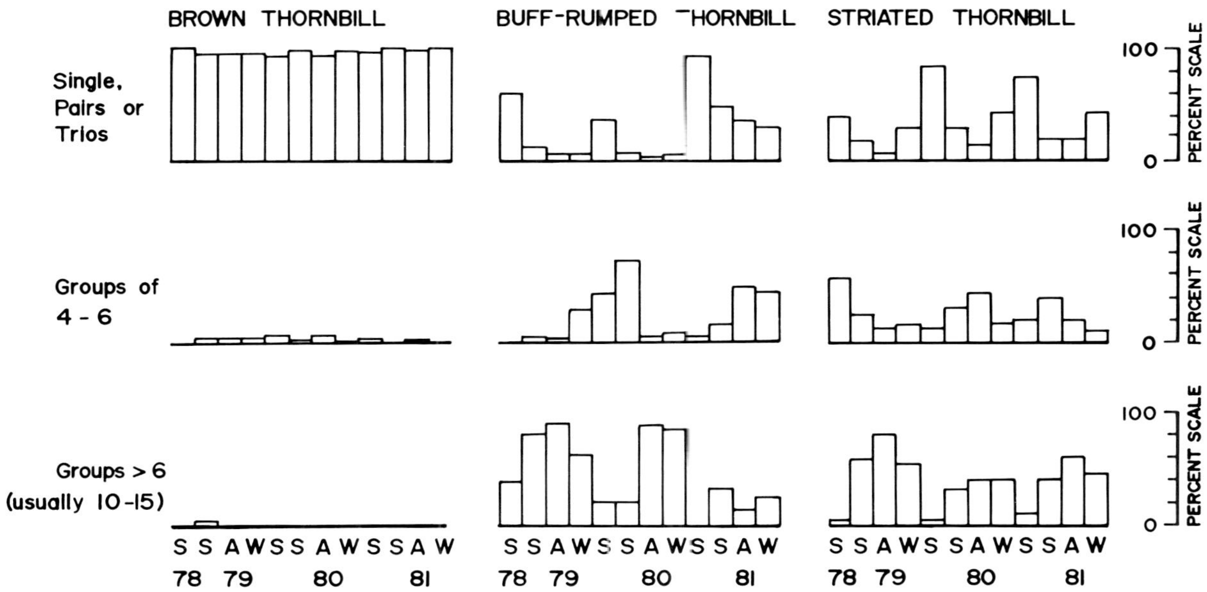

Fig. 1. Frequency of occurrences of Thornbills, by size of groups, by seasons of year, at Wollomombi 1978/81. (n = Brown Thornbills 4656, Buff-rumped Thornbills 4674, Striated Thornbills 5933). (Note that in 1980/81 insufficient Buff-rumped Thornbills survived to form groups > 6 birds).

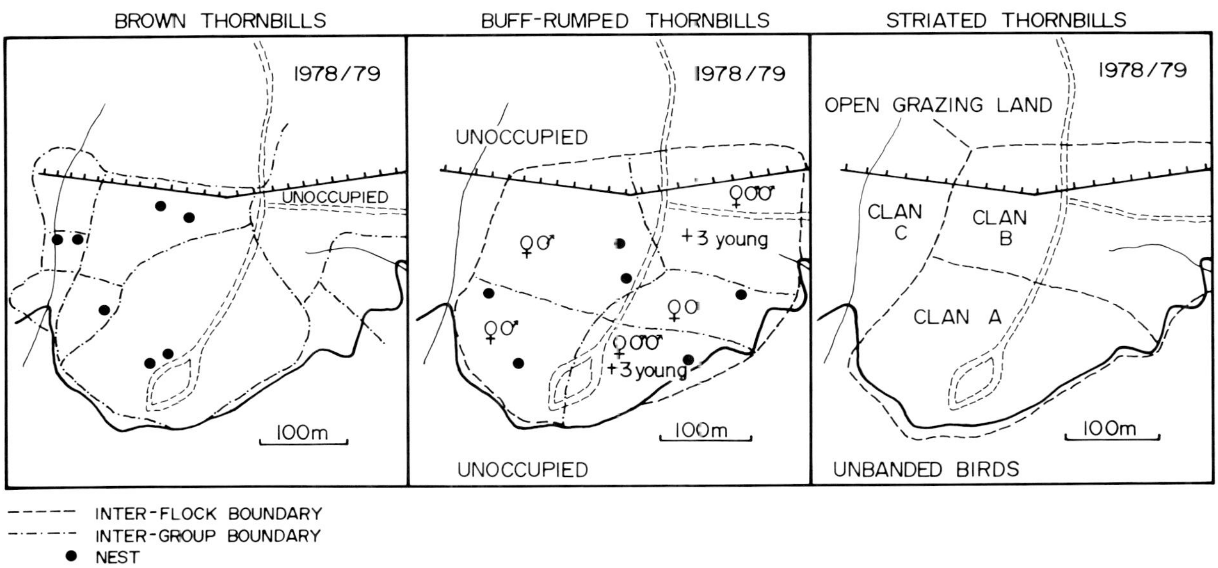

Fig. 2. Pair territories of Brown Thornbills, group territories within a clan of Buff-rumped Thornbills, and clan territories of Striated Thornbills, 1978/79. (Dots indicate nest sites).

of arthropods inside and on the surface of bark, from the studies of Noske (1983).

RESULTS

Social Organization

The three species had two types of social organization (Fig. 1). The Brown Thornbill occurred only as pairs, or pairs with attendant young. Each pair was permanent and occupied and defended its territory year-round. The young were evicted by their parents in autumn and the many unmarked birds seen at that period were probably also young evicted from elsewhere. These were prevented from settling by the residents. Sizes of territories differed greatly, from 0.23 to 3.7 ha (Fig. 2). Size of territory was correlated inversely with the percentage of a territory that was covered in understorey (Spearman Rank Correlation, r = -1.000, P < 0.01).

Buff-rumped and Striated Thornbills, in contrast to Brown Thornbills, occurred in clans of 10-20 birds, permanent in membership and which maintained a clan territory throughout the non-breeding period. The clan territory was ca. 13 ha for Buff-rumped and 5-9 ha for Striated Thornbills (Fig. 2). As shown in Figure 2 the clan territory was divided between breeding groups in spring. Each group consisted of one female and 1-3 males. Groups held the same breeding territory each year, defending it against the other groups with which they had been closely associated, within the clan, during the non-breeding period. When breeding concluded the clan re-formed. With Buff-rumped Thornbills, breeding groups that failed to fledge young helped to feed the fledglings of groups that were successful. During the non-breeding period the clan fed as a unit, always moving and roosting together.

Brown and Striated Thornbills bred from July to November and Buff-rumped Thornbills from September to December. Repeated attempts at nesting were made after failure but no pair or group nested again if young survived for at least two weeks after fledging. Only females incubated but both sexes, including helper males where present, shared equally in the feeding of nestlings and fledglings. Only females of Brown and Buff-rumped Thornbills built the nest. In some, but not all cases, the male Striated Thornbill participated in nest-building. Nesting success was low; only ten nests fledged young out of 51 in which eggs were found. Identifiable causes of failure were: wind (4), brood parasite (1), abandonment (adults still alive) (5), and predation (24). Abandonment may have been caused by brood parasites, because it usually followed the disappearance of one egg. Predation was witnessed four times. The Pied Currawong, *Strepera graculina* was seen to rob three nests (one with eggs, two with young) and an Olive-backed Oriole, *Oriolus sagittatus,* to rob a nest with young.

Almost all young of Buff-rumped Thornbills survived until the time of dispersal, in contrast to those of Brown Thornbills, the majority of which were predated a few weeks after fledging; with Striated Thornbills the picture was more obscure but survival was probably high.

Most young of the two communal species dispersed in spring, just before groups broke away from the clan to occupy breeding territories. With Buff-rumped Thornbills young females emigrated, some later being identified in neighbouring clans. Young males were added onto the parental breeding group. Recruitment of females into breeding groups was of individuals dispersing from other clans. However, after heavy mortality in the winter and drought of 1980, two young females replaced their (presumably) dead mothers in their parental group. There was no inter-change between groups within the clan, all immigration and emigration being outside the clan. However, one young male left his parental group, took as his mate a female recruited from outside the clan, and established a new territory within the clan. Although data for Striated Thornbills were less complete I believe that they behaved similarly to Buff-rumped Thornbills.

During the study, groups were more successful in breeding than were simple pairs. However, success seemed related entirely to freedom from predation. With Buff-rumped Thornbills, those groups with helpers were also those that nested in hollows and were not predated. Those that nested under bark on the sides of trunks were consistently predated and thus never raised any young males to add to the original pair.

Morphology

All three species weighed much the same. Brown Thornbills weighed 7.17 ± S.D. 0.58 g, Buff-rumped 7.12 ± S.D. 0.58 g and Striated 7.07 ± S.D. 0.44 g. This violates the theory of Hutchinson (1959), on the size-ratios of sympatric congeners, where co-existing species should differ, either in size or length of beaks, by at least a ratio of 1:1.2.

Striated Thornbills have significantly (χ^2, $P < 0.01$) blunter and shorter beaks than the other two species. The beaks of Brown and Buff-rumped Thornbills, approximately the same length, differ in that the Brown has a straight narrow needle-like beak, whereas that of the Buff-rumped has a pronounced curve. The tongues differ in that the Buff-rumped Thornbill's is more strongly bifurcated than that of the other two species, and the Striated Thornbill's tongue has a thick fringe at the rear.

The aerodynamic qualities also differ. The Brown Thornbill has a significantly lower (χ^2, $P < 0.005$) tail/wing ratio. The Striated Thornbill has the highest wing-loading and aspect ratio.

Striated Thornbills have significantly shorter (χ^2, $P < 0.05$) legs. Brown and Buff-rumped Thornbills have legs of similar length. Buff-rumped and Striated Thornbills have significantly more (χ^2, $P < 0.05$) curved claws than Brown Thornbills. Buff-rumped Thornbills have the significantly widest (χ^2, $P < 0.01$) foot-span. Tests made by me of grasping strength showed that Striated Thornbills had a much stronger grip (Wilcoxson's Sum of Ranks Test, $P < 0.002$) than the other species.

Diet

The composition of stomach samples was different from food seen taken by birds. Of insects seen taken by thornbills, 40% were lepidopteran larvae. Despite a diligent search for the mouth-parts of caterpillars, larvae of all kinds made up only 2-13% of items in stomachs. Striated Thornbills were seen to scrape their beaks on the blades of eucalypt leaves, apparently taking psyllids, but because lerps are quickly dissolved, few were found in stomachs.

The composition of stomach samples was similar for all species except that Buff-rumped Thornbills took more ants than the other species did. Also, Brown and Buff-rumped Thornbills took seeds, which Striated Thornbills did not. Thornbills discriminated in the size of prey taken. Most beetles in foliage samples were in the 1-2 mm size range whereas very few less than 3 mm in length were found in stomachs. Similarly, most ants in foliage samples were 3-4 mm long but very few below 5 mm were found in stomachs. During cooler months (i.e. in times of scarcity) insects of smaller sizes were taken and the variety of insects taken was greater than at times of abundance. Overlaps between species in the composition of stomach contents generally decreased in times of scarcity. Overlaps were, in warmer months: Brown/Buff-rumped 72%, Brown/Striated 83%, Buff-rumped/Striated 78%; in cooler months they were Brown/Buff-rumped 76%, Brown/Striated 61% and Buff-rumped/Striated 42%.

The increased diversity in items taken during cooler months, particularly of seeds and ants, suggests that there may be a need at such times to take less suitable or profitable items. Although ants made up no greater proportion of insects in wintertime compared with summer, they constituted a greater proportion of the prey taken in winter than in summer (for Brown Thornbills 20 vs 0%, for Buff-rumped Thornbills 22 vs 12%).

Foraging Behaviour

The heights and substrates from which thornbills foraged are depicted in Figure 3. The Brown Thornbill foraged mainly in the understorey, and chiefly off foliage. The Buff-rumped Thornbill foraged equally from the ground, understorey and main canopy, but foraged mainly off bark surfaces. Striated Thornbills foraged mostly in the tree canopy and almost entirely off leaves.

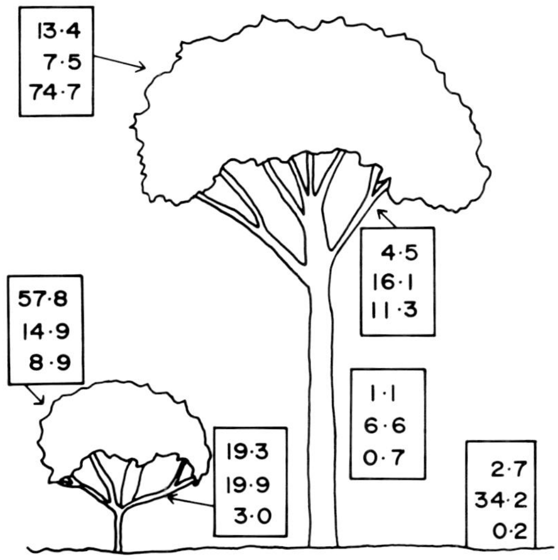

Fig. 3. Percentage foraging by Brown (top), Buff-rumped (centre) and Striated (bottom) Thornbills on substrates (no. of foraging moves: Brown 24,369; Buff-rumped 23,536; Striated 22, 232).

The thornbills foraged mainly by gleaning prey off a substrate on which they themselves were perched. They seldom sallied after aerial prey (ca. 1% of all moves; see Fig. 3 for total n). Striated Thornbills hovered significantly more (3% of all moves) than the other two species. Striated Thornbills hung more frequently (21%), Brown Thornbills less so (8%) and Buff-rumped Thornbills virtually never. Brown Thornbills snatched more (6%) than the other two. Although all species mostly gleaned, such differences in foraging method as they showed may have been sufficient to allow each species access to resources not so readily available to the others.

The three thornbills showed differences in the plant species on which they foraged, even when foraging at the same height above ground (Fig. 4). Striated Thornbills mainly frequented the canopy and thus foraged mainly on eucalypts, showing a slight preference for stringybarks and boxes over gums. They also fed on the taller shrubs like bipinnate *Acacia* and *Exocarpos*. Brown Thornbills, which foraged mainly in the shrub layer, foraged in bipinnate *Acacia* (40% of all foraging) and the very scarce *Exocarpos* to an extent much greater than the proportions of those species in the shrub layer. On the other hand, Buff-rumped Thornbills, when feeding in the shrub layer, foraged in *Jacksonia* and *Olearia* to a greater extent than the proportions of those species present.

Figure 5 shows that not only did the three species of thornbill forage substantially on different species of plants; they also used different substrates when foraging on the same species. Unlike the other two species the Buff-rumped Thornbill foraged more on bark than on foliage, except when foraging on *Jacksonia* and boxes. Only the Buff-rumped Thornbill foraged to a great extent on *Jacksonia* (10% of all foraging). However, when on *Jacksonia* it foraged off the leaves (in fact, cladodes), whereas on all other understorey shrubs it foraged mainly off bark. Buff-rumped Thornbills also foraged off the bark of stringybarks to a much greater degree than on other species of plants. The data of Noske (1983) showed that arthropods are abundant in the deeply fissured bark of stringybarks, particularly in winter. But why Buff-rumped Thornbills forage mostly off bark when on boxes, is unclear. Foliage samples showed similar abundances of insects on the leaves of both types of eucalypts. The other two species of Thornbills foraged on leaves of the different types of eucalypts in the same proportion as each type's occurrence.

Thornbills foraged more intensely, i.e. with less time between foraging moves, during periods when arthropods were scarce (Spearman Rank Correlation for intensity of foraging and abundance of arthropods; Brown $r = -0.509$, $P < 0.01$; Buff-rumped, $r = -0.501$, $P < 0.01$; Striated, $r = -0.423$, $P < 0.05$). They also moved further per move at times of scarcity (Spearman Rank Correlation; Brown $r = -0.510$, $P < 0.01$; Buff-rumped, $r = -0.458$, $P < 0.05$;

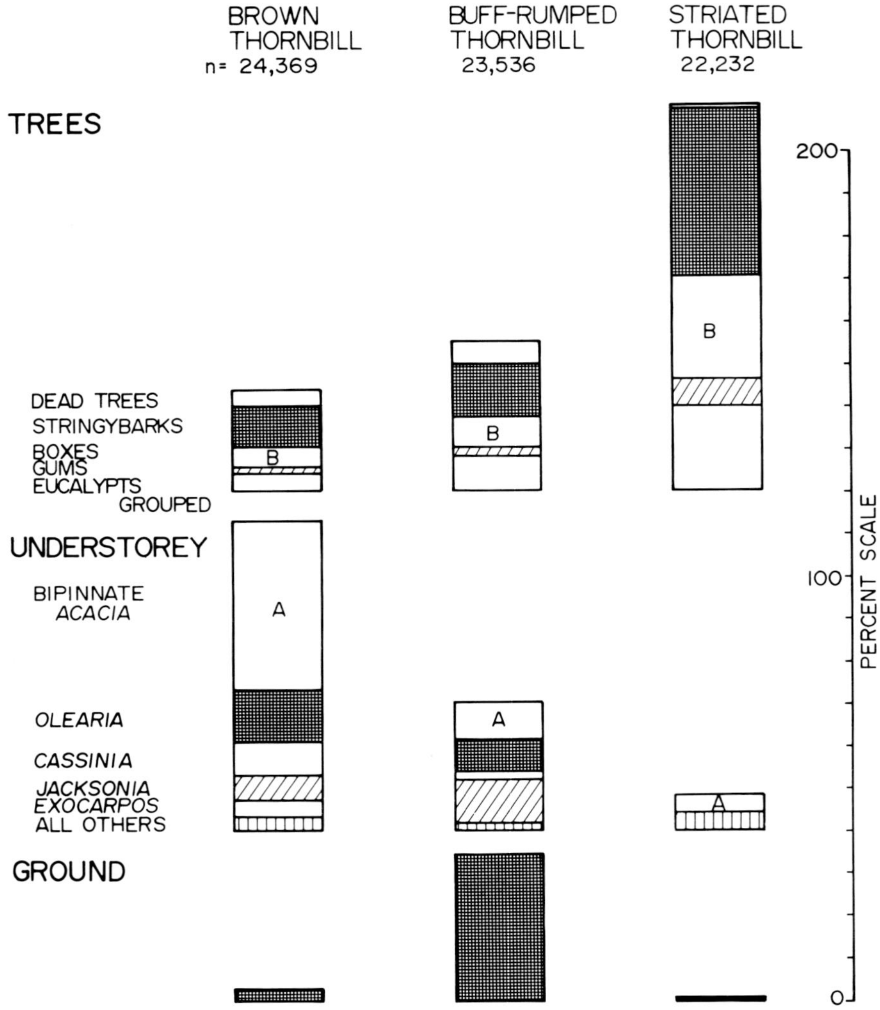

Fig. 4. Percentage foraging by thornbills on different plant species 1978/81. (Note that eucalypts were not differentiated in the first eight months, so are grouped).

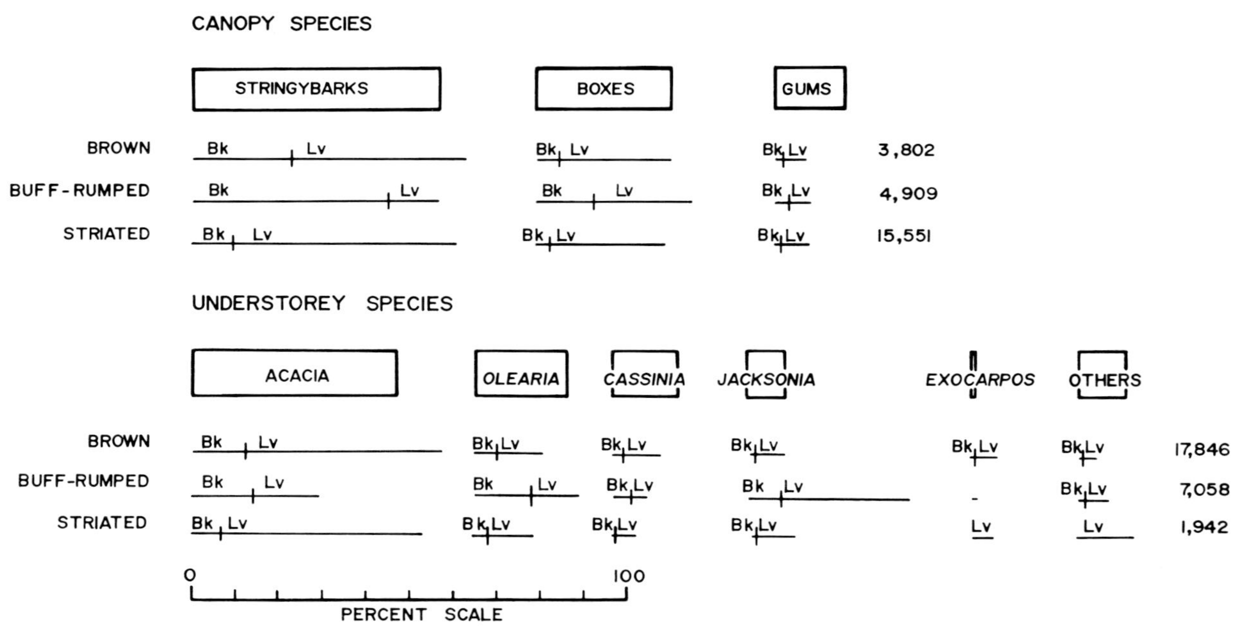

Fig. 5. Proportions of foraging, by substrate and plant species, of thornbills in canopy and understorey vegetation at Wollomombi (Bk: bark, Lv: foliage; numbers are foraging moves).

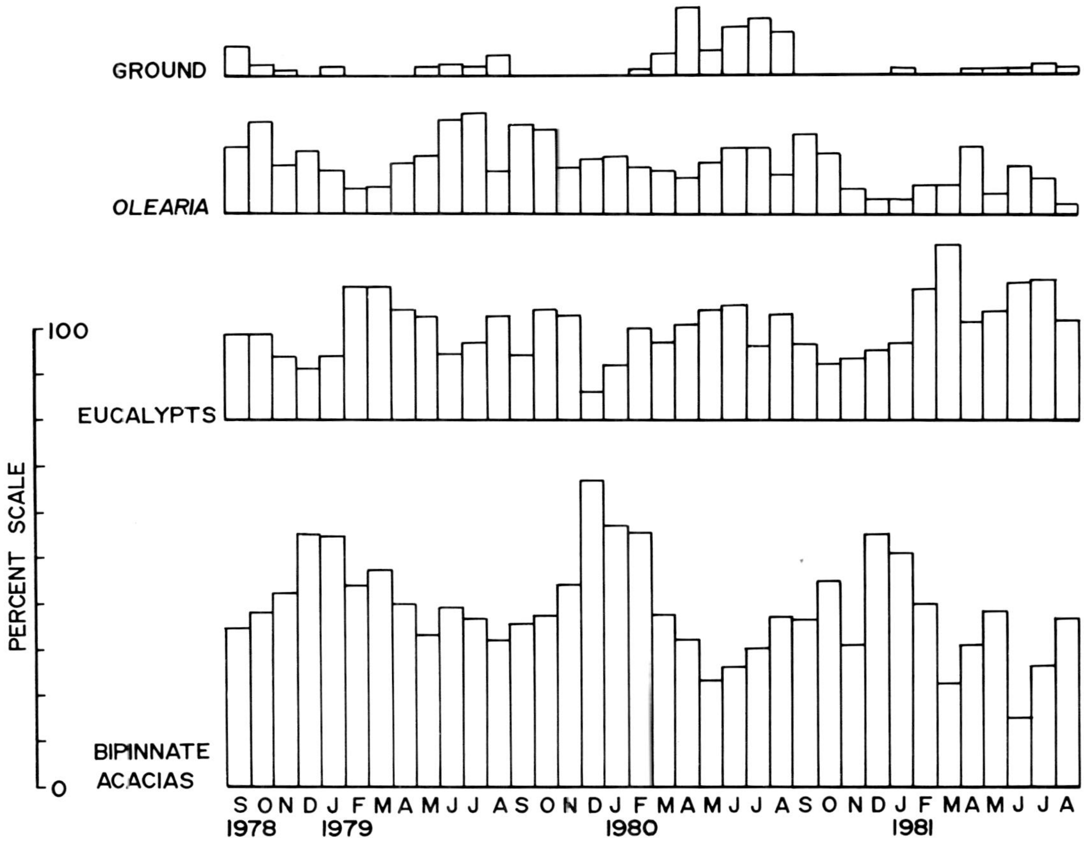

Fig. 6. Percentage frequency of foraging, by months, of Brown Thornbills on *Acacia,* eucalypts, *Olearia* and on the ground, 1978/81.

Striated r = -0.380, P < 0.05). Brown and Striated Thornbills, but not Buff-rumped, also foraged more intensely during mornings than at mid-day or afternoons. This may possibly be dictated by the need to quickly replace energy lost during the night and because insects, less active at cooler temperatures, were less easy to find in the mornings.

Seasonal Variations in Foraging Behaviour

The overall pattern of foraging by height, substrate, plant species and foraging method, for each species, was not necessarily the same year-round. Each species showed patterns of apparent seasonal variation in foraging behaviour. For example, Brown Thornbills changed to different plant species in autumn and winter, feeding less on bipinnate *Acacia* and more on *Olearia*, eucalypts and on the ground (Fig. 6). Samples of arthropods at such times showed no greater decline in abundance on *Acacia* than on other plant species; in fact, the greatest decline was on eucalypts. Perhaps it is harder to locate insects in the foliage of bipinnate *Acacia* at low light intensities. The marked switch to ground-feeding in the winter of 1980 may have been related to the drought-induced defoliation, then at its height.

Buff-rumped Thornbills showed an annual change from ground to arboreal foraging (Fig. 7). In autumn and winter Buff-rumped Thornbills foraged mostly on the ground but at the start of the breeding season (September) there was both a reversal from ground to arboreal foraging, and a bias towards leaf-foraging, rather than bark. This switch to leaf-foraging may relate to a spring flush of larvae, but no such flush occurred in 1980. A more likely cause of a change to arboreal foraging is the seasonal change in social organization, referred to earlier. At the start of the breeding season the clans of Buff-rumped Thornbills broke up into pairs or breeding groups. Therefore foraging by the species was in parties of four or fewer birds, compared to the 10-15, plus other species following, during the non-breeding period. Small parties tended to feed higher than large flocks (Spearman Rank Correlation, size of party and height above ground, r = -0.965, P < 0.01) (Fig. 3). The preference for ground-foraging in winter seems unrelated to abundance of food, because grass-sweeps showed a winter decline in arthropods similar to that on foliage. Other studies (e.g. Goldman 1980; Caraco *et al.* 1980) have shown experimentally that the larger a flock, the safer it is from predation. Thus, the break-up of the clan in spring may dictate a switch to arboreal foraging for reasons of safety.

Striated Thornbills showed a small but regular change to foraging in the understorey each spring

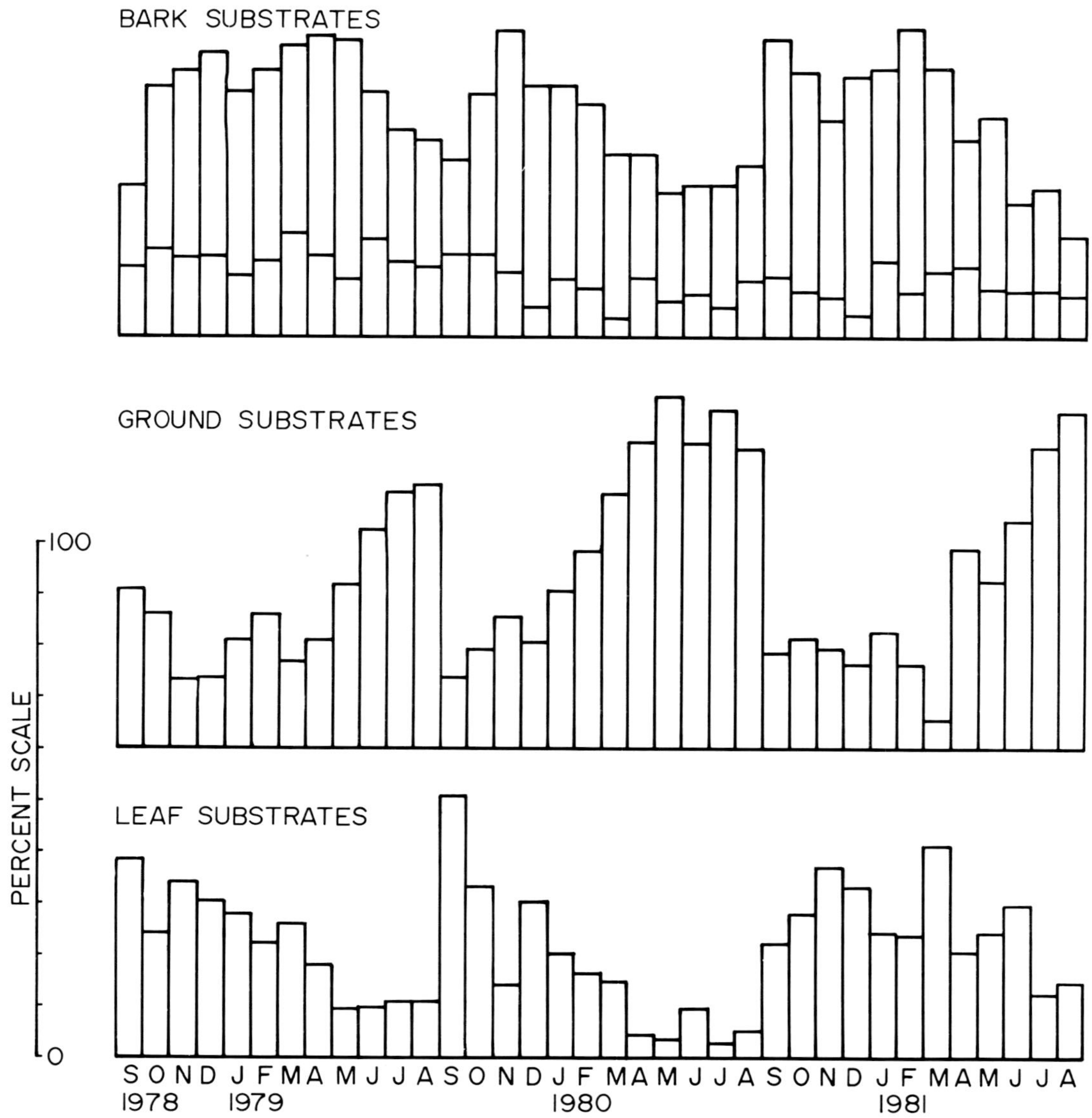

Fig. 7. Percentage frequency of foraging, by months, of Buff-rumped Thornbills on bark (branches-upper, trunk-lower), leaves and ground, 1978/81.

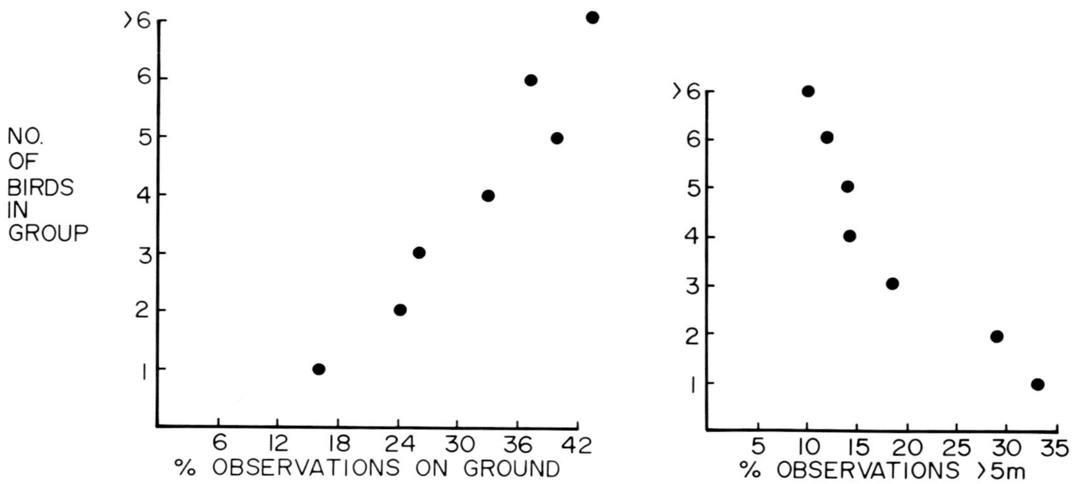

Fig. 8. Correlation between size of party of Buff-rumped Thornbills and amount of foraging and height from ground (n, individual moves: single birds 850, pairs 4426, trios 1176, quartets 2592, quintets 1406, sextets 1777, > six birds 9810).

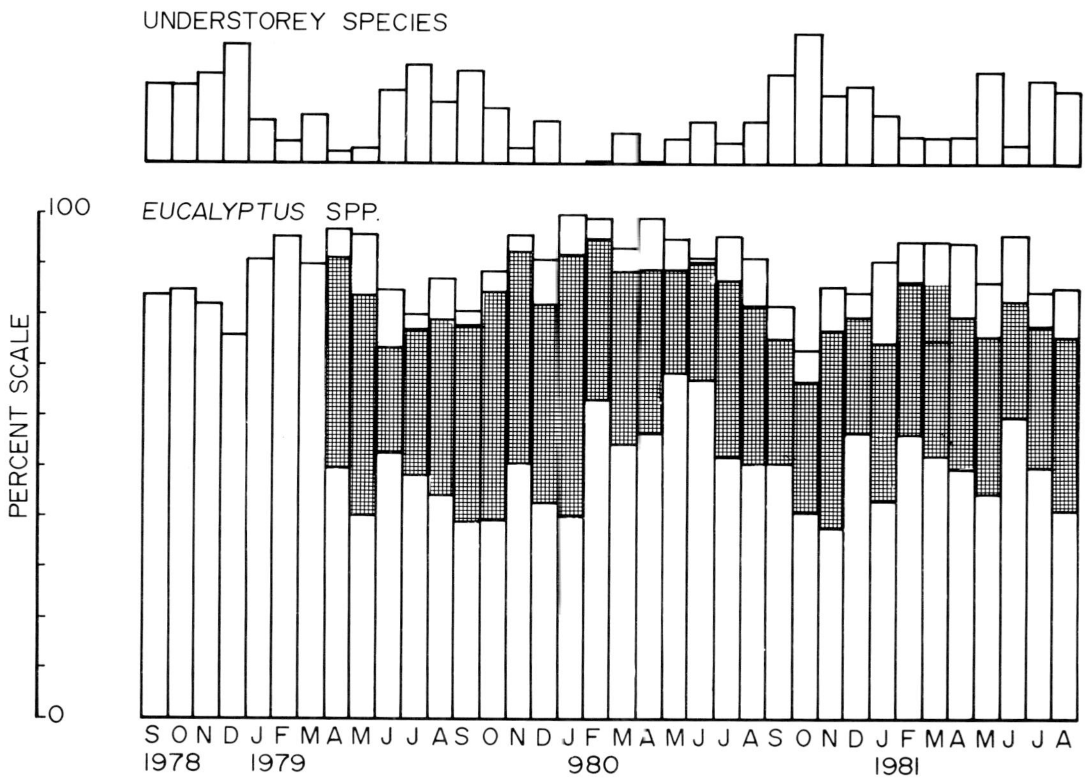

Fig. 9. Percentage frequency of foraging, by months, of Striated Thornbills on understorey plants and eucalypts, 1978/81. Eucalypts are separated into gums (top), stringybarks and boxes (bottom).

(Fig. 9). Smaller parties fed lower than larger parties (Spearman Rank Correlation, size of party and height from ground, r = 0.943, P < 0.05) (Fig. 10). The clans broke up into their breeding groups at this time and smaller groups are perhaps more vulnerable to predation in the extreme top of the eucalypt canopy, normally frequented by the large non-breeding clans of Striated Thornbills.

Relations between Changes in Foraging Behaviour and Abundance of Arthropods

The seasonal changes referred to earlier were probably, in many cases, related to changes in the abundance of food. Unfortunately, although the data on arthropods give values per unit of foliage, there are no data for the change, over time, in actual total volume of foliage. (One does not know of a drought until it is well-advanced.) Thus, values for abundance, over the three years, may not accurately reflect what food was available. However, in the first year, 1978/79, the pattern was 'normal', i.e. with summer peaks and winter troughs, since repeated after the drought broke.

Figure 11 shows the abundance of arthropods in 1978/79, and compares it (Spearman Rank Correlation) with various types of foraging which each species used significantly more than the others. Brown Thornbills foraged more in the dense understorey when food was scarcest (food with foraging at 0-2 m, r = -0.504 P < 0.01). Buff-rumped Thornbills foraged on trunks more when food was scarcest (r = -0.534, P < 0.01) and similarly on the ground (r = -0.566, P < 0.01). Striated Thornbills hovered more in times of scarcity (r = -0.543, P < 0.01) but hung significantly more in times of abundance (r = 0.644, P < 0.01).

In most cases, except for understorey foraging by Brown Thornbills, and ground-foraging by Buff-rumped, these specialized types of foraging were not a major part of each species feeding activity. However, each employed these specializations more than the others and used them most, except for hanging by Striated Thornbills, in times of scarcity.

Participation in Mixed-Feeding Flocks

Thornbills were regular participants in mixed-species flocks of insectivorous birds. Participation was year-round but fell off markedly during the breeding season. Buff-rumped and Striated Thornbills were nuclear species; that is, those around which other species gathered. Brown Thornbills, conversely, were usually followers. Buff-rumped Thornbills were followed more by other species (549 times vs 39 times following other species), than either Striated (218 vs 82 times), or Brown Thornbills (156 vs 228 times). Among the Thornbills, Brown Thornbills followed Buff-rumped and Striated, and Striated followed Buff-rumped Thornbills (Fig. 12).

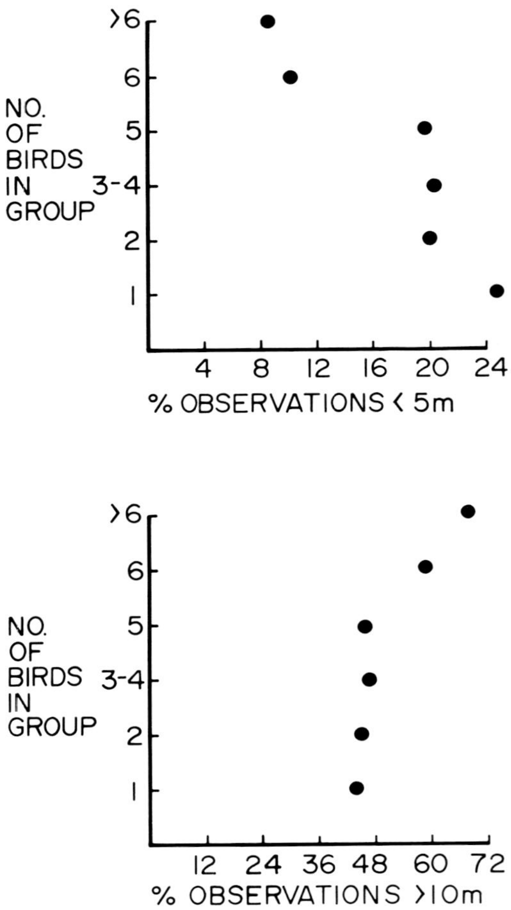

Fig. 10. Correlation between size of party of Striated Thornbills and height from ground (n, individual moves: single birds 1913; pairs 4004; trios and quartets 1960; quintets 682; sextets 687, > six birds 2392).

In some months up to 60% of Brown and Striated Thornbills and 80% of Buff-rumped Thornbills seen were in mixed-species flocks. Buff-rumped Thornbills tended to attract species that fed on the ground or in the understorey. Striated Thornbills attracted mainly species of the canopy or themselves followed Buff-rumped Thornbills or Varied Sittellas *Daphoenositta chrysoptera*. Brown Thornbills usually followed flocks of other Thornbills, Superb Fairy-wrens, *Malurus cyaneus,* or White-browed Scrub-wrens, *Sericornis frontalis*.

Thornbills, when in mixed-species flocks, often changed their foraging behaviour. This varied not only according to with what other species they associated, but also because of the numbers of individuals of each species involved (Fig. 13). Brown Thornbills foraged more on the ground when with the ground-foraging Buff-rumped Thornbill, but foraged higher when with canopy-foraging Striated Thornbills. The two communal species did not change their behaviour when with Brown Thornbills. Flocks of Buff-rumped Thornbills changed their behaviour little when with flocks of Striated Thornbills, but the Striated Thornbills foraged lower than normal during such associations. However, when only a few (1-3) Buff-rumped or Striated Thornbills joined a flock of the other species their foraging height conformed greatly to that species.

Some species appeared, when in mixed-species flocks, to differ in the plant species on which they foraged (Fig. 14). When feeding in the lower understorey, Brown Thornbills foraged off *Jacksonia* much more frequently when with Buff-rumped Thornbills than when on their own. Because *Jacksonia* was evenly distributed throughout the study area, and because only the Buff-rumped Thornbill forages off it to any great extent (see Fig. 4), the actions of the Brown Thornbill may be imitative foraging. Similarly, Striated Thornbills often joined sittellas, which foraged almost exclusively on stringybarks. Although overall, Striated Thornbills foraged in stringybarks in the same frequency as their proportion of the canopy (56%), when with sittellas they foraged 75% of the time in stringybarks.

DISCUSSION

The morphological differences between the three species seem related to those of their foraging habits. Striated Thornbills, foraging mostly off the blades of eucalypt leaves, have little need to probe crevices; thus they have stout and short beaks. Conversely, Brown Thornbills, which probe into densely-packed foliage such as bipinnate *Acacia, Cassinia* and *Exocarpos*, have longer needle-like beaks, more suitable for probing (Newton 1967). Of the three species only the Brown Thornbill has adapted to foraging in exotic conifers (e.g. Pawsey 1966). The bark-foraging Buff-rumped Thornbill also has a long, but curved beak, which enables it to probe places inside crevices which a straight beak might not reach. The differences in tongues are less easy to relate to foraging. However, of the three treecreepers *Climacteris* spp. at Wollomombi, (Noske 1979), that with the most deeply bifurcated tongue, the Brown Treecreeper, *C. picumnus,* is, like the Buff-rumped among thornbills, that which forages most on the ground and takes most ants.

The lower tail/wing ratio of the Brown Thornbill is considered an aid to manoeuvrability in flight (Hartman 1961), a necessary ability for the only one of the three species that forages amongst the dense foliage of the understorey. The high wing-loading and aspect ratio of the Striated Thornbill are considered to be aids to hovering (Partridge 1976). Striated Thornbills hovered more frequently than the other species and hovering ability would allow greater access to pendant leaves of eucalypts.

The Striated Thornbill has the shortest legs, to be expected of a mainly arboreal species, whereas the

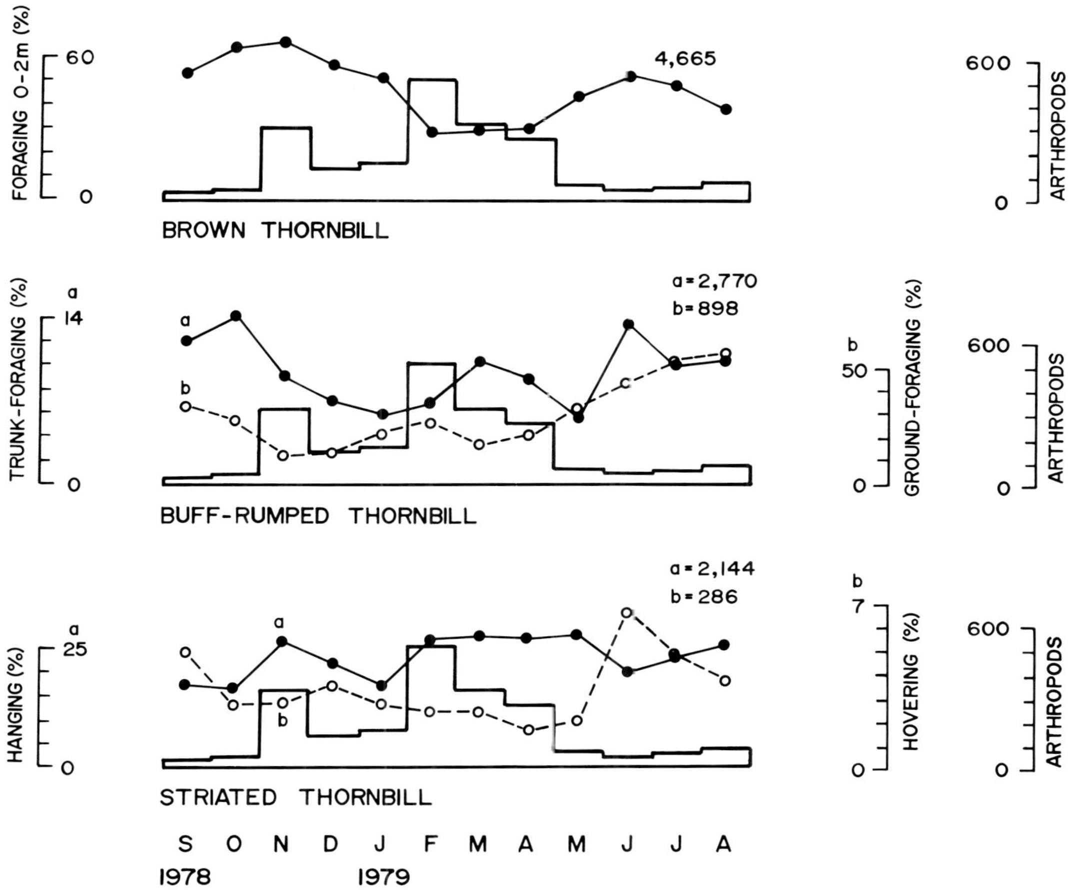

Fig. 11. Abundance of arthropods 1978/79 (mg wet weight/1000 g of leaves) compared with: Brown Thornbills, foraging in lower understorey 0-2 m; Buff-rumped Thornbills a. trunk-foraging b. ground-foraging; Striated Thornbills a. hanging b. hovering. (n: foraging moves; percentage is that of total monthly foraging moves).

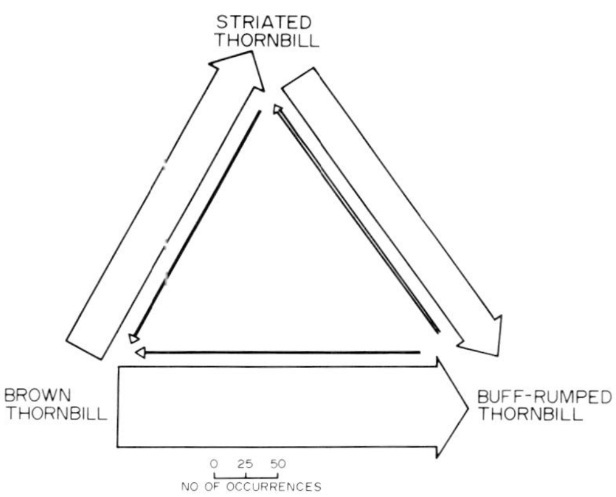

Fig. 12. Occurrences of species of thornbill following other species of thornbills at Wollomombi.

other two species had legs of similar length, even though one would expect the ground-foraging Buff-rumped Thornbill to have the longest (Promptov 1956).

The more curved claws of both Buff-rumped and Striated Thornbills seem related to the need of both species to cling to vertical surfaces, trunks in the case of Buff-rumped and pendant eucalypt leaves for Striated Thornbills (Norberg 1979). The wider foot-span of the Buff-rumped Thornbill is also an aid to perching on vertical surfaces (Norberg 1979), and the strong grasping-strength of the Striated Thornbill seems related to it hanging more than the other species.

However, the specialized foraging techniques, which are related to the morphological differences between species, are not the most used techniques. The three species are all mainly simple gleaners. They use their specialized techniques for only a small proportion of their total foraging. The important point is that each resorts to its specialized technique at times when food is scarcest. The only exception to this was the Striated Thornbill hanging most when food was abundant. I suspect, with no direct evidence, that the high incidence of scale insects on eucalypt leaves, during warmer months, prompts Striated Thornbills to hang onto the tips of leaves in order to gain access to the insects. Otherwise, it seems that each species, all of which basically forage in the same way, diverge in foraging habits at times of scarcity just sufficiently to gain access to additional food not so readily available to the other species. The increased diversity in the composition

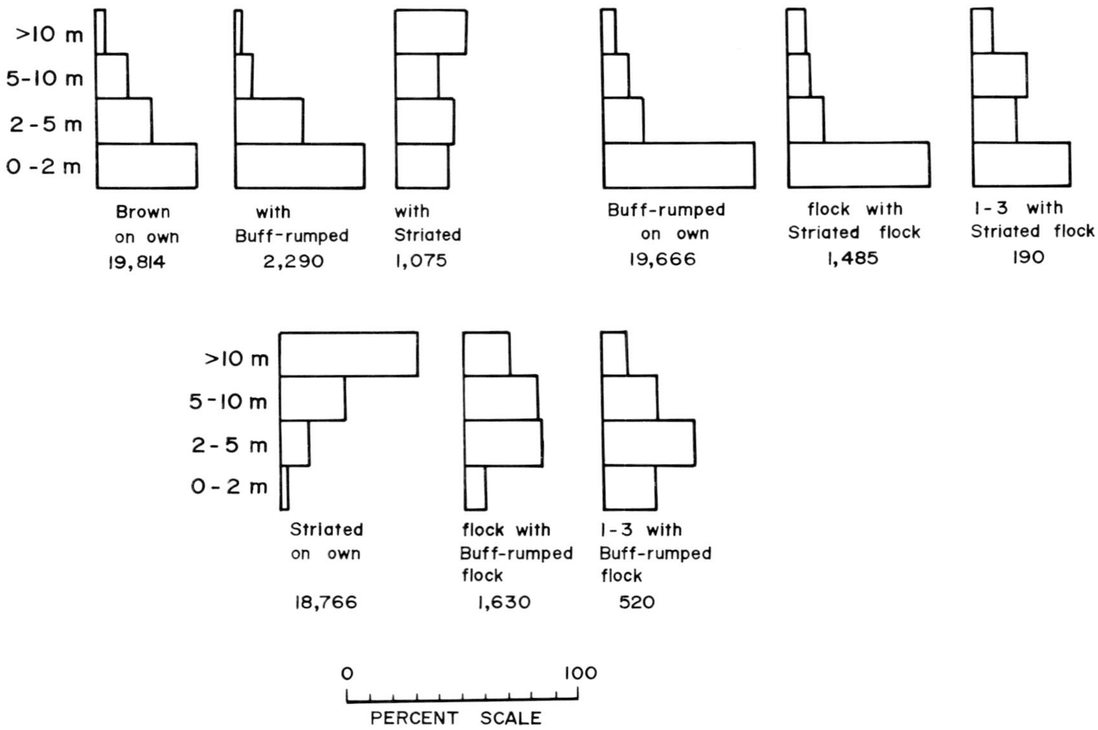

Fig. 13. Comparisons of foraging by thornbills when in and not in mixed-species flocks (n: foraging moves).

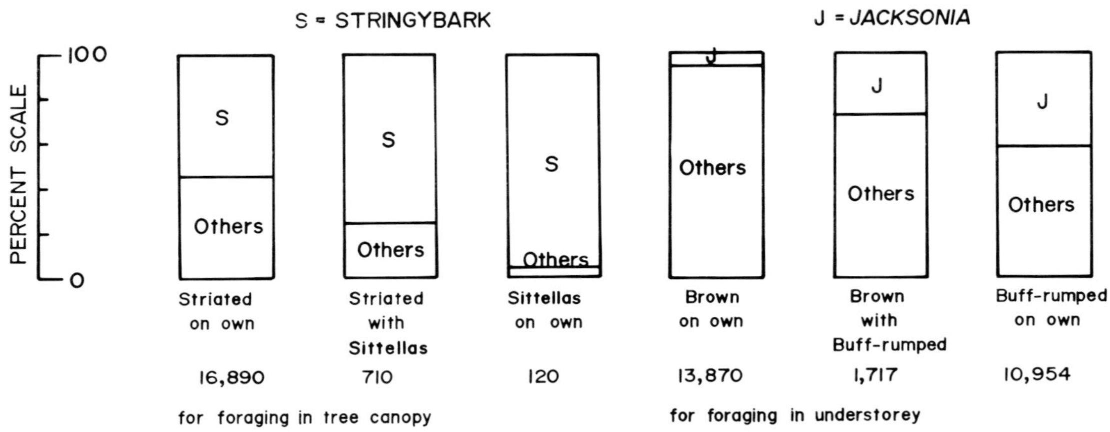

Fig. 14. Comparisons of foraging by thornbills on plant species when in and not in mixed-species flocks (n: foraging moves).

of diets and the reduced overlap in diets, in cooler months, i.e. times of scarcity, reinforces the evidence for specialization by each species at such times. This supports the contention of Schoener (1982) that 'during lean times, strong directional selection resulting from interspecific competition produces in each case adaptability most suitable for resources used relatively exclusively'.

The two-tiered social organization of the Buff-rumped and Striated Thornbills will probably be found in all gregarious species of thornbills, i.e. save the *pusilla* superspecies. Sufficient evidence was obtained in this study on the Yellow-rumped Thornbill, *A. chrysorrhoa*, to indicate that it too, had a social organization similar to Buff-rumped Thornbills. The advantages of the two systems is open to question. In the two gregarious species breeders have: advantages of helpers in the breeding group, but not too many to attract attention to the nest; further helpers when the clan re-forms; advantages from group-living of better detection of predators and access to clumped food resources; and the retention and survival of young until the following breeding season. However, such social organizations could inhibit colonization of new or ephemeral habitat, at which Brown Thornbills seem more successful. With Brown Thornbills the parental pair do not have to share resources with their young during winter; the young disperse early enough to find a vacant territory well before the next breeding season; a dispersing young needs to find only one mate in order to establish a breeding unit; the sex ratios of dispersed young are, unlike in the gregarious species, probably equal; and a 'new' pair needs less territory than a group in which to establish itself.

Only the monogamous *pusilla* group occurs in rainforest and may thus be the oldest species of the genus. Although there are many communal species of birds in rainforests one would imagine that group cohesion would be difficult to maintain in denser vegetation. Moreover, the communal species of rainforests do not have the multi-tiered social organization of *Acanthiza*. Additionally, Australian rainforests have probably long been fragmented, with even single pairs of Brown Thornbills occupying an isolated patch. The communal organization of Buff-rumped, Striated, and probably most other thornbills may be connected to the large areas of homogeneous habitat available, with little need to colonize or re-colonize isolated patches, and to the ease of maintaining social cohesion in more open habitat.

REFERENCES

BEADLE, N. C. W. AND COSTIN, A. B., 1952. Ecological classification and nomenclature. *Proc. Linn. Soc. NSW* 77: 61-82.

BELL, H. L., 1983. Resource-partitioning between three syntopic thornbills (Acanthizidae: *Acanthiza* Vigors and Horsfield). Unpubl. Ph.D. Thesis, University of New England, Armidale, NSW.

CARACO, T., MARTINDALE, S. AND PULLIAM, H. R., 1980. Avian flocking in the presence of a predator. *Nature* **285**: 400-01.

GOLDMAN, P., 1980. Flocking as a possible predator defense in Dark-eyed Juncos. *Wilson Bull.* **92**: 88-95.

HARTMAN, F. A., 1961. Locomotor mechanisms of birds. *Smiths. Misc. Coll.* **143**: 1-91.

HUTCHINSON, G. E., 1959. Homage to Santa Rosalia, or why are there so many kinds of animals? *Am. Nat.* **93**: 145-49.

NEWTON, I., 1967. The adaptive radiation and feeding ecology of some British finches. *Ibis* **109**: 33-98.

NORBERG, U. M., 1979. Morphology of the wings, legs and tail of three coniferous forest tits, the goldcrest and the treecreeper, in relation to the locomotor pattern and feeding station selection. *Proc. R. Soc. Lond.* **287**: 131-65.

NOSKE R. A., 1979. Co-existence of three species of treecreepers in northeastern New South Wales. *Emu* **79**: 120-28.

NOSKE R. A., 1983. Comparative behaviour and ecology of some Australian bark-foraging birds. Unpubl. Ph.D. Thesis, University of New England, Armidale, NSW.

PARTRIDGE, L., 1976. Some aspects of the morphology of Blue Tits (*Parus caeruleus*) and Coal Tits (*Parus ater*) in relation to their behaviour. *J. Zool. Lond.* **179**: 121-33.

PATON, D. C., 1979. The feeding ecology of the New Holland Honeyeater, *Phylidonyris novaehollandiae*. Unpubl. Ph.D. Thesis, Monash University, Melbourne.

PAWSEY C. K., 1966. Birds in relation to pine forests of the south-east of South Australia. *S. Aust. Orn.* **24**: 93-95.

PROMPTOV, A. N., 1956. [Essays on the problems of biological adaptiveness of the behaviour of passerine birds.] Moscow: Acad. Sci. USSR. [In Russian]; cited in Partridge 1976.

SCHOENER, T. W., 1982. The controversy over ecological competition: an inside view. *Am. Scientist* **70**: 586-95.

ZUG, G. R. AND ZUG, P. B., 1979. The Marine Toad, *Bufo marinus*: a natural history résumé of native populations. *Smithsonian Contributions to Zoology* **284**.

Spotted Pardalote, *Pardalotus punctatus*

CHAPTER 14

Foliage-Gleaners of the Treetops, the Pardalotes

John Woinarski[1]

The pardalotes are a family of four species of small birds which glean invertebrates from the foliage of eucalypts. The three species occurring in southeastern Australia are considered here. The diet of these species consists mostly of lerps, the sugary exudate of some psyllids. Lerp-forming psyllids are the most abundant arthropod group occurring on *Eucalyptus* foliage. They may also have dramatic population fluctuations. Pardalotes respond to these characteristics by restricting foraging to *Eucalyptus* species, being highly mobile, and having the potential for a high reproductive output relative to other Australian leaf-gleaning birds.

Lerps are also important in the diet of some honeyeaters (but not for most acanthizid warblers), and these may defend areas with high density of lerps. Pardalotes may lose a substantial amount of time and energy through being chased by such honeyeaters. Competition with the Striated Pardalote may be contributing to the recent rapid decline of the Tasmanian endemic species, the Forty-spotted Pardalote. The nesting habits of these two species and the Spotted Pardalote are compared.

INTRODUCTION

IN ALMOST ALL forests and woodlands throughout the world there occur species of small birds which forage mainly for insects among the foliage of trees and shrubs. In Australia these species come from two families — the Acanthizidae and Pardalotidae. Although some species of acanthizid warblers are found in New Zealand, New Guinea and Southeast Asia, most of the Acanthizidae, and all pardalotes, are Australian endemics. The leaf-gleaners in these families comprise four genera: the thornbills *Acanthiza*, the gerygone warblers *Gerygone* and the weebill *Smicrornis* (Acanthizidae) and pardalotes *Pardalotus* (Pardalotidae). These species are among the smallest Australian birds, with weights ranging from 6 to 12 g, and form an assemblage distinct from other, larger, leaf-gleaning species in Australia (mainly the honeyeaters (Meliphagidae) and whistlers). In this chapter I examine the resources available to these birds in forests of southeastern Australia, discuss how this availability influences the ecology of the pardalotes and acanthizid warblers, and compare the ecology of the pardalote species with that of acanthizids. Most of the information reported here comes from a three-year study based on 16 sites in Victoria (chosen to represent a broad range of forest and woodland habitats), and a three-month study in southeastern Tasmania.

PARDALOTES

1. *The group*. Pardalotes are of uncertain origin and their closest relatives are debatable. For long they have been placed with the flower-peckers (Dicaeidae) because of similarities in size, bright colours (unusual for small leaf-gleaners), nine (rather than the usual ten) primaries in the wing, and general body shape (Mayr and Amadon 1947; Salomonsen 1961a). More recent evidence, of egg-white protein and DNA hybridization (Sibley 1976; Sibley and Ahlquist 1980), suggests that these similarities are superficial, and that instead the pardalotes are part of an old Australian group including the honeyeaters and acanthizid warblers.

2. *The species*. The definition of species within the pardalotes has also been controversial (Hindwood and Mayr 1946; Serventy 1953; Salomonsen 1961b; Schodde 1975; Noske 1978). I believe that four species should be recognized — the Striated, *Pardalotus striatus*, which occurs in forests and woodlands throughout Australia in a series of colour morphs; the Red-browed, *P. rubricatus*, which is an Eyrean species typically found in trees lining watercourses in inland northern Australia; the Spotted, *P. punctatus*, of forests and woodlands of southern Australia and Tasmania, and which includes a representative in the drier country (the 'Yellow-rumped

[1]Zoology Department, Monash University, Clayton, Victoria 3168.

Pages 165-75 *in* BIRDS OF EUCALYPT FORESTS AND WOODLANDS: ECOLOGY, CONSERVATION MANAGEMENT ed. by A. Keast, H. F. Recher, H. Ford and D. Saunders. Royal Australasian Ornithologists Union and Surrey Beatty & Sons.

Pardalote, *P. xanthopygus*'); and the rare Forty-spotted Pardalote (*P. quadragintus*) now restricted to coastal forests of southeastern Tasmania and with a total population numbering less than 1000 individuals. In this discussion I will not consider *P. rubricatus* further, as it is not a forest species.

THE AUSTRALIAN ENVIRONMENT FOR LEAF-GLEANING SPECIES

The ecology of leaf-gleaning birds can be understood only in the light of what resources are available on foliage in Australian forests and woodlands. Unfortunately, little is known about the densities of invertebrates on such foliage, their community composition, or the way they vary seasonally and geographically. Most past research has concentrated on a few insect pest species, usually at abnormally high population levels. Based largely on this research, Morrow (1977) and Springett (1978) claimed that the density of foliage invertebrates in Australian forests was characteristically high (about an order of magnitude greater than that of comparable forests overseas) and that insect outbreaks are chronic. I found these claims difficult to reconcile with the sedentariness of most small insectivorous birds in Australia, and also with diversity and density estimates for such species in Australian forests, which were often low compared with estimates for similar small insectivores in northern temperate areas. So I examined base levels of invertebrate abundance in a range of forests and woodlands, taking more than 1400 foliage samples from plants in southeastern Australia and recording the number and types of arthropods on these. Samples were from 156 plant species, and were spread over all four seasons (Woinarski and Cullen 1984).

Generally, arthropod densities were low (averaging less than 10 medium-sized [> 3 mm] invertebrates per square metre of leaf area), but densities were affected by several environmental factors. One of the most important of these is shown in Figure 1a, which compares invertebrate abundance on the foliage of *Eucalyptus* species with that of other plant species. There were significantly more lerp-forming psyllids (Hemiptera: Psyllidae) and caterpillars on *Eucalyptus* foliage, and significantly more spiders, beetles, flies, and 'other sucking insects' on non-*Eucalyptus* foliage. Lerp-forming psyllids had by far the highest density of any arthropod category on *Eucalyptus* foliage. There were also pronounced differences between different *Eucalyptus* species (Fig. 1b). Eucalypts in the subgenus Symphyomyrtus (gums, boxes and ironbarks) had significantly higher invertebrate densities (especially of lerps) than eucalypts in the subgenus Monocalyptus (peppermints and stringybarks).

There was a complex array of seasonal changes in abundance of invertebrates. *Eucalyptus* and

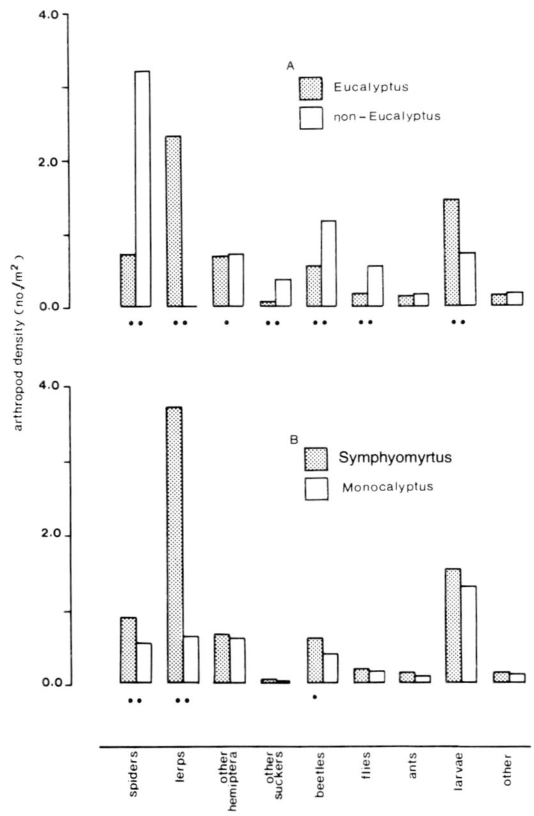

Fig. 1. Comparison of arthropod densities on foliage of different plant groups. Symbols below bars refer to significant differences between plant types (● = $0.01 < p < 0.05$; ●● = $p < 0.001$). 'Other Suckers' = *Psocoptera* + *Thysanoptera.* (a) *Eucalyptus* (shaded) versus NON-*Eucalyptus.* (b) *Symphyomyrtus* (shaded) versus *Monocalyptus.*

non-*Eucalyptus* species showed different patterns of seasonal variability and peaks, with non-*Eucalyptus* species less seasonal. Different invertebrate categories either peaked in different seasons or showed little seasonal variability. This variation was related to geographic position and forest type, with wetter sites (e.g. cool temperate rainforests and mountain *Eucalyptus* forests) having the lowest densities of arthropods but most pronounced seasonal changes. Overall the winter decline in arthropod density was not extreme at most sites, and seasonal variation at the drier sites was relatively minor. The distribution of different invertebrate categories also varied in relation to site and season, with lerp-forming psyllids a notable example. In summer these were distributed fairly evenly at sites across the entire moisture gradient sampled, whereas in winter and, less so, spring and autumn, wetter sites had significantly fewer lerps than drier sites.

LERPS

Lerps are a characteristic and important feature of *Eucalyptus* forests, and as such merit special consideration in any discussion of food available to

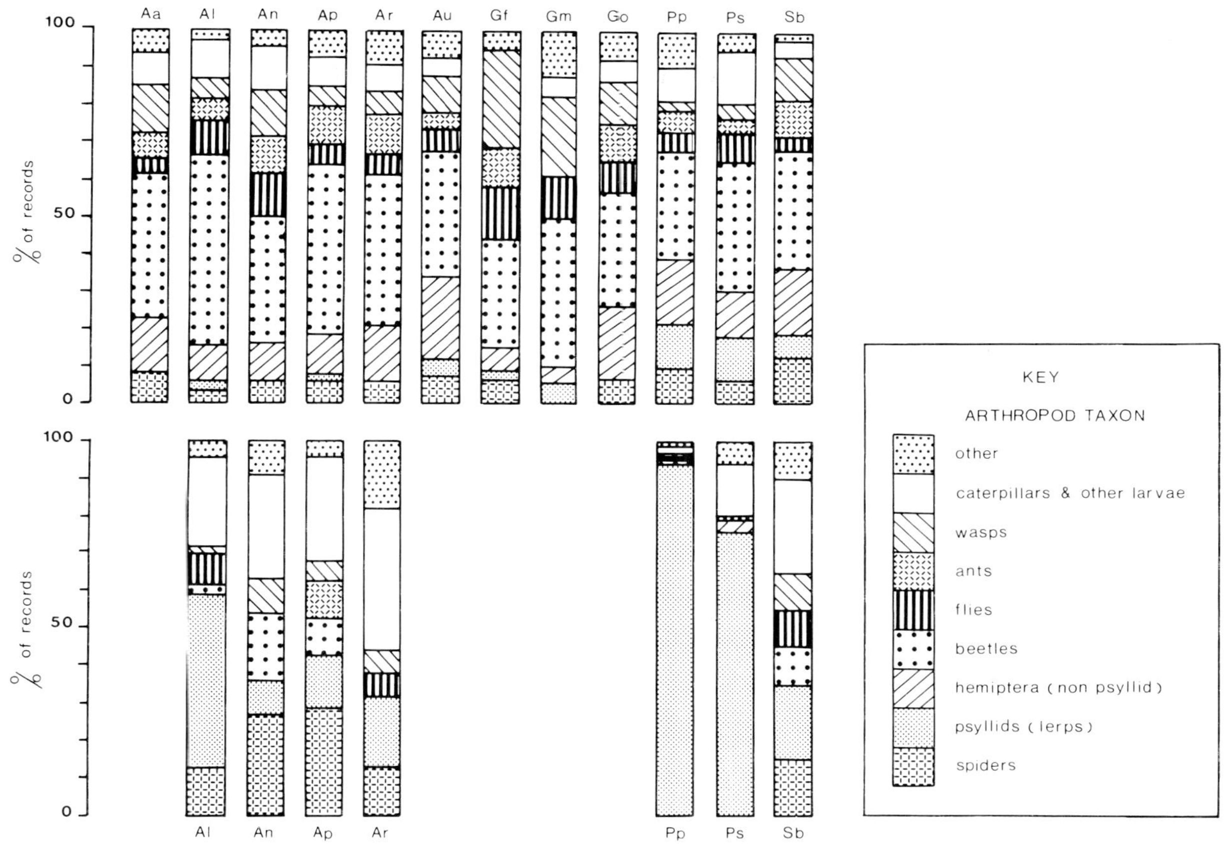

Fig. 2. Percentage of different arthropod taxa recorded in stomach analyses (top) and from feeding observations (bottom). Symbols for species are as in Table 1, with the addition of Gf = *Gerygone fusca*, Go = *G. olivacea and Px =P. xanthopygus*. Only species with at least ten observations of food items taken are shown.

Australian foliage-gleaning birds. They are formed as covering exudates by psyllids, which have radiated widely in Australia (Eastop 1978). Most psyllids, are relatively host-specific, many have high base levels of density and several frequently form massive outbreaks, often resulting in extensive loss of foliage or death of trees (e.g. Moore 1970; White 1971). As the lerp cover is larger than the actual psyllid nymph, and is rich in carbohydrates, lerps may constitute a valuable resource for birds capable of exploiting them (Paton 1980). Lerps possess several other characteristics that affect species foraging upon them: they are found only on leaves, and only on those of *Eucalyptus* species, they are stationary and may be difficult to remove, and their numbers may fluctuate dramatically. I believe that pardalotes specialize on lerps and have evolved a pattern of foraging based on these characteristics of lerps.

COMPARISON OF THE ECOLOGY OF PARDALOTES AND ACANTHIZIDS

1. *Diet.* In mainland southeastern Australia there are twelve species in the assemblage of small leaf-gleaners. Figure 2 presents a summary of all published information on diet of these species, along with data from some unpublished studies and my own observations. It is obvious from this figure that there is a vast difference between descriptions of diet based on data from stomach contents and that derived from foraging observations. This is mainly because food items are digested at different rates — for example lerps are absorbed quickly and leave little trace whereas the hard elytra of beetles remain distinctive and readily identifiable in most stomach samples. There are also biases in foraging observations towards recording larger and more obvious prey. Nonetheless, some dietary differences between groups of species are apparent. Psyllids occur appreciably only in the stomach samples of pardalotes and are the dominant food items in those species, based on foraging observations. Most thornbills have varied diets including beetles, bugs, ants, wasps and caterpillars and most gerygones include a high proportion of flying insects in the diet.

2. *Foraging behaviour.* Cody (1973) has emphasized the importance of foraging behaviour as:

> "perhaps the most basic and critical part of bird species' biology, upon which depend and to which are related most other characteristics (habitat choice, morphology, seasonality, etc.)"

I recorded types of foraging behaviour used at my study sites in southeastern Australia, and summarize the occurrence of different types of foraging in Table 1.

Table 1. Foraging behaviours of leaf-gleaning species in southeastern Australia.
(a) Percentage of different foraging acts.

sally		hover		hoversnatch		leafpeck		twigpeck		branchpeck		barkprobe		groundpeck		other	
Ar	0	Ar	0.8	Au	2.2	Ar	14.9	Pp	1.3	Pp	1.5	Pp	0.1	Al	0	Pp	0
Ps	0.1	Aa	0.8	Ar	3.0	Gm	22.0	Ps	1.4	Ps	3.4	Ps	0.2	Gm	0	Au	0
Pp	0.1	Ps	1.1	Ps	4.4	Au	23.5	Sb	2.1	Sb	3.8	Gm	0.6	Ap	0	Gm	0
Au	0.2	Au	1.9	Aa	4.7	Ap	33.5	Al	7.1	Gm	6.7	Sb	1.3	Ps	0.2	Sb	0.3
Al	0.2	Pp	3.2	Pp	6.4	Aa	34.2	Gm	8.2	Al	7.5	An	2.9	Pp	0.2	Aa	0.8
Sb	0.3	An	4.5	Ap	8.6	An	47.0	Ar	13.0	An	12.6	Aa	8.5	Sb	0.5	Ar	1.0
Aa	0.4	Ap	4.8	Al	8.6	Sb	48.8	Ap	16.5	Ap	18.9	Au	13.9	An	0.5	Ps	1.4
An	0.6	Al	7.0	An	12.0	Al	53.3	An	17.4	Aa	23.7	Ap	14.2	Aa	3.1	Al	2.1
Ap	0.6	Gm	19.6	Sb	12.6	Ps	87.7	Au	18.3	Ar	23.7	Al	14.2	Au	6.6	An	2.5
Gm	6.1	Sb	30.3	Gm	36.9	Pp	88.3	Aa	23.9	Au	33.5	Ar	15.9	Ar	27.8	Ap	3.1

(b) Percentage of substrates used for 'perches'.

air		leaf		twig		branch		ground	
Ar	3.8	Ar	2.0	Gm	23.7	Pp	1.4	Gm	0
Au	4.2	Ps	6.6	Ar	27.7	Sb	4.1	Ap	0
Ps	5.7	Gm	7.1	Sb	33.5	Ps	5.3	Al	0.1
Aa	6.1	Au	7.4	Au	34.9	Gm	6.8	Ps	0.2
Pp	10.3	Aa	8.1	Al	35.2	Al	12.6	Pp	0.2
Ap	13.9	Ap	8.8	Ap	44.4	An	16.0	Sb	0.5
An	17.0	Pp	15.5	Aa	49.9	Ap	32.4	An	0.6
Al	18.6	An	15.5	An	51.3	Aa	32.4	Aa	2.8
Sb	43.5	Sb	18.7	Pp	73.0	Ar	38.8	Au	6.6
Gm	63.1	Al	33.6	Ps	82.1	Au	45.7	Ar	27.6

(c) Feeding rates, etc. (No. per 100 sec.)

No. foraging acts		% time moving		No. intertree flights		No. intratree flights		Distance hopped (m)		% acts inverted		Diversity foraging acts		Diversity substrates	
Pp	18.7	Ps	28.7	Ps	2.2	Ar	2.5	Sb	5.3	Gm	3.9	Ps	0.3	Ps	0.4
Ps	18.7	Pp	28.8	Pp	2.3	Ps	2.8	Gm	5.8	Ar	7.6	Pp	0.3	Pp	0.5
Sb	22.2	Ar	31.7	Ap	2.7	Pp	3.5	Ps	6.7	Aa	11.1	Ar	0.8	Ar	0.5
Gm	25.8	Ap	32.3	Ar	3.3	Aa	4.8	Pp	7.3	Au	13.2	Al	0.9	Al	0.7
Al	29.0	An	35.4	Gm	3.4	Ap	5.5	An	9.0	Ap	13.9	Sb	0.9	Ap	0.7
Ar	29.6	Aa	35.4	Au	3.6	Au	5.7	Al	9.0	An	15.4	An	1.0	Gm	0.8
An	30.0	Al	35.6	An	3.9	An	6.5	Ar	10.6	Sb	16.1	Au	1.1	Au	0.8
Aa	30.8	Au	36.7	Al	4.2	Al	7.6	Ap	11.2	Pp	28.1	Ap	1.1	Aa	0.8
Ap	32.7	Gm	40.0	Aa	4.4	Sb	8.0	Aa	11.7	Ps	30.2	Aa	1.2	An	0.8
Au	33.8	Sb	40.3	Sb	5.7	Gm	10.7	Au	12.0	Al	32.4	Gm	1.2	Sb	0.9

Species symbols: Aa = *Acanthiza apicalis,* Al = *A. lineata,* An = *A. nana,* Ap = *A. pusilla,* Ar = *A. reguloides,* Au = *A. uropygialis,* Gm = *Gerygone mouki,* Pp = *Pardalotus punctatus,* Ps = *P. striatus,* Sb = *Smicrornis brevirostris.* Vertical lines join means which are not significantly different (Student-Neumann-Keuls test).

The distinctness of the pardalotes from the acanthizid warblers is manifest: pardalotes are relatively inactive and also relatively stereotyped in foraging actions and substrates used, with the bulk of feeding being pecking from leaf surfaces while the bird is perched on twigs. In comparison, *Acanthiza, Gerygone* and *Smicrornis* species have a wider repertoire of foraging behaviours. These also showed substantial shifts in foraging behaviours when foraging in different plant species, at different seasons, and at different study sites (Woinarski 1984a). This was not so apparent in the pardalotes. Nevertheless, pardalotes (and particularly Spotted Pardalotes) did show some foraging flexibility. There was some consistency between Spotted and Striated Pardalotes in the direction of seasonal foraging changes, with both feeding more rapidly and avoiding the foliage extremeties in winter. This may be related to shorter daylength, concentration of activity on feeding (as opposed to breeding), smaller prey in winter, and increased sheltering of prey, and birds, away from the exposed tree periphery. Elsewhere (Woinarski 1984a) I show that variation in the foraging of pardalotes on different plants is correlated with five characteristics of these tree species: average leaf area, leaf shape, tree height, and densities of psyllids and all invertebrates combined.

3. *Foraging sites.* I recorded the location (plant species, bird and tree height) of all birds observed along transects set up at every study site. The data on plant species used is summarized in Figure 3, which combines results from all study sites.

Within this assemblage there is a pronounced separation of species according to type of plant

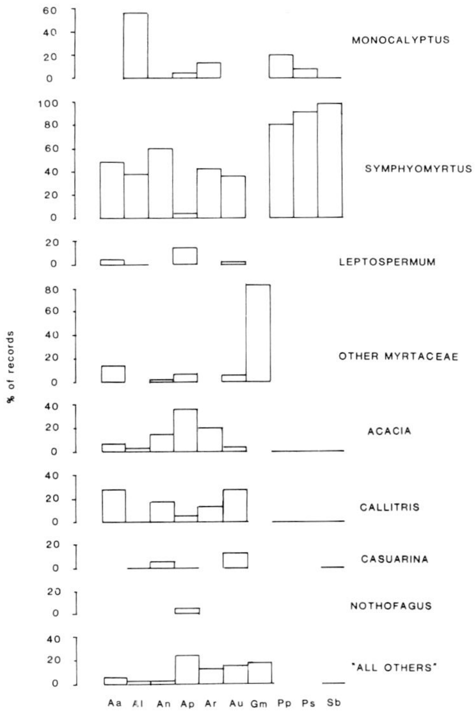

Fig. 3. Proportion of foraging records according to type of plant species used.

selected to forage in. Foliage in the Australian forest environment has major discontinuities, and bird species are using these to subdivide and partition resources. Striated Thornbills, Weebills, Striated Pardalotes and Spotted Pardalotes are *Eucalyptus* specialists, and of these Striated Thornbills alone feed extensively on Monocalyptus species. Other species direct feeding mainly at *Acacia* species (Brown Thornbill), species other than *Acacia* or *Eucalyptus* (Brown Gerygone), or show a substantial proportion of ground-feeding (Buff-rumped Thornbill). It follows that high diversity of small leaf-gleaners occurs only at sites where all these vegetation elements occur.

Related to these differences in type of plant species chosen, there are differences between species in foraging height. Segregation by foraging height is mainly secondary within this assemblage, and caused by differences in height between most *Eucalyptus* and most non-*Eucalyptus* species. Where non-*Eucalyptus* species form part of the canopy, 'shrub' specialists (e.g. Brown Thornbill) forage as high as 'tree' specialists (e.g. Striated Thornbill).

4. *Distribution, density, flock size, and movements.* Estimates of spring/summer and autumn/winter densities of foliage-gleaners at the 16 Victorian study sites are shown in Table 2. Several distributional factors are noteworthy: (i) both pardalote species occur at sites across the entire moisture gradient sampled, in contrast to all other leaf-gleaners which were more restricted (although the species-pair *A. apicalis*/*A. pusilla* jointly had nearly as wide a distribution); (ii) pardalotes occurred rarely at sites dominated by non-*Eucalyptus* species (and records at these sites were only of pardalotes feeding in the few scattered eucalypts); (iii) both pardalote species, but particularly *P. striatus*, moved down the moisture gradient towards drier sites in autumn and winter, whereas there was no such directional movement by *Acanthiza* or *Smicrornis* species; (iv) density peaks occurred at different sites and seasons for the two pardalote species, with Striated Pardalotes being less common than Spotted Pardalotes at wetter sites; (v) the number of co-occurring leaf-gleaning species is low and (vi) this number is lowest at wetter sites.

These species also vary in seasonal changes in size of foraging flocks (Fig. 4). The foraging of *A. apicalis*

Table 2. Distribution and densities of small leaf-gleaners at 16 Victorian sites.

Site	Species									
	Aa	Al	An	Ap	Ar	Au	Gm	Pp	Ps	Sb
A. Wyperfeld	1/2		3/5			2/6		.4/2	1/2	7/10
B. Bendigo	1/1		4/5		.4/1	1/1		9/3	2/17	9/7
C. Barmah		1/4	1/0		.1/1			0/1	7/9	5/10
D. Reedy Lake								0/.1	7/16	.4/.3
E. You Yangs		.2/.4	5/3	2/1	2/4			1/4	2/2	2/3
F. Tooborac		8/14	.3/2	1/2	7/10			6/4	1/3	1/3
G. Chiltern				0/.2				.4/2	1/6	0/1
H. Beechworth		6/10	3/2	8/12	12/8			2/3	.3/.1	6/1
I. Cranbourne		10/11		12/5				5/2	1/.2	
J. Sandy Point		1/2		11/12				10/6	.1/.1	
K. Lysterfield		12/20		8/6	.1/0			17/2	1/1	
L. Glenaladale		.2/.2	1/1	6/10			11/5	1/.2	.2/.1	
M. Tynong		15/32		14/12	0/.3			1/1	1/.1	
N. Bogong		8/19		9/7				3/3	3/1	
O. Coranderrk		8/16		11/11				1/.2	1/.1	
P. Donna Buang		6/2		13/11				1/0	1/.4	

Values shown are mean No. individuals per 10 ha. with spring-summer values numerator and autumn-winter denominator. Additionally, there were two rare species: *G. fusca* present at B (.1/.1) and C (.2/.1) and *G. olivacea* at F (0/.1), G (.1/.1) and H (.5/0). Brief vegetation descriptions of study sites: A — mallee, scrubpine + box woodland; B — mallee + box woodland; C, D and E — red gum woodland; F and G — box/ironbark open forest; H — box/ironbark and native pine open forest; I, J and K — coastal open forest; L — warm temperate rainforest; M, N and O — wet mountain forest; P — cool temperate rainforest. Sites are arranged from driest (A) to wettest (P).

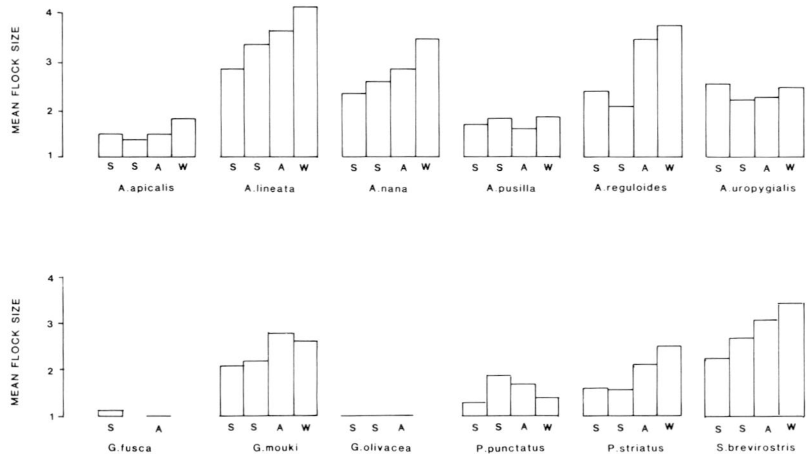

Fig. 4. Mean flock size in every season, averaged over study sites.

and *A. pusilla* is solitary or in pairs throughout the year. Most other species show distinct seasonal change in sociability, with flocks forming in the non-breeding season. Bell (1985) examines this in detail for *A. lineata* and *A. reguloides* in another chapter in this volume. The process in pardalotes is different again, with widespread dispersal in the non-breeding season a common phenomenon. Both pardalote species may form very large flocks; I have observed flocks of Spotted Pardalotes numbering about 1000 individuals (mostly females and juveniles) at three different study sites in late summer, and similar large flocks have been reported for Striated Pardalotes (e.g. Hindwood and Sharland 1964), although these are usually in autumn and winter. These flocks may wander widely, although most 'drift' towards warmer or drier areas for the winter. These movements may be extensive; for example, those Striated Pardalotes which breed in Tasmania move to New South Wales, Queensland and inland Australia for winter. Sadly, little of this movement has been documented through banding studies; however, a Spotted Pardalote banded near Adelaide in April was found breeding in Melbourne nine months later.

These mass movements of pardalotes are probably a response to a sharp winter decline in lerp numbers in wetter forests, with flocking nomadism a search for psyllid outbreaks which may persist throughout winter in drier areas. A hazard of this strategy is that the exact location of these outbreaks may be unpredictable or the occurrence and size of outbreaks reduced through inclement weather. This may lead to population crashes when psyllid outbreaks cannot be located, and indeed mass mortality of pardalotes has been recorded in association with winter flocking (Woinarski and McEvey 1983). This behaviour contrasts with that of most acanthizid warblers which are sedentary throughout the year, partly because their food sources fluctuate less dramatically and partly because they are less stereotyped in foraging behaviour, allowing them to switch diets or foraging sites to take advantage of whatever invertebrates are locally abundant or available.

5. *Breeding.* Typically, acanthizid warblers build domed nests in foliage or, for some species, behind hanging bark. Only the Chestnut-rumped Thornbill never nests in foliage, instead locating its nest behind bark or in natural crevices. This direction has been followed more extremely by the pardalotes, which are now obligate hole-nesters. As is usual with birds nesting in hollows, the clutch-size is larger and incubation period longer in pardalotes than in acanthizids (Table 3). There are few data available on breeding success of pardalotes. However, multiple nesting is common in Striated Pardalotes and I have recorded at least three broods per pair per season, and intervals of less than seven days between successful fledging of one brood and the commencement of the next. Striated Pardalotes have many attributes of 'boom-bust species', with high and prolonged reproductive output in favourable conditions. Other species in this assemblage may also have repeat broods, but the incidence of these is not documented and is probably less than that for Striated Pardalotes. Elsewhere (Woinarski 1985) I show that clutch size in all Australian leaf-gleaning species is considerably less than that of comparable species of northern temperate areas.

6. *Morphology.* I measured 13 body size variables for all species in this assemblage, and compared these simultaneously using a Principal Components Analysis. The results, summarized in Figure 5, show that pardalotes have longer, more pointed wings, shorter and stouter beaks, smaller rictal bristles, and are heavier than the other species. Most of these morphological features relate directly to their foraging behaviour: a short and powerful beak is particularly suitable for removing relatively large but inert lerps from broad leaf surfaces, whereas narrow beaks are more suitable for probing, and long beaks and rictal bristles for active, struggling prey. The longer wings and wingtips of pardalotes correlate

Table 3. Breeding data for all species.

	Species											
	Aa	Al	An	Ap	Ar	Au	Gf	Gm	Go	Pp	Ps	Sb
Clutch size	2.5	2.8	3.4	2.8	3.4	3.2	2.7	2.7	2.8	3.2	4.0	2.7
(No. records)	6	80	5	221	81	16	26	45	99	5	10	67
No. fledged	1.8	1.2	0.8	1.2	2.0	2.3	0.9	0.6	1.0	1.3	2.4	1.0
(No. records)	6	89	4	180	70	6	16	24	53	8	25	28
Incubation time	18-19	16-17	16-17	19	20-21	?	14-17	16-17	17	20-22	18-24	18-21
Fledging time	16-18	18-19	?	14-16	17-18	14-21	14-16	16-18	15-17	17-19	21-25	18-19

Data based on RAOU Nest Record Cards. Incubation + fledgling periods are not always precisely determinable, so ranges given.

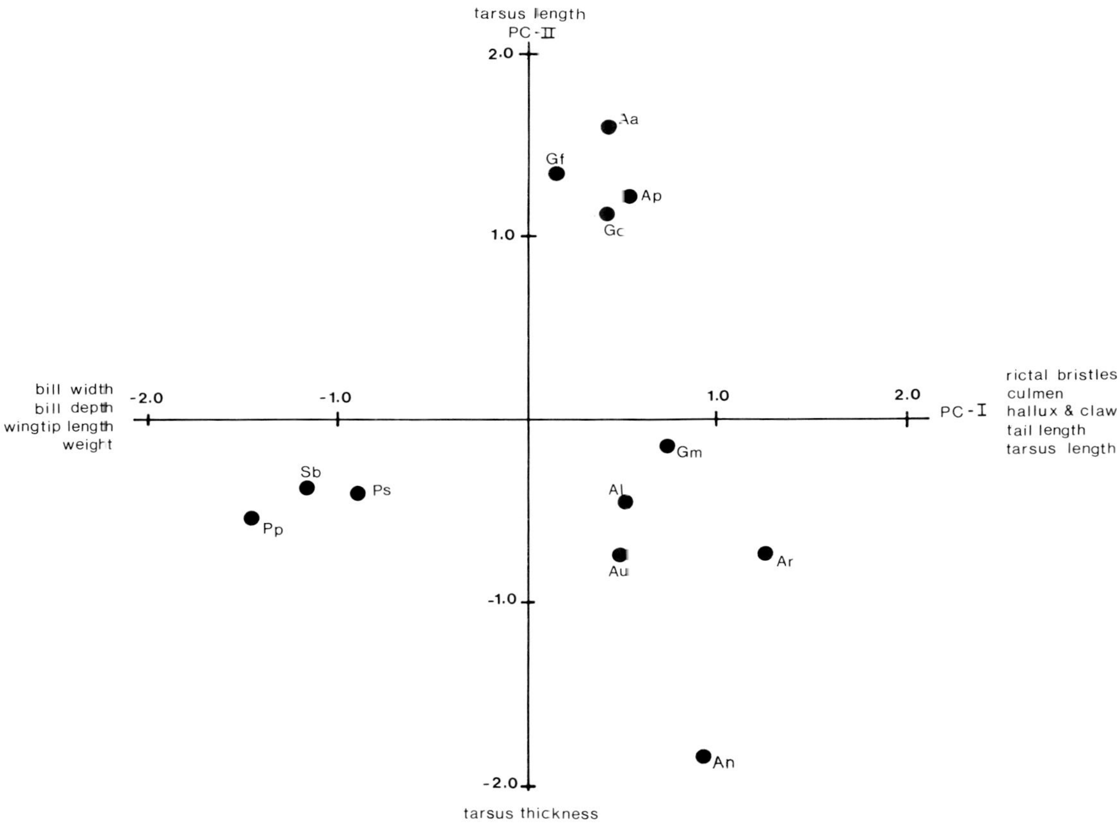

Fig. 5. The position of 12 species of small leaf-gleaners according to the first two principal components of 12 morphological measurements.

with their extensive dispersal movements; the short, round wings of *Acanthiza* and *Smicrornis* are typical for sedentary species. The pardalote morphological features are most closely approached by *Smicrornis* and, less so, *A. lineata,* and these may represent steps from a common acanthizid ancestral stock in the adaptive pathway to specialization on lerp-feeding.

PROBLEMS FOR PARDALOTES: THE DISADVANTAGE OF LERPS

Lerps are stationary and may occur in patches of very high density; as such they are defensible resources. Many studies have shown that honeyeaters may defend rich nectar sources from other birds. I have found (Woinarski 1984b) that they also defend lerp patches, attacking small insectivorous species and particularly pardalotes. By timing feeding observations I calculated the effects of such attacks on these small insectivores. Mean intervals between attacks are given in Table 4, and show that pardalotes are singled out. The attack rate is highest in winter and early spring when *Eucalyptus* flowers are least abundant (and hence lerps most important for honeyeaters), when pardalotes are flocking and nomadic (and hence may not have local knowledge of honeyeater-defended areas), and when honeyeaters may peak in aggression before the onset of breeding. I estimated that the amount of time pardalotes 'lose' through honeyeater attacks is 5-10 percent of daylight hours, and the energy expended 7-12 percent of daylight output in these seasons. This is a substantial cost for a small insectivorous bird, especially in winter when most of the day must be used for efficient foraging in order to survive. There

Table 4. Frequency of honeyeater attacks on small leaf-gleaners.

Species	No. foraging bouts observed	No. ending in chases	Mean interval between chases (sec.)
P. striatus	299	35	587
P. punctatus	323	18	1111
A. lineata	299	4	3460
A. nana	134	1	5990
A. pusilla	321	2	7668
S. brevirostris	176	1	9165
A. apicalis	35	0	>1400
A. reguloides	44	0	>1965
A. uropygialis	56	0	>2530
G. mouki	54	0	>2810

are also real risks of physical damage or death (Dow 1977) being inflicted. Honeyeater aggression may be a factor limiting the use of lerps by *Acanthiza, Gerygone* and *Smicrornis* species, and protection from attacks may have been an important factor in the development of hole-nesting by pardalote species.

There are other disadvantages in foraging predominantly from lerps. One of the most obvious is the restriction of lerp-forming psyllids to *Eucalyptus* species. As a result of this, pardalotes do not occur in forests composed of species other than eucalypts (e.g. rainforests, *Acacia* woodlands, and *Callitris*/*Casuarina* woodlands), and forage on non-*Eucalyptus* species only rarely and incidentally. Likewise, low psyllid densities on Monocalyptus species result in low densities of pardalotes in forests dominated by such species (for example, in the mountain areas of southeastern Australia). Another problem is the marked seasonality, patchiness and relative unpredictability in abundance of lerps. This may demand extensive nomadism and a high premium for opportunism in breeding. The effects of mass psyllid outbreaks on tree health may also help pardalotes locate areas of high psyllid density, as much canopy foliage becomes discoloured and obvious on trees suffering attacks. Nomadic pardalotes flying over such areas may be able to identify such characteristic damage and home in on psyllid outbreaks. There is evidence from studies on population dynamics of outbreaking psyllids that this indeed happens, with flocks of pardalotes moving into high density patches, mopping up psyllids, and then moving on (Clark 1964).

COMPARISONS BETWEEN PARDALOTE SPECIES

I have shown that there is a clear distinction between the ecology of pardalote species and the acanthizid warblers of southeastern Australia. But are there differences between the two pardalote species, how do they manage to co-occur and what happens when a third pardalote species, the Forty-spotted, is added in southeastern Tasmania?

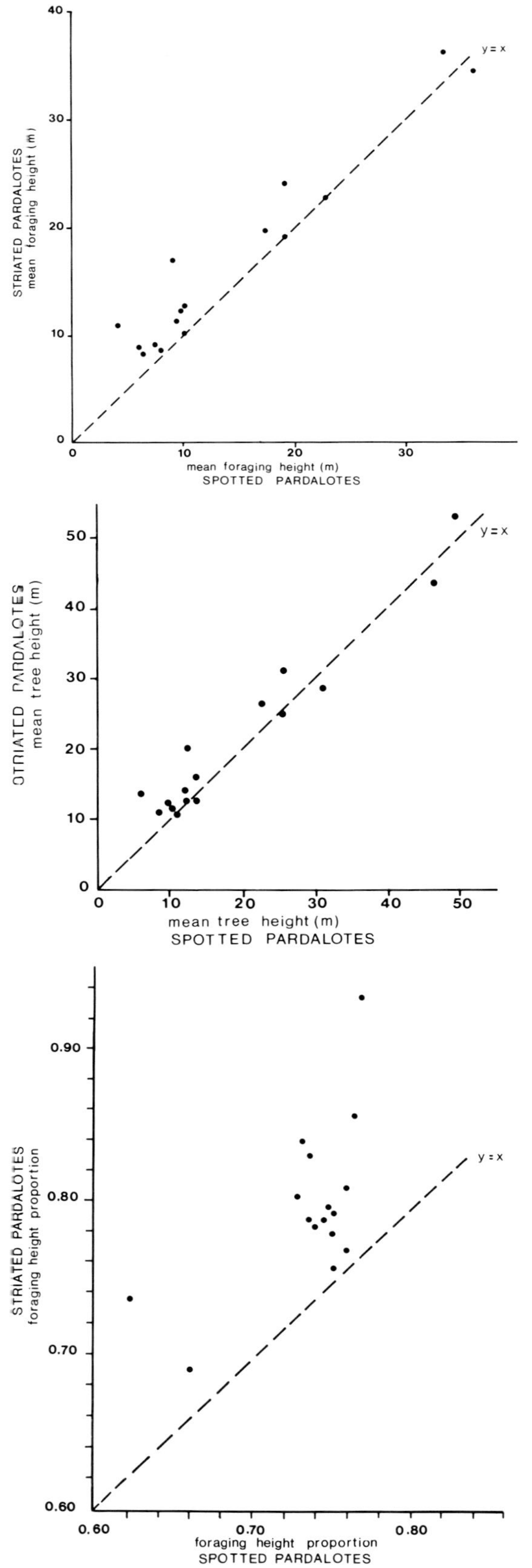

Fig. 6. Relative foraging heights of Spotted and Striated Pardalotes at 16 Victorian study sites: (a) mean height (in m) of foraging birds; (b) mean height (in m) of trees in which foraging occurred; (c) mean proportion of bird height:tree height.

Table 5. Tree species use by Spotted and Striated Pardalotes at particular study sites.

Site	Eucalyptus species	% of foraging records		χ^2
		P. striatus	*P. punctatus*	
A. Wyperfeld	*dumosa*	11.9	31.4	
	foecunda	0	22.9	
	incrassata	7.1	37.3	
	largiflorens	81.0	8.4	
	total records	42	83	67.9***
B. Bendigo	*behriana*	14.1	34.7	
	leucoxylon	43.1	24.2	
	microcarpa	37.1	18.7	
	polybractea	5.1	21.8	
	viridis	0.6	0.6	
	total records	453	326	119.9***
F + G. Tooborac + Chiltern	*blakelyi*	17.8	4.2	
	goniocalyx	6.7	13.1	
	macrorhyncha	11.8	11.3	
	melliodora	14.1	15.5	
	microcarpa	14.1	10.1	
	polyanthemos	13.3	30.9	
	sideroxylon	22.2	14.9	
	total records	135	168	29.8***
N. Bogong	*delegatensis*	15.1	28.2	
	dives	7.6	21.3	
	pauciflora	12.1	3.4	
	radiata	11.4	17.8	
	st. johnii	9.1	8.6	
	viminalis	43.3	18.4	
	other	1.5	2.3	
	total records	132	174	40.6***

Significance levels: *** $p < 0.001$.

I noted some differences between co-existing Spotted and Striated Pardalotes that were consistent across all study sites in Victoria. One of these was in foraging height, where Striated Pardalotes foraged higher than Spotted Pardalotes (Fig. 6a); this seems to be because Striated Pardalotes prefer higher parts of trees (Fig. 6c) rather than selecting taller trees (Fig. 6b). Another difference was in the proportion of some foraging methods used, with the smaller (9 g compared with 12 g) Spotted Pardalotes being more agile and aerial (Table 1); in contrast, the longer Striated Pardalotes appear to reach invertebrates on leaf extremities by perching on twigs and stretching, often while being upside-down. There are also differences in choice of habitat (Table 2), or tree species used in foraging at the same site (Table 5).

These differences seem sufficient to allow co-existence of two lerp-specialist species in most *Eucalyptus* forests and woodlands of mainland southeastern Australia. Co-existence may be facilitated by the wide diversity of lerp sizes, shapes and compositions, and the two pardalote species may also prefer distinctly different types of lerps.

The situation is more complicated in southeastern Tasmania where three pardalote species are found. Here all show preference for forests dominated by the Symphyomyrtus species *E. viminalis,* and for forests with a higher diversity of co-occurring *Eucalyptus* species (Woinarski and Rounsevell 1983; Woinarski and Bulman 1985). Habitat choice is also affected by the number of hollow-bearing trees (for Striated and Forty-spotted Pardalotes), and amount of disturbance. Forty-spotted Pardalotes appear to prefer relatively unmodified forests, Spotted Pardalotes slightly disturbed forests, and Striated Pardalotes show no preference. The foraging behaviour of Forty-spotted and Striated Pardalotes is very similar, and these species may compete directly for resources. Only at sites where there were relatively high densities of Forty-spotted Pardalotes (weight about 10.5 g) was that species dominant over the larger Striated Pardalotes, and was it able to forage without being harrassed. I believe that the resource requirements of these two species are too similar to permit stable co-existence, and that the rapid decline and the present parlous state of the Forty-spotted Pardalote is due to this competition which in turn may have become unbalanced through forest modification in historic times. The Forty-spotted Pardalote is distinctive in the very localized scale of its non-breeding dispersal, perhaps a legacy of the equitable climate of coastal forests. It still remains an intriguing puzzle why the Spotted and Striated Pardalotes are so widespread and abundant while the Forty-spotted Pardalote is perilously close to extinction and highly restricted. It may be that Forty-spotted Pardalotes are now too stereotyped in ecological requirements and behaviour to cope with recent changes in their forest environment.

It is in breeding behaviour that pardalotes show the most interesting variation. Spotted Pardalotes alone have marked sexual dimorphism, with male birds appreciably more colourful than females. In the breeding season males defend a relatively large territory (1-4 ha), are restricted in their foraging and other activities to this area, and exclude other males. In contrast, sexual dimorphism is minor in the Striated and Forty-spotted Pardalotes. Striated Pardalotes establish no feeding territory and only the nest site is defended. Forty-spotted Pardalotes may form large and dense colonies in areas where suitable breeding sites are abundant. The dispersion of breeding Forty-spotted Pardalotes differs from that of Striated Pardalotes, as the former appear to be obligatorily colonial (or loosely so) whereas nesting Striated Pardalotes may be solitary.

I believe these differences are tied into choice of nest sites. Spotted Pardalotes nest only in the ground, in a wide range of sites with loose soil (Woinarski 1974). This is a reasonably continuous resource and hence Spotted Pardalotes may be able to space themselves out more or less uniformly. Striated Pardalotes nest either in tree hollows or in fairly stereotyped locations of steep banks with relatively hard soil. The distribution of such sites tends to be more clumped, and hence breeding density will be maximized when all sites are used (and hollows suitable for breeding may be in short supply in most Australian forests). Almost all nests of Forty-spotted Pardalotes are in tree hollows. As these may also be suitable for the somewhat larger Striated Pardalotes, communal defence of grouped hollows occurs by Forty-spotted against Striated Pardalotes, and it seems that solitary pairs of Forty-spotted Pardalotes may be excluded from breeding sites by Striated Pardalotes. The risks of predation may also be important in spacing patterns; the less accessible nests of Striated and Forty-spotted Pardalotes may not need to be as widely separated as the more predator-prone nests of Spotted Pardalotes.

A further disadvantage of the nesting sites of Spotted Pardalotes is in their more ephemeral nature, with the soft soil around nests rarely preserving nesting tunnels for consecutive years. Hence pairs of Spotted Pardalotes must spend considerable time and energy constructing fresh burrows in most years. In contrast the nest sites of Striated and Forty-spotted Pardalotes may persist for several years and indeed may be used by pardalotes for at least 50 successive years (Rogers 1966). The use of an already suitable hollow or burrow is obviously energy efficient, and so there is considerable competition for these at the beginning of the breeding season. The relative rarity of such sites seems to be a pivotal feature in the ecology of both Striated and Forty-spotted Pardalotes, for, despite the widespread post-breeding dispersal of Striated Pardalotes, there is a strong degree of philopatry with most birds returning to their previous hollows in successive years. Likewise there is a strong traditional use of sites by colonies of Forty-spotted Pardalotes. Nesting requirements may also affect habitat choice, with a clear preference by both Forty-spotted and Striated Pardalotes for old forests containing hollow-bearing trees, and may exclude these species from young forests.

CONCLUSIONS

The ecology of the pardalotes is unique, differing from that of other Australian leaf-gleaning species in their adaptations to feeding on lerp-forming psyllids. As lerps are restricted to *Eucalyptus* species, there are no ecological equivalents of pardalotes in other continents. The richness and variety of lerps is sufficient to allow the co-existence of two pardalote species which differ in ecology but may not be sufficient to permit the stable co-existence of any more than two species.

REFERENCES

BELL, H., 1985. The social organization and foraging behaviour of three syntopic thornbills *Acanthiza* spp. Pp. 151-63 *in* Birds of Eucalypt Forests and Woodlands: Ecology, Conservation, Management ed. by A. Keast, H. F. Recher, H. Ford and D. Saunders. Royal Australasian Ornithologists Union.

CLARK, L. R., 1964. Predation by birds in relation to the population density of *Cardiaspina albitextura* (Psyllidae). *Aust. J. Zool.* **12**: 349-61.

CODY, M. L., 1973. Parallel evolution and bird niches. *In* "Mediterranean-type Ecosystems" ed. by F. di Castri and H. A. Mooney. Springer-Verlag: New York.

DOW, D. D., 1977. Indiscriminate interspecific aggression leading to almost sole occupancy of space by a single species of bird. *Emu* 7: 115-21.

EASTOP, V. F., 1978. Diversity of the Stenorrhyncha within major climatic zones. *In* "Diversity of Insect Faunas" ed. by L. A. Mound and N. Waloff. Blackwell: London.

HINDWOOD, K. A. AND MAYR, E., 1946. A revision of the striped-crowned pardalotes. *Emu* **46**: 49-67.

HINDWOOD, K. A. AND SHARLAND, M. S. R., 1964. The Swift Parrot. *Emu* **63**: 310-26.

MAYR, E. AND AMADON, D., 1947. A review of the Dicaeidae. *Amer. Mus. Novit.* **1360**: 1-32.

MOORE, K. M., 1970. Results from a study of the genus *Glycaspis* (Homoptera: Psyllidae). *Aust. Zool.* **15**: 343-76.

MORROW, P. A., 1977. The significance of phytophagous insects in the *Eucalyptus* forest in Australia. *In* "The Role of Arthropods in Forest Ecosystems" ed. by W. J. Mattson. Springer-Verlag: Berlin.

NOSKE, R., 1978. Comments on some of the scientific names used in the 'Interim List of Australian Songbirds'. *Aust. Birds* **13**: 27-35.

PATON, D. C., 1980. The importance of manna, honeydew and lerp in the diet of honeyeaters. *Emu* **80**: 213-26.

ROGERS, K., 1966. The Red-tipped Pardalote. *Vict. Nat.* **83**: 268.

SALOMONSEN, F., 1961a. Notes on the flowerpeckers (Aves, Dicaeidae). 5. The genera *Oreocharis, Paramythia* and *Pardalotus* (except the superspecies *Pardalotus striatus*). *Amer. Mus. Novit.* **2067**: 1-23.

SALOMONSEN, F., 1961b. Notes on the flowerpeckers (Aves, Dicaeidae). 6. The superspecies *Pardalotus striatus. Amer. Mus. Novit.* **2068**: 1-31.

SCHODDE, R., 1975. "Interim List of Australian Songbirds — Passerines". R.A.O.U.: Melbourne.

SERVENTY, D. L., 1953. Some speciation problems in Australian birds. *Emu* **53**: 131-45.

SIBLEY, C. G., 1976. Protein evidence of the origin of certain Australian birds. *Proc. XVI Int. Orn. Congr.* 66-70.

SIBLEY, C. G. AND AHLQUIST, J. E., 1980. The relationships of the "primitive insect-eaters" (Aves: Passeriformes) as indicated by DNA-DNA hybridization. *Proc. XVII Int. Orn. Congr.* 1215-20.

SPRINGETT, B. P., 1978. On the ecological role of insects in Australian eucalypt forests. *Aust. J. Ecol.* **3**: 129-39.

WHITE, T. C. R., 1971. Lerp insects (Homoptera: Psyllidae) on red gum (*E. camaldulensis*) in South Australia. *South Aust. Nat.* **46**: 20-3.

WOINARSKI, J. C. Z., 1974. A comparison of the ground-nesting of two species of pardalotes. *Emu* **74**: 224-8.

WOINARSKI, J. C. Z., 1984a. Ecology of Pardalotes in southeastern Australia. Ph.D Thesis: Monash University, Melbourne.

WOINARSKI, J. C. Z., 1984b. Small birds, lerp-feeding and the problem of honeyeaters. *Emu* **84**: 137-41.

WOINARSKI, J. C. Z., 1985. Breeding biology and life history of small insectivorous birds in Australian forests: response to a stable environment? *Proc. Ecol. Soc. Aust.* **14**: 159-68.

WOINARSKI, J. C. Z. AND BULMAN, C. E., 1985. Ecology and breeding biology of the Forty-spotted Pardalote and other species of pardalotes on North Bruny Island. *Emu* **85**: 106-20.

WOINARSKI, J. C. Z. AND CULLEN, J. M., 1984. Distribution of invertebrates on foliage in forests of southeastern Australia. *Aust. J. Ecol.* **9**: 207-32.

WOINARSKI, J. C. Z. AND MCEVEY, A. R., 1983. Mortality in pardalotes in Melbourne, 1960 to 1980. *Corella* **7**: 55-9.

WOINARSKI, J. C. Z. AND ROUNSEVELL, D. E., 1983. Comparative ecology of pardalotes, including the Forty-spotted Pardalote, *Pardalotus quadragintus* (Aves: Pardalotidae), in south-eastern Tasmania. *Aust. Wild. Res.* **10**: 349-59.

Grey Fantail, *Rhipidura fuliginosa*

CHAPTER 15

Habitat Usage and Foraging Behaviour of Three Fantails (*Rhipidura:* Pachycephalidae)

Elizabeth Cameron[1]

The habitat usage and foraging behaviour of three sympatric species of fantails in north-eastern New South Wales, are described. In a 25 ha area of diverse vegetation, Willie Wagtails and Rufous Fantails were almost entirely separated because they frequented different habitats. The former occurred in open grassy habitats with a sparse tree layer and the latter in forest habitats with a well-developed understorey, and sometimes in dense secondary growth. The Grey Fantails occupied a wide variety of habitats which included those frequented by the other two species. Wagtails concentrated activity close to, and on the ground, Rufous Fantails usually foraged within low strata of vegetation, and Grey Fantails used most strata but concentrated on the periphery of trees and shrubs and in the open air adjacent to them.

The Wagtail is primarily a "pursuer" predator, scanning for prey from low lookout perches and flycatching or gleaning single large insects. The Rufous Fantail is a typical "searcher" predator, almost continuously on the move through foliage in search of prey, hovering and gleaning insects as well as flycatching. The Grey Fantail employs both strategies and is the most skilful aerial forager, often flycatching many small gnats in a single roundabout flight. Morphological differences between species relate to their foraging strategies. The Grey Fantail has a low wing loading and is aerodynamically well suited to aerial foraging, but has a small weak bill. The large partly terrestrial Wagtail has proportionally longer legs and wings and a large strong bill, and the species which frequents undergrowth, the Rufous Fantail, is intermediate in limb and bill length and has short wings and tail.

The three species of fantails respond to changing seasonal conditions in different ways. The Rufous Fantail is a breeding summer visitor which migrates north in autumn; the populations of Grey Fantails and Willie Wagtails decline in autumn, probably as a result of local movements to lower altitudes. Habitat usage by the remaining Wagtails and Grey Fantails converges on open habitats in winter but their foraging strategies diverge. The Wagtail becomes more terrestrial, and the Grey Fantail, more aerial.

Although the Wagtail had the lowest density, its reproductive characteristics suggest it has the potential for rapid population increase if additional suitable habitat becomes available. The Rufous Fantail has a low potential rate of increase and although the Grey Fantail was the most abundant and widespread of the three species, its reproductive potential is probably lower than that of the Wagtail.

INTRODUCTION

THE MOUNTAIN ranges of northeastern New South Wales experience marked seasonal changes to which insectivorous birds respond in various ways. During the warm wet summer, eight flycatchers (genera *Monarcha, Myiagra* and *Rhipidura*) breed there but with the onset in autumn of cooler drier weather, five species migrate north to the tropics. How do these flycatchers share resources in summer? Do the habitats and foraging behaviour of migrants distinguish them from resident species, and do residents change their patterns of resource usage with the changing seasons? One genus will serve as an example.

The fantail genus *Rhipidura* is represented in the region by a breeding migrant — the Rufous Fantail (*R. rufifrons* (Latham 1801)) and by two resident species — the Grey Fantail (*R. fuliginosa* (Sparrman 1780)) and the Willie Wagtail (*R. leucophrys* (Latham 1801)). Fantails are characterised by

[1]Department of Zoology, University of New England, Armidale, New South Wales 2351. Present Address: Australian Museum, 6-8 College Street, Sydney, New South Wales 2000.
Pages 177-91 *in* BIRDS OF EUCALYPT FORESTS AND WOODLANDS: ECOLOGY, CONSERVATION, MANAGEMENT ed. by A. Keast, H. F. Recher, H. Ford and D. Saunders. Royal Australasian Ornithologists Union and Surrey Beatty & Sons.

long fan-shaped tails, short rounded wings, slender legs and feet and small, rather short broad bills surrounded by long rictal bristles; they catch much of their insect prey on the wing and can execute skilful aerial manoeuvres (Keast 1958). Sexes are indistinguishable in the field. The genus reaches its greatest diversity in the forests of New Guinea.

With an Australia-wide distribution the Grey Fantail is one of the most ubiquitous small birds of forest and woodland. The larger black and white Willie Wagtail is almost as widespread in woodland, scrubland and savanna, while the Rufous Fantail (slightly larger than the Grey Fantail) occurs in forests of the northern and eastern peripheries of the continent (Wilson 1976).

This chapter summarises the results of a study of the feeding ecology of sympatric populations of Grey and Rufous Fantails and Willie Wagtails. The aim of the study was to compare the feeding strategies of the three species and to look for correlations with habitat, food, morphology and residential status.

THE FIVE DAY CREEK STUDY

The study site was in the Five Day Creek valley, a tributary of the Macleay River east of the New England Plateau (30°36′S 152° 22′E; altitude 200 m). The deep valley was oriented SSE/NNW and in the middle of winter was sunlit only between 1000 and 1600 hours. Mean annual rainfall was about 1430 mm and 65% of this fell between October and March, with February the wettest month. Average summer maximum temperatures reached 25-35°C; in winter light frosts occurred between May and September and average maximum temperatures fell to 14-20°C. Originally forested, the site had been partially cleared for cattle grazing, resulting in a mosaic of vegetation in which twelve habitats were distinguished (Table 1). The closest comparable clearing in the forested valley was 2 km downstream. Characteristics of habitats were sampled in 0.05 ha plots and are listed in Table 2. The habitats separated into (i) "open" — generally fewer than 100 trees/ha with the dominant stratum less than 2 m high, and (ii) "closed" — generally more than 100 trees/ha with the dominant stratum trees or tall shrubs more than 2 m high.

Observations were made at fortnightly intervals over a three-year period and grouped into four seasons — early summer (October-December), late summer (January-March), early winter (April-June) and late winter (July-September). On each visit the distribution and number of fantails in a 25 ha area were recorded and the location of nests, song perches, aggressive encounters and flight paths were mapped. Spoken descriptions of foraging sequences, timed with a stop-watch, were recorded on cassettes and later transcribed. Data recorded for each sequence included duration, time of day, weather, habitat, estimated height and description of perches, and number and type of feeding manoeuvres. The fantails were almost continuously active and there was not time to record all details of behaviour and perches in each sequence. Each feeding manoeuvre was scored as a successful food capture unless an insect was seen to escape. The duration of uninterrupted surveillance of birds was influenced by the density of vegetation — recorded sequences of Rufous Fantails feeding in dense undergrowth were much shorter than those of Willie Wagtails on savanna grassland. Only the habitat in which a bird was initially encountered on a traverse of the site, was used to calculate the frequency of habitat usage.

Small samples of fantails near the study area were shot to obtain stomach samples and morphological data. Fresh carcasses were weighed, measured and sexed. To determine aerodynamic properties, wing area was calculated by tracing each wing on calibrated paper, and the lengths of flight feathers were measured. Stomach contents were stored in 75% alcohol and counted, measured and identified to the level of order or family with the aid of a stereomicroscope. Skeletons were cleaned in antiforminsol solution and their dimensions measured with dial calipers. To obtain standardised values for comparison of limb proportions between species, bone lengths were divided by the cube root of body weight of each bird.

An attempt was made to assess seasonal abundance and distribution of insects at the site by sweep netting and the use of sticky traps. Each fortnight 18 transparent plastic cylinders, each with a trapping surface area of 750 cm^2, were coated with an insect-trapping grease and exposed for 24 hours. Trapping stations ranged in height from ground level to 20 m above the ground in five habitats frequented by fantails. A 100 m^2 area of grass in the savanna grassland was swept twice during the same sampling period, at different times of day, Trapped insects were treated in the same way as stomach contents. Data up to 1971 from the Nest Record Scheme (N.R.S.) of the Royal Australasian Ornithologists Union were examined.

RESULTS

1. Habitat Requirements

1.1 *Vegetation*

None of the three fantails was randomly distributed in the study area. The Grey Fantail had the most uniform distribution and was seen in all habitats at the site. The habitats it "frequented" (those where more than 15% of sightings were made) were closed forest and three "open" habitats — bracken flat, savanna grassland and creek bank

Table 1. Habitats occupied by fantails at the study site. (Number of sightings shown in parentheses). *frequented habitat — where more than 15% of sightings were made.

(Left margin, reading down: DECREASING CANOPY COVER ↓↓; beside the habitat names: CLOSED … OPEN)

Habitat	% of 25 ha study site occupied by habitat	% Total Sightings: Grey Fantail Oct-Dec (199)	Grey Fantail Jan-Mar (246)	Grey Fantail Apr-Sep (149)	Rufous Fantail Oct-Dec (41)	Rufous Fantail Jan-Mar (123)	Willie Wagtail Oct-Dec (78)	Willie Wagtail Jan-Mar (105)	Willie Wagtail Apr-Sep (71)
Closed forest	6.3	18.6*	9.3	12.1	63.4*	30.9*			
Closed/open forest ecotone	1.1	3.5	4.1	2.7	12.2	11.4			
Layered open forest	11.9	9.5	13.0	2.0	12.2	13.8			
Tall open forest	7.9	5.0	8.5	7.4		2.4			
Grassy open forest	3.5		1.6	0.7				1.9	
Layered woodland	6.3	2.0	0.4	2.0			26.9*	8.6	2.8
Maclura thorn scrub	0.5	1.0	2.4	3.4				1.0	1.4
Solanum scrub (deciduous)	1.0	1.5	3.3	3.4		2.4			
Bracken flat *(Pteridium esculentum)* (winter dormant)	15.4	18.6*	17.5*	18.0*	9.8	27.6*	12.8	11.4	9.5
Shrubs, bracken and herbs	15.4	10.1	6.5	5.4		1.6	11.5	30.5*	18.3*
Savanna grassland	19.3	14.1	23.6*	16.1*		4.9	42.3*	41.0*	46.5*
Creek bank	11.4	16.1*	9.8	26.8*	2.4	4.9	6.4	5.7	22.5*

Table 2. Structural features of habitats frequented by fantails. (Values expressed as means ± standard deviations, followed by ranges).

	Grey Fantail	Rufous Fantail	Willie Wagtail
No. habitats	4	2	4
No. sample plots	11	6	13
% canopy cover 1.6 m above ground	38 ± 39 0-95	59 ± 37 5-95	9 ± 14 0-45
Vertical incident light 1.6 m above ground (% open sky reading)	82 ± 28 58-100	74 ± 27 58-100	95 ± 19 81-100
No. trees/ha	369 ± 544 0-1700	663 ± 601 60-1700	155 ± 296 0-1100
No. tall shrubs/ha (2-6 m high)	836 ± 1209 0-3200	1533 ± 1282 0-3200	262 ± 670 0-2400
No. low shrubs/ha (1-2 m high)	4818 ± 4554 0-12400	4133 ± 3602 0-9400	7015 ± 6188 200-22600
Ht. live ground cover (cm)	39 ± 46 0.5-160	63 ± 45 2-160	9 ± 15 0.5-76
% live ground cover (all species)	76 ± 23 40-100	68 ± 27 40-100	90 ± 11 70-100
% live ground cover (ferns)	35 ± 37 0-100	60 ± 32 30-100	9 ± 10 0-35
Litter depth (cm)	3 ± 2 0.5-10	4 ± 9 0.5-10	2 ± 2 0.5-14

(Table 1). These habitats varied greatly in structure (Table 2) but all contained some tall eucalypts and/or She-oaks, *Casuarina*, and some low shrubs. The use of these habitats by Grey Fantails changed during the year; after December (at the end of the breeding season), Grey Fantails were less abundant in closed forest and more abundant in open forest and savanna grassland. In winter (April-September) two-thirds of sightings were in open habitats.

Rufous Fantails were not seen in three of the habitats and 60% of sightings were in two habitats — closed forest and the bracken flat (Table 1); again a shift to open habitats occurred in late summer. Rufous Fantails were absent from the site in winter. The habitats frequented by Rufous Fantails had extensive canopy cover ($\bar{X}$ = 59%), dense tree and tall shrub strata ($\bar{X}$ = 663/ha and 1533/ha respectively) and tall discontinuous ground cover which

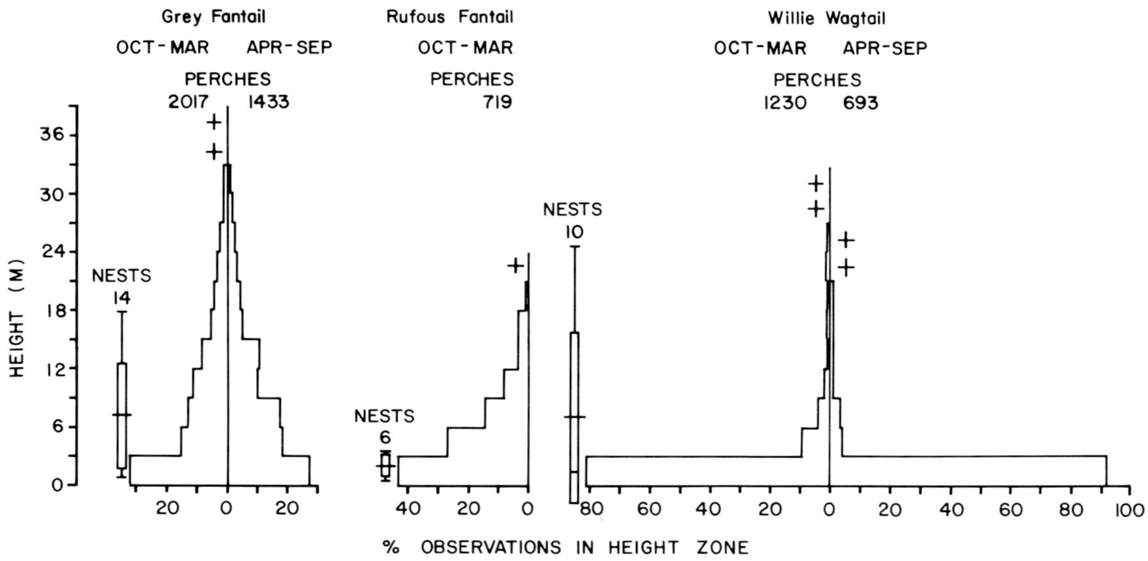

Fig. 1. Vertical distribution of nests and perches. (Very low percentages of sightings in the highest strata are represented by a +. For heights of nests, horizontal lines represent ranges and means and rectangular boxes represent one standard deviation either side of the means).

included a high proportion of ferns, particularly bracken (Table 2).

Willie Wagtails frequented open habitats, principally savanna grassland, the creek bank and the secondary growth of shrubs, bracken and herbs (Table 1). They were virtually absent from forest habitats and the densest habitat frequented — layered woodland — was used regularly only during the breeding season. From January onwards the proportion of sightings of Wagtails in open habitats increased steadily and reached 95% in winter. The habitats frequented by this species had few trees ($\bar{X}$ = 155/ha) or tall shrubs ($\bar{X}$ = 262/ha), and high light intensities; low spindly shrubs were abundant ($\bar{X}$ = 7015/ha) and there was an almost continuous ground cover of short grasses and forbs ($\bar{X}$ ht = 9 cm) (Table 2).

Although Rufous Fantails were not as widely distributed as Grey Fantails, the two often occurred together in forest habitats and on the bracken flat. Willie Wagtails also appeared regularly on the bracken flat but otherwise there was no overlap in the habitats frequented by Rufous Fantails and Wagtails. (In Kikkawa's 1974 analysis of avifauna of vegetation formations in eastern Australia, Rufous Fantails were highly predictable occupants of tall closed-forest and tall open-forest but were absent from all semi-arid forest, woodland and shrubland habitats, whereas Willie Wagtails were consistently present in the latter). Habitat overlap between Grey Fantails and Wagtails at the Five Day Creek site was highest in winter when 43% of Grey Fantail sightings and 70% of Willie Wagtail sightings were in savanna grassland and along the creek bank (Table 1).

1.2. *Foraging Zones*

Grey Fantails utilised most strata of vegetation. In summer nearly 50% of perches were below 6 m in height but foraging also occurred round the tops of trees 40 m tall (Fig. 1). There was a slight contraction of vertical foraging range in winter (April-September). Grey Fantails rarely entered the low shrub and fern strata in closed habitats or on the bracken flat, although they often foraged over the tops of shrubs and bracken. Less than 0.5% of all Grey Fantail perches were on the ground; ground foraging was usually associated with cold wet weather or autumn flocking.

Although Grey and Rufous Fantails both frequented the closed forest between October and December, there was little overlap in their foraging zones. The Rufous Fantails foraged lower and confined most activity to the understorey and shrub strata, sometimes descending to mossy logs but rarely alighting on the ground (about 1% of observations) (Fig. 1). Outside the forest Rufous Fantails often fed round dense bracken, perching on top of the fronds, then disappearing into their midst to forage in the dim humid interior.

The foraging zones frequented by Willie Wagtails differed from those of the other fantails. Most foraging occurred below 3 m (Fig. 1) and in summer 26% of perches were on the ground, often among short grass; between April and September the proportion of ground perches rose to 42% and towards the end of winter up to 90% of perches were less than 10 cm above the ground. Along the creek banks in winter, Wagtails usually perched on rocks and logs beside the water while Grey Fantails foraged in and around trees on the banks, only occasionally descending to perch at water level. On the bracken flat, Wagtails foraged in grassy clearings and in tall shrubs but rarely in the bracken stratum.

Song perches and perches used for maintenance activity by Grey and Rufous Fantails were well within their vertical foraging ranges. In contrast, perches at the top of the vertical range of Wagtails (Fig. 1) were used for maintenance and singing, and extended far above their normal foraging range. The highest perches used by Wagtails and Grey Fantails in winter were 6-8 m lower than in summer.

1.3. *Perch Sites*

Trees with sparse sclerophyllous foliage such as eucalypts and wattles, and she-oaks, provided 60% of perches for Grey Fantails; low inanimate perches such as logs and rocks were rarely used (Fig. 2). Grey Fantails used different types of perches for different foraging strategies; when "static searching" (see below), they often selected dead leafless twigs in the outer crown of a living tree, or a dead branch below the crown. When they progressed rapidly through vegetation in search of food, perches were usually live twigs and branches within the crown.

Most Rufous Fantail perches in forest habitats were thin live twigs of mesophyllous plants which formed an almost continuous network in the understorey. In layered open forest and along the creek bank, foraging occurred as often in the pine-like foliage of she-oaks as in eucalypt foliage (Fig. 2).

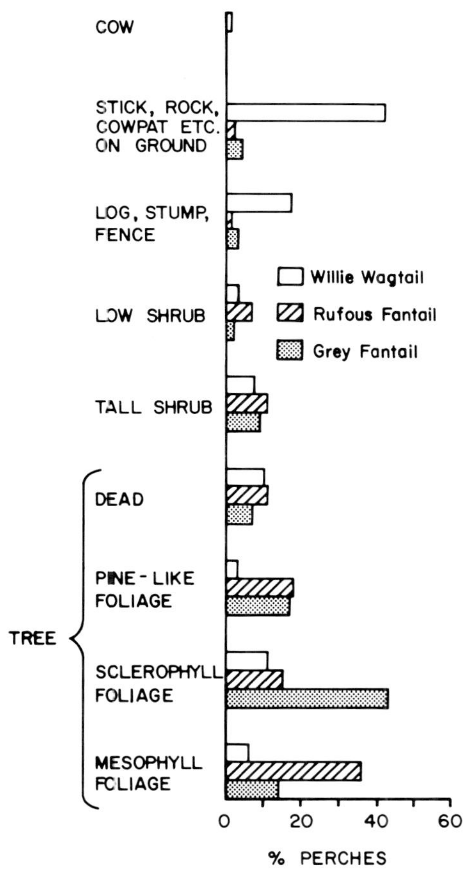

Fig. 2. Substrates used as perch sites throughout the year. (No. of perches: Grey Fantail — 5399; Rufous Fantail — 1866; Willie Wagtail — 1215).

When Willie Wagtails foraged round trees and shrubs, 35-42% of perches were exposed distal branches more than 5 mm thick, on which the birds might sit for several minutes while scanning their surroundings for food. This species also perched on stems of shrubs (and of isolated dead bracken ferns in winter), and foraged round the trunks of trees (about 20% of observations, Fig. 3); foraging occurred in the outer foliage only 20-27% of the time. In winter Wagtails rarely perched in trees and their typical perches were logs, fences, rocks, sticks, cowpats and the ground itself. These broad flat rigid surfaces contrasted with the thin flexible branches typically used by the other fantails.

Although 85% of trees and shrubs in which Grey Fantails foraged were alive, only 35% of their actual summer perches were alive (51% in winter). Ninety-five percent of plants in which Rufous Fantails foraged were alive, and 74% of their perches were alive. Living vegetation was the site of only 45-57% of foraging activity by Willie Wagtails and only 26-30% of their perches were alive.

The foraging stations of Grey Fantails were primarily in and adjacent to the outer foliage (46-55% of observations) and least often in proximal branches and around tree trunks (about 20% of observations) (Fig. 3). Foraging also occurred in open space beyond foliage and over the creek. Rufous Fantails foraged frequently in the outer foliage but as much as 41% of perches were on thin proximal branches and on shrub and bracken stems or tree trunks (Fig. 3). (Although all three fantails foraged around stems and tree trunks, less than 5% of their food captures were at these stations).

1.4. *Nest Sites*

Seventy percent of Grey Fantail nests found were built in forest habitats — five in closed forest and the ecotone between it and open forest, and five in open forest. Nests were also found in savanna grassland and along the creek bank. The tendency of Grey Fantails to nest in the densest or most mesic vegetation available has been recorded elsewhere (Wilson 1976; N.R.S.). In the breeding season parents foraged close to their nests, one reason for the high proportion of sightings in closed habitats between October and December. One nest was built as high as 17.5 m above the ground but the average height was 7 m, slightly above the median foraging height (Fig. 1). Most of the nests were built in trees with sclerophyllous foliage but four were in mesophyllous species.

Six Rufous Fantail nests were located, four in closed forest and two in open forest. Typical sites

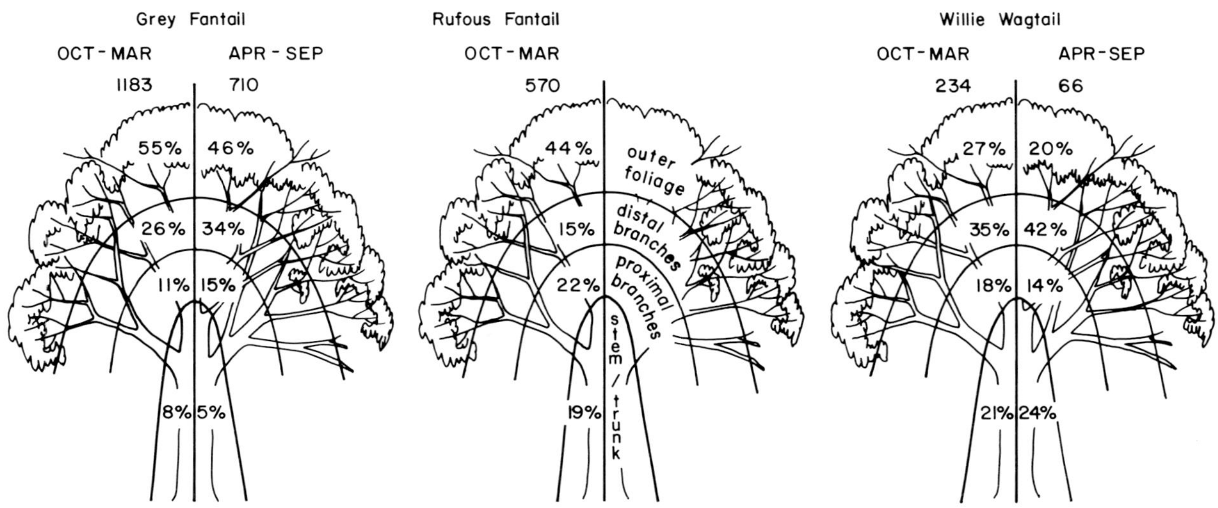

Fig. 3. Foraging stations in trees and shrubs. (No. of observations shown above the trees).

were shrubs or the lowest branches of mesophyllous trees, sometimes beside low cliffs bordering the creek. Maximum (3.7 m) and average (2 m) heights of nests were lower than those of Grey Fantails (Fig. 1). This species rarely nests higher than 3 m (N.R.S.).

Half of the 10 Wagtail nests were in grassy open forest and layered woodland and four more were built in eucalypts and thorny shrubs in savanna grassland. Like the other fantails, Wagtails usually nested in the densest habitats frequented. The highest Wagtail nests were 25 and 22 m above the ground in Blue Gums, *Eucalyptus saligna,* but the next highest was only 4.5 m up, illustrated by the high standard deviation in Figure 1. In the distribution of nest heights recorded in the N.R.S. there were two peaks — one around 3 m and a second smaller peak above 7 m (Marchant 1974).

2. Foraging Strategies

The feeding strategy of a bird may be defined as the sum of postures and locomotory movements involved in the location, capture and consumption of its food. Figure 4 illustrates components of the feeding strategy of fantails:

Static searching involved a bird remaining on a lookout perch for five seconds or more while surveying its surroundings for flying insects ("still-hunting" of Root 1967). Birds *searching progressively* were constantly on the move looking for (often stationary) prey as they progressed through vegetation. A form of locomotion used sometimes by progressively searching fantails was "spinning" (after Diamond 1972) in which a bird rotated through an arc of 180° as it hopped along a branch or up a trunk. Attack and capture tactics could be summarised as (i) flycatching: bird and prey both airborne; (ii) hovering (sometimes called flycatcher gleaning): bird airborne, prey on substrate such as a leaf or branch; (iii) gleaning (pecking): bird and prey on the same substrate.

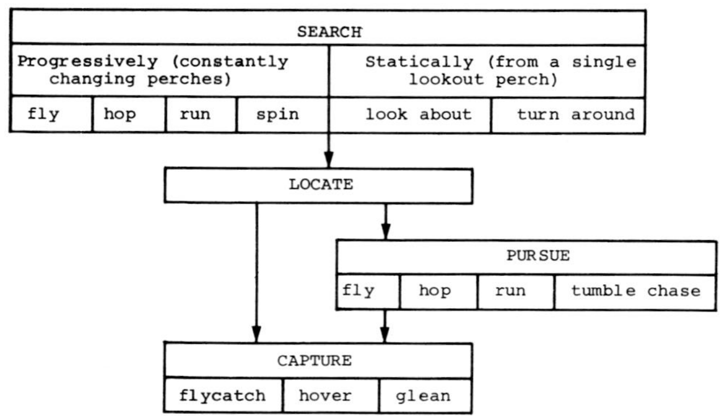

Fig. 4. Behavioural alternatives typical of the foraging sequence of a fantail.

2.1. *Static Searching and Pursuit*

Grey Fantails located most prey by static searching from a series of lookout perches, changing perches on an average of 7-9 times per minute (Fig. 5a). An alert posture was adopted with plumage sleeked, tail closed and wings slightly drooped as the bird looked about and frequently turned around on its perch. Three or four flights in pursuit of prey often followed in quick succession from the same perch. From low perches Grey Fantails performed dives, horizontal sweeps and cone flights over surrounding vegetation or water to catch a single insect. Also from lookouts, particularly in tall trees, they launched into long roundabout flights in the open air. Such flights often included five or six changes of direction (the maximum observed was 12) and lasted up to 20 seconds. Twelve percent of pursuit flights in early summer included these multiple manoeuvres and the proportion rose to 38% in late winter.

Static searching from a lookout perch was rarely performed by Rufous Fantails and was usually

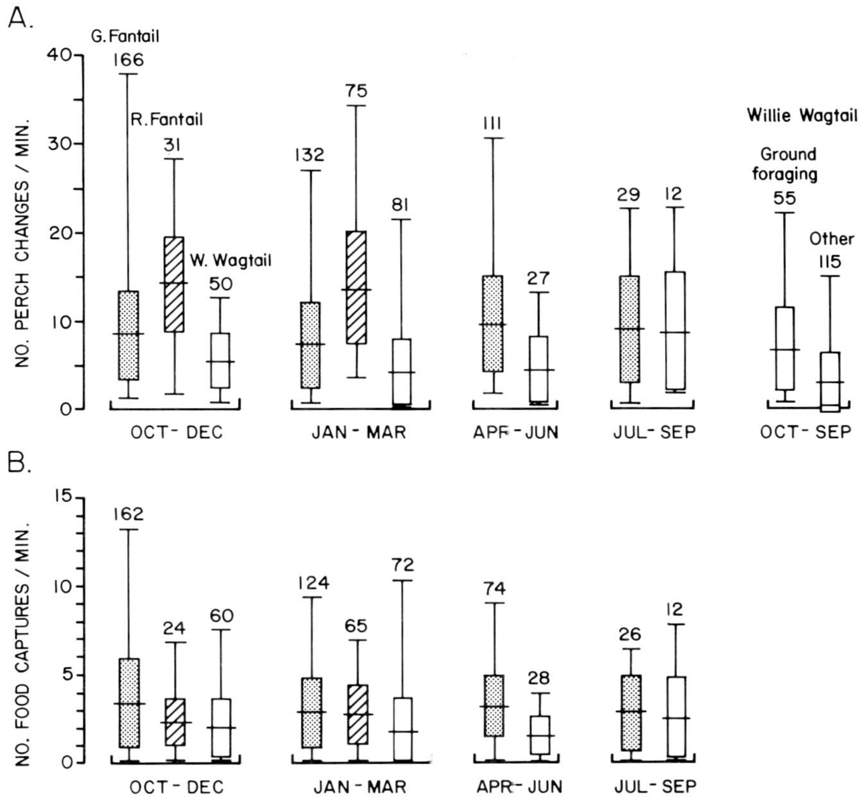

Fig. 5. Foraging intensity illustrated by A. frequency of movement between perches and B. frequency of food capture. (No. of feeding sequences in which foraging intensity measured, shown above columns. Horizontal lines represent ranges and means and rectangular boxes represent one standard deviation either side of the means).

followed by single manoeuvre pursuit flights within the canopy, or a horizontal zig-zag flight through bracken stems. Between 8 and 11% of prey pursuits involved multiple manoeuvres but Rufous Fantails did not perform the protracted aerobatics characteristic of Grey Fantails.

Of the three species, Wagtails most closely resembled "typical" flycatchers in their foraging strategies. They were principally static searchers from lookout perches, sometimes in trees or shrubs but more often on low elevations such as rocks, logs and dried cow dung. Wagtails sometimes remained almost motionless for up to 15 minutes on the same lookout perch and they returned many times to the same perch after insect captures. These low rates of perch turnover are illustrated in Figure 5a (means of 4-5 perches/minute between October and June).

For static searching the plumage was sleeked and the tail closed but the wings were not drooped; a tail-wag often accompanied visual scanning. Sometimes a Wagtail traced the flight of an insect visually but did not pursue it and sometimes Wagtails flew strongly 10-15 m in a straight line to catch a single insect. Pursuit flights from lookout perches included diagonal loops, dives and steep ascending cones; horizontal sweeps over water and grass were common. Multiple manoeuvres were most common in late summer (8.1% of pursuits) and dropped to 2.4% between July and September.

2.2 *Progressive Searching and Pursuit*

When Grey Fantails searched progressively through foliage they moved rapidly by short flights, hops and occasionally spinning, sometimes using more than 20 perches per minute; this behaviour has been described as "flitting" (Crome 1978). In progressive searching the wings were drooped very low and the tail was fanned wide and carried at an angle of 120° to the back, greatly increasing the effective size of the bird and its shadow. The adoption of this posture combined with spinning and wing flashing (wings flicked open and half closed again), resembled the progress of an agitated fan and appeared effective in dislodging and flushing concealed prey which the fantails pursued. The pursuit was often a tumble-chase (after Crome 1978) in which the bird dropped suddenly or fluttered down like a leaf, for a distance as great as 5 m; tumble chases composed 2-7% of all prey pursuits by Grey Fantails (but less than 2% of pursuits by Wagtails).

Rufous Fantails located most prey by progressive searching (flitting) through foliage and this correlates with the high proportion of live perches they used. They were almost constantly on the move, as illustrated by the high average turnover of 14 perches/minute (Fig. 5a). Hopping was the chief form of locomotion by Rufous Fantails, they did not fly as often as did Grey Fantails. Progressive searching in trees and shrubs usually followed a horizontal or ascending path, punctuated by tumble chases (10-15% of prey pursuits), steep dives or short attacking sorties at all angles within the crown. Rufous Fantails sometimes gleaned in clusters of dead leaves.

Progressive hunting by Wagtails was mostly confined to the ground and this is reflected in the faster perch turnover for ground foraging, which was most pronounced between July and September (Fig. 5a). Birds hunting progressively covered 20 m or more in zig-zagging series of runs, hops and low sweeping flights; wing-flashing and tail-fanning were associated with this foraging behaviour and Goodwin (1967) suggested they served to flush cryptic insects.

2.3. *Capture*

Grey Fantails caught most prey on the wing. Hovering at leaves, flowers or bark composed up to 20% of capture tactics in early summer but this species fed predominantly by sallying forth from perches in vegetation to flycatch (Fig. 6). In winter flycatching was used almost exclusively, and on the rare occasions that Grey Fantails foraged on the ground, they usually caught their food in the air. Gleaning accounted for less than 2% of foraging tactics throughout the year. Flycatching was also the principal foraging tactic of Rufous Fantails but hovering and gleaning together contributed up to 30% of their tactics (Fig. 6).

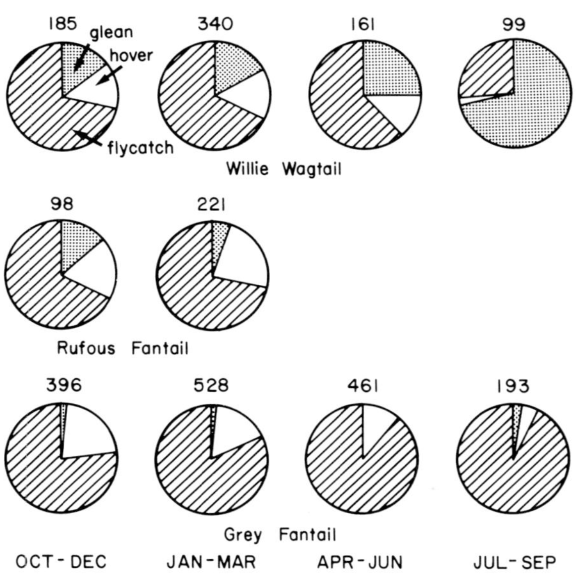

Fig. 6. Seasonal changes in capture tactics. (No. of feeding manoeuvres shown above pie charts).

In early summer Willie Wagtails caught most prey by flycatching, and hovering and gleaning each accounted for about 15% of tactics. However, from January onwards the proportion of flycatching steadily decreased and between July and September there was only 26% flycatching, compared with 72% gleaning, much of it among herbage on the ground. Thus, the foraging tactics of Grey Fantails and Willie Wagtails diverged widely in late winter.

Rate of food capture by Grey Fantails remained fairly constant throughout the year, averaging about three captures/minute with a maximum rate of 13/minute in early summer (Fig. 5b). Rufous Fantails had a slightly lower rate (about 2.5 captures/minute) and their maximum rate was only 7/minute. Willie Wagtails had the lowest mean rates of capture (1.5-2/minute) except in late winter (July-September) when their capture rates approached those of Grey Fantails. No consistent changes in rate of food capture through the day were observed for any of the fantails.

Most of the small prey of Grey and Rufous Fantails were swallowed whole; Willie Wagtails caught considerably larger prey items (see below) and often spent time subduing and dismembering insects or recapturing escapees before they were successfully eaten. They also travelled longer distances in pursuit of single prey items than did the other fantails.

2.4. *Foraging Associations*

Cattle were present in open habitats throughout the summer and between October and March 16% of feeding manoeuvres by Willie Wagtails were performed in their immediate vicinity. The foraging association was an active one — Wagtails used cattle as lookout perches and snatched insects which flew up in front of moving stock. Cattle also attracted arthropods such as flies, dung beetles and ticks which were in the Wagtails' diet (see below). The mean food capture rate by birds feeding with cattle was 2.13 insects/minute, close to the mean for all summer foraging (Fig. 5b). During the entire study Grey Fantails were only twice observed feeding around cattle although on five or six occasions they caught flies around the observer. Rufous Fantails were not observed feeding in association with mammals.

Grey Fantails appeared in mixed-species flocks of small insectivorous birds on 26 occasions, 21 of them between March and June. Rufous Fantails were seen in eight MSFs, four of them in March; Wagtails were not observed in such alliances. No records were kept of flocks which did not contain fantails but mixed species flocking appeared to be most common in autumn, with regular attendance by Grey Fantails. The nuclear species (according to Bell's terminology 1983) were Brown Thornbills and/or

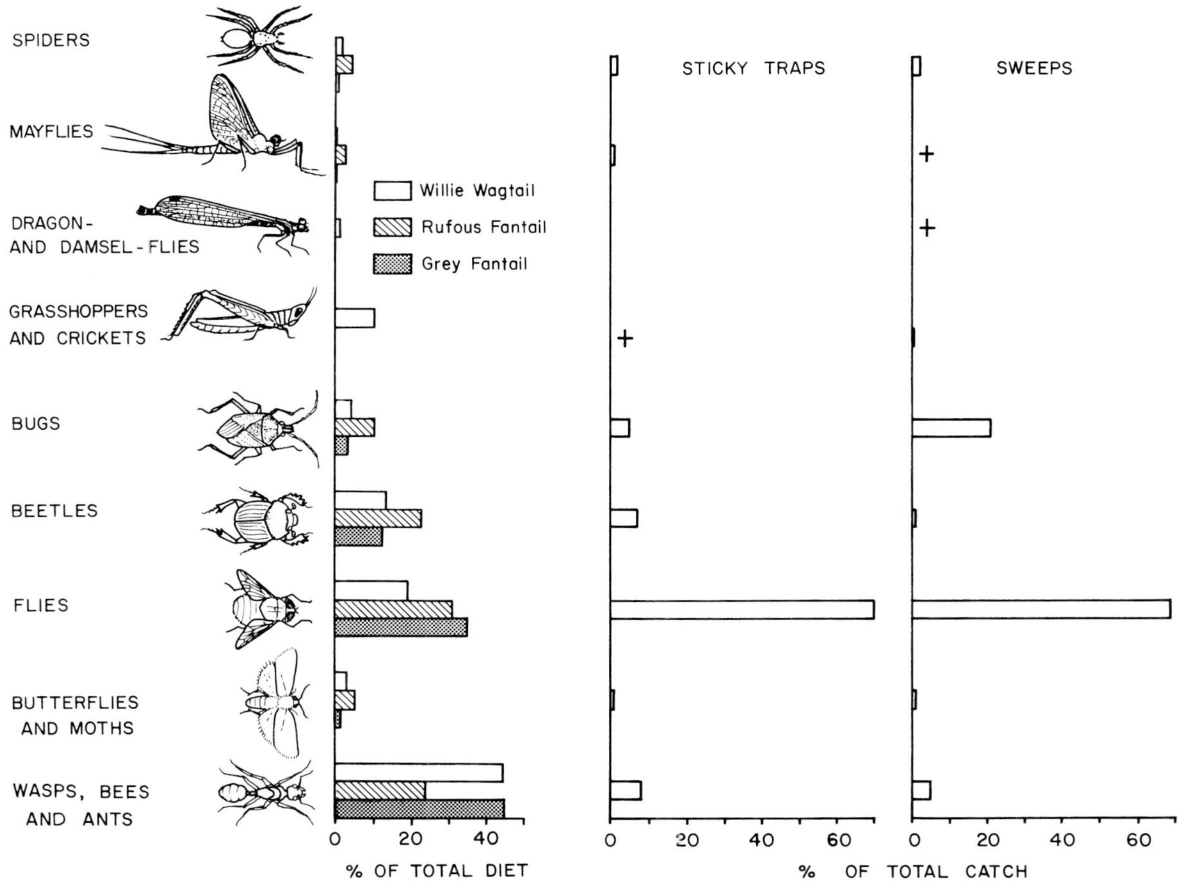

Fig. 7. Diet of fantails compared with the taxonomic composition of arthropods trapped by two methods. (Diet based on the following samples: Grey Fantail — 24 stomachs containing 853 arthropods; Rufous Fantail — 14, 272; Willie Wagtail — 18, 309).

Brown Warblers, and other regular attendants were White-throated Treecreepers and Lewin's Honeyeaters, while Eastern Yellow Robins, Rufous and Golden Whistlers and White-browed Scrubwrens, like Rufous Fantails, were less regular attendants. The association of fantails with other members of the MSFs did not appear to be random; on five occasions a fantail "shadowed" a treecreeper or warbler, maintaining a distance of less than 1 m from it, and catching insects flushed by it.

3. Diet

3.1. *Taxonomic Composition*

Between January and March 72-82% of the insects in the stomachs of Grey Fantails and Willie Wagtails were hymenopterans, including a high proportion of winged ants (Fig. 7). Other common prey of Grey Fantails were tiny flies (including sciarids, chironomid midges, mosquitoes and empidids), some larger flies (muscids, calliphorids, tabanids), small moths (but apparently not caterpillars) and plant-eating beetles and homopteran bugs.

In the diet of Rufous Fantails, salticid spiders and plant-eating homopteran bugs and beetles (relatively sedentary arthropods), were more prominent than in the diets of the other fantails. Crane Flies (tipulids) 8-10 mm long were the most conspicuous flies in the stomachs of this species and many fly families, including March Flies (tabanids) were common to Rufous and Grey Fantail samples. Moths and caterpillars were eaten regularly by Rufous Fantails, especially in late summer and autumn (February-April) when they were foraging extensively round bracken.

Willie Wagtails consumed the widest range of arthropod taxa (Fig. 7). Apart from the high proportion of wasps and winged ants, their diet included many muscid and tabanid flies, caterpillars, moths and butterflies. There was an increase in the proportion of beetles and lepidopterans in late winter samples (July-September) when Wagtail foraging consisted largely of gleaning on grassland and along the creek bank. The most abundant beetle was the dung beetle, *Onthophagus granulatus*, which occurred in 10 out of 24 Wagtail stomachs (and in 20% of Rufous Fantail stomachs); this diurnal beetle frequently occurs round cattle dung and carrion (Matthews 1972). One ixodid tick, an ectoparasite of cattle, was found in a Wagtail stomach.

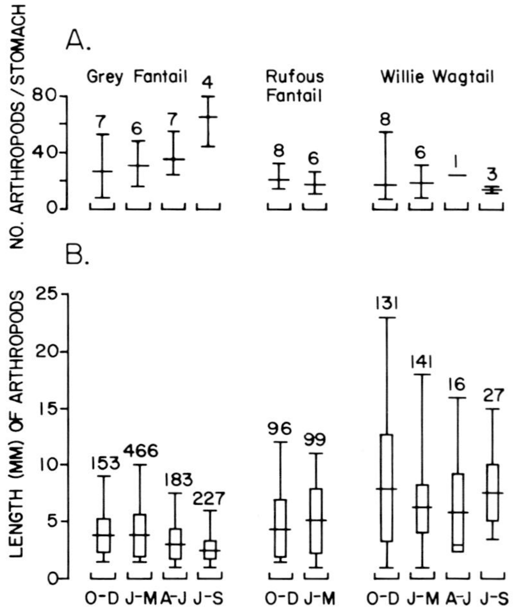

Fig. 8. Seasonal changes in A. number and B. size of prey in stomach samples. (No. of stomachs and arthropods, respectively, shown above columns in A. and B. Horizontal lines represent ranges and means and rectangular boxes represent one standard deviation either side of the means).

3.2. *Size and Diversity of Prey*

All three fantails consumed insects between 0.5 and 10 mm in length but increasing body size of birds corresponded with an increasing proportion of larger prey. The prey of Willie Wagtails was significantly larger than that of Grey Fantails (Fig. 8) and the largest insects recovered from Wagtail stomachs were damselflies and tabanid flies 20-23 mm long.

Stomachs of Grey Fantails collected in winter contained more, smaller insects than did summer samples (Fig. 8) and often a few species of flies or hymenopterans were present in large numbers; this is consistent with the increased tendency of Grey Fantails in winter to execute multiple manoeuvres among minute swarming insects. Rufous Fantail stomachs contained the most diverse and even array of prey and few species were represented by more than one individual. The number of prey decreased as prey size increased in late summer and the number of prey in Wagtail stomachs also varied inversely with average prey size (Fig. 8).

4. Food Availability

Both methods of insect sampling caught an overwhelming preponderance of flies (Fig. 7). It was not possible to extrapolate absolute food availability for fantails from these samples. Most insects trapped were smaller than 5 mm, considerably below the median length of insects in the stomachs, and although some insects 20 mm long were trapped, others 12-15 mm long escaped. The sticky traps favoured the capture of flying insects while much Rufous Fantail prey was sedentary. The sweep samples probably provided a reasonable assessment of food available to Wagtails foraging on grassland.

In Figure 9 the biomass of insect samples is overlaid by rainfall for the corresponding months, and periodic measurements of herbage height in the sweep plot. Rainfall generally corresponded with the abundance of arthropods on an annual scale (lowest between June and September) and to a lesser extent, from month to month, although there was often a time lag of 2-4 weeks between heavy rain and an increase in insect abundance (e.g. during October and November or February and March). Herbage and bracken were tall and lush in February and March but died back in winter. Large fluctuations in catches within habitats from fortnight to fortnight suggest that other weather conditions, too, had a strong influence on abundance and trapability of arthropods during the 24-hour sampling period.

Biomass of arthropods caught on sticky traps was highest between November and January (68-383 mg oven-dry wt/m^2/24 hours, in four out of five habitats) and lowest between June and August (13-145 mg/m^2/24 hours in the same habitats) (Fig. 9). Biomass of arthropods in sweep samples peaked later, in February and March (5.0-9.7 mg/m^2/hour) and was lowest in July and August. The sticky trap over the creek trapped up to eight times the biomass of insects trapped elsewhere, and samples from other open habitats — savanna grassland, and shrubs, bracken and herbs — were usually larger than those in the two forest habitats.

5. Comparative Morphology

The Willie Wagtail is one of the largest Australian flycatchers. Individuals at the study site were more than twice as heavy as Grey Fantails and usually 30-50% larger in linear dimensions, while Rufous Fantails were up to 15% larger than Grey Fantails in linear measurements (Table 3). When standardised against the cube root of body weight, the leg and particularly the wing bones of Wagtails were proportionally longer than those of the other two species (significant at the 5% rejection level in an analysis of variance). (Harrison (1976) found the leg bones of Wagtails were consistently longer than those of Grey Fantails when standardised against either cranial or coracoid length). Rufous Fantails had proportionally longer leg bones than did Grey Fantails, but shorter wing bones.

In most wing and tail measurements, Grey Fantails (the smaller species) equalled or exceeded Rufous Fantails (Table 3). Their flight feathers (primaries and rectrices) were 1-5 mm longer than the corresponding feathers of Rufous Fantails; they had longer tails and the wings of the two species were almost

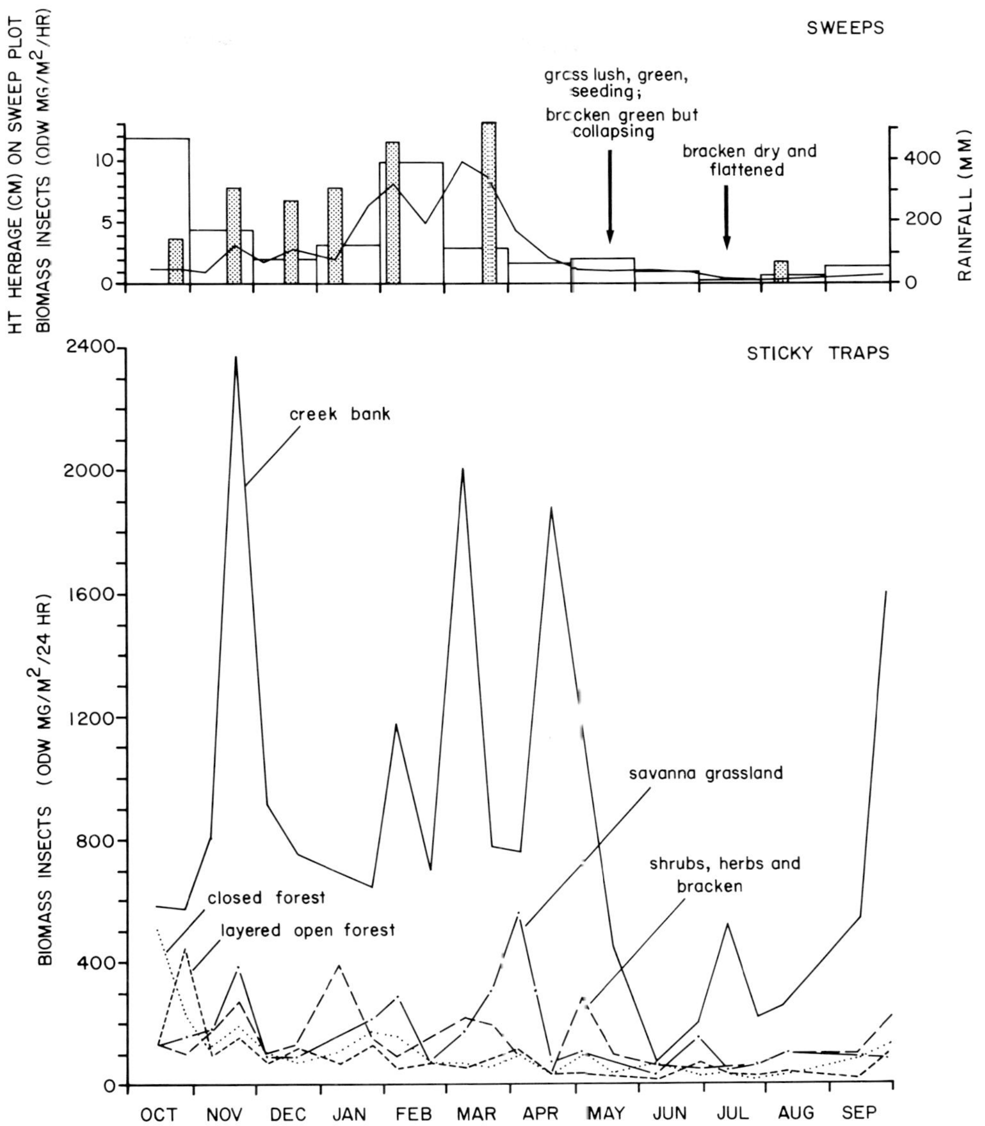

Fig. 9. Seasonal changes in biomass of arthropods trapped by two methods, compared with rainfall (broad open bars) and height of herbage (narrow stippled bars). (Values plotted are averages for each sampling period).

Table 3. Comparative dimensions of adult fantails. (Values expressed as means ± standard deviations).

Species	N	Fresh body weight (g)	Total length (mm)[1]	Leg length (mm)[2]	Wing length (mm)[3]	Tail length (mm)[4]	Wing loading (g/cm^2)[5]	Upper jaw length (mm)[6]
Grey Fantail	11-22	8.5 ± 0.2	160	49.6	86 ± 2	87 ± 2	0.108 ± 0.016	10.7 ± 0.2
Rufous Fantail	10-14	10.4 ± 0.6	160	56.4	87 ± 2	83 ± 2	0.141 ± 0.024	12.5 ± 0.2
Willie Wagtail	12-17	20.8 ± 1.0	200	73.5	120 ± 3	106 ± 1	0.144 ± 0.012	16.1 ± 0.3

[1] after Wilson (1976); [2] sum of mean lengths of femur, tibiotarsus and tarsometatarsus; [3] tip of longest primary on outstretched wing, to side of body; [4] tail tip to base of central rectrices; [5] body weight ÷ total wing area; [6] bill tip to base of feathers on culmen.

identical in length. As a consequence, the wing-loading of Grey Fantails was lower than that of Rufous Fantails and Willie Wagtails (significant at the 5% level of rejection in analysis of variance). In outline the wings of Grey and Rufous Fantails were rather blunt, while Wagtails had more sharply tapering wings.

Not only was the Grey Fantail's bill 15% shorter than that of the Rufous Fantail, it was less robust in structure (Fig. 10); the Wagtail's spear-shaped bill was 50% longer than that of the Grey Fantail. These size discrepancies in bill length were as large as any others measured on the birds. The fantails possess both upper and lower sets of mandibular bristles (which are characteristic of insectivores) but these are comparatively much longer on the Grey Fantail than on the Wagtail; Rufous Fantails are intermediate in this character.

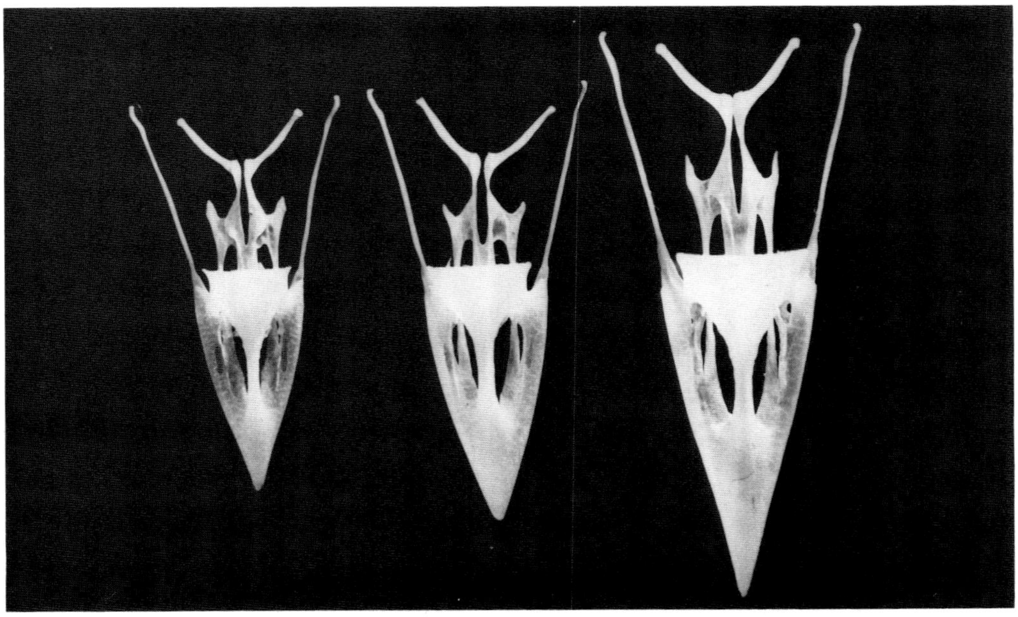

Fig. 10. Dorsal view of upper mandibles of mature female fantails. Left to right: Grey Fantail, Rufous Fantail, Willie Wagtail. Note the relatively smaller area of the bill of the Grey Fantail anterior to the nares.

6. Seasonal Fluctuations in Population Size

It was not possible to conduct an accurate census of unmarked birds at the study site. Grey Fantails were by far the most abundant species with a population during the breeding season of approximately 25 within an area of 25 ha, compared with fewer than 10 each of Rufous Fantails and Willie Wagtails (Fig. 11).

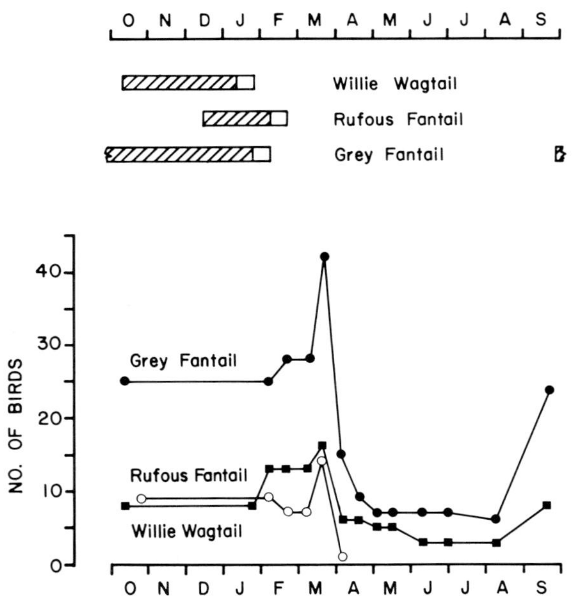

Fig. 11. Seasonal changes in the abundance of fantails in the 25 ha study area. The duration of breeding seasons is shown above the population graph. (Onset of egg-laying to last nestling — cross-hatched rectangles; fledged young still being fed by parents — open squares).

Densities during the breeding season calculated on the area covered by "frequented" habitats (those with more than 15% of sightings) were: Grey Fantails — 18/10 ha, Rufous Fantails — 16/10 ha and Willie Wagtails — 7/10 ha. On the south coast of New South Wales, Grey Fantail populations have been estimated at 8-14 birds/10 ha (Marchant *in* Ford 1981).

At the study site Rufous Fantails were totally migratory; the first individuals were recorded in mid-October and the last sightings were in late March or early April. This is the usual pattern for Rufous Fantails which breed in southeastern Australia and overwinter in northern Queensland (Wilson 1976). Although Grey Fantails and Willie Wagtails were resident throughout the year, winter censuses recorded only about one-third the numbers present in summer (Fig. 11) and some birds probably undertook local migrations. Local seasonal movements, including nomadism, have been recorded for some populations of these species (Wilson 1976).

Fluctuations in fantail populations were most pronounced in February and March. Numbers rose slightly in February with the recruitment of local young of the year, and in March there was a brief influx of Grey and Rufous Fantails. The latter appeared singly on their annual migration north and local Rufous Fantails followed in their wake. The Grey Fantail arrivals were much more conspicuous, flitting about the savanna grassland in flocks of three to six, continuously uttering "chip" calls and chasing one another. In 1973 these flocks appeared at Five Day Creek between 19th and 23rd of March; similar behaviour by a flock of 6-8 Grey Fantails was observed at Armidale (75 km west of the study site, altitude 1000 m) on 6th March 1972 (pers. obs.) and in autumn 1975 (R. Beeton, pers. comm.). (Grey Fantails are absent from Armidale in winter). The Grey Fantail population at Five Day Creek declined immediately after this time. No concerted movement of Wagtails was observed to explain the decline in their numbers at the study site.

7. Social Organisation and Breeding Success

Autumn flocking was not observed in Rufous Fantails or Willie Wagtails and all three species usually occurred singly or in pairs. In southeastern Australia Willie Wagtails appear to form stable pairs and defend their territories throughout the year (Janet Walker *in litt.*) and this was consistent with observations at Five Day Creek (where the average size of four Wagtail territories was 3.3 ha). Grey Fantails probably also form stable pairs (McLean 1912); they held mutually exclusive territories during the breeding season (mean area of 12 territories = 0.9 ha) but their territories overlapped broadly with those of Wagtails. Grey Fantail territories also overlapped the area occupied by Rufous Fantails but there were insufficient observations of the latter to determine the extent of their territories. A Willie Wagtail supplanted a Grey Fantail three times but the reverse action was not observed. In eight aggressive encounters between Grey and Rufous Fantails, the outcome favoured each species equally and on seven occasions the two foraged in close proximity without overt aggression. The distribution of Wagtails in one of their frequented habitats — layered woodland — was almost certainly restricted by aggression from a colony of Bell Miners (*Manorina melanophrys*) which mobbed and supplanted Wagtails on eight occasions.

The span of the nesting seasons of fantails at the study site is illustrated in Figure 11. Although the last fledglings of all three species left the nest within approximately three weeks of one another, Rufous Fantails commenced egg-laying approximately nine weeks after the other fantails and thus had a much shorter breeding season. In southeastern Australia the N.R.S. recorded a three month gap between onset of egg-laying by Wagtails and Grey Fantails, and egg-laying by Rufous Fantails; the last fledglings quit the nest between the second and third weeks in February.

Observations of breeding success at Five Day Creek were fragmentary. Grey Fantails deserted 26% of nests before egg-laying, while none of the Rufous Fantail nests located were deserted prior to egg-laying. Wagtails deserted 30% of nests but one Wagtail pair successfully reared two broods in the same nest. The success rates (i.e. fledglings left the nests) of those nests in which eggs were laid were: Grey Fantails, 59%; Rufous Fantails, 71% and Willie Wagtails, 67%. Another measure of breeding success was the number of free-flying young birds observed at the site. Over two years, 19 Grey Fantail pairs produced a minimum of 20 young and the comparable figures for Rufous Fantails and Willie Wagtails were 2/8 pairs and 19/8 pairs respectively. Average clutch size of Grey Fantails was 2.4 (5 clutches), of Rufous Fantails, 2.0 (3 clutches) and of Willie Wagtails, 3.0 (5 clutches). It was easier to monitor success of Wagtail nests because they were built in comparatively open habitats and parents defended them vociferously, and territories were large and easily distinguished. Grey and Rufous Fantails nested in denser vegetation, were less conspicuous in defence of nests and fledglings and their territories were smaller and more difficult to distinguish, so the above estimates of success must be considered biased.

DISCUSSION

Like many northern hemisphere flycatchers, the Willie Wagtail is primarily a type I predator or "pursuer" according to Schoener's definition (1969). The largest of the three fantails, it spends long periods on a lookout perch engaged in maintenance activities as well as scanning its surroundings for prey. It is sometimes selective as to the prey it pursues. It feeds less frequently and takes proportionally larger insects than do the two smaller fantails. This is consistent with the tendency for a negative correlation between body size and metabolic rate in birds (Kendeigh 1972). Wagtails make long direct flights across territories and in pursuit of single large airborne insects, but are also capable of hovering and fluttering in the air. Their versatility extends to gleaning stationary prey on the ground and on this substrate they frequently run, locomotion unusual in passerines which forage on the ground. Posturing (wing-flashing) may be used to startle prey into movement. Association with large grazing herbivores appears to be opportunistic and probably does not significantly increase the foraging success of Willie Wagtails.

The Rufous Fantail exhibits many characteristics of a type II predator or "searcher", and spends much time and energy engaged solely in searching for food within vegetation, and comparatively less energy in pursuit and capture of prey. This species appears to be an opportunistic feeder, taking any prey items it encounters within a certain size range. Prey of Rufous Fantails is more variable and proportionally larger than that of Grey Fantails and capture rate is generally lower. Progress during foraging is usually by hops and short flights but aerial manoeuvrability is displayed in hovering, tumble chases and occasional multiple capture flights; the two smaller fantails rarely fly far in a straight line. Posturing (fanning wings and tail and spinning) may be employed by Rufous and Grey Fantails to startle concealed prey into movement.

In its foraging behaviour the Grey Fantail displays some of the characteristics of a "pursuer" predator on the periphery of shrubs and trees, but forages with a considerably greater intensity than the Wagtail and appears to be an opportunistic feeder. Most of the Grey Fantail's prey are small flies and wasps caught in the air and it captures prey more frequently than do the other two fantails. Fluttering flight is the principle form of locomotion and the Grey Fantail efficiently utilises very small prey by making multiple manoeuvre flights within swarms of insects. Like the Willie Wagtail this species is a rather versatile forager and within vegetation can switch to a strategy of searching similar to that of Rufous Fantails. Participation in mixed-species flocks appears to be an integral part of the Grey Fantail's foraging strategy in autumn, and may also increase the foraging success of Rufous Fantails.

The foraging strategies of the Willie Wagtail are most effective in open habitats which provide extensive visibility, unimpeded flyways and a continuous stratum of herbage. Scattered open-crowned shrubs and trees or man-made structures (e.g. fences) are necessary to provide raised lookout perches as well as sites for nests and for the advertisment and defence of territory. Only short herbage is suitable for ground foraging by the short-legged Wagtails and they have adapted well to artificially created grasslands such as grazed pastures and lawns in parks and gardens.

Although Grey Fantails forage in structurally dissimilar habitats and in most strata of vegetation above the ground, their activity is concentrated on the periphery of vegetation where they can adopt pursuing or searching strategies. It is suggested the environmental features most important to Grey Fantails are extensive brightly-lit surfaces of sparse foliage extending from ground level to the tops of tall trees. Eucalypt woodland and forest edges along watercourses, roads and clearings, fulfil these criteria.

Of the three fantails, the Rufous Fantail has the most uniform foraging behaviour and its foraging and nesting are confined largely to the interior of low strata where light intensity is low. Although most habitats frequented by Rufous Fantails at Five Day Creek had a tree canopy, thickets of tall shrubs alone (*Casuarina, Melaleuca*), have provided suitable breeding habitat for this species (H. L. Bell, *in litt.*). It appears that a large volume of live vegetation which provides closely-spaced perches and has a continuous canopy, is important for the searching strategy of the Rufous Fantail. Secondary growth which results from clearing or burning, mesic vegetation left in gullies and along watercourses during timber clearance, and forests with well-developed understories, may all provide suitable conditions for Rufous Fantails.

Morphological differences between the three fantails correlate well with differences in their foraging strategies and the habitats they frequent. The largest species, the Willie Wagtail, is the most terrestrial and has proportionally the longest legs which enable it to run fast after prey. It also has proportionally the longest, most pointed wings and is a stronger, faster flier than the smaller fantails. Wagtails, which capture and manipulate large prey and often peck on the ground, have large strong bills which probably function like tongs (Lederer 1972), and short mandibular bristles.

The smallest species, the Grey Fantail, has the most aerial lifestyle, largest flight surface to volume ratio, lowest wing loading and shortest legs. In contrast to the Wagtail, it has a small weak bill surrounded by long bristles, which probably functions more like a funnel than a pair of tongs. The Rufous Fantail is slightly larger and more arboreal than the Grey Fantail; it has longer legs and shorter wings, which would appear to be advantageous in hopping and flying through dense vegetation. The bill of this species is considerably larger and more robust than that of the Grey Fantail and it takes larger prey and gleans more than does the latter.

It is suggested that the fantails at Five Day Creek respond to changes in their environment in the following ways.

In early summer the food supply is comparatively high in all habitats but fantails are concentrated in the denser habitats, perhaps because of better shelter and concealment afforded to nests and young. In late summer the food supply in closed habitats appears to decline in relation to that available in open habitats, and the arthropod biomass on the grassland reaches a peak; bracken and herbs are tall and lush. At this time Rufous Fantails feed extensively on the bracken flat and Grey Fantails and Wagtails frequent more open habitats.

The departure time of Rufous Fantails in March/April is very regular and occurs before food has become scarce and while undergrowth is still lush, so it may be determined by ultimate factors such as annual rainfall pattern or declining plant growth (Nix 1976). (By the middle of winter, conditions would almost certainly be unsuitable for Rufous Fantails, with very low arthropod biomass in closed habitats where light intensity and daytime temperatures drop very low, and with the senescence of undergrowth such as bracken in open habitats). The coincident decline in populations of Grey Fantails and Willie Wagtails in April may be related to the same ultimate factors.

Ford (1981) has reviewed the seasonal movements of the Grey Fantail species complex in Australia. In autumn the subspecies which breeds in northeastern New South Wales appears to disperse from the high altitudes of the Great Dividing Range westwards towards the foothills and eastwards towards the coast, and possibly northwards. Movement towards more open habitats from cold shaded valleys also occurs (Lamm and Calaby 1950). It is suggested that in autumn some Grey Fantails which have bred in the Five Day Creek valley join birds dispersing from the New England Plateau and move with them towards lower warmer altitudes near the coast. Thus competition for the remaining resources in the valley would be reduced.

For the birds which remain at Five Day Creek, food supplies are very low between June and August. Although Grey Fantails and Willie Wagtails converge on the same open habitats in winter, their feeding behaviour diverges and this may enable them to coexist if resources are limited. Wagtails become highly terrestrial and forage with greater intensity, gleaning and flycatching on the ground; herbage is low at this time and thus suitable for such a strategy. Grey Fantails seem to concentrate on aerial feeding, catching swarming gnats in multiple manoeuvre flights in the open air. This strategy may be adopted only where the winters are dry and sunny, because in damp dull weather Grey Fantails forage close to the ground (this study; McLean 1912). The congregation of insects round water may be important to both species in winter, when the food supply is universally low.

Differences in the reproductive biology of fantails may also contribute to differences in their adaptation to changing environments. In summer the population density of Willie Wagtails at Five Day Creek is about half that of Grey Fantails but their territories cover all suitable habitat. The factors limiting the breeding population appear to be lack of suitable habitat, large territory size and perhaps aggression from Bell Miners. However, the potential for population increase is high because the Wagtail's breeding season is long, clutch size is larger than that of the other fantails, and Wagtails appear to be more successful in rearing young. Such a potentially fertile population should be able to quickly colonise new areas of suitable habitat. Grey Fantails have the highest population density, perhaps because many more territories can fit into the large area of suitable habitat. Their breeding season is as long as that of the Wagtails but their clutch size is smaller and they have a higher rate of nest failure. Although their density in suitable habitat is high, Rufous Fantails have a much shorter breeding season and a small rather invariable clutch size so they do not have the same potential for population increase.

ACKNOWLEDGEMENTS

For assistance, encouragement and advice during the course of this study and its subsequent documentation, I would like to express my gratitude to the following colleagues:

The late Harry Bell, Walter Boles, Hal Cogger, Bob Harden, Hal Heatwole, Lorraine Hecker, Laurie Jessup, Wayne Longmore, Harry Recher and Pat Watters.

My father imbued me with a keen interest in natural history which led eventually to the fantails. The study was undertaken in fulfilment of an M.Sc. degree at the University of New England and was supported by a Commonwealth Postgraduate Research Award.

REFERENCES

BELL, H., 1983. A bird community of lowland rainforest in New Guinea. 5. Mixed-species feeding flocks. *Emu* **82** (Supplement): 256-75.

CROME, F. H. J., 1978. Foraging ecology of an assemblage of birds in lowland rainforest in northern Queensland. *Aust. J. Ecol.* 3: 195-212.

DIAMOND, J., 1972. Avifauna of the Eastern Highlands of New Guinea. *Publs Nuttall Ornithol. Club* **12**: 1-438.

FORD, J., 1981. Evolution, distribution and stage of speciation in the *Rhipidura fuliginosa* complex in Australia. *Emu* **81**: 128-44.

GOODWIN, D., 1967. Notes on behaviour of some Australian birds. *Emu* **66**: 237-51.

HARRISON, C. J. O., 1976. Some aspects of adaptation and evolution in Australian fan-tailed flycatchers. *Emu* **76**: 115-19.

KEAST, J. A., 1958. Variation and speciation in the Australian flycatchers (Aves: Muscicapinae). *Rec. Aust. Mus.* **24**: 73-108.

KENDEIGH, S. C., 1972. Energy control of size limits in birds. *Amer. Nat.* **106**: 79-88.

KIKKAWA, J., 1974. Comparison of avian communities between wet and semi-arid habitats of eastern Australia. *Aust. Wildl. Res.* **1**: 107-16.

LAMM, D. W. AND CALABY, J. H., 1950. Seasonal variation of bird populations along the Murrumbidgee in the Australian Capital Territory. *Emu* **50**: 114-22.

LEDERER R. J., 1972. The role of avian rictal bristles. *Wilson Bull.* **84**: 193-97.

MCLEAN, J. C., 1912. Bush birds of New Zealand. *Emu* **11**: 171-87.

MARCHANT, S., 1974. Analysis of nest-records of the Willie Wagtail. *Emu* **74**: 149-60.

MATTHEWS, E. G., 1972. A revision of the scarabaeine dung beetles of Australia. I, Tribe Onthophagini. *Aust J. Zool. Suppl.* **9**: 1-330.

NIX, H. A., 1976. Environmental control of breeding, post-breeding dispersal and migration of birds in the Australian region. Pp. 272-305 *in* "Proceedings of the 16th International Ornithological Congress" ed. by H. J. Frith and J. H. Calaby. Australian Academy of Sciences: Canberra.

ROOT, R. B., 1967. The niche exploitation pattern of the blue-gray gnatcatcher. *Ecol. Monogr.* **37**: 317-50.

SCHOENER, T. W., 1969. Models of optimal size for solitary predators. *Amer. Nat.* **103**: 277-313.

WILSON, S., 1976. in "Reader's Digest Complete Book of Australian Birds". Reader's Digest Services P/L: Sydney.

Red-browed Treecreeper, *Climacteris erythrops*

CHAPTER 16

Habitat Use by Three Bark-foragers of Eucalypt Forests

Richard A. Noske[1]

Three species of specialized bark-foraging birds commonly occur together in eucalypt forests of southeastern Australia. The foraging ecology and life histories of these three species — the White-throated Treecreeper *Cormobates (Climacteris) leucophaea*, Red-browed Treecreeper *Climacteris erythrops*, and the smaller Varied Sittella *Daphoenositta chrysoptera* — were studied over several years at Wollomombi Falls in northeastern New South Wales.

Sittellas foraged primarily on stringybarks, the most abundant eucalypt, where they spent about one-third of their time on dead branches. White-throated Treecreepers also preferred stringybarks, but spent proportionately more time foraging on the rough, fissured trunks and branches, and less time on dead substrates than Sittellas. The Red-browed Treecreeper foraged on smooth-barked eucalypts more than the other two species, but on stringybarks and boxes made greater use of the smooth or peeling bark on the outer branches.

Red-browed Treecreepers ate more spiders and small beetles than White-throated, but both species ate equally large quantities of ants, which they gleaned from the surface of the bark with their slightly decurved bills. By contrast, the Sittellas rarely ate ants; their specialized wedge-shaped bills were employed in excavating flakes of bark to uncover invertebrates living beneath the surface.

The White-throated Treecreeper mainly roosted on the surface of tree trunks and stumps and nested in knotholes, especially of stringybarks. Red-browed, on the other hand, preferred hollow dead branches of live boxes and gums, or entirely dead trees, for both nesting and roosting. Sittellas nested and roosted on dead branches in the canopy. Thus it is evident that dead trees and branches, and the hollows that form in them, constitute an important resource for all three forest-dwelling bark-foragers, especially the Red-browed Treecreeper.

Within its range, the White-throated Treecreeper is the most generalized of the three species in its habitat requirements, showing a remarkable capacity to colonize "new" habitats. In contrast, the more specialized foraging, nesting and roosting requirements of the Red-browed, and its large group-territories explain why this species typically occurs at lower population densities than the White-throated, where they coexist. The Sittella is apparently more specialized in its food requirements than the treecreepers, but more generalized in its nesting and roosting requirements.

INTRODUCTION

ONE OF THE most striking features of the genus *Eucalyptus* is the diversity of bark form displayed by different species, even in the same area. Indeed, most eucalypts can be conveniently classified according to the type of bark present on the trunk and branches. Stringybarks, for example, are characterized by vertically furrowed, coarsely fibrous bark persisting to the thinnest branches and twigs, which have a smooth surface. Gums, on the other hand, shed the whole of their outer bark in long strips ('ribbons') annually, leaving a powdery, smooth surface. Boxes are intermediate between these two extremes, but vary depending on the species. Most boxes possess flaky or scaly bark on the trunk and this persists to the small branches in some species, while in others, the larger limbs are smooth, often festooned with strips of peeling bark. Variation in bark structure undoubtedly influences the distribution of bark-dwelling invertebrates, and ultimately, the foraging tactics of birds preying upon them.

[1]Department of Biology, School of Technology and Science, Darwin Institute of Technology, Casuarina, Northern Territory 5792.

Pages 193-204 *in* BIRDS OF EUCALYPT FORESTS AND WOODLANDS: ECOLOGY, CONSERVATION, MANAGEMENT ed. by A. Keast, H. F. Recher, H. Ford and D. Saunders. Royal Australasian Ornithologists Union and Surrey Beatty & Sons.

Birds that forage on bark can be divided into two basic categories. First, there are those that regularly forage on other substrates (in particular, foliage) and fly or hop, rather than climb, to various bark surfaces. This group includes the Shrike-tit *Falcunculus frontatus* and several species of honeyeaters (e.g. Strong-billed and White-naped *Melithreptus validirostris* and *M. lunatus*, and White-eared and Yellow-throated Honeyeaters *Meliphaga leucotis* and *M. flavicollis*), which typically pry into loose or peeling bark, and in the case of the Shrike-tit, excavate decaying branches (Keast 1968; Ford and Paton 1976, 1977; Thomas 1980; pers. obs.). Other members of this group are the Buff-rumped Thornbill *Acanthiza reguloides* and Scarlet Robin *Petroica multicolor* which commonly glean or snatch insects hiding in fissures on the trunks and limbs of stringybarks and other rough-barked eucalypts (Bell 1983; Huddy 1979). The second group of bark-foragers have modified hind-limbs (viz. short legs, long toes and strongly curved claws) to enable them to climb steadily up vertical tree-trunks and in some cases, along the underside of horizontal branches.

Australia has two families of specialized bark-climbers: the Australasian treecreepers (Climacteridae) and sittellas (Neosittidae), both of which occur in New Guinea but are absent from Tasmania. Most of the six species of treecreepers are allopatric or prefer different habitats (Keast 1957). However, in the moist southeast of Australia, the Red-browed Treecreeper *Climacteris erythrops* invariably coexists with the White-throated *Cormobates (Climacteris) leucophaea*. The latter is found in almost all wooded habitats within its range, but the Red-browed is mainly restricted to eucalypt forests on the tablelands and adjacent coastal plains. These forests are also inhabited by the Orange-winged (Varied) Sittella *Daphoenositta c. chrysoptera*. This chapter compares the ways in which these three species use the eucalypt forests, focusing on their feeding, nesting and roosting requirements. These aspects of their life histories are related to their differing morphologies and social organizations.

STUDY SITE AND METHODS

Colour-banded populations of the three species were established at Wollomombi Falls, 40 km east of Armidale, New South Wales, mainly between 1977 and 1979. Few Sittellas were banded as these birds usually flew too high to be caught in mist-nets. The site comprised about 50 hectares of open forest, dominated by eucalypts, with few understorey trees (mainly *Acacia* spp.) and a generally sparse cover of shrubs (*Jacksonia, Olearie, Cassinia*). Most of the eucalypts were mature, with hollows and/or dead branches; deadfalls were abundant on the ground.

Live eucalypts consisted of four main types, arranged as follows (in order of decreasing abundance of trees, and increasing amount of smooth bark): stringybarks (*E. caliginosa, E. laevopinea*), rough boxes (*E. conica, E. bridgesiana, E.* sp. aff. *cypellocarpa*), yellow box (*E. melliodora*) and gums (*E. blakelyi, E. amplifolia, E. viminalis*). Previously, I combined *E.* sp. aff. *cypellocarpa* and *E. melliodora* under "smooth boxes" (Noske 1979) but smooth bark does not extend down to the trunk in the former as it commonly does in the latter. The site was subdivided into four sectors, each roughly equal in area. In each sector, the relative abundance of the four tree-types was determined by identifying the tree nearest to points at 20 m intervals along fixed transects covering the whole sector.

Foraging data were collected on fine days in all seasons from March 1978 to March 1980. The foraging location and behaviour of individual birds was described continuously at 10 second intervals or less using a tape-recorder for up to 10 minutes, rarely longer. Niche overlap was determined with the formula, $O_{ij} = \Sigma P_{min\ (i\ or\ j)}$ (Ricklefs and Cox 1977). Initially foraging diversity indices were calculated to indicate trends towards specialization, but as these were found to depend on the relative proportions of categories available, and thus often gave distorted pictures of degrees of specialization exhibited by the different species, they were abandoned.

Intersexual and temporal variation in foraging were examined using analyses of variance and discriminant analyses, but these results will be presented elsewhere.

Investigations of food availability were restricted to the lower trunks of trees because branches were generally inaccessible. Two methods were employed. Each month I counted the arthropods within marked grids (50 × 50 cm) on the trunks of 32 trees (eight each of the four main tree-types). Secondly, I removed bark from the live trunks and dead branches of unmarked trees. The bark was collected in plastic bags which were weighed and sorted thoroughly by hand for arthropods. The first method detected mainly surface-dwelling arthropods whereas the second included arthropods that hid or lived beneath the bark surface.

Specimens of treecreepers and sittellas were collected each month from areas two to 12 km outside the study site for analyses of diet and morphology. All three species were taken concurrently in the late afternoon to allow direct comparisons of diet and to minimize possible diurnal variation in stomach fullness. Additionally, faeces were collected and live prey noted at the study site. Stomach contents and faeces were examined under binocular microscope, and all items counted and identified (where possible to family level) using features such as those mentioned in Calver and Wooller (1982). The body length of items was

Table 1. Mensural characters of the three study species.

Species	Bill length (mm)	Wing length (mm)	Body weight (g)	Tarsus (mm)	Tail (mm)	Hind-claw curvature*
White-throated Treecreeper	14.3 ± 0.11	90.4 ± 0.46	22.1 ± 0.19	24.2 ± 0.13	63.3 ± 0.48	3.9 ± 0.05
Red-browed Treecreeper	14.1 ± 0.10	85.9 ± 0.26	22.8 ± 0.22	24.4 ± 0.16	64.1 ± 0.37	4.1 ± 0.06
Varied Sittella	11.6 ± 0.15	79.6 ± 0.51	11.3 ± 0.22	17.4 ± 0.10	37.7 ± 0.35	4.2 ± 0.06

Values represent mean ± standard error; vertical bar beside means indicates values not significantly different ($p>0.05$); sample size for all species is over 18. * measured as in Norberg (1979); the smaller the value, the greater the curvature.

estimated from comparison with entire specimens collected from bark and wet weight calculated from this using power equations computed for each major taxon (weight-length data supplied by G. Zug).

RESULTS

Morphology

As noted previously (Noske 1979), the White-throated and Red-browed Treecreepers are very similar in general morphology. Although the Red-browed is slightly (albeit significantly) heavier than the White-throated, it has shorter wings (Table 1). The bills of these two species do not differ in length, but the bill of the White-throated is thinner and straighter (Orenstein 1977). The White-throated is more specialized for climbing than the Red-browed, having shorter hind-limbs (relative to sternum length), longer toes, and smaller pads on the under-surface of the foot (Orenstein 1977; unpubl. data). Moreover, the hind-claw of the White-throated is significantly more curved than that of the Red-browed (Table 1), improving its clinging ability (Norberg 1979). Sittellas are approximately half the weight of the treecreepers, and much shorter-tailed. The bill of the Sittella is significantly shorter than that of either treecreeper (Table 1), and is slightly upturned.

Demography

The general social organization and demography of the two species of treecreepers has been described previously (Noske 1979, 1982a) and will be detailed elsewhere. White-throated Treecreepers held small territories ranging from 1.8 to 5.5 hectares, occupied by pairs or rarely unmated individuals. The average territory size for pairs was 3.9 hectares, but the sexes generally kept apart. The total population of the site varied between 26 and 30. Only four (15%) of the 27 adults banded up to 1980 survived into 1982. All young birds disappeared within 45 days of leaving the nest, most from 30 to 40 days post-fledging.

Red-browed Treecreepers had much larger territories than White-throated, ranging from 9.8 to 14.0 (averaging 11.6) hectares. These territories were occupied by a pair, which maintained close contact throughout the year, and in most cases, one or two of their male offspring.

The adult population of the site was almost one half that of the White-throated (15-17), but mortality was lower: eight (40%) of the 20 banded birds survived into 1982. Over a half of the young remained in their natal territories for at least two months. Yearling females that stayed usually dispersed just prior to the next breeding season, while males stayed for up to three years.

Sittellas occurred in groups of up to nine, the average group size (excluding juveniles) being about five. The demography of this species was impossible to determine as only seven adults were banded. The number of birds counted over the site varied from 17 to 28; thus, the population size was between that of the White-throated and Red-browed Treecreepers. As the groups to which banded birds belonged generally frequented exclusive areas and defended them against neighbouring groups whenever they came into contact, it is assumed that this species is territorial at times.

These territories were somewhat larger than those of the Red-browed Treecreeper. Observations at another study site (near Armidale) suggest that some groups amalgamate during the winter but split up again when breeding commences. Some non-breeding birds shared these breeding territories. Aggression was rarely seen within groups. There was some evidence that females dispersed more often than males.

Foraging Behaviour

The frequency of visits by each species to the four main tree-types was compared with the relative availability of these types in each of the four sectors (Fig. 1). As the resulting trends were consistent (except for the central sector), the data were lumped to yield an overall comparison. Of the three species, the Sittella was the most selective in its choice ($\chi^2 = 230.3$, 3df, $p<0.001$), showing a distinct preference for stringybarks. Red-browed were less discriminating than White-throated Treecreepers ($\chi^2 = 74.0$ and 158.3 respectively, both 3df, $p<0.001$), visiting boxes and gums more often than did either of the other two species. These interspecific differences are reflected in the total amount of time spent in all six tree-types (Fig. 2). Sittellas over-lapped more with the White-throated than with the Red-browed Treecreeper (84.5 and 65.6%), owing to greater use of stringybarks, as well as dead and "other" trees.

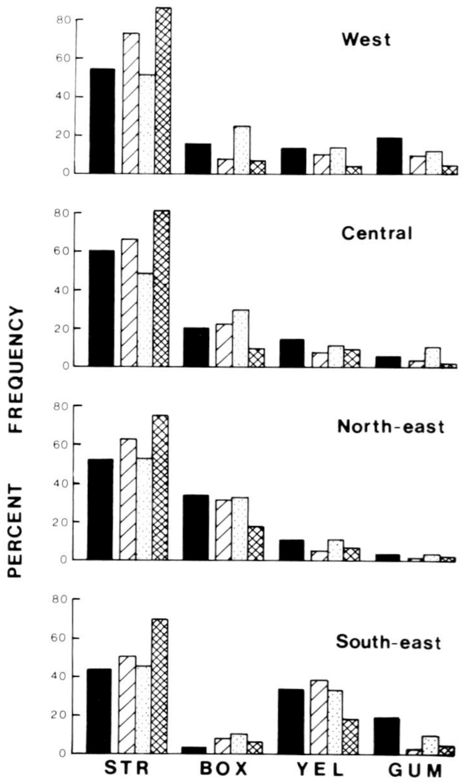

Fig. 1. Comparison of the frequency of use of the four main tree-types by birds and the availability of these tree-types in four sectors at Wollomombi Falls. Key: Black, trees; Diagonal lines, White-throated Treecreeper; Stippled, Red-browed Treecreeper; Cross-hatched, Varied Sittella. Over 300 observations in each sector for trees and both spp. of treecreepers; over 100 observations in each sector for Sittella.

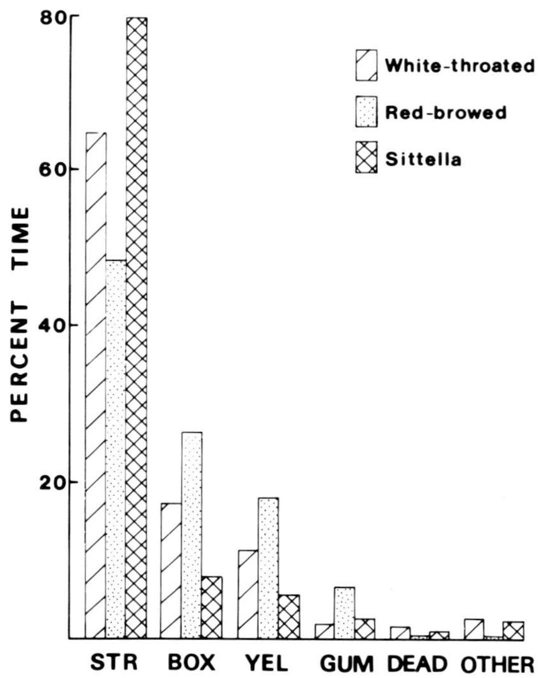

Fig. 2. Proportion of foraging time allocated to all tree-types by the three Wollomombi species. Total observation time: White-throated (1,854 min); Red-browed (1,737 min); Sittella (773 min). STR, Stringybark; BOX, Rough Box; YEL, Yellow Box.

The spatial distribution of stringybarks was significantly different from that of Yellow Boxes and gums, but not rough boxes (Fig. 3b). Since Red-browed Treecreepers visited Yellow Boxes and gums more frequently than the other species, one might expect this to be reflected in the distances they flew from one tree to the next. The direction of flights of this species might also be more random because they were less selective in their choice of trees. However, frequency distributions of flight distances revealed no significant differences between the three species (Fig. 3). Nor did the two species of treecreepers differ significantly in the angular orientation of their flights (Fig. 4).

The three species spent similar proportions of their time foraging at different heights (Fig. 5). None of the species differed significantly in their vertical distribution on stringybarks but in rough boxes, Red-browed spent more time than White-throated Treecreepers at upper levels (Kolmogorov-Smirnov test, $D = 21.9$, $p < 0.01$). Correspondingly, vertical overlap between these two species was least in rough boxes (78.1%) and greatest in stringybarks

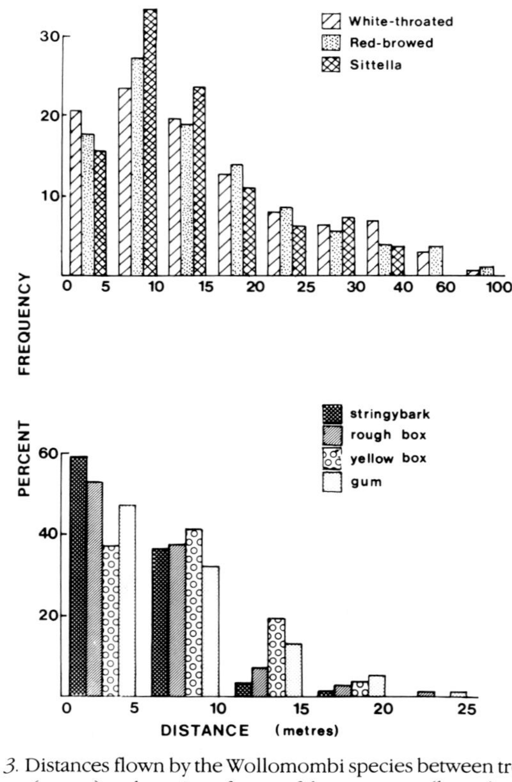

Fig. 3. Distances flown by the Wollomombi species between trees (upper) and spacing of trees of the same type (lower). Distances measured during foraging observations (for birds) and during transects (for trees); number of observations as follows: Upper (a): White-throated, 467; Red-browed, 498; Sittella, 111. Lower (b): stringybark, 436; rough box, 143; Yellow Box, 164; gum, 93. (a) no significant differences between two treecreepers, nor between Sittella and two treecreepers (Kolmogorov-Smirnov test, $D = 2.9$, 8.2, 8.7, respectively; $p > 0.05$); (b) stringybarks significantly different from Yellow Boxes and gums ($D = 22.7$, $p < 0.001$ and $D = 16.8$, $p < 0.05$) but not from rough boxes ($D = 6.7$, $p > 0.05$).

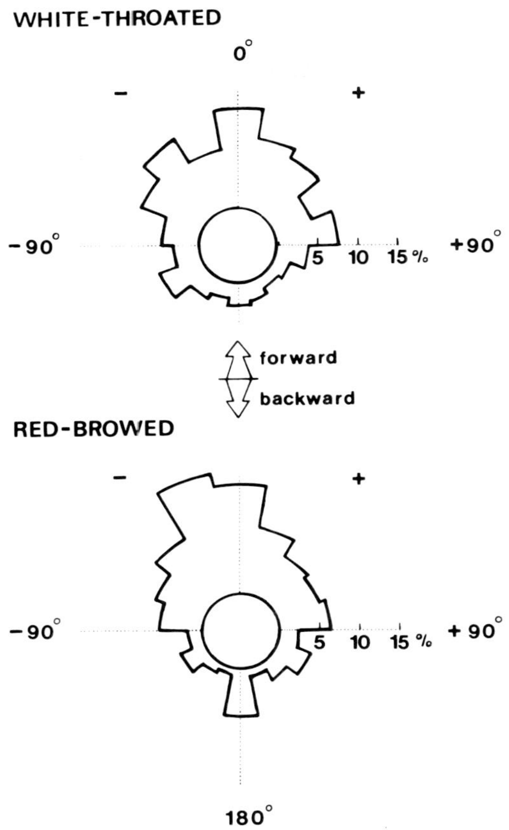

Fig. 4. Frequency (percent) of angular deviations for successive flights of the two Wollomombi treecreepers. Positive values indicate deviation to the right; negative to the left; only distances exceeding five metres were used; intervals, 20 degrees; total number of observation for White-throated — 284, Red-browed — 246. No significant difference between the two species when 20 degree intervals compared ($\chi^2 = 8.57$, 8df, $p > 0.2$).

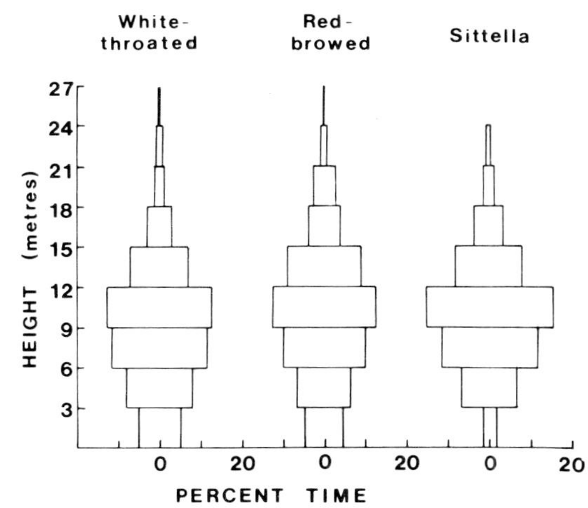

Fig. 5. Vertical distribution of the three Wollomombi species, based on the proportion of time* spent at different levels.

	Observation time (Mins.)	N (trees)
White-throated	1082	1090
Red-browed	820	950
Sittella	360	335

	Overlap Indices White-throated	Red-browed
Red-browed	92.3	–
Sittella	91.3	90.8

* Kolmogorov-Smirnov tests (in text) based on number of trees, in which heights were recorded at each interval.

(93.2%). Despite significant differences in the heights of the four main tree-types, the two species of treecreepers chose trees of similar heights (2-way ANOVA, p>0.5). Tree height had a significant effect on the total time birds spent foraging in each tree (ANACOVA, p<0.001 for both treecreepers in all tree-types).

Though the three species overlapped extensively in their vertical distribution on stringybarks, their use of substrates on these trees differed markedly. White-throated Treecreepers devoted most of their time to the fibrous bark on the trunk and limbs, while the other species made greater use of the smooth bark on the outermost branches and twigs (Fig. 6a). Sittellas spent almost a third of their time on dead branches, whereas Red-browed Treecreepers generally avoided these substrates. These patterns of substrate use were also evident in the other three main tree-types (Fig. 6b). White-throated and Red-browed Treecreepers preferred rough (mostly scaly) and smooth bark respectively, while Sittellas concentrated on dead branches. The Sittella rarely used thick smooth substrates because they rarely used Yellow Boxes and gums. Both this species and the Red-browed Treecreeper exhibited marked sexual differences in foraging substrates.

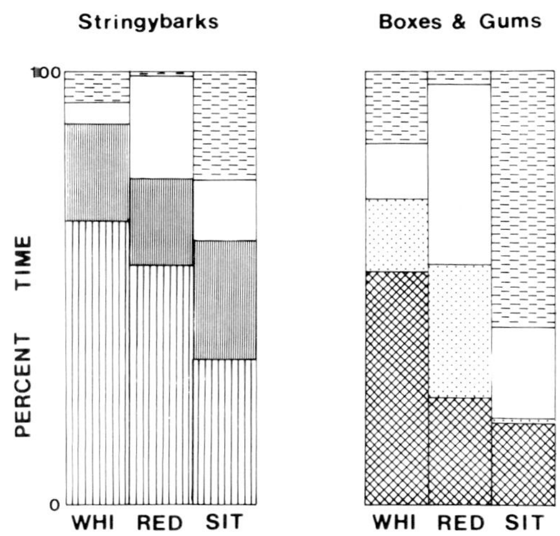

Fig. 6. Foraging substrates of the three bark-foraging species at Wollomombi Falls in (a) stringybarks and (b) boxes (including Yellow Box) and gums. Key: Broad vertical lines — Fibrous (rough), thick; Narrow vertical lines — Fibrous (rough), thin; Crosshatched — Scaly/shaggy (rough); Stippled — Smooth, thick; White — Smooth, thin; Dashed — Dead. N.B. Substrates were called "thin" if similar in diameter, or smaller, than a treecreeper, "thick" if greater in diameter.

In contrast to treecreepers, Sittellas spent about one-third of their time climbing downwards or in towards the trunk (Fig. 7) and did not climb along the underside of branches. They also climbed faster, switched from branch to branch more frequently, and spent more time foraging in one spot ("drilling"), than treecreepers (Noske 1982b). Red-browed were quicker climbers than White-throated Treecreepers on stringybarks, but slower on rough boxes. Red-browed also switched from branch to branch more frequently than White-throated Treecreepers, but the distance involved in these "switches" was shorter presumably due to the greater use of closely-spaced thin branches. White-throated spent proportionally more time "drilling" in each stringybark than Red-browed Treecreepers, with longer "drills", on average, on fibrous substrates. However, the average duration of drills by the two species was similar when on smooth branches.

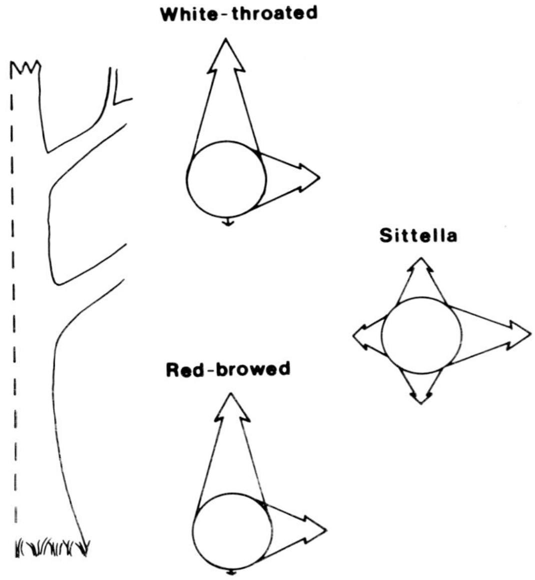

Fig. 7. Proportion of time spent climbing in each of four directions (upwards, downwards, outwards, and in towards trunk) by the three Wollomombi species. Total observation time: White-throated Treecreeper (970 min); Red-browed Treecreeper (870 min); Varied Sittella (465 min).

Food

The relative availability of different taxa observed on the surface of tree-trunks differed dramatically from that found beneath the bark. Ants were the most numerous and conspicuous arthropods on the surface of trunks of all tree-types, but were commonest on gums and least abundant on stringybarks (Fig. 8a). Over three-quarters of the ants recorded were small *Iridomyrmex* spp. (3-4 mm long), which streamed up and down the trunks between their food sources and ground nests. Other important insect groups on the surface were flies, moths and wasps (Fig. 8b), which merely rested on tree-trunks. These transient taxa were largely absent from beneath the surface, where sedentary groups such as beetles, larvae and cockroaches predominated (Fig. 8c). Spiders were prevalent on both the surface and sub-surface, though different families were represented.

Ants (Formicidae) comprised about 90% of the arthropods found in the stomachs of both species of treecreepers (Fig. 9) and were present in every faecal sample examined. Red-browed ate significantly more large ants (over 4 mm long) than White-throated Treecreepers but the size of other prey did not differ between these species. Red-browed stomachs contained more spiders and beetles than White-throated Treecreepers (χ^2 = 35.6 and 20.4 respectively, both 1df, $p < 0.001$). Red-browed were also observed handling spiders more often than White-throated Treecreepers ($\chi^2 = 4.35$, 1df, $p<0.05$)

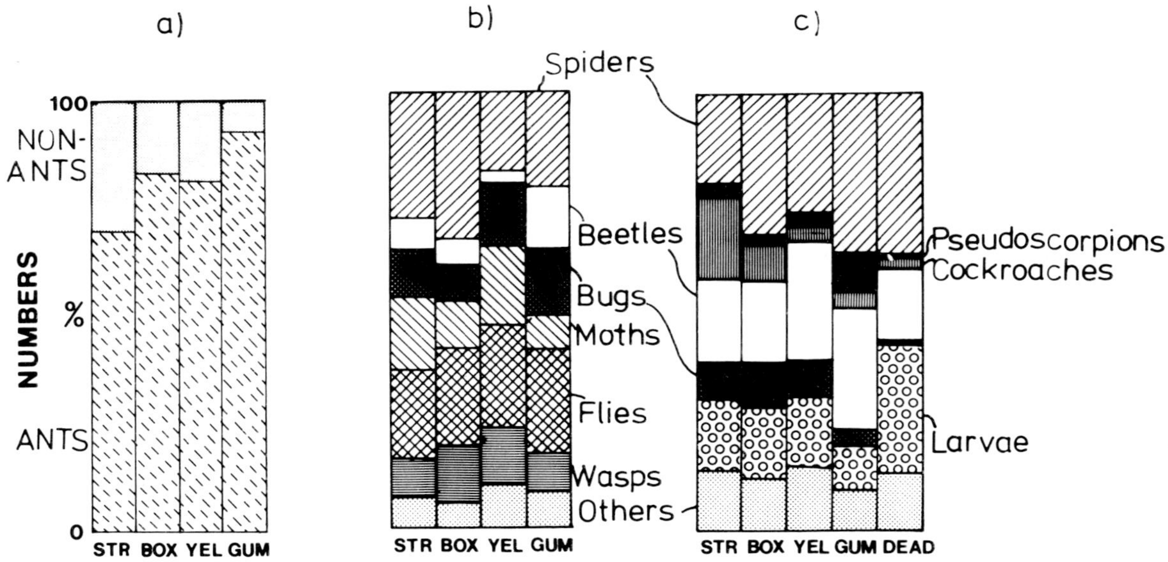

Fig. 8. Taxonomic composition of arthropods found on the surface of tree-trunks (including ants, a; excluding ants, b) and under bark (c).

	No. of arthropods (excluding ants)	
	Surface (b)	Sub-surface (c)
Stringybark	349	261
Rough Box	324	226
Yellow Box	288	213
Gum	265	235
Dead tree	–	236

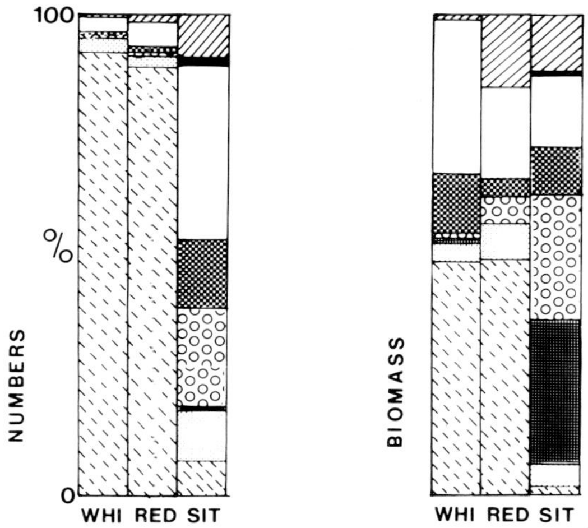

Fig. 9. Taxonomic composition of arthropods found in stomachs of the three bark-foraging species near Wollomombi Falls, according to numbers (left) and wet weight (right). No. of arthropods: White-throated Treecreeper 4446; Red-browed Treecreeper 4038; Varied Sittella 487. Key: as in Fig. 8; cross-hatched, crickets.

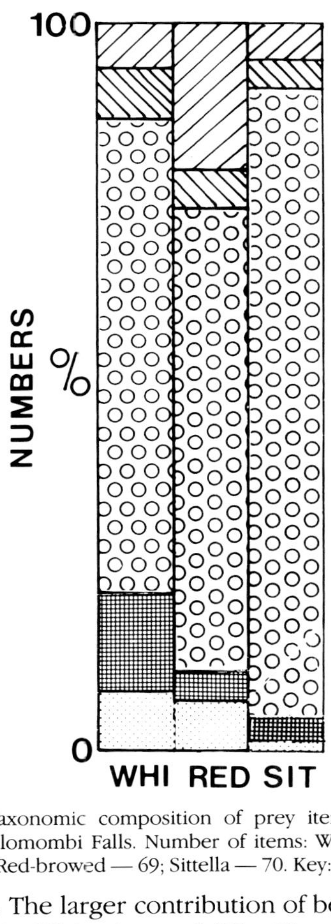

Fig. 10. Taxonomic composition of prey items observed at Wollomombi Falls. Number of items: White-throated — 47; Red-browed — 69; Sittella — 70. Key: as in Fig. 8.

(Fig. 10). The larger contribution of beetles to total biomass in White-throated (Fig. 8) was due to the greater number of large Chrysomelids (subfamily Chrysomelinae) in this species compared to the Red-browed Treecreeper. Over half of the food items observed for both species were larvae (Fig. 10), reflecting the ease with which this group can be identified in the field.

Ants formed only a minor proportion (7%) of arthropods found in the stomachs of Sittellas (Fig. 8). In addition, only one of the ten faecal samples of this species contained ants. Thus, in view of the abundance of ants on bark, the Sittella must be considered more specialized in its food requirements than the treecreepers. Sittellas took significantly smaller beetles and spiders than Red-browed Treecreepers (Fig. 11) as might be expected from their shorter bill (cf. Table 1). However, the size of larvae and bugs did not differ between these two species. Larvae comprised a greater proportion of the total numbers of arthropods in stomachs and observations of Sittellas, than in treecreepers (Figs. 9, 10). Coleopteran (beetle) larvae were more numerous than Lepidopteran (moth) larvae in stomachs, but the latter predominated in sightings, presumably because of their generally larger size. The large contribution of crickets to total biomass in stomachs resulted from four large Gryllacridids (20-30 mm).

Despite the lack of variation between tree-types in their arthropod fauna, some correlations were found between the diets of the three species and their preferred tree-types. Spiders were more prevalent under the bark of boxes and gums than under stringybark, and Red-browed apparently ate more spiders than White-throated Treecreepers. Psocids (Psocoptera) were present on all tree-types except stringybark, and Red-browed took more of these insects than White-throated (28 vs 6). Clerids and Tenebrionids figured prominently in the beetle larvae identified from the stomach contents of Sittellas, and these families mainly occurred on stringybarks and dead substrates, respectively (Fig. 12). Three of the eight larvae in Red-browed Treecreeper stomachs were Elaterids, which were commonest on boxes and gums.

Nesting and Roosting

The single-brooded White-throated Treecreeper had a shorter breeding season than the other two species, which sometimes attempted second broods. Egg laying peaked in early September for the Red-browed and almost a month later in the White-throated Treecreeper. The last White-throated clutches were laid in mid-November but some Red-browed Treecreeper clutches were recorded in December. In the Sittellas, incubation was noted as late as mid-January, but the greatest number of concurrent nests with incubating birds occurred in late September.

Like all Australasian treecreepers, the White-throated and Red-browed Treecreepers nest in holes. Each species preferred different types of hollows. The majority of White-throated Treecreeper nests were built in knotholes on the trunks or main limbs of live trees, particularly stringybarks (Table 2). Few dead trees were used. In contrast, most Red-browed Treecreeper nests were situated in hollow dead branches of live trees or entirely

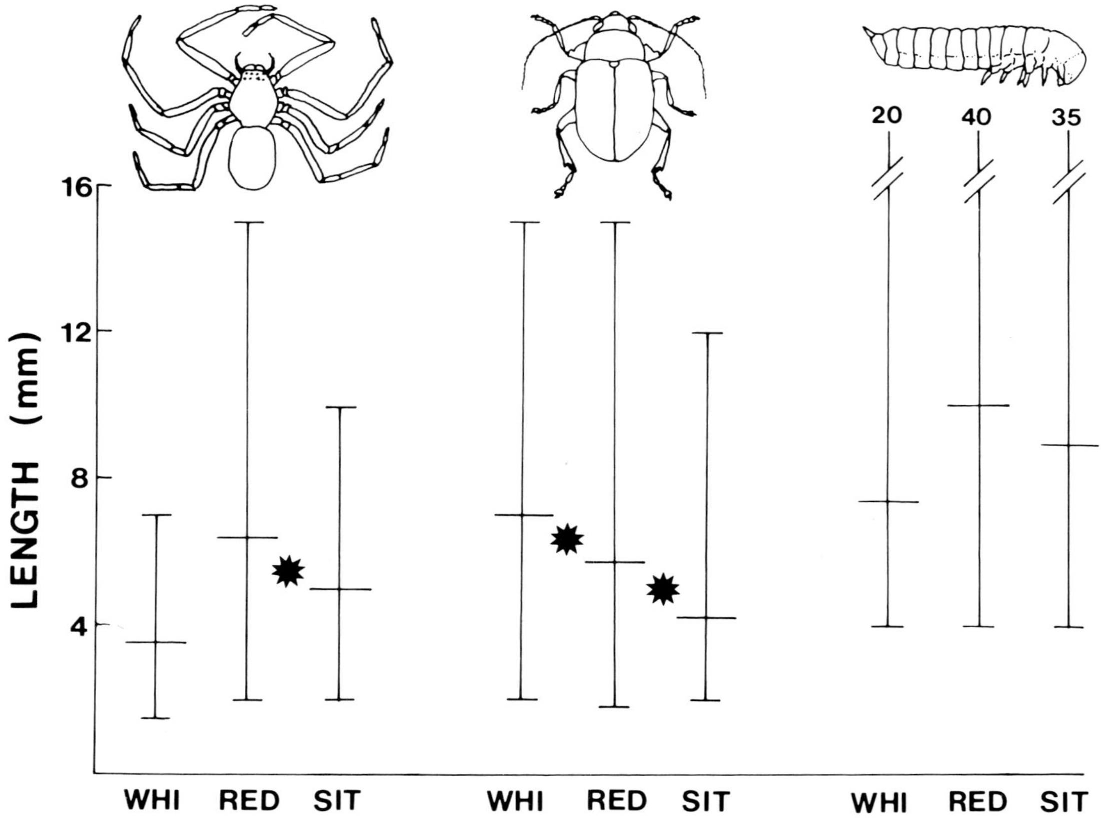

Fig. 11. Estimated lengths (mean and range) of spiders, beetles and larvae found in stomachs of the three bark-foraging species near Wollomombi Falls. Stars beside means indicate significant differences (using t-tests, $p < 0.05$). Sample sizes are follows:

	Spiders	Beetles	Larvae
WHI	7	49	42
RED	144	209	172
SIT	21	31	102

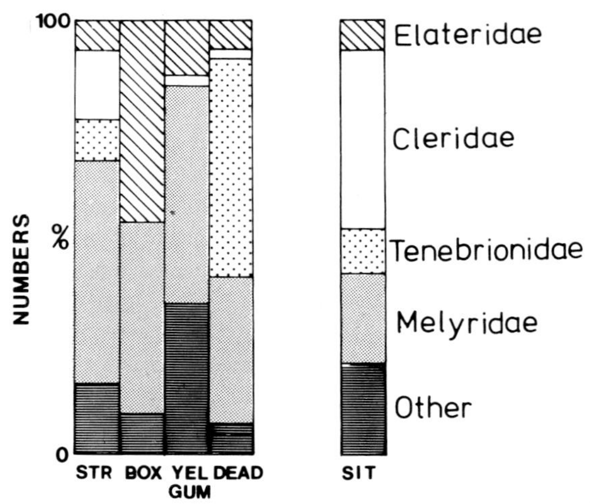

Fig. 12. Family composition of beetle larvae found under bark of four tree-types (left) and in stomachs of Sittellas (right). Number of larvae: Stringybark — 31; Rough Box — 32; Yellow Box/gum — 32; Dead tree — 47; Sittella — 29.

dead trees. Stringybarks were rarely used. The nest-sites of Red-browed were significantly higher than those of the White-throated Treecreeper (Table 3). Both species re-used nest-sites: six (30%) of the 20 different White-throated sites and five (24%) of the 21 Red-browed Treecreeper sites were used in two or more seasons.

Table 2. Nest-sites* of treecreepers at Wollomombi Falls.

	White-throated		Red-browed	
Tree-type	Knothole	Spout	Knothole	Spout
Stringybark	12	4	1	–
Box†	3	3	3	8
Gum	3	–	–	6
Dead	2	1	1	10
Total	20	8	5	24

* includes all nesting attempts; † includes Yellow Box.

Table 3. Heights (in metres) of nest-sites of treecreepers and sittellas at Wollomombi Falls.

	White-throated Treecreeper	Red-browed Treecreeper	Varied Sittella
Minimum	4.0	5.3	6.5
Maximum	13.4	20.0	19.0
Mean ($\bar{X}$)	8.1	10.6	13.1
Standard error	0.56	0.81	1.08
N	25	27	14

Both species of treecreepers roosted solitarily but they chose different roost-sites. The White-throated slept in external situations, with part or whole of the body exposed, whereas Red-browed Treecreeper roost-sites were enclosed. Most White-throated Treecreepers roosted either on burnt surfaces ("fire-boxes") at the base of live stringybarks and boxes, or in depressions at the junction of cleft tree-trunks (stringybarks only). Others roosted in shallow cavities or directly beneath swellings on the trunks of boxes and gums. Crevices on the exterior wall of dead trees were also used. All but two of the Red-browed Treecreeper roost-sites were in the hollow tops of broken tree-trunks (mostly in live boxes) or in hollow spouts of dead trees. Again, stringybarks were rarely used (Table 4). Like their nests, Red-browed roost-sites were significantly higher than those of the White-throated (Table 5).

Table 4. Roost-sites of treecreepers at Wollomombi Falls.

	Tree-type				Burnt	
	Stringybark	Box	Gum	Dead*	Stump	Total
White-throated	9	7	2	2	3	23
Red-browed	2	7	1	5	–	15

* unburnt.

Table 5. Heights (in metres) of roost-sites of treecreepers (at Wollomombi Falls) and sittellas (near Armidale).

	White-throated Treecreeper	Red-browed Treecreeper	Varied Sittella
Minimum	0.7	2.5	7.6
Maximum	6.2	12.0	16.8
Mean ($\bar{X}$)	3.0	6.6	14.0
Standard error	0.39	0.76	0.29
N	21	15	120*

* nights of observation, rather than individual sites (as in treecreepers).

The open cup nests of the Sittellas were invariably built in the crotch of a vertical forked dead branch but occasionally on a leaning single-stemmed dead branch or live fork. All but four of the 24 nests at Wollomombi were in stringybarks, the exceptions in boxes. Unlike treecreepers, Sittellas roost communally, the members of each group huddling together along a thin horizontal dead branch (McGill 1967; Beruldsen 1978). Few roost-trees were found at Wollomombi but near Armidale, a colour-banded group was known to use 15 sites over 18 months (unpubl. data). Stringybarks were the most commonly used tree-type. The average height of the roost-sites over 120 nights was more than twice the height of treecreeper roost-sites (Table 5).

DISCUSSION

Feeding ecology

As I found earlier (Noske 1979), White-throated Treecreepers at Wollomombi foraged in rough-barked stringybarks more often than Red-browed. Similarly, White-throated Treecreepers in the Blue Mountains near Sydney visited rough-barked mahoganies (*E. notabilis*) and peppermints (*E. piperita*) more than Red-browed, which concentrated on gums (*E. deanei*) (Orenstein 1977). White-throated also used dead substrates more than Red-browed Treecreepers. In the Blue Mountains, they spent 28% of their time on dead substrates, compared to only 4% for the Red-browed (Table 5, Orenstein 1977). At Wollomombi, White-throated Treecreepers and Sittellas spent a greater proportion of time on both dead trees and dead branches of live trees than Red-browed. They also made greater use of "other" trees (Fig. 1) such as *Acacia diphylla* and *Jacksonia scoparia*, both of which possess rough bark similar to stringybarks. That the White-throated Treecreeper and Sittella are predisposed towards rough-barked trees is consistent with their occurrence in forests of ironbarks (eucalypts with rough, corrugated bark) and Cypress Pine *Callitris*, as well as *Casuarina* thickets (River Oak, Belar). These habitats are avoided by the Red-browed Treecreeper.

Both species of treecreepers obtain much of their food by gleaning (pecking the bark surface). The other main foraging technique of the White-throated is excavating, while the Red-browed relies more on probing (Orenstein 1977; Noske 1979). The bill of the White-throated Treecreepers is straighter than that of the Red-browed, and is thus, better adapted for excavating (see Bock 1966). The Sittella appears to excavate and probe more than glean; this is also inferred by the fact that it rarely eats ants, which occur mainly on the bark surface and do not hide in fissures or under loose bark. Its laterally-compressed, wedge-shaped bill is better suited to scaling (inserting the bill under loose pieces of bark and using it as a lever to dislodge them) rather than hammering and tugging bark like the White-throated Treecreeper (Orenstein 1977). The extensive use of friable rough bark and decaying wood by both the White-throated Treecreeper and Sittella accords with their excavating techniques. By contrast, the probing techniques of the Red-browed Treecreeper are more appropriate for extracting insects from ribbons of peeling bark.

Although the numerical proportion of ants in the stomachs of the two species of treecreepers corresponded with the availability of ants relative to other prey types, their importance in the diets of these birds is probably overestimated. Because of their tough exoskeleton, ants persist longer than most other insects in the digestive tract of birds; indeed, many ants in the stomachs of treecreepers were intact. Ants could be a source of carbohydrate, since the vast majority of those eaten by treecreepers belong to species which are nectarivorous, collecting sugar solutions from various arboreal sources (e.g. honeydew) to feed their larvae (R. Taylor, pers.

comm.). Yet if ants were the favourite food item or main source of nutrition of these birds, one might expect them to perch beside ant trails for long periods during the day, rather than climb continuously, visiting hundreds of trees. On the other hand, little energy is expended in gleaning ants while searching for better items.

Closely-related species are invariably found to differ in some aspect of their foraging behaviour, and these differences are usually interpreted as the result of interspecific competition for food. Although competition almost certainly occurs among nectarivorous birds (e.g. Ford 1979), there is still little direct evidence that insectivores commonly compete. Connell (1975) considered that aggressive displacement of one species by another was convincing evidence of competition. Subordinate species typically decrease their niche breadth (i.e. specialize) in the presence of a dominant species (Morse 1974). As Red-browed Treecreepers were always dominant to White-throated in aggressive encounters (Noske 1979, 1982b) it could be argued that they prevented White-throated from foraging on smooth-barked trees, forcing the latter to specialize on stringybarks. However, at Swan Vale (100 km north-west of Armidale) where Red-browed were absent, White-throated Treecreepers still visited gums less often than expected according to their availability (Noske 1982b). Indeed, the larger Brown Treecreeper, *Climacteris picumnus,* frequently aggressed the White-throated here, but as the former was semi-terrestrial it is doubtful if these two species competed to any great extent.

Orenstein (1977) proposed that the Red-browed Treecreeper concentrated on smooth-barked trees because it was less well adapted for climbing than the White-throated, having proportionally longer legs and broader pads on the underside of the toes. He argued that the peeling bark of gums provides discrete patches containing insects which even non-climbing species, such as the Shrike-tit, can exploit. The absence of such patches on rough-barked trees required a bird to search the entire surface to locate food. This hypothesis is supported by my observations of the foraging behaviour of these birds and data on the abundance of invertebrates on these substrates (Noske 1982b). Although Red-browed were less specialized than White-throated Treecreepers in their choice of trees, they were apparently more specialized in their selection of substrates, spending a large proportion of their time on the smooth bark that was restricted to the outermost branches of stringybarks and rough boxes.

Orenstein (1977) also related the less specialized hind-limbs of the Red-browed to greater use of the ground, but he observed foraging of this species for only 96 minutes. My studies clearly indicate that the Red-browed Treecreeper is just as arboreal as the White-throated. In most life history and plumage characteristics, the Red-browed resembles the Brown Treecreeper more closely than the White-throated (Noske 1982a, b, 1983), so it seems more likely that the relatively unspecialized hind-limbs of the Red-browed are due to its closer affinity to the semi-terrestrial members of the family.

Nesting and Roosting Requirements

The White-throated was more generalized than the Red-browed Treecreeper in its nesting and roosting requirements. Over a half of the White-throated nests were in stringybarks which were the commonest tree-type at the study site, while over a third of Red-browed nests were in dead trees, which were the scarcest. Hyem (1936) also noted that the White-throated nests in "any sort of hole . . . in large hollows in the trunks of dead trees, behind loose pieces of bark and in crevices in old, rotting stumps", while the Red-browed Treecreeper usually chooses "upright, hollow branches". In addition, the exposed situations where the White-throated Treecreeper sleeps give it great flexibility in its roosting habits. This is reflected in its use of human constructions, including the public toilet at Wollomombi, a chimney (Dickison 1933), a garage (Bryant 1934) and a house (Noske 1977). It has also been recorded roosting on the ceiling of a cave (Mackness 1976). By contrast, the Red-browed Treecreeper apparently requires hollow spouts.

There was little evidence of interspecific competition for nest-holes at the study site. This contrasts with the situation in the north temperate region (Short 1979; Alerstam and Högstedt 1981), where there is severe competition for holes, despite the fact that many birds (e.g. woodpeckers) can excavate their own hollows. Although a high proportion of the nest-sites in both species of treecreepers were used more than once, there did not appear to be any shortage of suitable hollows. Treecreepers did not take long to locate new hollows and build their nests. One Red-browed, for example, laid a second clutch only 20 days after her first brood fledged from another nest-site 270 metres away. Near Armidale, a White-throated Treecreeper began to lay only seven days after she had abandoned her previous clutch in a different tree. Saunders *et al.* (1982) showed that less than half of the available hollows in a patch of eucalypt woodland in Western Australia were used by hole-nesting species.

The two treecreepers preferred different nest-sites, so were unlikely to compete with each other for holes. Of the five other common arboreal hole-nesters at Wollomombi, four (one parrot and three Coraciiformes) were larger, and the other (Striated Pardalote, *Pardalotus striatus*) smaller than treecreepers, and probably favoured larger and smaller holes respectively. Saunders *et al.* (1982) found a correlation between the body size of four species of Psittaciformes and the size of their nest-holes. Naturally, large species are restricted to large holes, whereas smaller species can use large holes but risk displacement or predation by large species. Indeed, one White-throated Treecreeper nest-site at Wollomombi was subsequently occupied by Dollarbirds, *Eurystomus orientalis*, but this was the only known case of a hole being used by two species. The Red-browed Treecreeper may compete with some mammals (e.g. possums, bats) for roost-holes, as these were generally larger than those chosen for nests.

CONCLUSIONS

The fundamental difference between the life history strategies represented by the White-throated Treecreeper on the one hand, and the Red-browed Treecreeper and Sittella on the other, lies in their treatment of offspring. In the White-throated, the complete eviction of young as soon as they are independent leads to widespread dispersal. In the other species, the tolerance and retention of a substantial proportion of young results in reduced dispersal and ultimately in communal breeding (Noske 1980a, b). White-throated Treecreepers suffered higher adult mortality than the Red-browed during the last two years of the study, but raised more young than the latter on a *per capita* basis (Noske 1982b). Thus, compared to the White-throated, the Red-browed Treecreeper has a low rate of dispersal, high survival (longevity) and low fecundity, typical attributes of communal breeders (Brown 1974; Emlen 1978; Gaston 1978; Dow 1980).

In the White-throated Treecreeper, dispersal is undoubtedly facilitated by its generalized habitat requirements. Its ability to colonize "new" habitats is exemplified by its frequent occurrence in small isolated patches of forest (Howe 1982), as well as in plantations of exotic pine *Pinus radiata* (Stevens 1975; Gepp 1976; Driscoll 1977; Debus 1983). The Red-browed Treecreeper has a more localized distribution than the White-throated in New South Wales (cf. Anon 1979; Noske 1982b), reflecting its more specialized habitat requirements. Howe (1982) found that this species was generally absent from small forest islands in cleared areas, despite their presence in larger tracts of forest nearby. The large territories of Red-browed Treecreepers at Wollomombi suggest that they require more space than White-throateds to obtain sufficient resources. This is consistent with their more specialized foraging, nesting and roosting site requirements, and probably explains why they typically occur at lower population densities than White-throated (e.g. Hyem 1936; Disney and Stokes 1976).

Dead wood was an important resource for all three bark-foragers. The White-throated Treecreeper and Sittella used dead substrates extensively for foraging, and the latter invariably built its nests and roosted on dead branches. The Red-browed Treecreeper nested mainly in hollow dead branches of live trees or in dead trees, and also roosted in similar situations. Most eucalypts do not develop hollows until they are at least 50 years old (Disney and Stokes 1976; Saunders *et al.* 1982). Culling of mature or dead trees therefore greatly reduces the breeding opportunities for both species of treecreepers. The retention of over-mature eucalypts is particularly important for the Red-browed Treecreeper, which depends on them for both nesting and roosting.

ACKNOWLEDGEMENTS

I am indebted to my wife, Susan, for her patience and help in all aspects of this study. I also thank Drs Hugh Ford and Harry Bell for their advice and encouragement, and Mr R. Edgar for permission to work on his property adjoining the reserve. The NSW National Parks and Wildlife Service issued me with authorities to trap and collect birds, and Mr D. Purchase (Division of Wildlife Research, C.S.I.R.O.) supplied bands. This study was undertaken while in receipt of a Commonwealth Postgraduate Research Award at the Department of Zoology, University of New England, whose facilities are gratefully acknowledged. I thank Mrs M. Brown for typing the final draft.

REFERENCES

ALERSTAM, T. AND HÖGSTEDT, G., 1981. Evolution of hole-nesting in birds. *Ornis. Scand.* **12**: 188-93.

ANON, 1979. Royal Australasian Ornithologists Union Pilot Atlas Scheme. *Corella* **2**: 97-162.

BELL, H. L., 1983. Resource partitioning between three syntopic thornbills. Ph.D. Thesis, University of New England, Armidale.

BERULDSEN, G. R., 1978. Roosting of sittellas. *Sunbird* **9**: 14.

BOCK, W. J., 1966. An approach to the functional analysis of bill shape. *Auk* **83**: 10-51.

BROWN, J. L., 1974. Alternate routes to sociality in jays — with a theory for the evolution of altruism and communal breeding. *Am. Zool.* **14**: 63-80.

BRYANT, J. J., 1934. Bird notes from Toolern Vale. *Emu* **34**: 113-20.

CALVER, M. C. AND WOOLLER, R. D., 1982. A technique for assessing the taxa, length, dry weight and energy content of the arthropod prey of birds. *Aust. Wildl. Res.* **9**: 293-301.

CONNELL, J. H., 1975. Some mechanisms producing structure in natural communities. *In* "Ecology and Evolution of Communities" ed. by M. L. Cody and J. M. Diamond. Belknap Press, Cambridge, Mass.

DEBUS, S. J. S., 1983. Birds in a pine plantation at Armidale, New South Wales. *Corella* **7**: 14-16.

DICKISON, D. J., 1933. Notes on the White-throated Treecreeper. *Emu* **32**: 279-81.

DISNEY, H. J. de S. AND STOKES, A., 1976. Birds in pine and native forests. *Emu* **76**: 133-38.

DOW, D D., 1980. Systems and strategies of communal breeding in Australian birds. *Proc. XVII Int. Orn. Congr.* Pp. 875-81.

DRISCOLL, P. V., 1977. Comparison of bird counts from pine forests and indigenous vegetation. *Aust. Wildl. Res.* **4**: 281-88.

EMLEN, S. T., 1978. The evolution of co-operative breeding in birds. *In* 'Behavioural ecology: An evolutionary approach" ed. by J. Krebs and N. Davies. Blackwell, Oxford.

Ford, H. A., 1979. Interspecific competition in Australian honeyeaters — depletion of common resources. *Aust. J. Ecol.* **4**: 145-64.

Ford, H. A. and Paton, D. C., 1976. Resource partitioning and competition in honeyeaters of the genus *Meliphaga*. *Aust. J. Ecol.* **1**: 281-87.

Ford, H. A. and Paton, D. C., 1977. The comparative ecology of ten species of honeyeaters in South Australia. *Aust. J. Ecol.* **2**: 399-407.

Gaston, A. J., 1978. The evolution of group territorial behaviour and co-operative breeding. *Am. Nat.* **112**: 1091-1100.

Gepp, B. C., 1976. Bird species distribution and habitat diversity in an exotic forest in South Australia. *Aust. For.* **39**: 269-87.

Howe, R. W., 1982. Biogeography and ecology of birds in small forest habitat islands of Australia and North America. Ph.D. Thesis, Univ. of Wisconsin, Madison.

Huddy, L., 1979. Social behaviour and feeding ecology of Scarlet Robins (*Petroica multicolor*). B.Sc. (Hons.) Thesis, Univ. of New England, Armidale, NSW.

Hyem, E. L., 1936. Notes on the birds of Mernot, Barrington, NSW *Emu* **36**: 109-27.

Keast, A., 1957. Variation and speciation in the genus *Climacteris* Temminck. *Aust. J. Zool.* **5**: 474-95.

Keast, A., 1968. Competitive interactions and the evolution of ecological niches as illustrated by the Australian honeyeater genus *Melithreptus* (Meliphagidae). *Evolution* **22**: 762-84.

Mackness, B., 1976. White-throated Treecreeper roosting in a cave. *Sunbird* **7**: 48.

McGill, A. R., 1967. Roosting of the Black-capped Sittella. *Emu* **66**: 388.

Morse, D. H., 1974. Niche breadth as a function of social dominance. *Am. Nat.* **108**: 818-30.

Norberg, U. M., 1979. Morphology of the wings, legs and tail of three coniferous forest tits, the Goldcrest, and the Treecreeper in relation to locomotor pattern and feeding station selection. *Phil. Trans. R. Soc. Lond.* **287**: 131-65.

Noske, R. A., 1977. Roosting of treecreepers. *Sunbird* **8**: 41-43.

Noske, R. A., 1979. Co-existence of three species of treecreepers in northeastern New South Wales. *Emu* **79**: 120-28.

Noske, R. A., 1980a. Co-operative breeding by treecreepers. *Emu* **80**: 35-36.

Noske, R. A., 1980b. Co-operative breeding and plumage variation in the Orange-winged (Varied) Sittella. *Corella* **4**: 45-53.

Noske, R. A., 1982a. The private lives of treecreepers. *Aust. Nat. Hist.* **20**: 419-24.

Noske, R. A., 1982b. Comparative behaviour and ecology of some Australian bark-foraging birds. Ph.D. Thesis, Univ. of New England, Armidale.

Orenstein, R. I., 1977. Morphological Adaptations for bark foraging in the Australian Treecreepers (Aves; Climacteridae). Ph.D. Thesis, Univ. of Michigan, Ann Arbor.

Ricklefs, R. E. and Cox, G. W., 1977. Morphological similarity and ecological overlap among passerine birds on St. Kitts, British West Indies. *Oikos* **29**: 60-66.

Saunders, D. A., Smith, G. T. and Rowley, I., 1982. The availability and dimensions of tree hollows that provide nest-sites for cockatoos (Psittaciformes) in Western Australia. *Aust. Wildl. Res.* **9**: 541-56.

Short, L. L., 1979. Burdens of the Picid hole-nesting habit. *Wilson Bull.* **91**: 16-28.

Stevens, J. H., 1975. Exotic pine forest as a bird habitat. *Aust. Bird Watcher* **6**: 77-79.

Thomas, D. G., 1980. Foraging of honeyeaters in an area of Tasmanian eucalypt forest. *Emu* **80**: 55-58.

CHAPTER 17

The Helmeted Honeyeater and Related Honeyeaters of Victorian Woodlands

B. J. Wykes[1]

Co-existing populations of Yellow-tufted, Helmeted and other short-billed honeyeaters were studied in a variety of southern Victorian habitats. They generally forage on insects and associated sugars of eucalypt bark and foliage during their spring-summer breeding season while avoiding winter shortages in their breeding habitat by switching to nectarivory. Three distinctive life-styles are described for which adaptations to foraging in the breeding season and winter are compatible. Breeding Yellow-tufted and Helmeted Honeyeaters were interspecifically territorial and monopolised riparian habitat rich in carbohydrates and arthropods of bark and foliage, and areas of auriferous forest where there was year-round nectar. They sought local rich nectar sources in winter which were also aggresively defended. Similar life-styles were probably possessed by the Fuscous, White-plumed and Yellow-plumed Honeyeaters, except that each was adapted to different habitats. The White-eared Honeyeater was a specialist bark-foraging species which was excluded from much of the food-rich habitat occupied by generalised, energy-profligate relatives but also occurred in many habitats unattractive to these species. It was also generally excluded from winter nectar but its year-round staple of bark-dwelling arthropods was probably less depleted than were those of more exposed sites.

The Bellminer was a honeyeater specialised to taking psyllid-produced lerps from foliage of riparian eucalypts. Areas rich in these carbohydrates were monopolised by dense colonies throughout the year through indiscriminate interspecific aggression. Bellminers have taken over much of the habitat reserved for the rare Helmeted Honeyeater since the trees have become chronically infested by psyllids. Ways in which the reserve can be managed to favour Helmeted Honeyeaters are suggested.

INTRODUCTION

In the following chapter by David Paton (1985) compares the ecologies of highly nectarivorous, long-billed honeyeaters. This paper concerns some short-billed honeyeaters of the genus *Lichenostomus* (formerly part of Meliphaga; Schodde 1975). The genus is a reasonably uniform group of species that are said to alternate seasonally between insectivory and nectarivory, and occur in separate habitats (Keast 1976; Ford and Paton 1976b). However, up to five species may be sympatric in Victoria. The primary subject of my study was the ecology of the rare Helmeted Honeyeater, *L. melanops cassidix,* and another more common subspecies of Yellow-tufted Honeyeater, *L. melanops meltoni,* in Victoria. I compared the habitats and resources used by these subspecies and co-occurring congeners, plus the ecologically similar Bellminer, *Manorina melanophrys.* The aims of the research were to assess the role of competition in moulding the species' niches and whether interspecific competition contributes to the scarcity of Helmeted Honeyeaters.

The study was conducted in three study sites, each of which supported *L. melanops* populations and differing arrays of congeners. It was hoped to assess the effects of competition on *L. melanops* ecology by studying populations in these differing habitat and competitor environments. Densities, distributions, foraging behaviour, breeding characteristics and aggressive interactions were recorded in all seasons. Year-round arthropod and nectar availabilities to the honeyeaters were monitored as an aid to interpreting the ecological observations.

[1]School of Biology, Western Australian Institute of Technology, Kent Street, Bentley, Western Australia 6102

Pages 205-17 *in* BIRDS OF EUCALYPT FORESTS AND WOODLANDS: ECOLOGY, CONSERVATION, MANAGEMENT ed. by A. Keast, H. F. Recher, H. Ford and D. Saunders. Royal Australasian Ornithologists Union and Surrey Beatty & Sons.

STUDY SITES

The three study sites chosen were within 150 km of Melbourne (Fig. 1). The Yellingbo site was of 0.7 km^2 in a small faunal reserve for Helmeted Honeyeaters. In contrast, the sites at the Brisbane Ranges National Park and in State Forest near Tarnagulla were each of approximately 6 km^2, within more extensive forest.

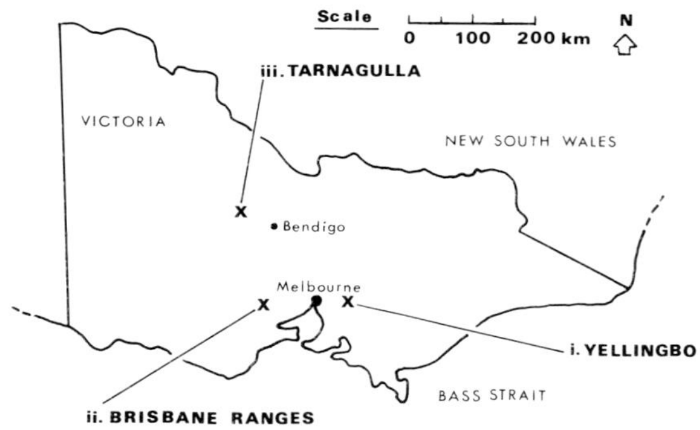

Fig. 1. Locations of the study areas. The honeyeater species studied at each area were: i. Yellingbo — Helmeted H.E., White-eared H.E., Bellminer; ii. Brisbane Ranges — Yellow-tufted H.E., White-eared H.E., Yellow-faced H.E. Fuscous H.E. (non-breeding)., White-plumed H.E. (auxiliary site); iii. Tarnagulla — Yellow-tufted H.E. Fuscous H.E., White-eared H.E., Yellow-plumed H.E. (non-breeding)., Yellow-faced H.E. (non-breeding).

Yellingbo is in the foothills of the Great Divide where the annual rainfall is high (about 1200 mm), with a winter maximum. The Helmeted Honeyeater reserve consists primarily of riparian vegetation among farmland. The small study site comprised a swampy river-flat on the Cockatoo Ck, dominated by Swamp Gum, *Eucalyptus ovata,* over a mixture of open reeds and dense, tall sclerophyllous shrubs, such as Tea-tree, *Leptospermum juniperinum.* The adjacent slope supported woodland of various trees, including *E. ovata,* Scent-bark, *E. aromamphloia,* Peppermint, *E. radiata* and Messmate, *E. obliqua,* with a dense, diverse shrub-layer. Foraging observations were also collected 2 km downstream at an auxiliary site of about 20 ha where well-drained river-flat was co-dominated by Manna Gum *E. viminalis* and *E. ovata.* Both sides were surrounded by farmland.

The Brisbane Ranges site was chosen as an area with similar riparian habitat to Yellingbo in which the *meltoni* race of *L. melanops* occurred, but, unlike Yellingbo, within extensive forest. The region has a moderate annual rainfall of about 700 mm, with an early spring maximum. The study site was on the eastern edge of a 180 m scarp that rises from undulating volcanic plain. It was bisected by the deep, narrow Stony Creek valley and subsidiary gullies, where *E. viminalis, E. ovata* and *E. radiata* occupied the river-flats. Adjacent slopes and an upper, sandy plateau supported a further eight eucalypt species including Iron Bark, *E. sideroxylon,* Brown Stringybark, *E. baxteri,* and Long-leaf Box, *E. goniocalyx.* The shrub-layer on the slopes was of relatively sparse, woodland species, while a dense, diverse heath shrub-layer occurred on the plateau. Foraging observations were also gathered 5 km away in an auxiliary area of more extensive riparian woodland, again dominated by *E. viminalis* and *E. ovata.*

Tarnagulla differed in climate and vegetation from the other sites but also supported a *L. melanops meltoni* population. It was west of Bendigo in a region of low annual rainfall (about 450 mm), which falls predominantly in winter. Yellow Gum, *E. leucoxylon,* predominated along the water-courses, which only flowed in winter, over a generally sparse shrub-layer among alluvial gold-mining, mulloch heaps. Open woodland of *E. sideroxylon* and Grey Box, *E. microcarpa,* occurred on the adjacent, shallow slopes, over a low, dense shrub-layer dominated by Flame Heath, *Astroloma conostephioides.* A dense, low woodland of *E. microcarpa* and Red Box, *E. polyanthemos,* dominated the ridges, where patches of low shrubs and 3 m-tall *Acacia calamifolia* formed an open understorey. Green Mallee, *E. viridis,* thickets with little understorey occurred on dry, northerly slopes, dissected by gullies of *E. sideroxylon* and dense shrubs.

THE HONEYEATER SUBJECTS

The distributions, habitat preferences and morphologies of the *Lichenostomus* honeyeaters that occurred in the study areas are summarised in Figure 2. The Yellow-tufted Honeyeater is an eastern Australian species of which three races are recognised in Victoria. The Helmeted Honeyeater, *L. melanops cassidix,* occurs along foot-hill streams of the southeast and was represented at Yellingbo. *L. m. meltoni* is a small, inland race of auriferous woodland that was represented at the Brisbane Ranges and Tarnagulla. *L. m. gippslandica* occurs in Gippsland in similar habitat to the Helmeted Honeyeater and was not represented in the study.

Five other species of *Lichenostomus* honeyeater and the Bellminer were represented on the study sites. The White-eared, Yellow-faced and Fuscous Honeyeaters overlap in distribution with the Yellow-tufted Honeyeater along the east coast. The White-plumed and Yellow-plumed Honeyeaters are species of similar size and appearance to the Fuscous Honeyeater and these three are generally complementary in their ranges, and differ in choice of habitat and vegetation. The Bellminer is an eastern Australian species of riparian eucalypts. It was included as a principal species in the research because it was numerous at Yellingbo and has been considered a competitor of the Helmeted Honeyeater (Cooper 1967).

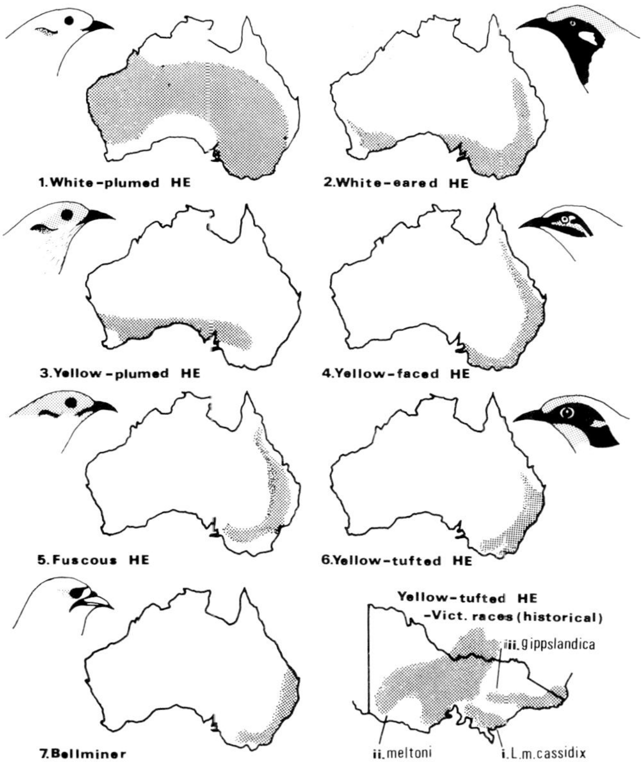

Fig. 2. Distributions of the honeyeater species studied (after Reader's Digest 1976; Crome 1976). These species have the following body measurements (weight, g; total length, mm) (study populations, Wykes 1982) and habitat characteristics (Reader's Digest 1976): 1. *Lichenostomus penicillatus* — 23 g; 160 mm; riparian eucalypts, eastern woodlands. 2. *L. leucotis* — 26 g; 178 mm; woodland and forest. 3. *L. ornatus* — 19 g; 156 mm; mallee, mulga, savannah-woodland. 4. *L. chrysops* — 18 g; 150 mm; forest and dense gullies. 5. *L. fuscus* — 19 g; 152 mm; open forest west of the divide. 6. *L. melanops* — 34 g, 194 mm, riparian forest (*cassidix*); 23 g, 175 mm, riparian forest, auriferous woodland (*meltoni*). 7. *Manorina melanophrys* — 32 g; 179 mm; riparian forest east of the divide.

MATERIALS AND METHODS

The research was conducted between September 1976 and December 1978 by making short, consecutive visits to the three areas at two- to three-month intervals.

1. *Population Densities and Distributions*

A strip-transect method was employed to record densities of all bird species. The Tarnagulla and Brisbane Ranges transects were approximately 11 km long. A third of a transect was covered on each of three mornings of a census. A 2376 m transect was covered on each of three mornings for a census at the small Yellingbo site. The strips were 40 m wide, giving a total coverage of 44 ha at the larger sites, and 9.5 ha (x 3) at Yellingbo.

The strip-transects were subdivided into 91 m sections (120 in the larger strips, 26 at Yellingbo) in order to record distributions of bird sightings. Distributions of the honeyeater populations in relationship to plant-species composition and structure of micro-habitats within the study areas were analysed using these strip-transect data (Wykes 1982). However, micro-habitat preferences are only presented generally here.

2. *Food Supply*

Arthropods were collected using sweep-nets in the shrub-layer, placing sticky-traps in the tree-layer and clipping small branches from trees. The sweep-net samples were obtained by making 50 sweeps per transect-section in 13 sections of the strip-transect during each visit. Foliage-dwelling arthropods were sampled from ten trees, representative of species at five strip-transect markers, by randomly examining 50 leaves from each of four branches per tree. Ten plastic grids of 40 × 40 cm, covered on one face with 'Tangle trap', formed the sticky-traps. Two were hung in separate trees within the lower canopy at each of five transect-markers, for 72-hour periods. All collecting was conducted in the absence of strong winds and rain. Biomass of the catches was estimated as the sum of lengths by widths of all individuals of all taxa.

Availability of nectar was estimated by counting flowering plants (of species attracted to nectarivores) along line-transects which followed the total strip-transect route at Yellingbo, and a third of each of the other, larger routes. Degree of flowering was visually estimated as percentage of total reproductive structures that were fresh, open flowers. The counts were then adjusted by weightings of x2 and x3 for plants with 20-50% and 50% flowering, respectively.

3. *Foraging*

Observations were systematically recorded when traversing the study areas by noting details of up to five attempts to obtain food items per bird at one time. Foraging stratum, plant species, posture and station were noted. Stomach contents were obtained from birds caught in mist-nets in order to analyse diet. Nestling diets were also investigated by placing rings of pipe-cleaner around necks of nestlings to restrict swallowing (Orians 1966).

4. *Aggressive Interaction*

All aggressive interaction (chases and displacement of one bird by another) that were observed between and within species over the research period were recorded.

5. *Breeding Seasons and Nesting Requirements*

All observations of nesting by the honeyeaters were recorded in order to compare breeding season and nest-site characteristics of the species.

Fig. 3. Changes in honeyeater density and food resources. Changes in density of the honeyeater populations at each study area are shown for the research period, with means and standard variations noted below. The results of One-way Analysis of Variance and Student-Newman-Keuls tests to compare the mean densities ($\alpha = 0.05$) are indicated. The figure also shows changes in relative abundance of arthropod populations that were sampled by sweep-net (SN), leaf-clippers (LC) and sticky-traps (ST). These are presented in units of "biomass index", which was the total sum of dorsal lengths-by-widths of the arthropod individuals captured. Each SN value is the mean of twenty 50-sweep samples, each LC value is the mean of five 200-leaf samples, and each ST value the mean per trap of 10 samples. Nectar availability estimates are shown for plant species which produced blossom that attracted honeyeaters. "Nectar index" values were obtained by totalling the number of flowering plants that were intercepted on a line-transect, with weightings of x2 and x3 for 'medium' and 'rich' blossom (see methods).

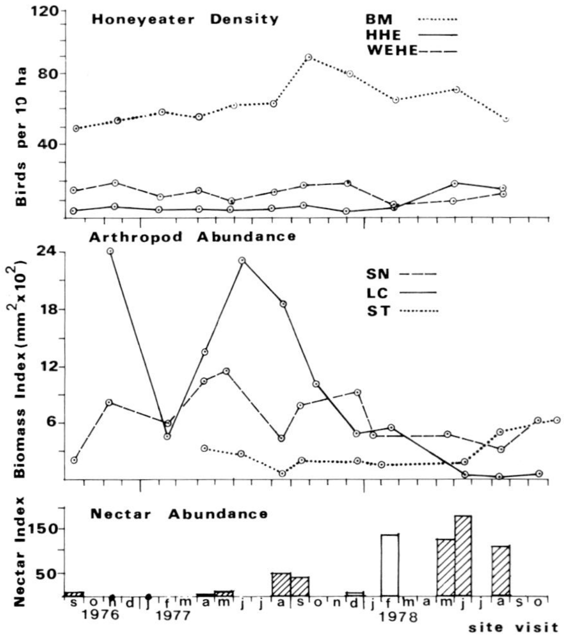

Fig. 3(i).

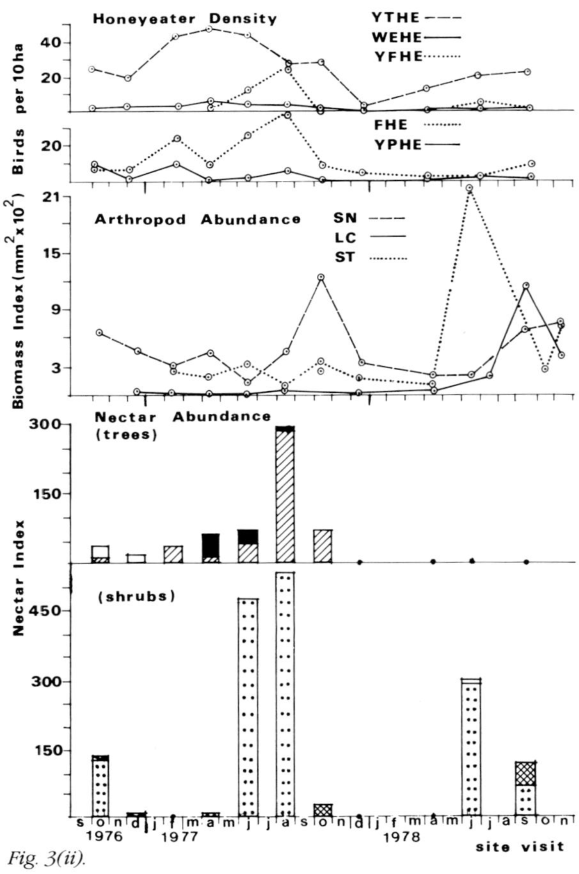

Fig. 3(ii).

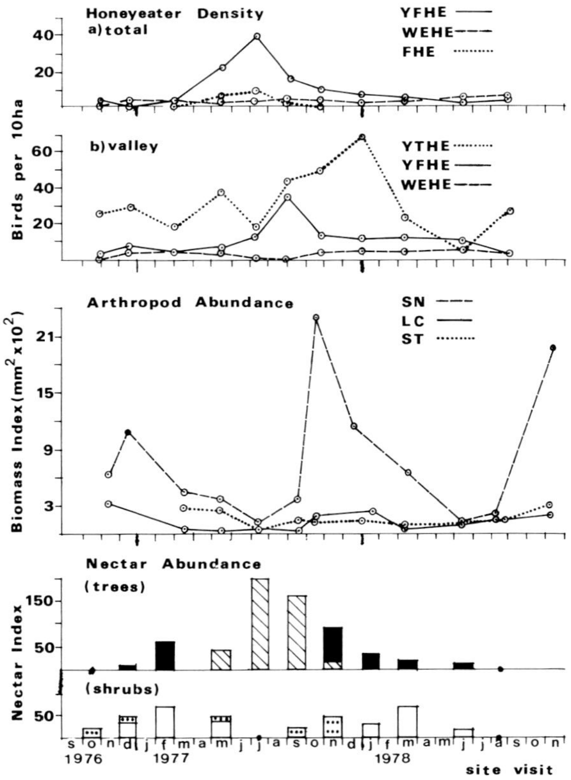

Fig. 3(iii).

Fig. 3(i). Yellingbo. Honeyeater mean densities per 10 ha were: Bellminer — 64±12.4; Helmeted HE — 6±5.6; White-eared HE — 13±4.2. A.O.V.:BM > (WEHE=HHE). Key to nectar categories: hatching = *Eucalyptus ovata*; clear = *E. obliqua*, *E. radiata* and *E. aramophloia*.

Fig. 3(ii). Brisbane Ranges. Honeyeater densities are given for a) the total 44 ha census-area (Yellow-tufted HE excepted), and for b) the 7 ha section of river valley (see text). Mean densities per 10 ha were: a) Yellow-faced HE — 10±11.8, White-eared HE — 3±0.8, Fuscous HE — 1±2.8. A.O.V.: WEHE=YFHE. b) Yellow-tufted HE — 31±17.6; YFHE — 10±8.9; WEHE — 3±1.5. A.O.V.: YTHE > (YFHE = WEHE). Key to nectar categories: hatching = *Eucalyptus sideroxylon*; black = *E. viminalis, dives, radiata, polyanthemos, baxteri, goniocalyx*; clear = *Amyema* spp.; stippling = *Grevillea* spp.

Fig. 3(iii). Tarnagulla. Mean honeyeater densities per 10 ha were: Yellow-tufted HE — 26±12.7; Fuscous HE — 12±11.2; Yellow-plumed HE — 2±3.7; White-eared HE — 1±1.1; Yellow-faced HE — 4±7.8. A.O.V.: YTHE > FHE > (YPHE = WEHE). Key to nectar categories: hatching = *Eucalyptus sideroxylon*; clear = *E. leucoxylon*; black = *E. microcarpa*; stippling = *Astroloma conostephioides*; cross-hatching = *Grevillea alpina*.

POPULATION DENSITIES, DISTRIBUTIONS AND FOOD SUPPLY

1. *Yellingbo*

Helmeted Honeyeaters bred in only two groups, 1 km apart, within the Yellingbo study area. Colour-banding indicated a total population of around 10 birds. Other small groups arrived for the 1977 winter and there was a large influx in May 1978. These winter influxes were in response to flowering of *E. ovata,* as indicated by the estimates of nectar abundance (Fig. 3(i)). Sixty-five birds were estimated to be present in June 1978, based on ratios of unmarked to 26 dyed birds in the population. This was an extremely high proportion of known Helmeted Honeyeater numbers. For comparison, I organised two surveys, conducted by teams of 25 and 36 observers, to cover all known Helmeted Honeyeater habitat in the 1978-79 summer. These gave counts of 59 and 71 individuals, respectively (Wykes 1981).

Breeding Helmeted Honeyeaters in the study area were restricted to *E. ovata* river-flat. This was also the principal habitat of the Bellminer, which greatly out-numbered Helmeted Honeyeaters — mean densities of 64 and 6/10 ha, respectively (Fig. 3(i)). Bellminers were associated with trees suffering psyllid infestation of the foliage (48% of variance in distribution for the research period was explained by correlation with areas of defoliation). They had a mean density of 104/10 ha for the 16 (of 26) transect-sections in which they were recorded. Infestation of *E. ovata* and other eucalypt species in some side-creeks, by a lerp-forming *Glycaspis* psyllid (Hemiptera), was similar to that described in *E. saligna* of NSW (Moore 1961). The infested trees were on low-lying ground prone to waterlogging in winter, whereas adjacent patches of healthy *E. ovata* were generally on higher ground. The Helmeted Honeyeater groups bred in this healthy vegetation.

There were no seasonal fluctuations in density of the colonial, sedentary Bellminer (Smith and Robertson 1978). However, distributions altered when groups moved into areas of fresh psyllid infestation and deserted trees that partially recovered or were defoliated during the research period. Such Bellminer movements were reflected in changes in density along the strip-transect, while dramatic fluctuations in psyllid abundances (which dominated the leaf-clip samples) were recorded in the trees that were sampled (Fig. 3(i)).

Segregation between breeding Helmeted Honeyeaters and Bellminers was also recorded at the auxiliary study site. Dense groups of Bellminers occupied swampy areas dominated by psyllid-infested *E. ovata,* whereas better drained patches in which *E. viminalis* was more abundant than *E. ovata* were exclusively occupied by a total of 204 Helmeted Honeyeaters. Segregation of the two species was not maintained at either study site in winter. Helmeted Honeyeaters and other nectarivores aggregated at flowering *E. ovata* including trees in psyllid-infested areas inhabitated by Bellminers. Movements of up to 3 km between breeding sites and the winter aggregation at *E. ovata* blossom in the main study site in 1978 were recorded for colour-banded Helmeted Honeyeaters.

The third species studied at Yellingbo, the White-eared Honeyeater, had a moderate mean density of 12/10 ha (Fig. 3(i)). It was recorded in most transect-sections and all habitat-types over the research period but there was an overall negative correlation with areas of psyllid infestation (25% of variance explained). Thus, White-eared Honeyeaters utilised the river-flat vegetation except for psyllid-infested areas dominated by Bellminers, but also occurred in hillside woodland from which Helmeted Honeyeaters and Bellminers were absent.

Changes in White-eared Honeyeater density followed a seasonal pattern. Density increased in early spring, when breeding territories were established, and peaked with addition of offspring around November. The population then dispersed for the winter, with a wider distribution within the study area and a lower total density (Fig. 3(i)). The changes in density matched seasonal fluctuations in arthropod abundance (Fig. 3(i)). This was most clear for the total sweep-net values but most taxa from all sampling methods followed this pattern except for aseasonal fluctuations in psyllids on the foliage and winter abundances of small flies in the sticky-trap catches.

2. *Brisbane Ranges*

Yellow-tufted Honeyeaters were confined to river-flats of the Stony Creek Valley during the breeding season, in similar habitat to that of the Helmeted Honeyeater at Yellingbo. The densities of the honeyeaters are presented separately to the total transect for the 2 km section which was routed along this valley (Fig. 3(ii)). Yellow-tufted Honeyeater breeding occurred in the spring and summer, coincident with peak arthropod abundances as measured by all three sampling techniques (Fig. 3(ii)). The distribution of the breeding population along the valley was uneven, with all territories at peak density in 1977. No particular floristic or structural attributes were found for this section but the river-flats were broader there than elsewhere (Wykes 1982).

Many of the Yellow-tufted Honeyeaters foraged throughout the study area for nectar during the autumn and winter seasons of arthropod poverty (Fig. 3(ii)). They were recorded singly and in small parties scattered among valley, gully and ridge vegetation during the 1977 winter, when *E. sideroxylon* blossom was abundant (Fig. 3(ii)). The valley population then reached a maximum for the study period in the 1977 breeding season. Very little nectar was available during the following winter and Yellow-tufted Honeyeater density dropped to a minimum when nectar was presumably sought outside as well as throughout the study area.

The two other *Lichenostomus* species that were resident at the Brisbane Ranges were of significantly lower density to the Yellow-tufted Honeyeater in the Stony Creek valley. However, unlike the Yellow-tufted Honeyeaters, both also occurred at these same densities throughout the study area (Fig. 3(ii)). The White-eared Honeyeater had a mean density of 3/10 ha, which varied little with season. Most breeding territories were established along the valley and gullies and in dense, heath-woodland of the plateau. Foraging ranges broadened in winter to include the open woodlands of the slopes during this season of low arthropod abundance. As a result, there was a reduction in density along the main valley during each winter and early spring (Fig. 3(ii)), in the same way as White-eared Honeyeater density fluctuated in the small Yellingbo study area (Fig. 3(i)).

The third resident species was the Yellow-faced Honeyeater, with a mean density of 10/10 ha. This was not significantly different to that of the White-eared Honeyeater because density was low except for up to 40 birds/10 ha during a 1977 winter influx (Fig. 3(ii)). Breeding pairs occurred along the main valley and in gullies where the shrub-layer was tall. Nectar was sought throughout the study area in the autumn and winter of 1977 when nectar was plentiful. Immigrant flocks of up to 50 were associated with another *Lichenostomus,* the Fuscous Honeyeater, and the White-naped Honeyeater, *Melithreptus lunatus.* This was the first known record of Fuscous Honeyeaters for the National Park. No influx of nectarivorous honeyeaters occurred in 1978 and Yellow-faced Honeyeater density decreased after the breeding season, when nectar was presumably sought away from the study area.

3. *Tarnagulla*

The Yellow-tufted Honeyeater was broadly distributed within the Tarnagulla study area, being recorded in 91 of the 120 transect-sections, at a mean density of 26/10 ha for the research period. This was in contrast to the localised, riparian populations of *L. melanops* (Helmeted Honeyeater at Yellingbo) at the other study sites. Late winter/spring breeding densities were consistently around 25/10 ha (Fig. 3(iii)), with pairs most numerous in slope vegetation of *E. sideroxylon* and *E. microcarpa* with dense, low shrub-layer. Breeding occurred during the brief spring periods of peak arthropod abundances, as measured by all sampling methods except that small midges (Diptera) were also numerous on sticky-traps in winter (Fig. 3(iii)).

Post-breeding and winter densities of Yellow-tufted Honeyeaters varied greatly over the research period. An influx in early 1977 took density to 45/10 ha, when family groups aggregated at eucalypt nectar sources throughout much of the study area. Nectar was still abundant when the Yellow-tufted Honeyeater population had returned to its breeding distribution and lower density (Fig. 3 (iii)). The Tarnagulla district suffered a drought from autumn 1977 through to the following year. The Yellow-tufted Honeyeaters had a brief early spring breeding season in that year and then deserted the study area over the 1977/1978 summer. These may have been responses to relatively low arthropod abundances (except for an abundance of a beetle species in the sweep-net catches), lack of nectar (Fig. 3(iii)), or directly to absence of surface water for drinking. The Yellow-tufted Honeyeater population and arthropod abundances recovered by the 1978 breeding season while sources of eucalypt nectar remained low.

A variety of other *Lichenostomus* species, less abundant than the Yellow-tufted Honeyeater, occurred at Tarnagulla, including some not found at the other study sites. The most numerous was the Fuscous Honeyeater, with a mean density of 12/10 ha.

Fuscous Honeyeater breeding density was consistently at around 8/10 ha, with breeding groups confined to water-course vegetation of *E. leucoxylon-E. microcarpa* where the shrub-layer was sparse. They were more widely distributed in autumn and winter of 1977, when immigrant flocks sought out the abundant nectar sources. Fuscous Honeyeaters bred in late spring and aggregated in high densities at nectar sources through the 1977 winter, prior to deserting the site during the drought, without breeding. They did not return until the spring of 1978 (Fig. 3(iii)).

The third resident species, the White-eared Honeyeater, was scarce and restricted in distribution, in contrast to widespread populations at the other sites. Breeding pairs were confined to ridge habitat of low *E. microcarpa* and tall shrub-layer. The population was more widely distributed in the 1977 winter but density remained low throughout the research period, with a mean of 1/10 ha (Fig. 3(iii)).

Yellow-faced Honeyeaters were winter immigrants to Tarnagulla. Large flocks arrived in June 1977 and were then absent from October until a smaller influx the following winter. These nectarivorous flocks were widespread at the abundant nectar sources in 1977 but most numerous in ridge habitat where there was a tall, dense shrub-layer.

They were virtually restricted to this habitat when *Astroloma* was the only nectar source in the following winter (Fig. 3(iii)).

The Yellow-plumed Honeyeater was at the southern limit of its range at Tarnagulla and never numerous (mean density of 2/10 ha). Specimens taken from late-winter groups, which occurred in dry, Green Mallee habitat with sparse understorey, were males with enlarged testes. However, few birds were present in spring and breeding was not recorded. Most other records were of post-breeding individuals or small groups that aggregated with other honeyeaters at nectar sources in summer and winter, 1977. Yellow-plumed Honeyeaters were virtually absent in 1978.

Occasional post-breeding, immature White-plumed Honeyeaters occurred at Tarnagulla (mean 0.4/10 ha), and one pair bred in watercourse vegetation. These were out-liers of populations resident in Yellow-gum, *E. meliodora* and Red-gum, *E. camaldulensis,* along nearby, permanent streams.

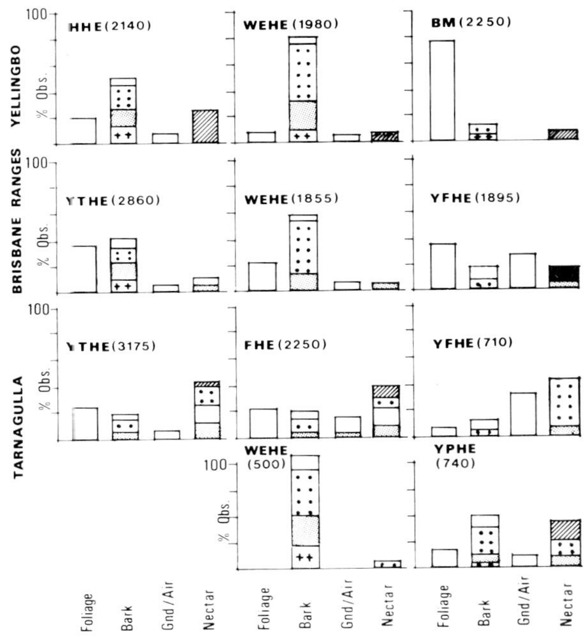

Fig. 4. Use of foraging stations. Summaries of foraging station usage by the honeyeater populations are given as means of the visit by visit frequencies. Visit samples of less than 45 obs. were not included. Total numbers of observations are noted. Yellingbo samples were only from the main site. Key — Bark sub-classes: twig-clear, branch-coarse stipple, limb-fine stipple, trunk-cross; Gnd/air: ground-fine stipple, air-clear; Nectar: (Yellingbo) *Eucalyptus ovata*-hatch; (Brisbane Ranges) *E. sideroxylon*-clear, *Amyema* spp.-cross; (Tarnagulla) *E. sideroxylon*-fine stipple, *E. leucoxylon*-clear, *E. microcarpa*-hatch, *Astroloma conostephioides*-course hatch.

FORAGING BEHAVIOUR AND DIET

1. *Riparian L. melanops Populations*

Foraging by the Helmeted Honeyeaters at Yellingbo and Yellow-tufted Honeyeater at the Brisbane Ranges was confined to riparian gums, except for some nectarivory in winter. Thus, 89% per visit of Helmeted Honeyeater non-nectar foraging records (total n=1400) were in *E. ovata* at the main site, while *E. viminalis* (68%) and *E. ovata* (21%) were utilised at the auxilliary site (n=1995). *E. viminalis* (42%) and *E. ovata* (38%) were also favoured by the Yellow-tufted Honeyeater at the Brisbane Ranges.

Foraging was by gleaning at foliage and all bark size-classes (Fig. 4). The adult (no Helmeted Honeyeater samples) and nestling diets included bark- and foliage-dwelling beetles, ants and bugs. However, there are also carbohydrates that are gleaned from these stations by honeyeaters which are generally undetected in diet samples. The three major types are the lerp coatings produced by leaf-dwelling psyllids (Hemiptera), honeydew excreted by psyllids and other Hemiptera of foliage and bark, and manna as a fluid or in lumps, exuded by plants at points of injury (Paton 1980). These carbohydrates are most abundant where there is decorticating bark since this gives insects access to sap and provides them with shelter. Riparian Gums, such as the aptly named Manna Gum, *E. viminalis,* have decorticating bark on all structures and are prolific producers of manna in spring/summer (Paton 1980). Others, such as *E. ovata,* are particularly susceptible to infestation by lerp-forming psyllids.

Non-nectar carbohydrates at Yellingbo and the Brisbane Ranges formed a large but unquantified portion of the *L. melanops* food supplies. Close observation showed that honeydew and manna were frequently licked or plucked from bark and leaf surfaces and from under strips of bark. Futhermore, the birds regularly foraged at the same sites to obtain these renewing sugars. Lerps occurred in adults and nestling food samples of Yellow-tufted Honeyeaters at the Brisbane Ranges but, surprisingly, considering the high abundances, not in nestling samples from Helmeted Honeyeaters at Yellingbo. However, *Coprosma hirtella* berries, presumably rich in carbohydrates, were common in the diets of some Helmeted Honeyeater nestlings. This was the only frugivory that was recorded for *Lichenostomus* honeyeaters.

Large portions of the riparian populations of *L. melanops* exhibited further foraging-station versatility by turning to nectarivory in autumn and winter. The most abundantly flowering eucalypt species during the research period — *E. ovata* at Yellingbo and *E. sideroxylon* at the Brisbane Ranges — were the main nectar sources. Other species might be more abundant in other years and exploited by Yellow-tufted Honeyeaters at the Brisbane Ranges but the Helmeted Honeyeater was restricted in its access to other nectar sources since surrounding habitats at Yellingbo have been cleared. However, grevillea and banksia blossom in local gardens formed a novel source of nectar for Helmeted Honeyeaters at the auxiliary site. The *L. melanops* populations supplemented their winter nectar intake with small flies that were hawked among and above the tree tops. This was probably a case of protein being obtained at an energy loss by honeyeaters with access to sugars but little protein (Ford and Paton 1976a).

2. *Potential Competitors of Riparian L. melanops Populations*

White-eared Honeyeaters at Yellingbo and the Brisbane Ranges were specialised bark foragers which gleaned and probed at decorticating bark of branches, limbs and trunks in many eucalypt species (Fig. 4). However, foraging was confined to the branches on species that have persistant bark on larger structures. Beetles and ants dominated adult diets, while beetles, spiders, flies and psyllids from bark and some foliage were fed to nestlings. Bark was the main foraging site even during the winter periods of minimum arthropod and non-nectar carbohydrate availability. Nectar was utilised when extremely abundant or in small patches and consequently poorly utilised by other species. Honeydew was seen to be taken from under bark of riparian gums, and lerps occurred in diet samples from Yellingbo, but most foraging appeared to be for arthropods.

A species of higher potential for conflict with the Helmeted Honeyeater at Yellingbo was the numerous Bellminer. It foraged in *E. ovata* (79% per visit of 2085 non-nectar observations) at the main study area and in both *E. ovata* (37%) and *E. viminalis* (41% per visit of 1890 obs.) at the auxilliary site. In contrast to the Helmeted Honeyeater, which utilised these same eucalypts, the body and beak of the stocky Bellminer were specifically adapted to one foraging technique — hanging from twigs and leaves to pluck items from the pendulous leaves (Wykes 1982). Psyllid larvae with whole and partially digested, 2-4 mm diameter lerps dominated the adult and nestling diets. Only larvae that adhered to their lerps were seen to be taken but, with a mean of 56 (maximum 241) per digestive tract, they were an important nutrient source. Psyllids did not infest *E. viminalis* at the auxiliary site and, instead, lumps of manna on the foliage and twigs attracted Bellminers to this species. Lerps and manna were the principal food source of Bellminers throughout the year and nectar was only taken when it occurred in their normal territories.

The Yellow-faced Honeyeater was a potential competitor of Yellow-tufted Honeyeaters at the Brisbane Ranges. Its spring/summer diet was of Mistletoe (*Amyema*) nectar and arthropods snatched and hawked from foliage and blossom. Foliage-dwelling beetles, spiders, lerps and flies formed the chick-meals. The winter diet was also of nectar and insects hawked in the shrub-layer. The nectar was primarily from *E. sideroxylon* and the abundant flowering of this species in 1977 attracted the large migrant influx. Small flies formed a median of 82% of adult food items, with a maximum of 67/ stomach.

The limited observations of White-plumed Honeyeaters at the auxilliary Brisbane Ranges site indicated very similar foraging habits to the Yellow-tufted Honeyeater and thus a high potential for competition. They gleaned in *E. viminalis* (45% per visit, n = 220) and *E. ovata* (43%) at foliage and all bark stations for arthropods and carbohydrates. The extent of nectarivory in winter was not determined at the Brisbane Ranges but White-plumed Honeyeaters were attracted to nectar in the Tarnagulla area.

3. *Non-riparian Yellow-tufted Honeyeater Population*

Yellow-tufted Honeyeaters at Tarnagulla occurred in open forest where persistent-barked eucalypts, *E. sideroxylon* (mean 58% per visit, n = 1578) and, to a lesser extent, co-dominant *E. microcarpa* (15%), were the main sites of non-nectar foraging in the breeding season from foliage, twigs and branches. Beetles and ants were predominant in the adult diet samples while lerps, psyllid larvae and diptera were most abundant in nestling meals. Nectar, particularly that of the ubiquitous shrub, *Astroloma conostephioides,* was also utilised by breeding pairs. Family groups aggregated at eucalypt blossom shortly after chicks fledged in late spring and fed on nectar throughout autumn and winter. This was supplemented by hawking for small flies and snatching items such as moths close to the ground.

These Yellow-tufted Honeyeaters were thus distinctive in not being associated with rich carbohydrate sources of riparian gums. Instead, they were widespread in forest where nectar was available throughout the year and sugary lerps were abundant on foliage in the breeding season.

4. *Potential Competitors of Yellow-tufted Honeyeaters at Tarnagulla*

Fuscous Honeyeaters often foraged in the same tree species as did the Yellow-tufted Honeyeater — mean 30% per visit (n = 1223) in *E. microcarpa* and 21% in *E. sideroxylon* — but most was along the water-courses rather than on the slopes, where *E. leucoxylon* also occurred (27%). Foraging habits were similar to those of the Yellow-tufted Honeyeaters, with high frequencies at twigs, branches, foliage, air and blossom. Lerps and eucalypt blossom were carbohydrate sources in the breeding season, while ants, beetles and a variety of other taxa occurred in stomach samples (Wykes 1982). *E. leucoxylon* has decorticating bark on all woody parts but this sheds in fine flakes, rather than strips, so that trunks and limbs have no foot-holds for honeyeaters or shelter to harbour insects and associated carbohydrates. Consequently, there was little foraging at limbs or trunks of *E. leucoxylon* by the Fuscous or other honeyeaters. Fuscous Honeyeaters utilised eucalypt and *Astroloma* nectar along watercourses and in other habitats during the non-breeding seasons, supplemented by small flies, which were hawked, and beetles and ants from around blossom.

Yellow-plumed Honeyeaters that occurred in their typical mallee habitat at Tarnagulla foraged in a similar manner to the Yellow-tufted and Fuscous Honeyeaters except that branches of the mallee *E. viridis* were particularly favoured (Fig. 4). The small population utilised the same nectar sources as did the other honeyeaters in winter.

The White-eared Honeyeater at Tarnagulla was a specialist gleaner and prober at bark sites, as elsewhere. Beetles and ants dominated the adult diet (nestling meals were not obtained). Most foraging was in the two widespread eucalypts, *E. microcarpa* (mean per visit 54%, total n = 505) and *E. sideroxylon* (26%). It was therefore surprising that the breeding population was very small and restricted in distribution to one habitat-type, *E. microcarpa*-dominated ridges.

Yellow-faced Honeyeaters were an abundant winter immigrant to Tarnagulla with the potential to compete with resident species for nectar. However, *E. sideroxylon* nectar that was favoured by other species was virtually only utilised when it was particularly abundant in 1977. In that year and the following winter, Yellow-faced Honeyeater flocks primarily foraged on *Astroloma* nectar and small flies (up to 60/stomach) among the tall shrub-layer of the ridge habitat. Here they were frequently the nuclei for mixed species honeyeater flocks.

INTERSPECIFIC AGGRESSION AS A FACTOR IN DIVISION OF RESOURCES

Interspecific aggression often plays a role in the structure of nectarivorous bird communities (Case and Gilpin 1974; Dow 1979). To understand the role of aggression in *Lichenostomus* communities completely would involve detailed behavioural research. Here I have merely compiled frequencies of aggression (Fig. 5).

Intraspecific aggression was relatively infrequent within the small riparian Helmeted and Yellow-tufted Honeyeater populations but common within the more widely distributed Yellow-tufted Honeyeater population at Tarnagulla and within the large Bellminer population at Yellingbo. All of these

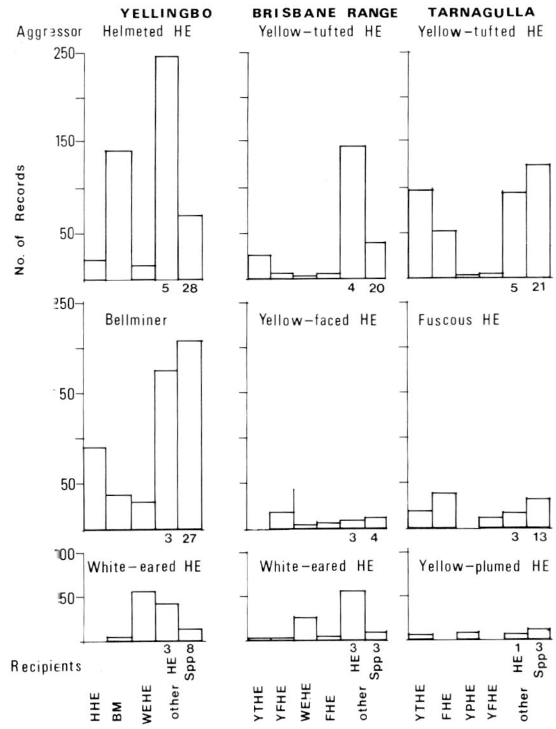

Fig. 5. Aggressive interactions between honeyeaters. Total frequencies of aggressive interactions that were recorded involving the honeyeaters are shown for each study population. Frequencies of intraspecific attack and attack against other *Lichenostomus* species, the Bellminer, other honeyeaters (pooled as 'other HE') and 'other bird species' are shown for each aggressor species. The numbers of species in the pooled categories are noted. Also recorded but not shown was aggression against the study populations by the Red Wattlebird (all sites) and the White-naped Honeyeater (Yellingbo and Brisbane Ranges).

populations formed co-operative/communal breeding groups (Wykes 1982; Dow 1980) which monopolised the tree-layers of their preferred habitats through indiscriminate aggression against all other bird species that intruded. Breeding groups of Fuscous and White-plumed Honeyeaters were also often aggressive towards many bird species.

White-eared and Yellow-faced Honeyeaters bred in pairs which generally defended territories only against conspecifics and some smaller honeyeater species. The breeding populations of these species at Yellingbo and the Brisbane Ranges were relatively widespread but excluded from the mosaics of territories occupied by their pugnacious, co-operative congeners. The high abundances of Yellow-tufted and Fuscous Honeyeaters throughout much of the Tarnagulla site probably contributed to the restricted distribution of that White-eared Honeyeater population.

The dominance relationships that were observed in the breeding season were also maintained at winter nectar sources. Helmeted, Yellow-tufted and Fuscous Honeyeaters defended small, feeding territories at eucalypt blossom against other honeyeater species. Yellow-faced Honeyeater flocks were attacked by territorial species but gained nectar from unattended trees when large influxes arrived at Tarnagulla and the Brisbane Ranges in years when *E. sideroxylon* blossom was particularly abundant. Furthermore, *Astroloma* nectar in the shrub-layer was an undefended resource at Tarnagulla that was particularly favoured by Yellow-faced Honeyeaters. White-eared Honeyeaters opportunistically utilised undefended blossom that was encountered within their ranges.

Some breeding populations of Helmeted and Yellow-tufted Honeyeaters occurred in the same, or similar, habitat to that utilised by co-existing species. Bellminers monopolised lerp-infested patches of *E. ovata,* river-flat vegetation at Yellingbo which would probably otherwise have been utilised by Helmeted Honeyeaters. Helmeted Honeyeaters excluded Bellminers from healthier *E. ovata* and areas in which *E. viminalis* predominated but Bellminers evicted resident Helmeted Honeyeaters if these became infested. Bellminers continued to defend lerp-infested areas through the winter but persistent attacks against Helmeted Honeyeaters and other honeyeater species seeking nectar from flowering *E. ovata* did not always discourage the intruders.

Yellow-tufted Honeyeaters and White-plumed Honeyeaters at the Brisbane Ranges auxilliary site, and Yellow-tufted and Fuscous Honeyeaters at Tarnagulla, defended mutually-exclusive territories within the same or similar habitat-types. The reasons why one species did not exclude the other from the region were not fully apparent from their foraging or behavioural characteristics.

BREEDING AND NESTING CHARACTERISTICS AS FACTORS IN DIVISION OF RESOURCES

Most of the *Lichenostomus* species bred in the spring and summer, coincident with the period of peak arthropod abundance. The White-eared Honeyeater, Bellminer, and riparian populations of Helmeted and Yellow-tufted Honeyeaters had prolonged breeding seasons, as might be expected for gleaners of bark and foliage. The Tarnagulla Yellow-tufted Honeyeater had a late-winter, early-spring season, related to a short spring peak in arthropods at this site and occurrence of *Astroloma* and *E. sideroxylon* nectar at this time in their breeding habitat. Peak Fuscous Honeyeater breeding was significantly different in being a month later (Wykes 1982), coincident with blooming of *E. leucoxylon* in their water-course habitat (Fig. 3(iii)). This asynchrony may reflect greater differences in foraging ecologies and food resources than was apparent from the foregoing foraging analysis.

Migrant Yellow-faced flocks were still wintering at Tarnagulla when Yellow-tufted Honeyeaters were breeding and their breeding season was significantly later than that of the Yellow-tufted Honeyeater at the Brisbane Ranges (Wykes 1982). This characteristic, early-summer, breeding period (Ford 1981) coincided with blossoming of *Amyema* and high abundance of the active foliage-dwelling insects on which they fed.

Thus, breeding season characteristics of the honeyeaters did not promote co-existence of species through staggered use of resources (Cody 1974) but some differences in season reflected differences in foraging habits. However, nest-site requirements of the Helmeted and Yellow-tufted Honeyeater did affect their distributions and their relationships with congeners. Helmeted, Yellow-tufted and White-eared Honeyeaters built relatively bulky, well-lined nests, attached by the sides and rims. These were generally placed in dense foliage and twigs of shrubs. Median nest heights for the populations were: HHE — 1.5 m (n = 19), YTHE (BR) — 0.75 m (n = 35), YTHE (T) — 0.5 m (n = 29); WEHE (Y) — 0.5 m (n = 22), WEHE (BR) — 0.5 m (n = 11). Yellow-faced Honeyeaters also built relatively bulky nests, in tall shrubs (median height 2.3 m, n = 10). Bellminer nests were flimisier, and attached by the rims at sites ranging from low shrubs to saplings and trees, at a median height of 2 m (n = 40). In contrast, Fuscous Honeyeaters built light, flexible cups of fine grass and cobweb which they suspended from twigs and leaves of saplings and trees (median height 10 m, n = 34).

Tree-foraging passerines might be expected to nest in the tree-layer in order to be inaccessible to ground-foraging predators, inconspicuous when travelling to the nest and efficient in carrying food (MacArthur 1958). However, a low diversity of native mammalian and reptilian nest predators and the sparse foliage of eucalypts in southern Australia may be factors which would promote building well-insulated nests in the shrub-layer. This would not be advantageous for the Fuscous Honeyeater since it is a species of open savannah where shrub-layer vegetation is often sparse. Similarly, the White-plumed Honeyeater and Yellow-plumed Honeyeater are species which occur in habitats where the shrub-layer is often sparse and suspend fine nests among eucalypt foliage (Reader's Digest 1976).

A consequence for Helmeted and Yellow-tufted Honeyeaters of requiring patches of dense shrubs in which to nest was their absence from habitat in which they might otherwise have occurred. The Helmeted Honeyeater was absent from some areas of river-flat at Yellingbo in which they were present prior to removal of shrub-layer for grazing and which were not infested by psyllids and associated Bellminers. Similarly, grazing and flooding at the auxiliary Brisbane Ranges site had removed the shrub-layer in areas of the *E. viminalis-E. ovata* woodland. These appeared suitable for Yellow-tufted Honeyeaters, and were adjacent to Yellow-tufted Honeyeater territories in habitat with a dense bracken shrub-layer, but were occupied by White-plumed Honeyeaters. Clearing of the shrub-layer for picnic areas along the stream in the main Brisbane Ranges study site also appeared to reduce the area of broad river-flat in which the small population held territories.

Differences in nest site requirements was also likely to have been one of a number of factors which contributed to habitat selection by Yellow-tufted Honeyeaters and co-existence with Fuscous Honeyeaters at Tarnagulla. The shrub-layer of the water-course habitat in which Fuscous Honeyeaters bred was very patchy and sparse due to past alluvial-gold diggings, whereas the hill-side slopes that were the principle breeding habitat of Yellow-tufted Honeyeaters had a dense, low shrub-layer. Yellow-tufted Honeyeater pairs bred in the water-course vegetation where the shrub-layer of *Astroloma* remained, and utilised it for both nest sites and to obtain nectar.

DISCUSSION

1. *Population Regulation and Competition*

A general problem for birds in temperate climates is that arthropod resources fluctuate seasonally. Any bird populations that are dependent on these resources are likely to be regulated by mortality during the period of minimum food availability (Fretwell 1972). However, many species avoid this shortage to some degree by seasonal movement between habitats and/or by switching to alternative food supplies (Morse 1971). The solution to this problem for most *Lichenostomus* honeyeaters was to forage for various food sources on bark and foliage of eucalypts when they were abundant in the spring-summer breeding period and then to turn to nectar during the winter.

In some years nectar was scarce in the study areas and segments of most populations departed and possibly suffered high winter mortality. As nectar sources were often defended, then this mortality would be density-dependent. However, in years when nectar was abundant or when the search for nectar away from the study areas was successful, survival was likely to be good. Consequently, relatively constant breeding densities for many of these populations suggested that the populations were primarily limited by availability of habitat. Intraspecifically territorial, breeding groups probably occupied all suitable breeding habitat and the excess population was likely to survive as auxilliaries to breeding pairs (Dow 1980), occupy suboptimal habitat and seek suitable habitat elsewhere. Since a number of *Lichenostomus* species occurred at each study area, all foraging in relatively similar manners in the breeding season and winter, interspecific competition also had the potential to regulate the densities and distributions of the populations. Indeed, interspecific aggression at blossom and breeding territories showed that interspecific competition was likely to be a major factor in the evolution and maintenance of each species' niche.

2. *Resource Partitioning*

Each of the species solved the problem of obtaining food in both the breeding season and winter in different ways. The White-eared Honeyeater primarily foraged for arthropods at bark on vertical surfaces and morphological specialisation to this niche probably restricted its ability to efficiently obtain other resources such as nectar. It would therefore probably suffer relatively higher mortality than its nectarivorous congeners in winter but arthropod populations under bark are likely to fluctuate less than at other stations (Noske 1985). The White-eared Honeyeater occurred in a wider variety of habitats than did other species since most eucalypts provide branch foraging sites but was excluded from areas monopolised by interspecifically aggressive congeners.

Helmeted Honeyeaters were associated with breeding habitat rich in various carbohydrates which were obtained by a variety of foraging techniques and monopolised by indiscriminate, interspecific territoriality. The larger body size and flight manoeuvreability of the Helmeted Honeyeater (Wykes 1982) were possibly adaptations for aggressive interaction (Carpenter 1978; Feinsinger and Colwell 1978) which would involve sacrifices in foraging efficiency. Compatibility of breeding and winter adaptations was achieved through winter aggregations at rich nectar sources within the general breeding region which were often defended against other honeyeaters.

The Fuscous Honeyeater was similar to the Yellow-tufted Honeyeater at Tarnagulla in foraging habits and in being interspecifically aggressive. Co-existence of these populations may have been promoted through differing nest-site requirements. The Fuscous Honeyeater also differs in being migratory, at least for some populations (Keast 1968) and migrant influxes occurred at the study areas when blossom was abundant in winter.

The Yellow-faced Honeyeater breeds in many forest habitats, particularly in montane areas, on active insects of eucalypt and shrub foliage which would be highly seasonal in abundance. Nectar was also utilised in the summer at the Brisbane Ranges but may not be generally available to breeding populations. The Yellow-faced Honeyeater was not interspecifically aggressive but is a migrant (Keast

1968; Ford 1977) that wintered in the study areas in high numbers only when there was superabundance of blossom. The nectar was supplemented by small flying insects captured in a similar way to arthropods in the breeding season.

The Bellminer was a specialised leaf-gleaner adapted to exploit the rich carbohydrate sources of lerp and manna that are a year-round resource in riparian habitats where some populations of eastern Australia form relatively long-lived colonies (Swainson 1970; McCulloch and Noelher 1974). However, the population has erupted at Yellingbo in response to psyllid infestation of *E. ovata* swamps. All other bird species were excluded from utilising the Bellminer territories through indiscriminate interspecific aggression.

3. *Interspecific Aggression*

Overt, interspecific aggression is unusual (Dow 1979) and generally considered maladaptive (Orians and Willson 1964). Orians and Willson suggest that it is prevalent in nectarivorous communities because nectar is a seasonal resource around which precise partitioning mechanisms cannot evolve. More importantly, however, nectar is a readily obtained, easily digested, rapidly renewed resource that often occurs in patches which can be aggressively defended at a low cost/benefit ratio (Case and Gilpin 1974). The resource can often be partitioned through pugnacious species monopolising rich, 'economically defendable' (Brown 1964) clumps, while less aggressive species efficiently exploit dispersed or less-productive blossom (Feinsinger and Colwell 1978; Ford 1979). Nectar remains a common resource in the evolution of these communities, and pugnacious species must remain so in order to obtain rich sources.

The interspecific aggression shown by sedentary honeyeaters such as the primarily 'insectivorious' species of *Manorina* (Dow 1977; Smith and Robertson 1978) is superficially different from that of nectarivores. However, contrary to Dow's (1977) unusual suggestion that the aggression centres around common resources of small insects, Paton (1980) argues that these honeyeaters basically feed on carbohydrates that are similar to nectar. The breeding colonies of the interspecifically aggressive honeyeaters of this study were certainly associated with habitat in which abundant manna, honeydew and lerps were exploited. The Tarnagulla population of Yellow-tufted Honeyeater was exceptional in relying on year-round nectar instead.

I suggest that the alternative carbohydrates, as for nectar, are readily obtained and digested, renewable energy sources for many bird species, and economically defendable in some habitats. However, the interspecific aggression of the Yellow-tufted Honeyeaters and *Manorina* species is remarkably indiscriminate (this study; Dow 1979). Interference is favoured when the cost of the aggression is small, when it is effective and when there is high overlap of food resources with the target species. Many characteristics of the social systems of these honeyeaters appear adapted to reducing costs of aggression while maximising its effect (Wykes 1982). Furthermore, foraging overlap with other bird types is higher than might be expected. Many birds are known to feed on lerps, and probably also take other sugars from bark and leaves (Woinarski 1985). The honeyeaters might also benefit by excluding all insectivores because they would eat larval and adult Hemiptera that initiate lerp, manna and honeydew production, although the honeyeaters themselves eat these insects (Campbell and Moore 1957; Wykes 1982).

4. *Helmeted Honeyeater Conservation*

The results of this study show that the rare Helmeted Honeyeater is a form of Yellow-tufted Honeyeater which is adapted to monopolizing rich sources of carbohydrates in riparian gums through overt aggression. The ultimate threat to its continued existence is habitat destruction through clearing or fire. Within the reserved habitat it is likely that some areas are unsuitable because shrub-layer nest-sites are lacking. A further difficulty is likely to arise in winters when the usual source of nectar, *E. ovata,* flowers poorly, as during one of the study years. Alternative nectar sources are no longer available in countryside surrounding the reserves since these have been largely cleared for farming. However, the establishment of shrub-layers for nesting and planting of winter-flowering nectar-producers are simple remedies to these problems.

A more subtle, less easily rectified threat is that changes to the habitat are favouring a competitor species, the Bellminer. Psyllid infestations probably result from water stress to trees in particular soil types (Moore 1961) under certain climatic conditions (White 1971; Landsberg and Wylie 1983). These occur naturally but may be chronic in areas such as Yellingbo where the seasonal water-table regime has been altered by factors such as clearing in the catchment (Wykes 1982). Bellminers exploit these psyllid outbreaks (Campbell and Moore 1956) and since both these and Helmeted Honeyeaters defend mutually exclusive territories, Bellminer eruption at Yellingbo has been at the expense of the Helmeted Honeyeaters.

However, the psyllid infestation appears confined to *E. ovata* on low-lying swampy, river-flats and was not recorded in *E. viminalis,* which occurs on higher ground. A short-term means of providing additional habitat for Helmeted Honeyeaters would be to improve present stands of *E. viminalis* by providing shrub nest-sites and winter nectar sources, while re-establishing *E. viminalis* and *E. ovata* on the generally cleared areas of well-drained river-flats.

Further research should be directed at discovering means to control psyllid infestation and Bellminer proliferation in the region.

ACKNOWLEDGEMENTS

The research was financed by a scholarship and field grant from the Victorian Ministry for Conservation, administered by K. Dempster of Fisheries and Wildlife Division. I am grateful for this support. I thank Elaine Wykes for unstinting support during the research, and owe a great debt to my supervisor, Alan Lill. I am also indebted to L. Willaby of Fisheries and Wildlife, M. Kewish of National Parks and K. Jerone of the Forestry Commission for their co-operation at the study sites. The C.S.I.R.O. bird-banding scheme co-ordinated the banding programme. I also thank Ms A. Hoar and M. Wilson for typing and preparation.

REFERENCES

BROWN, J. L., 1964. The evolution of diversity in avian territorial systems. Wilson Bull., 76: 160-69.

CAMPBELL, K. G. AND MOORE, K. M., 1956. An investigation of the food of the Bellminer, *Manorina melanophrys* Latham. *Proc. R. Zool. Soc. NSW (May, 1957)* Pp. 72-73.

CARPENTER, F. L., 1978. A spectrum of nectar-eater communities. *A. Zool.,* **18**: 809-19.

CASE, T. J. AND GILPIN, M. E., 1974. Interference competition and niche theory. *Proc. Nat. Acad. Sci., U.S.A.* **71**: 3073-77.

CODY, M. L., 1974. "Competition and the Structure of Bird Communities". Princeton University Press, Princeton.

COOPER, R. P., 1967. Is the Helmeted Honeyeater doomed? *Aust. Bird Watcher* **3**: 1-14.

CROME, F. H. J., 1973. The relationship of the Helmeted and Yellow-tufted Honeyeaters. *Emu* **73**: 12-18.

DOW, D. D., 1977. Indiscriminate interspecific aggression leading to almost sole occupancy of space by a single species of bird. *Emu* **77**: 115-21.

DOW, D. D., 1979. Agonistic and spacing behaviour of the Noisy Miner, *Manorina melanocephala*, a communally breeding honeyeater. *Ibis* **121**: 431-36.

DOW, D. D., 1980. Communally breeding Australian birds, with an analysis of distributional and environmental factors. *Emu* **80**: 121-40.

FEINSINGER, P. AND COLWELL, R. K., 1978. Community organisation among neotropical nectar-feeding birds. *Am. Zool.* **18**: 779-95.

FORD, H. A., 1977. The ecology of honeyeaters in South Australia. *S. Aust. Ornithol.* **27**: 199-203.

FORD, H. A., 1979. Interspecific competition in Australian Honeyeaters — depletion of common resources. *Aust. J. Ecol.* **4**: 145-64.

FORD, H. A., 1981. Analysis of nest-record cards for three species of honeyeater. *R.A.O.U. Newsletter* **47**: 11.

FORD, H. A. AND PATON, D. C., 1976a. The value of insects and nectar to honeyeaters. *Emu* **76**: 83-84.

FORD, H. A. AND PATON, D. C., 1976b. Resource partitioning and competition in honeyeaters of the genus *Meliphaga. Aust. J. Ecol.* **1**: 281-87.

FRETWELL, S. D., 1972. "Populations in a Seasonal Environment". Princeton University Press, Princeton.

READER'S DIGEST, 1976. "Complete Book of Australian Birds". Reader's Digest Services Pty. Ltd., Sydney.

KEAST, J. A., 1968. Seasonal movements in the Australian honeyeaters (Meliphagidae) and their ecological significance. *Emu* **67**: 159-210.

KEAST, J. A., 1976. The origins of adaptive zone utilisations and adaptive radiations, as illustrated by the Australian Meliphagidae. *Proc. 16th Int. Ornithol. Congr.* Pp. 71-82.

LANDSBERG, J. AND WYLIE, F. R., 1983. Water stress, leaf nutrients and defoliation; a model of dieback of rural eucalyptus. *Aust. J. Ecol.* **8**: 27-41.

MACARTHUR, R. H., 1958. Population ecology of some Warblers of northeastern coniferous forests. *Ecology* **39**: 599-619.

MCCULLOCH, E. M. AND NOELKER, F., 1974. Bellminers in the Melbourne area. *V.O.R.G. Report No. 38. Victorian Nat.* **91**: 288-303.

MOORE, K. M., 1961. The biology and occurrence of *Glycaspis baileyi* Moore in New South Wales. *Proc. Linn. Soc. NSW* **86**: 185-200.

MORSE, D. H., 1971. The insectivorous bird as an adaptive strategy. *Ann. Rev. Ecol. Syst.* **2**: 177-200.

NOSKE, R. A., 1985. Habitat Usage by Three Bark-foragers of Eucalypt Forests. Pp. 193-204 *in* Birds of Eucalypt Forests and Woodlands: Ecology, Conservation, Management ed. by A. Keast, H. F. Recher, H. Ford and D. Saunders. Royal Australasian Ornithologists Union.

ORIANS, G. M., 1966. Food of nestling Yellow-headed Blackbirds, Caribou Parklands, British Columbia. *Condor* **68**: 321-37.

ORIANS, G. M. AND WILLSON, M. F., 1964. Interspecific territories in birds. *Ecology* **45**: 736-45.

PATON, D. C., 1980.The importance of manna, honeydew and lerp in the diets of honeyeaters. *Emu* **80**: 213-26.

PATON, D. C., 1985. Food Supply, Population Structure and Behaviour of New Holland Honeyeaters *Phylidonyris novaehollandiae* in Woodland near Horsham, Victoria. Pp. 219-30 *in* Birds of Eucalypt Forests and Woodlands: Ecology, Conservation, Management ed. by A. Keast, H. F. Recher, H. Ford and D. Saunders. Royal Australasian Ornithologists Union.

SHODDE, R., 1975. "Interim List of Australian Song-birds: Passerines". R.A.O.U., Melbourne.

SMITH, A. J. AND ROBERTSON, B. I., 1978. Social organization of Bellminers. *Emu* **78**: 169-78.

SWAINSON, G. W., 1970. Co-operative rearing in the Bellminer. *Emu* **70**: 183-88.

WHITE, T. C. R., 1971. Lerp insects (Homoptera: Psyllidae) on Red Gum *Eucalyptus camaldulensis,* in South Australia. *S. Aust. Nat* **46**: 20-23.

WOINARSKI, J., 1985. Foliage-Gleaners of the Treetops, the Pardalotes. Pp. 165-75 *in* Birds of Eucalypt Forests and Woodlands: Ecology, Conservation, Management ed. by A. Keast, H. F. Recher, H. Ford and D. Saunders. Royal Australasian Ornithologists Union.

WYKES, B. J., 1981. "The ecology of the Helmeted Honeyeater and its relationships with potential competitors". Report to Fisheries and Wildlife Division, Victoria. 291 pp.

WYKES, B. J., 1982. Resource partitioning and the role of competition in structuring *Lichenostomus* honeyeater (and *Manorina melanophrys*) communities in southern Victoria. Ph.D. Thesis, Monash University, Melbourne. (Available in the R.A.O.U. microfiche series).

New Holland Honeyeater, *Phylidonyris novaehollandiae*

CHAPTER 18

Food Supply, Population Structure, and Behaviour of New Holland Honeyeaters *Phylidonyris novaehollandiae* in Woodland near Horsham, Victoria

David C. Paton[1]

In woodland in central Victoria, New Holland Honeyeaters depended on nectar or honeydew for energy and their abundance, breeding and physical condition depended on the distribution and abundance of these resources. Nectar production was usually about 1500 kJ ha^{-1} d^{-1}, with a peak of 7000 kJ $ha^{-1}d^{-1}$ in late spring when *Callistemon* flowered. The lowest level occurred in autumn, when nectar production dropped to about 400 kJ $ha^{-1}d^{-1}$ or less. Counts of honeyeaters on the study area were highest in late spring and lowest in autumn 1978. Breeding did not occur during the peak in nectar production but occurred in late winter and spring, when *Astroloma* and *Grevillea* flowered. These shrubs provided the most profitable foraging during the afternoon.

Throughout the year the daily nectar production was consumed by birds and insects suggesting that the area was at carrying capacity with respect to nectarivorous animals. There were always surplus or transient birds moving through the area consistent with the area being at carrying capacity. Juveniles were more likely to be transients than adults, and females more likely than males. Adult males usually outnumbered adult females by about 2 to 1, and this biased sex ratio was due to differential mortality.

INTRODUCTION

THERE have been numerous studies on the ecology of honeyeaters in southern Australia. Most studies have examined resource partitioning (e.g. Keast 1968a; Recher and Abbott 1970; Ford and Paton 1976, 1977, 1982; Wykes 1982) or the foraging behaviour and diets of honeyeaters (e.g. Halse 1978; Paton 1979, 1980, 1982a; Pyke 1980; Thomas 1980). These studies agree that honeyeaters in southern Australia can be divided broadly into two groups, those that feed largely on nectar and those that do not (Keast 1976; Ford and Paton 1977). The mainly nectarivorous *Phylidonyris, Anthochaera* and *Acanthorhynchus* have received most attention (e.g. Recher 1971, 1977; Ford 1979; Ford and Paton 1982; Ford and Pursey 1982; Paton 1982a; Pyke 1983). The inference from these studies is that many aspects of the ecology and behaviour of honeyeaters can be related to the distribution and abundance of nectar and that these honeyeaters are in some way limited by this resource.

Several studies have shown correlation between numbers of honeyeaters in an area and flowering levels of native plants (e.g. Ashton 1975; Keast 1968b; Ford 1979; Ford and Pursey 1982; Collins and Briffa 1982). Recently, Pyke (1983) has questioned this assumed relationship between nectar abundance and honeyeater abundance. Certainly in some areas nectar production far exceeds the requirements of nectar-feeding birds, the surplus nectar accumulating in flowers and often dripping to the ground (e.g. Recher 1981). Such an observation led Carpenter (1978) to speculate that honeyeater communities may be limited by the abundance of insects rather than nectar.

No species of honeyeater feeds entirely on nectar, since nectar is only a simple sugar solution, almost

[1]Department of Zoology, University of Adelaide, G.P.O. Box 498, Adelaide, South Australia 5001.

Pages 219-30 *in* BIRDS OF EUCALYPT FORESTS AND WOODLANDS: ECOLOGY, CONSERVATION, MANAGEMENT ed. by A. Keast, H. F. Recher, H. Ford and D. Saunders. Royal Australasian Ornithologists Union and Surrey Beatty & Sons.

devoid of any other nutrients (Baker and Baker 1975; Paton 1982a). All species of honeyeater take at least a few insects in their diet and these provide the birds with nutrients and protein (Pyke 1980). In the only detailed study, Paton (1982a) showed that New Holland Honeyeaters, *Phylidonyris novaehollandiae,* depended on nectar or an equivalent source of carbohydrate for energy, and caught small flying insects, sometimes at an energetic loss, to satisfy their protein requirements. The small size of the insects made them inadequate as a source of energy. Sufficient insects to provide daily requirements of protein could be caught in as little as 10 minutes at any time in the year. To obtain sufficient energy for one day the birds foraged at nectar sources for several to many hours. Thus nectar is more likely to limit the abundance and behaviour of New Holland Honeyeaters, than insects.

In this chapter, I summarize a study in which nectar abundance and the abundance and behaviour of New Holland Honeyeaters were recorded simultaneously. I conclude that the quantity as well as the quality of the nectar supply largely determined the abundance and behaviour of these birds.

STUDY SITE

The study site was situated on Golton Vale, a farming property owned by Mr Howard Pipkorn of Horsham. It adjoined the Grampians State Forest and was on the Western Highway, midway between Stawell and Horsham, and approximately 265 km WNW of Melbourne. The habitat was dry sclerophyll forest, predominantly *Eucalyptus obliqua,* with scattered clumps of *E. camaldulensis, E. melliodora* and *E. viminalis.* There was a low (<1 m), dense heath layer of mainly epacrids, predominantly *Astroloma conostephioides,* and scattered larger bushes of *Callistemon macropunctatus, Xanthorrhoea australis, Banksia marginata, Grevillea aquifolia* and *Acacia* species.

METHODS

I visited Golton Vale for six days about every two months from October 1976 to September 1978. During each visit I measured the abundance of nectar and insects, determined the condition of the birds and estimated their abundance. Prior to, and since this detailed study, I have made irregular trips to the study site to collect information on flowering seasons.

Population Counts of Honeyeaters

The study site was essentially a broad strip of roadside vegetation that varied in width from 250 to 400 metres. A series of short 200, 300 or 400 m transects, totalling 2000 m was laid out across the strip at 100 m intervals, using the roadside fence as a baseline. I estimated population sizes of honeyeaters by recording all honeyeaters seen or heard within 50 m of either side of these marked transects. Essentially an area of 20 ha was censused. The censuses began about an hour after dawn and took about two hours.

Condition of Honeyeaters

I operated up to six mistnets for one to three days during each visit. All trapped birds were banded, weighed to the nearest half gram, aged (Disney 1966; Paton 1982b, c) and sexed (Disney 1966; Rooke 1976; Paton 1982b and unpubl.). Birds were also scored for breeding condition (cloacal protuberance or brood patch), moult (Paton 1982b, c) and levels of fat. Subcutaneous fat was visible under the skin of the abdomen (and neck) of many honeyeaters and was scored on a scale of 0 (none) to 4 (bulging with fat, Paton 1979).

Behaviour of Honeyeaters

Mistnetted birds that were recaptured were individually colourbanded and studied intensively using time budgets. This involved recording to the nearest second the time individuals spent feeding, preening, sitting, flying, nestbuilding, incubating, feeding nestlings, and chasing intruders during one hour periods spread evenly over the day. Time budgets were converted to energy budgets using standard techiniques (see Paton 1979, 1982a). The methods used to measure foraging rates and calculate food intake and energy expenditures have been explained elsewhere (Paton 1979, 1982a). In addition I kept records of colour-banded birds sighted during each visit, noting where they were sighted and whether they were defending a territory.

Survival Rates and Population Structure

From the trapping data and observations, I compiled a "trapping and observation" record for each bird. I assumed that birds were present in the population between captures and sightings, and determined the number of males, females, adults, immatures (7-12 months of age) and juveniles (1-6 months of age) known in the population during each period (month or season) by adding the number actually caught or sighted to those assumed to be present. Sex ratios and the proportion of adults, immatures and juveniles were then calculated.

Many of the birds were caught only once and probably most these were moving through the study area. The rate at which such birds were caught was used as an estimate of the number of transient birds in the population.

Survival rates were calculated from the trapping and observation records. Annual rates of survival were calculated as the proportion of birds that remained in the records for a twelve month period. Only birds trapped by February 1977 were used.

This meant that there was at least a six month period after February 1978, in which I could trap or resight the marked birds. Estimates of seasonal survival over three month periods were made only on birds caught by December 1977 for similar reasons.

Abundance of Nectar and Honeydew

a) Nectar

On each visit I measured rates of nectar production, standing crops of nectar, and rates of consumption of this food supply by birds and insects. Standing crops of nectar were measured at or just before dawn and at 3 to 4 h intervals during the day till dusk. I took samples of 40 to 100 flowers (or 5 to 10 inflorenscences) from several plants each time and measured the volume of nectar in each flower with a 10 μl capillary tube and its sugar concentration (% wt/wt) with a hand refractometer compensated to 20°C. I then calculated the energy content of the nectar by converting volumes and concentrations to milligrams of sugar, taking into account the density of different sugar solutions and assuming that one milligram of sugar gave 16.7 J.

Rates of nectar production were measured by bagging 120 to 300 flowers with fine nylon netting (0.7 mm mesh) and measuring the accumulated nectar in flowers 24 h later. Production was calculated by subtracting the amount of nectar present in flowers at the time of initial bagging from that accumulated in bagged flowers.

Estimates of production of nectar per hectare per day were calculated by multiplying the average daily production of a flower by the average number of flowers per plant and the density of the plant. Plant densities were determined from 200 5 m × 5 m quadrats in which the number of individual plants of each species had been counted, while the average number of flowers per plant was based on counts of flowers made on samples of 20 to 500 plants (depending on the species of plant and its density) during each visit. Estimates of nectar production on territories of individual birds were determined by counting all flowers on a territory and multiplying these by measured rates of nectar production per flower.

A measure of the level of exploitation of nectar by nectarivorous animals (i.e. birds and insects combined) was obtained by comparing standing crops with production rates, and determining what proportion of daily nectar production had been consumed. I also determined the periods (both seasonal and diurnal) when insects were likely to be affecting nectar abundance by counting the number of bees and other insects feeding on the nectar of a sample of flowers at about 3 h intervals, commencing at dawn and ending at dusk. I also covered flowers with 1.2 cm bird wire or 3.1 cm mesh nylon net to exclude birds but not insects and then sampled some of these flowers over a 24 h cycle.

b) Honeydew

Honeydew of psyllids and eriococcids was measured by counting the number of nymphs living under sections of bark, estimating the amount of suitable bark, and measuring the volume and concentration of the honeydew in the same way as for nectar. I also estimated the production of honeydew by psyllid nymphs by sampling honeydew from the same nymphs at dawn and dusk, keeping the insects covered. For more details see Paton (1980).

RESULTS

Seasonal Production of Nectar and Honeydew

The abundance of nectar within the study area followed a seasonal pattern (Fig. 1). Over winter and spring productivities were usually about 1500 kJ ha^{-1} d^{-1}, with peaks of 4000-7000 kJ ha^{-1} d^{-1} in late spring. The lowest levels of nectar production occurred over summer and autumn when rates dropped to 400 kJ ha^{-1} d^{-1} or less.

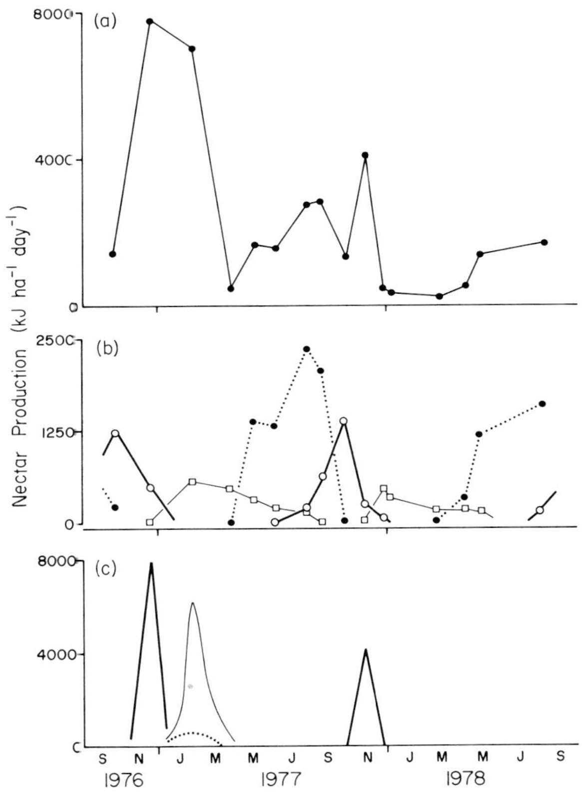

Fig. 1. Production of nectar (kJ ha^{-1} d^{-1}) within the 20 ha study area at Golton Vale. a) total production; b) production by *Banksia marginata* (□), *Astroloma conostephioides* (●) and *Grevillea aquifolia* (○); c) production by *Callistemon macropunctatus* (thick line), *Eucalyptus obliqua* (thin line) and *Eucalyptus melliodora* (dotted line).

Four shrubs with staggered flowering seasons produced nectar regularly, and in sufficient quantities to be important to the birds. *Astroloma conostephioides* produced over 1200 kJ ha^{-1} d^{-1} between

June and September, *Grevillea aquifolia* 400 to 1200 kJ ha^{-1} d^{-1} during spring and *Banksia marginata* about 400 kJ ha^{-1} d^{-1} during summer and autumn. *Callistemon macropunctatus* flowered for about six weeks in November and December and produced about 4000 kJ ha^{-1} d^{-1} and up to 7000 kJ ha^{-1} d^{-1} at peak flowering. Each of these plants produced fewer flowers and flowered approximately one month earlier in 1977-78 (July-June) than 1976-77.

Three eucalypts also supplied important quantities of nectar to the honeyeaters but did not flower consistently. *Eucalyptus obliqua* produced over 6000 kJ ha^{-1} d^{-1} during February 1977, but in the following year produced few flowers and supplied only about 40 kJ ha^{-1} d^{-1}. *E. melliodora* flowered between September and May, but individuals flowered asynchronously and only a few plants were in flower at one time, except in February 1977, when about a third of the plants were in flower and supplied about 500 kJ ha^{-1} d^{-1}. In 1977, the nectar of *E. obliqua* and *E. melliodora* greatly supplemented that of *Banksia marginata* within the study plot, but not in 1978 (Fig. 1.). *E. leucoxylon* did not occur within the study area, but over 20 plants grew in paddocks adjacent to the study site, and in the relatively lean autumn of 1978, these plants produced about 2500 kJ tree^{-1} d^{-1} of nectar at peak flowering. In the previous autumn they did not flower.

Other sources of carbohydrates were also important to the birds, particularly the secretions of eriococcids (Coccoidea) living in the bark of *Eucalyptus obliqua* and those of psyllids (Psyllidae), living under the bark of *E. viminalis* (Paton 1980). Psyllid nymphs produced about 600 to 750 kJ of honeydew daily from July to September in the two hectare area where *E. viminalis* occurred. Averaged over the whole study area, their contribution to total carbohydrate production was small, being about 33 to 37 kJ ha^{-1} d^{-1}. I have no estimates of production of honeydew by eriococcids. Small amounts of eriococcid honeydew were present throughout the year, but quantities were noticeably more abundant in May 1977, December 1977 and May 1978, when branches of *E. obliqua* were sticky to touch. In May 1977, Red Wattlebirds defended large trees for several days, and production rates of 200 kJ tree^{-1} d^{-1} would have been required to sustain them (Paton 1979, 1980). Given that there were 160 *E. obliqua* trees ha^{-1} and using a conservative estimate of 20 kJ tree^{-1} d^{-1} (smaller trees), honeydew production at these peak times may have exceeded 3000 kJ ha^{-1} d^{-1}.

Seasonal and Diurnal Variation in Profitability of the Carbohydrate Supply

The profitability of nectar-feeding (rate of net energy gain) varied seasonally and diurnally (Fig. 2) with changes in the species of plant, the density of flowers and the amount of nectar in flowers.

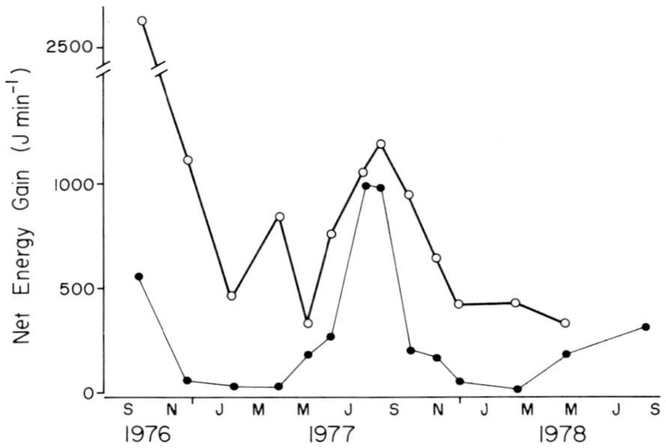

Fig. 2. Seasonal and diurnal changes in the profitability of nectar sources at Golton Vale. The figure shows the best net energy gains (J min^{-1}) for a New Holland Honeyeater foraging at dawn (○) and in the afternoon (●) at different times in the year.

Profitability was highest at dawn when the quantities of nectar "standing" in flowers were highest. Rates of net energy gain for New Holland Honeyeaters were usually above 400 J min^{-1} at dawn. Profitability then declined during the day since the birds and insects harvested nectar (and honeydew) at a rate faster than its was being produced. At night nectar accumulated in flowers because few, if any, nocturnal animals consumed the nectar while production continued. Afternoon rates of net energy gain showed a marked seasonal pattern, exceeding 200 J min^{-1} during winter and spring, but dropping to as little as 40 J min^{-1} during summer and autumn.

Exploitation of Nectar by Birds and Insects

The quantities of nectar consumed per day by birds and insects closely matched the amounts being produced per day throughout the year (Fig. 3a). Insects, particularly *Apis mellifera*, consumed a large proportion of this nectar, except during winter. Seasonal changes in the number of bees exploiting the nectar of flowers used by birds are shown in Figure 3b. During the warmer months, insects typically removed 80% or more of the nectar in flowers that had been covered with netting to exclude birds (Fig. 3c). This estimate is likely to be an overestimate of the amounts actually consumed by insects for two reasons. First, insects usually commenced foraging several hours after dawn by which time the birds would have removed some of the nectar. Second, because flowers which were covered with netting had higher nectar levels than uncovered flowers, insects may have preferentially fed on them, at least initially.

A more precise estimate of the quantities of nectar consumed by birds and insects can be made. This uses information on the diurnal foraging patterns of the two animals and diurnal changes in nectar levels (see Paton 1979 for details). An example of such an estimate is given in Table 1 for birds and insects

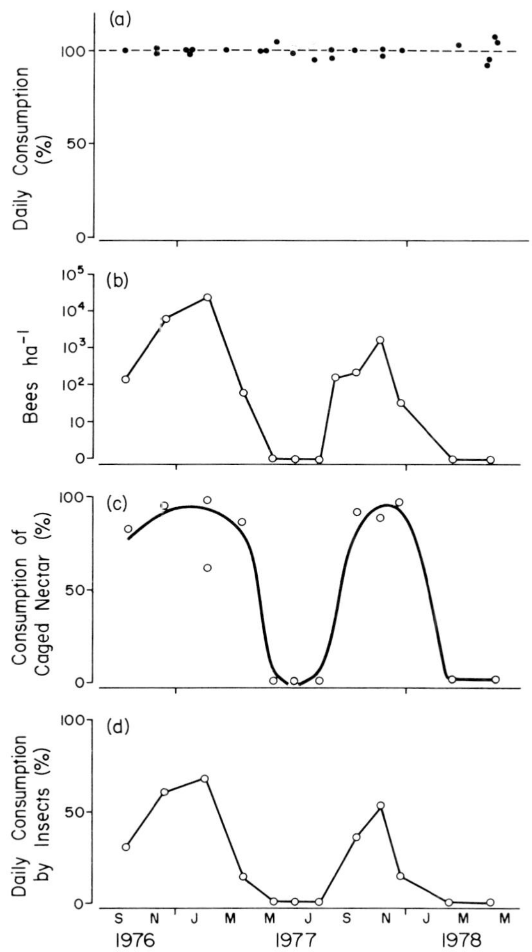

Fig. 3. Seasonal changes in nectar consumed by all visitors and by bees at Golton Vale. a) percentage of daily production consumed each day; b) maximum number of bees feeding at one time on flowers used by honeyeaters; c) percentage of nectar removed from caged flowers; d) percentage of daily production consumed by bees (e.g. see Table 1).

exploiting the nectar of *Grevillea aquifolia. Apis mellifera* consumed about 36% of the nectar and the birds 64%. By making similar calculations at other times in the year, insects accounted for typically 30-60% of the nectar produced during the warmer months (Fig. 3d).

The Honeyeater Community at Golton Vale

Honeyeaters were the predominant birds in the avian community at Golton Vale and usually accounted for two-thirds of all the birds counted during a census (Fig. 4). Only in March 1978, when there was little carbohydrate food, did honeyeaters account for less than half of the birds counted on the study site. Seventeen species of honeyeater were recorded, of which eight were present throughout the year, five were present seasonally and four were seen on one or two occasions only (Table 2). One species, the New Holland Honeyeater predominated. Approximately 50% of all birds and 60-90% of all honeyeaters counted at Golton Vale were New Holland Honeyeaters (Fig. 4). Changes in the abundance of honeyeaters at Golton Vale largely reflected changes in the abundance of New Holland Honeyeaters (Figs 4, 5a).

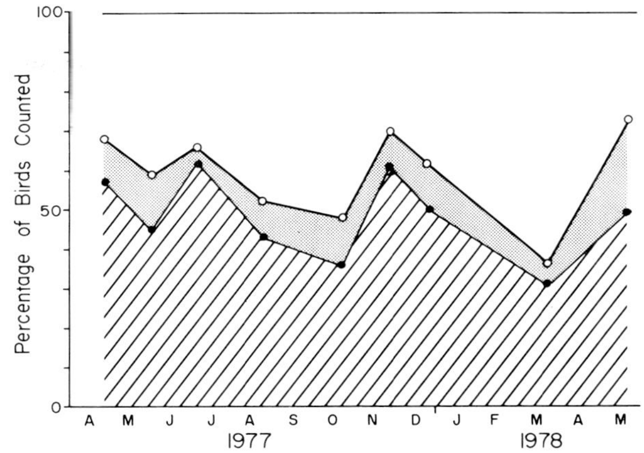

Fig. 4 Proportion of honeyeaters in the avian community at Golton Vale. New Holland Honeyeaters (hatched), other honeyeaters (stippled) and other species (blank).

Seasonal Changes in Population Size, Population Structure and Mortality of New Holland Honeyeaters

Densities of New Holland Honeyeaters at Golton Vale were usually 6 birds ha^{-1}, but the population doubled in late spring, when *Callistemon macropunctatus* flowered (Fig. 5a). Most changes in population density reflected changes in the number of juveniles and immatures, with the density of adults remaining at about 4 adults ha^{-1} throughout the year (Fig. 5a). Breeding occurred during late winter and spring, and following this there were large numbers of juveniles in the population.

Table 1. Number of visits and nectar consumption by New Holland Honeyeaters and *Apis mellifera* feeding on the nectar of *Grevillea aquifolia* in October 1977 at Golton Vale. In this example, birds and insects are assumed to harvest all the nectar available during each period and in proportion to the number of visits to flowers that each makes.

Period	Nectar available (J)	Visits $flower^{-1}$ birds	Visits $flower^{-1}$ bees	Nectar consumed per flower (J) birds	Nectar consumed per flower (J) bees	% collected by birds
0530-0900	27.7	2.34	0	27.7	0	100
0900-1230	7.2	1.40	5.43	1.5	5.7	20
1230-1600	7.2	1.24	9.25	0.8	6.4	12
1600-1930	7.2	2.82	5.43	1.4	5.8	25
Total	49.3	7.8	20.1	31.4	17.9	64

Table 2. Number of honeyeaters counted along 2000 m of transect (20 ha) at Golton Vale between October 1976 and May 1978.

Species of honeyeater	Number of honeyeaters counted:											
	Oct	Dec	Feb	Apr	Jun	Jul	Aug	Oct	Nov	Dec	Mar	May
New Holland*	137	245	137	138	139	133	113	130	298	180	71	192
Brown-headed*	13	5	5	8	19	+	9	10	13	9	2	10
White-plumed*	3	10	3	9	15	2	2	4	1	3	1	4
Fuscous*	4	5	2	1	+	+	2	15	10	3	+	9
Tawny-crowned*	2	+	1	2	4	1	2	7	11	13	1	2
Black-chinned*	11	7	2	+	2	+	3	3	5	5	2	8
Yellow-faced*	3	+	2	1	2	+	1	+	6	2	1	21
Red Wattlebird*	2	7	2	+	2	3	3	1	1	9	3	9
Eastern Spinebill*		+	+	1	+						+	1
White-fronted*			+					1	1	2		
White-eared*				3	2	1			+			2
White-naped*	+			+	1	1	2				1	30
Crescent*			+								1	1
Painted*										+		
Little Wattlebird*			+									
Yellow-tufted*			+			+						
Noisy Miner*						+						

+ seen on study site, but not during census. *See Appendix 1 for Scientific names.

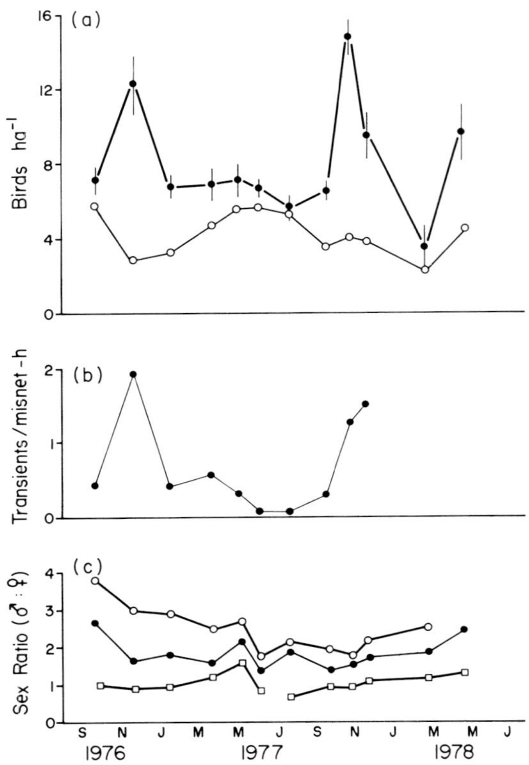

Fig. 5. Seasonal changes in the structure of the New Holland Honeyeater population at Golton Vale. a) mean number (● ± s.e.) of New Holland Honeyeaters counted per 100 m of transect (≡ 1 ha) and number of adults (○); b) number of transients caught per mistnet-hour (10 m mistnet, at least 50 mistnet-hours per sample); c) sex ratios (male:female) of all adults (○), resident adults only (●) and juveniles and immatures (□).

Although many individuals, particularly adults, were resident in the area throughout the year, a proportion of the birds present at any one time were "transients". The number of transients in the population showed a marked seasonal pattern (Fig. 5b). During winter only 0.1 transients were caught per mistnet-hour, while in late spring and early summer ten times this number were caught per hour. Most transients were juveniles and adult females were more often transients than adult males (Table 3). The flush of transients following spring breeding suggests post-breeding dispersal of juveniles. To confirm that a large number of birds were transients in late spring, the first 50 New Holland Honeyeaters caught in November 1977 were dyed red and their presence monitored. On the following day, only 19 of these dyed birds were still present on the study site (Paton 1979). Thus at this time of the year transient birds remained in the area for only a few days or less.

Table 3. Level of transiency and annual survival of resident New Holland Honeyeaters differing in age and sex at Golton Vale.

Age	Sex	Proportion of birds (n)		
		Transients	Surviving 12 months a	b
Adult	male	0.63 (178)	0.68 (39)	0.51
Adult	female	0.71 (163)	0.42 (19)	0.25
Juvenile	male	0.82 (308)	0.19 (16)	0.16
Juvenile	female	0.84 (378)	0.10 (21)	0.02

a, survival of individuals over 12 month period; b, seasonal survival rates multiplied together.

Table 4 provides estimates of seasonal survival of New Holland Honeyeaters for all birds and for 'resident' birds only. The figures based on all birds include transients and will underestimate true survival. Furthermore, seasonal variations in these figures will partly reflect seasonal variations in the number of transients. Survival of resident birds was lowest over summer and highest over autumn (Table 4). The figures for autumn, however, are based solely on survival over autumn 1977, and do not include autumn 1978. Autumn 1977 was a period of plentiful food and the birds were fat, so survival should have been high. In autumn 1978, however,

Table 4. Seasonal survival of New Holland Honeyeaters at Golton Vale.

Season	Proportion of birds surviving (n): All birds	Resident birds: Adult male	Adult female	Juvenile male	Juvenile female	Total
Spring (ASO)	0.40 (270)	0.74 (72)	0.66 (35)	0.85 (26)	0.55 (20)	0.71
Summer (NDJ)	0.14 (557)	0.91 (57)	0.68 (28)	0.42 (45)	0.16 (45)	0.62
Autumn (FMA)	0.41 (157)	0.94 (36)	0.82 (17)	0.77 (13)	0.50 (12)	0.82
Winter (MJJ)	0.33 (213)	0.81 (48)	0.67 (27)	0.58 (12)	0.55 (11)	0.71

food was scarce; there were fewer birds and they were in poor condition, hence survival would have been lower than in autumn 1977.

Adults survived better than juveniles and males better than females (χ^2, $p < .05$, Tables 3 and 4). The better survival of males is supported by the change in sex ratio from close to unity in juveniles to a 2:1 ratio in favour of males in adults (Fig. 5c).

Changes in Physical Condition of New Holland Honeyeaters

New Holland Honeyeaters generally maintained some subcutaneous fat, but had least during spring when the birds were breeding and during autumn 1978 when food was scarce (Fig. 6). The low weights of autumn 1978 contrasted with the higher weights of the previous autumn, when nectar and honeydew were abundant. The fattest birds averaged 4 to 5 grams of fat or 20 to 25% above lean body weight. Generally juveniles and immatures, though structurally the same size as adults, weighed less than adults of the same sex (Fig. 6). Females are approximately 10% smaller and lighter than males (Disney 1966; Rooke 1976; Paton 1979).

Annual Cycle of New Holland Honeyeaters at Golton Vale

New Holland Honeyeaters commenced breeding in early July at Golton Vale, when pairs began to re-establish breeding territories. Breeding then continued until late November or early December when the last fledglings reached independence. Adults underwent a complete moult following cessation of breeding, with males starting two weeks earlier than females (Paton 1982b). Most adults had replaced all their flight feathers by the end of summer (Paton 1982b). No breeding occurred during autumn. Juveniles commenced an incomplete moult during summer that continued through autumn. Many juveniles arrested this moult or showed faulty feather growth during autumn 1978 (Paton 1982c).

Seasonal Changes in Foraging

New Holland Honeyeaters maintained a predominantly carbohydrate diet throughout the year, and simply switched from one nectar source to another as each became available (Paton 1982a, Fig. 7). The major sources of nectar were *Banksia*

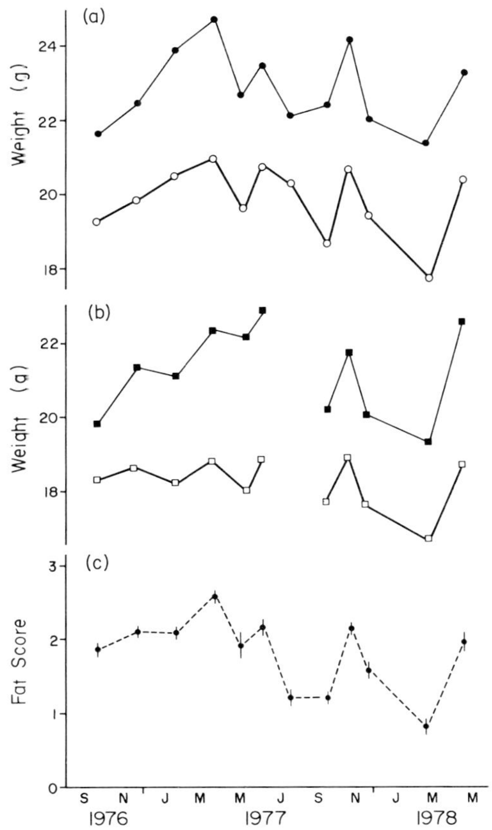

Fig. 6. Seasonal changes in weight and fat of New Holland Honeyeaters at Golton Vale. a) weight of adults (males (●), females (○)); b) weight of juveniles and immatures (males (■), females (□)); c) mean fat score ± s.e. (sample sizes range from 23 to 197).

marginata, Grevillea aquifolia, Astroloma conostephioides and *Callistemon macropunctatus.* Nectar of *Eucalyptus obliqua, E. melliodora* and *E. leucoxylon* was used in autumn and honeydew excreted by eriococcids living on *E. obliqua* was an important carbohydrate source during summer and autumn. The importance of psyllid honeydew to honeyeaters in winter is over-represented in Figure 7. My observations during winter were concentrated in an area where psyllid honeydew occurred. Only some birds had access to this resource (Paton 1980). In other parts of the study site, the nectar of *Astroloma conostephioides* was used almost exclusively.

Insect-feeding by New Holland Honeyeaters remained more or less constant throughout the year

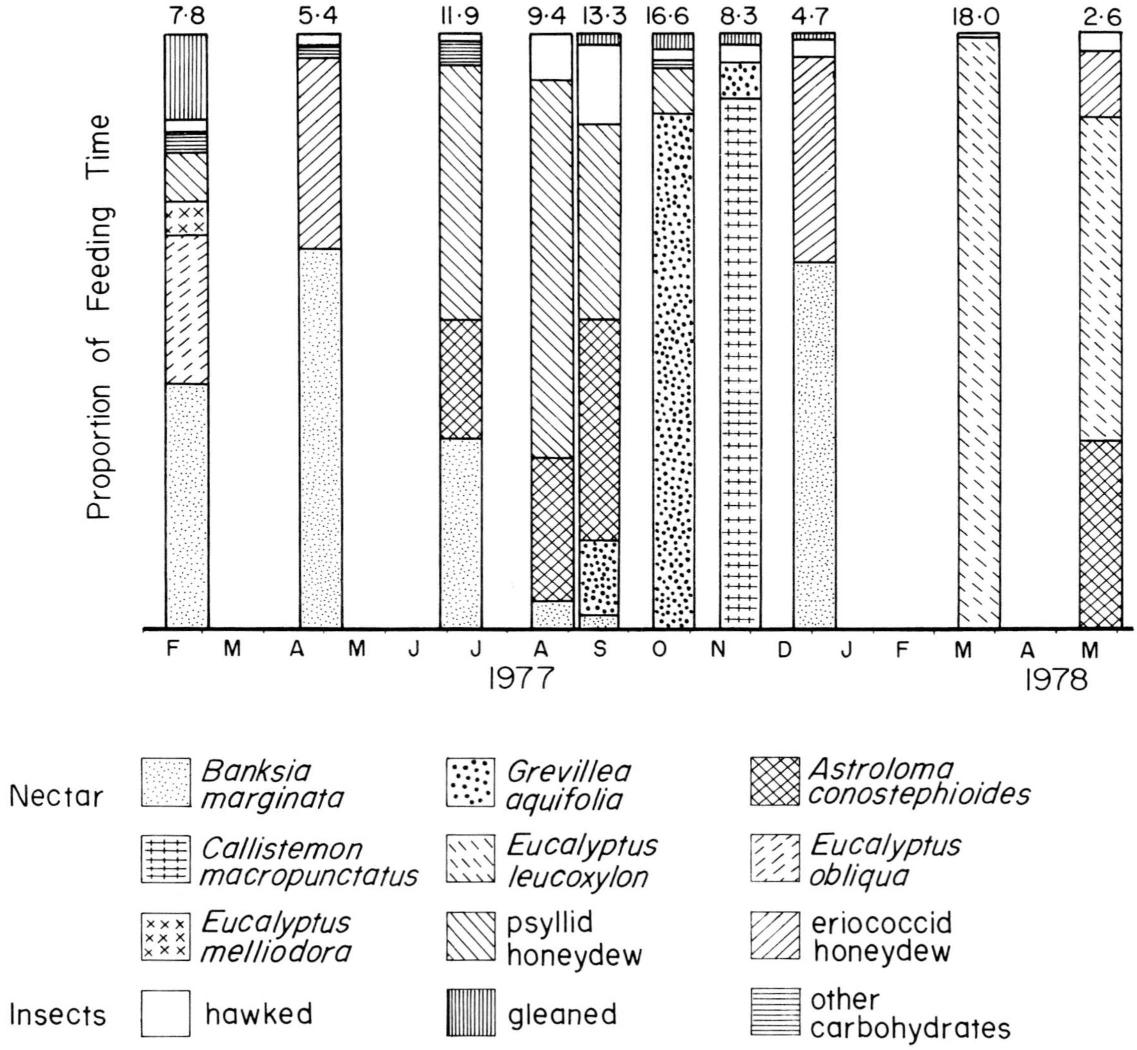

Fig. 7. Seasonal changes in foraging by New Holland Honeyeaters at Golton Vale. The figure shows the proportion of foraging time spent collecting nectar, honeydew and insects at different times in the year. Hours of foraging time given for each period at top of figure.

with non-breeding birds spending less than 20 min d^{-1} catching small flying insects (Fig. 8b). Breeding birds, however, spent up to 80 min d^{-1} hawking small flying insects to meet the protein requirements of nestlings or fledglings (Paton 1979, 1982a).

The amount of daylight spent foraging varied seasonally (Fig. 8a). Non-breeding New Holland Honeyeaters usually spent about 50% of the day foraging except during autumn when over 80% of the day (and more than 10 h) was spent foraging. This increase in foraging time reflected a decrease in the profitability of the carbohydrate sources available to the birds. During autumn the birds used the nectar of various eucalypts and *Banksia marginata* (Fig. 7). These plants only provided a small energy profit to the birds for most of the day. The honeydew of eriococcids, though abundant, was thinly spread and never provided high energy returns. Despite spending more time foraging in autumn, New Holland Honeyeaters still harvested less food than at other times of the year (Fig. 8c). Furthermore, the amounts consumed were less than their estimated energy requirements (Fig. 8c) and the birds lost weight (Fig. 6).

Territorial Behaviour of New Holland Honeyeaters

New Holland Honeyeaters defended sources of nectar at all times in the year, but the intensity and duration of this defence varied seasonally. Defence consisted of attacking and chasing intruders off territories. Only intruders that attempted to rob nectar or feed on honeydew were attacked (Paton 1980). Conspecifics and smaller species of honeyeater were usually quickly removed, but the larger wattlebirds (*Anthochaera*) and Musk Lorikeets, *Glossopsitta concinna,* were not easily driven away. Territorial New Holland Honeyeaters also had difficulty in driving off flocks of Brown-headed Honeyeaters, *Melithreptus brevirostris,* which occasionally swamped territories. Adult and juvenile New

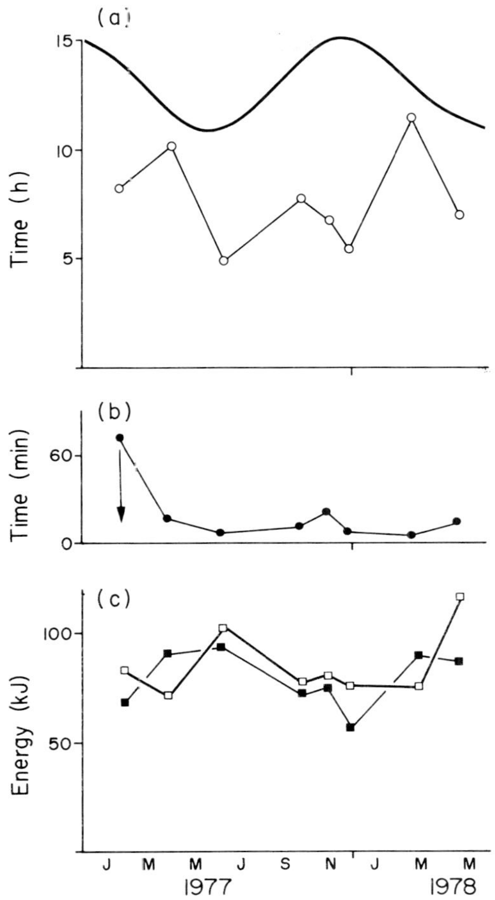

Fig. 8. Seasonal changes in foraging time, energy intake and energy expenditure for non-breeding New Holland Honeyeaters at Golton Vale. a) total foraging time (○, hours d^{-1}) and hours of daylight (thick line); b) time spent insect feeding per day. In February 1977 one juvenile spent a lot of time catching insects (see Paton 1982a). This raised the average daytime spent insect feeding. Other birds spent much less time catching insects in February 1977 (indicated by arrow); c) estimated energy intake (□, kJ d^{-1}) and estimated energy expenditure (■, kJ d^{-1}) for an individual New Holland Honeyeater.

Holland Honeyeaters of both sexes were all capable of defending individual feeding territories.

Territorial behaviour was least prominent during May 1977, December 1977 and May 1978 when the honeydew of eriococcids was abundant. During these times and to a lesser extent in December 1976 when *Callistemon* was in peak flower, individuals only defended small areas for short periods of time (1-5 h). The area defended was usually too small to provide sufficient food to support the birds throughout the day. When not using and defending this area, the birds foraged in nearby trees and shrubs. Often the same area was defended by the same individual for several consecutive days, but only for part of each day.

At all other times in the year, individuals defended a feeding area for the whole day and for several consecutive days at least, and had almost exclusive use of the defended resources. These territories produced sufficient nectar to satisfy the daily energy requirements of the owners (Table 5). Typically, most sources of nectar on the study site were controlled by territorial individuals. Territorial behaviour was most intense during March 1978. This was a time when food was scarce, and many birds had moved into a neighbouring area that contained flowering *Eucalyptus leucoxylon*. As many as 20 New Holland Honeyeaters and the occasional Red Wattlebird, *Anthochaera carunculata,* defended individual feeding territories in the one tree, and territory owners averaged over 24 intrusions per hour. At other times in the year the birds had less than five intrusions per hour (Fig. 9).

New Holland Honeyeaters did not attempt to defend nectar sources against insects. Instead when insects were consuming a large proportion of the nectar, they defended larger territories that produced more nectar than they required. That surplus usually accounted for the estimated losses of nectar to insects (Table 5).

DISCUSSION

Comparison of Nectar Productivities with other Areas

Nectar production at Golton Vale was usually about 1200 to 1500 kJ ha^{-1} d^{-1}, sufficient to support 15 to 20 New Holland Honeyeaters or their equivalent per hectare. These rates of nectar production are similar to those found in other areas. Paton (1979) found that nectar production in coastal heath near Cranbourne varied seasonally from about 400 to 1500 kJ ha^{-1} d^{-1}. Ford and Pursey (1982) estimated nectar production at New England National Park at 26000 kJ ha^{-1} d^{-1}, when *Banskia collina* was at the peak of flowering. Seasonal peaks of nectar production of the order of 10000 kJ ha^{-1} d^{-1} occur in areas of woodland and heath in South Australia (Ford 1979; Ford and Paton 1982) and along the coast near Sydney (Pyke 1983). However, in all of these studies, the study sites were selected because nectar-producing plants or nectar-feeding birds were abundant. The figures quoted may not be typical of all woodland and heathland in south-eastern Australia.

The Quantity of Nectar as a Limiting Resource for Honeyeater Populations

At Golton Vale, nectar production exceeded the estimated energy requirements of the honeyeater population for most of the year (Fig. 10a). There was often a surplus, even after the consumption of nectar by insects had been subtracted (Fig. 10b). These results are inconsistent with: (1) the presence of territorial behaviour throughout the year (Table 5, Fig. 9); (2) measurements of complete or nearly complete consumption of the nectar supply (Fig.

Table 5. Nectar production, nectar availability and size of non-breeding feeding territories of New Holland Honeyeaters at Golton Vale at different times in the year.

Date	Energy require-ments of birds (kJ)	Major nectar source	Territory parameters: Area (m^2)	Nectar production (kJ)	Nectar lost to insects (kJ)	Nectar available to birds (kJ)
Feb. 77	67	*Banksia* + Euc. obliqua	100-2000	158	79	79
Jul. 77	92	*Astroloma*	2100-2700	104	0	104
Oct. 77	72	*Grevillea*	20	119	43	76
Nov. 77	75	*Callistemon*	30	166	86	80
Mar. 78	90	*Euc. leucoxylon*	4-12*	173	73*	100

* m^3 of canopy, nectar lost to lorikeets.

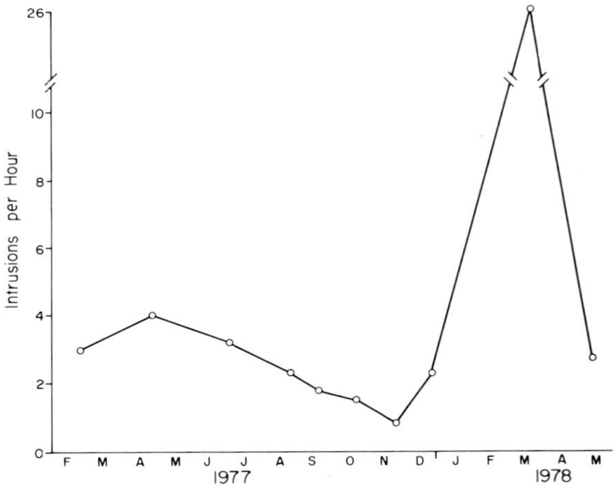

Fig. 9. Seasonal changes in intrusion rates onto territories of New Holland Honeyeaters at Golton Vale.

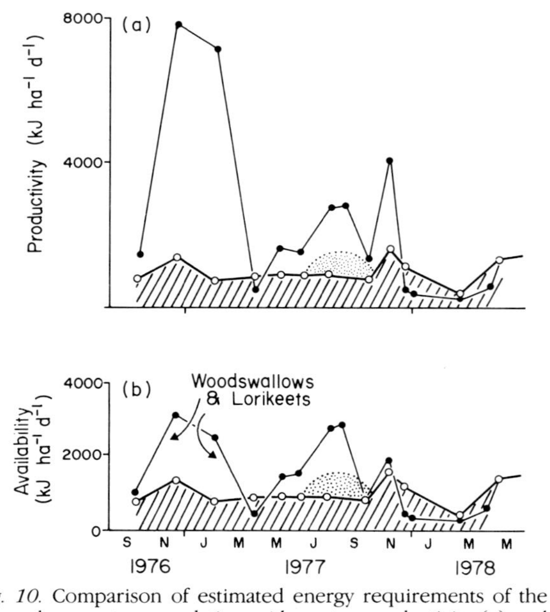

Fig. 10. Comparison of estimated energy requirements of the honeyeater population with nectar productivity (a) and nectar availability (b). Hatched area represents energy requirements of free-flying honeyeaters, stippled area represents additional energy required during breeding. Nectar available is production less consumption by insects (see Fig. 3d). Energy requirements of the honeyeater population were estimated using the per gram energy requirements of New Holland Honeyeaters (approx. 4 kJ $g^{-1} d^{-1}$), the average weight for each species of honeyeater, and their abundance from transect counts (Table 2). When the birds were breeding their energy requirements were approximately doubled (Paton 1979, 1982a).

3a); and (3) the knowledge that most of the flowers in the study site were being defended by territorial individuals (Paton 1979). Territoriality is unlikely to develop when there are surplus resources (Carpenter 1978). Furthermore, there were always surplus birds moving through the area (transients). These transients presumably would have remained in the study area if sufficient resources existed for them. These inconsistencies, however, are easily explained.

During summer 1976-1977 there were many woodswallows (*Artamus* spp.) and lorikeets (*Glossopsitta* spp.) also taking the nectar of *Callistemon* and to a lesser extent *Eucalyptus* during the early morning. These species probably accounted for most of the surplus nectar during this period. Furthermore, honeyeaters gained weight over this period (Fig. 6), indicating that some of the surplus food was being consumed and stored by them (Fig. 8). This storage of food may simply be a response to surplus food being available, but could equally be part of a longer-term strategy, where the birds deliberately gained weight during summer so that they could survive possible food shortages during autumn. During autumn the honeyeater population exceeds the carrying capacity of the nectar-producing plants (Fig. 10). Although honeydew from eriococcids is available during this period, it is thinly spread and time-consuming to collect, even when abundant (i.e. of low quality, Paton 1980, Fig. 8). A weight gain also occurred during the period of nectar surplus in early winter, prior to the energy demands of breeding.

Estimates of population sizes based on counts along transects might be out by 25% during August when most birds were breeding and had eggs. Approximately 40% of the population were females and given that they spend 65% of their time on the nest, about 25% of the free-flying population may not have been counted. Thus, the apparent surplus of nectar shown in Figure 10 may not exist.

I have confidence in my population counts of honeyeaters at other times in the year, since the number of birds counted along a section of transect, closely matched the number known to be living in that area. Such a statement may seem incongruous with the large numbers of transient birds reported (Fig. 5b). However transients, even when abundant, make up a very small proportion of the population at any one instance in time. For example, territorial individuals are on their territories for about 57 minutes of every hour, while other individuals trespass on the territory for a combined total that rarely exceeds five minutes per hour. Thus, transients account for less than 10% of the bird-time on an area, and probably much less, since some of the trespassers came from neighbouring territories.

Although I do not have sufficient quantitative data to be convincing I believe the honeyeater population at Golton Vale is limited by the quantity of nectar being produced. The following sequence of events adds support to my contention and suggests a mechanism by which honeyeaters can maintain their numbers at the carrying capacity of the nectar-producing plants.

Most flowers at Golton Vale were controlled by territorial New Holland Honeyeaters whose territories contained sufficient flowers to provide them with their daily energy requirements and a little surplus (Table 5). When resources declined, dominant individuals (adults, males) expanded their territories and maintained sufficient flowers to meet their energy requirements (Paton 1979). In so doing, subordinates (juveniles, females) were often displaced from their territories, or forced to abandon them because of inadequate nectar supply. Such birds joined the pool of transients and presumably fed on scattered undefended resources or robbed territories on the study site and elsewhere. Any increase in resource availability or spaces for territorial individuals should be quickly taken by transients, so enabling honeyeaters to track changes in nectar supply and maintain numbers close to carrying capacity. Such a mechanism is consistent with differences in the survival and mobility of New Holland Honeyeaters that differed in age and sex (Table 3).

My hypothesis can be tested experimentally by manipulating the nectar supply and recording responses in the honeyeater population. Such a manipulation has been carried out in heathland, near Cranbourne, Victoria (Paton 1979 and unpubl.). When nectar was added to inflorescences of *Banksia marginata,* the territorial New Holland Honeyeater reduced the size of her territory and allowed an unmarked immature to use the undefended part.

The Quality of the Nectar Supply as a Determinant of Breeding Seasons

At Golton Vale the nectar levels in flowers and hence foraging profitability (quality) declined during the day, particularly in summer (Fig. 2). Often during the middle of the day and afternoon the foraging profitability for New Holland Honeyeaters approached zero. New Holland Honeyeaters could potentially avoid the afternoon shortage by foraging mainly at dawn. However, they did not do this, and foraged throughout the day (Paton 1979, e.g. Table 1). Breeding birds need to collect food not only for themselves but also for their offspring and need profitable energy sources throughout the day to meet these added energy requirements (Paton 1979, 1982a). It is not surprising, then, that breeding only occurred during winter and spring when the net energy gains from nectar feeding (quality) remained high throughout the afternoon (Fig. 2, 11b).

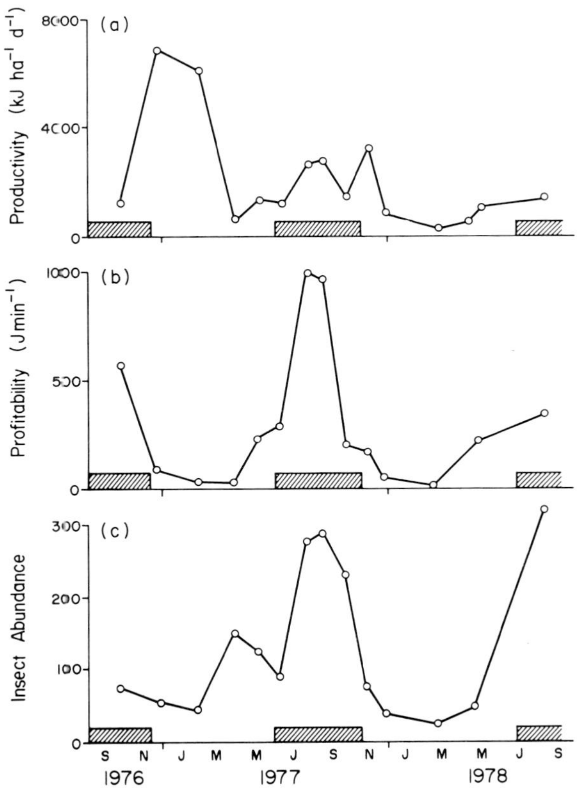

Fig. 11. Food supply and breeding seasons of New Holland Honeyeaters at Golton Vale. Breeding seasons (hatched) are shown against nectar production (a, from Fig. 1), foraging profitability in the afternoon (b, from Fig. 2) and insect abundance (c). Insect abundance is average number of flying insects caught in fifteen sweep samples (approx. equivalent to 150 m^3) made above heath (see Paton 1979 for more details).

Breeding was not correlated with peaks in nectar production or with insect abundance (Fig. 11a, c). Although breeding coincided with spring peaks of small flying insects in 1977 and 1978, New Holland Honeyeaters bred successfully with low insects abundance in 1976. Since New Holland Honeyeaters caught insects at about the same rate, irrespective of insect abundance (Paton 1982a), their breeding should not be restricted to periods when flying insects were abundant.

The Importance of the Shrub Layer to Honeyeaters at Golton Vale

The abundance and predominance of the New Holland Honeyeater in the avian community at Golton Vale is almost certainly due to the shrubs in the understorey rather than the eucalypt overstorey. *Astroloma, Grevillea, Callistemon* and *Banksia* flowered in all six years that I examined them (1975, 76, 77, 78, 80, 81) and with just *Astroloma, Grevillea* and *Banksia* in the plant community, a year round supply of nectar can be obtained (Fig. 1b). By comparison the eucalypts were unreliable sources of nectar, since each species of *Eucalyptus* failed to flower in one or more years. Although *Eucalyptus obliqua* housed the eriococcids that produced the honeydew over autumn, this honeydew was of low quality. Furthermore, the shrubs *Astroloma* and *Grevillea* produced the 'quality' resources that enabled New Holland Honeyeaters to breed and it was in shrubs that they built their nests (Paton 1979). The importance of the shrub layer can be illustrated with the following example. When the flowering of *Banksia marginata* was poor in autumn 1978 the number of honeyeaters on the study site declined (Fig. 4, 5a). Those that remained lost weight (Fig. 6). Once *Astroloma conostephioides* began flowering in May, the numbers of honeyeaters increased and their physical condition improved.

ACKNOWLEDGEMENTS

I am grateful to Howard Pipkorn for allowing me to work on his property and the late Douglas Dorward for his supervision and encouragement. Penny Paton assisted with field work. Hugh Ford, Allen Keast, Penny Paton and Harry Recher provided comments on earlier drafts. Financial support from a Monash Graduate Scholarship is gratefully acknowledged.

REFERENCES

Ashton, D. H., 1975. Studies on the flowering behaviour in *Eucalyptus regnans* F. Muell. *Aust. J. Bot.* **23**: 399-411.

Baker, H. G. and Baker, I., 1975. Studies of nectar constitution and plant-pollinator co-evolution. Pp. 100-40 *in* "Plant and animal co-evolution" ed. by L. E. Gilbert and P. H. Raven. Univ. of Texas Press: Austin.

Carpenter, F. L., 1978. A spectrum of nectar-eater communities. *Amer. Zool.* **18**: 809-19.

Collins, B. G. and Briffa, P., 1982. Seasonal variation of abundance and foraging of three species of Australian honeyeater. *Aust. Wildl. Res.* **9**: 557-70.

Disney, H. J. deG., 1966. Bird in the hand. R.A.O.U. No. 631. New Holland Honeyeater *(Meliornis novaehollandiae). Aust. Bird Bander* **4**: 14-5.

Ford, H. A., 1979. Interspecific competition in Australian honeyeaters — depletion of common resources. *Aust. J. Ecol.* **4**: 145-64.

Ford, H. A. and Paton, D. C., 1976. Resource partitioning and competition in honeyeaters of the genus *Meliphaga. Aust. J. Ecol.* **1**: 281-7.

Ford, H. A. and Paton, D. C., 1977. The comparative ecology of ten honeyeaters in South Australia. *Aust. J. Ecol.* **2**: 399-408.

Ford, H. A. and Paton, D. C., 1982. Partitioning of nectar sources in an Australian honeyeater community. *Aust. J. Ecol.* **7**: 149-59.

Ford, H. A. and Pursey, J. F., 1982. Status and feeding of the Eastern Spinebill *Acanthorhynchus tenuirostris* at New England National Park, northeastern NSW. *Emu* **82**: 203-11.

Halse, S. A., 1978. Feeding habits of six species of honeyeater in SW Australia. *Emu* **78**: 145-8.

Keast, J. A., 1968a. Competitive interactions and the evolution of ecological niches as illustrated by the Australian honeyeater genus *Melithreptus* (Meliphagidae). *Evol.* **22**: 762-84.

Keast, J. A., 1968b. Seasonal movements in the Australian honeyeaters (Meliphagidae) and their ecological significance. *Emu* **67**: 159-210.

Keast, J. A., 1976. The origins and adaptive zone utilizations and adaptive radiations as illustrated by the Australian Meliphagidae. *Proc. XVI Int. Orn. Congr.* Pp. 71-82.

Paton, D. C., 1979. "The behaviour and feeding ecology of the New Holland Honeyeater, *Phylidonyris novaehollandiae,* in Victoria". Ph.D. Thesis: Monash University, Melbourne.

Paton, D. C., 1980. The importance of manna, honeydew and lerp in the diets of honeyeaters. *Emu* **80**: 213-26.

Paton, D. C., 1982a. The diet of the New Holland Honeyeater, *Phylidonyris novaehollandiae. Aust. J. Ecol.* **7**: 279-98.

Paton, D. C., 1982b. The moult of New Holland Honeyeaters, *Phylidonyris novaehollandiae* (Aves: Meliphagidae) in Victoria. I. Moult of adults. *Aust. Wildl. Res.* **9**: 332-44.

Paton, D. C., 1982c. The moult of New Holland Honeyeaters, *Phylidonyris novaehollandiae* , (Aves: Meliphagidae) in Victoria. II. Moult of juveniles. *Aust. Wildl. Res.* **9**: 345-56.

Pyke, G. H., 1980. The foraging behaviour of Australian honeyeaters: a review with some comparisons with hummingbirds. *Aust. J. Ecol.* **5**: 343-69.

Pyke, G. H., 1983. Seasonal pattern of abundance of honeyeaters and their resources in heathland areas near Sydney. *Aust. J. Ecol.* **8**: 217-33.

Recher, H. F., 1971. Sharing of habitat by three congeneric honeyeaters. *Emu* **71**: 147-52.

Recher, H. F., 1977. Ecology of co-existing White-cheeked and New Holland Honeyeaters. *Emu* **77**: 136-42.

Recher, H. F., 1981. Nectar-feeding and its evolution among Australian vertebrates. Pp. 1637-48 *in* "Ecological biogeography in Australia" ed. by A. Keast. W. Junk, The Hague.

Recher, H. F. and Abbott, I. J., 1970. Some differences in use of habitat by White-eared and White-cheeked Honeyeaters. *Emu* **70**: 117-25.

Rooke, I. J., 1976. A measurement for sexing New Holland Honeyeaters. *Aust. Bird Bander* **14**: 72.

Thomas, D. G., 1980. Foraging of honeyeaters in an area of Tasmanian sclerophyll forest. *Emu* **80**: 55-8.

Wykes, B. J., 1982. "Resources partitioning and the role of competition in structuring *Lichenostomus* honeyeater (and *Manorina melanophrys*) communities in southern Victoria". Ph.D. Thesis: Monash University, Melbourne.

CHAPTER 19

Breeding of the Eastern Yellow Robin *Eopsaltria australis*

S. Marchant[1]

A population of colour-banded Eastern Yellow Robins *Eopsaltria australis* was studied at Moruya from 1975 to 1982 in two woodland and forest plots. The birds nested solitarily (nests usually 50 m apart) throughout the areas. Territories ranged from about 0.8 to 2 ha in size; established pairs maintained them all year. Boundary disputes were mild, consisting mostly of chases from perch to perch. Maintenance of the pair-bond by allopreening was not noticed but courtship feeding became frequent and regular as breeding started.

Almost one quarter of primary breeding pairs had auxiliaries, probably all first-year males and usually reared by the pair concerned. Part of the population (young birds) dispersed in the first three months of the year, when breeding was over, but dispersal also occurred as late as June-July.

Laying was at intervals of 27 hours approximately. Females alone incubated and were fed on the nest by males and auxiliaries. The incubation period was 15 days 20 hours on average; the nestling period varied from 10 to 14 days. The female incubated for about 80% of the time and brooded the young for a similar amount of the time for at least the first eight days but less later, though she sometimes sheltered them from the sun up to the time they fledged. Nestlings fledged at 82% of the adult weight, retained some juvenile plumage till 90 days old and hid in the undergrowth for the first 30 days or so out of the nest. Males and auxiliaries did some 75% of feeding of the young, the auxiliaries, when present, relieving the males of about half of this burden. Females performed injury feigning when the young had hatched.

Clutch size averaged 2.3 and was significantly smaller in three years after a fire than in the preceeding five. It was also significantly smaller early in the season than late, and for simple pairs as opposed to those with auxiliaries. Three broods could be reared during a season, which lasted from August-September to the end of December. Breeding success was about 30%, not differing annually but significantly better in nests with auxiliaries. The start of breeding could not be correlated with the weather but its end was probably controlled by drought and heat in December-January.

The birds fed mostly on the ground, dropping on to prey from vantage points on horizontal perches or vertical tree-trunks, that choice being apparently dictated merely by the type of vegetation in the territory or locality. They ate a variety of invertebrates in larval or adult form but also took small skinks. Life expectancy of birds that reached independence was about 3.5 years.

Fire did not affect the population immediately, except that the birds tended to nest high in trees more often after than before the fire, but it declined a year later because regeneration of grasses had deprived the birds of foraging areas; destruction of nesting sites was probably less important.

INTRODUCTION

ROBIN-LIKE birds that feed largely on insects, taken on the ground, are an important component of Australian avifauna in forests and woodlands. They differ from other members of the avifauna that live in the foliage in adaptations, food and population dynamics. Little attempt has so far been made, however, to compare these elements. The social organization, pattern of dispersion, annual cycle, territorialism and feeding of the 'robins' are poorly known, as is the way that they are adapted to live in the forests.

[1]Box 123, Moruya, New South Wales 2537.

Pages 231-40 *in* BIRDS OF EUCALYPT FORESTS AND WOODLANDS: ECOLOGY, CONSERVATION, MANAGEMENT ed. by A. Keast, H. F. Recher, H. Ford and D. Saunders. Royal Australasian Ornithologists Union and Surrey Beatty & Sons.

There are about six genera and 15 species of these 'robins' in Australian woodlands, belonging not to the Turdinae but to the endemic Australian fly-catchers. Most feed on the ground, at least in part, either by dropping on their prey from above or by hopping over the ground in search of it. As far as is known they are predominantly insectivorous.

This chapter is concerned with a typical member of the assemblage, the Eastern Yellow Robin, *Eopsaltria australis*, which is one of the most familiar birds of the bush in eastern Australia. In Western Australia its direct ecological counterparts are the Western Yellow Robin, *E. griseogularis,* and the White-breasted Robin, *E. georgiana*; inland, the Hooded Robin, *Melanodryas cucullata*; in the north, the Mangrove Robin, *E. pulverulenta*; and others elsewhere.

Only general accounts of the habits and breeding of the Eastern Yellow Robin have been published, none improving much on the information given by Campbell (1900) and North (1912). However, Dow (1980a) reported co-operative breeding in the species and Littlejohns (1933) recorded more than two birds attending a nest. Brown and Brown (1980) recorded co-operative breeding in *E. griseogularis*.

Since 1975, I studied the breeding of *E. australis* in open forest about five km north-west of Moruya, NSW (lat. 35°53′S, 150°03′E), in two plots, separated by about 50 m , on either side of a north-south ridge about 100 m asl, along which ran a road. Plot A on the western slope covered just over 9 ha of mixed eucalypt woodland to about 25 m high, with an understorey of acacias and casuarinas about 6 m high, a ground cover of Burrawang, *Macrozamia communis,* in places and thickets of *Melaleuca* Paper-bark and Tick Bush, *Kunzea ambigua,* elsewhere. Plot B of 10 ha was in State forest on the eastern slope in a stand of Spotted Gum, *Eucalyptus maculata,* and Grey Ironbark, *E. paniculata,* to 30-35 m high, with patchy understorey of acacias and casuarinas to 6-8 m high, some shrubberies along the gullies and areas of Burrawang.

From 1975 to March 1979 rainfall was above average (1000 mm), coming mostly as periodic heavy falls. Drought set in after March 1979 and became serious by the end of 1982. Plot A was burned out on 11 August 1980, except for two small patches of about 0.5 ha each. The burned area was bare until some regeneration started in January 1981. Since then, the paperbarks and Burrawangs regenerated well and acacias grew thickly from seed but the thickets of Tick Bush and other shrubs were mostly destroyed and grasses have grown much more thickly than before in their place. The western part of Plot B near the road was also burned out on 11 August 1980 and at about the same time much of the higher ground along the northern and southern edges of the plot was logged for pitprops, which ruined the area for systematic work on the Robins.

TERRITORIES

The robins nested as simple pairs or as primary pairs with auxiliaries throughout the woods, their nests being usually about 50 m or more apart; on three occasions I found them separated by only 15-20 m. The size of territories ranged from 0.8 to 2.0 ha (n = 37) for those that fell within my plots but was probably larger on the drier ridges outside the plots where there was less undergrowth and understorey. Once established as breeders, birds stayed in the same area for a long time (Marchant 1982) with the same mate until one or other died, though by reason of a gradual shift a territory at the end of three or four years might differ from that at the beginning. I knew only two females that moved more than 150 m after they had once bred, and that certainly or probably changed mates.

Boundary disputes, in which both members of a pair and auxiliaries took part were not common. The birds chased one-another from perch to perch with agitated piping and scolding in such a rapid and confusing way that I could not determine the roles of individuals. The chased bird gave way to the chaser a few seconds before it arrived so that I did not see contact between contestants in any dispute. The chaser landed on or near the abandoned perch and slowly and exaggeratedly cocked its tail, piping or scolding at the same time. Before starting a chase, birds sometimes sleeked their plumage and pointed aggressively at rivals.

Outside the breeding season I did not record these boundary disputes and saw no behaviour that seemed directed at maintaining the pair-bond, e.g. allopreening. Members of a pair were not then regularly in company. As the breeding season approached in July and August, they associated more closely and courtship feeding became regular. Females spread their wings, fluttering them rapidly in a begging manner, often several moments before males appeared, whether they were on or off the nest. The males quickly fed the females and flew away without any demonstration. I witnessed only two brief attempts at copulation, neither preceded by display. In July and August males began to give their territorial song (a loud emphatic 'chop-chop' call) persistently about dawn and dusk but seldom during daylight hours, though I did not try to quantify this.

Throughout the period of my study, the occupants of 35 territories bred as simple pairs; ten primary pairs had one auxiliary and two had two. Eight of the auxiliaries had been reared during the previous breeding season, five being the off-spring of the pair being helped and three having been reared in an adjacent territory. Two others were almost certainly year-old birds. Three of the fourteen auxiliaries bred in later years and were males; the others disappeared.

NESTING

The bulky cup-shaped nests of the Eastern Yellow Robin, decorated with lichen and strips of bark, are well known. They were placed in the forks of many sorts of bushes or of trees in the understorey, sometimes where dead leafy twigs or strips of bark had fallen and been caught, and in the forks of branches high in gum trees. Table 1 lists the sites in broad categories and Figure 1 shows their distribution in intervals of one metre from the ground. The average height from the ground was 301.5 cm. The occurrence of few nests between five and seven metres correlates with a lack of nesting sites between the top of the vegetation of the understorey and the lower branches of the forest trees. Before the fire in 1980 13 nests were more than seven metres from the ground and 147 less; after the fire 17 were high (>7 m) and 63 low (<7 m). The difference is significant ($\chi^2 = 7.23$, 1 df, $P < 0.01$).

Table 1. Nest sites of the Eastern Yellow Robin at Moruya.

Small live trees and shrubs (acacias, gum saplings, casuarinas, *Melaleuca, Kunzea, Persoonia, Helichrysum* spp. *et al.*)	138
Small dead trees and shrubs	31
Large trees (*Eucalyptus, Angophora* spp.)	42
On bark, twigs fallen into shrubs, small trees	22
Unusual sites (Burrawangs, fallen dead trees, etc.)	9
	242

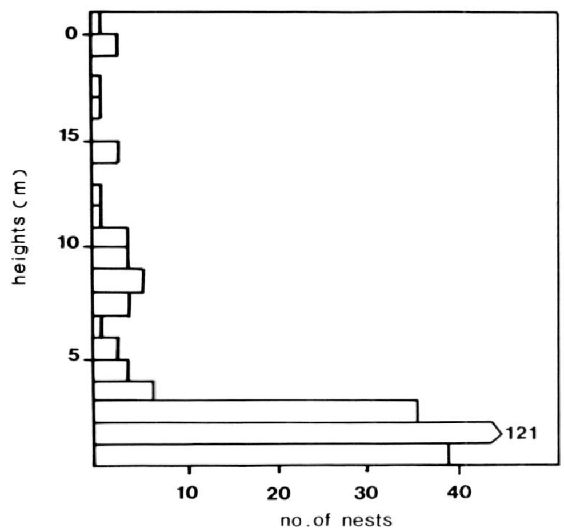

Fig. 1. Distribution of nests of Eastern Yellow Robins at Moruya, in intervals of one metre above the ground.

Males accompanied or preceded females when hunting for nest sites and showed them to her. For the first nest of a season a site was usually chosen and abandoned after a few shreds of material had been laid; later nests were built without this preliminary. New nests were made for each attempt at breeding but twice nests from which a clutch had been lost were used again in the same season (different females in different years) and one female laid a single egg in a nest from which her previous nestling had been lost and which still contained an infertile egg. First nests of a season took 10-15 days to be built (by females alone) but lay empty and apparently complete for 2-3 days; later nests took 8-10 days to build.

The laying interval was 27 hours approximately and incubation started partially after the laying of the second egg so that in a clutch of three the first two eggs hatched within about six hours of each other and the third up to 18 hours later. Females alone incubated and were fed on the nest by males and by auxiliaries. They spent about 80% of daylight hours on the nest, sitting in spells that ranged from 1 to 98 minutes, and were fed by attendants once or twice an hour (Table 2). The incubation period from laying of the last egg to its hatching was 15 days 20 hours on average.

The nestling period from hatching of the last egg until the last young left varied from 10 days 2 hours to 14 days 1 hour. The nestling that hatched first, from the first egg laid, weighed slightly more than later ones, at least till the eighth day and fledged first, at dawn or up to four hours later. Other members of the brood left 2-6 hours later but from two nests the last young to fledge did so 22 and 44 hours after the first. Figures 2a and b show the growth curves of nestlings and demonstrates that their heaviest weight was achieved about Day 9 and that they lost weight during their last 2-3 days in the nest. The average weight of 21 birds at least a year old was 19.9 g (range 17.5-22.5); so the nestlings fledged at about 82% of the adult weight. Linear regressions of the growth curves, excepting data from the flat tops, were highly significant, accounting for 95% of the variance ($P < 0.001$, data transformed $\sqrt{x}$). There was no significant difference between the slopes of the regressions for broods reared by simple pairs as opposed

Table 2. Periods of incubation and rates of feeding during incubation by Eastern Yellow Robins at Moruya.

♀	Total watch (mins)	♀ on nest (mins) (%)	n	Spells on nest Range (mins)	Average	♂	Feeds by aux.	unident.
1	176	158(90)	3	16-33	22.0	7		
2*	714	631(88)	27	1-98	17.9	12	10	4
3	686	588(86)	19	3-45	24.4	8		
4*	868	664(77)	44	1-49	13.5	4	5	9
5	620	450(73)	46	1-30	9.4	4		
	3064	2491(81)	139	1-98	14.7	35	15	13

* with auxiliary. Females 1 and 2 had nested in area for 4 years; Females 4 and 5 were nesting in area for first time; history of Female 3 not certain.

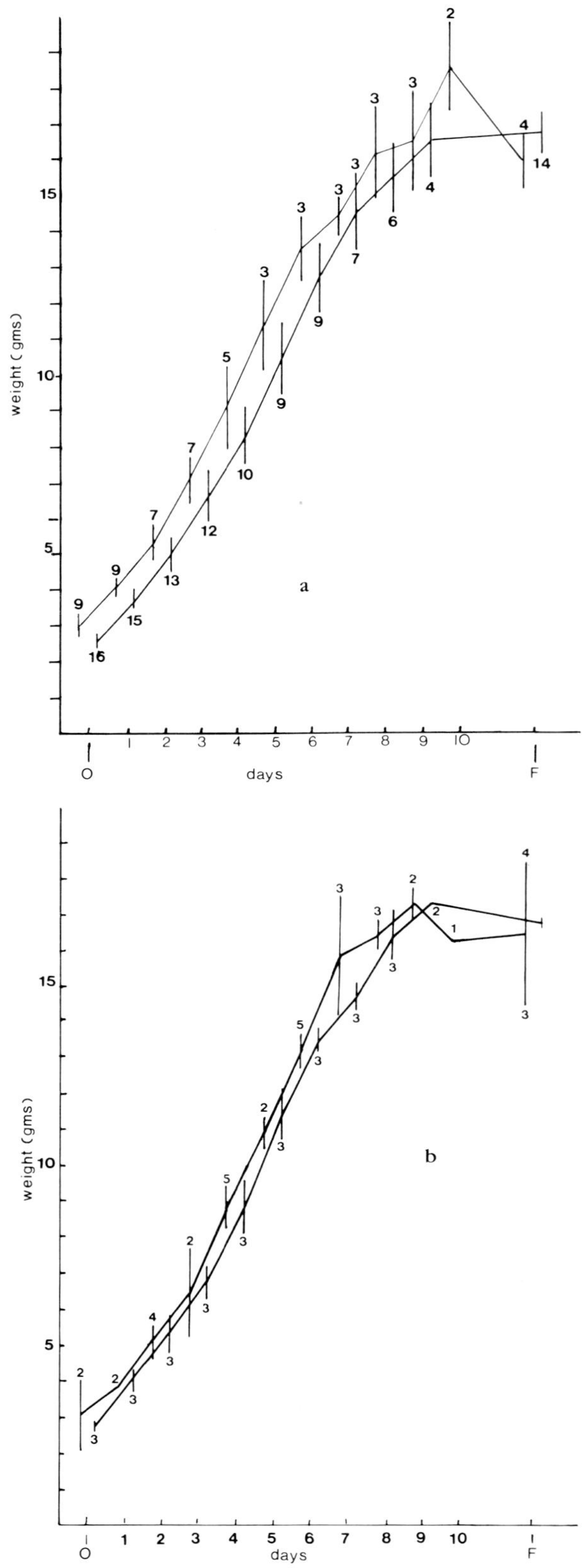

Fig. 2. Growth-curves of weight (g) for nestling Eastern Yellow Robins at Moruya, shown as means ± 2 s.e. from day of hatching (0) to fledgling (F). a: Broods of two; left-hand or upper curve of pairs with auxiliaries, n in sample above; right-hand or lower curve, of simple pairs, n in sample below. Offest for clarity. b: Broods of three. Same conventions.

to those reared by pairs with auxiliaries, for broods of both two and three. Broods of two nestlings, reared with or without auxiliaries, had different intercepts for weight ($P < 0.05$). This was confirmed by a *t* test for the weights on Day 0 ($P < 0.05$). Thus nestlings in broods of two reared by pairs with auxiliaries were heavier on hatching than those reared by simple pairs and remained so till fledging.

On leaving the nest, fledgelings fluttered to the ground and hopped into thick cover where they hid for about three to four weeks until full grown and depended on adults and auxiliaries. They emerged and became active about 30 days after leaving the nest, still in their brown plumage but soon afterwards began the body moult into adult plumage. Three siblings, when 38, 38 and 41 days out of the nest, had moulted about half their juvenile plumage, which showed that this moult takes longer than suggested by Lane (1976); other fledgelings retained some juvenile feathers till 92 days out of the nest and could be recognized as immature birds till at least one year old by the yellowish soles of their feet.

Table 3 shows the amount of brooding daily by females of simple pairs and of those with auxiliaries. A Wilcoxon matched-pair test showed no difference between the performances. The amount of brooding declined slowly but remained high till Day 8, when the nestlings were already well-feathered; thereafter it fell drastically but occurred even until the young fledged, though in the last few days some sessions on the nest were evidently an attempt to shade the young from hot sun rather than brooding. Table 4 gives the rates of feeding at nests with and without auxiliaries and with different sizes of brood. On average, each nestling was fed 3-4 times an hour and, though the role of the female remained much the same for all categories, auxiliaries relieved the male of half his duties. When the female was on the nest, food was usually given to her and she passed it on to the young or ate it herself. Feeding visits occurred in bursts; once the young were satisfied, none occurred for long periods. Each attendant removed faecal sacs and dropped them at a regularly used perch 20-30 m from the nest where they accumulated into a conspicuous pile; this provided a certain clue that there was or had been a nest nearby. Females removed hatched egg-shells and dropped them at the same spot as the faecal sacs.

When disturbed from a brood, especially in the later part of the nestling period, the female feigned injury by flopping or fluttering to the ground and there puffing out her body feathers, waving and spreading her wings to display the whitish wing-bar and hopping or creeping over the ground, all the time silently. I saw a male perform similarly at only one nest, of a simple pair, without any ostensible cause.

Table 3. Amount of brooding by Eastern Yellow Robins at Moruya.

Day	Simple pairs Total watch	Simple pairs Total brooding (mins)	%	Pairs with aux. Total watch	Pairs with aux. Total brooding	%	Total Total watch	Total Total brooding	%
0	464	386	83	220	206	94	684	592	87
1	365	314	86	284	213	75	649	527	81
2	141	124	88	154	122	79	295	246	83
3	181	149	82	332	294	88	513	443	86
4	288	213	74	297	258	87	585	471	80
5	309	244	79	72	39	54	381	283	74
6	285	189	66	165	106	64	450	295	65
7	212	115	54	190	100	53	402	215	53
8	169	115	68	161	122	76	330	237	72
9	213	45	21	416	60	14	629	105	17
10	250	43	17	300	172	57	550	215	39
11	389	115	30	380	127	33	769	242	31
12	289	92	32						
13	243	35	14						

Day 0 = day of hatching.

Table 4. Rates of feeding of young and roles of attendants by Eastern Yellow Robins at Moruya.

Brood Size	No.	Hours watched	Feeds by ♀	♂	aux.	unident.	Rate per hr
Simple pairs							
1	1	8.45	7	32			4.5
2	4	31.22	60	150			6.7
3	1	21.54	68	179			11.3
		62.00	135 (27%)	361 (73%)			8
Pairs with auxiliaries							
2	3	20.42	9	33	45	25	5.4
3	2	29.53	76	98	109	21	10.2
		50.35	85 (23%)	131 (35%)	154 (42%)	46	8.3

Other species of birds were tolerated except very close to the nest and thornbills *Acanthiza* spp. were not molested even at the nest. I recorded that females chased away from the nesting bush seven species of honeyeater, even as large as Lewin's Honeyeater, *Meliphaga lewinii*, Golden and Rufous Whistlers, *Pachycephala pectoralis,* and *P. rufiventris*, Grey Shrike-thrush, *Colluricincla harmonica,* and Grey Fantail, *Rhipidura fuliginosa*. If a visitor came close to her when sitting on the nest, she puffed out her feathers and gaped widely at it until it left.

CLUTCH SIZE

Table 5 shows the clutch size annually and for the whole period of study. Only those clutches that I followed through from laying or saw repeatedly with the same number of eggs are included. There was no significant difference in the average clutch size between the years. However, when the total number of clutches and counts of eggs for the years 1975-79 were compared with those for the years 1980-82, there were significantly smaller clutches and fewer eggs for the later period (P = 0.018, Mann-Whitney U test).

Table 5. Clutch sizes of Eastern Yellow Robin at Moruya annually.

	Clutch size 1	2	3	Total eggs	Average
1975	0	15	6	48	2.3
1976	2	17	7	57	2.2
1977	0	15	8	54	2.35
1978	0	10	14	62	2.6
1979	0	17	13	73	2.4
1980	0	12	4	36	2.25
1981	0	13	6	44	2.3
1982	2	8	2	24	2.0
Totals	4	107	60	398	2.3

Table 6. Clutch sizes of Eastern Yellow Robin at Moruya during the season.

	Clutch size 1	2	3	Eggs laid	Average
Before 1 October	3	58	11	152	2.1
During October	0	22	17	95	2.4
During November	0	13	22	92	2.6
After 30 November	1	14	10	59	2.4
Totals	4	107	60	398	2.3

Table 7. Clutch sizes of simple pairs and pairs with auxiliaries of the Eastern Yellow Robin at Moruya.

	Clutch size 1	2	3	Eggs laid	Average
Simple pairs	0	50	24	172	2.3
Pairs + auxiliary	2	10	15	67	2.5

Table 6 shows the clutch size during the breeding season, month by month. Clutches laid before 1 October were significantly smaller than those laid during October (χ^2= 9.3, 1 df, P < 0.01) and so were those laid before 1 November compared with those laid later (χ^2 = 12.3, 1 df, P < 0.01).

Table 7 compares the clutch sizes of simple pairs with those of pairs with auxiliaries. The difference was significant (χ^2 = 11.36, 1 df, P < 0.001).

Once a pair started to breed, it went on attempting to do so, successfully or unsuccessfully, until conditions became unsuitable in December and January,

Table 8. Breeding success of Eastern Yellow Robins at Moruya.

	Eggs Laid	Eggs Hatched	%	Young Fledged	%	Total success %	Production per nest
1975	45	28	62	10	36	22	0.5
1976	55	35	64	24	69	44	0.8
1977	55	42	76	14	33	25	0.6
1978	60	27	45	22	81	37	1.0
1979	76	40	53	15	37	20	0.5
1980	41	15	37	7	47	17	0.4
1981	40	28	70	11	39	27	0.6
1982	25	16	64	10	62	40	0.8
Total	397	231	58	113	49	28	0.7
C/1	5	3	60	2	67	40	0.4
C/2	206	115	56	54	47	26	0.5
C/3	186	113	61	57	50	31	0.9

Table 9. Breeding success of simple pairs and pairs with auxiliaries of Eastern Yellow Robins at Moruya.

	Eggs Laid	Eggs Hatched	%	Young Fledged	%	Total success %	Production per nest
Simple pairs							
C/2	100	56	56	26	46	26	0.5
C/3	72	40	55	20	50	31	0.8
	172	96	55	46	46	26	0.6
Pairs with auxiliaries							
C/1	2	1	50	1	100	50	0.5
C/2	20	16	80	9	56	45	0.9
C/3	45	33	73	20	61	44	1.3
	67	50	74	30	60	45	1.1

i.e. there was a period of 100-150 days during which the birds nested (see Breeding Season below). A successful nesting cycle took about 40 days and theoretically a pair could have 2-4 broods annually according to conditions. Many pairs raised two broods during a season but, because so many nests failed, I knew only one pair that successfully raised three broods in a season, between September 1976 and 12 January 1977. On the other hand some pairs made five or six unsuccessful attempts to breed during a season.

BREEDING SUCCESS

As a rough guide to breeding success, 75 nests produced some young and 156 failed to produce any (32% success). Simple pairs (38 successful v. 82 unsuccessful; 32%) were little more successful than pairs with auxiliaries (13 v. 31: 30%). Table 8 shows detailed success (eggs laid, eggs hatched, young fledged) year by year. There was no significant difference in success for the years 1975-1979 compared with 1980-1982 (Mann-Whitney, U = 3, one-tailed test). Table 9 shows the division between the success of simple pairs and that of pairs with auxiliaries. Here the survival to fledging was significantly better in nests with auxiliaries ($\chi^2 = 11.36$, 1 df, $P < 0.001$).

Of 106 nests that were totally lost, eight failed during laying, 24 in the first four days of incubation and 38 in the first week of the nestling period. So the greatest risk was probably soon after a nest was started and in the first week of the nestling period. The only positive evidence of loss was that two broods of two succumbed to heat within 48 hours of hatching and another of three died thus when 8-9 days old and fully feathered. Losses of eggs early in the nesting cycle occurred often at night and were probably due to nocturnal mammals.

BREEDING SEASON

Figure 3 presents the start of laying (in periods of five days) in all nests during the study to which I could give an exact date. In general, laying started at the beginning of September and reached a peak in the middle of that month for first nests. Later peaks at the end of October and in late November to early

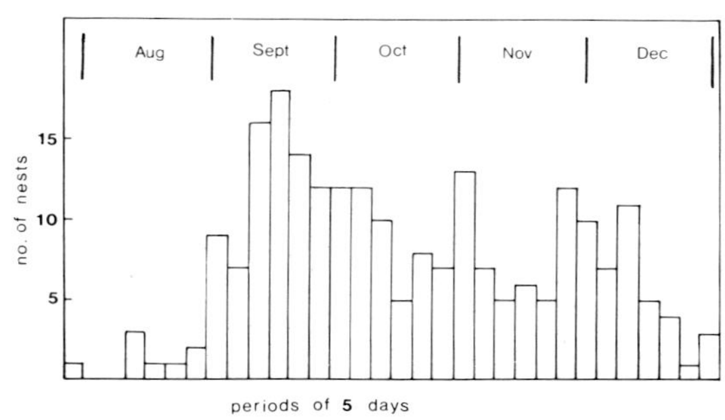

Fig. 3. Start of laying by Eastern Yellow Robins at Moruya, 1975-1982, shown as number of nests started in each period of five days from 1 August.

Table 10. Period of laying of Eastern Yellow Robins at Moruya, with rainfall and temperatures before and during the period.

	Start of laying			Rainfall (mm)		Degree days	
	First nest	Last nest		May-Aug	Sept-Dec	May-Aug	Sept-Dec
1975	4 Sept	22 Dec		No data	243	no data	
1976	7 Sept	2 Jan		91	988	2473	2955
1977	3 Sept	1 Dec		119	162	2527	3323
1978	4 Sept	19 Dec		582	385	2292	3002
1979	15 Aug	24 Dec		177	189	2375	3306
1980	22 Aug	20 Dec		106	200	2258	3146
1981	27 July	1 Jan		223	399	2113	2899
1982	13 Sept	10 Dec		69	236	2497	3136
			$\bar{x}$	195	350	2362	3110

December no doubt represented second and third broods but laying quickly fell off after mid-December and rarely occurred in January, when the birds began to moult. On 2 April, 1981, I found a pair building a nest, which did not receive eggs; this was the only breeding attempt noticed outside the breeding season shown in Figure 3.

Table 10 gives the limits of breeding annually, with rainfall and temperatures during the four months before breeding started and during the breeding season. There is no obvious correlation between the start of laying and the weather. The three years with an early start (1979-81) were moderately wet but rather cool and in 1981 the earliest start occurred in the coldest year. In the wettest year (1978) it was also rather cool and the start was late. The latest start (1982) occurred in the driest year, which was also warm. In 1977 a drop in average maximum temperatures of 10°C between 29 August and 2 September apparently did not inhibit the start of laying, as might have been expected. The end of the breeding season is more clearly associated with weather. Both 1976 and 1981 when laying went on into early January were wet and cool, whereas, when laying stopped early (1977 and 1982), rainfall had been poor and temperatures high.

FEEDING

I did not try to quantify the feeding behaviour of the robins. Most of their food is collected on the ground. The birds perched within about three metres of the ground and flew or dropped down to catch their prey; they dealt with it where it was caught or hopped up onto a log where they knocked it about, ate it or flew away with it for courtship feeding. Having eaten an item, birds then might catch others nearby by pouncing down from the log or low perch but generally they flew up to another perch and repeated the performance. The use of tree-trunks as perches seemed to occur as often as the use of horizontal perches on lower branches in the understorey and is probably dictated by the structure of the vegetation; in a thicket of *Melaleuca* there are few horizontal perches and the birds perch often on the trunks; in thickets of *Kunzea* or among *Macrozamia* they avail themselves of horizontal perches as often as of tree-trunks. Their requirement seem to be a good view of the ground and even in areas of thick *Macrozamia* this is provided by the dead fronds at the base of the plants. I noticed no seasonal variation in foraging habits. However, some food is collected well above the ground. Birds will fly to a tree-trunk from their perches and pick off an insect in much the same way as Laughing Kookaburras *Dacelo novaeguineae* and occasionally indulge in brief hawking sallies.

Rose (1973) recorded spiders, ground-living Hemiptera, Coleoptera and ants of three species in stomach contents. While chopping wood, which attracts Robins, I noticed birds taking termites, the larvae of cossid moths, centipedes 50-60 cm long and the larvae of beetles that looked like mealworms. When nestlings were being fed, items large enough to be identified were recorded as follows: worms (4), caterpillars (9), cricket (1), tabanid fly (1), imago of the Common Brown butterfly *Heteronympha merope*, moths (7), spider (1) and tail-less skinks Scincidae or their wiggling tails (9).

DISPERSAL

As has been said, once birds were established in a territory they remained there for the rest of their lives. At the end of the breeding season (January onwards) I netted my plots to catch and colour-band surviving juveniles, banded as pulli in the previous breeding season. Apart from such known birds, in the first four months of 1978, 1979 and 1980 I caught 41 others, all juveniles except perhaps six. Two of these stayed and bred together in 1980; three others stayed just outside my plots and one stayed in my plot for at least six months; the other 35 disappeared within a couple of weeks at most of being caught. This showed that there is a period of about three months at the start of the year when a proportion of the young birds of the year wander; the rest, of course, stay in their natal area and either breed in the following season or become auxiliaries to established pairs.

There was also probably another period of dispersal before breeding started about June-July. Two birds that settled to breed in or near my plots arrived in June and July. Three immature birds or birds that had been acting as auxiliaries for at least one year were last seen in July 300-400 m from their habitual haunts.

MORTALITY

The annual survival of banded birds in my plots from the beginning to the end of four years was:

1976	32 out of 39	— (82%)
1977	28 out of 42	— (66%)
1978	23 out of 31	— (74%)
1979	32 out of 41	— (78%)

Of the 39 birds known to be alive at the end of 1975, 32 (82%) survived to the end of 1976; 20 (51%) to the end of 1977; 15 (38%) to the end of 1978; 11 (28%) to the end of 1979; 8 (20%) to the end of 1980; and 4 (10%) to the end of 1982.

If one takes the annual mortality ($\underline{m}$) as 0.25 and uses the formula $2\text{-}\underline{m}/2\underline{m}$ (e.g. Gibb 1961), the expectancy of life of adult robins that reach independence is 3.5 years. One bird that fledged on 7 October 1975 was still alive on 3 September 1983; a male bred in Plot A for each of eight seasons and a female for seven.

EFFECT OF FIRE

The fire of 11 August 1980 was of moderate intensity. It entirely destroyed the ground cover, shrubs, *Melaleuca* and *Kunzea* thickets, except for two small patches that miraculously were untouched, and all acacias and casuarinas in the understorey with few exceptions. The *Macrozamia* was burned out, leaving only the spherical black tops of the stems. The flames, though up to 8-10 m high, did not ignite the canopy; all the foliage of the gum trees, however, was scorched and turned brown within a week or so. The trunks of the big trees were all singed or scorched and the bark of Spotted Gums quickly turned orange and fell off by the end of the year.

The fire had little immediate effect on the robins. As far as I knew all the marked birds in the plot before the fire survived. Four pairs had nested in the area during 1979 and in 1980, immediately after the fire, three of these same pairs bred again; indeed one of them must have started to build within a day or two of the fire, in the unburned patch of *Melaleuca*. Three other pairs had nests in my plot after the fire but most of their territories lay outside it.

As already pointed out, after the fire nests were generally higher and clutches smaller but breeding success did not decrease. No doubt the fire by destroying many low possible nesting sites forced the birds to nest higher but the smaller clutches and lower breeding success probably resulted from increasing drought rather than from the fire directly.

The major effect of the fire was delayed till 1981. By that time there had been no regeneration of the thickets that covered about half of Plot A; the stems of old *Kunzea* bushes had begun to collapse and grasses had grown thickly over most of the area formerly occupied by them. In consequence only three pairs of robins nested in the plot, all birds that had held territories for 3-6 years — two in the eastern part of the block where the woodland had been less disturbed in the past and had large areas of *Macrozamia* and one in a particularly barren area that never supported much ground cover. Though there were still plenty of nest sites and perches from which the birds could forage in the rest of the block, I assumed that the dense ground-cover of grass made it unsuitable because it prevented the birds from foraging.

CONCLUSION

Eastern Yellow Robins live in a variety of habitats, some of which differ in the structure of the vegetation from the woodlands at Moruya. Therefore, their habits, as observed at Moruya, may not be the same everywhere but probably they always need a shady environment with a rather sparse low ground cover and plenty of perches below about three metres from the ground whence they can spot their prey. In my opinion their habit of clinging to upright trunks is opportunistic.

Although the birds spend most of their time below the top of the understorey in forest, they also fly up into the tree-tops, where I could not follow them for long enough to see what they did. As a consequence of living at low levels, they ought to be endangered or at least seriously disturbed by any sort of fire but this apparently did not happen at Moruya immediately in August 1980. This suggests that ground fires even of moderate intensity do not seriously deplete their supply of food, though they do destroy low nest sites and perhaps induce the birds to nest higher than usual. If, however, such a fire occurred during breeding rather than in the non-breeding season or at its start, some nests would be destroyed and the effect probably more drastic. The chief effect of a ground fire seems delayed until regeneration of the ground cover and, if grasses and other plants grow up thick and tall, the feeding area of the robins is no longer available to them and the population falls. This probably happens only in areas where the woodlands have been disturbed by previous clearing. As might be expected, drought adversely affected clutch size but not apparently breeding success.

Use of the upper levels of forest by the robins is shown by their ability to nest high but their habit of nesting both low and high is shared by several other species of woodland birds, e.g. Rufous Whistlers, Grey Shrike-thrushes, Grey Fantails, Yellow-faced

Honeyeaters, *Lichenostomus chrysops*, Eastern Spinebills, *Acanthorhynchus tenuirostris*, Striated Thornbills, *Acanthiza lineata,* and Red-browed Firetails, *Emblema temporalis,* (pers. obs.). It is hard to get comparative data on this matter, which needs investigation, especially in the robins whose low nests are often absurdly conspicuous, precariously placed and unprotected. All the same, the robins have an uncanny ability to place their low nests where they are easily overlooked, at least by people. The siting of nests conspicuously and even precariously suggests that the sites are selected because potential predators hunt by smell, not by sight. They may be nocturnal mammals and certainly I had no evidence for predation by day.

In general, the pattern of nesting cycle by the robins conforms to that of most small passerines that lay in open cup-shaped nests, though the incubation period is about a maximum for such birds. Two aspects need mention, The laying interval of about 27 hours is unusual; in my experience small passerines lay at intervals of about 24 hours about dawn (or at intervals of 48 hours, as in the quite exceptional acanthizids and climacterids and some tyrannids) but why the robins adopt this strange interval could be determined only by careful study of the availability of food at the critical period of laying and the physiology of the birds. The nestling period is short and curtailed by the nestlings leaving the nest well before they are full grown and hiding in the ground cover for about a month. This suggests that it is safer for them to hide away from the nest and again raises the question of when predation occurs and what are the potential predators. I had no direct evidence on this point. This adaptation of early departure by nestlings is common to some other species of birds in the lower levels of the forest, e.g. Golden Whistlers and Eastern Whipbird, *Psophodes olivaceus,* but their young leave even more prematurely than those of the robins, at about 60-75% of adult weight (pers. obs.). The development of a cryptic juvenile plumage, totally different from that of the adult, by robins and Golden Whistlers is an obvious adaptation for this period of their lives, not needed by the drab Whipbirds.

Being sedentary, long-lived, with poor breeding success and inhabiting an environment without strongly marked seasons, the robins seem closer to tropical passerines in their nesting habits than to species of northern temperate zones but they rank low as possible communal breeders according to the criteria listed by Dow (1980b). In fact, only about a quarter of breeding pairs have auxiliaries, which are probably only males bred one or two years previously. Thus pairs, if they survive from year to year, are not likely to have an auxiliary one year unless they successfully reared male offspring in the year before. A proportion of the juveniles wander away from their natal areas soon after they reach independence. The sex of these birds and the distances that they travel need studying.

An obvious benefit of auxiliaries at nests is to relieve the primary male of some of his duties. Their part in courtship feeding may bring about the larger average clutch size in nests of pairs with auxiliaries than in those of simple pairs by enabling the female to lay down more reserves before and during the critical laying period. Similarly, the nestlings could be fed and grow better, resulting in the better production from nests of pairs with auxiliaries. On top of that, the auxiliaries probably gain experience in a breeding role against the time when they get the chance to enter fully into the breeding population. However, communal breeding in the robins is less thorough and more simple than it is among such birds as fairy-wrens, miners and babblers among others (as described by Rowley (1965), Dow (1970) and King (1980)) (see Dow 1980a for more refs) and probably arose in a different way from those other systems.

The breeding season of the robins is broadly from the beginning of September to the end of December. Its end is readily explained by the combination of heat and dryness that desiccates the environment sooner or later from early December onwards. The proximal factors controlling the start are less easily recognised. No doubt they are a subtle combination of climatic factors that control the supply of food and other requirements; a much more detailed study than I could make would be needed to decide the matter. However, maybe the age of breeding birds plays a part, because in none of the years in which I found early nests was there a burst of general laying before 1 September. I did not have enough breeding birds of known age to determine this.

ACKNOWLEDGEMENTS

I am grateful to the Forestry Commission of NSW for permission to work in the State Forest; to Drs R. Traynier and M. Guppy for supplying the statistical analysis: to Drs D. D. Dow and P. J. Fullagar for helpful discussions on many aspects of the breeding of the robins; and to the editors for instruction and advice on how to write this chapter.

REFERENCES

Brown, R. J. and Brown, M. N., 1980. Co-operative breeding in robins of the genus *Eopsaltria*. *Emu* **80**: 89.

Campbell, A. J., 1900. Nests and Eggs of Australian Birds. Sheffield: privately.

Dow, D. D., 1970. Communal behaviour of nesting Noisy Miners. *Emu* **70**: 131-34.

Dow, D. D., 1980a. Communal breeding Australian birds with an analysis of distributional and environmental factors. *Emu* **80**: 121-40.

DOW, D. D., 1980b. Systems and strategies of communal breeders in Australian birds. Proc. XVII Int. Orn. Congr.: 375-81.

GIBB, J. A. 1961. Bird populations. *In* "Biology and Comparative Physiology of Birds, 2" ed. by A. J. Marshall. New York: Academic Press.

KING, B. R., 1980. Social organization and behaviour of the Grey-crowned Babbler, *Pomatostomus temporalis*. *Emu* **80**: 59-76.

LANE, S. G., 1976. Results from banding Southern Yellow Robins. *Aust. Bird Bander* **14**: 63-66.

LITTLEJOHNS, R. T., 1933. In quest of unusual bird photographs. *Emu* **32**: 259-62.

MARCHANT, S., 1982. The sedentary nature of passerine birds at Moruya, New South Wales. *Corella* **6**: 87-88.

NORTH, A. J., 1912. Nests and Eggs of Birds found breeding in Australia and Tasmania. Spec. Cat. 1 Aust. Mus. Sydney.

ROSE, A. B., 1973. Food of Some Australian Birds. *Emu* **73**: 177-83.

ROWLEY, I., 1965. The life history of the Superb Blue Wren *Malurus cyaneus*. *Emu* **64**: 251-97.

CHAPTER 20

Habitat Requirements of a Rare Species, The Rufous Scrub-bird

Simon Ferrier[1]

The Rufous Scrub-bird, *Atrichornis rufescens,* is a small ground-dweller, at present confined to upland forests of northeastern New South Wales and far southeastern Queensland.

The habitat requirements of the species were studied in subtropical rainforest at Wiangarie and open eucalypt forest at Gloucester Tops. In each area, multivariate habitat comparisons were made between plots centred on male song territories and plots located at random throughout unoccupied forest. The critical factors determining whether or not habitat was suitable appeared to be identical in the two areas, despite broad differences between the forest types occupied. The major requirements were extremely dense cover 2-50 cm above the ground, moderately dense cover 50-100 cm above the ground, a moist microclimate at ground level, and abundant leaf litter. These requirements closely resembled those of the Noisy Scrub-bird, *A. clamosus,* in Western Australia.

Selective logging increased habitat suitability in the rainforest, yet lowered suitability in the eucalypt forest. Suitable habitat within euclaypt forest seemed to be associated with an intermediate seral stage of post-fire succession. The effects of logging and prescribed burning on habitat suitability need to be considered in planning the management of this species.

INTRODUCTION

THE SCRUB-BIRDS (family Atrichornithidae) represent one of the most interesting and scientifically important elements of the Australian forest avifauna. Although these distant relatives of the lyrebirds were probably more widespread during the Tertiary and Pleistocene, they are today restricted to a few small relic populations (Smith 1977).

The Noisy Scrub-bird, *Atrichornis clamosus,* is at present known to occur only at Two Peoples Bay in southwestern Australia. This population has been the subject of intensive research and management (Smith and Robinson 1976; Davies *et al.* 1982). The Two Peoples Bay area was declared an 'A' Class Reserve in 1966, primarily for the purpose of conserving the Noisy Scrub-bird.

The Rufous Scrub-bird, *A. rufescens,* is confined to moist forests of northeastern New South Wales and far southeastern Queensland. Although more abundant than the western species, the Rufous Scrub-bird has nevertheless been the cause of considerable concern. Since European settlement it has become extinct in all lowland areas of its range and is now restricted to five isolated upland populations (Ferrier 1982, 1983). Both species of scrub-bird are included in the Red Data Book list of endangered birds compiled by the International Council for Bird Preservation (King 1981).

The following study of Rufous Scrub-bird habitat requirements is presented as an example of the research needed to ensure preservation of these fascinating species. Further information concerning research on the Rufous Scrub-bird is given in Ferrier (1984), while the results of research on the Noisy Scrub-bird can be found in Smith (1976a, 1977, 1979), Smith and Robinson (1976), Robinson and Smith (1976), Smith and Forrester (1981), and Davies *et al.* (1982).

The impression obtained from previous descriptions of Rufous Scrub-bird habitat (e.g. Smith 1976b; Pizzey 1980) is that the bird is restricted to areas of dense undergrowth within rainforest or contiguous sclerophyll forest. While qualitative descriptions

[1]Department of Zoology, University of New England, Armidale, New South Wales 2351. Present Address: C/- National Parks and Wildlife Service, P.O. Box 402, Armidale, New South Wales 2350.
Pages 241-48 *in* BIRDS OF EUCALYPT FORESTS AND WOODLANDS: ECOLOGY, CONSERVATION, MANAGEMENT ed. by A. Keast, H. F. Recher, H. Ford and D. Saunders. Royal Australasian Ornithologists Union and Surrey Beatty & Sons.

such as this provide a valuable starting point for the study of habitat, detailed quantitative information is desirable for the purposes of habitat management (see Conner and Adkisson 1976; Rotenberry 1981). Those factors determining whether or not habitat is suitable for the Rufous Scrub-bird need to be precisely identified.

Previous research has suggested that birds often select habitats on the basis of factors not immediately connected with their survival. These proximate factors are probably used as easily recognized guides to ultimate factors directly connected with survival (Partridge 1978; Rotenberry 1981). For example, vegetation structure may be used as a proximate factor indicating the availability of food. However, vegetation structure may also act as an ultimate factor by providing cover for protection from predators. In most cases careful experimentation is needed to distinguish between proximate and ultimate factors. For the purposes of this chapter, both will be viewed simply as "critical habitat factors"; those variables determining, either proximately or ultimately, the suitability of habitat for the Rufous Scrub-bird.

METHODS

Details of the methods used are given in Ferrier (1984). Most of the habitat research was conducted at two localities: 1. Wiangarie State Forest (28°25′S, 153°05′E) 20 km northeast of Kyogle, and 2. Gloucester Tops (32°05′S, 151°35′E) 35 km west of Gloucester, lying partly within Barrington Tops State Forest and partly within Barrington Tops National Park. Altitudes ranged between 700 m and 1100 m for the Wiangarie study area and between 900 m and 1300 m for the Gloucester Tops study area. These areas were chosen because they represented the extremes of both latitudinal range and habitat type for the Rufous Scrub-bird. The Wiangarie area consisted mostly of subtropical rainforest with a few small patches of cool temperate rainforest dominated by *Nothofagus moorei*. The Gloucester Tops area consisted mainly of open forest dominated by *Eucalyptus fastigata* and *E. obliqua*, interspersed with cool temperate rainforest. Scrub-birds at Wiangarie were found mostly within subtropical rainforest whereas those at Gloucester Tops occurred mostly within open forest.

Eighteen kilometres of transect line were mapped in each area. The transects were walked 18 times between August 1981 and October 1982. Male scrub-birds heard singing within 150 metres either side of the transect line were mapped by triangulation of compass bearings taken on the source of sound. The song produced by males is loud and penetrating, and provides the only useful means by which scrub-birds can be detected.

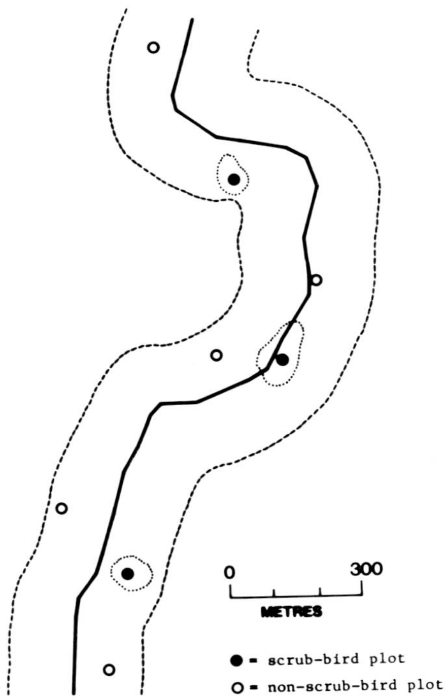

Fig. 1. A typical section of transect strip used in the study of Rufous Scrub-bird habitat (Wiangarie State Forest). The boundaries of scrub bird territories are indicated by dotted lines. Each symbol indicates the centre of a habitat plot.

Accumulated detections of singing males after 18 transect walks formed distinct clusters. Detailed study has suggested that these clusters represent the territories of individual singing males.

In order to compare the habitat in male scrub-bird territories with habitat in nearby, unoccupied areas, two types of plots (30 metre radius) were established (see Fig. 1). Plots of the first type were centred on the activity centres of male territories (22 scrub-bird plots in each study area). Plots of the second type were located at random throughout the 300 metre wide transect strips; the only condition was that these plots did not overlap scrub-bird plots (40 non-scrub-bird plots in each study area).

Seventy habitat variables were measured in each of the 124 plots. These variables covered a wide variety of habitat features including vegetation structure and floristics, topography, soil surface characteristics, microclimate, and proximity to neighbouring scrub-bird territories.

Less detailed investigation of Rufous Scrub-bird habitat was also made at subsidiary sites scattered throughout the range of the species: Mt. Barney, Lamington Plateau, Gibraltar Range, Ebor-Dorrigo area, Hastings Range. A total of 174 territories was located during the study.

RESULTS AND DISCUSSION

Critical Habitat Factors

Did the habitat of scrub-bird plots differ from that of non-scrub-bird plots? Because many of the habitat variables were correlated with one another, it was difficult to assess overall habitat differences by looking at variables individually. For this reason, discriminant analysis was used to combine all of the habitat variables into a single index, the discriminant function, in a way that maximized the separation between scrub-bird and non-scrub-bird plots. Details regarding the use of discriminant analysis in studying avian habitats can be found in Conner and Adkisson (1976), MacKenzie and Sealy (1981), and Titus and Mosher (1981).

In Figure 2, scrub-bird and non-scrub-bird sites are plotted along the discriminant functions for Wiangarie and Gloucester Tops. By combining all of the variables in this way, complete separation of scrub-bird and non-scrub-bird habitat was achieved for both study areas.

What were the critical factors determining whether or not habitat was suitable for the Rufous Scrub-bird? A backward elimination procedure was used to reduce the number of variables in each discriminant function. This involved eliminating, one by one, those variables that did not contribute significantly to the separation of scrub-bird and non-scrub-bird sites, over and above the separation achieved using all other remaining variables. The functions were thereby reduced to include only those variables that were most important in the separation of scrub-bird and non-scrub-bird habitat. These were regarded as critical habitat variables. While such variables may not represent the exact factors used by scrub-birds in selecting habitats, they are nevertheless the best approximations available based on the variables measured.

The critical habitat variables identified at Wiangarie and Gloucester Tops are summarized in Table 1. Note that different variables have been identified for the two areas. This does not necessarily mean that scrub-birds in these areas have different habitat requirements. The availability of each requirement within an area appears to have played a large part in determining whether or not that requirement was identified. For example, dense cover 2-50 cm above the ground was quite common at Gloucester Tops and therefore failed to be identified as a critical component of scrub-bird plots. It was only at Wiangarie, where ground cover was generally sparse, that this variable emerged as an important factor limiting the suitability of habitat.

A similar situation was revealed for the variables relative humidity and leaf litter volume, which were identified as critical factors only at Gloucester Tops.

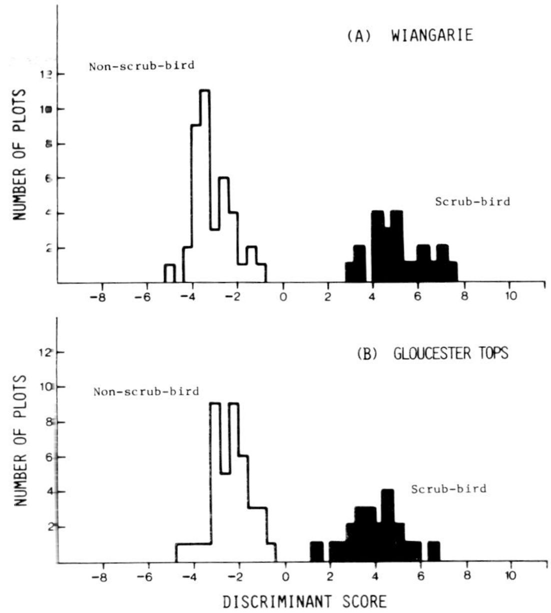

Fig. 2. Frequency distribution of discriminant scores for scrub-bird plots (solid histograms) and non-scrub-bird plots (open histograms) at Wiangarie and Gloucester Tops. The discriminant scores are derived from a combination of 70 habitat variables.

At Wiangarie, both requirements were satisfied throughout most of the rainforest, and therefore failed to be identified.

The distance between a plot and the nearest neighbouring scrub-bird territory was identified as a critical factor at Gloucester Tops but not at Wiangarie. Suitable habitat at Wiangarie occurred as small, isolated patches beneath gaps in the rainforest canopy. Social spacing mechanisms appeared to be of little importance as patches of habitat were already well spaced and each patch was usually too small to accommodate more than one territory. Suitable habitat at Gloucester Tops generally occurred as long strips within eucalypt forest fringing the edge of rainforest. Here, social spacing mechanisms seemed to play a critical role in preventing scrub-birds from taking up territories close to existing territories in what may otherwise have been ideal habitat.

The Rufous Scrub-bird appeared to have the same habitat requirements at Wiangarie and Gloucester Tops despite broad differences between the forest types occupied in these two areas. The basic requirements of dense ground cover, moist microclimate, and deep leaf litter, closely resemble those of the Noisy Scrub-bird in Western Australia (Smith 1979; Davies *et al.* 1982).

Guidelines for the identification of Rufous Scrub-bird habitat are included in Table 1. Suitable habitat occurs in a variety of situations within a diversity of forest types including subtropical, warm temperate

Table 1. Critical factors determining the suitability of habitat for the Rufous Scrub-bird.

Habitat Factor	Significance[a] Wiangarie	Significance[a] Gloucester Topes	Minimum Requirement (30 metre radius plot)
1. Cover 2-50 cm above ground[b]	***		55% projected cover
2. Cover 50-100 cm above ground[b]	***	***	40% projected cover
3. Relative humidity 5 cm above ground[c]		**	Moss index = 1.4
4. Leaf litter volume		***	8,000 cm^3m^{-2}
5. Distance to nearest neighbouring scrub-bird territory (plot centre to territory centre)		**	250 metres

[a] Significance of the variable in a discriminant function after removal of non-significant habitat variables, ** $p < 0.01$; *** $p < 0.001$, in all cases scrub-bird means were greater than non-scrub-bird means.

[b] Cover was measured using the light meter technique described by Fox (1979). For the purposes of rough assessment the required minimum has been converted to a percentage projected cover.

[c] This variable required complex standardization in the present study. The required minimum has therefore been given in terms of a moss index found to be closely correlated with the humidity variable. The index is an average rating of the amount of moss below 50 cm on tree trunks greater than 20 cm diameter. O = moss absent or inconspicuous, 1 = moss very conspicuous, 2 = thick mossy covering to most of lower trunk.

and cool temperate rainforests, and adjoining eucalypt forests. The required ground cover may consist of either living material such as ferns, shrubs, sedges and vines, or non-living material such as logs and other debris.

Over the last few decades the Rufous Scrub-bird has rarely been found below an altitude of 600 metres, despite the fact that it was once relatively abundant within lowland rainforests such as the Big Scrub near Lismore (Chisholm 1951; Smith 1977). The cause of this upward retreat is unknown. Both the Wiangarie and Gloucester Tops study areas lay above 600 metres, and therefore the present study could not determine the influence of altitude-related factors.

The results discussed above have concerned the habitat within a 30 metre radius of each male's territory centre. In order to estimate the area of habitat required by a pair of scrub-birds, information is needed on the size of a male's territory and the location of the female relative to the male. Detailed mapping of selected territories at Wiangarie and Gloucester Tops has shown that males do 95% of their singing within an average area of 1.13 hectares (based on 22 territories and a total of 417 mapped singing points). Territory shape varies, but is generally elongate. Males usually roost just outside the singing territory (up to 30 metres beyond its boundary). Females vocalize much less than males and therefore yield little information. They seem to occupy a distinct home range which only partly overlaps the male's territory. If we assume the female's area to be no larger than the male's, then the total area of habitat required by a pair is probably in the order of two hectares.

Effects of Forest Management Practices

The effects of some land use practices on scrub-bird habitat are obvious. For example, there is little doubt that the conversion of forest to farmland by clearfelling creates habitat totally unsuitable for the species. However, the influence of other activities may not be as obvious and can only be determined through careful research.

1. Logging

Both the Wiangarie and Gloucester Tops study areas included a mixture of logged and unlogged forest.

Multiple regression analysis was used to investigate relationships between logging and scrub-bird habitat suitability on the 124 sample plots described earlier. Each plot's discriminant function score was used as a measure of habitat suitability for that plot. The basal area of trees greater than 20 cm diameter at breast height was used as a measure of logging intensity. (Basal area is the sum of the sectional areas of trees at breast height). Logging was the major cause of variation in basal area within the two study areas. The time since logging and other factors such as altitude, aspect, proportions of tree types in the canopy, and proximity to important features (e.g. gullies, forest type boundaries, neighbouring scrub-bird territories) were also included in the analysis.

Within the rainforest at Wiangarie, basal area and time since logging explained over 40% of the variation in habitat suitability. Basal area was negatively correlated with habitat suitability suggesting that logging probably promoted the development of scrub-bird habitat (see Fig. 3). However, it should be emphasized

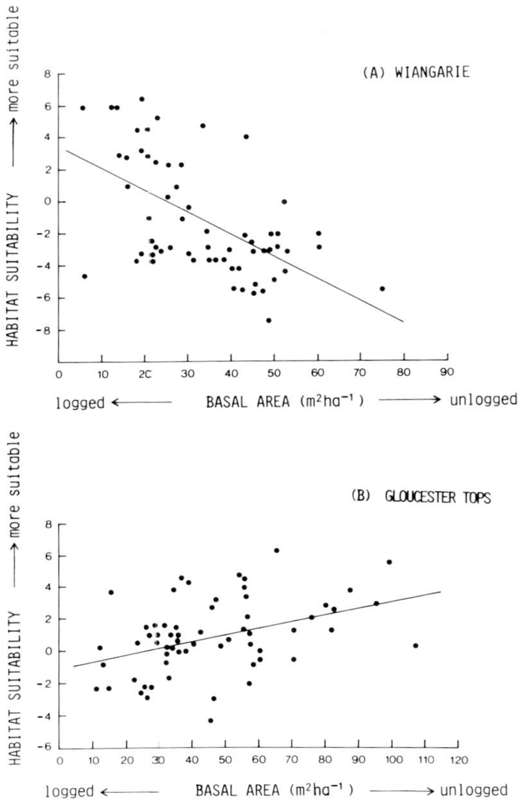

Fig. 3. The relationship between habitat suitability and basal area of trees over 20 cm d.b.h., in rainforest (Wiangarie) and eucalypt forest (Gloucester Tops). Basal area provides an indication of logging intensity. Discriminant scores are used as a measure of habitat suitability. These have been adjusted to remove the influence of other variables included in the multiple regressions. Each point represents a habitat plot.

Regression line for Wiangarie: Y = 3.5-0.14X: r = -0.57. Regression line for Gloucester Tops: Y = -1.0 + 0.04X: r = 0.39.

that the forest at Wiangarie was selectively logged (50% canopy reduction) and that the observed relationship between logging and habitat suitability cannot be extrapolated to heavier types of logging. Selective logging created gaps in the canopy, allowing more light to reach the ground and thereby promoting the development of dense ground cover, which was otherwise scarce within the rainforest. The effects of selective logging resembled those of natural tree falls, with which scrub-birds were also associated. The effects of heavier rainforest logging on habitat suitability are not known. While heavier logging also creates dense ground cover (Pattemore and Kikkawa 1975), it may have detrimental effects on the availability of leaf litter and the microclimate at ground level, both of importance to the Rufous Scrub-bird.

Habitat suitability at Wiangarie declined with time since logging. The regeneration of trees gradually reduced the amount of light reaching the ground and thereby the ground cover. The most recently logged sites studied had been logged five years previously and therefore little is known about habitat suitability less than five years after logging. Pattemore and Kikkawa (1975) have shown that one year after logging at Wiangarie the density of ground cover is low, resembling that of unlogged forest. Therefore it seems that habitat suitability is probably low immediately after logging, but increases at some stage within five years following logging, and then gradually declines. The suitability of selectively logged rainforest after five years can be predicted from the regression equation:

Habitat suitability = 3.47–0.14 (basal area of trees over 20 cm d.b.h. in m^2ha^{-1}) + 126 (Years since logged)$^{-2}$

Plots with suitability scores greater than 1.0 can be considered suitable for the Rufous Scrub-bird. In order to predict long term habitat changes following removal of a known basal area, the subsequent recovery of basal area over time needs to be taken into account.

At Gloucester Tops, logging explained less than 10% of variation in the suitability of habitat. This suggested that logging had less influence on habitat suitability within eucalypt forest than in rainforest. The effects of logging at Gloucester Tops were also quite different from those described for Wiangarie. Basal area was positively correlated with habitat suitability, suggesting that logging probably had a detrimental effect on scrub-bird habitat (see Fig. 3). This effect appeared to result not from changes in ground cover density, but rather from a reduction in leaf litter volume and humidity at ground level. The suitability of habitat increased with time since logging, in close correlation with the recovery of basal area. Logging at Gloucester Tops was relatively selective. The effects of heavier logging on habitat suitability within eucalypt forest are not known. While such practices would probably promote regeneration of dense ground and shrub vegetation (Recher *et al.* 1980), this influence may be counteracted by a reduction in litter volume and humidity due to loss of overstorey vegetation.

The overall impact of logging on the Rufous Scrub-bird will be a product not only of long term habitat changes to the general areas involved, but also of short term effects on existing territories. Logging within an existing territory is likely to have deleterious effects on the occupants of that territory, regardless of the long term influence on habitat. Management of the Rufous Scrub-bird in relation to logging therefore needs to fulfil two aims. Firstly, the protection of existing territories from logging. Secondly, the management of remaining areas in order to provide adequate habitat for the future maintenance, if not expansion, of present populations. The first aim has already been fulfilled by the Forestry Commission of New South Wales, which has undertaken to exclude all known scrub-bird territories from logging. The importance of the second aim should not be underestimated. Many existing territories are susceptible to habitat changes caused by natural phenomena such as fire and forest succession. The provision of suitable habitat in currently unoccupied areas may provide insurance against the loss of existing territories.

2. Fire

Fire seems to have played a major role in shaping the Rufous Scrub-bird's present distribution and, in particular, its close association with rainforest. However, the influence of current fire management policies on the species and its habitat are difficult to ascertain. As in the case of logging, two aspects of the problem need to be considered: 1. Short term effects on the occupants of existing territories, and 2. Long term effects on the suitability of habitat.

Almost a third of the 174 territories located during the present study were situated in eucalypt forest. Most of these displayed evidence of past fire (e.g. charred tree trunks) suggesting that there was at least some likelihood of burning in the future. The fire management policy most commonly adopted throughout the range of the Rufous Scrub-bird is one of low intensity prescribed burning aimed at preventing the occurrence of high intensity wildfires. Part of the Gloucester Tops study area was subjected to prescribed burning in September 1981, following a period of below average rainfall. Three territories were situated in eucalypt forest within the burnt area. The fire burnt up to, but not beyond, the boundary of each territory. All three males were heard singing from within their territories during and after the fire. The territories were apparently moist enough to escape burning. However, high intensity wildfires have been known to destroy territories elsewhere in the Gloucester Tops area (E. L. Hyem pers. comm.). While prescribed burning probably has little direct influence on existing territories it may provide a surrounding buffer zone against the effects of potentially destructive wildfires. In many cases, territories are also naturally buffered from such fires by surrounding patches of rainforest. Rufous Scrub-bird territories in eucalypt forest are rarely found further than 300 metres from the nearest rainforest.

The long term influence of prescribed burning on habitat suitability is a complex problem. Suitable habitat in eucalypt forest is probably associated with a particular seral stage of post-fire succession (Hodgson 1969; Ashton 1981; Christensen *et al.* 1981). The frequency of fire must therefore be an important determinant of habitat availability. Research on the Noisy Scrub-bird in Western Australia has suggested that burning at intervals of less than three to five years may not allow regeneration of suitable habitat, whereas burning at intervals greater than 30 years may allow successional changes to render areas unsuitable for that species (Smith 1977). The relationship between fire frequency and habitat suitability is probably similar for the Rufous Scrub-bird, although details concerning optimal burning intervals are yet to be established. At Gloucester Tops, habitat suitability 10 months after prescribed burning was found to be extremely low in terms of cover available, leaf litter volume, and humidity at ground level. At the other end of the successional scale, eucalypt forest adjacent to rainforest in New South Wales tends to develop into rainforest if fire is excluded for a long period (Turner 1976; Smith and Guyer 1983). Habitat suitability is probably lowered during this transition due to a reduction in ground cover density.

Fire management of eucalypt forest in relation to the Rufous Scrub-bird needs to ensure sufficient availability, at all times, of an intermediate seral stage suitable for the species.

GENERAL BIOLOGY: A SUMMARY OF CURRENT KNOWLEDGE

The following information on the general biology of *Atrichornis clamosus* and *A. rufescens* has been summarized from Chisholm (1951), Smith (l976a, l976b, 1979), Smith and Robinson (1976), Davies *et al.* (1982), and Ferrier (1984).

1. *Social Organization*

Breeding males occupy permanent territories defended with loud and penetrating song. Singing activity is most intense during the breeding season. Ephemeral (non-breeding) males may defend short-term territories at the onset of the breeding season, usually within sub-optimal habitat. In general, only one female is associated with each breeding male.

2. *Territories*

Males do over 80% of their singing from within an area of one hectare. Roosting sites are usually situated near the boundary of this core singing area. The female occupies a small nesting area on the periphery of the male's territory.

3. *Density*

Within areas of optimal habitat: 10 male territories per km^2 for *A. clamosus* and six male territories per km^2 for *A. rufescens*.

4. *Feeding*

Scrub-birds are incapable of sustained flight and spend most of their time feeding on the ground. Their diet consists mainly of leaf litter invertebrates.

5. *Breeding Season*

From April to October for *A. clamosus* and from September to December for *A. rufescens.*

6. *Breeding Biology*

The female alone is responsible for nest-building, incubation, and feeding the chicks. The nest is a domed structure lined on the inside with a cardboard-like substance. Clutch size is one for *A. calmosus* and two for *A. rufescens.* The incubation period is about 36-38 days. The young leave the nest three to four weeks after hatching, but continue to be fed by the mother for at least a further three weeks. Breeding generally takes place only once a year.

7. *Longevity*

Territory turn-over data for *A. clamosus* suggest that territorial males may live for at least 10 years.

THE FUTURE

The scrub-birds at present fulfil only minor roles in the ecology of Australian forests. They are nevertheless of great importance both as records of the past and as alternative components providing ecosystems with the flexibility needed to cope with future environmental change (see Main 1982).

In discussing the problem of scrub-bird conservation, Smith (1977) wrote:

> "If they (the species) are restricted in distribution and rare because changes in the climate and vegetation continue to take place then their preservation will need major management efforts involving large sums of money, which society may not consider justified. On the other hand, decline due to man-made effects can be halted and sometimes reversed with very little cost to society."

The rarity of the scrub-birds has resulted from a combination of natural and human causes. Both species have probably been declining in status since the Tertiary as a result of long-term climatic trends towards a drier environment. Habitat alteration by Europeans has simply increased the rate of this decline. Conservation efforts should aim to halt and, where possible, reverse this human influence.

Halliday (1978) has described two approaches to the conservation of rare species. Passive conservation entails the reservation of land and the prevention of harmful influences. Active conservation involves direct intervention by manipulating various elements of a species' biology. The Noisy Scrub-bird has required active management in order to ensure its future survival. Efforts have been made to breed birds in captivity (Smith 1979; Davies *et al.* 1982) and recently both males and females were successfully translocated from the Two Peoples Bay population to a nearby area of unoccupied habitat (D. V. Merton pers. comm.). The Rufous Scrub-bird is not sufficiently endangered to warrant this type of management. Passive conservation is more appropriate. Information on the species' habitat requirements, such as that presented in this chapter, can help in the identification and selection of areas requiring reservation. Over 90% of the territories found to date already lie within State Forests and National Parks. Unfortunately, reservation of land does not necessarily guarantee preservation of suitable habitat. The species generally occupies forest at an intermediate successional stage following disturbance through fire or opening of the canopy. In order to maintain sufficient areas of scrub-bird habitat, authorities need to consider the influence that prescribed burning and logging have on a forest's natural disturbance regime. This chapter has presented only preliminary answers to these problems. Future research will hopefully provide a better understanding of the relationship between forest management practices and scrub-bird habitat suitability.

ACKNOWLEDGEMENTS

The research described in this chapter was funded by the Commonwealth Government (as a post-graduate research award) and the National Parks Foundation of New South Wales. Information on logging and fire history was kindly provided by the Forestry Commission of New South Wales and the National Parks and Wildlife Service of New South Wales. I thank Glenn Holmes and Dr G. T. Smith for valuable discussion concerning scrub-birds. The assistance of numerous other individuals during the course of this study is also gratefully acknowledged.

REFERENCES

Ashton D. H., 1981. Fire in tall open forests. Pp.339-366 *in* Fire and the Australian biota ed. by A. M. Gill, R. H. Groves and I. R. Noble. Australian Academy of Science: Canberra.

Chisholm, A. H., 1951. The story of the scrub-birds. *Emu* **51**: 88-112, 258-97.

Christensen, P., Recher, H. and Hoare, J., 1981. Responses of open forests to fire regimes. Pp. 367-394 *in* Fire and the Australian biota ed. by A. M. Gill, R. H. Groves and I. R. Noble. Australian Academy of Science: Canberra.

Conner, R. N. and Adkisson, C. S., 1976. Discriminant function analysis: possible aid in determining the impact of forest management on woodpecker nesting habitat. *Forest Sci.* **22**: 122-27.

Davies, S. J. J. F., Smith, G. T. and Robinson, F. N., 1982. The Noisy Scrub-bird in Western Australia. Pp. 117-127 *in* Species at risk: research in Australia ed. by R. H. Groves and W. D. L. Ride. Australian Academy of Science: Canberra.

Ferrier, S., 1982. The Rufous Scrub-bird. Pp. 209 *in* Species at risk: research in Australia ed. by R. H. Groves and W. D. L. Ride. Australian Academy of Science: Canberra.

Ferrier, S., 1983. The Rufous Scrub-bird: a small bird with a big voice. *Forest and Timber* **19**: 9-11.

FERRIER, S., 1984. The status of the Rufous Scrub-bird, *Atrichornis rufescens*: habitat, geographical variation and abundance. Unpubl. Ph.D. Thesis: University of New England, Armidale.

FOX, B. J., 1979. An objective method of measuring the vegetation structure of animal habitats. *Aust. Wildl. Res.* **6**: 297-303.

HALLIDAY, T., 1978. Vanishing birds: their natural history and conservation. Holt, Rinehart and Winston: New York.

HODGSON, A., 1969. Historical and present role of low intensity fires in eucalypt forests. *Proc. 1st Fire Ecology Symposium*, Pp. 1-16. For. Comm. Vict. and Monash Univ.: Melbourne.

KING, W. B., 1981. Endangered birds of the world: the ICBP Red Data book. Smithsonian Inst. Press: Washington.

MACKENZIE, D. I. AND SEALY, S. G., 1981. Nest site selection in Eastern and Western Kingbirds: a multivariate approach. *Condor* **83**: 310-21.

MAIN, A. R., 1982. Rare species: precious or dross? Pp. 163-174 *in* Species at risk: research in Australia ed. by R. H. Groves and W. D. L. Ride. Australian Academy of Science: Canberra.

PARTRIDGE, L., 1978. Habitat selection. Pp. 351-376 *in* Behavioural ecology: an evolutionary approach ed. by J. R. Krebs and N. B. Davies. Blackwell: Oxford.

PATTEMORE, V. AND KIKKAWA, J., 1975. Comparison of bird populations in logged and unlogged rainforest at Wiangarie State Forest NSW. *Aust. For.* **37**: 188-98.

PIZZEY, G., 1980. A field guide to the birds of Australia. Collins: Sydney.

RECHER, H. F., ROHAN-JONES, W. AND SMITH, P., 1980. Effects of the Eden woodchip industry on terrestrial vertebrates with recommendations for management. For. Comm. NSW Res. Note No. 42.

ROBINSON, F. N. AND SMITH, G. T., 1976. The Noisy Scrub-bird — fact and fiction. *Western Australian Naturalist* **11**: 119-23.

ROTENBERRY, J. T., 1981. Why measure bird habitat? Pp. 29-32 *in* The use of multivariate statistics in studies of wildlife habitat ed. by D. E. Capen. USDA Forest Service General Technician Report: Fort Collins: Col.

SMITH, G. T., 1976a. Ecological and behavioural comparisons between the Atrichornithidae and Menuridae. Proc. 16th International Ornithological Congress, Pp. 125-36.

SMITH, G. T., 1976b. Rufous Scrub-bird. Pp. 334 *in* Complete book of Australian birds. Readers Digest Services: Sydney.

SMITH, G. T., 1977. The effect of environmental change on six rare birds. *Emu* 77: 173-79.

SMITH, G. T., 1979. The Noisy Scrub-bird. Pp. 117-121 *in* The status of endangered Australasian wildlife ed. by M. J. Tyler. Royal Zoological Society of South Australia: Adelaide.

SMITH, G. T. AND FORRESTER, R. I., 1981. The status of the Noisy Scrub-bird, *Atrichornis clamosus*. *Biological Conservation* **19**: 239-54.

SMITH, G. T. AND ROBINSON, F. N., 1976. The Noisy Scrub-bird: an interim report. *Emu* **76**: 37-42.

SMITH, J. M. B. AND GUYER, I. J., 1983. Rainforest-eucalypt forest interactions and the relevance of the biological nomad concept. *Aust. J. Ecol.* **8**: 55-60.

TITUS, K. AND MOSHER, J. A., 1981. Nest-site habitat selected by woodland hawks in the Central Appalachians. *Auk* **98**: 270-81.

TURNER, J. C., 1976. An altitudinal transect in rainforest in the Barrington Tops area, New South Wales. *Aust. J. Ecol.* **1**: 155-74.

CHAPTER 21

A Synthesis of the Foraging Ecology and Behaviour of Birds in Eucalypt Forests and Woodlands

Hugh A. Ford[1]

IN THIS chapter I synthesise the ways in which Australian forests influence the birds that live in them. The main features of the forest that influence birds are the structure and composition of the vegetation. These provide refuges, foraging and breeding sites and food, on which birds depend for survival and breeding.

One of the most exciting developments of the last few years has been the disentangling of the relationships of the Australian passerine birds to those on other continents. Australian forests have treecreepers, wrens, warblers, flycatchers, thrushes and robins just like Eurasian, African and American forests. However in Australia the treecreepers are related to bowerbirds and lyrebirds, wrens and warblers to honeyeaters and the other three groups to crows, rather than to birds bearing the same names on other continents (Sibley and Ahlquist 1985). Thus a study of the types of niches occupied by birds here and the ways in which relatives partition their resources provides a test of the ideas about avian ecology that have been developed in the Northern Hemisphere.

The structure of the vegetation limits the variety of foraging sites offered by a forest and consequently the diversity of bird species that it can support. Major foraging sites are exploited by ecologically similar groups or guilds of species. In eucalypt forest, pardalotes, thornbills, gerygone warblers and honeyeaters are the main components of the foliage-gleaning guild (Recher and Holmes 1985; Woinarski 1985a; Bell 1985a and Wykes 1985). Treecreepers, the sittella and shrike-tit are the major bark-foraging birds (Noske 1985). Few species feed primarily in the shrub layer but honeyeaters (Wykes 1985; Paton 1985) exploit nectar and Brown Thornbills, *Acanthiza pusilla*, (Bell 1985a) insects in this stratum. Ground feeders include rosellas (Wyndham and Cannon 1985), the Eastern Yellow Robin, *Eopsaltria australis*, that pounces on its prey (Marchant 1985), the gleaning Brown Treecreeper, *Climacteris picumnus*, (Noske 1985) and the secretive Rufous Scrub-bird, *Atrichornis rufescens*, (Ferrier 1985). Aerial feeders include the fantails (Cameron 1985) and monarch flycatchers. Some honeyeaters also take the insect part of their diet from the air (Paton 1985).

Several other important groups have not been mentioned in the chapters here but have been studied previously. These include fairy-wrens (Rowley 1965, 1981), choughs (Rowley 1978), butcherbirds, magpies (Carrick 1972), currawongs (Recher 1976) and woodswallows (Recher and Schulz 1983). Another important group includes the whistlers and cuckoo-shrikes, which snatch their prey from the foliage (Bridges 1980; Bell 1985b; Recher *et al.* 1985; Ford *et al.* 1986). Few of the forest and woodland raptors have been studied (Debus 1984; Baker-Gabb 1984) and even less is known about the owls (Schodde and Mason 1980).

The structure of the vegetation also determines whether safe refuges from predators and nesting sites are present. The low breeding success of thornbills (Bell 1985a) and Yellow Robins (Marchant 1985) indicates how important predation is in eucalypt forests. The Rufous Scrub-bird requires dense understorey or debris below 2 metres, perhaps principally to provide cover, but also for the microclimate that it produces (Ferrier 1985). The Yellow-tufted Honeyeater, *Lichenostomus melanops*, requires shrubs for breeding, though much of its foraging is in the canopy (Wykes 1985).

[1]Department of Zoology, University of New England, Armidale, New South Wales 2351.

Pages 249-54 *in* BIRDS OF EUCALYPT FORESTS AND WOODLANDS: ECOLOGY, CONSERVATION, MANAGEMENT ed. by A. Keast, H. F. Recher, H. Ford and D. Saunders. Royal Australasian Ornithologists Union and Surrey Beatty & Sons.

Treecreepers and parrots require nest-holes in trees (Noske 1985; Wyndham and Cannon 1985). Although eucalypts appear to provide plenty of these, many are unsuitable and others will be occupied by mammals. Spotted Pardalotes, *Pardalotus punctatus,* build their own nest-holes and soil type, topography and even the impact of people, rather than vegetation structure, determine where they can nest (Woinarski 1985a).

Generally eucalypt forests do not suffer from extremes of heat or cold or frequent high winds. They do experience frequent fires though, which many birds appear to survive (Marchant 1985). In addition they are evergreen and hence provide some protection when cold conditions are experienced. Shelter may not be as important as in the deciduous or even coniferous forests of northern Europe or North America.

The range of foods available strongly influences the bird species that live in a forest, their foraging behaviour and population dynamics. Seasonal changes in the abundance of different foods may cause birds to change their diet through the year, alter their foraging behaviour or even to leave the forest for a while. Breeding should also be geared to periods of abundance of food. Populations may have to respond to years of plenty and years of scarcity, by altering reproductive effort, moving locally or nomadically.

Nectar is a far more important food source in Australian forests than it is in other temperate parts of the world. This is exploited by honeyeaters (Paton 1985; Wykes 1985) and lorikeets (Wyndham and Cannon 1985a). Lerps, manna and honeydew, which are like nectar in being principally carbohydrate, are almost uniquely important to birds in eucalypt forests. These are favoured foods for pardalotes (Woinarski 1985a), and rosellas (Wyndham and Cannon 1985), as well as for honeyeaters (Wykes 1985). Large aggressive honeyeaters, like Bell Miners, *Manorina melanophrys,* can defend rich or persistent sources of nectar or lerps, and become sedentary. Small species, such as Yellow-faced Honeyeaters, *Lichenostomus chrysops,*may have to be highly mobile so that they can find unexploited or superabundant resources. Whether these movements are nomadic or show regular patterns is one of the major unanswered questions of honeyeater biology. Detailed studies of single species, such as that of David Paton (1985), on the New Holland Honeyeater, reveal both sedentary and mobile segments of the population. Lorikeets, despite being large and aggressive, seem not to defend nectar sources, but to be highly mobile. Christine Cannon's studies in southeastern Queensland will, in time, provide an interesting comparison with the rather better studied honeyeaters (1985).

New Holland Honeyeaters in Victoria breed in late winter, spring and sometimes autumn, when nectar is most abundant. Moult and dispersal of young birds follows as nectar becomes scarce. In contrast the short-billed honeyeaters studied by Boyd Wykes breed in spring and early summer, coinciding with abundant insects, rather than nectar. New Holland Honeyeaters appear strictly monogamous, whereas some of the short-billed honeyeaters (e.g. miners, Yellow-tufted Honeyeaters) breed co-operatively. How phylogeny, habitat and food supply influence breeding systems in honeyeaters is another challenge for the future.

Fruit is a minor source of food for birds in eucalypt forest. It is mentioned briefly in the chapters on short-billed honeyeaters (Wykes 1985) and parrots (Wyndham and Cannon 1985). Mistletoes provide fleshy fruits for Mistletoebirds, *Dicaeum hirundinaceum,* and some honeyeaters (Reid 1985), but were not the subject of any chapter in this book.

Seeds consumed by birds in forests fall into two classes, grass-seeds and those of woody shrubs and trees. Grass-seeds are exploited by finches such as the Red-browed and Diamond Firetails, *Emblema temporalis* and *E. guttata,* whose ecology has so far received little attention. Rosellas, and other parrots, consume both grass-seeds and seeds from shrubs and trees. Eastern Rosellas, *Platycercus eximius,* exploit a wide range of seeds whereas Pale-headed Rosellas, *P. adscitus,* feed more on the ground and Crimson Rosellas, *P. elegans,* more in the trees (Wyndham and Cannon 1985). They appear to time their breeding to anticipate abundant food as the young leave the nest. As in the New Holland Honeyeater, the Eastern Rosella has sedentary (core) and mobile members of the population. Perhaps nest-sites rather than food limits the breeding population.

Arthropods are the principal food of forest birds in Australia. As discussed above these are captured in a variety of sites and in a variety of ways. Their abundance and diversity permit fine subdivision of niches. Thornbills, fantails and treecreepers have been compared at single locations (Bell 1985; Cameron 1985; Noske 1985) and over a broader area the whole foliage-foraging guild has been examined (Woinarski 1985a). The short-billed honeyeaters also depend on arthropods and their diets and methods and sites of capture have been compared (Wykes 1985; see also Keast 1968; Ford and Paton 1976).

A decade ago there had been virtually no studies comparing the foraging behaviour of congeneric Australian birds. In recent years this deficiency has been redressed and data have been presented in preceding chapters on no less than seven genera. In three cases a pair of congeners has been compared,

Table 1. Differences between congeners of birds in eucalypt forests and woodlands.

	Morphology	Food	Foraging Method	Foraging Site	Habitat	Plant Species	Height	References
Rosellas *Platycercus*	—	+	—	+++	–	—	++	Wyndham and Cannon (1985)
Lorikeets *Trichoglossus*	—	+	—	—	+	+	—	Wyndham and Cannon (1985)
Thornbills *Acanthiza*	+	+	+	+++	—	++	++	Bell (1985)
Pardalotes *Pardalotus*	+	—	+	—	+	+	+	Woinarski (1985)
Fantails *Rhipidura*	+	+	++	++	+++	+	+	Cameron (1985)
Treecreepers *Climacteris*	+	—	++	++	—	++	+	Noske (1985)
Honeyeaters *Lichenostomus*	+	++	+	++	++	+	+	Wykes (1985)
Honeyeaters *Meliphaga* *(Lichenostomus)*	+	+	+	+–	+++	+	+	Ford and Paton (1977)
Honeyeaters *Melithreptus*	+	+	+	+–	++	+	—	Keast (1968)
Honeyeaters *Phylidonyris*	—	—	+	+	++	+	—	Recher (1977)
Robins *Petroica*	+	?	++	++	++	+	++	Fleming (1980)
Whistlers *Pachycephala*	+	—	++	+	+–	+	—	Bridges (1980)

— negligible; + slight; ++ moderate; +++ major.

in three a trio of species and in one no less than seven species have been studied. I have summarized the degree and types of difference displayed by congeners in seven characteristics (Table 1). I have also added the results of five other studies by; Keast (1968) on *Melithreptus,* Ford and Paton (1977) on *Meliphaga,* Recher (1977) on *Phylidonyris,* Fleming (1980) on *Petroica* and Bridges (1980) on *Pachycephala.* As thirteen observers with different precision in subdividing characteristics were involved, my classification is a qualitative rather than a quantitative one. Also the degree of difference does not necessarily indicate the importance of this difference to the species concerned.

Overall it seems that congeners do not differ markedly in morphology or food. Lack of morphological differences may of course be the reason why taxonomists have placed the species in the same genus Similarity in diet may be related to similar morphology or may simply show that most species are generalized in what they eat. The more strictly insectivorous species often show moderate differences in foraging method.

The fantails show varying amounts of gleaning and flycatching, treecreepers of pecking and peering (Noske 1979), robins of pouncing and hawking and whistlers of gleaning and snatching. Differences in foraging site and habitat account for moderate to major differences in five genera.

The three thornbills separate into a canopy-feeder, a shrub-feeder and a bark/ground specialist. The two rosellas differ markedly in the time they spend feeding in trees and on the ground. The fantails differ in their amount of ground, aerial and vegetation feeding and use habitats of differing density. One of the three treecreepers includes a significant degree of ground-feeding in its repertoire. Among the short-beaked honeyeaters there is wide variation in the extent to which bark is used as a substrate for foraging, in addition most species show specific habitat preferences. The robins too also show some separation by habitat with the Rose Robin, *Petroica rosea,* at the dense extreme and the Flame Robin, *P. phoenicea,* in open habitats. The former mostly snatches insects from vegetation or the air while the latter pounces on terrestrial prey. The Scarlet Robin, *P. multicolor,* is intermediate. In the Whistlers, the Golden, *Pachycephala pectoralis,* is a forest dweller whereas the Rufous, *P. rufiventris,* occupies woodland.

Plant species and height are correlated with foraging site (e.g. eucalypt leaf-gleaners feed high, bark feeders lower, with ground feeders obviously lower still), though they may account for additional differences. For instance both Brown and Striated Thornbills, *A. lineata,* are leaf-gleaners, but the former feeds low on shrubs and the latter high on eucalypts. The two species of pardalotes, both glean from eucalypt leaves, but they show slightly different preferences for tree species in some sites.

The pardalotes and lorikeets emerge as the genera whose species are most similar, with no differences even being classed as moderate or major. It is interesting that both feed largely on carbohydrate foods, lerps and nectar respectively. Two of the highly nectarivorous *Phylidonyris* honeyeaters are also very similar in foraging behaviour (Recher 1977). All three groups have been described as nomadic. Perhaps the foods of these three groups, which are chiefly carbohydrate ones, are often superabundant, though patchy and rather unpredictable in distribution.

It is impressive to see so many attempts to measure seasonal changes in the abundance of arthropods (Bell 1985a; Woinarski 1985a; Cameron 1985; Noske 1985; Wykes 1985; see also Woinarski and Cullen 1984; Bell 1985c). This is a notoriously tedious procedure, where clear patterns are plagued by the diversity of organisms and variety of short- and long-term influences. Any generalizations about patterns of abundance must be very tentative at this stage, though some common threads do emerge. First, arthropods are more abundant in the warmer than the cooler months. This is the typical pattern of temperate climates. However the spring-summer peaks seem less impressive in Australia than in Europe and North America. Perhaps the evergreen trees and equable climate of eucalypt forest permit good survival of arthropods and even some activity during the winter. Both larvae and midges may be plentiful on winter days, though ants and beetles tend to be more available in summer than winter. Many eucalypts are rough-barked or shed their bark. This provides a variety of rich substrates on which birds can forage, especially for overwintering insects. Another obvious pattern is one of differences in abundances between years. It is widely believed that droughts can decimate birds populations in Australia. Boyd Wykes in Victoria (1985) and Harry Bell in New England (1985a) have shown how nectar and arthropods do indeed become scarce after persistent dry weather. Reductions in bird populations are perhaps less dramatic than might have been expected and it is hard to assess the degree to which these reflect mortality or emigration. Possibly the major effect is on reduced breeding success rather than increased mortality.

Most insectivorous birds breed in spring, although some start very early, e.g. Striated Thornbills in July. Many species have a long breeding season, e.g. Yellow Robins August-January. This is perhaps a response to the fact that there is not a dramatic peak in abundance of arthropods, yet there is much year to year variation. Most of the insectivorous or nectarivorous species discussed here have clutches of two or three eggs (Woinarski 1986). Compare this with the three to nine eggs of rosellas (Wyndham and Cannon 1985), and generally higher clutches in Europe and North America. Yom-Tov (ms) has pointed out that it is the families with a long history in Australia rather than the recent arrivals that have small clutches. Both Yom-Tov (ms) and Woinarski (1985b) attribute this low clutch-size to the equable climate, past and present, in Australia.

A feature of many Australian birds is the tendency to breed co-operatively. This ranges from the presence of occasional first year birds that assist their parents, to the complex social organization of the miners (Dow 1980). Of the groups discussed here, co-operative breeding is found in such diverse birds as treecreepers, thornbills, Bell Miners, Yellow Robins and Striated Pardalotes, *Pardalotus striatus*. Other detailed studies of co-operative breeders include those on fairy-wrens (Rowley 1965, 1981), choughs (Rowley 1978), babblers (Counsilman 1979; King 1980) and Kookaburras, *Dacelo novaeguineae,* (Parry 1973). On the other hand some thornbills and treecreepers show a different breeding biology from their co-operative congeners. They breed strictly in territorial pairs. The relationship between habitat, foraging behaviour and phylogeny and social organization in Australian birds is one of the exciting areas of research for the future. Of equal interest is the question of why Australia appears to have so many co-operatively breeding species.

On the whole the insectivorous species described in this section are sedentary, or at least have sedentary components of their population. Only Rufous Fantail, *Rhipidura rufifrons,* White-throated Gerygone, *Gerygone olivacea,* and Yellow-faced Honeyeater are true migrants. This low incidence of regular migration may reflect the lack of marked seasonality in abundance of food, particularly arthropods. However, Recher *et al.* (1983) found that migrants made up a significant component of the bird community in an upland site in southern New South Wales.

Although species whose whole populations depart in autumn are few, many insectivorous species are partial migrants. These include Grey Fantails, *Rhipidura fuliginosa,* where most of the population leaves, and the pardalotes and honeyeaters whose complex movements have been described as nomadic.

In many species there is extensive dispersal of the young at the end of the breeding season (Yellow Robin, White-throated Treecreeper, New Holland Honeyeater, Eastern Rosella, and Brown Thornbill). Such dispersal often results in these species turning up in areas outside their breeding range, which has been regarded as migration or nomadism. In species that breed co-operatively dispersal may take place just before the next breeding season (Red-browed *C. erythrops* and Brown Treecreepers — Noske 1983, Buff-rumped Thornbills, *Acanthiza reguloides* — Bell 1983, Superb Fairy-wren, *Malurus cyaneus* — Rowley 1965), which gives the juveniles the benefit of the parental territory for the autumn and winter. Two periods of dispersal may be shown by Yellow Robins (Marchant 1985) and Superb Fairy-wrens (Nias pers. comm.).

It is also impressive to see estimates of survival for small passerines in Australia. Fry (1980) previously suggested that these are higher than in the Northern Hemisphere, and data presented here support this view. Adults on established territories show annual survival rates of 75% in Yellow Robins (Marchant 1985), 70-80% in treecreepers (Noske 1983), and 66% in Superb Fairy-wrens (Rowley 1965). Even the tiny thornbills showed 72-88% overwinter survival and over 50% of adults survived a year of severe drought (Bell 1983). Interestingly in both New Holland Honeyeaters and Splendid Fairy-wrens, *Malurus splendens,* (Rowley 1981) territorial males survive better than do females (68% vs 42%, and

71% vs 43%). Differential mortality is presumably the reason for imbalanced sex ratios in many Australian passerines. Estimates of survival for juveniles and dispersing or transient birds are harder to make. Paton (1985) estimated that about 10-19% of transient New Holland Honeyeaters survive for one year. Although this is probably an underestimate, transient birds probably suffer greater mortality than do territory holders. I suggest that insectivorous birds in eucalypt forests experience a sufficiency of food most of the year and in most years. Perhaps times of great abundance are as rare as times of scarcity. The birds have responded by staying put, if they can, and by modest reproductive efforts, even to the extent of "helping" others rather than breeding themselves. Although data are scanty I predict that populations will prove to be rather stable and resources often heavily exploited.

One of the main aims of the research reported here is to ensure the conservation of Australia's forest birds. Three taxa that have been discussed are rare and endangered. These are the Rufous Scrub-bird (Ferrier 1985), Forty-spotted Pardalote, *Pardalotus quadragintus,* (Woinarski 1985a) and the Helmeted Honeyeater, *Lichenostomus melanops cassidix,* (Wykes 1985).

Simon Ferrier recommends both passive and active management of the Rufous Scrub-bird. The fact that most known territories occur in State Forest or National Park makes management easier, though logging influences the suitability of the habitat for scrub-birds. Logging is detrimental to scrub-birds in eucalypt forest but in some circumstances may be beneficial in rainforest. Fire may also affect habitat suitability though exactly how this will influence management is less obvious.

There are similarities between the history and present plight of the Helmeted Honeyeater and Forty-spotted Pardalote. Both are isolated populations of more widespread taxa. Both have suffered from habitat loss or modification over the last century. However there is evidence that both are currently suffering from interspecific competition. Striated Pardalotes may dominate Forty-spotted Pardalotes where the latter is rare, but the dominance is reversed where it is common (Rounsevell and Woinarski 1983; Woinarski and Wykes 1983; Woinarski and Bulman 1985). Bell Miners may aggressively exclude Helmeted Honeyeaters from lerp-infested areas (Wykes 1981).

Now that we have basic information on so many groups, research on birds in eucalypt forests and woodlands over the next decade should perhaps concentrate on the following topics:

1. long term studies on breeding biology and population dynamics, in relation to changes in conditions. The studies of Rowley (1965, 1981) and Marchant (1985) should be taken as examples.

2. measurement of food resources, the extent to which they are exploited and how birds respond to abundance and scarcity.

3. the relationship between a species' ecology and its social organization, both within and between species.

Experiments that manipulate the abundance of food and other resources (e.g. nest-sites) and the different components of the population (sex ratios, helpers etc.), should play a part in these studies. However they must complement rather than replace long-term investigations. Only when we have results from both can we be confident that we understand how bird-populations are controlled. When we have this knowledge we are in a position to manage Australia's eucalypt forests and woodlands to ensure a diverse and healthy avifauna.

REFERENCES

BAKER-GABB, D. J., 1984. The breeding ecology of twelve species of diurnal raptors in northwestern Victoria, Australia. *Aust. Wildl. Res.* 7: 433-54.

BELL, H. L., 1983. Resource-partitioning between three syntopic thornbills (Acanthizidae: *Acanthiza* Vigors and Horsfield). Ph.D. Thesis, University of New England, Armidale, New South Wales.

BELL, H. L., 1985a. The social organization and foraging behaviour of three syntopic thornbills *Acanthiza* spp. Pp. 151-63 *in* Birds of Eucalypt Forests and Woodlands: Ecology, Conservation, Managment ed. by A. Keast, H. F. Recher, H. Ford and D. Saunders. Royal Australasian Ornithologists Union.

BELL, H. L., 1985b. Sexual differences in the behaviour of wintering Golden Whistlers, *Pachycephala pectoralis,* at Wollomombi, New South Wales, *Emu* (in press).

BELL, H. L., 1985c. Seasonal variation and the effects of drought on the abundance of arthropods in Savanna woodland on the Northern Tablelands of New South Wales. *Aust. J. Ecol.* **10**: (in press).

BRIDGES, L., 1980. Some examples of the behaviour and feeding ecology of the Rufous *(Pachycephala rufiventris)* and Golden *(Pachycephala pectoralis)* Whistler. B.Sc. (Hons) Thesis, University of New England, Armidale, New South Wales.

CAMERON, E., 1985. Habitat usage and foraging behaviour of three fantails *(Rhipidura:* Pachycephalidae). Pp. 177-91 *in* Birds of Eucalypt Forests and Woodlands: Ecology, Conservation, Managment ed. by A. Keast, H. F. Recher, H. Ford and D. Saunders. Royal Australasian Ornithologists Union.

CARRICK, R., 1972. Population ecology of the Australian Black-backed Magpie, Royal Penguin and Silver Gull. Population Ecology of migratory birds: A symposium. U.S. Dept. of Interior. Wildlife Research Report 2.

COUNSILMAN, J. J., 1979. Notes on the breeding biology of the Grey-crowned Babbler. *Bird Behaviour* **2**: 114-24.

DEBUS, S. J. S., 1984. Biology of the Little Eagle on the Northern Tablelands of New South Wales. *Emu* **84**: 87-92.

DOW, D. D., 1980. Communally breeding Australian birds, with an analysis of distributional and environmental factors. *Emu* **80**: 121-40.

FERRIER, S., 1985. Habitat requirements of a rare species, the Rufous Scrub-bird. Pp. 241-48 *in* Birds of Eucalypt Forests and Woodlands: Ecology, Conservation, Managment ed. by A. Keast, H. F. Recher, H. Ford and D. Saunders. Royal Australasian Ornithologists Union.

FLEMING, P., 1980. The comparative ecology of four sympatric robins. B.Sc. (Hons) Thesis, University of New England, Armidale, NSW.

FORD, H. A. AND PATON, D. C., 1976. Resource partitioning and competition in honeyeaters of the genus *Meliphaga*. *Aust. J. Ecol.* **1**: 281-87.

FORD, H. A. AND PATON, D. C., 1977. The comparative ecology of ten species of honeyeaters in South Australia. *Aust. J. Ecol.* **2**: 399-407.

FORD, H. A., NOSKE, S. AND BRIDGES, L., 1986. Foraging behaviour of birds in eucalypt woodland in northeastern New South Wales. *Emu* **86**: (in press).

FRY, C. H., 1980. Survival and longevity among tropical land-birds. Proc. IV Pan-Afr. orn. Congr. 333-43.

KEAST, J. A., 1968. Competitive interactions and the evolution of ecological niches as illustrated by the Australian honeyeater genus *Melithreptus* (Meliphagidae). *Evolution* **22**: 762-84.

KING, B. R., 1980. Social organization and behaviour of the Grey-crowned Babbler, *Pomatostomus temporalis*. *Emu* **80**: 59-76.

MARCHANT, S., 1985. Breeding of the Eastern Yellow Robin, *Eopsaltria australis*. Pp. 231-40 *in* Birds of Eucalypt Forests and Woodlands: Ecology, Conservation, Managment ed. by A. Keast, H. F. Recher, H. Ford and D. Saunders. Royal Australasian Ornithologists Union.

NOSKE, R. A., 1979. Co-existence of three species of treecreepers in northeastern New South Wales. *Emu* **79**: 120-28.

NOSKE, R. A., 1983. Comparative behaviour and ecology of some Australian bark foraging birds. Ph.D. Thesis, University of New England, Armidale, NSW.

NOSKE, R. A., 1985. Habitat use by three bark-foragers of eucalypt forest. Pp. 193-204 *in* Birds of Eucalypt Forests and Woodlands: Ecology, Conservation, Managment ed. by A. Keast, H. F. Recher, H. Ford and D. Saunders. Royal Australasian Ornithologists Union.

PARRY, V., 1973. The auxiliary social system and its effect on territory and breeding in kookaburras. *Emu* **73**: 81-100.

PATON, D. C., 1985. Food supply, population structure and behaviour of New Holland Honeyeaters, *Phylidonyris novaehollandiae*, in Woodland near Horsham, Victoria. Pp. 219-30 *in* Birds of Eucalypt Forests and Woodlands: Ecology, Conservation, Managment ed. by A. Keast, H. F. Recher, H. Ford and D. Saunders. Royal Australasian Ornithologists Union.

RECHER, H. F., 1976. Reproductive behaviour of a pair of Pied Currawongs. *Emu* **76**: 224-27.

RECHER, H. F., 1977 Ecology of coexisting White-checked and New Holland Honeyeaters. *Emu* **77**: 136-42.

RECHER, H. F. AND HOLMES, R. T., 1985. Foraging ecology and seasonal patterns of abundance in a forest avifauna. Pp. 79-96 *in* Birds of Eucalypt Forests and Woodlands: Ecology, Conservation, Managment ed. by A. Keast, H. F. Recher, H. Ford and D. Saunders. Royal Australasian Ornithologists Union.

RECHER, H. F., GOWING, G., KAVANAGH, R., SHIELDS, J. AND ROHAN-JONES, W., 1983. Birds, resources and time in a tablelands forest. *Proc. Ecol. Soc. Aust.* **12**: 101-23.

RECHER, H. F. AND SCHULZ, M., 1983. Observations on the breeding of White-browed Woodswallows. *Corella* **7**: 1-6.

RECHER, H. F., HOLMES, R. T., SCHULZ, M., SHIELDS, J. AND KAVANAGH, R., 1985. Foraging patterns of birds in eucalypt forest and woodland on the tablelands of southeastern Australia. *Aust. J. Ecol.* **10**: (in press).

REID, N., 1985. Mistletoes and birds. *In* The dynamic partnership: birds and plants in southern Australia ed. by H. A. Ford and D. C. Paton. Flora and Fauna Board of South Australia: Adelaide (in press).

ROUNSEVELL, D. E. AND WOINARSKI, J. 1983. Status and conservation of the Forty-Spotted Pardalote, *Pardalotus quadragintus*, (Aves: Pardalotidae). *Aust. Wildl. Res.* **10**: 343-49.

ROWLEY, I., 1965. The life history of the Superb Blue Wren, *Malurus cyaneus*. *Emu* **64**: 251-97.

ROWLEY, I., 1978. Communal activities among White-winged Choughs, *Corcoras melanorhamphus*. *Ibis* **120**: 178-97.

ROWLEY, I., 1981. The communal way of life in the Splendid Wren, *Malurus splendens*. *Z. Tierpsychol.* **55**: 227-67.

SCHODDE, R. AND MASON, I. J., 1980. Nocturnal birds of Australia. Melbourne: Lansdowne.

SIBLEY, C. G. AND AHLQUIST, J. E., 1985. The phylogeny and classification of the Australo-Papuan passerine birds. *Emu* **84**: 1-14.

WOINARSKI, J., 1985a. Foliage gleaners of the tree tops, the Pardalotes. Pp. 165-75 *in* Birds of Eucalypt Forests and Woodlands: Ecology, Conservation, Managment ed. by A. Keast, H. F. Recher, H. Ford and D. Saunders. Royal Australasian Ornithologists Union.

WOINARSKI, J., 1985b. Breeding biology and life history of small insectivorous birds in Australian forests: response to a stable environment? *Proc. Ecol. Soc. Aust.* **14**: 159-68.

WOINARSKI, J. AND WYKES, B., 1983. Decline and extinction of the Helmeted Honeyeater at Cardinia Creek. *Biol. Conserv.* **27**: 7-22.

WOINARSKI, J. AND CULLEN, J. M., 1984. Distribution of invertebrates on foliage in forests of southeastern Australia. *Aust. J. Ecol.* **9**: 207-32.

WOINARSKI, J. AND BULMAN, C., 1985. Ecology and breeding biology of the Forty-spotted Pardalote and other pardalotes on North Bruny Island. *Emu* **85**: 106-20.

WYKES, B. J., 1981. The ecology of the Helmeted Honeyeater and its relationships with potential competitors. Report to Fisheries and Wildlife Division, Victoria.

WYKES, B. J., 1985. The Helmeted Honeyeater and related honeyeaters of Victorian woodlands. Pp. 205-17 *in* Birds of Eucalypt Forests and Woodlands: Ecology, Conservation, Managment ed. by A. Keast, H. F. Recher, H. Ford and D. Saunders. Royal Australasian Ornithologists Union.

WYNDHAM, E. AND CANNON, C. E., 1985. Parrots of eastern Australian forests and woodlands: the genera *Platycercus* and *Trichoglossus*. Pp. 141-50 *in* Birds of Eucalypt Forests and Woodlands: Ecology, Conservation, Management ed. by A. Keast, H. F. Recher, H. Ford and D. Saunders. Royal Australasian Ornithologists Union.

YOM-TOV, Y., 1985. The reproductive rates of Australian birds. (ms).

Human Impact: The Response of Forest and Woodland Bird Communities

Co-ordinator D. A. Saunders

White-tailed Black Cockatoo, *Calyptorhynchus baudinii*

INTRODUCTION

THE environments in which animals live have always been subject to change. In fact, it is the dynamics of this change which provides the impetus for evolution with those organisms that fail to adapt to the changes in their environment becoming extinct. With the arrival and settling of Europeans in Australia nearly two hundred years ago, a distinct new period of human induced changes was initiated. Since 1788 the pace of this change has varied but it has become much more rapid in recent years. For example, in the southwest of Western Australia only 500 km^2 of land had been cleared by 1890 but by 1968 the total was nearly 130,000 km^2 with over half of it being cleared since 1945. Since the widespread application of the internal combustion engine a single person has been given the potential to wreak greater change on his or her surroundings in far less time than ever before. Unfortunately our knowledge of the consequences of these changes to the environment is negligible and has not kept pace with our capacity to bring about the changes. This fact has been obvious for many years. Graeme Smith (Chapter 26) quotes a passage by Tom Carter published in 1924 in which Carter describes the change in coastal southwestern Australia from impenetrable native vegetation to bare sand drifts brought about by frequent fires which were set to improve the country for cattle grazing.

The aim of this section of the book is to examine the effects on our avifauna of some changes brought about by European settlement and to discuss methods for ameliorating these effects. Some of the more controversial changes are as a result of clear-felling areas of forest for saw timber and woodchips, the bulk of which is exported. Peter Smith (Chapter 22) and Kavanagh *et al.* (Chapter 23) examine the effects of these activities on bird populations in the southeast of New South Wales and discuss the implications for successful management to ensure the survival of the forest avifauna.

Throughout Australia, Europeans changed the fire regime practiced by the Aborigines before them. There is very little information about the effects of fire on our avifauna or of the consequences of our fire regimes on forest and woodland ecosystems in the long term. Fire can occur as an uncontrolled event (wildfire) or as a controlled event (prescribed burning) and two of the chapters address these issues. The effects of wildfire and logging on bird populations in southeastern Australia are dealt with by Recher *et al.* (Chapter 24) while Christensen *et al.* (Chapter 25) present data on the effects of controlled fire for litter reduction on bird populations in southwestern Australia.

One drastic effect of altered fire management is the extinction of populations or even whole species. The Noisy Scrub-bird, *Atrichornis clamosus,* was believed to have become extinct about 1890, until its rediscovery nearly 70 years later on one isolated headland in the southwest of Western Australia. Graeme Smith (Chapter 26) discusses the role of the changed fire regime in the disappearance of the Scrub-bird and the reasons for its survival at Two Peoples Bay. He then presents options for the successful managment of this species based on his long term study of its biology.

One of the most obvious changes to the Australian landscape has been the massive alteration brought about by the clearing of native vegetation for agricultural pursuits. This has had a dramatic effect on the avifauna and Saunders *et al.* (Chapter 27) document the changes in the distribution of cockatoos in the southwest of Western Australia as a result of clearing for agriculture and discuss those aspects of the biology of species which have been affected by the changes.

When land is cleared for agriculture, this leaves isolated patches of native vegetation. This network of patches provides the only refuges for much of our

Pages 255-58 *in* BIRDS OF EUCALYPT FORESTS AND WOODLANDS: ECOLOGY, CONSERVATION, MANAGEMENT ed. by A. Keast, H. F. Recher, H. Ford and D. Saunders. Royal Australasian Ornithologists Union and Surrey Beatty & Sons.

flora and fauna. The role of these patches in the conservation of birds in the Gippsland area of Victoria is discussed by Loyn (Chapter 28) who notes that "even small patches can be useful if they contain habitats the value of which can be maintained in the long term".

Considerable areas of forest and woodland throughout Australia are affected by a phenomenon widely called "Eucalypt Dieback". This is a generic name which covers a number of different causes. Trees in the Northern Tablelands of New South Wales have suffered from New England Dieback which may be a result of depredations by leaf-eating insects. Ford (Chapter 29) discusses the effect of dieback on birds and the possible role of birds in the control of dieback.

Forests in Australia are often managed for a range of uses. In the southwest of Western Australia forests are scarce and the different demands for their use often conflict. Much of the northern Jarrah, *Eucalyptus marginata,* forest is used as water catchment yet bauxite is being mined, with much of the forest being covered by mining leases. One of the biggest conservation problems facing the bauxite miners is the rehabilitation of mined lands. Collins *et al.* (Chapter 30) have been studying how birds recolonize bauxite mines that have been restored to native vegetation. By examining mine sites of different rehabilitation histories they have been able to evaluate the efficacy of each technique and so suggest suitable management.

Each of the chapters in this section of the book provides new data which can be used to manage our environment to ensure the survival of our rich and diverse avifauna.

D. A. Saunders

CHAPTER 22

Woodchip Logging and Birds near Bega, New South Wales

Peter Smith[1]

WHEN THE First Fleet arrived in 1788 practically all of the eastern half of New South Wales, some 40 million hectares, was covered by eucalypt forest or woodland. Today only about a third of that area remains. For the first hundred years the forests were regarded as a hinderance to settlement, to be cleared as quickly as possible, and as an inexhaustible supply of wood. The first steps to conserve the forest resources of the State were only taken in the 1870's and 1880's and it was not until the Forests Act of 1909 that the Forestry Commission of NSW was established to implement a definite policy of forest protection and management. Then, as now, the major objective of the Commission was to provide for the present and future timber requirements of the State (Forestry Commission of NSW 1976).

The 1960's saw an increasing public concern for the environment. The NSW National Parks and Wildlife Service was established in 1967 to consolidate the Government's nature conservation activities under one department. At the same time, the management of Crown timber-lands began to be directed more towards multiple objectives: maintaining their value for recreation, education, scientific research, catchment protection and wildlife conservation, in addition to their value for timber (Forwood 1974; Carron *et al.* 1979).

These developments coincided with an expanding demand for wood. The 1970's saw the growth of a woodchip industry in New South Wales and the replacement of large areas of eucalypt forest with plantations of exotic pines. There were many bitter confrontations between conservationists on the one side, and foresters and the timber industry on the other (Routley and Routley 1975). One such confrontation occurred in 1977 when woodchip logging began in the forests near Bega on the far south coast of the State. As a consequence, State Cabinet directed the National Parks and Wildlife Service to survey the forest fauna of the area and investigate the effects of woodchip logging on that fauna. This chapter outlines the results of the study in relation to birds.

WOODCHIP LOGGING

Woodchips are flat pieces of wood about 4 cm in diameter and 0.5 cm thick. They are the raw material from which paper is made. In 1968 the NSW Government entered into a 20-year agreement with Harris-Daishowa Pty Ltd to provide 530,000 tonnes of logs a year for export to Japan as woodchips. A chipmill was constructed at Eden (Fig. 1) and began operating in 1969. Although wood from Tasmanian and Victorian eucalypt forests had been chipped and made into paper locally since the late 1930's, the Eden project was the first to export woodchips from Australia. It was soon followed by export projects in Tasmania and Western Australia and more are being developed (Forestry Commission of NSW 1979; Australian Forestry Council 1979).

Wood for the chipmill comes mainly from the forests within a radius of 80 km around Eden (Fig. 1). Within this concession area are 297,000 ha of Crown timber-lands and 53,000 ha of national parks and nature reserves. A further 76,000 ha national park is located on the northwestern periphery (Wadbilliga National Park). Approximately 74 percent of the Crown timber-lands are to be intensively logged over a period of 40 to 45 years that began in 1970. The remainder is too steep or rocky for logging or will be reserved for stream protection, recreation, preservation of scenic qualities, and conservation of flora and fauna. Some 20,000 ha of the logged forest will be converted to pine plantation (*Pinus radiata*), leaving about 200,000 ha on which eucalypts will be regenerated (Forestry Commission of NSW 1982; Bridges 1983).

The logging operations are termed 'integrated' logging, since they involve simultaneous harvesting of logs for the chipmill and for local sawmills. The logging is carried out by contractors who are engaged by the chipmill and sawmills but work under the supervision of the Forestry Commission. The trees are felled by chainsaw and the crowns removed. The logs are then dragged by tractor to a

[1]New South Wales National Parks and Wildlife Service, P.O. Box N189, Grosvenor Street, Sydney, New South Wales 2000. Present Address: 44 Hawkins Parade, Blaxland, New South Wales 2774.
Pages 259-71 *in* BIRDS OF EUCALYPT FORESTS AND WOODLANDS: ECOLOGY, CONSERVATION, MANAGEMENT ed. by A. Keast, H. F. Recher, H. Ford and D. Saunders. Royal Australasian Ornithologists Union and Surrey Beatty & Sons.

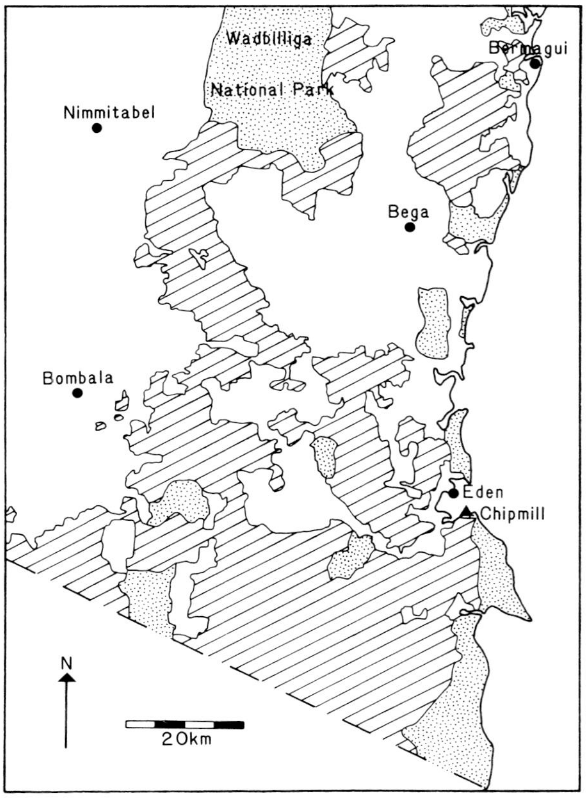

Fig. 1. The Eden woodchip management area. Hatched areas are Crown timber-lands. Stippled areas are national parks and nature reserves. The National Parks and Wildlife Service study was carried out in the forests between Bega and Bermagui.

central log dump where their bark is removed and they are loaded onto trucks. The better quality logs go to the sawmills to provide sawn timber. The poorer quality logs go to the chipmill and are cut up into woodchips (Forestry Commission of NSW 1979).

Since 1969 there have been continual changes in the planning and management of integrated logging. At the time of the study the logging areas, or 'coupes', were 10-20 ha in area. All of the merchantable trees were felled in each coupe but young trees that were likely to develop into good sawlogs were left and there were several tree species that were unsuitable even for chipping. Logged coupes were allowed to regenerate naturally without burning, seeding or felling of the unmerchantable trees, but the heavily compacted soil of the log dumps was ripped and planted out with eucalypt seedlings (Forestry Commission of NSW 1979). Logging procedures have since been changed to reduce the potential impact of wildfires. Additional trees are now retained in each logging area to provide a source of seeds in case of fire, and there is a more intensive programme of hazard reduction burning. Because of the burning programme, coupes are now 30-50 ha in area and some are up to 100 ha (Forestry Commission of NSW 1982; Bridges 1983).

By the end of the first 40 to 45-year logging cycle the older regeneration will be suitable for logging again. What form that logging will take is uncertain. Current predictions are that higher yields of wood will be obtained because the young trees, although smaller, will be denser than in the original forest. This should allow a greater flexibility in logging procedures during the second cycle (Bridges 1983).

THE BIRDLIFE OF THE BEGA FORESTS

The National Parks and Wildlife Service study was carried out in the forests between Bega and Bermagui at the northern end of the woodchip management area (Fig. 1). These form a strip of forested land about 40 km long by 20 km wide that is bordered on the east by the Pacific Ocean and on the west by cleared farmlands. Altogether, 116 bird species were recorded in eucalypt forest or rainforest in this area (Table 1). Twenty-two species were transients, while 94 species were regular inhabitants of the forest. Of the latter, two-thirds were resident species and one-third were migratory or nomadic species.

There is a variety of eucalypt forest-types in the area, with different tree species compositions and different understoreys. The greatest differences, in both vegetation and birds, are between the dry forest of the ridges and the taller, more moist forest of the gullies (Smith 1984). These two forest-types are closely intermingled, as the distance from ridge to gully is generally only 200-800 m. The gullies are richer in birds than the ridges (Fig. 4) and there are also differences in species composition (Table 1). Characteristic birds of the gullies include the Wonga Pigeon, Eastern Whipbird and Brown Warbler. Characteristic birds of the ridges include the Black-faced Cuckoo-shrike, Rufous Whistler and Spotted Quail-thrush (see Table 1 for Scientific names).

Flowering eucalypts and banksias attract many honeyeaters and lorikeets to the area over winter. The major species involved are Coast Banksia, *Banksia integrifolia,* Spotted Gum, *Eucalyptus maculata,* Woollybutt, *E. longifolia* and Red Ironbark, *E. sideroxylon.* The intensity of flowering of the eucalypts fluctuates from year to year and honeyeater and lorikeet numbers fluctuate accordingly. Similar movement patterns have been reported elsewhere in New South Wales (Keast 1968).

Although eucalypts are the dominant forest vegetation, patches of low rainforest (5-20 m tall) occur in the most sheltered gullies, particularly towards the coast. The dominant tree species in these patches is usually Lilly Pilly, *Acmena smithii.* Bird species that are closely associated with eucalypts, such as the Striated Thornbill, White-throated Treecreeper and Spotted Pardalote, are rare in the rainforest, which is consequently less rich in birds than the adjacent

Table 1. Bird species recorded in eucalypt forest and rainforest near Bega, 1977-1982. The status of each species is the likely status of most adult birds seen within the forest. Double listings indicate that there are two distinct populations of the species. Asterisks indicate evidence of breeding. sa — Silvertop Ash, *Eucalyptus sieberi* forest; wb — Woollybutt, *E. longifolia* forest; mg — Mountain Grey Gum, *E. cypellocarpa* forest; pm — Peppermint, *E. elata/E. smithii* forest. The ages are the ages of regeneration after intensive logging. The figures are the number of survey sites at which the species was recorded (10 sites per habitat). The + signs indicate records outside the survey sites. Birds that were flying over have only been included if they were raptors or aerial feeders.

		Eucalypt forest							
		ridges				gullies			
		mature		9-17	0-4	mature		9-17	
Species	Status	sa	wb	year	year	mg	pm	year	Rainforest
White-faced Heron *Ardea novaehollandiae*	transient					+			
Square-tailed Kite *Lophoictinia isura*	transient				+				
Whistling Eagle *Haliastur sphenurus*	resident	1	+	+	2	2	+	+	+
Australian Goshawk *Accipiter fasciatus*	resident	1	+	1	+	+	1	+	+
Collared Sparrowhawk *A. cirrhocephalus*	resident	+	+	+	+	+	+	+	+
Grey Goshawk *A. novaehollandiae*	resident		+			1	1	1	+
White-breasted Sea Eagle *Haliaeetus leucogaster*	transient		+		+				
Wedge-tailed Eagle *Aquila audax*	*resident	1	2	+	3	3	1	1	2
Australian Little Eagle *Hieraaetus morphnoides*	resident	+	+	+	+	+	1	2	+
Swamp Harrier *Circus aeruginosus*	transient						+		
Peregrine Falcon *Falco peregrinus*	transient		+						
Little Falcon *F. longipennis*	resident		+		+		+		+
Brown Hawk *F. berigora*	transient	+		+	+				
Nankeen Kestrel *F. cenchroides*	transient	+							
Brown Quail *Coturnix australis*	transient	+							
Painted Quail *Turnix varia*	resident		+						
Top-knot Pigeon *Lopholaimus antarcticus*	transient						+		+
Brown Pigeon *Macropygia amboinensis*	resident		+	+		2	+	1	1
Peaceful Dove *Geopelia placida*	transient		+						
Common Bronzewing *Phaps chalcoptera*	transient			+					
Brush Bronzewing *P. elegans*	resident		+	1	+	1		1	
Wonga Pigeon *Leucosarcia melanoleuca*	resident	1	4	2		4	5	2	7
Glossy Black Cockatoo *Calyptorhynchus lathami*	resident	3	1		2	1	+		
Yellow-tailed Black Cockatoo *C. funereus*	resident	+	1	+	+	4	+	2	1
Gang-gang Cockatoo *Callocephalon fimbriatum*	resident	9	2	4	3	7	2	4	
Rainbow Lorikeet *Trichoglossus haematodus*	nomad		+		+		+		
Musk Lorikeet *Glossopsitta concinna*	nomad		+		1		+		
Little Lorikeet *G. pusilla*	nomad		+				+	+	
King Parrot *Alisterus scapularis*	*nomad	3	2	1	2	5	5	6	5
Crimson Rosella *Platycercus elegans*	*resident	8	8	6	5	10	6	7	
Eastern Rosella *P. eximius*	transient	1							
Pallid Cuckoo *Cuculus pallidus*	transient		+						

Table 1. Continued.

		Eucalypt forest							
		ridges				gullies			
		mature		9-17	0-4	mature		9-17	
Species	Status	sa	wb	year	year	mg	pm	year	Rainforest
Brush Cuckoo *C. variolosus*	summer migrant		+	+		1	3	1	+
Fan-tailed Cuckoo *C. pyrrhophanus*	*summer migrant	6	7	5	3	5	3	7	1
Golden Bronze-cuckoo *Chrysococcyx lucidus*	summer migrant	1	3	2	1	3	4	5	
Powerful Owl *Ninox strenua*	resident						+		+
Boobook Owl *N. novaeseelandiae*	resident	+	+		+	+	+		+
Sooty Owl *Tyto tenebricosa*	resident					+		+	+
Tawny Frogmouth *Podargus strigoides*	resident	+	+		+	+	+		
Owlet-nightjar *Aegotheles cristatus*	resident	+	+		+	+	+		
White-throated Nightjar *Caprimulgus mystacalis*	*summer migrant	+	+		+	+	+		
Spine-tailed Swift *Hirundapus caudacutus*	summer migrant	3	2	2	4	1	+	1	1
Azure Kingfisher *Ceyx azurea*	resident					+	+		
Laughing Kookaburra *Dacelo novaeguineae*	*resident	5	4	2	+	1	1	2	
Sacred Kingfisher *Halcyon sancta*	*summer migrant	+	+		2	1	1		
Dollar-bird *Eurystomus orientalis*	transient						+		
Superb Lyrebird *Menura novaehollandiae*	*resident	1	3	4		3	3	5	6
Welcome Swallow *Hirundo neoxena*	resident		1	1	1		1	1	2
Tree-martin *Cecropis nigricans*	summer migrant			1	1				1
Black-faced Cuckoo-shrike *Coracina novaehollandiae*	*summer migrant	7	5	4	1	1	1	1	
Little Cuckoo-shrike *C. papuensis*	*resident	2	2				+		
Cicada-bird *C. tenuirostris*	summer migrant	3	2	+		1	+		
Australian Ground-thrush *Zoothera dauma*	*resident			1		2	1	1	2
Rose Robin *Petroica rosea*	*resident	5	5	8	4	9	7	10	5
Flame Robin *P. phoenicea*	winter migrant	+		1	1	1			
Scarlet Robin *P. multicolor*	*resident	4	2	1	3	1			
Eastern Yellow Robin *Eopsaltria australis*	*resident	9	10	10	8	10	10	9	10
Jacky Winter *Microeca leucophaea*	*resident	7	5		7		1		
Shrike-tit *Falcunculus frontatus*	*resident	4	4	3		4	3	4	1
Olive Whistler *Pachycephala olivacea*	winter migrant					2		+	
Golden Whistler *P. pectoralis*	*resident	10	10	10	8	10	10	10	10
Rufous Whistler *P. rufiventris*	*summer migrant	6	9	2	6	1	3	2	
Grey Shrike-thrush *Colluricincla harmonica*	*resident	10	10	10	10	10	10	10	2
Black-faced Flycatcher *Monarcha melanopsis*	*summer migrant	4	4	6	2	6	6	9	8
Leaden Flycatcher *Myiagra rubecula*	*summer migrant	1	2	5		1	1	1	
Satin Flycatcher *M. cyanoleuca*	summer migrant		+						

Table 1. Continued.

Species	Status	Eucalypt forest ridges mature sa	Eucalypt forest ridges mature wb	Eucalypt forest ridges 9-17 year	Eucalypt forest ridges 0-4 year	Eucalypt forest gullies mature mg	Eucalypt forest gullies mature pm	Eucalypt forest gullies 9-17 year	Rainforest
Restless Flycatcher *M. inquieta*	transient	+	+		+		+		
Rufous Fantail *Rhipidura rufifrons*	*summer migrant	1	3	5	1	5	4	7	8
Grey Fantail *R. fuliginosa*	*summer migrant/ winter migrant	10	10	10	10	10	10	10	8
Eastern Whipbird *Psophodes olivaceus*	*resident	3	+	2	1	8	7	9	10
Spotted Quail-thrush *Cinclosoma punctatum*	resident	1	3	+	+				
Superb Blue Wren *Malurus cyaneus*	*resident	10	9	8	10	8	8	9	5
Variegated Wren *M. lamberti*	A rare resident in another forest-type (Spotted Gum, *E. maculata*).								
Pilotbird *Pycnoptilus floccosus*	resident	+	1	1	1	5	4	6	3
Large-billed Scrubwren *Sericornis magnirostris*	*resident		1			3	+	1	7
Yellow-throated Scrubwren *S. citreogularis*	resident								+
White-browed Scrubwren *S. frontalis*	*resident	9	8	8	9	10	10	10	10
Brown Warbler *Gerygone mouki*	*resident	3	5	6	1	9	8	10	9
White-throated Warbler *G. olivacea*	transient		+						
Brown Thornbill *Acanthiza pusilla*	*resident	10	10	10	10	10	10	10	10
Buff-rumped Thornbill *A. reguloides*	resident	2	4		+	1			
Striated Thornbill *A. lineata*	*resident	10	10	10	10	10	10	10	3
Orange-winged Sittella *Daphoenositta chrysoptera*	*resident	5	7	3	4	3	5	5	
White-throated Treecreeper *Climacteris leucophaea*	*resident	10	10	10	10	10	10	10	1
Red-browed Treecreeper *C. erythrops*	*resident	8	3	2	4	9	4	3	
Red Wattlebird *Anthochaera carunculata*	winter migrant	4	5	3	5	+	1	2	
Little Wattlebird *A. chrysoptera*	transient	1							
Noisy Friarbird *Philemon corniculatus*	nomad	1	1						
Bell Miner *Manorina melanophrys*	*resident	1	2	+	1	1	4	3	7
Lewin Honeyeater *Meliphaga lewinii*	*resident	4	6	7	4	9	9	10	10
Yellow-faced Honeyeater *Lichenostomus chrysops*	*resident/winter migrant	10	10	10	10	10	10	10	6
White-eared Honeyeater *L. leucotis*	winter migrant	6	5	4	4	5	5	4	1
Yellow-tufted Honeyeater *L. melanops*	transient		+				+		
Fuscous Honeyeater *L. fuscus*	winter migrant	+					+		
Brown-headed Honeyeater *Melithreptus brevirostris*	resident	6	3	4	2	7	1	4	
White-naped Honeyeater *M. lunatus*	*resident/winter migrant	7	9	8	7	8	10	8	3
Crescent Honeyeater *Phylidonyris pyrrhoptera*	*resident/winter migrant	2	1	2	2	8	2	7	1
New Holland Honeyeater *P. novaehollandiae*	*resident	3	2	2	5	7	5	6	3
Eastern Spinebill *Acanthorhynchus tenuirostris*	*resident	10	10	10	10	10	10	10	9
Scarlet Honeyeater *Myzomela sanguinolenta*	nomad		+				1		1

Table 1. Continued.

		Eucalypt forest							
		ridges				gullies			
		mature		9-17	0-4	mature		9-17	
Species	Status	sa	wb	year	year	mg	pm	year	Rainforest
Mistletoebird *Dicaeum hirundinaceum*	*nomad	7	5	3	6	10	4	7	
Spotted Pardalote *Pardalotus punctatus*	*resident	10	10	9	9	10	9	10	+
Yellow-tipped Pardalote *P. striatus*	winter migrant	6	5	4	3	4	1	3	
Grey-breasted Silvereye *Zosterops lateralis*	*summer migrant/ winter migrant	8	5	8	8	8	6	7	6
Goldfinch *Carduelis carduelis*	transient				1				
Red-browed Finch *Emblema temporalis*	*resident	4	4	6	10	8	10	7	5
Olive-backed Oriole *Oriolus sagittatus*	*nomad	4	4	1	1	1	4	1	
Spangled Drongo *Dicrurus hottentottus*	transient		+						
Satin Bowerbird *Ptilonorhynchus violaceus*	nomad	2	3	3		5	5	7	8
Magpie Lark *Grallina cyanoleuca*	transient		+						
White-browed Woodswallow *Artamus superciliosus*	transient		+						
Dusky Woodswallow *A. cyanopterus*	*summer migrant	3	2	2	2	+	+		1
Grey Butcherbird *Cracticus torquatus*	resident	2	4	1	2	1	1	3	
Pied Currawong *Strepera graculina*	resident	4	4	3	3	8	1	5	4
Grey Currawong *S. versicolor*	winter migrant	+				+			
Australian Raven *Corvus coronoides*	resident	1	+	1	1	3	+	1	+

eucalypt forest (Table 1). The most characteristic rainforest bird is the Large-billed Scrubwren. Other birds that are common in the rainforest are also common in eucalypt forest (Smith 1984).

THE BIRDLIFE OF REGENERATING FOREST

In the forests near Bega two stages of regeneration after intensive logging were available for study: 0-4 year regeneration in areas logged for woodchips, and 9-17 year regeneration in areas logged in a timber stand improvement (t.s.i.) programme. In the t.s.i. operations all of the merchantable sawlogs were removed, the area was burnt, and then most of the remaining trees were felled or ringbarked to prevent them suppressing regeneration. Some trees were left standing to provide a source of seeds. The object was to promote the growth of a dense stand of vigorous, straight and even-aged eucalypts (E. Dowton, former Forest Foreman, pers. comm.). The t.s.i. operations occurred before the development of the woodchip industry but represent a similar intensity of logging. However, the t.s.i. logged areas are continuous, while the woodchip logged areas are 10-20 ha patches separated by similar-sized patches of mature forest that is destined for logging in the future.

Ten sites were selected in 0-4 year regeneration on ridges, ten in 9-17 year regeneration on ridges, and ten in 9-17 year regeneration in gullies. Comparable gully sites were not available in 0-4 year regeneration because of the pattern of the woodchip logging. For regeneration on ridges, half the sites were in Silvertop Ash, *Eucalyptus sieberi,* forest and half in Woollybutt forest. For regeneration in gullies, half the sites were in Mountain Grey Gum, *E. cypellocarpa,* forest and half in River Peppermint, *E. elata*/Gully Peppermint, *E. smithii,* forest. For comparison, ten sites were selected in mature forest of each of these four forest-types.

Each site was surveyed for birds once each season for a year. Each survey was for an hour and all bird species seen or heard in the habitat were listed. A species list for each site was obtained by combining the records from all four surveys. These lists were then used to analyse the similarity in bird species composition between habitats, weighting presence/absence differences over differences in frequency (Smith 1984).

The bird species composition of the 0-4 year regeneration was the most distinctive of the seven habitats (Fig. 2). Many species were less frequent than in mature forest and few species were found breeding (Table 1). In the 9-17 year regeneration

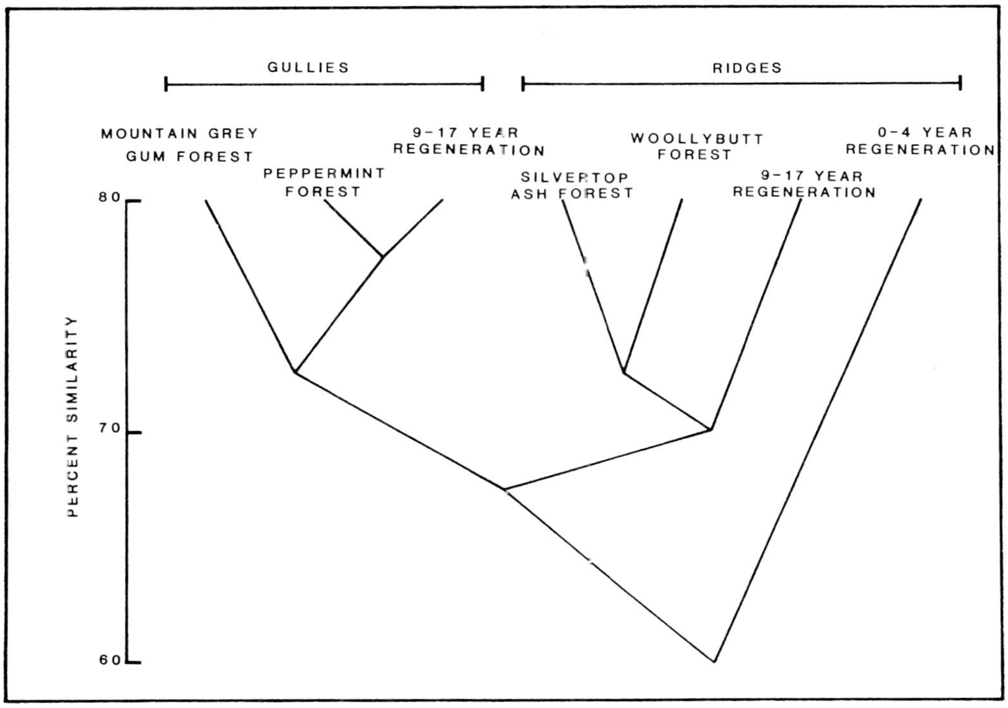

Fig. 2. Similarity in composition between the bird communities of mature forest and logging regeneration near Bega. The level at which the lines from two habitats join is the measure of the similarity of their bird communities (Smith 1984).

the eucalypt saplings had grown sufficiently to form a low forest (6-12 m tall; Fig. 3). The composition of the bird community at this stage of regeneration more closely resembled that of mature forest (Fig. 2).

Similar results have been found for woodchip areas in Tasmania (Pattemore 1980), southern Victoria (Loyn 1980) and near Eden (Recher *et al.* 1980). Each study recorded an immediate reduction in bird numbers and diversity after logging, but after ten years of regeneration the bird community was more like that of mature forest. All three studies also reported some colonization of the early stages of regeneration by birds of open country or heath that were absent from mature forest. After ten years most of these additional species had disappeared. In the United States, bird numbers in the early stages of regeneration after intensive logging commonly exceed the numbers in mature forest because of colonization by non-forest birds (Hagar 1960; Conner and Adkisson 1975; Webb *et al.* 1977; Titterington *et al.* 1979). No such colonization was noted at Bega (Table 1), probably because the logged areas were too small to attract and support non-forest birds. Also, one of the species that colonizes logged areas at Eden, the Jacky Winter (Recher *et al.* 1980), was already common in the mature forest at Bega, where there has been more of a history of intensive logging.

To examine more closely how the bird community of the older regeneration differed from that of mature forest and which species had been most affected by the logging, bird censuses were carried out on 32 circular 1 ha plots (Smith 1985). There were eight plots in each of four habitats: mature Silvertop Ash forest on ridges, Silvertop Ash forest 10-15 years after intensive t.s.i. logging, mature Mountain Grey Gum forest in gullies, and Mountain Grey Gum forest 10-15 years after logging (Fig. 3). Each plot was censused on three mornings each season for a year. Each census was a 30 minute count, from the centre, of all birds seen or heard within the plot. The location of each bird was marked on a map of the plot and the maps were used to test whether the more distant birds were less detectable in the denser than in the more open habitats. For those species for which there was a significant difference in detectability, subsequent analyses were restricted to records within 28 m of the centre of each plot; that is, 0.25 ha plots were used instead of 1 ha plots.

The number of bird species recorded per census in each habitat and season is shown in Figure 4. No statistically significant difference was found between the values for 10-15 year regeneration and mature forest in the gullies, but on the ridges, which supported fewer birds, values for 10-15 year regeneration were significantly less than those for mature forest

Analyses of the data for each species showed that for 17 species in the gullies there were significant differences in their abundance between 10-15 year regeneration and mature forest. Eight species were

Fig. 3. Mature forest and logging regeneration near Bega. A. Mature Silvertop Ash, *Eucalyptus sieberi,* forest on ridges. B. Mature Mountain Grey Gum, *E. cypellocarpa,* forest in gullies. C. Silvertop Ash forest 15 years after intensive logging. D. Mountain Grey Gum forest 13 years after logging.

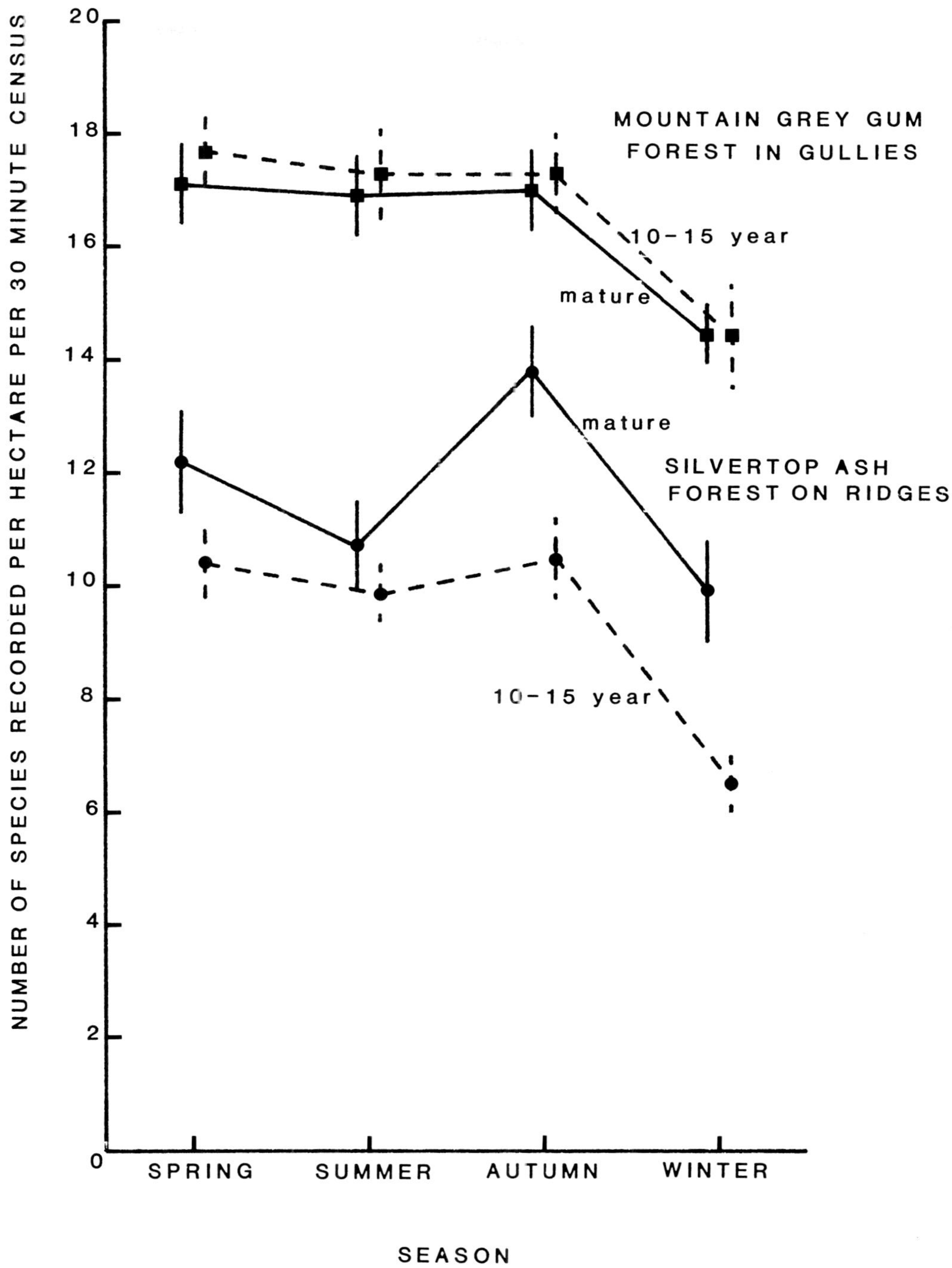

Fig. 4. Number of bird species recorded per census in mature forest and 10-15 year logging regeneration near Bega. Vertical lines are two standard errors long (N = 24).

less abundant in regeneration (mean difference in abundance 60%), while nine species were less abundant in mature forest (mean difference 66%). On the ridges there were significant differences in abundance for 14 species. Twelve were less abundant in regeneration (mean difference 77%), while only two were less abundant in mature forest (mean difference 76%). Basically, 10-15 year regeneration on ridges supported a smaller bird community than mature forest, while 10-15 year regeneration in gullies supported a bird community of similar size but different composition to that in mature forest.

SPECIES AFFECTED BY LOGGING

The species that were more abundant in the 10-15 year regeneration were mostly birds that feed, nest or shelter in dense vegetation: Superb Lyrebird, Eastern Whipbird, Pilotbird, Bell Miner, Lewin's

Honeyeater, New Holland Honeyeater and Grey-breasted Silvereye (Smith 1985). These were generally birds of the gullies and their greater abundance in the regeneration corresponded with the particularly dense and extensive layer of shrubs and saplings in that habitat (Fig. 3).

Differences in the abundance of some species could be related to differences in the abundance of particular food-plants. The Satin Bowerbird was more numerous in regenerating gully forest in winter than in other habitats or at other seasons. This corresponded with the fruiting of the vine *Cissus hypoglauca*, a plant of the gullies that was more common in the regeneration. Similarly, the Red-browed Finch was common in regenerating gully forest in summer, corresponding with the seeding of the grasses *Entolasia stricta* and *Tetrarrhena juncea*. On the ridges, however, the finches were feeding on the seeds of *Casuarina littoralis* trees and were more numerous in mature forest, where *Casuarina* trees were also more numerous.

Bird species that were less abundant in the 10-15 year regeneration were mostly those that feed in mature eucalypts. These species were the Gang-gang Cockatoo, Orange-winged Sittella, Red-browed Treecreeper, White-eared Honeyeater, Brown-headed Honeyeater, Eastern Spinebill, Mistletoe-bird and Spotted Pardalote (Smith 1985). However, not all bird species that feed in eucalypts were as affected by the logging. For example, the Striated Thornbill and White-throated Treecreeper feed mainly in eucalypts but do not require mature trees. Both were common in the regeneration, feeding mainly in the eucalypt saplings.

The White-throated Treecreeper is an insectivorous species that obtains its food from the bark of eucalypts. It was less dependent on mature trees than were three other species of similar foraging habits, the Red-browed Treecreeper, Orange-winged Sittella and Shrike-tit, which were either less abundant in regenerating forest or dependent on the retained trees. These differing requirements may be explained by differences in foraging behaviour (Fig. 5). White-throated Treecreepers obtain most of their food from the surface of rough bark on live trunks and branches. This type of bark is found on eucalypts of all ages, but the characteristic foraging sites of the other three species are associated mostly with older trees. Red-browed Treecreepers often probe behind large, loose flakes and sheets of bark; Orange-winged Sittellas often forage on the dead branches of live eucalypts; and Shrike-tits, which have heavier and stronger bills than the other species, often prise open and probe curled ribbons of bark.

Other species that were less abundant in the 10-15 year regeneration were the Large-billed Scrubwren, Rufous Whistler, Jacky Winter and Scarlet Robin (Smith 1985). The Large-billed Scrubwren was closely associated with rainforest (Table 1) and many of the small patches of rainforest on the plots in the regeneration had been disturbed during the logging of the adjacent eucalypt forest. The other three species are characteristic of woodland and forest with an open understorey. Although scarce in the dense vegetation of the 10-15 year regeneration, all three were relatively common in the open vegetation of the first few years after logging (Table 1).

Intensive logging reduces the number of trees with hollows suitable for nesting. Little is known about the formation of hollows in eucalypts but they are certainly a feature of old trees rather than young trees. Mackowski (1984) estimates that Blackbutt, *Eucalyptus pilularis,* in northeastern NSW does not contain hollows large enough for nesting cockatoos until at least 190 years old. In a forest subject to intensive logging at intervals less frequent than this, most trees will be felled before they can develop suitable hollows, particularly if management practices involve deliberate felling of large old trees to foster regeneration (Cowley 1971; McIlroy 1978; Recher *et al.* 1980).

In the regeneration at Bega many of the retained trees were ones that contained hollows and there was little indication that bird numbers were being limited by a shortage of nest-hollows. Although two hollow-nesting species, the Gang-gang Cockatoo and Red-browed Treecreeper, were rare in the 10-15 year regeneration, another two species, the Crimson Rosella and White-throated Treecreeper, were common and were nesting in the retained trees. Similar results were obtained by Franzreb and Ohmart (1978) in a study of the effects of logging of similar intensity on birds in coniferous forest in Arizona. Birds that foraged on the trunks, branches and foliage of trees were the group most reduced in numbers by logging. The effect on hollow-nesting birds depended largely on their foraging behaviour, since retained live and dead trees provided sufficient hollows for nesting. In the long term, however, a shortage of hollows can be expected in the Bega forests unless steps are taken to maintain this resource. Wombat State Forest in central Victoria has been intensively logged for over a century and trees with hollows are now rare in that forest and their shortage is affecting the numbers and breeding success of hollow-nesting birds (Golding 1979; Manderson 1979).

WILDLIFE MANAGEMENT IN A WOODCHIP FOREST

Traditionally, forest management has aimed at a maximum, sustained production of timber, with other forest values such as wildlife being a minor consideration. At the other extreme, there has been a widely held view among conservationists that

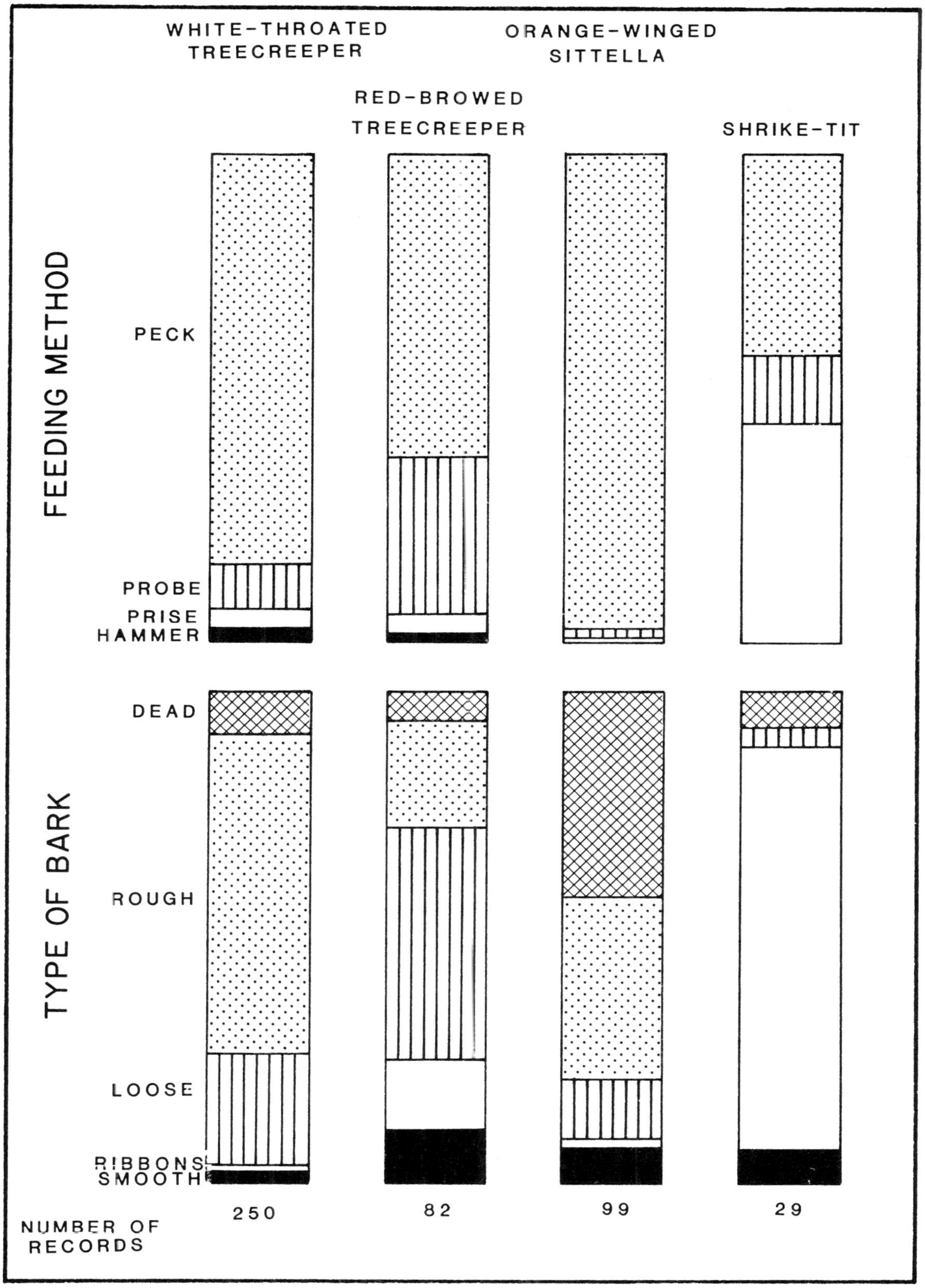

Fig 5. Differences in the foraging behaviour of four bird species that forage on eucalypt bark. Foraging observations were made independently of each other at sites throughout the Bega-Bermagui forests over a two-year period.

wildlife conservation can only be achieved through the reservation of land in national parks and nature reserves. Commercial forestry and wildlife conservation are not mutually exclusive. It is possible to manage forests for both timber and wildlife, but this requires compromise. Management for timber production may result in lower wildlife populations, while management for wildlife may involve a lower yield of timber (Carron *et al.* 1979; Day 1981; Recher and Rohan-Jones 1981).

Regenerating forest at Bega contained fewer birds than mature forest, despite the fact that some species were more common in the regenerating areas (Smith 1984, 1985). Similar results have been reported elsewhere in southeastern Australia (Loyn 1980; Pattemore 1980; Recher *et al.* 1980). The oldest regeneration at Bega was 17 years old but in Gippsland, Victoria, older ages occur and bird populations in even-aged forest 40 to 100 years after logging are also lower than in mature forest of mixed age (Loyn 1980; Loyn *et al.* 1980).

In the Eden woodchip management area the intensity of logging is much greater now than before the woodchip industry started. The total area of mature forest has decreased and the area of regenerating forest has increased. Thus the forests of the region now support fewer birds than before, and will continue to do so while they continue to be intensively logged. However, the impact of woodchip logging, where the native forest is allowed to regenerate, is less severe than the impact of clearing for pine plantations or for agriculture (e.g. Disney and Stokes 1976; Driscoll 1977; Friend 1982; Saunders 1982). Also, because the logging operations are under the control of the Forestry Commission, it is possible to lessen the adverse effects. Two management procedures with this aim are outlined below.

i. Retention of trees at each logging site. Retained trees in the 10-15 year regeneration at Bega were used intensively by birds. The mean density of these trees (live eucalypts over 18 m tall) was 5 per ha on the ridge plots and 12 per ha on the gully plots, compared with densities on plots in mature forest of 61 per ha on the ridges and 43 per ha in the gullies. Of the 40 most frequently recorded bird species in the regeneration, 14 (35%) were feeding mainly in the retained trees (Smith 1985). Several species were using hollows in retained trees as nest sites. Dead trees were also a source of nest-hollows and, in fact, some bird species are known to use hollows in dead trees in preference to those in live trees (Saunders *et al.* 1982). There would have been many fewer birds in the regeneration if no live or dead trees had been left. Although the more trees that are left, the better, retaining even a few trees at each logging site is effective in reducing the adverse effects of logging on birds, particularly where those trees include ones with mistletoes (whose fruits and flowers are important food sources for a number of birds) and ones with hollows. Special attention should be given to retaining trees with hollows, and also those that are likely to develop hollows, so that there is a continuity of supply as the older trees die and fall.

ii. Reduction of logging effects in gullies. Birds were not uniformly distributed through the Bega forests but were concentrated in the gullies (Fig. 4), where moisture and nutrient levels are higher and fires are less frequent. The distribution of birds in relation to ridges and gullies was studied on one 13 ha plot of mature forest, three-quarters of which were dry Woollybutt and Spotted Gum ridge forests and one-quarter Gully Peppermint forest and rainforest in a deep, moist gully. Breeding birds were concentrated in the gully forests. Of 22 species found nesting on the plot, 18 nested in the gully forests, compared with only nine in the ridge forests.

The gullies also act as refuge areas after fires. The study plot was burnt by a low intensity wildfire in October 1981. Eighty percent of the understorey in the ridge forests was heavily burnt, but only 4% of the understorey in the gully forests. Although the fire was a light one, with little damage to the tree layer, there was a marked shift of birds out of the ridge forests into the gully forests in the year after the fire. This pattern was statistically significant in ten species. Not surprisingly, most were birds of the understorey. Before the fire many birds had been caught and individually marked with colour-bands and their territories mapped. Adult White-browed Scrub-wrens, Brown Thornbills and Striated Thornbills that had occupied territories in those parts of the plot that were most heavily burnt were displaced by the fire and their territories were still vacant a year after. Some of these birds disappeared from the plot but others set up new territories among the established pairs in the gully forest.

Not all forest birds live in gullies and a number of those that do are more numerous at some stages of regeneration than in mature forest. Nevertheless, because of the importance of the gullies to breeding birds and to birds seeking refuge after fires, reduction of logging effects in gullies should be given priority over reduction on ridges.

In the Eden woodchip management area both of the above wildlife management procedures have been introduced to some extent. Additional trees are now retained in logging areas as 'wildlife habitat' trees and a number of areas of gully forest have been reserved from logging specifically for wildlife. Wildlife also benefits from similar management procedures used for other purposes, namely the retention of seed trees in logged areas and the exclusion of logging machinery from the vicinity of major watercourses to prevent erosion and siltation (Forestry Commission of NSW 1982; Bridges 1983).

The integration of wildlife conservation with wood production will need, however, to be a continuing and evolving process, involving both further research and the application of research results to forest management.

ACKNOWLEDGEMENTS

This research project formed part of the Five Forests Study of the NSW National Parks and Wildlife Service. I would like to thank all who helped during the work, in particular Dan Lunney, the Team Leader of the Five Forests Study; Brian Cullis and Tony Underwood, for help with statistical analyses; and John Barker, Ian Bevege, Ross Dobbyns, Dan Lunney, Denis Saunders and Jim Shields, for their invaluable comments on the manuscript. Staff of the Forestry Commission of NSW at Bega and Eden provided facilities and assistance and their co-operation is appreciated, particularly Regional Forester Ross Dobbyns and District Forester John Reynolds.

REFERENCES

AUSTRALIAN FORESTRY COUNCIL, 1979. The Australian Pulpwood Story. Tasmanian Forestry Commission: Hobart.

BRIDGES, R. G., 1983. Integrated logging and regeneration in the Silvertop Ash-Stringybark forests of the Eden Region. For. Comm. NSW. Res. Paper No. 2.

CARRON, L. T., SHEPHERD, K. R., FLORENCE, R. G. AND BACHELARD, E. P., 1979. Australia's forests: A perspective on environmental management. *Aust. For.* **42**: 63-109.

CONNER, R. R. AND ADKISSON, C. S., 1975. Effects of clearcutting on the diversity of breeding birds. *J. For.* **73**: 781-85.

COWLEY, R. D., 1971. Birds and forest management. *Aust. For.* **35**: 234-49.

DAY, M. T. (ed.), 1981. Australia's Forests: their role in our future. Australian Academy of Science: Canberra.

DISNEY, H. J. de S. AND STOKES, A., 1976. Birds in pine and native forests. *Emu* **76**: 133-38.

DRISCOLL, P. V., 1977 Comparison of bird counts from pine forests and indigenous vegetation. *Aust. Wildl. Res.* **4**: 281-88.

FORESTRY COMMISSION OF NSW, 1976. Indigenous Forest Policy. Forestry Commission of NSW: Sydney.

FORESTRY COMMISSION OF NSW, 1979. Woodchips from Eden. Forestry Commission of NSW: Sydney.

FORESTRY COMMISSION OF NSW, 1982. Eden Native Forest Management Plan. Forestry Commission of NSW: Sydney.

FORWOOD, 1974. Report of Panel 3 — multiple use of the forest. Forestry and Wood-based Industries Development Conference: Canberra.

FRANZREB, K. E. AND OHMART, R. D., 1978. The effects of timber harvesting on breeding birds in a mixed-coniferous forest. *Condor* **80**: 431-41.

FRIEND, G. R., 1982. Bird populations in exotic pine plantations and indigenous eucalypt forests in Gippsland, Victoria. *Emu* **82**: 80-91.

GOLDING, B. G., 1979. Use of artificial hollows by mammals and birds in the Wombat State Forest, Daylesford, Victoria. M. Env. Sc. Thesis: Monash University, Victoria.

HAGAR, D. C., 1960. The interrelationships of logging, birds and timber regeneration in the Douglas-fir region of northwestern California. *Ecology* **41**: 116-25.

KEAST, A., 1968. Seasonal movements in the Australian honeyeaters (Meliphagidae) and their ecological significance. *Emu* **67**: 159-209.

LOYN, R. H., 1980. Bird populations in a mixed eucalypt forest used for production of wood in Gippsland, Victoria. *Emu* **80**: 145-56.

LOYN, R. H., MACFARLANE, M. A., CHESTERFIELD, E. A. AND HARRIS, J. A., 1980. Forest utilisation and the flora and fauna in Boola Boola State Forest in southeastern Victoria. Forests Commission Victoria Bulletin No. 28.

MACKOWSKI, C. M., 1984. The ontogeny of hollows in Blackbutt (*Eucalyptus pilularis*) and its relevance to the management of forests for possums, gliders and timber. Pp. 553-67 *in* "Possums and Gliders" ed. by A. Smith and I. Hume. Australian Mammal Society, Sydney.

MANDERSON, A. D., 1979. Aspects of forest management for natural hollows in the Wombat State Forest. M. Env. Sc. Thesis: Monash University, Victoria.

MCILROY, J. C., 1978. The effects of forestry practices on wildlife in Australia: a review. *Aust. For.* **41**: 78-94.

PATTEMORE, V., 1980. Effects of the pulpwood industry on wildlife in Tasmania. Part 3: Succession in bird communities and their habitats. Tasmania N.P.W.S. Tech. Rep. 80/1.

RECHER, H. F., ROHAN-JONES, W. AND SMITH, P., 1980. Effects of the Eden woodchip industry on terrestrial vertebrates with recommendations for management. For. Comm. NSW Res. Note No. 42.

RECHER, H. F. AND ROHAN-JONES, W., 1981. Forests and wildlife management: conflict or challenge? *Living Earth*, June **1981**: 11-14.

ROUTLEY, R. AND ROUTLEY, V., 1975. The Fight for the Forests: the takeover of Australian forests for pines, woodchips and intensive forestry. 3rd Ed. A.N.U. Research School of Social Sciences: Canberra.

SAUNDERS, D. A., 1982. The breeding behaviour and biology of the short-billed form of the White-tailed Black Cockatoo, *Calyptorhynchus funereus*. *Ibis* **124**: 422-55.

SAUNDERS, D. A., SMITH, G. T. AND ROWLEY, I., 1982. The availability and dimensions of tree hollows that provide nest sites for cockatoos (Psittaciformes) in Western Australia. *Aust. Wildl. Res.* **9**: 541-56.

SMITH, P., 1984. The forest avifauna near Bega, New South Wales. I. Differences between forest types. *Emu* **84**: 200-10.

SMITH, P., 1985. Effects of intensive logging on birds in eucalypt forest near Bega, New South Wales. *Emu* **85**: 15-21.

TITTERINGTON, R. W., CRAWFORD, H. S. AND BURGASON, B. N., 1979. Songbird responses to commercial clear-cutting in Maine spruce-fir forests. *J. Wildl. Manage.* **43**: 602-09.

WEBB, W. L., BEHREND, D. T. AND LAISORN, B., 1977. Effects of logging on songbird populations in a northern hardwood forest. *Wildl. Monogr.* 55, 35 pp.

Yellow Robin, *Eopsaltria australis*

CHAPTER 23

Bird Populations of a Logged and Unlogged Forest Mosaic at Eden, New South Wales

R. P. Kavanagh[1], J. M. Shields[1], H. F. Recher[2] and W. G. Rohan-Jones[1, 3]

Bird counts were made in alternate logged (clearfelled) and unlogged small coupes in the Eden Forestry District. Populations of some bird species were greatly altered as a result of logging and remained so for the duration of the study (1976-1980). The differences in the avifauna were greatest immediately after logging. There were fewer bird species and individuals in the logged compared to the unlogged coupes. As vegetation on the logged plots regenerated, the number of species and the number of birds increased. Four years after logging 78% of the species present in unlogged forest occurred on the logged plots.

The adverse effects of logging were greatest on birds which require tree hollows for nesting, on species which forage in the forest canopy, and on birds which forage in litter on the ground. In general, non-passerines were affected more seriously than passerines. Species favoured by logging were those normally inhabiting open disturbed situations or requiring dense ground or shrub vegetation. The dense eucalypt regeneration following logging provided additional nesting and foraging sites for some honeyeaters.

Coupe size in an alternate coupe logging operation may be an important factor affecting bird populations. The bird density of some species was lower in the small unlogged coupes than in nearby unlogged forest.

INTRODUCTION

THE Eden Woodchip Management Area includes 297,000 ha of Crown timber land in southeastern New South Wales, approximately 77,000 ha of which has been reserved from logging (Bridges 1983). An additional forest area of about 53,000 ha has been dedicated as National Parks or Nature Reserves in the region and a further 76,000 ha (Wadbilliga National Park) is located on its periphery. Nearly half of the public forest land in the region will not be logged. The remaining forests are managed for the production of sawlogs and hardwood pulp in an intensive operation which often approaches clearfelling (see Smith 1985 for a description of logging operations). Little is known of the effects of this type of forest management on wildlife yet such information is required to ensure that the diversity of forest values is maintained (McIlroy 1978).

The impact of intensive logging is likely to vary according to a number of factors: the original value of the forest as habitat for wildlife; the logging intensity; the duration of the logging cycle; and the proximity of mature forest. The current practice at Eden is to clearfell alternate small coupes creating a mosaic of logged and unlogged forest. (The term 'coupe' describes an area felled in a single operation). During the first logging cycle (20-25 years) half of the mature forest cover will be adjacent to the logged areas providing potential for recolonization. During the second logging cycle the unlogged coupes will be harvested leaving mature forest only in reserves, corridors, and areas unsuitable for logging. Without special provisions to ensure the retention of species dependent on mature forest, it could be expected that the majority of the future forests in the Eden Region will support only fauna typical of a 0-50 year-old forest.

Current management prescriptions now provide for the retention of some mature trees within the coupe as a seed source and others specifically for wildlife

[1]Forestry Commission of New South Wales, Forest Research Division, P.O. Box 100, Beecroft, New South Wales 2119.
[2]Australian Museum, P.O. Box A285, Sydney South 2000.
[3]Present Address: National Parks and Wildlife Service, New South Wales, P.O. Box 22, Parramatta, New South Wales 2150.
Pages 273-81 *in* BIRDS OF EUCALYPT FORESTS AND WOODLANDS: ECOLOGY, CONSERVATION, MANAGEMENT ed. by A. Keast, H. F. Recher, H. Ford and D. Saunders. Royal Australasian Ornithologists Union and Surrey Beatty & Sons.

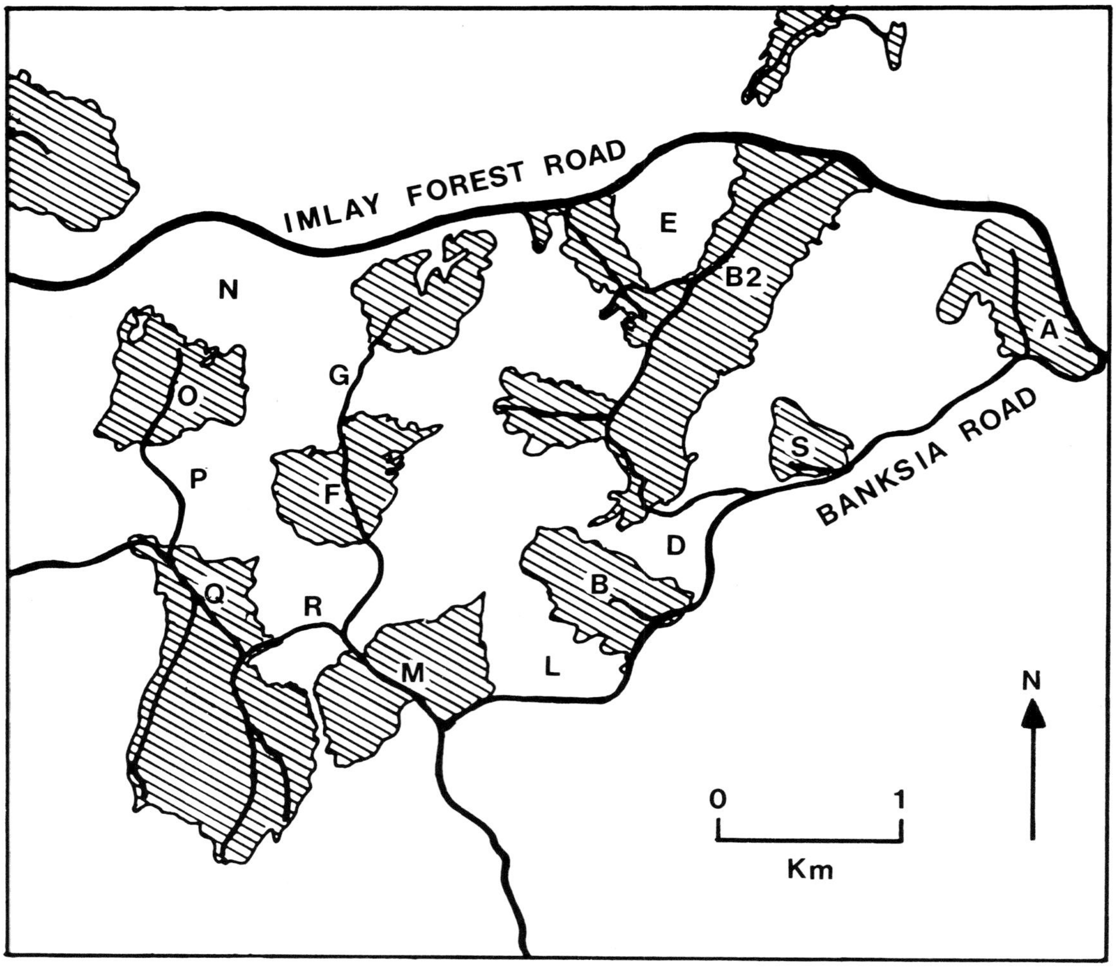

Fig. 1. The pattern of alternate small (~ 15 ha) coupe logging in the Banksia Road study area. Logged coupes are shaded. Minor logging roads follow the ridgelines. Letters indicate the coupes in which bird counts were made.

habitat. Our study reports on operations occurring between 1976 and 1980 when few or no mature trees were retained in logged coupes and coupe size averaged 15 ha (range 5-25 ha). Coupes logged before 1976 were much larger being of the order of 200-800 ha but were reduced in size to take advantage of perceived wildlife, aesthetic and hydrological benefits (Humphreys 1978; Ovington and Thistlethwaite 1976). The aim of this study was to examine the short-term effects of an intensive logging operation on bird populations. The role of coupe size as a management variable influencing the overall impact of logging operations on bird populations is discussed.

METHODS

1. *The Study Areas.* The study area was located in open forest on ridges dominated by Silvertop Ash, *Eucalyptus sieberi,* Blue-leaved Stringybark, *E. agglomerata,* and White Stringybark, *E. globoidea*. The alternate logged and unlogged plots were located in an area bounded by Banksia Road and the Imlay Forest Road in East Boyd State Forest about 30 km south of Eden, New South Wales (Fig. 1). The topography was irregular with slopes averaging about 15° and aspects ranging from northwest through to east. The area was burnt during wildfires in 1952 and 1972 which resulted in the understorey of the unlogged coupes being sparse and dominated by *Acacia longifolia, A. mucronata* and *A. terminalis*. A dense growth of *Goodenia ovata* developed immediately after logging although many bare patches were visible where mineral soil was exposed. Logs, branches and tree stumps made up a large portion of the ground cover. In the last two years of the study the vegetation consisted of a dense layer of shrubs dominated by

Acacia longifolia, A. mucronata, A. terminalis, G. ovata and *Daviesia latifolia* with emergent eucalypt regeneration 3-6 m high. Logging was completed in late 1976 with no mature seed trees or habitat trees being retained. No immediate pre- or post-logging burning was carried out on the study coupes. Bird counts made in 1977 and 1978 on four large (> 100 ha) logged and four large unlogged coupes in a similar forest type nearby were compared with the small coupe logging results.

2. *Census Method.* Birds were counted on one hectare plots in the small logged and unlogged coupes in November of each year from 1976 to 1980. The number of coupes censused varied from three in 1976 to 14 in 1980. All study plots were censused at least twice, a number were censused three times, but only one coupe was censused every year from 1976 (Table 1). All birds seen and heard within 60 m of a point during one hour were counted. Four one hour counts were made on different mornings on each plot by four observers during each sample period. A census consisted of the average number of birds recorded during the four one hour counts. In 1979 and 1980, the birds present during each ten minute period were also scored separately. This allowed analyses of the census results based on the number of ten minute periods in which a species was recorded as well as the total number of birds in each one hour count. Individuals scored in more than one ten minute period were identified on the census sheets but were counted only once for the one hour count. Adjacent logged and unlogged coupes were treated as matched pairs for analysis. Because of the inherent bias in the bird census results favouring the more open logged plots, a 10% level of significance has been used throughout this chapter instead of the more conventional 5% significance level.

RESULTS

1. *Differences in Bird Numbers.* The effect of logging was a significant reduction in the mean number of species ($P < 0.002$, $n = 16$ matched comparisons) and the mean number of birds ($P < 0.002$, $n = 15$) per one hour census, and a significant reduction in the total number of species present ($P < 0.002$, $n = 15$) on the logged compared to the unlogged coupes during 1977-1980 (Wilcoxon's matched-pairs signed ranks test, Siegel 1956) (Table 1). The impact of logging was most severe during the first two years after logging when counts of individual birds and number of species present in logged coupes were about 50% lower than unlogged coupes. Forest regeneration three to four years after logging appeared to provide conditions more suitable to some birds. However, the differences between logged and unlogged coupes were still significant with an average of 20% fewer individuals and 30% fewer species per one hour census recorded in logged compared to unlogged coupes. The total number of species recorded during each complete census period in 1979 and 1980 was reduced by an average of 25% in logged coupes (Table 1).

The comparisons given above should be regarded as a minimum estimate of differences between logged and unlogged coupes due to the census method. The number of birds recorded during a census depends on the number present and their detectability. Since detectability increases in open habitats, it could be expected that birds were more detectable in the logged coupes. In addition, birds recorded during censuses may have been passing from one unlogged coupe to another so that any resulting bias would be in favour of logged coupes.

2. *Differences in Bird Species Composition.* Within any one year, about 20% of the total number of species recorded were found only on logged coupes. Species exclusive to small logged coupes were open country and non-forest birds such as the Welcome Swallow, *Hirundo neoxena*, Australian Pipit, *Anthus novaeseelandiae,* Jacky Winter, *Microeca leucophaea,* Willie Wagtail, *Rhipidura leucophrys,* and Chestnut-tailed Heathwren, *Hylacola pyrrhopygius* (Table 2). One year after logging, 35% of the total number of species recorded were found only in unlogged coupes but this figure declined to 22% after four years. The total number of species common to both logged and unlogged coupes increased from 45% to 60% over years 1 to 4 after logging. When the open country and non-forest bird species are excluded from this analysis, an average of 55% of species recorded in unlogged coupes during the first and second years after logging also occurred on logged coupes. This percentage increased to 66% in the third year and to 78% in the fourth year after logging. Species which were found only on unlogged coupes included Collared Sparrowhawk, *Accipiter cirrhocephalus,* Forest Bronzewing Pigeon, *Phaps chalcoptera,* Wonga Pigeon, *Leucosarcia melanoleuca,* King Parrot, *Alisterus scapularis,* Musk Lorikeet, *Glossopsitta concinna,* Brush Cuckoo, *Cuculus variolosus,* Restless Flycatcher, *Myiagra inquieta,* Leaden Flycatcher, *M. rubecula,* Red-browed Treecreeper, *Climacteris erythrops,* and Striated Pardalote, *Pardalotus striatus* (Table 2).

Most bird species were found to be relatively uncommon on the plots during this study. During the years 1979 and 1980, for which most data are available, about half of the total number of species in the study area were observed in less than 5% of the 10 minute counts. These data reflect the generally poor habitat for birds provided by dry ridge forest compared to moist gully forests in the Eden District (Recher *et al.* 1980). The effect of logging on the density of birds is best illustrated then by changes in the abundance of the most common species.

Table 1. Summary of bird census results on the small coupes 1976-1980. All logging took place in 1976.

Logging treatment	Adjacent matched small coupes	x̄ indivs/ha/hr					x̄ species/ha/hr					Total species/ha/4 hrs				
		76	77	78	79	80	76	77	78	79	80	76	77	78	79	80
Years after logging		0	1	2	3	4	0	1	2	3	4	0	1	2	3	4
Logged	B	1.0	2.0	4.0	9.5	14.0	1.0	1.8	2.5	5.5	4.8	2	7	6	14	12
Unlogged	D	—	14.5	20.0	19.5	20.0	—	6.0	9.5	9.0	9.0	—	10	16	17	14
Logged	B2	12.5	22.0	—	—	22.0	5.3	9.3	—	—	9.5	12	17	—	—	22
Unlogged	E	—	17.0	12.0	—	23.0	—	9.0	7.0	—	11.5	—	17	14	—	25
Logged	F	—	3.0	—	—	12.0	—	2.8	—	—	4.3	—	6	—	—	9
Unlogged	G	—	9.0	16.5	—	16.0	—	4.5	—	8.0	8.5	—	10	15	—	20
Logged	O	—	—	—	8.0	24.0	—	—	—	4.3	9.3	—	—	—	10	22
Unlogged	N	—	—	—	24.3	29.0	—	—	—	11.5	13.8	—	—	—	24	25
Logged	Q	—	—	—	28.5	26.5	—	—	—	10.3	11.0	—	—	—	22	24
Unlogged	P	—	—	—	29.5	27.0	—	—	—	14.8	11.5	—	—	—	29	20
Logged	M	—	—	—	30.3	30.0	—	—	—	11.0	12.0	—	—	—	20	22
Unlogged	R	—	—	—	32.5	30.0	—	—	—	13.3	12.5	—	—	—	27	25
Logged	S	—	—	—	16.5	20.0	—	—	—	6.3	5.5	—	—	—	12	13
Unlogged	L	—	—	—	23.3	24.5	—	—	—	11.5	9.0	—	—	—	25	19
Logged	A	17.0	*4.5	—	—	—	7.0	*3.3	—	—	—	14	*9	—	—	—
Logged Plots x̄		13.6	7.9	4.0	18.6	21.3	4.4	4.3	2.5	7.5	8.0	9.3	9.8	6.0	15.6	17.7
Unlogged Plots x̄		—	13.5	16.2	25.8	24.2	—	6.5	8.2	12.0	10.8	—	12.3	15.0	24.4	21.1

*Plot was burnt and later discontinued.

Species found to be significantly more abundant in unlogged coupes compared to logged coupes were: in 1977, Grey Fantail, *Rhipidura fuliginosus* ($P < 0.05$), Striated Thornbill, *Acanthiza lineata* ($P < 0.1$), and White-throated Treecreeper, *Climacteris leucophaea* ($P < 0.01$); in 1979, Yellow Robin, *Eopsaltria australis* ($P < 0.05$), Rufous Whistler, *Pachycephala rufiventris* ($P < 0.05$), Striated Thornbill ($P < 0.1$), White-throated Treecreeper ($P < 0.002$) and Spotted Pardalote, *Pardalotus punctatus* ($P < 0.1$); in 1980, Fan-tailed Cuckoo, *Cuculus pyrrhophanus* ($P < 0.05$), Black-faced Cuckoo-shrike, *Coracina novaehollandiae* ($P = 0.1$), Scarlet Robin, *Petroica multicolor* ($P = 0.05$), Yellow Robin ($P < 0.05$), Rufous Whistler ($P < 0.002$), Grey Shrike-thrush, *Colluricincla harmonica* ($P < 0.002$), Rufous Fantail, *R. rufifrons* ($P = 0.1$), Striated Thornbill ($P < 0.05$) and White-throated Treecreeper ($P < 0.002$) (Wilcoxon's matched-pairs signed-ranks test, Siegel 1956), based on the data from census replicate counts for each treatment).

Species found to be significantly more abundant in logged coupes compared to unlogged coupes were: in 1977, nil; in 1979, Superb Blue Wren, *Malurus cyaneus* ($P < 0.05$), and White-eared Honeyeater, *Lichenostomus leucotis* ($P = 0.1$); in 1980, Superb Blue Wren ($P < 0.1$), Pilot Bird, *Pycnoptilis floccosus* ($P < 0.01$), Yellow-faced Honeyeater, *L. chrysops* ($P < 0.002$) and White-eared Honeyeater ($P = 0.1$).

More species were found to show significant differences between treatments in 1979 and 1980 compared to 1977. This was because the statistical tests used could be more discriminating due to the increased replication of study plots in 1979 and 1980. However, trends were evident for other species which were not numerous enough to make statistical comparisons. Kookaburra, *Dacelo gigas*, Welcome Swallow, Australian Pipit, Black-faced Cuckoo-shrike, Jacky Winter, Spotted Quail-thrush, *Cinclosoma punctatum,* White-browed Scrub-wren, *Sericornis frontalis,* Olive-backed Oriole, *Oriolus sagittatus,* Dusky Woodswallow, *Artamus cyanopteris* and Pied Currawong, *Strepera graculina,* were relatively common on the recently logged coupes but numbers of these birds declined by the second year after logging. This group includes a few large forest dwelling species which spent some time foraging in cleared areas. In the third and fourth year after logging, with increasing amounts of eucalypt and shrub regeneration, a different suite of birds became prominent on the logged coupes. These species were Superb Blue Wren, Pilot Bird, Yellow-faced Honeyeater, White-eared Honeyeater and Yellow-tufted Honeyeater, *L. melanops,* together with an increasing component of 'forest' species such as Crimson Rosella, *Platycercus elegans,* Yellow Robin, White-throated Treecreeper, Brown-headed Honeyeater, *Melithreptus brevirostris,* White-naped Honeyeater, *M. lunatus,* Eastern Spinebill, *Acanthorhynchus tenuirostris* and Spotted Pardalote,

Table 2. Number of birds per hectare per hour counted near Eden in small (~ 15 ha) and large (> 100 ha) logged and unlogged areas. (In brackets are the number of plots censused for each year. The density of birds is rounded to one decimal place).

Species	Small Coupes										Large Coupes			
	Logged					Unlogged					Unlogged		1-6 year logged	
	(3) 76	(4) 77	(1) 78	(5) 79	(7) 80	(0) 76	(3) 77	(3) 78	(5) 79	(7) 80	(4) 77	(3) 78	(4) 77	(2) 78
Collared Sparrowhawk*							+		0.1	+				
Brown Goshawk*		+								+				
Wedge-tailed Eagle*									+					
Swamp Harrier*														0.1
Brown Falcon*														0.1
Nankeen Kestrel*													0.1	
Forest Bronze-winged Pigeon*										+				
Wonga Pigeon*									+					
Yellow-tailed Black Cockatoo*												0.2		
Gang-gang Cockatoo*				0.1					0.1	0.1	+	+		0.1
Musk Lorikeet*									0.2		+			
Little Lorikeet*													0.3	
King Parrot*					+				0.2	0.2	+	+		
Crimson Rosella*				0.1	0.5			0.6	0.4	0.4	0.1	0.3	0.1	
Brush Cuckoo*							+				+	0.4		0.1
Pallid Cuckoo*													+	0.3
Fan-tailed Cuckoo*	0.3	0.1		0.4	+			+	0.8	0.4		0.3	0.1	
Golden-bronze Cuckoo*				+					0.3	+	0.1	+	0.1	0.1
Spine-tailed Swift*	1.8											+		0.4
Laughing Kookaburra*	0.3	+												
Sacred Kingfisher*				+					+					
Superb Lyrebird*				0.2					0.2	0.3				0.1
Welcome Swallow*	0.9													
Australian Pipit*		0.2												
Black-faced Cuckoo-shrike*	1.3	0.4		0.1	+		+	0.6	0.4	0.3	0.3	+	+	0.1
Cicada Bird*														0.1
Flame Robin*	+	0.1			+		0.3		+			+	+	
Scarlet Robin*		+			+		0.2		0.2	0.4	0.3	+		
Eastern Yellow Robin*				0.3	0.2		0.3	1.0	1.0	0.7	0.4	0.8	+	0.1
Jacky Winter*	0.3	0.4			+								+	0.1
Shrike-tit*											+			
Rufous Whistler*	+			+	+		0.4	1.1	0.5	1.0	1.2	1.8	0.2	1.8
Golden Whistler*				0.4	0.7			+	0.9	0.6	0.4	+	+	0.3
Grey Shrike-thrush*		0.1		0.3	+		+	1.2	0.5	0.5	0.7	+	0.2	0.5
Black-faced Flycatcher*					+			+	0.2	0.2		+		0.5
Restless Flycatcher*									0.1					
Leaden Flycatcher*							+		+	0.1	0.7	0.5	+	
Rufous Fantail*					0.1			+	+	0.4	0.1			
Grey Fantail*		0.1	0.3	1.1	1.3		0.8	2.2	1.3	1.8	1.4	1.8	1.8	1.3
Willie Wagtail*				+										
Spotted Quail-thrush*			1.0					+				0.3		
Eastern Whipbird*				0.4	0.1				0.3		+			
Superb Blue Wren*		0.4		1.5	0.8			0.2	0.5	0.2			1.0	
Pilotbird*				0.4	0.7			+	0.1	0.1				
White-browed Scrubwren*	0.8	1.0	0.3	0.3	0.6		0.3	0.4	0.3	0.6	+	+	1.1	0.1
Chestnut-tailed Heathwren*		0.1											+	
Brown Thornbill*	0.7	0.8	0.5	1.3	1.4		1.2	1.3	2.0	1.5	0.7	0.9	2.4	1.8
Buff-tailed Thornbill*		0.1		0.2					0.4	+	0.1	0.4		
Striated Thornbill*		1.1	0.5	3.6	3.5		4.2	2.8	5.5	6.0	4.4	2.0	0.6	1.0
Brown Warbler*					+					+				
Orange-winged Sittella*	+				+		0.2	+			0.6	0.5	0.1	
White-throated Treecreeper*		+		0.3	0.1		1.3	0.3	2.1	1.1	1.3	0.8	0.2	0.1
Red-browed Treecreeper*									0.2			0.2		
Red Wattlebird*	0.2				+				0.2		0.1		0.2	
Bell Miner*					+					+				
Lewin's Honeyeater*				0.1					0.2				+	
Yellow-faced Honeyeater*	0.4	1.3	1.5	4.5	4.2		1.7	2.2	3.2	2.4	2.1	2.8	1.5	2.4
Yellow-tufted Honeyeater*					0.2									
White-eared Honeyeater*				0.6	0.3				0.2	+				0.1
Brown-headed Honeyeater*					0.4		0.6		0.2	0.2	0.6			
White-naped Honeyeater*		0.2		0.9	1.9		0.2	0.4	0.7	2.3	2.3	1.7		
New-Holland Honeyeater*					+					+		+		

Table 2. Continued.

Species	Small Coupes										Large Coupes			
	Logged					Unlogged					Unlogged		1-6 year logged	
	(3) 76	(4) 77	(1) 78	(5) 79	(7) 80	(0) 76	(3) 77	(3) 78	(5) 79	(7) 80	(4) 77	(3) 78	(4) 77	(2) 78
Crescent Honeyeater*				0.4	+				0.4			0.2	+	
Eastern Spinebill*	+	0.1		0.5	0.6		0.8	0.3	0.4	0.6	1.6	1.1	0.4	
Mistletoebird*				+			0.2		0.1		0.3	0.4		
Spotted Pardalote*	0.2	+		0.3	0.4		0.4	0.3	0.9	0.6	0.4	0.9	0.2	0.1
Striated Pardalote*								0.2	0.3	+	+	0.2		0.1
Silvereye*				0.2	0.2				0.2	0.3			0.1	
Red-browed Finch*		0.1			0.3		+						0.1	0.6
Olive-backed Oriole*	0.9	0.1						+			0.4	+		0.1
Satin Bowerbird*					+				0.2	0.1			+	
White-browed Woodswallow*					1.4									
Dusky Woodswallow*	0.6	0.8		+					+			0.2		
Grey Butcherbird*	+													
Australian Magpie*														0.3
Pied Currawong*	0.8			+	0.1			0.3	0.3	0.3		+	0.1	0.1
Australian Raven*				0.2	+			0.3		+	+	+	0.1	
Total No. of birds/ha/hr	9.5	7.5	4.1	18.6	21.5		13.3	15.7	26.3	24.5	20.6	18.7	11.0	12.9

+ = < 0.1 bird/ha/hr. *See Appendix 1 for Scientific names.

which were previously restricted to the unlogged plots (Table 2).

During 1979 and 1980, data were gathered in ten minute intervals to provide a measure of the time spent on each study plot by birds. We were interested to know whether birds really spent much time on the study plots in addition to simply being recorded as present. These data were also expected to reduce bias caused by difficulties in counting numbers of abundant species. Species recorded frequently and which consistently contributed most to the differences between treatments in Wilcoxon's signed ranks test analyses were considered as species advantaged or disadvantaged by logging. The abundance of the most common or frequently occurring species are given in Table 3 as percentages of the number of ten minute periods in which a species was recorded. These analyses yielded similar results to those presented above for bird counts per ha per hour of censusing.

Data on avian biomass, derived from the mean number of individuals counted per hour in each treatment and from bird weights as obtained from banding schedules and museum records, were also examined to describe differences in the avifauna. Logged plots had between one-half and two-thirds of the total avian biomass of the unlogged plots at three to four years after logging. In 1979, the average bird biomass per ha per hour was 605 g on the small logged coupes compared to 1015 g on unlogged coupes. The figures were 498 g and 875 g, respectively, in 1980. Most of the differences could be attributed to fewer numbers of the largest birds, including all of the non-passerines, on the logged coupes.

3. *Effect of Coupe Size*. In 1977 and 1978, the effect of logging in large coupes (> 100 ha) compared to large unlogged coupes was a reduction in the mean number of species ($P < 0.1$, $n = 6$ matched comparisons) and the mean number of birds per one hour census ($P < 0.1$, $n = 6$) and a reduction in the total number of species present ($P < 0.1$, $n = 6$) (Table 2). Only moderate significance levels were achieved because of small sample sizes but the trends are evident. The pattern of change in species abundance between large logged and large unlogged coupes was similar to that obtained for small coupe logging. Similar species were found to be disadvantaged by logging: Yellow Robin ($P < 0.05$), Rufous Whistler ($P < 0.01$), Leaden Flycatcher ($P < 0.05$), Striated Thornbill ($P < 0.01$), White-throated Treecreeper ($P < 0.002$), White-naped Honeyeater ($P = 0.01$), Eastern Spinebill ($P < 0.05$) and Spotted Pardalote ($P < 0.05$). Species advantaged by logging were Superb Blue Wren ($P < 0.01$), White-browed Scrubwren ($P = 0.05$) and Brown Thornbill, *Acanthiza pusilla* ($P < 0.05$).

No significant differences were found in comparisons between small logged and large logged coupes in terms of their overall number of birds in 1977 ($P > 0.1$). However, in species comparisons, Grey Fantail ($P < 0.002$) and Brown Thornbill ($P < 0.1$) were significantly more abundant in large logged coupes. Black-faced Cuckoo-shrike was more abundant ($P = 0.1$) in small logged coupes.

Similarly, no significant differences were found in comparisons between small unlogged and large unlogged coupes in terms of their overall number of

Table 3. Differences in abundance for selected species. Figures are rounded percentages concerning the number of 10 minute periods in which a species was observed. (1979, n = 120: 1980, n = 168).

	1979 (3rd Year Post-logging)		1980 (4th Year Post-logging)	
Species Advantaged by Logging	Unlogged	Logged	Unlogged	Logged
Yellow-faced Honeyeater*	75	88	59	79
Pilotbird*	3	13	5	20
Superb Blue Wren*	6	17	2	11
White-eared Honeyeater*	3	18	1	7
Species Disadvantaged by Logging	1979 (3rd Year Post-logging)		1980 (4th Year Post-logging)	
Striated Thornbill*	71	41	62	29
White-throated Treecreeper*	53	5	46	3
Rufous Whistler*	13	1	36	5
Spotted Pardalote*	22	6	20	11
Grey Shrike-thrush*	13	6	19	2
Yellow Robin*	23	6	29	6
Scarlet Robin*	7	0	7	1
Fan-tailed Cuckoo*	30	10	11	1
Grey Fantail*	42	35	57	33

*See Appendix 1 for Scientific names.

birds in 1977 ($P > 0.1$). However, Rufous Whistler ($P < 0.1$), Grey Shrike-thrush ($P < 0.01$), Leaden Flycatcher ($P < 0.1$) and White-naped Honeyeater ($P < 0.05$) were significantly more abundant in large unlogged coupes. Brush Cuckoo, Orange-winged Sittella, *Daphoenositta chrysoptera,* Yellow-tailed Black Cockatoo, *Calyptorhynchus funereus,* Shrike-tit, *Falcunculus frontatus,* and Mistletoebird, *Dicaeum hirundinaceum,* may have been disadvantaged by the size of small unlogged coupes (Table 2) but numbers of these species were too low for statistical analysis. No species were more abundant in small unlogged coupes.

Large logged areas contained some open country species absent from the small logged coupes. These birds, which are not typical of forested environments, included the Swamp Harrier, *Circus aeruginosus,* Brown Falcon, *Falco subniger,* Nankeen Kestrel, *Falco cenchroides,* Pallid Cuckoo, *Cuculus pallidus* and Australian Magpie, *Gymnorhina tibicen* (Table 2).

DISCUSSION AND CONCLUSIONS

Clearfelling replaces the structurally complex vegetation of a mature forest with a single layer of regrowth. As expected from the correlation between bird species diversity and foliage height diversity (Recher 1971), the number of bird species present in early logging regeneration is less than in mature forest, although individual species may be more abundant.

Differences in bird species composition between the logged and unlogged forest patches studied, can best be explained in terms of the foraging behaviour of the birds (Recher *et al.* 1985). The species most affected by logging were: those birds which nest in tree hollows and which forage extensively among bark substrates, e.g. White-throated Treecreeper, Grey Shrike-thrush; birds which feed among the foliage of canopy and understorey trees, e.g. Striated Thornbill, Spotted Pardalote, Rufous Whistler, White-naped Honeyeater; birds which catch their insect prey from the air close to the edge of the tree canopy, e.g. Leaden Flycatcher, Grey Fantail, Rufous Fantail; species which forage among the moist layer of litter on the forest floor, e.g. Eastern Yellow Robin, Scarlet Robin; and the non-passerines, e.g. Fan-tailed Cuckoo. Non-passerines were two to four times more abundant on unlogged coupes.

Bird species favoured by logging appeared to fall into two groups. One group appeared immediately after logging, presumably in response to the highly disturbed ground conditions and sparse cover provided by logging debris, only to disappear within two years of logging when these conditions ceased to exist. The second group rose to prominence in three to four year old forest regeneration. This group included species which forage, nest and shelter in a clumped or thick shrubby understorey with an open canopy, e.g. Superb Blue Wren, Pilotbird and species which can nest and forage among the foliage of a 3-6 metre layer of eucalypt regrowth, e.g. Yellow-faced Honeyeater and White-eared Honeyeater. Similar results have been reported from a Victorian study in essentially the same ridge forest type (Loyn 1980).

The proximity of mature forest in this study must have contributed significantly to the rapid rate of recolonization by forest birds following clearfelling. Four years after logging, 78% of bird species recorded in the study area were observed on logged areas. Many of these species, however, were still in low numbers because they required resources (food, nest sites) which were only available in the adjacent unlogged coupes. It is these species and the remaining 20% of the avifauna which are dependent on mature forest that are of most concern to forest managers.

The important question is whether birds which depend on mature forest can fully recolonize logged areas within the 40-50 year span of each complete logging rotation. Only two studies have compared bird populations over a range of age classes following pulpwood logging operations in Australia. Both Loyn (1980) and Pattemore (1980), found that 40 year old forest regeneration was still unsuitable for many species of birds requiring mature forest. The present study is supportive of the work by Loyn (1980) and Pattemore (1980) who found that most of the common and generalist bird species had reached pre-logging densities by 10 years after clearfelling. However, it is not until trees begin to senesce at about 150 years of age (Mackowski 1984), that cavities develop which can be used as nest sites by hole-nesting species. Other species such as Leaden Flycatcher, Orange-winged Sittella and Shrike-tit do not require tree hollows for nesting but forage on substrates normally found on or around larger trees.

No bird species were excluded from the study area as a result of small coupe logging operations. However, the original diversity of birds was clearly maintained by the presence of unlogged forest on adjacent coupes. The size of these coupes was identified as a management variable which could be used to manipulate bird populations. Few differences existed between numbers of birds in large and small logged coupes, but evidence was found to suggest that some species dependent on mature forest were disadvantaged in unlogged coupes by small coupe logging. An increase in coupe size to approximately 50-100 ha would ensure that viable populations of all species were maintained in unlogged coupes for the duration of the first logging rotation (40-50 years). Nevertheless, it is likely that populations of birds dependent on mature forest will decline markedly in the Eden Region after the first logging rotation unless special provisions are made.

One solution is to retain an adequate number of mature trees at the time of logging to ensure that species dependent on mature forest can use logged areas. Mature trees provide specialized foraging substrates and nesting sites compared to young trees. Retained trees also act as stopover points for birds that may otherwise fly over logged areas. Current logging prescriptions (post 1980) in the Eden District specify the retention of five trees per ha for seed purposes and five trees per fifteen ha specifically for wildlife habitat. The effectiveness of this measure for conserving populations of forest dependent species has yet to be tested. Work in similar forest near Bega has shown that many bird species dependent on mature forest were present in clearfelled forest where an average of nine mature trees per ha were retained at the time of logging (Smith 1985). Another study has shown that at least three hollow-bearing trees per ha were required to maintain populations of possums and gliders in Blackbutt, *Eucalyptus pilularis*, forest in north-eastern New South Wales (Mackowski 1984). Whether mature trees should be retained in clumps or scattered evenly throughout logged areas still needs to be determined. Moreover potential habitat trees must be identified and recruited for use in subsequent logging rotations if this strategy is to be successful.

The most important solution to maintaining the diversity of wildlife within an intensive logging operation, given the current state of knowledge, is to ensure that adequate areas of mature forest are retained as reserves and connecting corridors throughout the productive forest. To be effective, such reserves need to be located in areas which are known to be favourable habitats for wildlife. The forest habitats in this study were poor or marginal for many bird species compared to the adjacent gullies and creek flats (Recher *et al.* 1980). Wildlife management should not therefore be concentrated in the dry ridge forests examined in this study but based around a system of "wildlife priority areas" in the gullies, creek flats and other more suitable habitats. A conservation strategy based on the mapping of such areas has recently been developed for the Eden district (Dobbyns and Ryan 1983). In addition research is needed to determine the effects of logging on other bird species which were only poorly represented in the study area. In particular, these studies should be situated in habitats more favourable to birds and where questions regarding the optimal number of trees that should be retained for wildlife and their configuration can be more closely examined.

ACKNOWLEDGEMENTS

Support and encouragement for this research was provided by the Forestry Commission of NSW and the Australian Museum. The involvement of the Australian Museum was supported by a grant from Harris-Daishowa Pty. Ltd. Renée Levy, David Milledge, Martin Shulz and Peter Smith provided field assistance with bird censuses. Lisa Strickland kindly assisted with statistical analyses. Drs. Denis Saunders and Peter Smith provided helpful comments on the manuscript.

REFERENCES

Bridges, R. G., 1983. Integrated logging and regeneration in the Silvertop Ash-Stringybark Forests of the Eden Region. Forestry Commission of New South Wales. Research Paper No. 2.

DOBBYNS, R. AND RYAN, D., 1983. Birds, Glider Possums and Monkey Gums: The Wildlife Reserve System in the Eden District. *Forest and Timber* **19**: 12-15.

HUMPHREYS, N., 1978. Determination of costs and benefits of different coupe sizes in the Eden Working Circle. *Appita* **32**(6): 450-59.

LOYN, R. H., 1980. Bird populations in a mixed eucalypt forest used for production of wood in Gippsland, Victoria. *Emu* **80**: 145-56.

MACKOWSKI, C. M., 1984. The ontogeny of hollows in Blackbutt (*Eucalyptus pilularis*) and its relevance to the management of forests for possums, gliders and timber. Pages 553-67 *in* Possums and Gliders ed. by A. P. Smith and I. D. Hume, Australian Mammal Society, Sydney.

MCILROY, J. C., 1978. The effects of forestry practices on wildlife in Australia: A review. *Aust For.* **41**(2): 78-94.

OVINGTON, J. D. AND THISTLETHWAITE, R. J., 1976. The Woodchip Industry: Environmental effects of cutting and regeneration practices. *Search* 7(9): 383-92.

PATTEMORE, V., 1980. Effects of the pulpwood industry on wildlife in Tasmania. 3. Succession in bird communities and their habitats. Wildlife Division Technical Report 80/1. National Parks and Wildlife Service, Tasmania.

RECHER, H. F., 1971. Bird species diversity: a review of the relationship between species number and environment. *Proc. Ecol. Soc. Aust.* **6**: 135-52.

RECHER, H. F., HOLMES, R. T., SHULZ, M., SHIELDS, J. AND KAVANAGH, R., 1985. Foraging patterns of birds in eucalypt forest and woodland on the tablelands of southeastern Australia. *Aust. J. Ecol.* **10**: (in press).

RECHER, H. F., ROHAN-JONES, W. AND SMITH, P., 1980. Effects of the Eden Woodchip Industry on Terrestrial Vertebrates with Recommendations for Management. Forestry Commission of NSW Research Note 42.

SIEGEL, S., 1956. Non-parametric Statistics for the Behavioural Sciences. McGraw-Hill, Tokyo.

SMITH, P., 1985. Woodchip Logging and Birds near Bega, New South Wales. Pp. 259-71 *in* Birds of Eucalypt Forests: Ecology, Conservation, Management ed. by A. Keast, H. F. Recher, H. Ford and D. Saunders. Royal Australasian Ornithologists Union.

Gang Gang, *Callocephalon fimbriatum*

CHAPTER 24

The Impact of Wildfire on Birds in an Intensively Logged Forest

Harry F. Recher[1], David Allen[1] and Greg Gowing[1]

Fire is a frequent and important ecological event in eucalypt forests and woodlands throughout Australia. Through its effects on the structure and species composition of the vegetation, fire plays a major role in determining the distribution and abundance of forest birds. One year after a wildfire in forests managed for the production of pulpwood (woodchips) near Eden in southeastern New South Wales the numbers of birds and species richness was significantly lower in burnt than in unburnt forest. The difference occurred despite what appeared to be greater numbers of insects in the burnt forests. At Eden the combined effect of fire and logging was greater than either alone and such interactions must be taken into account when developing plans for management of fauna in forests managed principally for wood production.

INTRODUCTION

FIRE is an important ecological event in eucalypt forests and woodlands throughout Australia. The frequency of fire, its intensity and season of occurrence are major factors affecting the structure and species composition of plant communities (Gill 1975; Specht 1981) and through that the distribution and abundance of forest and woodland birds. In order to understand why bird communities differ between types of forest and change through time, the ways in which fires affect birds and the reasons for these effects need to be considered.

In November 1980 wildfire burnt 44,000 ha of forest near Eden in southeastern New South Wales. Much of the area burnt included State Forests managed for the production of hardwood pulp (woodchips) and included areas of regrowth following clear-felling (logging) (see Smith 1985, and Kavanagh *et al.* 1985, for details for logging procedures at Eden). Wildfires have been recurring events in the Eden district with major fires in 1939, 1952, 1964 and 1972 affecting all or parts of the forest burnt in 1980. Reduction in the severity and frequency of wildfires at Eden is therefore a major goal of forest management. To achieve this it will be necessary to use fuel-reduction burns as a pre-or post-logging treatment coupled with regular controlled burning of unlogged forest and logging regeneration (Anon. 1982; Sneeuwjagt 1981, Bridges 1983). The effects of fire on fauna therefore need to be considered in developing plans of management for these forests.

The 1980 fire provided an opportunity to measure the impact of wildfire on forest birds and to compare the effects of fire to those of logging. One of the authors (H.F.R.) noted that after the 1972 Eden fire birds rapidly recolonized burnt forests and were possibly more abundant in fire regeneration during the 1973 breeding season than they had been before the fire. The apparent abundance of birds in 1973 coincided with rapid post-fire revegetation and an outbreak in the numbers of foliage-eating insects. None of these observations was quantified, but they provided a model of post-fire events. By releasing nutrients from the litter and woody shrubs, fire stimulates the growth of vegetation. The vegetation in turn is rich in nutrients and particularly palatable. This leads then to a build-up in insect numbers and through that to the rapid recolonization of burnt forests by birds with an increase in the abundance of the avifauna relative to unburnt forests of a similar type.

The objective after the 1980 fire was to compare the numbers of birds in burnt and unburnt forest and to relate the numbers of birds to the amount of vegetation and the abundance of insects in each category. Logged and unlogged areas were censused to obtain a measure of the combined and separate effects of fire and logging on birds and insects. Bird and insect populations were censused and the vegetation measured one year after the fire during the breeding season (November/December) for forest birds in southeastern New South Wales (Recher *et al.* 1983; Marchant 1981). These results

---

[1]The Australian Museum, P.O. Box A285, Sydney South, New South Wales, Australia 2000.

Pages 283-90 *in* BIRDS OF EUCALYPT FORESTS AND WOODLANDS: ECOLOGY, CONSERVATION, MANAGEMENT ed. by A. Keast, H. F. Recher, H. Ford and D. Saunders. Royal Australasian Ornithologists Union and Surrey Beatty & Sons.

are presented and discussed in this chapter in respect to the response of forest birds to fire and logging and how this affects the management of the forest avifauna.

METHODS

Birds were censused, insect abundance measured and vegetation density estimated on 12 burnt and 12 unburnt plots. Eight were located along creeks in tall open-forest, four were in low open-forest and twelve were in open-forest. The forests at Eden are logged on an alternate coupe system and except in tall open-forest where pulpwood logging does not occur, we used pairs of adjoining logged and unlogged coupes. The area burnt by the fire and the way in which logging has proceeded at Eden prevented the location of equal numbers of plots having the same logging and fire histories in each type of forest. The four plots (i.e. two pairs of adjacent logged and unlogged coupes) in low open-forest were burnt while eight plots in open-forest and four plots in tall open-forest were unburnt and the remainder burnt. The eight plots in tall open-forest were on different creeks or drainages.

An interval count procedure was used to census birds (Pyke and Recher 1984; Recher 1981) between 1 November and 15 December 1981. Twelve points were located on each plot and all birds seen or heard within 60 m of the observer were recorded. Points were approximately 120 m apart and each was censused four times for five minutes. Two of the counts were conducted early in the morning (0600-0800) and two at mid-morning (0830-1030). Two observers were used and each did an equal number of morning and mid-morning counts on each plot. Weather permitting, pairs of plots were censused on four consecutive mornings.

Projected foliage cover was estimated for ground, shrub, subcanopy and canopy layers at each of 10 points on the bird census plots. A sampling procedure was followed in which the percentage cover of each layer was estimated in the four quadrats at each point as established by north-south and east-west compass directions. The 40 values obtained were used to provide an estimate of the average cover values for the coupe. At the same points the 10 nearest canopy and subcanopy trees and the 20 nearest shrubs within a 10 m radius were identified. The ground vegetation was described and the dominant species recorded.

The abundance of insects was measured in two ways. First, the numbers of small flying insects were estimated using 10 cm square, white plastic sticky traps set at 1 m intervals above the ground to a height of 10 m. The traps were located on a rope suspended from a tree branch and left in position for the four days of the bird census. One rope with 10 traps was set centrally on each coupe. Second, the abundance of foliage dwelling insects was estimated from visual inspection of samples of the foliage on each of the dominant species of trees and shrubs on the coupes. Each sample consisted of 40 leaves growing in one area of the plant within 2 m of the ground. One hundred such samples were taken for each species of plant.

The length of all insects scored on the foliage or taken on the sticky traps was measured and the length converted to weight using appropriate regressions of length on weight (Gowing and Recher 1984). In addition to the bird census plots, insects were sampled on two additional unburnt plots in tall open-forest and eight unburnt coupes in open-forest. These plots were sampled primarily to obtain additional information on the effects of logging on insect numbers.

RESULTS

Vegetation

Low open-, open- and tall open-forests differed in plant species composition, height of the canopy, subcanopy and shrub layers and in the projected cover values for each layer. Cover values and possibly the plant species composition of the shrub and ground vegetation were affected by logging and fire.

Low open-forest was dominated by *Eucalyptus consideniana* and *E. globoidea* with *Banksia serrata, Casuarina littoralis* and eucalypts in the subcanopy. All of the low open-forest was burnt and the shrub layer was poorly developed (Table 1). Ground vegetation was dominatd by *Pteridium, Haloragis, Lomandra* and seedlings or root sprouts of *Leptospermum, Eucalyptus, Casuarina* and *Acacia*. In unlogged forest the canopy varied between 12 and 20 m in height. The topography of low open-forest was flat to undulating. Open-forest occurred in more rugged situations with steep ridges and slopes grading into moist gullies or creek flats on which tall open-forest grew. Open-forest was dominated by *E. muellerana, E. sieberi* and *E. cypellocarpa* with *E. agglomerata, E. obliqua, E. globoidea* and *E. consideniana* being abundant on some sites. Eucalypts dominated the subcanopy with *Bursaria spinosa, Daviesia buxifolia, Leucopogon lanceolatus, Eucalyptus* spp. and *Acacia* spp. typical of the shrub layer. *Daviesia, Pteridium* and *Poa* were common ground plants in unburnt and unlogged forest. Canopy heights in open-forest varied between 15 and 35 m. In tall open-forest the canopy varied between 20 and 50 m in height with trees generally larger and more widely spaced than in open-forest. Tall open-forest was dominated by *E. cypellocarpa, E. longifolia, E. muellerana* and *E. elata*. Eucalypts, *Casuarina torulosa* and *Acacia mearnsii* dominated the subcanopy. Shrub layers were complex with ferns, *Pomaderis, Elaeocarpus,*

Table 1. Projected cover values (% cover) ($\bar{x}$ ± s.e.) for vegetation layers in forests affected by logging and fire at Eden, NSW.

Forest type treatment	Vegetation layer			
	Canopy	Subcanopy	Shrub	Ground
1. Unburnt and unlogged				
low open-forest	—	—	—	—
open-forest	33 ± 2	25 ± 2	18 ± 3	24 ± 2
tall open-forest	24 ± 1	40 ± 2	26 ± 2	43 ± 2
2. Unburnt and logged				
low open-forest	—	—	—	—
open-forest	0.1 ± 0.1	5 ± 1	27 ± 3	48 ± 5
tall open-forest	—	—	—	—
3. Burnt and unlogged				
low open-forest	4 ± 1	9 ± 2	13 ± 2	31 ± 5
open-forest	2 ± 1	5 ± 1	11 ± 1	36 ± 2
tall open-forest	10 ± 1	15 ± 1	23 ± 2	54 ± 2
4. Burnt and logged				
low open-forest	5 ± 6	6 ± 1	10 ± 1	38 ± 2
open-forest	0.1 ± 0.1	6 ± 2	18 ± 3	45 ± 3
tall open-forest	—	—	—	—

Table 2. Numbers ($\bar{x}$ ± s.e.) of invertebrates sampled by sticky traps and recorded on foliage in forests at Eden affected by logging and fire. Data from low open- and open-forest have been combined.

Forest type and treatment (N)	Invertebrates		
	Individuals	Taxa (Orders)	Biomass (mgms)
A. *Sticky Traps* (numbers per 10 traps)			
burnt open-forest, logged (4)[a]	145 ± 32	6 ± 1	148 ± 90
burnt open-forest, unlogged (4)	185 ± 43	7 ± 1	463 ± 268
unburnt open-forest, logged (8)	254 ± 27	7 ± 1	270 ± 138
unburnt open-forest, unlogged (8)	159 ± 15	7 ± 1	83 ☆ 22
burnt tall open-forest, unlogged (4)	288 ± 18	6 ± 1	1027 ± 868
unburnt tall open-forest, unlogged (6)	234 ± 37	7 ± 1	175 ± 47
B. *Foliage Counts* (numbers per 4000 leaves)			
burnt open-forest, logged (6)[b]	277 ± 108	8 ± 1	2375 ± 887
burnt open-forest, unlogged (8)	96 ± 5	9 ± 1	578 ± 191
unburnt open-forest, logged (22)	63 ± 8	7 ± 1	76 ± 14
unburnt open-forest, unlogged (17)	63 ± 12	7 ± 1	315 ± 168
burnt tall open-forest, unlogged (13)	349 ± 71	9 ± 1	1853 ± 751
unburnt tall open-forest, unlogged (8)	90 ± 25	8 ± 1	120 ± 30

[a] number of plots sampled with 10 traps per site
[b] total number of tree and/or shrub species sampled on the plots within each treatment category. The number of plots is the same as for the sticky trap samples.

Goodenia, Bedfordia, Acacia and tree ferns dominant elements. The ground vegetation was mainly ferns with *Gahnia, Lomandra, Pimelea oxiflora* and *Goodenia* among the more conspicuous plants.

Cover values in burnt and logged forest differed significantly from unburnt and unlogged forest (Table 1). Logged coupes had a sparse canopy and subcanopy but developed a dense ground and shrub vegetation. Burnt plots were similar in having sparser canopy and subcanopy vegetation than unlogged forest and developing a dense ground vegetation. The shrub layer in burnt forest tended to be less dense than in unburnt plots. Unburnt and unlogged open- and tall open-forest also differed in cover values for the different vegetation layers. Open-forest tended to have a denser canopy than tall open-forest, but less dense subcanopy, shrub and ground vegetation. The relatively sparse canopy in the tall open-forest was probably the result of logging during the 1940s which removed many of the larger trees.

Invertebrates

Analysis of variance was used to test differences between plots with different logging and fire histories in the taxa of invertebrates, number of individuals and total biomass (Table 2). There were no differences between low open- and open-forest plots and the samples from these types of forest have been combined as open-forest in Table 2.

Within forest types, flying insects were more abundant in unlogged tall open-forest than in unlogged open- or low open-forest ($p = 0.01$). For open- and low open-forest, unburnt logged sites had significantly greater numbers of flying insects than other treatments ($p = 0.03$). There were no significant differences in the number of taxa or in the biomass of insects sampled with sticky traps. The differences in biomass shown in Table 2 are largely the result of chance captures of a few large insects (e.g. moths, butterflies).

Greater differences were found in the abundance, taxa and biomass of invertebrates on foliage (Table 2). However, as with the sticky traps, the presence of a few very large insects affected some calculations of biomass. For example, although the number of insects on foliage in logged, burnt open-forest was less than on foliage in unlogged, burnt, tall open-forest, these few large insects resulted in higher average biomass for the open-forest plots. Generally, foliage invertebrates tended to be more abundant, larger and represented by more taxa in burnt unlogged tall-open forest than in any other treatment or forest type. In tall open-forest invertebrates were significantly more abundant on burnt than unburnt plots ($p = 0.01$), but the differences in the number of taxa ($p = 0.06$) and biomass ($p = 0.09$) were not significantly different. In open- and low open-forest burnt plots had significantly more individuals, more taxa and larger invertebrates on foliage than unburnt plots ($p < 0.05$). Logging had no effect on the number of invertebrates or taxa, but burnt logged plots had significantly larger invertebrates than other treatments ($p < 0.05$).

Birds

A total of 80 species of birds were recorded during censuses. Sixty-three species were recorded in tall open-forest, 60 species in open-forest and 48 species in low open-forest. These numbers are comparable with the results obtained by Recher *et al.* (1980) who censused unburnt and unlogged forest in south-eastern New South Wales between 1975 and 1978. They recorded 79 species in tall open-forest, 71 species in open-forest and 61 species in low open-forest (see also Milledge and Recher 1985). The species not recorded in 1981 were uncommon or transient during the earlier censuses (e.g. Little Falcon, *Falco longipennis,* and Scarlet Honeyeater, *Myzomela sanguinolenta*).

There was a significant effect of fire and logging on the numbers of birds. Burnt and logged plots had lower numbers and fewer species of birds than unburnt or unlogged plots. The combined effect of logging and fire was greater than either treatment alone (Table 3). Fire had a greater impact on the numbers of individuals than on the number of species. In unlogged open-forest the average number of individuals recorded per census point was reduced from an average of 5.5 birds per census point in unburnt to 2.1 in burnt forest while the number of species was reduced from 3.7 to 3.2 respectively. In unlogged tall open-forest there was a similar effect with the numbers of individuals reduced from 11.3 per point in unburnt forest to 6.1 in burnt forest with the number of species reduced from 6.2 to 4.4 respectively. On logged plots in open-forest the numbers of birds was reduced from 4.6 individuals per point on unburnt to 1.8 on burnt plots while numbers of species decreased from 2.1 to 1.3 respectively.

The initial impact of logging on birds is a significant reduction in the numbers of individuals and species (Recher *et al.* 1980; Loyn 1980; Smith 1985; Kavanagh *et al.* 1985). However, numbers recover rapidly as revegetation proceeds. The plots censused in 1981 had been logged from three to five years previously and the differences between logged and unlogged areas were less than if the plots had been logged in the preceeding 12 months. In open-forest the number of individuals per point on logged plots was 4.6 and the number of species was 2.1 compared with 5.5 individuals and 3.7 species per point on unlogged plots. Plots in open-forest which had been logged and burnt had only 1.8 individuals and 1.3 species per point. In low open-forest 1.9 individuals and 1.4 species were recorded per point in burnt and logged forest compared with 8.5 individuals and 3.5 species on plots which had been burnt but not logged.

Species responded differently to the effects of fire and logging. In logging regeneration which was three to five years old, ground and shrub foraging species such as Brown Thornbill, *Acanthiza pusilla,* and Yellow-faced Honeyeater, *Meliphaga chrysops,* were more abundant than in unlogged forest (Table 4) (see also Smith 1985; Kavanagh *et al.* 1985). In contrast, canopy species such as Striated Thornbill, *A. lineata,* and bark foragers such as White-throated Treecreeper, *Climacteris leucophaea,* were reduced in abundance (Table 4).

In general, species were less abundant in burnt than unburnt forest. The numbers of Brown Thornbill, Striated Thornbill, White-throated Treecreeper and Grey Fantail, *Rhipidura fuliginosa*, for example, were less in burnt than unburnt habitats (Table 4). However, in some instances species were more abundant on burnt plots than on unburnt plots. For example, in tall open-forest Musk Lorikeet, *Glossopsitta concinna*, Yellow-faced Honeyeater, Yellow-tufted Honeyeater, *M. melanops*, and Spotted Pardalote, *Pardalotus punctatus,* were more abundant in burnt than unburnt forest. Musk Lorikeet and Yellow-faced, Yellow-tufted and White-naped, *Melithreptus lunatus*, honeyeaters were also more abundant in burnt than unburnt, unlogged open-forest (Table 4). All these species feed on nectar or take other carbohydrates (e.g. manna, lerp and honeydew) and may have responded to an increase in carbohydrates

Table 3. The number of individuals and species of birds recorded during the census in forests at Eden affected by logging and fire.

	Fire and logging history			
Forest type	Unburnt and unlogged	Unburnt and logged	Burnt and unlogged	Burnt and logged
A. Low Open-Forest				
No. individuals (x̄/pt.)	—	—	8.5	1.9
No. species (x̄/pt.)	—	—	3.5	1.4
No. species (x̄/census)	—	—	17.6	8.6
Total species	—	—	40	23
B. Open-Forest				
No. individuals (x̄/pt.)	5.5	4.6	2.1	1.8
No. species (x̄/pt.)	3.7	2.1	3.2	1.3
No. species (x̄/census)	14.3	12.5	14.0	7.4
Total species	42	39	33	17
C. Tall Open-Forest				
No. individuals (x̄/pt.)	11.3	—	6.1	—
No. species (x̄/pt.)	6.2	—	4.4	—
No. species (x̄/census)	26.1	—	19.1	—
Total species	55		59	

Table 4. Abundance of 12 common species of birds in forests at Eden affected by logging and the 1980 wildlife. Numbers are the average number of individuals recorded per point × 100 (± s.e.).

	Fire and logging history			
Species/type of forest	Unburnt and unlogged	Unburnt and logged	Burnt and unlogged	Burnt and logged
Musk Lorikeet				
Open-Forest	1 ± 1	2 ± 2	58 ± 11	2 ± 1
Tall Open-Forest	45 ± 8	—	76 ± 11	—
White-throated Treecreeper				
Open-Forest	27 ± 5	2 ± 1	29 ± 6	6 ± 4
Tall Open-Forest	41 ± 6	—	20 ± 5	—
Grey Fantail				
Open-Forest	40 ± 4	46 ± 4	24 ± 4	8 ± 2
Tall Open-Forest	61 ± 6	—	44 ± 7	—
Brown Thornbill				
Open-Forest	33 ± 5	50 ± 6	7 ± 2	12 ± 3
Tall Open-Forest	52 ± 6	—	30 ± 5	—
Striated Thornbill				
Open-Forest	100 ± 14	20 ± 6	8 ± 4	5 ± 2
Tall Open-Forest	45 ± 11	—	17 ± 5	—
Spotted Pardalote				
Open-Forest	38 ± 5	3 ± 1	36 ± 6	21 ± 4
Tall Open-Forest	17 ± 3	—	24 ± 4	—
Bell Miner				
Open-Forest	0	0	0	0
Tall Open-Forest	266 ± 41	—	1 ± 1	—
Yellow-tufted Honeyeater				
Open-Forest	0	0	32 ± 9	10 ± 4
Tall Open-Forest	8 ± 3	—	47 ± 7	—
Yellow-faced Honeyeater				
Open-Forest	75 ± 8	135 ± 9	116 ± 11	48 ± 7
Tall Open-Forest	45 ± 6	—	73 ± 7	—
Rufous Whistler				
Open-Forest	11 ± 2	5 ± 2	30 ± 4	2 ± 1
Tall Open-Forest	47 ± 4	—	39 ± 5	—
Golden Whistler				
Open-Forest	25 ± 4	26 ± 4	2 ± 1	0
Tall Open-Forest	45 ± 6	—	16 ± 3	—
White-naped Honeyeater				
Open-Forest	56 ± 7	23 ± 5	120 ± 19	3 ± 2
Tall Open-Forest	107 ± 13	—	76 ± 9	—

such as lerp and manna on the burnt plots. If carbohydrates were more abundant in burnt than unburnt forest, it was not detected in surveys of foliage below 2 m. Any increase of carbohydrate abundance on burnt plots may have been an effect of fire promoting insect populations or may be related to the absence of Bell Miners, *Manorina melanophrys*, on the burnt plots (Table 4) (see Loyn *et al.* 1983, Loyn 1985 and Wykes 1985 for a discussion of the effects of Bell Miners on the populations of forest birds). The insectivorous Rufous and Golden Whistlers, *Pachycephala rufiventris* and *P. pectoralis*, were possibly more affected by the changed structure of the forest after burning than by changes in insect abundance. The shrub-foraging Golden Whistler decreased in abundance on burnt plots while the Rufous Whistler, which favours open habitats, increased in abundance in the burnt open forest, but was less abundant in burnt tall open-forest relative to unburnt plots (Table 4).

DISCUSSION

Recher and Christensen (1981) suggested that the response of birds to fire was predictable. In their model the numbers of birds decreased initially after fire, but then increased rapidly as re-vegetation proceeded. Observations by one of the authors (H.F.R.) after a wildfire in southeastern New South Wales and by Christensen and Kimber (1975) following prescribed burns in Jarrah, *Eucalyptus marginata,* forest in Western Australia seemed to indicate that the numbers and species of birds in burnt forest were greater the first year ater burning than in comparable unburnt habitats. Wooller and Brooker (1980) also recorded more birds a year after a prescribed burn in a Karri, *E. diversicolor,* forest in Western Australia, but considered the difference in total numbers was insignificant. In both the Jarrah and Karri forest some species decreased in abundance after the fires while others increased.

Smith (1985) compared the abundance and distribution of birds on a 13 ha plot in mixed open- and tall open-forest near Bega in southeastern New South Wales a year before and a year after a low intensity wildfire. Twelve species were less abundant the year after the fire and none were more abundant. The effect of the fire was greatest on species which fed or sheltered in dense ground and shrub vegetation. These were the layers most affected by the fire. Cowley (1974), Christensen and Kimber (1975) and Wooller and Brooker (1980) also found that the greatest impact of fire on birds was on species which normally frequented ground and shrub vegetation.

The fire studied at Eden was more intense than those studied to date with the canopy and subcanopy as well as the ground and shrub layers being burnt. This resulted in a reduction in the numbers of most species of birds. However, some nectar-feeding birds (lorikeets and honeyeaters) were more abundant in burnt than unburnt forest (Table 4). Smith (1985) also recorded an increase in the abundance of some nectarivorous species a year after fire burnt his plot at Bega, but attributed this to the blossoming of eucalypts and not to the effects of the fire. The differences in the abundance of nectar-feeders between burnt and unburnt forest at Eden was probably the result of a combination of factors including possible differences in the abundance of nectar-rich flowers, availability of other carbohydrates and the presence or absence of Bell Miners.

Considering the frequency with which fires occur in eucalypt forests and woodlands and their dramatic impact on the landscape, there is remarkably little information about the effects of fire on fauna or the long-term consequences of burning on forest ecosystems. The species of bird affected by any given burn varies according to the intensity of the fire and amount and layers of vegetation burnt. It is probable that changes in the abundance of birds after a fire occur in response to changes in the structure of the vegetation. In most instances the impact will be greatest on birds which forage or nest in ground and shrub vegetation while those favouring more open habitats may benefit (Christensen and Kimber 1975; Wooller and Brooker 1980). How long these effects last after a fire has not been determined, but recolonization and an increase in the abundance of birds probably proceeds in parallel with the rate of post-fire re-vegetation.

The model tested here proposed that the release of nutrients by burning stimulated the growth of nutrient-rich vegetation and resulted in an increase in the abundance of insects and other invertebrates. The increased abundance of food would then lead to an increase in the abundance of insect-eating birds. There is evidence that small, ground-dwelling mammals (e.g. *Rattus fuscipes, Antechinus stuartii*) increase in numbers relative to pre-fire densities from two to six years after wildfire (Recher *et al.* 1974; Newsome *et al.* 1975; Recher and Christensen 1981). Presumably the increased abundance of small mammals is related to an increased food supply, but similar changes have yet to be convincingly demonstrated for forest birds. Cowley (1974) and Christensen and Kimber (1975) reported an influx of ground-foraging birds onto burnt habitats and we observed Diamond Dove, *Geopelia cuneata*, Common Bronzewing, *Phaps chalcoptera*, Forest Bronzewing, *P. histrionica,* Brown Quail, *Coturnix australis,* and Painted Quail, *Turnix varia,* congregating in burnt forest where the fire had stimulated the release of seed from wattles, banksias and hakeas. The same species along with Red-browed Finch, *Emblema temporalis,* and Beautiful

Firetail, *E. bella,* also responded to the growth of grasses after the 1980 fire and congregated where these plants were seeding. However, in each instance the birds moved into the burnt forest from outside and unlike the small mammals, breeding was not demonstrated.

Although insects were found to be more abundant on the foliage of post-fire vegetation, no increase in the numbers of insect-eating birds was recorded. Birds were less abundant in burnt than unburnt forest. The best explanation for the apparent failure of birds to respond to an increased food supply is the much lower density of vegetation on the burnt plots relative to the unburnt sites. The difference in the density of foliage may have affected birds in two ways. Firstly there was inadequate cover and birds which frequented post-fire vegetation may have been exposed to an increased risk of predation. Secondly, although the numbers of insects per leaf were greater on post-fire vegetation, the smaller amount of foliage on the burnt plots may have meant that the total amount of food available to birds was actually less than on unburnt sites. The converse situation may have occurred in unburnt logged and unlogged open-forest. Both treatments had similar numbers of invertebrates per leaf and although the biomass of invertebrates was greater in the unburnt forest this was the result of relatively small numbers of very large insects and spiders. As the density of foliage in the ground and shrub layers on the logged sites was twice that in unlogged forest, the total amount of food available to birds foraging in these layers was probably greater on the logged plots. The dense foliage also provided more cover. Birds which frequented ground and shrub vegetation were more abundant on the unburnt logged plots than the unlogged coupes (e.g. Brown Thornbill, Table 4).

The lack of precise information on the long-term effects of fire on forest birds or the reasons for changes in abundance between burnt and unburnt habitats identifies an important gap in our knowledge of forest avifaunas. In order to understand why bird communities differ between types of forest and change through time, the ways in which fires affect birds and the reasons for these effects need to be considered. To obtain this information it will be necessary to study fires of different intensities and to monitor the response of the avifauna for a number of years after a burn occurs.

The results from Eden have shown that an intense wildfire significantly reduces the abundance of birds and that recolonization of burnt forest probably occurs over a number of years. They have also shown that the combined effect of logging and fire is greater than either treatment alone. The effects of both can be related to the ways in which they change the structure of the forest.

The control of wildfires by the use of pre- or post-logging burns and the use of prescription burning to reduce fuel loads in unlogged forest and logging regeneration will have its greatest effect on ground and shrub foraging species. As most of these species appear to benefit from the effects of logging, the immediate consequences of prescription burning in the Eden forests may not be significant. On logged areas it will probably delay the colonization of burnt coupes by shrub birds rather than prevent it. In unlogged forest there will probably be a decrease in the abundance of such species as Brown Thornbill and an increase of such species as Buff-rumped Thornbills, *Acanthiza reguloides,* which forage more on the ground than in shrubs (see Recher and Holmes 1985; and Bell 1985 for a discussion of thornbill foraging behaviour). What can not be predicted with certainty are the long-term consequences of repetitive burns on the various components of forest ecosystems including birds. The rich avifauna which persisted in tall open-forest despite burning emphasizes the importance of these habitats to forest birds and their value as a source for recolonizing forest affected by logging or fire. In the absence of information on the long-term effects of fire or on the rate of recolonization of burnt forest, fire should be excluded from tall open-forest along creeks and in gullies within the woodchip management area. Additionally where unlogged forest is burnt as part of a fuel reduction programme or where it forms part of a fire break, patches of unburnt ground and shrub vegetation must be retained to provide cover and nesting sites for those species affected by low intensity burns.

ACKNOWLEDGEMENTS

Sandy Gilmore and David Milledge conducted the bird censuses and assisted with measuring the vegetation. The Forestry Commission of New South Wales kindly allowed us to work in the State Forests at Eden and local foresters were generous with their support. We greatly appreciate the time they made available to us and their assistance in locating suitable plots. The research at Eden was funded by a grant to the Australian Museum from Harris-Daishowa Pty. Ltd.

REFERENCES

ANON, 1982. Fire management policy — Eden Region. For. Comm. NSW Internal Report. Sydney.

BELL, H., 1985. The social organization and foraging behaviour of three syntopic thornbills *Acanthiza* spp. Pp 151-63 *in* Birds of Eucalypt Forests and Woodlands: Ecology, Conservation, Management ed. by A. Keast, H. F. Recher, H. Ford and D. Saunders. Royal Australasian Ornithologists Union.

BRIDGES, R. G., 1983. Integrated logging and regeneration in the Silvertop Ash-Stringy-bark forests of the Eden region. Research Paper No. 2, Forestry Commission of NSW, Sydney.

CHRISTENSEN, P. AND KIMBER, P., 1975. Effect of prescribed burning on the flora and fauna of south-west Australian Forest. *Proc. Ecol. Soc. Aust.* 9: 85-107.

COWLEY, R. D., 1974. Effects of prescribed burning on birds in mixed species forests of west central Victoria. Third Fire Ecology Symposium, Forests Commission, Victoria. 58-65.

GILL, M. A., 1975. Fire and the Australian Flora. *Aust. Forestry* **38**: 4-25.

GOWING, G. AND RECHER, H. F., 1984. Length-weight relationships for invertebrates from forest in southeastern New South Wales. *Aust. J. Ecol.* **9**: 5-8.

KAVANAGH, R. P., SHIELDS, J. M., RECHER, H. F. AND ROHAN-JONES, W. G., 1985. Bird populations cf a logged and unlogged forest mosaic in the Eden Woodchip Area. Pp 273-81 *in* Birds of Eucalypt Forests and Woodlands: Ecology, Conservation, Management ed. by A. Keast, H. F. Recher, H. Ford and D. Saunders. Royal Australasian Ornithologists Union.

LOYN, R. H., 1980. Bird populations in a mixed eucalypt forest used for production of wood in Gippsland, Victoria. *Emu* **80**: 145-56.

LOYN, R. H., 1985. Ecology, distribution and density of birds in Victorian Forests. Pp 33-46 *in* Birds of Eucalypt Forests and Woodlands: Ecology, Conservation, Management ed. by A. Keast, H. F. Recher, H. Ford and D. Saunders. Royal Australasian Ornithologists Union.

LOYN, R. H., RUNNALLS, R. G., FORWARD, G. Y. AND TYERS, J., 1983. Territorial Bell Miners and other birds affecting populations of insect prey. *Science* **221**: 1411-13.

MARCHANT, S., 1981. The breeding season at Moruya, New South Wales. *Corella* **5**: 19-25.

MILLEDGE, D. R. AND RECHER, H. F., 1985. A comparison of forest bird communities on the New South Wales, South and mid- North Coasts. Pp 47-52 *in* Birds of Eucalypt Forests and Woodlands: Ecology, Conservation, Management ed. by A. Keast, H. F. Recher, H. Ford and D. Saunders. Royal Australasian Ornithologists Union.

NEWSOME, A. E. MCILROY, J. AND CATLING, P., 1975. The effects of an extensive wildfire on populations of twenty ground vertebrates in south-east Australia. *Proc. Ecol. Soc. Aust. Vol.* **9**: 107-23.

PYKE, G. H. AND RECHER, H. F., 1984. Censusing Australian birds: A summary of procedures and a scheme for standardization of data presentation and storage. Pp. 55-63 *in* Methods of censusing birds in Australia ed. by S. J. J. F. Davies. R.A.O.U. Report No. 7.

RECHER, H. F., 1981. Report of working group on the need of standardised census methods. Studies in Avian Biology No. **6**: 580-81.

RECHER, H. F. AND CHRISTENSEN, P., 1981. Fire and the evolution of the Australian biota. Pp. 137-162 *in* Ecological Biogeography of Australia ed. by A. Keast. Dr. W. Junk, The Hague.

RECHER, H. F., GOWING, G., KAVANAGH, R. P., SHIELDS, J. M. AND ROHAN-JONES, W. G., 1983. Birds, resources and time in a tablelands forest. *Proc. Ecol. Soc. Aust.* **12**: 101-23.

RECHER, H. F. AND HOLMES, R. T., 1985. Foraging ecology and seasonal patterns of abundance in a forest avifauna. Pp. 79-96 *in* Birds of Eucalypt Forests and Woodlands: Ecology, Conservation, Management ed. by A. Keast, H. F. Recher, H. Ford and D. Saunders. Royal Australasian Ornithologists Union.

RECHER, H. F., LUNNEY, D., AND POSAMENTIER, H., 1974. Effects of wildfire on small mammals at Nadgee Nature Reserve, NSW. Proc. Third Fire Ecology Symposium, Forests Commission Victoria.

RECHER, H. F., ROHAN-JONES, W. G. AND SMITH, P., 1980. Effects of integrated logging on wildlife with recommendations for management. Forestry Commission of NSW Research Note No. 42, For. Comm. (NSW), Sydney.

SNEEUWJAGT, R. J., 1981. The Eden Fire and its implication to fire protection. I.F.A. Newsletter **22**(3).

SMITH, P., 1985. Effects of Fire on birds in eucalypt Forest. *Emu* **85**: (in press).

SMITH, P., 1985. The effects of woodchip logging on birds at Bega, New South Wales. Pp. 259-71 *in* Birds of Eucalypt Forests and Woodlands: Ecology, Conservation, Management ed. by A. Keast, H. F. Recher, H. Ford and D. Saunders. Royal Australasian Ornithologists Union.

SPECHT, R. L., 1981. Ecophysiological principles determining the biogeography of major vegetation formations in Australia. Pp. 301-331 *in* Ecological Biogeography of Australia ed. by A. Keast. Dr. W. Junk, The Hague.

WOOLLER, R. D. AND BROOKER , K. S., 1980. The effects of controlled burning on some birds of the understorey in Karri Forest. *Emu* **80**: 165-66.

WYKES, B., 1985. The Helmeted Honeyeater and related honeyeaters of Victorian woodlands. Pp. 205-17 *in* Birds of Eucalypt Forests and Woodlands: Ecology, Conservation, Management, ed. by A. Keast, H. F. Recher, H. Ford and D. Saunders. Royal Australasian Ornithologists Union.

CHAPTER 25

Birds and Fire in Southwestern Forests

P. E. S. Christensen[1], G. Wardell-Johnson[1] and P. Kimber[2]

Number of individuals, species, and diversity of birds were increased in spring, six months after a fire burnt a study site in open Jarrah forest east of Manjimup, W.A. Numbers of species and individuals remained greater for the area burnt than in a control for three years after the fire. The effect of the fire on individual species was in general inversely proportional to their main foraging height in the foliage profile. Transects through upland sites in Jarrah forest of different fire history support this conclusion and indicate that the effect of a fire on bird populations is largely dependent on the intensity of the fire. Other studies of birds and fire in Australia are reviewed in this light. Inadequate knowledge of fire regimes makes long term effects on bird communities more difficult to predict.

INTRODUCTION

FIRE is an integral part of the forest environment in Australia (Mount 1969; Gill 1975; Recher and Christensen 1981) and is considered an important aspect of forest management (Shea *et al.* 1981; Underwood and Christensen 1981). However, it is only relatively recently that prescribed fire has been widely used in Australian forests as a management tool. Its use has been almost entirely confined to operations to reduce hazards and to obtain regeneration. Technology and methodology of prescribed burning has improved markedly over the past 20 years. Hence a far greater range of options in the use of fire for management is now available. Nevertheless the introduction of broad scale prescription burning has preceded a complete evaluation of the effects of fire in the forest environment.

Vegetation floristics and structure are changed at least temporarily through the effects of a single fire event, although these changes will depend on the season and intensity of the fire (Noble and Slatyer 1977; Posamentier *et al.* 1981; Recher and Christensen 1981). These changes will affect the bird community since the numbers and kinds of birds in a forest environment are affected primarily by the vegetation structure (MacArthur and MacArthur 1961; MacArthur *et al.* 1962; Recher 1971). The effects of fire on birds in Australia and in southwestern Australia in particular, however, is not extensively researched.

In this chapter we present our findings from two studies of fire and birds in the dry sclerophyll forest of southwestern Australia where Jarrah, *Eucalyptus marginata* Sm., is the dominant tree species. We consider three groups of resident species occupying different foraging heights in the vegetation profile when presenting our data. The results from these studies are collated and reviewed with other published data on the effects of fire on birds in Australian forests with the intention of highlighting certain broad ecological principles which emerge from the data.

AREAS AND METHODS

The two studies were carried out in open Jarrah forest in the Perup 40 km east of Manjimup, Western Australia, 34° 10′S, 116° 40′E (Fig. 1). The topography of the area consists of broad valleys and low ridges.

Study 1

The main study was carried out in two sites of 4.5 ha in broad valleys in the Perup. The forest on the two sites had an understorey ranging from low xeric shrubs to thickets of Heartleaf Poison, *Gastrolobium bilobum* R.Br. Both areas were burnt under cool conditions in spring, nine years prior to the commencement of the study. The Heartleaf thickets in both blocks had not burnt in this fire and were assessed through ring counting to have originated from a wildfire in 1951.

[1]W.A. Forests Department, Brain Street, Manjimup, Western Australia 6258.
[2]W.A. Forests Department, 50 Hayman Road, Como, Western Australia 6152
Pages 291-99 *in* BIRDS OF EUCALYPT FORESTS AND WOODLANDS: ECOLOGY, CONSERVATION, MANAGEMENT ed. by A. Keast, H. F. Recher, H. Ford and D. Saunders. Royal Australasian Ornithologists Union and Surrey Beatty & Sons.

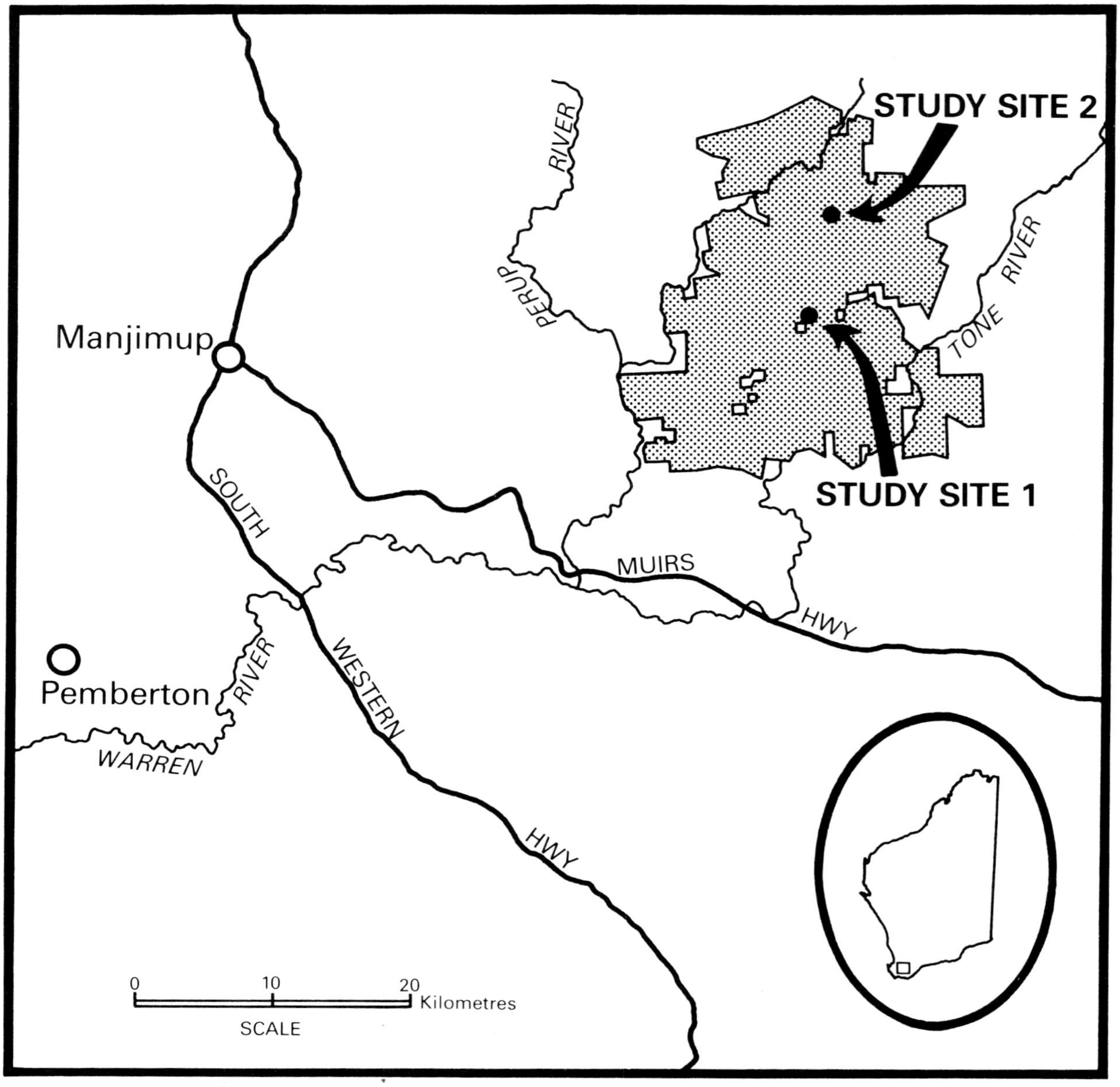

Fig. 1. Map of the southwest of Western Australia showing study sites in relation to major towns in the area.

Bird counts were made along three lines in each of the two study sites. Six assessment points were located at 50 m intervals along each transect giving a total of 18 points in each area. At each count, an observer spent 10 minutes at each point recording the location of observed birds on a map of the site. Counts were made in the morning in fine, still conditions five times each spring for five years, commencing in 1979.

An area of 900 ha including one of the study sites was prescribed burnt under dry conditions in autumn 1981 after two years of censusing. This study site was 250 m from the edge of unburnt forest. The intensity of the fire in the study site ranged up to 10,000 kWm^{-1} but was generally between 1000 and 5000 kWm^{-1}. Complete scorching and pockets of defoliation occurred in some areas of the study site and no unburnt patches remained. The unburnt control site was approximately 1.5 km from the edge of the burn. Bird counts were continued at both sites for three years after the fire.

Study 2

Bird counts using transects were carried out in three upland areas of Jarrah forest with different fire history in the Perup during February and March 1983. One area had not been burnt for 14 years. A second area had been subject to a mild fire in spring six months prior to the study. The third area had been burnt in a severe wildfire one year previously.

Overstorey vegetation and the position of the site in the landscape appeared to be very similar between the three areas. The area burnt under mild conditions displayed the greatest variety of understorey vegetation as approximately half of the study area remained unburnt following the controlled fire. The area burnt by a wildfire had an open ground layer at the time of the census. However, considerable sprouting of epicormic shoots from the trees and saplings formed a dense layer of vegetation in the lower to upper canopy.

Birds were counted along point transects after Reynolds *et al.* (1980) and Recher and Pyke (pers. comm.). The three study areas were each traversed by a transect of 10 points, 120 m apart. Each bird seen or heard within 60 m radius of each point during a 5 min. interval was recorded. Four counts were made along the unburnt and wildfire transects and six along the mild fire transect. All counts were made within four hours of sunrise by the same observers.

We have considered three groups of resident species occupying different foraging heights in the vegetation profile when presenting our data from both studies. Robins (White-breasted Robin, *Eopsaltria georgiana*, Western Yellow Robin, *Eopsaltria griseogularis,* and Scarlet Robin, *Petroica multicolor*) are an important group of ground foragers in the study areas. Thornbills (Inland Thornbill, *Acanthiza apicalis,* and Western Thornbill, *Acanthiza inornata*) occupy the shrub and midstorey of the forest. Overstorey leaf gleaners are represented by the Pardalotes (Spotted Pardalote, *Pardalotus punctatus*, Striated Pardalote, *P. striatus*) and Western Gerygone *(Gerygone fusca)*. Results of differences and changes in Golden Whistler *(Pachycephala pectoralis)* detections are also presented. Enough detections were obtained to suggest an influence of fire on the population density of this species. Each of these species are common in the study areas and will be referred to as study group birds. Study group birds made up 49.57% of the detections in Study 2 suggesting that these species are important indicators of the overall effect of a fire on the bird community in this area.

RESULTS

Fire and Numbers of Individuals and Species

Study 1

Changes in number of detections and number of species is shown following fire, over five years of bird counts in valley sites in Jarrah forest (Fig. 2). The number of detections per survey was significantly greater ($p < 0.05$) for the three years following the burn than before it. There were, however, year to year fluctuations and high variability between individual counts carried out in the same season. The total number of species detected also increased

Table 1. Diversity of birds (H′) following burning in autumn in open Jarrah forest (Study 1). All detections are included and the Shannon-Wiener index used. The area used for the study was burnt in autumn 1981.

	1979	1980	1981	1982	1983
Burn	1.1	1.0	1.3	1.3	1.2
Control	1.1	1.2	1.1	1.2	1.0

during spring six months after the burn but remained constant for the control. Thirteen species appeared after the fire that had not been detected before it. Numbers of species recorded per survey remained significantly greater ($p < 0.05$) in the area burnt than the control for three years after the fire. These changes in numbers of individuals and species are summarised by a diversity index (the Shannon-Wiener Index, $H' = \Sigma_i\, pi \log pi$, where pi is the proportion of the ith species) which takes into account both the number of species and their relative abundance (MacArthur and MacArthur 1961). Prior to the fire (1979 and 1980) bird species diversities (H′) on the control and burn plots were similar (Table 1). Following the burn bird species diversity (H′) increased on the burnt plot and in the three years following the burn (1981-83), was significantly greater ($p < 0.05$) on the burnt plot relative to the unburnt control.

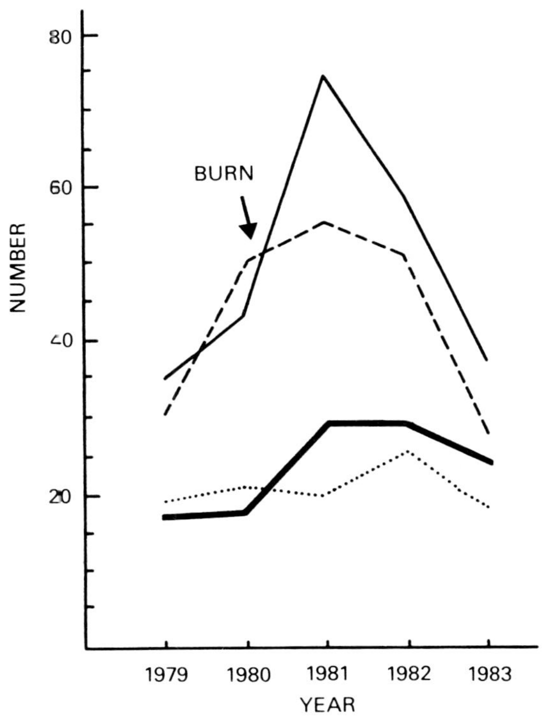

Fig. 2. The number of birds per survey (burn; ——— and control; - - - - -) and total bird species (burn; ▬▬▬ and control; ··········) following burning in open Jarrah forest at Perup (Study 1).

Study 2

Differences in numbers of detections and numbers of species in areas of different fire history

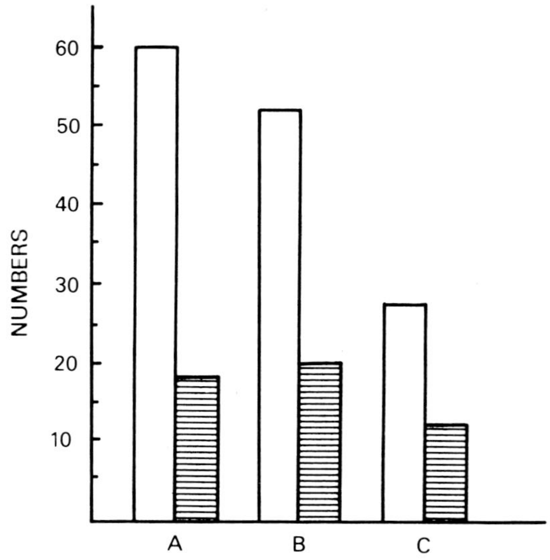

Fig. 3. The number of birds per transect ☐ and bird species per transect ▤ related to fire history (A — unburnt 14 years; B — burnt under mild conditions 6 months prior to study and C — burnt during a wildfire 1 year prior to study) in open Jarrah forest (Study 2).

is shown through transect work in upland Jarrah forest (Fig. 3). Numbers of detections and numbers of species per survey were similar between a transect through forest that had been burnt under mild conditions in spring of the previous year and a control that had not been burnt for fourteen years. Fewer individuals and species per survey ($p < 0.01$) were recorded following a severe wildfire than in either the control or the area burnt under mild conditions.

Fire and Individual Species

Study 1

Changes which relate to foraging height were observed in the species composition of the bird community following the burn in autumn 1981. Ground or low understorey foragers and birds of the open country were found to be most affected by the burn. The White-breasted Robin and White-browed Scrubwren, *Sericornis frontalis,* did not reappear until two years after the burn. Neither species was common in the study area before the burn and both depend on a dense shrub layer. Colonisers of the area burnt included the White-winged Triller, *Lalage sueurii,* Australian Magpie, *Gymnorhina tibicen,* and Australian Raven, *Corvus coronoides.* The Dusky Woodswallow, *Artamus cyanopterus,* and Tree Martin, *Cecropis nigricans,* increased in numbers after the burn.

Of the study group birds, the Scarlet Robin and Western Yellow Robin increased substantially in numbers after the burn (Fig. 4a). Two species common in the shrub or lower canopy layer increased in number following the burn (Fig. 4b). The Inland Thornbill and Western Thornbill increased in the first year after the burn. Detections of both species dropped to slightly above the pre-fire levels after three years. Leaf-gleaning birds in the canopy were largely unaffected by the fire (Fig. 4c). For the Spotted Pardalote, Striated Pardalote and Western Gerygone, the effect of the fire was of less importance than year to year variability caused by other factors.

The Golden Whistler did not react to the fire according to a generalised foraging height class. It was the only species in which enough detections were obtained to show a significant ($p < 0.01$) negative effect of the fire (Fig. 5). Numbers of detections of Golden Whistlers continued to drop three years after the burn but remained stable (except in 1981) throughout the study period in the control.

Non-resident species reacting to short term availability of food resources are affected only secondarily by fire and did not respond according to a generalised foraging height class. The flowering of the Jarrah trees accounts for the appearance of several species in the area following the burn. The effect of flowering plants producing nectar has a major effect on the numbers of species and individuals observed over and above the effects of the fire (Keast 1968; Recher and Christensen 1981), although fire affects overstorey flowering (Kimber 1978). These include the Purple-crowned Lorikeet, *Glossopsitta porphyrocephala*, Silvereye, *Zosterops lateralis*, Red Wattlebird, *Anthochaera carunculata,* and Western Spinebill, *Acanthorhynchus superciliosus.*

Study 2

The results from transects in upland Jarrah Forest suggest variable affects following burns of different intensities (Fig. 6). These effects can also be summarised according to the foraging heights of the most common species. Scarlet and Western Yellow Robins were most common in the control and least common in the area burnt by a mild fire. Thornbills were most common after a mild fire and least common after a wildfire. The two species reacted differently to fire. Inland Thornbills were most common in the control and least common in the area burnt by a wildfire. Western Thornbills were most common after the mild fire and least common in the control. Proportionally Western Thornbills were most common after the wildfire and least common in the control. The numbers of leaf-gleaning birds that forage in the canopy were very similar in each of the three transects. The numbers of each of these three groups of most common birds encountered in transects were most similar in the control. Thornbills were a dominant group following the wildfire. Robins were poorly represented after the wildfire. These results show different trends from

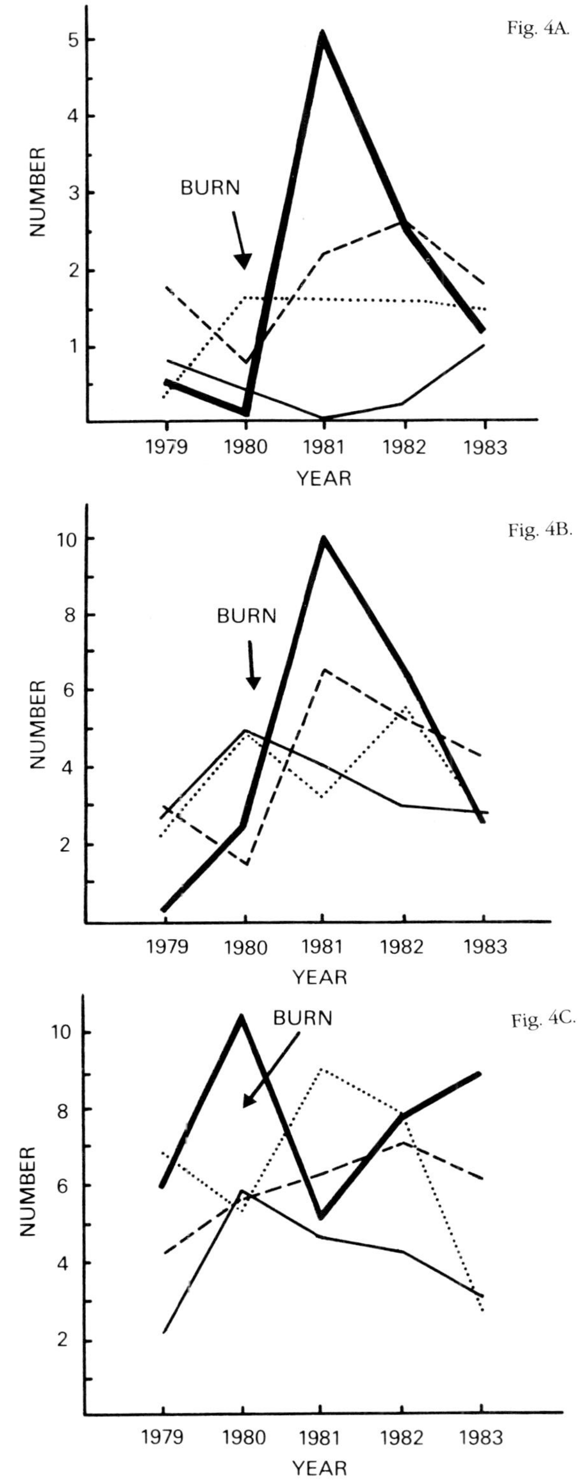

Fig. 4. The numbers of birds per survey following burning in open Jarrah forest at Perup. 4A. Ground foragers (Robins; Scarlet Robin, burn ━━━; Scarlet Robin, control ·········· Yellow Robin, burn - - - - - -; Yellow Robin, control ————): 4B. Scrub and mid-storey foragers (Thornbills; Inland Thornbill, burn ━━━; Inland Thornbill, control ··········; Western Thornbill, burn - - - - - -; Western Thornbill, control ————): 4C. Overstorey leaf gleaners (Pardalotes and Western Gerygone; Spotted and Striated Pardalotes, burn ━━━; Spotted and Striated Pardalotes, control ··········; Western Gerygone, burn - - - -; Western Gerygone, control ————).

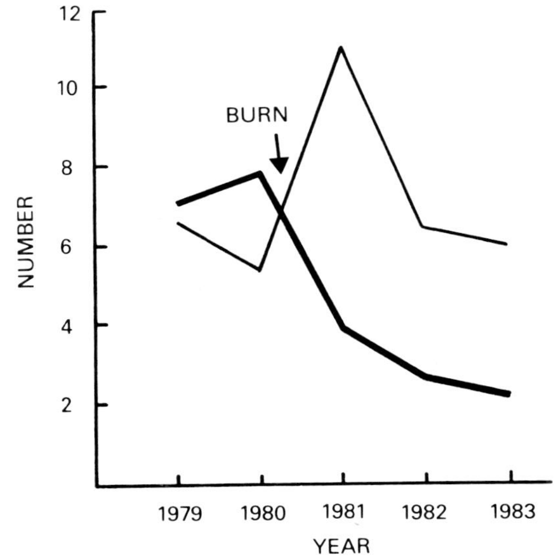

Fig. 5 The number of Golden Whistlers per survey following burning in open Jarrah forest; Study 1 (Burn ━━━, control ————).

those in study one. However, the numbers of Golden Whistlers declined with increasing intensity of burn in transects in upland sites, supporting the results of a study one. Numbers of detections per survey were 4.8, 3.3 and 1.5 for the transects through the control, mild fire and wildfire respectively.

Changes were observed in the species composition of the bird community following the burn in autumn 1981 in Study 1. The composition of the bird communities also varied between the three transects of different fire history in Study 2. These differences were not always consistent. However, seasonal and site differences may be more important in comparing Study 1 and 2 than the effects of burning. Several species that were not detected in the transect through a severe wildfire area in upland Jarrah forest were present or common in a valley site after the same wildfire.

DISCUSSION

Although information on the effects of fire on bird populations in eucalypt forests is limited, the data available suggest some basic trends in the response of bird populations and communities to fire. The trends observed appear similar in both wet and dry sclerophyll forests and are related to the intensity of the fire event. Following a fire of a given intensity, the response of individual species depends on their versatility and their main foraging height in the foliage profile.

Fire and Bird Numbers

The current study demonstrated an increase in the number of individuals and species within one year

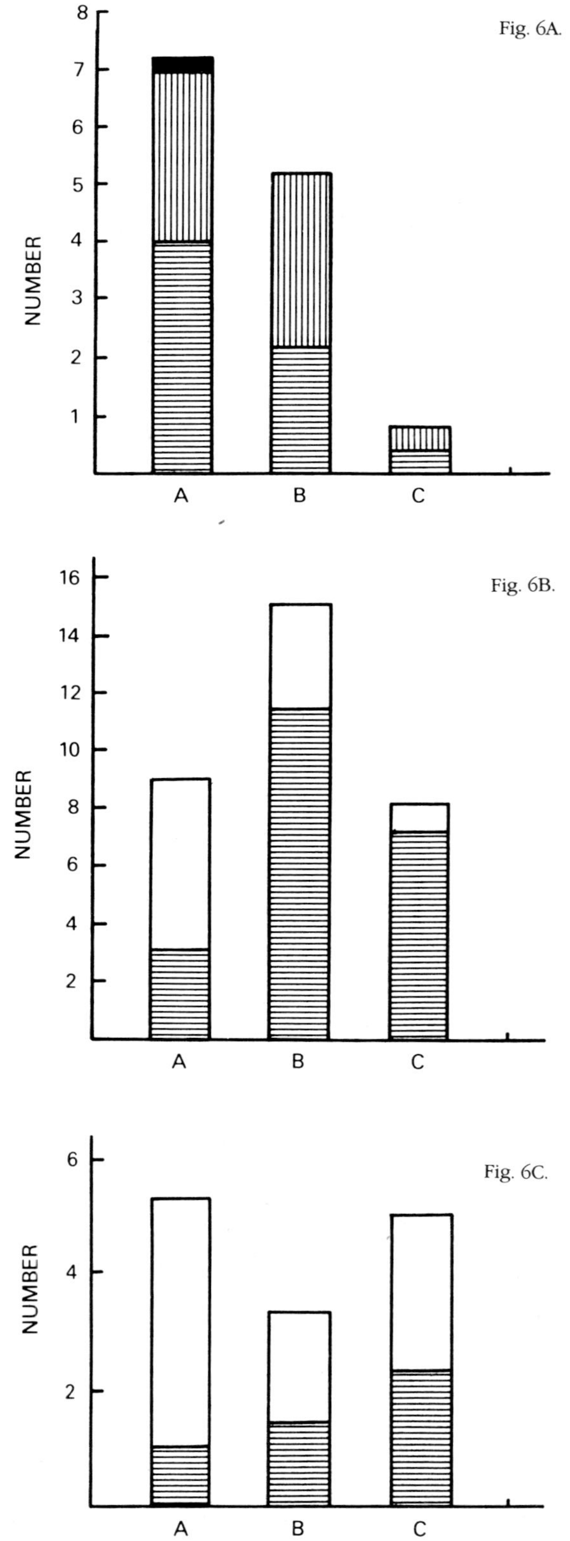

Fig. 6. The number of birds per transect related to fire history (A — unburnt 14 years; B — burnt under mild conditions six months prior to study and C — burnt in wildfire one year prior to study) in open Jarrah forest; 6A. Ground foragers (Robins; White-breasted Robin, Scarlet Robin and Yellow Robin); 6B. Scrub and mid-storey foragers (Thornbills; Inland Thornbill and Western Thornbill); 6C. Overstorey leaf gleaners (Western Gerygone , Spotted and Striated Pardalotes).

after a fire of moderate to high intensity in dry sclerophyll Jarrah forest near Manjimup. Kimber (1974) noted little change in numbers the year after a controlled burn in spring in Jarrah forest near Dwellingup, but found an increase within three years. In eastern Australia, Catling and Newsome (1981) reported data collected by Wilson using mist nets in *Eucalyptus dalrympleana* Maiden. forest in mountains west of Canberra. One year after an intense summer fire the numbers of bird captures in mist nets was similar to the year before the fire. An increase was recorded in the second year following the fire and captures returned to pre-fire levels after three years.

A similar pattern of an increase in numbers and in species following fire has been reported in wet sclerophyll forests. Christensen and Kimber (1975) working in Karri, *Eucalyptus diversicolor* F. Muell., forest found the number of individuals to be little changed one year after a fire of moderate to high intensity. Two years after the fire numbers were much higher than in an area of forest used as an unburnt control.

Limited studies using mist nets indicate that many birds survive fires of low to moderate intensity, suggesting that the changes noted are not all due to movements of birds into and away from the burnt area. Thus Cowley (1974) working in Victoria in dry sclerophyll forest found that 18 of 27 banded birds continued to use the same areas after a controlled fire of moderate intensity burnt the study area. Wooller and Brooker (1979) report similar numbers of captures in mist nets before and one year after a low intensity prescribed burn in Karri forest. Recaptures were obtained of birds inhabiting the understorey in the year following the fire.

A pattern of an increase in bird numbers and sometimes also of species in the second and third year following fires of low to moderate intensity is emerging. The trend appears to hold for both dry and wet sclerophyll forest. It is not clear whether the increase is due to increased survival of young in the seasons following the fire or whether it is a result of immigration from surrounding unburnt forest.

Birds have been observed to move easily from danger in fires of low to moderate intensity but it appears that they may be killed by suffocation or heat in fires of very high intensity. High intensity wildfires cover large areas, demonstrate violent physical behaviour and often result in tree canopy defoliation over extensive areas (Luke and MacArthur 1978). Immediately following such fires bird numbers usually decrease.

Following the huge Tasmanian fires in 1967 (Hemsley 1967), individuals of 59 of the 72 species found in the sclerophyll forests in southwest

Tasmania were collected, washed up on beaches. A similar phenomenon was observed following the Nadgee fire in New South Wales where individuals of 42 species were found washed up on beaches (Recher and Christensen 1981). Fox (1978) recorded a decline in bird numbers immediately following this fire.

Recovery of bird populations following wildfires of high intensity is slow. The observed losses during such fires undoubtedly have some influence on recovery but there appears to be more fundamental causes for the slow recovery. Structural changes are greater following high intensity wildfires and their extent may influence the rate of immigration. However, the delayed recovery observed in the comparatively small wildfire in this study suggests that factors other than fire area and structural changes may be the cause.

It is suggested that a delayed recovery in the invertebrate population of the area may be responsible. The degree to which invertebrate activity is influenced following fire is not adequately known, although Campbell and Tanton (1981) present evidence suggesting little effect following a single mild fire. It is likely that more intense fires have a more significant impact. Many of the woodland and forest birds of Australia are insect eaters and would therefore be affected by any delayed recovery in invertebrates following fire.

Site may have a localised effect on the recovery of bird populations following fire. It has been observed, for example, that recovery is more rapid in valley areas than on ridges. This appears to be due to the more rapid recovery of the vegetation on the moister and more favourable valley sites.

Fire and Bird Species

The kind of fire which burns the forest has been shown to have a profound influence on bird numbers. Suites of birds and populations of individual species are likewise influenced by the kind of fire which burns their habitat.

As a general rule the effect of fires of mild to moderate intensity on birds is inversely proportional to their main foraging height in the foliage profile. The greatest degree of disturbance following fire is experienced by the birds of the ground and understorey layers, the layers most influenced by such fires. Birds of the mid and upper canopy layers, which are minimally affected by such fires, show little change in numbers and composition.

The effect of fire on bird species in this study follows a trend of decreasing effect with increasing height from the ground. Kimber (1974) and Christensen and Kimber (1975) also demonstrated this trend in Jarrah forest similar to this study. As in the present study overstorey leaf-gleaning birds in the canopy were largely unaffected by the fire. Catling and Newsome (1981), in summarising Wilson's data, also demonstrated this trend in eastern Australia.

The influence on ground foragers is most noticeable following fires of medium to high intensity. The Scarlet Robin increased greatly in numbers after fire in this study and also appeared following a medium to high intensity burn in Karri forest (Christensen 1974). It was also captured in mist nets more frequently following a mild spring burn in Karri forest (Wooller and Brooker 1980). Wilson reported by Catling and Newsome (1981) found that ground foragers, although not a high proportion of total captures, quadrupled in the first year following an intense fire in dry sclerophyll forest near Canberra. Wilson's data also demonstrate substantially increased numbers of upper shrub and lower tree-dwellers. Inland Thornbills and Brown Thornbills, *Acanthiza pusilla,* which normally made up the largest numbers of residents in these layers changed little in the first year but increased substantially in numbers two years later.

There are exceptions to this general trend. The Golden Whistler, a species dwelling in the mid-storey was adversly affected by a fire of moderate intensity to a greater degree than would have been expected from the changes in habitat structure. The Grey Fantail, *Rhipidura fuliginosa,* also a species dwelling in the mid-storey was on the other hand largely unaffected by the same fire. These exceptions to the general trend highlight a problem involved in concluding that the changes in numbers of individuals of common species represent the effect of fire on all species. While these species may show changes for the community as a whole, rare or uncommon species may be more sensitive to changes in habitat structure through the effects of fire. For these species there are too few detections in studies such as these to enable any conclusions to be drawn. Yet in terms of the conservation of a diverse avian community, rare or uncommon species may be extremely important.

CONCLUSION

The limited information currently available supports the view that fire intensity is the major factor determining the effect of fire on birds in sclerophyll forests in Australia. A single forest fire causes temporary short term changes in species composition and numbers of residents. Such changes depend on the extent of the changes in habitat structure which relate primarily to fire intensity. Such changes are never permanent (unless the total available colonised habitat is rendered unsuitable) and recovery is usually rapid. The season of burning in itself may be of less importance than the influence which it has on fire intensity. This view is contrary to popular belief where it is often suggested that spring burning is especially detrimental to some communities because of an alleged disruption to the nesting of some species.

Although the breeding season of birds is lengthy in southwestern Australia, the majority of nesting coincides with the spring fuel reduction burning programme (Davies 1979). Undoubtedly many birds have their breeding disturbed. However, this may be of relatively minor biological significance in itself for many birds have high mobility and high reproductive potential. Furthermore, the inherent mozaic pattern of fires ensures that patches of unburnt habitat remain. Populations of common species and general community structure appear not to be changed to any significant degree in the long term by a single fire event.

Many species also nest out of reach of the flames of mild fires. We have recorded nests which have escaped fire and in which broods were subsequently reared. Individuals of species which nest close to the ground are however more severely affected.

The observed rapid recovery in both numbers and species composition following fires of mild to moderate intensity during all seasons supports the view that season of burning as such is of only minor significance to bird communities in fire prone environments. More intense fires burning during summer or autumn which may kill many birds and which cause greater changes to forest structure and perhaps also to invertebrate populations cause more extensive change. The process of recovery after such fires is less rapid.

While most bird species and communities show a relatively rapid response to the changes induced by a single fire in a fire prone environment, very little is known of the long term effects of certain fire regimes. It has been demonstrated in Jarrah forest that a long interval without fire leads to a lowering of the diversity of the avian community (Christensen and Kimber 1975). The use of fire for a single purpose over a large area may also lead to a simplication of the structure and floristics of the forest community and thus the bird community. Long term studies are needed to determine the changes induced by fire regimes. Individual species of birds will react differently to fire regimes of varying intensity, frequency and season. Hence a diverse burning pattern may be the most appropriate to ensure the conservation of diverse avian communities in the forest environment. This will of necessity include areas left unburnt as biological reference areas as well as areas burnt regularly to reduce the accumulation of fuels.

ACKNOWLEDGEMENTS

We would like to thank Graeme Liddelow, John Serventy and Chris Vellios for work in the field, Julie O'Connor for statistical guidance, Harry Recher and Graham Pyke for use of a standard census method proposed for Australia and Dennis Saunders for comments on an earlier draft of the chapter. Thanks also to Liz Bartuccio for drafting the figures and Fiona Ward for typing the manuscript.

REFERENCES

Campbell, A. S. and Tanton, M. T., 1981. Effects of fire on the invertebrate fauna of soil and litter of a eucalypt forest. Pp. 215-42 *in* Fire and the Australian Biota. ed. by A. M. Gill, R. H. Groves and I. R. Noble. Australian Academy of Science.

Catling, P. C. and Newsome, A. E., 1981. Responses of the Australian vertebrate fauna to fire; an evolutionary approach. Pp. 273-310 *in* Fire and the Australian Biota. ed. by A. M. Gill, R. H. Groves and I. R. Noble. Australian Academy of Science.

Christensen, P., 1974. The concept of fauna priority areas. Proc. Third Fire Ecology Symp., Forests Comm. Vic., Melbourne, Pp. 66-73.

Christensen, P. E. and Kimber, P. C., 1975. Effect of prescribed burning on the flora and fauna of south-west Australian forests. *Proc. Ecol. Soc. Aust.* **9**: 85-106.

Cowley, R., 1974. Effects of prescribed burning on birds of the mixed species forests of west central Victoria. Proc. Third Fire Ecology Symp., Forests Comm. Vic., Melbourne, Pp. 58-66.

Davies, S. J. J. F., 1979. The breeding seasons of birds in south-western Australia. *J. Roy. Soc. of Western Australia* **62**: 53-64.

Fox, A., 1978. The '72 fire of Nadgee Nature Reserve. *Parks and Wildl.* **2**: 5-24.

Gill, A. M., 1975. Fire and the Australian flora; a review. *Aust. For.* **38**: 4-25.

Hemsley, J., 1967. Bushfire — S.E. Tasmania — Feb. 1967. National Parks and Wildlife Service (Tas.).

Keast, A., 1968. Seasonal movements in the Australian Honey-eaters (Meliphagidae) and their ecological significance. *Emu* **67**: 159-210.

Kimber, P., 1974. Some effects of prescribed burning on jarrah forest birds. Proc. Third Fire Ecol. Sym., Forests Comm. Vic., Melbourne, Pp. 49-57.

Kimber, P., 1978. Increased girth increment associated with crown scorch in jarrah. For. Dept. W. Aust. Res. Pap. No. 37.

Luke, R. H. and MacArthur, A. G., 1978. Bushfires in Australia. Dept. of Primary Industry, C.S.I.R.O. Div. of For. Res., Aust. Govt. Publishing Service.

MacArthur, R. H. and MacArthur, J., 1961. On bird species diversity. *Ecology* **42**: 584-98.

MacArthur, R. H., MacArthur, J. W. and Preer, J., 1962. On bird species diversity. II Prediction of bird censuses from habitat measurements. *Am. Nat.* **96**: 167-74.

Mount, A., 1969. Eucalypt ecology as related to fire. *Proc. Tall Timbers Fire. Ecol. Conf.* **9**: 75-108.

Noble, I. R. and Slatyer, R. O., 1977. Post fire succession of plants in Mediterranean ecosystems. Pp. 27-36 *in* Proceedings of the Symposium on the Environmental Consequences of Fire and Fuel Management in Mediterrenean Ecosystems ed. by H. A. Mooney and C. E. Conrad. USDA For. Serv. Gen. Tech. Rep. W0-3.

POSAMENTIER, H. G., CLARK, S. S., HAIN, D. L. AND RECHER, H. F., 1981. Succession following wildfire in coastal heathland (Nadgee Nature Reserve, NSW). *Aust. J. Ecol.* **6**: 165-75.

RECHER, H. F., 1971. Bird species diversity: A review of the relation between species number and environment. *Proc. Ecol. Soc. Aust.* **6**: 135-52.

RECHER, H. F. AND CHRISTENSEN, P. E., 1981. Fire and the evolution of the Australian Biota. Pp. 135-62 *in* Ecological Biogeogaphy of Australia ed. by A. Keast. Junk, The Hague.

REYNOLDS, R. T., SCOTT, J. M. AND NUSSBAUM, R. A., 1980. A variable circular-plot method for estimating bird numbers. *Condor* **82**: 309-13.

SHEA, S. R., PEET, G. B. AND CHENEY, N. P., 1981. The role of fire in forest management. Pp. 443-70 *in* Fire and the Australian Biota ed. by A. M. Gill, R. H. Groves and I. R. Noble. Australian Academy of Science.

UNDERWOOD, R. J. AND CHRISTENSEN, P. E. S., 1981. Forest fire management in Western Australia. Special Focus No. 1, Forests Dept. of Western Australia.

WOOLLER, R. D. AND BROOKER, K. S., 1979. The effects of controlled burning on some birds of the understorey in karri forest, *Emu* **80**: 165-66.

Noisy Scrub-bird, *Atrichornis clamosus*

CHAPTER 26

The Noisy Scrub-bird *Atrichornis clamosus.* Does It's Past Suggest a Future?

G. T. Smith[1]

In the period 1842 to 1889 the Noisy Scrub-bird, *Atrichornis clamosus,* was found in a number of coastal and near coastal localities between Drakesbrook and Albany in Western Australia. After 1889 there were no further sightings and the bird was thought to be extinct until it was discovered at Two Peoples Bay in 1961.

The Scrub-bird is a species whose ecological traits were established under more stable conditions than those operating in the last million years and in particular since the arrival of Europeans. Its terrestrial and territorial behaviour, low reproductive potential and low dispersive ability reduced its ability to recolonize after local extinctions, while its adaptation to the linear and patchy distribution of the ecotone between swamp and forest placed further constraints on its ability to survive.

Its survival at Two Peoples Bay is a consequence of the area having no agricultural value and the topography which provides natural fire breaks which have prevented the whole area being burnt at one time. Management of the reserve since its creation in 1966 has concentrated on fire prevention and this is the main reason for the increase in population from 45 in 1970 to 138 in 1983. The expansion of the population has effectively created two separate populations which increases the species chance of survival. Translocation of ten males and six females to the Mt Manypeaks reserve and the establishment of the feasibility of captive breeding have added further support for the long term survival of the species.

INTRODUCTION

THE FIRST account of the discovery of the Noisy Scrub-bird, *Atrichornis clamosus,* was given by the botanist James Drummond in a letter to the Perth Inquirer on 15 February 1843, in which he related his observations while travelling to Augusta with John Gilbert in November 1842. In the following 46 years the Noisy Scrub-bird was recorded in a number of coastal areas between Perth and Albany (Fig. 1). Nothing more was seen or heard of the bird, despite numerous searches by ornithologists, until its rediscovery at Two Peoples Bay, 40 km east of Albany by Webster (1962).

The limited data from these initial observations of the bird raise questions as to its status at the time of European settlement, the reasons for its decline to near extinction and its prospects for a secure future. The aim of this chapter is to briefly outline the biology of the species as background to a discussion of these questions.

LIFE HISTORY OF THE NOISY SCRUB-BIRD

Noisy Scrub-birds are small (♀ 35 g, ♂ 50 g) solidly built birds with strong pointed bills, long powerful legs, long graduated tails, short round wings and cryptic brown plumage. These are ideal attributes for birds which live close to the ground in dense vegetation. They are fast and agile runners using their limited flying ability to move from shrub to shrub without having to go to the ground. While their colour and ground hugging behaviour in their dense habitat makes them difficult subjects to observe, they are not as secretive as casual observations would suggest; in fact they are inquisitive birds, quickly moving to observe the cause of some disturbance before moving off, often unheard and unseen. Various aspects of the biology of the Scrub-bird have been given in Smith (1976), Smith and Robinson (1976), Robinson and Smith (1976), Smith (1977), Smith (1978), Smith and Forrester (1981), and Smith and Calver (1984).

[1]Division of Wildlife and Rangelands Research, CSIRO, LMB No. 4, P.O. Midland, Western Australia 6056.
Pages 301-8 *in* BIRDS OF EUCALYPT FORESTS AND WOODLANDS: ECOLOGY, CONSERVATION, MANAGEMENT ed. by A. Keast, H. F. Recher, H. Ford and D. Saunders. Royal Australasian Ornithologists Union and Surrey Beatty & Sons.

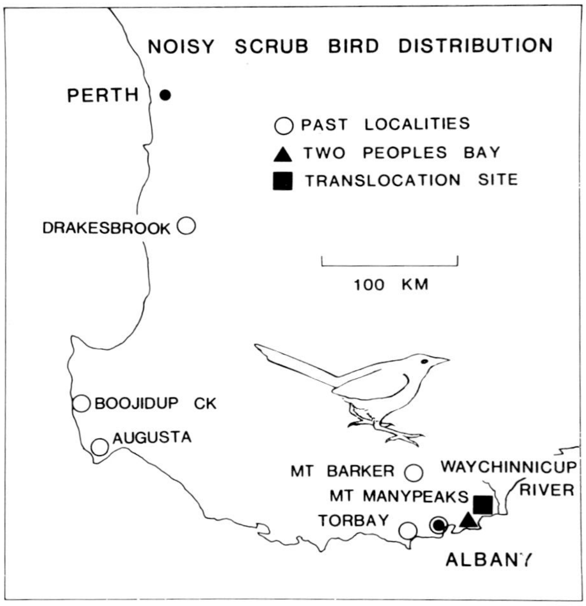

Fig. 1. Locations where Noisy Scrub-birds have been recorded and the location of the translocation area.

Habitat

Examination of the habitat at Two Peoples Bay and at locations where Scrub-birds were known to have occurred indicates that they were confined to the wetter areas within the distribution of the Marri, *Eucalyptus calophylla*, Jarrah, *E. marginata* forest, in particular, to the ecotone between forest and swamp vegetation which appears to have been the most common habitat in the localities recorded in the last century. The dense layer of rushes and shrubs at the edge of the swamp provides cover as well as the sites and material for nests. The reduced moisture regime uphill from the swamp allows the development of middle and upper storey vegetation which maintains the low light intensity at ground level with a reduced density of the ground storey. The latter allows the development of a thick layer of litter. Broadly, the swamp can be thought of as providing the breeding habitat while the forest provides the feeding habitat. At Two Peoples Bay such a situation is uncommon. In the wetter creeks the breeding habitat is confined to a narrow zone (2-5 m) either side of the creek bed, while the feeding habitat is above this zone. In the drier drainage lines without a well defined creek bed suitable habitat is found in a small scale mosaic of breeding and feeding habitat (see Fig. 2).

Territory

The males are territorial, defending their territories with a loud, directional song which they sing throughout the year. The location, size and use

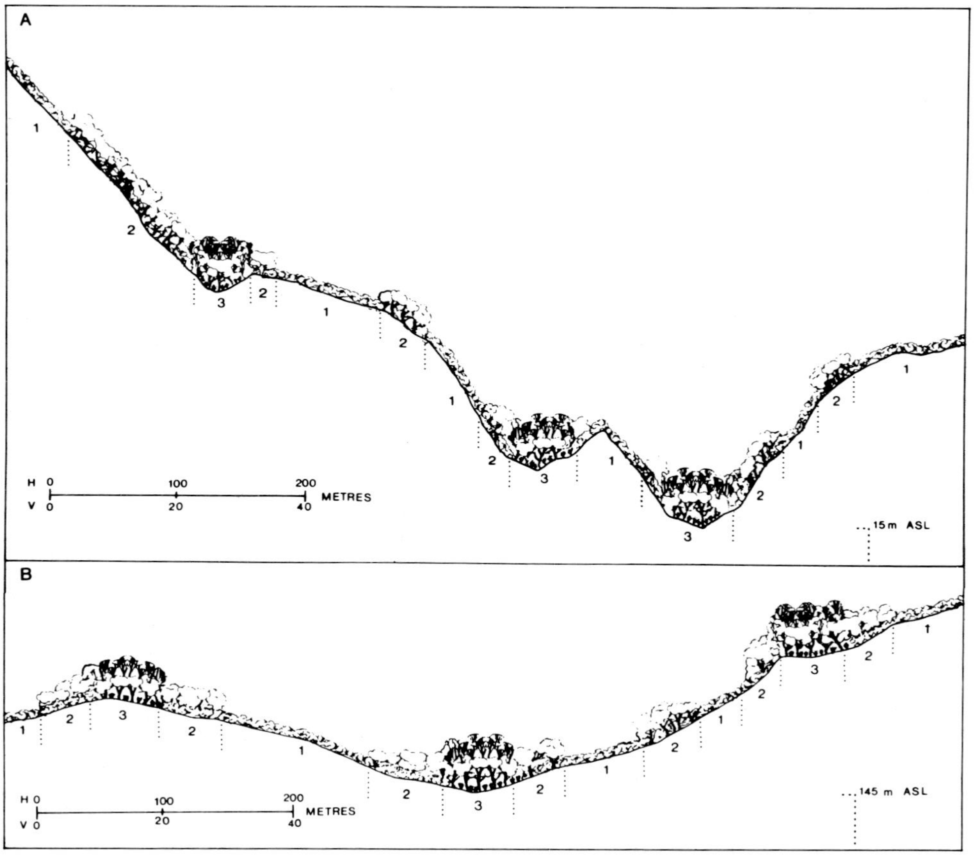

Fig. 2. Vegetation profiles in an area of deep gullies A, and shallow gullies B. Location of profiles shown on Figure 4.

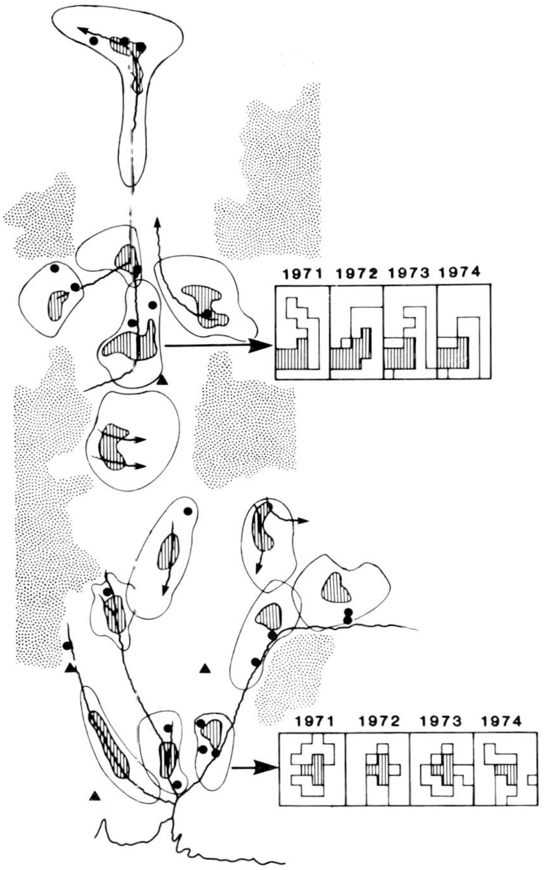

Fig. 3. Territories in study area showing core areas -vertical lines; nesting areas, ●; location of short term territories ▲. The boxes show the territory outline and core areas for two territories from 1971 to 1974. Stippled area are rock outcrops, continuous lines are streams.

of the 14 territories in the study area during the period 1973 to 1976 were determined by plotting the location of singing males (Fig. 3). The average size of the territories was 6 ha (range 4 to 9 ha), and within this area 80% of the observations were made within a core area that averaged 1.25 ha (range 0.75 to 2.25 ha). Detailed observation of ten territories from 1971 to 1976 showed that the core areas remained virtually constant throughout the year and from year to year (Fig. 3). Less detailed observations from 1977 to 1983 showed that these same core areas were still being used. The core areas are characterized by having a dense understorey, low light intensity and a thick layer of litter with abundant invertebrates, that is they are high quality feeding areas. Marked changes in the area used by a male were only observed when a neighbouring male stopped singing, the remaining male then spent some time most days during the breeding season in his former neighbour's territory, until it was reoccupied.

Territories may be divided into two classes (Fig. 3). Firstly, long term territories in which the male sings throughout the year and in which there is adequate feeding and breeding habitat. Evidence of breeding was found in most territories of this type, but the difficulty of finding nests made it impossible to verify breeding in all. Secondly, short term territories in which the male sings infrequently during the breeding season, and in which the feeding and/or breeding habibat was inadequate. No evidence of breeding has been found in these territories and they are probably occupied by immature males. The increasing population since 1976 has tended to blur these distinctions; there are now an increasing number of males in localities with poor and/or small areas of breeding and feeding habitat which sing throughout the year. It is possible that breeding may occur in these territories because a nest has been found in similar habitat adjacent to two long term territories, where at least one of the breeding attempts was successful. In addition there are more males occupying heath and thicket, vegetation formations that can best be described as "survival" habitat. In 1970 three males (7%) were located in such areas while in 1983, 28 (23%) males were found in these vegetation formations.

During the breeding season the male moves slowly around his territory each day, the pattern of movement changing from day to day. The only predictable movements are those to and from the roosting areas which are usually outside the core area. The roost sites are at the top of shrubs or small trees and may be used continuously for up to at least three months. Little is known of the daily movements of the female, however, a limited number of observations suggest that as with the male, the female spends most of her time in a small area around the nest area. This area is usually on the periphery or outside the core area of the male's territory (Fig. 3), but some breeding females have been located in areas away from male territories (Fig. 3).

Breeding

The period from the start of nest building to the last chick leaving its nest may extend from May to November although it more commonly ranges from June to October. Males who defend their territories with song all year begin to sing more frequently in April reaching a peak in May/June. A high but variable level of singing is maintained until October, after which it declines (Smith and Robinson 1976). From April to June males in short term territories begin singing and this may lead to boundary disputes with adjacent breeding males. Such confrontations between males with long term territories and those with short term territories may occur as often as two to three times a week compared with a maximum of one or two disputes per breeding season between males with long term territories (Smith 1976).

Egg laying begins in late May with the peak in late June and ending in early October. Little is known of the factors controlling the breeding season. Decreasing photoperiod is probably the proximate factor, while the interaction of rainfall and temperature on the production and decay of litter and its invertebrate fauna are probably important factors controlling the timing of events.

Males do not build nests, incubate or feed the chicks. The nest is globular (approximate diameter 18 cm) with a small side entrance, the bottom half of the cavity is lined with a papier mâché like substance. The nest is constructed of a variety of rushes with some leaves and twigs and usually sited about 20 cm above the ground in a dense rush clump or shrub. Building takes two to three weeks and the egg is laid ten to 16 days later. Second nests may be built in a week presumably because the female spends more time building per day, also, the nests have fewer pieces of rush and the lining is not as extensive.

There is no direct evidence for polygamy in the field, however, observations of two males in the study area showed that they regularly visited areas with breeding females outside their territories. In one case these visits ceased after the area was re-occupied by a male. Only one active nest has ever been found within a male's territory, although breeding females have occasionally been found in suitable areas outside the territories, the extent of this extraterritorial breeding is unknown. Females start breeding in their first year and males in their third year (Smith 1978) and it is likely that males are opportunistically polygamous, mating with the female in their territory as well as females on the periphery or outside their territory. This conclusion is supported by the fact that in captivity a single male successfully mated with three females in consecutive years (Smith *et al.* 1983).

The single egg is laid one to two weeks after the nest is lined, incubation lasts 36 to 38 days and the chick fledges three to four weeks after hatching. The chick stays with its mother for some time after fledging, probably until it has finished its post-fledging moult at about three months old (Smith *et al.* 1983). The female tends to build her nest in the same area year after year, often within 5 m of the previous nest. If the egg is lost the female will build another nest, usually some distance (20-50 m) away from the first nest and re-lay; a second clutch is not attempted if the chick is lost (Smith 1976).

Food

Scrub-birds spend most of their time feeding on or near the ground, foraging in litter, the bases of rush clumps, dense shrubs and decaying wood. The diet of adults has not been documented in detail but it appears to be mainly insectivorous. The diet fed to nestlings has been obtained from nest observations and the analysis of the contents of faecal sacs. Seventeen orders of invertebrates and vertebrates were identified, the most common items were Araneida, Orthoptera, Blattodea, Lumbricidae, Myriapoda and various insect larvae (Smith and Calver 1984).

Population

The loud, directional territorial song of the male (females do not sing) provides the only practical means of doing a census, the statistic being the number of singing males. Census methods and population data for the period 1962 to 1977 are given in Smith and Forrester (1981) while the number of singing males found in the censuses from 1970 to 1983 are shown in Table 1. In the period 1962-66 the number of singing males was probably 40 to 45, rising to about 50 in 1968. By the time of the first census in 1970 the number was 45, falling to 44 in the following year. Thereafter there was a steady increase to 138 in 1983.

Table 1. Numbers of singing males located during censuses from 1970 to 1983 (A) and the number of males located in sub-optimal habitat (B).

	70	71	72	73	74	75	76	77	79	80	82	83
A	45	44	50	59	66	69	74	74	107	116	130	138
B	3	2	4	6	8	11	13	12	22	26	29	28

From a management point of view, the population can be considered as consisting of two sub-populations; the original population around Mt Gardner and the population around Lake Gardner and its outlet creek (Fig. 4). The former has increased from 45 singing males in 1970 to 120 in 1983, while 12 territories in this area have been established in areas burnt between 1962 and 1964; the rest of the increase has been in areas that have not been burnt for at least 35 years. Qualitative observations in a small number of areas indicates that some males have occupied areas that have had suitable habitat since 1970 while others have occupied areas that have become suitable since 1970. There are also an increasing number of males occupying areas of unsuitable habitat (Table 1) which suggest that most of the suitable areas are occupied. Areas around Lake Gardner and its creek were burnt a number of times in the 1960s and by 1970 there were no Scrub-birds in the area. The first bird along the creek was recorded in 1973 and the first around the lake in 1979. In 1983 there were 18 birds in the area, with ample habitat between lakes (Fig. 4) into which the population can expand.

CHANGES IN STATUS

The Noisy Scrub-bird and Rufous Scrub-bird, *A. rufescens,* are survivors of the old endemic avifauna and may pre-date the Eocene (Schodde and Calaby

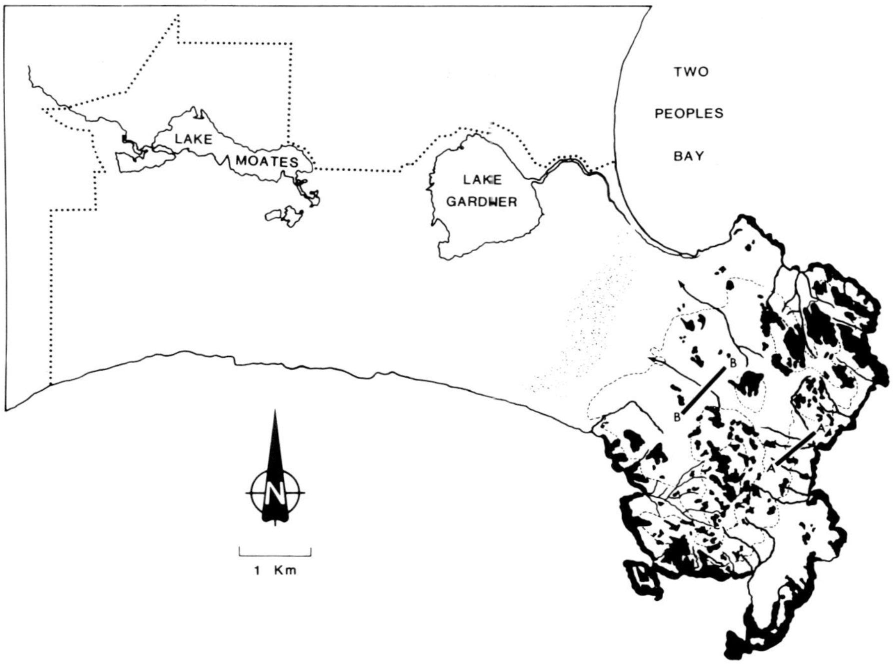

Fig. 4. Two Peoples Bay reserve showing areas of rock outcrop (black) and major drainage lines ——, stippled area, control burn blocks.

1972; Crook 1981; Feduccia and Olsen 1982). A scenario of their possible evolutionary history (Smith 1977) suggests that they are evolutionarily robust and that their present small populations and disjunct distribution represents just one of many low points during the climatic changes in the Pleistocene. In considering their past history and their prospects for the future it is important to take into account their past evolutionary history which has shaped their present relict status.

The Noisy Scrub-bird was first recorded by John Gilbert in a swampy creek valley near Drakesbrook on the western side of the Darling Range in November 1842 and later on the same journey at Augusta. In 1843 he travelled from York to Albany and noted that he first heard the Scrub-bird around Mt Barker, the numbers increasing as he approached Albany (locations on Fig. 1). His general conclusion was that the species was most common around Albany and was not found more that 60 km from the coast. George Masters in 1866 and 1868/69 collected a number of specimens in the Albany area as did William Webb, a local collector. In a letter to the Albany Mail in 1895 Webb referred to the Noisy Scrub-bird as " . . . not, strictly speaking a rare bird, but is rarely seen . . . ". In 1889 A. J. Campbell collected a single male at Torbay and later in the year recorded having heard it at Boojidup Creek near Margaret River (Whittell 1943) (Fig. 1).

If we assume that Scrub-birds were at Two Peoples Bay at the time of European settlement, then there were a minimum of six coastal or near coastal localities. While positive data are limited, it would appear that the main population was around Albany with numbers decreasing to the east, north and west. The populations at Augusta, Boojidup Creek and Drakesbrook were probably representative of a larger number of small isolated populations in the coastal districts to the west and north of Albany, while to the east, Scrub-birds were probably as far east as Mt Manypeaks. Such a status is in keeping with the abundance of habitat which can be deduced from topographic and vegetation maps. By the end of the eighteenth century the only extant population was probably that at Two Peoples Bay. What were the causes of this relatively rapid decline in status? The most obvious agent of change was the arrival of Europeans.

The development of the agricultural and pastoral industries and the growth of urban centres in Western Australia was slow, the population in 1891 being only 53,000. While the area used for agriculture was small, pastoral activities were extensive with pastoral leases being taken up on most of the coastal land (Anon 1979), much of this activity being concentrated on the most fertile areas between Perth and Albany, the area encompassing the entire distribution of the Noisy Scrub-bird. The area was also favourable for Aborigines and given that fire has been considered the main cause of the decline (Smith 1977), it is worthwhile considering the differences in the fire regimes of the Aborigines and Europeans. Hallam (1975) gives a general account of the Aboriginal use of fire in the southwest, while the only specific accounts for an area with Scrub-birds are those of Scott and Nind from Albany (Green 1979). Aborigines from a wide area visited Albany during the summer. In this period the largest fires were those lit by the men while hunting for kangaroos in the forests, but these fires probably had little effect on Scrub-birds. On the other hand, the women used to light smaller fires in the denser vegetation around the creeks and swamps to hunt for small mammals and lizards. These fires would have been in or close to Scrub-bird habitat and would have destroyed all or parts of some territories. However, the small area affected by such fires and the likelihood of a patchy burn would have minimized the damage. Given the fact that Scrub-birds appear to have been relatively common in the area it is likely that such fires did not adversely affect them and may have been beneficial in that they regenerated the habitat and increased its heterogeneity.

The arrival of Europeans resulted in a rapid breakdown in the traditional hunting methods of the Aborigines in the areas of contact with the result that the prevailing fire regime was changed. Europeans altered the fire regime in two ways. Firstly, in coastal areas the practice of burning the heath and thicket vegetation every two to three years to provide new growth for stock became established. The effects on the vegetation in an area close to Boojidup Creek are described by Carter (1924), " . . . where there had been impenetrable scrub there was mostly bare sand drifts caused by fire made to improve the country for cattle grazing". This change took place between 1902 and 1916. The results of this type of fire regime can be seen at Two Peoples Bay. In the wetter gullies the vegetation takes four to six years to grow to a condition suitable for Scrub-birds, while in drier areas the regeneration may take up to 10 years. It is clear that fire frequencies of less than four years and probably more will eventually change the habitat and its fauna. Although the evidence that this burning practice affected Scrub-birds is only circumstantial, there is little doubt that the frequent and uncontrolled coastal burning must have destroyed the habitat of many birds, especially in the drier areas which would constitute their feeding areas. Grazing stock in such areas would have resulted in further deterioration of the area. This latter effect was clearly demonstrated at Boojidup Lake where Campbell heard the bird in 1889. In 1974 when I first visited the area, the forest areas fringing the swamp vegetation had been destroyed, but the swamp had all the elements of an excellent breeding area except that it was too open because of the cattle tracks. When I returned in 1979, cattle had been excluded for about three years and the vegetation had regenerated and was excellent nesting habitat.

Forest areas also were subjected to a different fire regime. European settlers quickly discovered that their axes were unsuitable for large scale felling of eucalypts and resorted to the use of fire for clearing, resulting in an epidemic of summer fires. It was not until 1847 that burning in summer was prohibited (Cameron 1979). Subsequent attempts to limit fires in the absence of the Aboriginal cycle of fires would have led to fewer but more destructive fires. These fires would have had the greatest effect on Scrub-birds living in isolated valleys such as Drakesbrook and Boojidup Creek. In such areas, all the feeding areas and some if not all the breeding areas would have been destroyed and recolonization made more difficult.

The effect of clearing the forests on the slopes around the coastal swamps can be seen, even today, at Boojidup Creek and at certain localities in the Albany area. Essentially, the areas favoured for nesting are still intact, but the feeding areas have either been destroyed or the trees have been thinned so that the ground storey and the litter layer are no longer suitable for feeding. Clearing would have affected relatively small areas at any one time, but its cumulative effect in conjunction with fire must have made an important contribution to the decline, especially as Europeans also favoured areas suitable for Scrub-birds.

Climate has been suggested as another cause, because of the below average rainfall in the period 1880-1890 (Smith 1977). However, rainfall in this period was similar to that over the last decade when the population has been increasing. While below average rainfall may adversely affect reproduction, its most serious consequence may be in making the habitat more vulnerable to fire. Nothing is known of the effect of introduced predators such as the cat or fox. The vegetation is probably too dense for the fox and cat; certainly in the period 1971 to 1976 none of the 150 new or recent nests examined showed any evidence of having been damaged by these predators, even though foxes were in the area.

While the above discussion is speculative, there are sufficient clues to suggest that the outline is correct and therefore of use in helping to shape the framework for the future management of the species.

MANAGEMENT IMPLICATIONS

Management of the Noisy Scrub-bird can be considered under three main headings; management of the Two Peoples Bay reserve, translocation and captive breeding.

The survival of the species at Two Peoples Bay suggests that there is something different about the area. The major difference is the topography, a combination of a hilly and rocky headland backed by lakes and sand dunes which make the area unsuitable for agricultural and pastoral activities. Further, the stronghold of the species is on the hilly headland around Mt Gardner which has a series of bare granite ridges lying across the prevailing winds (Fig. 4). These ridges act as natural firebreaks and prevent the whole area from being burnt at any one time. The effectiveness of this system of natural firebreaks is illustrated by aerial photography taken in 1946, which show that there were numerous small fires in the preceeding years but that none had burnt a significant percentage of the area. In the last 13 years fire has been excluded from Two Peoples Bay and the growth of the vegetation over this period has provided additional habitat for a percentage of the increased population. While the increased density of the vegetation has benefited the Scrub-bird it also has created a potential hazard. In 1946 the combined effect of the granite ridges and numerous small fires reduced the probability of a severe fire. Today, the vegetation, especially on the ridges, is sufficiently dense to cast doubt on whether the ridges will any longer act as effective fire breaks. The only effective means of improving the ridges as firebreaks is thinning the vegetation by hand, a large but not impossible job. Fortunately the population can be considered as two populations separated by a number of tracks, firebreaks, the main road and a strip of controlled burn blocks (Fig. 4). The population around Lake Gardner and its outlet creek has 18 singing males with sufficient habitat to the west to allow this population to double or even treble in size. While it is possible to protect the area from fire, it is almost impossible to stop a fire once it has started, nevertheless the growth of this population has provided the species with a further safeguard.

Since the creation of the reserve at Two Peoples Bay the policy has been to protect it from fire. While this protective management practice has worked well in the past it may not be the best policy in the future. While some idea of the minimum fire frequency is available, there are no data on how long the habitat will remain suitable for Scrub-birds after fire. All territories show signs of having been burnt in the past, most in the period 39 to 44 years ago; possibly only nine territories have escaped fire for more than 42 years. While no areas that have had long term territories have become unsuitable for Scrub-birds, there have been obvious changes in the vegetation over the last 13 years. Perhaps the most disturbing is the death of the rushes *Lepidosperma* spp. and shrubs in nesting areas in a number of territories. In two territories these areas have remained unsuitable for nesting for the last 10 years. Because fire is not a practical management tool in this area, it is important to understand the types and rates of change in the vegetation so that methods for maintaining areas suitable for Scrub-birds can be devised and applied when necessary.

A single population in a restricted area is always at risk, one method that can decrease the risk of extinction is translocation. In May 1975, I surveyed the area between the Waychinicup River and Mt Manypeaks (Fig. 1) and found that there were a number of areas suitable for Scrub-birds. Acting on this information the State Department of Fisheries and Wildlife asked for the area to be set aside for the translocation of Noisy Scrub-birds. The area was finally set aside for this purpose in April 1979. Unfortunately the area was almost entirely burnt out during the summer 1978/79.

In 1983 the Ranger at Two Peoples Bay found an area that had been only partially burnt and that was suitable for Scrub-birds. Ten males and four females were caught (males are easier to locate and hence trap) and moved to this area in June and July 1983. At least four males survived and established territories. A further two females were moved in November 1983 (G. Folley, pers. comm.).

The 1978/79 fire showed the area's vulnerability to fire and although it was the first fire in the area in at least 15 years, more frequent fires in the future cannot be excluded, especially as the area is difficult to protect. While it is unlikely that Mt Manypeaks and Two Peoples Bay will be burnt at the same time, translocation to other areas such as Torbay and Boojidup Creek would provide greater protection for the species.

The final option is captive breeding, often considered a last resort (Conway 1977). In the period 1975 to 1983 I ran a small captive breeding programme using hand-reared birds (three female, one male) collected when they were 16 to 19 days old. Twenty eggs were laid and four chicks were hatched in the two breeding seasons in which the male was sexually mature. The main problem was that the chicks suffered from calcium deficiency and only one chick survived (Smith *et al.* 1983). Unfortunately, financial restrictions led to the programme closing before this problem could be solved. This small, part-time programme showed that captive breeding, while feasible, was very expensive (approximately nine man years and $40,000 for maintenance and equipment). A programme to re-establish birds in the wild would cost many times these figures, a cost that may be better spent in translocation or protection of areas with Scrub-birds.

CONCLUSION

The Noisy Scrub-bird is a species whose ecological traits were established under more stable conditions than those operating in the last million years and in particular since the arrival of Europeans. Its terrestrial and territorial behaviour, low reproductive potential and low dispersive abilities reduce its ability to recolonize after local extinctions, while its adaptation to the ecotone between forest and swamp placed further constraints on its survival. Such habitat is limited in area and has a linear and patchy distribution, factors which compound the poor colonizing ability of the species. The species has only managed to survive the changes to the environment caused by Europeans because of the fortuitous coincidence of factors at Two Peoples Bay. At the present its future looks secure with a healthy and expanding population at Two Peoples Bay and the nucleus of another population at Mt Manypeaks. Further translocations, which are relatively cheap, would further secure its future and perhaps do away with the necessity of expensive and time consuming management of a single reserve.

REFERENCES

ANON, 1979. Western Australia. An atlas of human endeavour 1829-1979. Western Australia Government Printer.

CAMERON, J. M. R., 1979. Ambitions Fire. The agricultural colonization of pre-convict Western Australia. University of Western Australia Press, Perth.

CARTER, T., 1924. Birds of the Broome Hill district. *Emu* **23**: 306-18.

CONWAY, G. W., 1977. Breeding endangered birds in captivity: The last resort. Pp. 225-30 *in* Endangered birds — Management techniques for preserving threatened species ed. by S. A. Temple. University of Wisconsin Press, Madison.

CROOK, K. A. W., 1981. The break-up of the Australian-Antarctic segment of Gondwanaland. Pp 1-14 *in* Ecological biogeography of Australia ed. by A. Keast. Junk. The Hague.

FEDUCCIA, A. AND OLSON, S. L., 1982. Morphological similarities between the Menurae and the Rhinocryptidae, relict passerine birds of the Southern Hemisphere. Smithsonian Institute Press, Washington.

GREEN, N., 1979. Nyungar, the people. Aboriginal customs in the southwest of Australia. Creative Research, Perth.

HALLAM, S. J., 1975. Fire and Hearth. Canberra. Aust. Inst. Aborig. Studies.

ROBINSON, F. N. AND SMITH, G. T., 1976. The Noisy Scrub-bird — fact and fiction. *West. Aust. Nat.* **13**: 119-22.

SCHODDE, R. AND CALABY, J. H., 1972. The biogeography of the Australo-Papuan bird and mammal faunas in relation to Torres Strait. Pp. 257-300 *in* Bridge and Barrier. The Natural and cultural history of Torres Strait ed. by D. Walker. Department of Biogeography and Geomorphology, A.N.U., Canberra.

SMITH, G. T., 1976. Ecological and behavioural comparisons between the Atrichornithidae and Menuridae. Proc. 16th Internat. Ornithol. Congr. 125-36.

SMITH, G. T., 1977. The effect of environmental change on six rare birds. *Emu* **77**: 173-79.

SMITH, G. T., 1978. The Noisy Scrub-bird, Pp. 117-21 *in* The status of endangered Australasian wildlife ed. by M. J. Tyler. The Royal Society of South Australia.

SMITH, G. T. AND CALVER, M. C., 1984. The diet of the nestling Noisy Scrub-bird, *Atrichornis clamosus. Aust. Wildl. Res.* **11**: 553-8.

SMITH, G. T. AND FORRESTER, R. I., 1981. The status of the Noisy Scrub-bird *(Atrichornis clamosus). Biol. Conserv.* **19**: 239-54.

SMITH, G. T., NICHOLLS, C. A., MOORE, L. A. AND DAVIS, H., 1983. The results of a breeding programme for the Noisy Scrub-bird (*Atrichornis clamosus*) in captivity. *West. Aust. Nat.* **15**: 151-7.

SMITH, G. T. AND ROBINSON, F. N., 1976. The Noisy Scrub-bird, an interim report. *Emu* **76**: 37-43.

WEBSTER, H. O., 1962. Rediscovery of the Noisy Scrub-bird. *West. Aust. Nat.* **8**: 57-9.

WHITTELL, H. M., 1943. The Noisy Scrub-bird (*Atrichornis clamosus*). *Emu* **42**: 217-34.

CHAPTER 27

The Effects of Clearing for Agriculture on the Distribution of Cockatoos in the Southwest of Western Australia

D. A. Saunders[1], Ian Rowley[1] and G. T. Smith[1]

With the arrival of the Europeans in the southwest of Western Australia in 1827 native vegetation was cleared to create land for agriculture. The cumulative effect of this clearing has resulted in considerable changes to the ecology of the area. This chapter examines how these changes affected the distribution of the endemic cockatoos. The Galah, Little Corella, Long-billed Corella and Inland Red-tailed Black Cockatoo all feed on cereal crops and agricultural weeds and have increased their pre-settlement ranges, while the distribution of the Forest Red-tailed Black Cockatoo and Long-billed White-tailed Black Cockatoo which feed on native vegetation and nest in forest reserves (of which there are still large areas) have not changed. The Short-billed White-tailed Black Cockatoo and Major Mitchell's Cockatoo have undergone a reduction in range through reduction in food supply and nest sites. The Sulphur-crested Cockatoo was introduced by Europeans and has established breeding populations near Perth and Pinjarra.

The continued survival of the cockatoos depends on an adequate supply of trees with hollows. In the agricultural area of Western Australia most of the remnant woodland is on private land and is rapidly deteriorating with no regeneration to replace dying trees. It is suggested that there should be a ban on further clearing of native vegetation and release of land for agriculture until there is sufficient knowledge to plan such releases so that all species present before release are represented in perpetuity. Failure to do this or to rehabilitate and regenerate existing remnant woodland will lead to a total loss of native vegetation in the agricultural area, with a consequent loss of wildlife.

INTRODUCTION

THE ARRIVAL of Europeans in Western Australia in 1827 heralded a period of rapid changes to the landscape. The area which was most affected is the major agricultural region of the State which lies between the mouth of the Murchison River in the north and Esperance in the east (Fig. 1). Nine different cockatoos occur in this region; eight of them being present before 1827, while one, the Sulphur-crested Cockatoo, *Cacatua galerita*, was introduced sometime after 1930. All of these species have been influenced by the environmental changes that have followed European settlement. This chapter looks at some of these changes and considers those aspects of the biology of cockatoos that have determined the birds' responses to these alterations. We also discuss the implications for the conservation of these spectacular endemic animals.

HUMAN IMPACT ON THE ENVIRONMENT

The impact of Aborigines on the environment of the southwest is poorly known, but there is little doubt that it was spread over thousands of years whereas the Europeans brought about massive changes over a very short time. Initially the changes were small, but by 1850 the colonists had some idea of what crops could be grown and what domesticated animals could be farmed successfully. Farming areas started to evolve but the little land that was developed was mainly around towns and homesteads, and by 1890, only 500 km^2 of the southwest had been cleared (Fig. 1). Surrounding areas were largely unfenced and used to graze stock with shepherds moving the animals round the countryside. During the present century, farming spread to all suitable areas of the southwest. The government released land for agriculture as transport links from the ports were established and as the settlers moved

[1]C.S.I.R.O. Division of Wildlife and Rangelands Research, LMB No. 4, P.O. Midland, Western Australia 6056.

Pages 309-21 *in* BIRDS OF EUCALYPT FORESTS AND WOODLANDS: ECOLOGY, CONSERVATION, MANAGEMENT ed. by A. Keast, H. F. Recher, H. Ford and D. Saunders. Royal Australasian Ornithologists Union and Surrey Beatty & Sons.

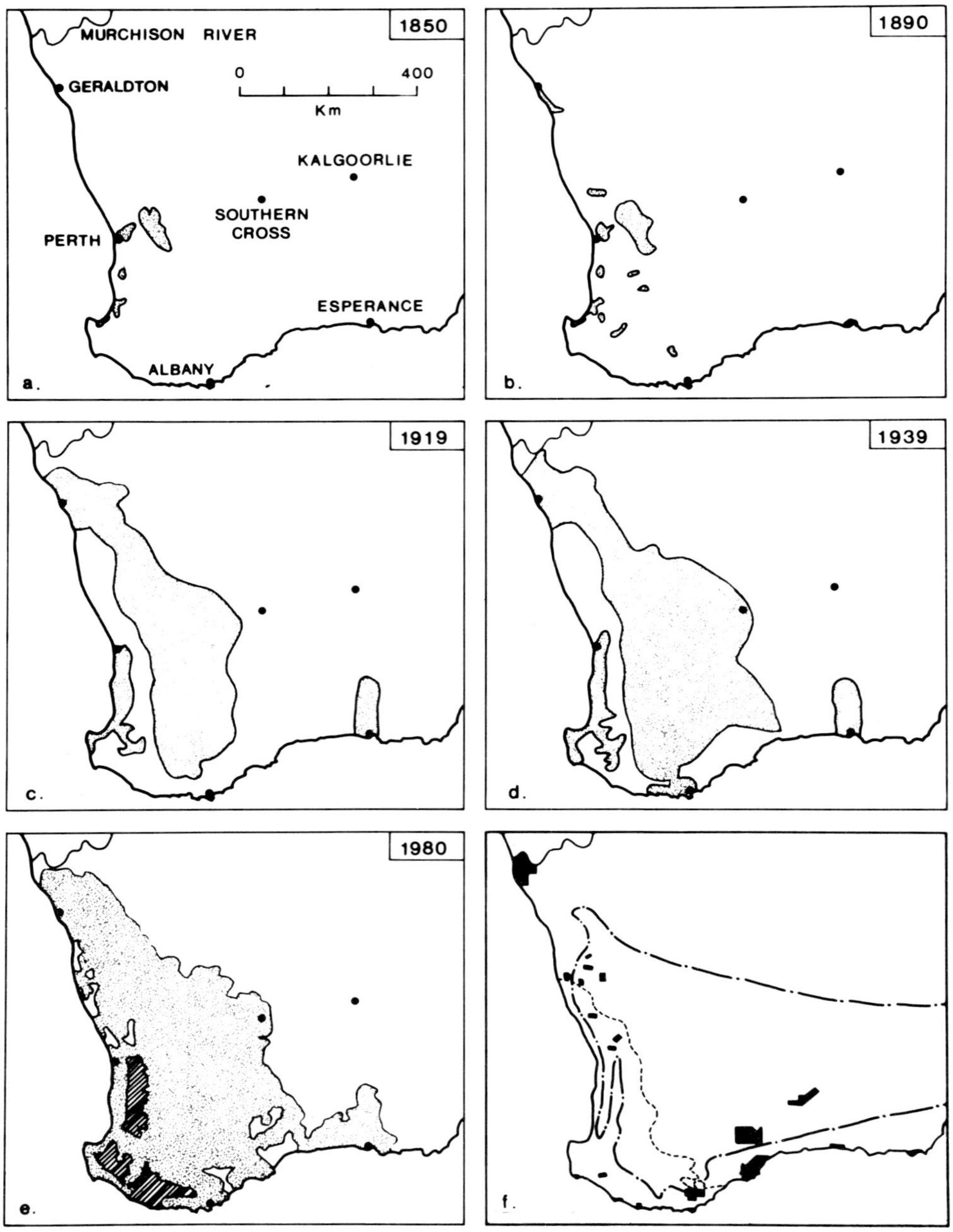

Fig 1. The spread of intensive agriculture as shown by area of land released for agriculture (modified from Anon 1979). Obviously land released for agriculture does not represent land cleared for agriculture but it does show the extent of the areas over which land was cleared. Areas of State Forest are shown in cross hatching in e. Figure f shows the area in which forest (Jarrah, Marri and Karri) formerly occurred (- - - - -), the area in which woodland (Wandoo and Salmon Gum) occurred (·—·—·—·) and the areas now designated as National Parks or Nature reserves.

in, more land was cleared. By 1968 nearly 130,000 km^2 had been cleared.

Western Australia is one of the few places in the Western world to undergo large scale land development in recent years. Since 1945 more than 3300 new farms have been created in the southwest (Anon 1979) and 54 percent of all land developed for agriculture in Western Australia was cleared between 1945 and 1982. The spread of agriculture from 1850 until 1982 is shown in Figure 1 and the cumulative area of land cleared between 1890 and 1970 is shown in Figure 2.

With the exception of coastal areas, a few National Parks and the State Forests, the southwest of Western Australia is now extensively farmed (see Fig. 1) and is generally known as "the wheatbelt". In fact, National Parks and Nature Reserves account for just over 2% of this area, State Forests about 4% and about 18% is vacant Crown land. The remainder is taken up by rural holdings most of which have been

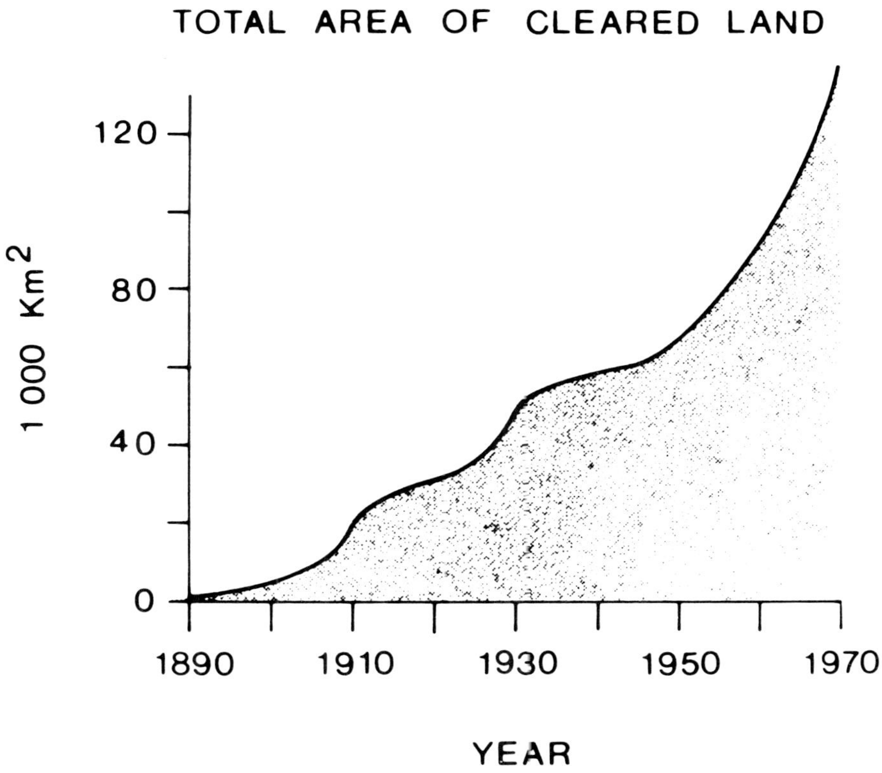

Fig. 2. Cumulative area of land cleared in the southwest of Western Australia from 1890.

cleared and are now farmed (Kitchener *et al.* 1980). The clearing of this large area has drastically altered the landscape with large areas of forest and woodland being replaced by plains clothed in exotic annual vegetation. There is no doubt that the ecology of the region has changed dramatically as a result (Main 1967).

While most of the native vegetation has disappeared, remnant patches of woodland have been left as shelter around buildings and paddocks, along roads and as isolated individual trees. The cockatoos which occur in this area are obligate hole nesters (Saunders *et al.* 1982) and trees are the only source of such holes in the region. In other parts of Australia some species do nest in holes in cliffs (Forshaw 1981) but there are few such sites in the southwest of Western Australia. Some cockatoo species feed predominantly on native vegetation (Table 1) but throughout the wheatbelt there is little or no native vegetation under or around the remaining trees. At present there are many nest sites available (Saunders *et al.* 1982) but for some species of cockatoo, food may be the critical limiting resource.

CHANGES IN COCKATOO DISTRIBUTION

The cockatoos which occur in the southwest of Western Australia are: the Galah, *Cacatua roseicapilla assimilis*; Major Mitchell's Cockatoo, *C. leadbeateri*; the Sulphur-crested Cockatoo; the Little Corella, *C. pastinator gymnopis*; the Long-billed Corella, *C. p. pastinator* [Corella nomenclature follows Schodde *et al.* (1979)]; the Forest Red-tailed Black Cockatoo, *Calyptorhynchus magnificus naso*; the Inland Red-tailed Black Cockatoo, *C. m. samueli* [nomenclature follows review by Saunders and Schodde (in prep.)]; the Long-billed White-tailed Black Cockatoo, *C. baudinii*; and the Short-billed White-tailed Black Cockatoo, *C. funereus latirostris* [nomenclature follows Saunders (1979a)]. The distribution of each of these cockatoos at the turn of the century and their present distributions are shown in Figure 3. The diet of each species or subspecies in the southwest is shown in Table 1, while clutch size, breeding seasons, attachment to nesting hollow during the year, abundance and a statement of population trends are given in Table 2. The productivity of those species for which data are available is given in terms of fledglings per nesting attempt and fledging weights expressed as a percentage of the mean weight of the adult female are given in Table 3. These data will be discussed in the following sections in relation to each cockatoo and its responses to changes in the environment.

i. *The Galah.* The Galah is the great success story of the cockatoo world and dramatically expanded its range as a result of agricultural development (Fig. 3a). This species has been studied at several

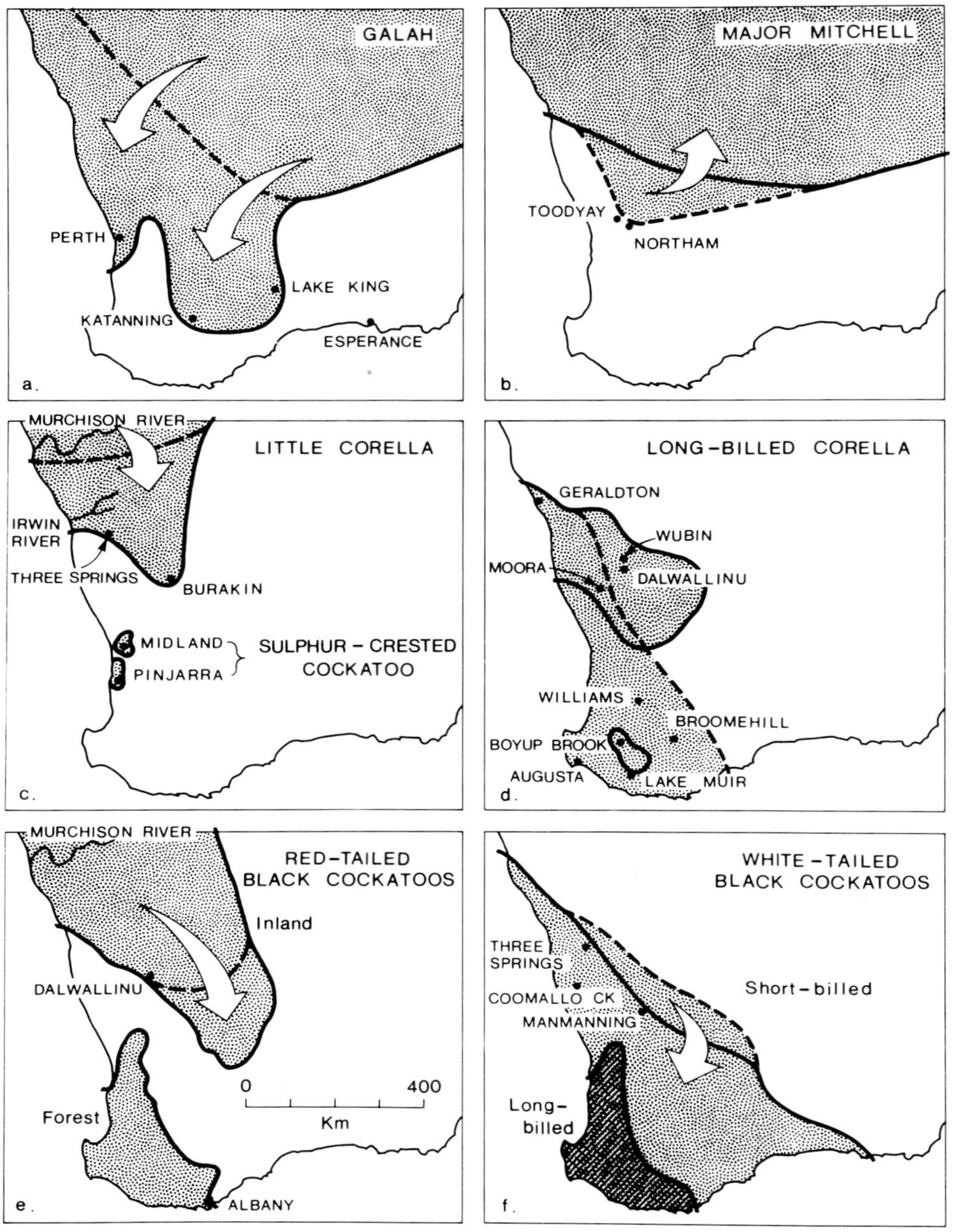

Fig. 3. Distributions of the cockatoos found in the southwest of Western Australia at the turn of the century and present. The dotted line indicates limit of distribution around 1900 and the solid line indicates distribution at present. Arrows indicate direction of change in range. Place names mentioned in the text are shown also.

locations in the wheatbelt since 1970 and the results of these studies are presented in Buckland *et al.* (1983), Rowley (1980, 1983), Saunders and Smith (1981), Saunders *et al.* (1982), Smith and Saunders (1986). It is predominantly a ground feeder, foraging for a variety of seeds (Table 1) which, before the advent of agriculture, derived from grasses, *Acacia* spp. and other arid or semi-arid shrubs. Being both a seed eater and an obligate hole nester, it occurred along the water-courses of the arid and semi-arid zones, where both resources were concentrated. Before the natural vegetation was removed, the Galah does not appear to have penetrated the sandplain heath which occupied much of what is now known as the wheatbelt. In clearing this area, man created a "riverine plain", with accessible water throughout the area and with nest sites and food (cereal grain and weed seeds, Table 1) in abundance. This suited the Galah which was probably already in an expansive phase following long-term climatic changes (Serventy and Whittell 1976).

The spread of the Galah was largely undocumented and by the 1970's it had spread throughout the wheatbelt to the southernmost

Table 1. Major food items of the cockatoos in the southwest of W.A. (++ indicates that item comprised the bulk of the diet. In the case of Major Mitchell's Cockatoo the three items so indicated all appear to be important).

	Agricultural Grains	Grass or Herb Seed	Double Gee Seed	Marri Seed	Other Eucalypt Seed	Proteaceae Seed	*Acacia* Seed	*Callitris* Seed	Wild Melon Seed	*Pinus* Seed	*Erodium* Seed	Insect Larvae	Digging tubers from ground
Galah	++	+				+	+				+	+	
Major Mitchell's Cockatoo	++	+	++			+	+	+	++			+	
Sulphur-crested Cockatoo	++	+		+					+	+		+	
Little Corella	++	+	+									+	
Long-billed Corella	++	+	+								+	+	+
Forest Red-tailed Black Cockatoo				++	+								
Inland Red-tailed Black Cockatoo			++						+				
Long-billed White-tailed Black Cockatoo				++	+	+					+	+	
Short-billed White-tailed Black Cockatoo				+	+	++				+	+		

extremity, approximately coincidental with the railway lines and silos serving farmers between Katanning and Lake King. It has not yet moved into the Esperance area, but probably will in time. There has also been a slow spread from the north, down the coastal plain, west of the forest on the Darling Scarp (see Fig. 1f for areas of forest) and throughout the Perth metropolitan area. This spread has been aided by the escape or deliberate release of aviary birds as there are numerous birds of the eastern subspecies *C. r. roseicapilla* (with pale crowns and red periophthalmic rings) in this population. This coastal expansion has been helped by the great increase in the popularity of horses over the last 10 years or so and the consequent trend towards establishing "farmlets" of about two hectares where these animals are kept for riding. Such holdings are too small to maintain horses and they are usually fed grain, much of which passes through undigested; the Galah is not a fussy feeder and thrives on these "left-overs".

Recent studies (Rowley 1983) suggest that it is the juvenile element of the population that is most mobile and acts as colonizers. Once established in a new area, the young birds range as a flock over an area of about 10 km^2. From within this flock, pairs form, select and take up suitable hollows, thus developing into new sedentary local breeding populations which in turn bud off colonizers to continue the range expansion. Data from the resighting of birds tagged as nestlings (Buckland *et al.* 1983) and from the return of their leg bands (Rowley 1983) suggest that 12% of these birds dispersed more than 20 km from their birth place. Since the Galah seldom breeds before it is three years old, this dispersal allows a rate of spread in excess of 7 km per year which accounts for the colonization of the wheatbelt since the 1920's.

The Galah has a decided advantage over most cockatoos because the breeding pair maintains an interest in the nest hole all year, returning to it at least once a day during the non-breeding season. The other endemic cockatoos have a period outside the breeding season when they take no interest in the hollow (Table 2) while the introduced Sulphur-crested Cockatoo has a similar advantage to the Galah. The Galah pair are able to hold their hollow against "take-over" bids from other species under most conditions. At Manmanning, an area where the Galah and Short-billed White-tailed Black Cockatoo both breed, a pair of the latter species laid two eggs in a hollow which had been used by a Galah pair in earlier seasons. When the Black Cockatoo laid her eggs, the Galah pair were also preparing the hollow, dropping in sprays of green leaves. For some days there was fighting between the two pairs and the Galah pair did enter the hollow when the Black Cockatoo female was absent. Both Black Cockatoo eggs were broken and the Galah pair proceeded to nest in the hollow.

The Galah lays more eggs in a clutch than any other cockatoo (Table 2) but it produces roughly the same numbers of young per nesting attempt as other members of the genus *Cacatua* (Table 3). The difference between potential productivity and actual productivity results mainly from high nestling mortality. While the generally abundant food and nest sites have fostered a large population, up to 60% of the nestlings in some populations may be killed by a disease which is always present although its incidence appears to vary with both season and area

Table 2. Aspects of the biology of the cockatoos in the southwest of Western Australia. In the first five species listed, clutches of one are probably incomplete. Abundance is ranked in a crude way based on our knowledge of these species in the southwest. The Inland Red-tailed Black Cockatoo has two egg-laying periods each year; autumn shown as A and spring shown as S.

	Usual clutch size (range)	Egg-laying period	Interest by adults in nest hollow during year	Abundance of birds in their range in the southwest	Population trend at present
Galah	4 (2-4)	August-October	All year	Very common	Increasing
Major Mitchell's Cockatoo	3 (1-4)	August-September	Breeding season and up to 6 months before	Uncommon	Decreasing
Sulphur-crested Cockatoo	2 (1-3)	July-August	All year	Total of 300-500 birds	Increasing when not subject of control
Little Corella	3 (1-4)	August-October	Breeding season	Uncommon	Increasing
Long-billed Corella	3 (1-4)	August-September	Breeding season and up to 6 months before	Uncommon	Increasing
Forest Red-tailed Black Cockatoo	Probably 1	Probably August-October	Breeding season	Uncommon	Stable
Inland Red-tailed Black Cockatoo	1	A. February-April S. July-October	Breeding season	Common	Increasing
Long-billed White-tailed Black Cockatoo	Probably 2 (1-2)	August-October	Breeding season	Uncommon	Stable
Short-billed White-tailed Black Cockatoo	2 (1-2)	July-October	Breeding season	Common	Decreasing

(Rowley 1980; Smith and Saunders 1986). However the Galah renests readily after its first nesting attempt has failed (80 of 648 nesting attempts listed in Table 3 were re-lays) and this raises the figure for productivity per pair to 2.16 fledglings per year.

ii. *Major Mitchell's Cockatoo*. This species is distributed patchily throughout the semi-arid and arid areas of Australia and has been studied in the northeastern wheatbelt since 1978. Most populations number 50-150 and these birds coalesce into one or two feeding flocks outside the breeding season.

Historical records document Major Mitchell's Cockatoo breeding at Toodyay and Northam (Fig. 3b), but nowhere to the south of the Great Eastern Highway except for the coastal strip near Esperance. This south coastal population is probably contiguous with populations further east along the coast into South Australia. Major Mitchell's Cockatoo is a versatile feeder, consuming insect larvae and a variety of seeds from cones, nuts and fruits. In the wheatbelt, spilt grain and weed seeds such as Doublegee, *Emex australis,* and Wild Melon, *Citrullus vulgaris,* provide the major part of its diet (Table 1). With abundant food and ample nest sites, this species should not be under any threat but although its annual productivity and survival of fledged young are the equal of the much more common Galah, its range appears to be shrinking. Part of the explanation for this decline concerns the availability of nest sites. While there are enough nest sites *in toto* the distribution in blocks of woodland isolated by clearing limits their availability. This species is intolerant of breeding conspecifics within 2 km of an active nest and a breeding pair will deny other pairs access to potential nest sites within this area. Thus while a 10 ha block of residual woodland may accommodate one pair of breeding Major Mitchell's Cockatoo, it may have 10 or more breeding pairs of the Galah. Over a 450 km^2 study area with its patchwork of woodland admidst large wheat paddocks, only 16 pairs of Major Mitchell's Cockatoo had breeding territories while a further 100 birds formed a non-breeding flock. The corresponding figures for the Galah would be approximately 350 breeding pairs and a non-breeding flock of more than 3000 birds. As the land is cleared to create bigger paddocks so more clumps of woodland are destroyed, displacing breeding pairs which may not be able to establish themselves in the area.

Another reason for the decline of Major Mitchell's Cockatoo is its popularity as an aviary bird and its high price in relation to more common cockatoos. The locations of nests were known to land owners who tended to "harvest" them every year for the avicultural trade. This has been a common practice, which, when combined with clearing has resulted in the decline of this cockatoo.

iii. *The Sulphur-crested Cockatoo*. Although found naturally in the Kimberley Region of northwestern Australia, this species did not occur in the southwest before the arrival of the Europeans. There are now two populations in the southwest, one centred on Guildford-Midland on the outskirts of Perth, and the other at Pinjarra, further south on the Coastal Plain (Fig. 3c). These populations almost certainly became established as a result of aviary escapes or deliberate releases, the first reports being from near Pinjarra in 1935 (Long 1981). By the 1950's, the number of

Table 3. Productivity of those cockatoos studied in some detail.

	Area	No. fledglings per nesting attempt	No. nests studied (year)	Fledgling weight as a % of mean adult ♀ weight in brackets	
Galah	Three Springs) Burakin)	1.35	65 (1979)	92	(317 g)
	Manmanning	1.89	648 (1970-1977)	98	(317 g)
Major Mitchell's Cockatoo	Mukinbudin	1.93	83 (1977-1982)	88	(390 g)
Little Corella	Three Springs	1.20	5 (1979)	85	(490 g)
Long-billed Corella	Coomallo Creek	1.87	15 (1972-1976)	94	(610 g)
	Three Springs) Burakin)	1.51	37 (1979)	84	(610 g)
Inland Red-tailed Black Cockatoo	Three Springs	0.38	428 (1975-1980)	76	(660 g)
Short-billed White-tailed Black Cockatoo	Coomallo Creek	0.65	400 (1970-1976)	89	(650 g)
	Manmanning	0.35	94 (1970-1976)	76	(650 g)
	Three Springs	0.23	13 (1975, 1976, 1979)	65	(650 g)

reports had increased markedly and the birds were known to have bred successfully in the 1970's; by 1982 the combined population was of the order of 300-500 birds. The area where this species occurs is occupied mainly by small holdings and as with the Galah, the Sulphur-crested Cockatoo obtains food from grain supplied to stock (Table 1).

The presence of this species is viewed with concern by the Department of Fisheries and Wildlife and by the Agriculture Protection Board. The former is concerned because of the possibility that the Sulphur-crested Cockatoo will compete with the endemic cockatoos, particularly for nest sites. In this context it may have an advantage because of its larger size and the habit of the breeding pairs returning to their nest site at least once a day throughout the year (Table 2). The Agriculture Protection Board is concerned because the Sulphur-crested Cockatoo is known to be a pest of various crops, including grapes which are grown commercially around Perth.

Several attempts at control have been made. The first was to declare an open season and to encourage the trapping and shooting of the birds. It was hoped that licensed trappers would remove birds for the aviary trade. This did not work because the best economic return was from nestlings and the trappers concentrated on them and few free-flying birds were trapped. Between 1977 and 1980, only 21 birds were taken. During the same period, 36 birds were exported from the State and 109 imported, a net gain of 94 within the State. At present, there is a steady campaign of shooting the birds as opportunity presents.

iv. *The Little Corella.* This species originally extended from the Pilbara south to the Murchison River (Fig. 3c). Within this area it occurred along watercourses, nesting in the River Gums, *Eucalyptus camaldulensis* and foraging on the riverine plains where it fed on the seeds of shrubs or grasses and wasp larvae from galls on *Acacia spp*. Provision of permanent water for stock in the pastoral area probably allowed the Little Corella to expand away from the watercourses although it may have been extending its range already following long-term climatic changes. Where it met the wheatbelt the creation of suitable habitat and the provision of abundant food from grain (Table 1) allowed it to expand south. By 1928 it had been recorded from the lower reaches of the Irwin River and recently it has been noted breeding just north of Burakin and near Three Springs (Saunders *et al.* 1982). There is no reason why the Little Corella should not expand its range further. It occurs in sympatry with the Galah although both have similar diets in the wheatbelt (Table 1). Food is not a limiting factor and competition over nest sites is minimised because the Galah can use a hollow with a smaller entrance (Saunders *et al.* 1982). The Little Corella does overlap the range of the Long-billed Corella and the two subspecies feed and flock together. In those areas where both subspecies are known to breed sympatrically there is no evidence of interbreeding and collections from the area of overlap provide no evidence of hybrids. Both subspecies have the same hollow requirements (Saunders *et al. loc. cit.*) and this may lead to competition for nest sites and the eventual extinction of the northern population of the Long-billed Corella.

v. *The Long-billed Corella.* This subspecies has been studied at several locations and the results are presented in Saunders (1977a), Schodde *et al.* (1979), Saunders and Smith (1981), Saunders *et al.*

(1982), and Smith and Saunders (1986) (Table 3). Historically it appears that the Long-billed Corella inhabited the woodlands surrounding the forest areas of the southwest (Fig. 3d and Fig. 1f) as well as some of the more open valleys within the forest and there is evidence that it visited the coastal plains around Perth and Albany during the non-breeding season. By 1912 there seemed to be a marked reduction in numbers, especially in the area to the south and southeast of Perth (Carter 1923). Since these birds have been reported eating grain and damaging ripening crops it is likely that poisoning was one cause of this decline.

By the 1920's this cockatoo had two distinct populations. The larger northern population was located in the area bounded by Moora, Geraldton and the Coast. The southern population was centred round Lake Muir and the area to the north where there were about 100 birds, with smaller groups recorded as far as Williams, Augusta and Broome Hill (Fig. 3d). Since the 1930's the range of this southern population has contracted but is still centred round Lake Muir where most of the population breeds. After breeding, the population moves north to the east of Boyup Brook and spends the summer in one or two large flocks. In 1978 this population numbered about 1000, which, although small, could pose problems if farming practices in the area changed. Sunflower is not grown in any significant quantity at present, but these birds will become a nuisance if oil seeds are grown commercially. This situation has already developed in Victoria, where the Slender-billed Corella, *Cacatua tenuirostris,* a rare species poses threats to an agricultural enterprise and needs to be managed (Forshaw 1981).

Since the 1920's, the northern population of the Long-billed Corella has expanded to the east following agricultural development. By the mid 1930's there were small numbers at Dalwallinu, beyond the pre-settlement range, and Wubin had been reached by the early 1940's. By 1982 the bulk of the population was in the western two-thirds of the range, while in the eastern third flocks were small and widely dispersed. Throughout this area, cereal grain is the most important food for the Long-billed Corella (Table 1), and given the productivity of this bird and the abundance of food (Table 3) there appears to be no reason why this expansion of range should not continue until all suitable habitat is colonized. This bird is raucous and noisy so that many farmers do their best to discourage it from settling in their neighbourhood.

vi. *The Forest Red-tailed Black Cockatoo*. Two subspecies of Red-tailed Black Cockatoo occur in the southwest: the Forest Red-tailed Black Cockatoo and the Inland Red-tailed Black Cockatoo. Morphological differences exist between the two, the major ones being the larger mandible and maxilla of the Forest Red-tailed Black Cockatoo and the brighter striping and spotting of that female's body feathers.

The distribution of the Forest Red-tailed Black Cockatoo is shown in Figure 3e. This bird is restricted to the Jarrah, *E. marginata*, Marri, *E. calophylla* and Karri, *E. diversicolor* forests of the extreme southwest (from Perth south to Albany). There are extensive areas of uncleared native forest throughout this region (Fig. 1f), and while the forest has been altered by logging, Marri (the main food item, Table 1) is widespread and common and nest sites are still present in sufficient quantities to ensure the continued survival of this cockatoo.

The mandible and maxilla of the Forest Red-tailed Black Cockatoo are enlarged, presumably as an adaptation to Marri nuts. The powerful bill enables it to chew out the base of the mature seed capsule and extract the seed. While not common this bird is widely distributed and the population may be stable. Nothing is known about breeding, except that birds have been seen entering and leaving hollows in very tall trees in spring.

vii. *Inland Red-tailed Black Cockatoo*. This cockatoo is widely distributed through central Australia and has been studied at several localities in the southwest since 1975 and the results have been published in Saunders (1977b), Saunders and Smith (1981), Saunders *et al.* (1982) and Smith and Saunders (1986). Before settlement it occurred on the watercourses of the arid and semi-arid areas, probably having its southern limits around Dalwallinu. It fed on the seeds of grasses and *Acacia* spp., *Casuarina* spp. and *Hakea* spp., as well as insect larvae infesting various species of plant. This species is the only cockatoo known to breed in the autumn and spring (Saunders 1977b) and some individuals may breed at six month intervals.

At present, throughout the range of this species south of the Murchison River it feeds mainly on seeds of the Doublegee (Table 1). Of 237 observations of feeding flocks, 219 were seen feeding on Doublegee. The remainder were seen on other agricultural weeds (Safron Thistle, *Carthamus lanatus* 2; Wild Radish, *Rhaphanus raphanistrum* 1; Wild Melon 5); native vegetation (*Acacia acuminata* 1; *A. costata* 1; *Grevillea paniculata* 4; *Hakea recurva* 2); and insect larvae (2). Analysis of the crop contents of 16 dead birds showed the same dependence on Doublegee. All 16 contained Doublegee seed and there was Wild Radish seed in one and Wild Turnip, *Brassica tournefortii,* in another. Doublegee was introduced to the southwest in the 1830's and areas of heavy infestation (mainly the central and northern wheatbelt) may contain nearly 9000 seeds per m^2 (Gilbey n.d.). The seeds remain viable in the soil for several years. As each bird needs no more than 3500 seeds per day there is ample food available for this cockatoo which is gradually increasing its range.

Productivity is low with slightly more than one-third of all nesting attempts giving rise to free flying young and the mean weight of the young at fledging is about 76% of the mean weight of the adult females. These figures are much lower than those of the other cockatoos we have studied and indicate that the Inland Red-tailed Black Cockatoo may be experiencing difficulty in breeding successfully. Their productivity and fledging weights are comparable with those of the Short-billed White-tailed Black Cockatoo from Manmanning, an area where the population suffered from shortage of food and subsequently became extinct (Saunders 1982). The Inland Red-tailed Black Cockatoo at Three Springs has a hatching success of 64.7% (428 nests 1975-1980). This is less than the Short-billed White-tailed Black Cockatoo from Manmanning which had a hatching success of 70.7% (157 eggs) while that species at Coomallo Creek (an area where food was not limiting) had a much higher hatching success of 76.5% (712 eggs). The Short-billed White-tailed Black Cockatoo has a usual clutch of two eggs, while the Inland Red-tailed Black Cockatoo has only one egg per clutch. The Galah and Long-billed Corella with larger clutches, four and three respectively, also have greater hatching success than the Inland Red-tailed Black Cockatoo [77.2%; n = 246: 82.1%; n = 95: Smith and Saunders (1986)]. The low hatching success and poor fledging weights of the Inland Red-tailed Black Cockatoo indicate that the population in the wheatbelt is being affected by some detrimental factor or factors which result in lowered productivity.

Food quality rather than quantity may be one cause of the problem. It is possible that the seed of the Doublegee is not providing the nestlings with a suitable diet. Another explanation of this low productivity involves agricultural treatment of Doublegee and the other weed species on which the Inland Red-tailed Black Cockatoo feeds. All of these weeds are sprayed with herbicides as part of the recommended method of control (Gilbey *loc. cit.*). Ingestion of these herbicides or their metabolites by the birds is inevitable and they may be having some effect on the incubation or brooding behaviour of the females resulting in a reduction in the number of chicks hatching and fledging.

viii. *Long-billed White-tailed Black Cockatoo*. This cockatoo is regarded as a species separate from the Short-billed White-tailed Black Cockatoo (Saunders 1974a, 1979a; Campbell and Saunders 1976) and occupies the same range as the Forest Red-tailed Black Cockatoo (Fig. 3f). This bird also relies on the seeds of the Marri (Table 1) and uses its elongate maxilla to hook the seeds out of the capsule without having to tear the capsule apart. It does cause occasional damage to orchards when the birds attack the fruit to extract the seeds. The range has not altered a great deal since settlement, for the same reasons set out for the Forest Red-tailed Black Cockatoo.

ix. *Short-billed White-tailed Black Cockatoo*. This cockatoo has been studied since 1969 and the results of this work may be found in Saunders (1974b, 1974c, 1977c, 1979b, 1980, 1982), Saunders and Smith (1981) and Smith and Saunders (1986). It feeds primarily on the seeds of native vegetation, particularly those of the Proteaceae (Table 1). Unlike most of the other cockatoos it does not feed on cereal crops and in the wheatbelt it is entirely dependant on native vegetation. The breeding biology of this cockatoo was studied at two sites over seven years (Fig. 3f). At Coomallo Creek there were extensive areas of uncleared vegetation while at Manmanning (towards the limit of its range) clearing began in the 1920's and by the 1970's there was little native vegetation remaining. The difference in productivity between these two areas is shown in Table 3 with only one-third of the breeding pairs at Manmanning producing young, compared with nearly twice that at Coomallo Creek. Females at Manmanning often foraged while they had eggs in the nest in contrast to females at Coomallo Creek where they depended on the males for food and were never seen away from the immediate area of the nest during incubation. In addition females at Manmanning ceased brooding the nestlings during the day when the nestlings were about 10 days old while those at Coomallo Creek never left their nestlings alone for long periods until they were at least 21 days old. The birds at Manmanning often did not return to the nest between dawn and dusk whereas the birds at Coomallo Creek invariably returned at mid-morning to feed their nestlings. Weight of nestlings increased more slowly at Manmanning (9.8 g per day) than those at Coomallo Creek (13.0 g per day) and nestlings leaving the nest were lighter at Manmanning (mean fledging weight 501 g; range 280-650 g compared with 584 g; range 480-750 g). These differences can be related to the poor food supply at Manmanning and the need to forage over greater areas to obtain sufficient food. The population at Manmanning was less than one-fifth the size of that at Coomallo Creek yet they foraged over an area nearly four times greater. By 1978 the Short-billed White-tailed Black Cockatoo had become extinct at Manmanning. A study at Three Springs, also towards the limit of the birds range, showed that what had happened at Manmanning was also taking place at Three Springs for the same reasons (Table 3).

The contraction in the range of this cockatoo is a recent event and will continue for some time.

PROBLEMS POSED BY AVICULTURE

Parrots have long been popular as pets and aviary birds. The cockatoos in particular have been favourites all over the world and there is a ready market for them. The Sulphur-crested Cockatoo is one of the most popular of all cage birds (Forshaw 1981) yet there is little attempt to breed them in captivity. Demand for this species exceeds the ability of aviculture to provide aviary-bred birds and so the majority are being taken from the wild. This harvesting may not have any effect on species like the Galah, Long-billed Corella, Little Corella or Sulphur-crested Cockatoo but as mentioned earlier, may be hastening the decline of Major Mitchell's Cockatoo. The Short-billed White-tailed Black Cockatoo has only been recorded as breeding in captivity at one institution (Saunders 1976) and virtually all of the birds held in aviaries have been taken from the wild. Fortunately this species is not a popular aviary bird. The Inland Red-tailed Black Cockatoo does not breed readily in captivity, but like the former species is not common in aviaries and most of these birds have been wild caught. The Long-billed White-tailed Black Cockatoo and Forest Red-tailed Black Cockatoo are almost unknown in captivity because they are very difficult to catch. They both nest in tall trees (greater than 20 metres) and they are beyond the reach of all but the most foolhardy.

Most wild caught birds are taken as nestlings because hand-reared birds make more tractable aviary birds. This practice poses more problems in damage to nest sites than it does to direct regulation of the populations of most species. In Western Australia there is a flourshing illegal trade in wild caught nestlings. Many of these birds are taken from the wheatbelt where nest sites are between 2 and 16 metres from the ground (Saunders *et al.* 1982). A hole is often cut in the side of the tree close to the floor of the hollow to gain access to the nestlings; this is frequently left uncovered and cockatoos will not use such a hollow again. In some cases the tree itself is cut down to get the nestlings, if they survive the fall. A study of the Short-billed White-tailed Black Cockatoo was discontinued at one site because of human predation (Saunders 1979b).

The keeping of aviary birds and the movement of these birds to other areas poses other problems and these are illustrated in the sections on the Sulphur-crested Cockatoo and Galah. In the first case a feral population has become established in an area where they may compete with endemic species. In the second case there are birds from the eastern subspecies interbreeding with the local subspecies. Both cases are a direct result of the escape or deliberate release of aviary birds. Species like the Sulphur-crested Cockatoo and the Corellas are very noisy and complaints from neighbours often results in the birds being released. There are populations of both the Little Corella and the Long-billed Corella in the Perth metropolitan area and they occasionally occur in flocks with the Sulphur-crested Cockatoo; all of these populations are certainly the result of aviary escapees. Within these flocks there were at least two Slender-billed Corellas both of which originated in South Australia where they were trapped as part of a control programme, banded with South Australian National Parks and Wildlife Service bands and made available to the avicultural trade. They were transported to Perth where they were subsequently released or escaped. This practice is not restricted to the Perth area as a Slender-billed Corella was seen by G. T. Smith in a flock of the Long-billed Corella near Yerecoin (140 km north-east of Perth) during 1977. The result of these unwelcome additions to wild populations is the potential for changes to the gene pools of the endemic cockatoos as a result of breeding with the escapees.

FLOCKING AND THE POTENTIAL FOR CONTROL

One aspect of the biology of these cockatoos which has an important bearing on their conservation is their flocking behaviour during the non-breeding season. These flocks can represent the total population of the species from a very large area. The Short-billed White-tailed Black Cockatoo is the best documented example of this phenomenon (Saunders 1980). Following breeding the birds leave the nesting areas and cockatoos from many breeding populations amalgamate to form large flocks the size of which depends on the vegetation on which they are feeding. These flocks are concentrated in the coastal parts of the range of this species north of Perth and on the south coast (Fig. 1e). It is these areas which have the largest tracts of uncleared native vegetation. In addition there are several large pine *Pinus* spp. plantations (Gnangara planted area 7992 ha; Pinjar, 7559 ha; and Yanchep 4830 ha) around the Perth metropolitan area and these also provide food during the non-breeding season (Saunders 1974c). When feeding on native vegetation where food is dispersed, the flocks average 170 birds (range 11-400; N = 80) while it is not unusual to see flocks of over 1000 birds in pine plantations where food is concentrated and flocks of 5000-6000 birds have been reported (Perry 1948). The movements of individuals marked with wing-tags (Rowley and Saunders 1980) have shown that birds from about 25% of the total breeding range of the species concentrate around Perth in the non-breeding season. Estimates of the total population of the species, based on the distribution of birds during the breeding season, range from 11,000 to 60,000 birds, thus one flock of 5000 birds in a pine plantation may represent 8-45% of the total population. The formation of these flocks have lead to gross overestimates of population numbers. In the past, where poisoning was used as a form of control it was these large aggregations which were subject to control. As a result it is possible for a significant proportion of the population to be eliminated quickly. This has happened within the last 10 years when about 1000 Corellas were poisoned on one farm. Sightings of tagged individuals have shown that a flock of this size represents the population from an area of up to 6000 km^2.

AVAILABILITY OF NEST SITES

The continued existence of the cockatoos depends on an adequate supply of trees with hollows in them. In the southwest of Western Australia there are two regions to be considered; forest areas of the high rainfall regions of the southwest corner and the woodland areas of the drier agricultural region (Fig. 1f). The problems posed by the development of these regions are different and must be considered separately.

The Long-billed White-tailed Black Cockatoo and the Forest Red-tailed Black Cockatoo both occur in the southwest forests and the birds will be affected by accelerated harvesting of the forests. Mining for bauxite in the northern Jarrah forest (Collins *et al.* 1985) involves clearfelling with the loss of both food and nest sites Mining has been carried out since 1963 and will probably continue until the turn of the century. Only 7-8% of the forest can be mined and much of this is subject to various environmental constraints (Collins *et al. loc. cit.*). The clearfelling of large tracts of forest for woodchipping is of much more immediate concern. The Forests Department are aware of the problem of loss of nest sites and they state that "for some time those species which frequent the tree canopy and those needing holes in trees for breeding will remain severely reduced. This includes all the parrot family, possums and some other animals" (Anon. 1973). They say that "it appears that although numbers of some species will be temporarily reduced, no species will be wiped out and most will eventually recolonize cut-over country as it grows to maturity". The stated management policy is to convert mature forest stands to fully stocked regrowth stands which will result in increased productivity. Thus in "the foreseeable future the saw-milling industry will change from using a predominantly mature log resource to one comprising mainly regrowth logs. This change will come regardless of a woodchip industry. Except in that a woodchip industry will make the regrowth sawmill resource larger and be available sooner, . . ." (Anon. *loc. cit*). Basically the forest will not be allowed to grow to maturity and hollows are unlikely to develop in these areas. The existence of all hole-nesting species will depend on those areas preserved in an uncut state, along streams, around swamps and along road reserves. In an example of how forest blocks will be managed for woodchipping (Anon. 1973), areas reserved from cutting amount to 19.5% of the block, but 4.8% of the area is not forest. Hole nesting species must then depend on 14.7% of the block for nest sites and for food.

At present there is no information on the distribution of hollows in different species of tree within the forest, nor on the distribution of hollows within the forests, nor on how long it takes to form a hollow in any species. Without this basic knowledge there is no way of knowing if the management plans of the Forests Department are adequate to preserve the two Black Cockatoos. It will be a shortage of nest sites which will limit the populations, not food as the regenerating forest will provide ample sustenance. There is no doubt that the total numbers of both species of Black Cockatoo will fall as large areas of nest sites are removed from the forest. At present there is no way of knowing if either species will survive in the forests of the future.

Throughout the wheatbelt, the majority of hollows occur in the Salmon Gum, *E. salmonophloia* and Wandoo, *E. wandoo*, with the former being more important in the drier areas. The continued survival of the Galah, Major Mitchell's Cockatoo, Little Corella, Long-billed Corella, Inland Red-tailed Black Cockatoo and Short-billed White-tailed Black Cockatoo are linked to the survival of these species of tree. As agriculture developed it was recognized that the soils carrying Salmon Gum would produce good cereal crops and this indication of good soils meant that woodland areas were quickly taken up and cleared. The result is that very little woodland is represented on Nature Reserves or National Parks. Most of the remaining woodland is on private land, or along roads and railway lines. Because of the importance of woodland on private land, a study was carried out to examine the characteristics of hollows selected by various cockatoos using Salmon Gum woodland, the size composition of the woodland and its health (Saunders *et al.* 1982). The area studied was near Three Springs and was a 15 ha block left standing after the surrounding land had been cleared of native vegetation and developed for mixed wheat and sheep farming. Sheep had access to the block and grazed and sheltered under the trees.

The block contained 173 trees which had 241 hollows large enough for cockatoos. During the spring of 1978 47% of the hollows were occupied [Galah (48); Long-billed and Little Corella (28); Inland Red-tailed Black Cockatoo (16); Short-billed White-tailed Black Cockatoo (1); Port Lincoln Parrot, *Barnardius zonarius* (13); Mountain Duck, *Tadorna tadornoides* (5); Nankeen Kestrel, *Falco cenchroides* (2); and Cockatiel, *Nymphicus hollandicus* (1)]. There were differences in the sizes of hollows used by the cockatoos and there was a trend for hollow size to decrease in the order Black Cockatoos, Corellas and the Galah.

Examination of the size composition of the woodland revealed that there were few trees in the smallest size class. The smallest Salmon Gum had a circumference at breast height of 0.11 m and predates the introduction of sheep in 1929 indicating that there had been no recruitment since then. The smallest tree to provide a hollow for use by any cockatoo had a circumference at breast height of 0.70 m. From the limited data on growth rates available, this tree was in excess of 130 years old.

During May 1978 each tree was rated according to its health and this was repeated in July 1981. Of the 194 (173 hollow trees plus those with no apparent hollows) trees in the sample, nearly 23% were healthy and 19% were dead in 1978. By 1981, less than 5% were healthy and over 40% were dead.

These findings typify what is happening throughout the wheatbelt and the long-term prospect for any species depending on trees for their survival is not good. At this stage it is essential to place a total ban on any further release of vacant Crown land for extension of the agricultural areas and consequent clearing of native vegetation. There are proposals to release more land for farming but these plans should be held in abeyance until such time as there is sufficient knowledge to ensure that every species can be maintained in such areas after development. Of vital concern at present is the rehabilitation and regeneration of native vegetation in the wheatbelt. No expertise is available to guide such a project and a failure to come to grips with the problem will lead to the total loss of native vegetation in the wheatbelt and the consequent loss of wildlife.

ACKNOWLEDGEMENTS

We are grateful to Dr E. M. Russell and Dr R. D. Wooller for commenting on an earlier draft of this chapter; to Mr C. P. de Rebeira who drew the figures; and to Miss P. Large and Miss C. Nota who typed the manuscript.

REFERENCES

ANON, 1973. Focus on the Marri Woodchip Project. *Forest Focus* **12**: 3-15. Forests Department of Western Australia.

ANON, 1979. Western Australia. An atlas of human endeavour 1829-1979. Western Australian Government Printer.

BUCKLAND, S. T., ROWLEY, I. AND WILLIAMS, D. A., 1983. Estimation of survival from repeated sightings of tagged galahs. *J. Anim. Ecol.* **52**: 563-73.

CAMPBELL, N. A. AND SAUNDERS, D. A., 1976. Morphological variation in the White-tailed Black Cockatoo, *Calyptorhynchus baudinii*, in Western Australia. *Aust. J. Zool.* **24**: 589-95.

CARTER, T., 1923. Birds of the Broome Hill District. *Emu* **23**: 125-42, 223-35.

COLLINS, B. G., WYKES, B. J. AND NICHOLS, O. G., 1985. Re-colonization of restored bauxite minelands by birds in southwestern Australian forests. Pp. 341-54 *in* Birds of the Eucalypt Forest and Woodlands: Ecology, Conservation, Management ed. by A. Keast, H. F. Recher, H. Ford and D. Saunders. Royal Australasian Ornithologists Union.

FORSHAW, J. M., 1981. Australian Parrots. Second (revised) edition. Lansdowne Editions. Melbourne.

GILBEY, D. J. (no date). The Doublegee problem in Western Australia. Bulletin of the Western Australian Department of Agriculture. 3pp.

KITCHENER, D. J., CHAPMAN, A., DELL, J. AND MUIR, B. G., 1980. Lizard assemblage and reserve size and structure in the Western Australian wheatbelt — some implications for conservation. *Biol. Conserv.* **17**: 25-62.

LONG, J. L., 1981. Introduced birds of the world. A. H. and A. W. Reed, Sydney.

MAIN, B. Y., 1967. Between Wodjil and Tor. Jacaranda Press, Melbourne and Landfall Press, Perth.

PERRY, D. H., 1948. Black Cockatoos and pine plantations. *West. Aust. Nat.* **1**: 133-35.

ROWLEY, I., 1980. Parent-offspring recognition in a cockatoo, the Galah, *Cacatua roseicapilla*. *Aust. J. Zool.* **28**: 445-56.

ROWLEY, I., 1983. Mortality and dispersal of juvenile Galahs, *Cacatua roseicapilla*, in the Western Australian wheatbelt. *Aust. Wildl. Res.* **10**: 329-42.

ROWLEY, I. AND SAUNDERS, D. A., 1980. Rigid wing-tags for cockatoos. *Corella* **4**: 1-7.

SAUNDERS, D. A., 1974a. Subspeciation in the White-tailed Black Cockatoo, *Calyptorhynchus baudinii*, in Western Australia. *Aust. Wildl. Res.* **1**: 55-69.

SAUNDERS, D. A., 1974b. The function of displays in the breeding of the White-tailed Black Cockatoo. *Emu* **74**: 43-46.

SAUNDERS, D. A., 1974c. The occurrence of the White-tailed Black Cockatoo, *Calyptorhynchus baudinii*, in *Pinus* plantations in Western Australia. *Aust. Wildl. Res.* **1**: 45-54.

SAUNDERS, D. A., 1976. Breeding of the White-tailed Black Cockatoo in captivity. *West. Aust. Nat.* **13**: 171-72.

SAUNDERS, D. A., 1977a. Breeding of the Long-billed Corella at Coomallo Creek, W.A. *Emu* **77**: 223-27.

SAUNDERS, D. A., 1977b. Red-tailed Black Cockatoo breeding twice a year in the south-west of Western Australia. *Emu* **77**: 107-10.

SAUNDERS, D. A., 1977c. The effect of agricultural clearing on the breeding success of the White-tailed Black Cockatoo. *Emu* **77**: 180-84.

SAUNDERS, D. A., 1979a. Distribution and taxonomy of the White-tailed and Yellow-tailed Black Cockatoos, *Calyptorhynchus* spp. *Emu* **79**: 215-27.

SAUNDERS, D. A., 1979b. The availability of tree hollows for the use as nest sites by White-tailed Black Cockatoos. *Aust. Wildl. Res.* **6**: 205-16.

SAUNDERS, D. A., 1980. Food and movements of the Short-billed form of the White-tailed Black Cockatoo. *Aust. Wildl. Res.* **7**: 257-69.

SAUNDERS, D. A., 1982. The breeding behaviour and biology of the Short-billed form of the White-tailed Black Cockatoo, *Calyptorhynchus funereus*. *Ibis* **124**: 422-55.

SAUNDERS, D. A. AND SMITH, G. T., 1981. Egg dimensions and egg weight loss during incubation in five species of cockatoo, and the use of measurements to determine the stage of incubation of birds' eggs. *Aust. Wildl. Res.* **8**: 411-19.

SAUNDERS, D. A., SMITH, G. T. AND ROWLEY, I., 1982. The availability and dimensions of tree hollows that provide nest sites for cockatoos (Psittaciformes) in Western Australia. *Aust. Wildl. Res.* 9: 541-56.

SCHODDE, R., SMITH, G. T., MASON, I. J. AND WEATHERLY, R. G., 1979. Relationships and speciation in the Australian corellas (Psittacidae). *Bull. B.O.C.* 99: 128-37.

SERVENTY, D. L. AND WHITTELL, H. M., 1976. Birds of Western Australia. 5th Edition. University of Western Australia Press; Perth.

SMITH, G. T. AND SAUNDERS, D. A., 1986. Clutch size and productivity in three sympatric species of (Psittaciformes) in the southwest of Western Australia. *Aust. Wildl. Res.* 13: (in press).

Long-billed White-tailed Black Cockatoo, *Calyptorhynchus baudinii*

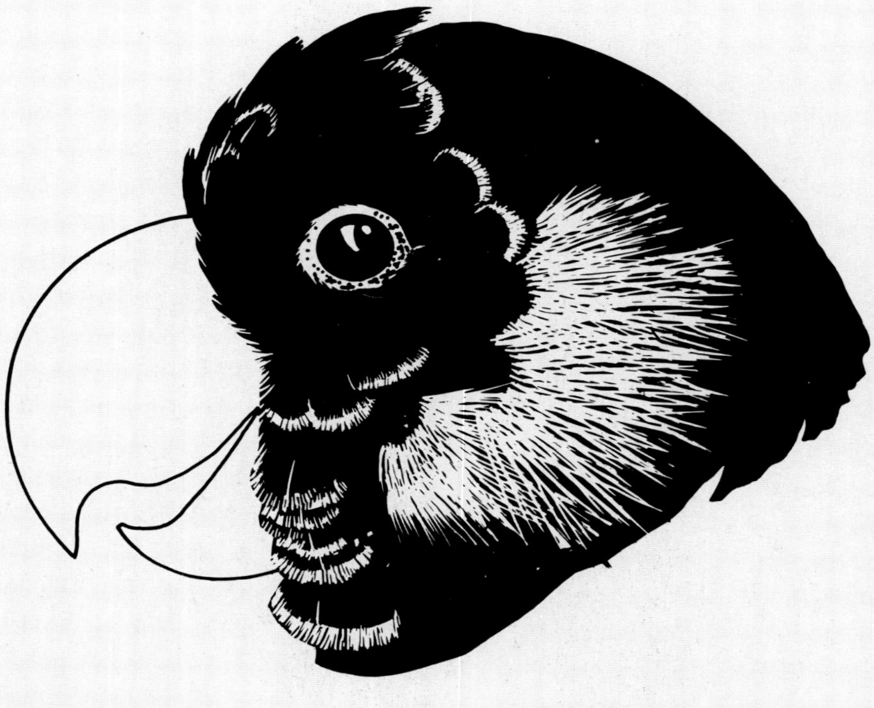

Short-billed White-tailed Black Cockatoo, *Calyptorhynchus funereus latirostris*

CHAPTER 28

Birds in Fragmented Forests in Gippsland, Victoria

Richard H. Loyn[1]

Birds were studied in 56 forest patches (from 0.1 to 1771 ha) isolated by clearing for agriculture or pine plantations. Most patches of less than 10 ha were seriously degraded and had been taken over by farmland birds including Noisy Miners, *Manorina melanocephala,* which aggressively excluded other species. Larger patches supported a selection of forest birds depending on the habitats represented; numbers of species increased with size of patch. The only species confined to the two largest patches was the Powerful Owl, *Ninox strenua.*

INTRODUCTION

VICTORIA still has large tracts of continuous forest but many special habitats can occur as isolated patches including stands of a particular tree species or age. Hence, it is important to know the effects of fragmentation when establishing a system of reserves. To this end 56 patches of mixed-species eucalypt forest have been studied in the Latrobe Valley near Traralgon in Central Gippsland (Victoria), isolated by clearing for agriculture and pine plantations.

The mammals of these patches were studied by Suckling (1980, 1982), and he estimated that in 1978, 34% of the 82,000 ha study region remained as fragmented native forests or woodlands, with 52% as agricultural land and 14% as pine plantations. The vegetation of the patches included stands of Narrow-leaf Peppermint, *Eucalyptus radiata,* with some Manna Gum, *E. viminalis,* and stands of Yertchuk, *E. consideniana.* The understorey was dominated either by heathy shrublands, by swards of Bracken, *Pteridium esculentum,* or by introduced grasses in heavily grazed patches. Thickets of Burgan, *Leptospermum phylicoides,* occurred frequently and some gullies contained tall gully shrubs, though generally the region was too low (60 to 360 m above sea level) and dry (about 750 mm average annual rainfall at 180 m), for gully vegetation to be well developed.

The patches studied ranged in size from 0.1 to 1771 ha and were scattered through the region (Fig. 1). They had been isolated for periods from 7 to 100 years and most agricultural clearing occurred between 1870 and the turn of the century. Extensive pine plantations were established from 1950 and some clearing still continues. Three patches (M7, T29 and F10) have been partly cleared for use as shire rubbish tips, and T29 now consists of a ring of forest round the main tip for Traralgon.

Most of the patches are on private property though a few are shire reserves and extensive uncleared public land has been reserved nearby.

This chapter presents a preliminary analysis from a study of birds on these patches.

METHODS

All patches were visited in each of three spring/summer seasons from November 1980 to March 1983. Most were visited five times or more, in time-periods from November 1980-April 1981, October-December 1981, January-March 1982, November-December 1982 and December 1982 to March 1983. All birds observed were recorded, along with numbers seen or heard on 20-minute counts. On the smallest patches one or two counts were sufficient to cover the whole area on each visit, while on the larger patches up to 19 counts were made on a visit with the aim of searching all habitats. In the later visits counts were continued until few additional species could be found in any patch and extra visits were made where necessary to ensure that all parts of each patch had been searched. Altogether the final visit to each area added a mean of 0.9 species to the list for that area and most of the species added at that stage were vagrants rather than resident species.

---

[1]Forests Commission, Victoria (now Department of Conservation, Forests and Lands). Mountain Research Station, Sherbrooke, Victoria 3789.

Pages 323-31 *in* BIRDS OF EUCALYPT FORESTS AND WOODLANDS: ECOLOGY, CONSERVATION, MANAGEMENT ed. by A. Keast, H. F. Recher, H. Ford and D. Saunders. Royal Australasian Ornithologists Union and Surrey Beatty & Sons.

Fig. 1. Distribution of study patches and other habitats in the Latrobe Valley.

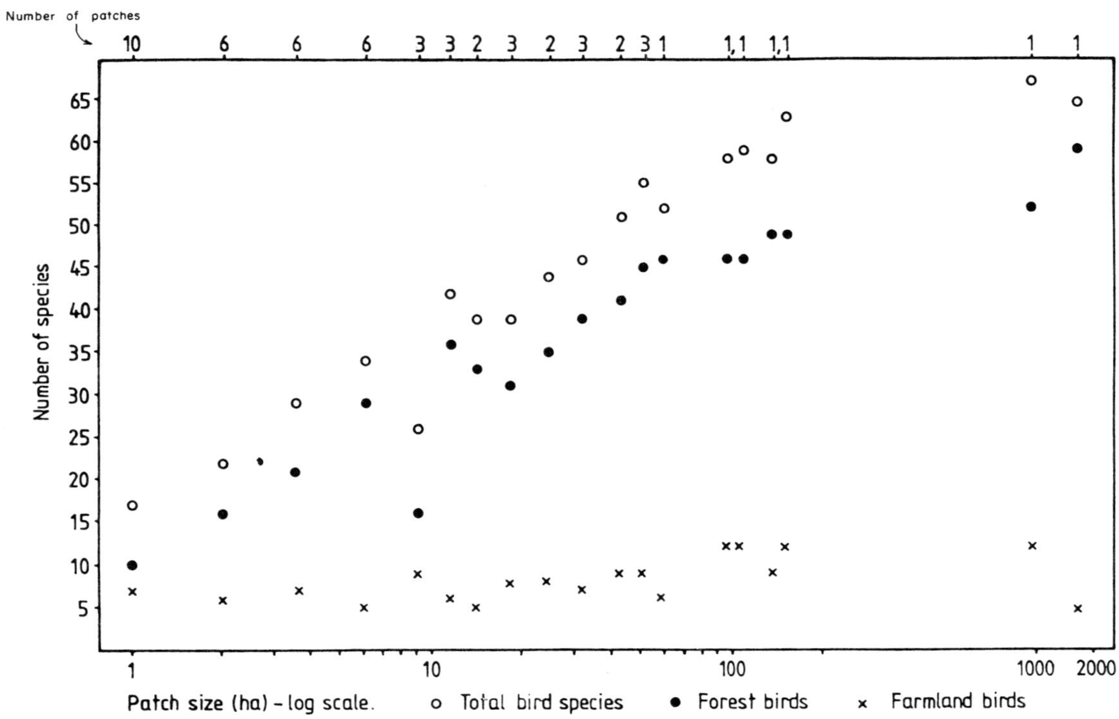

Fig. 2. Mean number of bird species in forest patches of different size (i.e. 0-1, 2, 3-4, 5-7, 8-10, 11-12, 13-15, 17-19, 22-26, 29-34, 40-44, 48-50 ha and single larger areas). Numbers of patches used to calculate means are shown above.

RESULTS

A list of species observed in or around the study patches is given in Appendix 1 with information on their abundance in patches of different size and occurrence in different habitats. Numbers of species in patches of different size are shown in Figure 2 and they are divided into forest, farmland and water birds in Table 1.

Table 1. Numbers of bird species observed on 26 counts in forest patches and fragmented combinations of patches each totalling about 50 ha.

	Number of patches (= degree of fragmentation)	Sizes of patches (ha)	Total area (ha)	Forest bird species	Farmland bird species	Water bird species	Total bird species
A	23	3 × 0.5, 7 × 1, 6 × 2, 1 × 3, 4 × 4, 2 × 5	49.5	44	13	1	58
B	7	1 × 5, 2 × 6, 2 × 7, 1 × 9, 1 × 10	50	43	14	4	61
C	4	11, 12, 12, 13	48	41	7	0	48
D	3	15, 17, 19	51	43	9	0	52
E	2	22, 26	48	43	10	0	53
F	1	48	48	43	10	2	55
G	1	49	49	46	11	1	58
H	1	50	50	40	4	0	44

No single patch contained all the habitats in the region or all of the bird species. The larger patches contained more species than the smaller ones and the presence or absence of birds in patches larger than about 10 ha could usually be explained by habitat availability. Some species thought to be dependent on large forest tracts (e.g. Red-browed Treecreeper, *Climacteris erythrops,* Varied Sitella, *Daphoenositta chrysoptera*) were found to breed in patches as small as 12 ha. Other species were local and only occurred in a few patches with particular vegetation. For example, Emus, *Dromaius novaehollandiae,* were found only in adjacent areas with 134, 57 and 26 ha of heathy woodland next to a stream and pasture where they grazed, although they were widespread in nearby pine plantations. The only species confined to the two largest patches was the Powerful Owl, *Ninox strenua.* It and some other large owls have large territories, need hollow trees for nesting and prey on arboreal mammals. Hence, they require extensive forest reserves. The most notable absentee from the region was the Superb Lyrebird, *Menura novaehollandiae.* Lyrebirds survive well in managed continuous forests elsewhere (Loyn 1980) and the region under study was on the drier limit for this species. The typical birds of wet foothill gullies were scarce in the region and the extent of suitable habitat for them was much less than the amount of forest indicated in Figure 1.

Most patches smaller than 10 ha had been grazed heavily by stock and little understorey vegetation remained. They had been taken over by farmland birds including Common Starling, *Sturnus vulgaris,* Eastern Rosella, *Platycercus eximius,* Australian Magpie, *Gymnorhina tibicen,* Australian Magpie-lark, *Grallina cyanoleuca,* and Noisy Miner, *Manorina melanocephala,* all of which were rare in larger patches (Appendix 1). All except Noisy Miner fed mainly in pasture. The Noisy Miners fed both in pasture and the tree canopy, and drove out other small birds. Usually the only species that remained were a few hole-nesting Striated Pardalotes, *Pardalotus striatus,* and Tree Martins, *Cecropis nigricans.* Hence, few birds except Noisy Miners and the occasional pardalote fed on insects from the canopy (see also Dow 1977) and a parallel situation may exist to that described for Bell Miners, *Manorina melanophrys,* where Bell Miners effectively defend high populations of insect prey which other birds could have reduced (Loyn *et al.* 1983). Severe defoliation by insects and subsequent dieback was common in the small patches occupied by Noisy Miners.

In an attempt to distinguish effects of fragmentation from effects of habitat reduction, data were pooled from groups of various numbers of patches with total areas of about 50 ha (Table 1). As more total time was spent on counts in combinations of areas than in single areas, information was used for only the first 26 counts in each area or combination. The combinations of many small areas (A and B, representing highly fragmented systems) contained slightly more farmland species than the single areas or combinations of fewer larger areas, reflecting the greater amount of farmland boundary and the degradation of the small areas. Numbers of forest species were similar in all combinations, though numbers of individuals observed in the most fragmented combinations (A and B) were very low (Appendix 1) and consisted mainly of transient birds except in the few patches where understorey remained intact. Most of the forest birds in the less fragmented combinations (C, D and E) were resident there and mean numbers of forest birds per count in patches of 11-20 ha were 81% of those in the two largest patches. Hence, it is concluded that in this situation the major effect of forest clearing has been to reduce habitat and the degree of fragmentation of remaining habitat has been of lesser importance.

DISCUSSION

The pattern of bird species found in this study is broadly similar to that found in forest patches in the wheatbelt of Western Australia (Kitchener *et al.* 1982) and by Robert Howe on the New England Tablelands of New South Wales (Ford 1981).

When considering design of reserves for fauna, it is nothing new to say the bigger the better and this is supported by the present results as well as by analogy with biogeographic theory from oceanic islands (Macarthur and Wilson 1967; Diamond 1975). Controversy has concerned the relative values of several small fragmented reserves or fewer large reserves of equivalent total area. Simberloff and Abele (1982) reviewed the literature and found no evidence that large reserves supported more animal species than fragmented systems of equal area in any taxonomic group, refuting the contrary assertions of several authors. Similar conclusions were reached in the present analysis of 50 ha combinations, showing that habitat reduction had been more detrimental than fragmentation. This was true even when forest bird species were considered separately from farmland species, despite low populations of forest birds in the most fragmented combinations. In Western Australia, Kitchener *et al.* (1982) found that when passerine birds were classified into groups corresponding to forest and farmland species, the latter responded positively to fragmentation (as in this study) but the numbers of forest bird species were reduced by fragmentation (Humphreys and Kitchener 1982). In the Western Australian study distances between reserves were greater, there was no extensive uncleared vegetation nearby and the areas were larger than those included in the present analysis. Effects of parameters such as these will be examined in subsequent analyses.

There is little doubt that both fragmentation and habitat reduction could affect particular species and this must be taken into account when designing reserves. The important point is to reserve the right habitats and this study has shown that even small patches can be useful if they contain habitats the value of which can be maintained in the long term. As patches of less than 10 ha had often been degraded by grazing and contained low populations of forest birds, it appears difficult to maintain the value of such small patches on farmland. Exclusion of grazing to protect understorey and regeneration would greatly enhance their value and provide cover for insectivorous birds that benefit the health of the trees. Reserves surrounded by productive eucalypt forest are more likely to be viable as they would not be subjected to the same pressures and the intervening habitat would allow easier movement and colonisation by forest birds. The needs of individual species and their territory sizes and mobility should be considered in designing reserve systems.

ACKNOWLEDGEMENTS

I thank my colleague Dr Graeme Suckling who selected the study area and encouraged this work. Several people helped with fieldwork or analysis, specially Barry Traill, David Eades, John Martindale, Mark Cavill, Peta O'Donahue, Steve Smith and Andrew Moore; I am grateful to them all.

REFERENCES

Diamond, J. M., 1975. The island dilemma: lessons of modern biogeographic studies for the design of natural reserves. *Biol. Conserv.* 7: 129-46.

Dow, D. D., 1977. Indiscriminate interspecific aggression leading to almost sole occupancy of space by a single species of bird. *Emu* 77: 115-21.

Ford, H. A., 1981. Birds and eucalypt dieback. Pp. 112-20 *in* Eucalypt dieback in forests and woodlands ed. by K. M. Old, G. A. Kile and C. P. Ohmart. C.S.I.R.O. Melbourne, 285 pp.

Humphreys, W. F. and Kitchener, D. J., 1982. The effect of habitat utilisation on species-area curves; implications for optimal reserve area. *J. Biogeog.* 9: 391-96.

Kitchener, D. J., Dell, J. and Muir, B. G., 1982. Birds in Western Australian wheatbelt reserves — implications for conservation. *Biol. Conserv.* **22**: 127-63.

Loyn, R. H., 1980. Bird populations in a mixed eucalypt forest used for production of wood. *Emu* **80**: 145-56.

Loyn, R. H., Runnalls, R. G., Forward, G. Y. and Tyers, J., 1983. Territorial Bell Miners and other birds affecting populations of insect prey. *Science* **221**: 1411-13.

MacArthur, R. H. and Wilson, E. O., 1967. The theory of island biogeography. Prineton Univ. Press, New Jersey, 203 pp.

Raou, 1978. Recommended English Names for Australian Birds. *Emu* 77, Supplement : 245-313.

Simberloff, D. and Abele, L. G., 1982. Refuge design and island biogeographic theory: effects of framentation. *Am. Nat.* **120**: 41-50.

Suckling, G. C., 1980. The effects of fragmentation and disturbance of forest on mammals in a region of Gippsland, Victoria. Ph.D. Thesis, Monash Univ. Clayton, Vic. 303 pp.

Suckling, G. C., 1982. Value of reserved habitat for mammal conservation in plantations. *Aust. For.* **45**: 19-27.

APPENDIX 1.

Bird list for forest patches in the Latrobe Valley: F = forest bird, O = farmland or open country bird, W = water bird, S = mainly summer visitor to the forest patches, P = mainly passage migrant or winter visitor, I = introduced species and H = nesting in tree hollows. These categories are somewhat arbitrary. For example, swifts could be considered as F or O. Data from counts are presented for all areas except F4, T19, and T29 which were atypical. The list includes all species observed in the region during the study other than uncommon water birds. Names and order follow RAOU (1978).

Common name	Scientific name	Main habitat	Nesting in tree hollows	Seasonal occurrence	Number of patches species occurred (total 56)	Number of individuals per 100 counts in patches of size: 0-5 ha	6-10 ha	11-22 ha	26-57 ha	94-144 ha	973-1771 ha	Comments
Emu	*Dromaius novaehollandiae*	F			3	0	0	0	4	18	0	Widespread in pine plantations and resident in heathy woodland (F12, M23E, M23W) next to river flats where they grazed.
Hoary-headed Grebe	*Poliocephalus poliocephalus*	W			0							
Australasian Grebe	*Tachybaptus novaehollandiae*	W			1							
Little Pied Cormorant	*Phalacrocorax melanoleucos*	W			1							
Pacific Heron	*Ardea pacificus*	W			0							
White-faced Heron	*A. novaehollandiae*	W			10	11	0	3	1	3	2	Widespread, feeding in swampy paddocks or beside dams. Nesting in trees mainly in patches 0-5 ha.
Great Egret	*Egretta alba*	W			0							
Sacred Ibis	*Threskiornis aethiopica*	W			3	0	0	0	0	3	22	Small numbers in swamps and wet paddocks, occasionally roosting in trees on edge of M11 (973 ha).
Straw-necked Ibis	*Threskiornis spinicollis*	W			1	0	0	0	0	0	15	Irregular visitor to wet and dry paddocks, occasionally roosting in trees with Sacred Ibis on edge of M11 (973 ha).
Yellow-billed Spoonbill	*Platalea flavipes*	W			2	3	1	0	0	0	0	Small numbers feeding beside farm dams. A pair bred successfully T28 (1 ha), 500 m from nearest dam.
Black Swan	*Cygnus atratus*	W			0							
Australian Shelduck	*Tadorna tadornoides*	W			1							
Pacific Black Duck	*Anas superciliosa*	W			1							
Grey Teal	*A. gibberifrons*	W			0							
Maned Duck	*Chenonetta jubata*	W	H		1	0	105	0	0	0	0	Common near farm dams, nesting in trees. Only regular MD1 (10 ha) but elsewhere breeds in patches of many different sizes.
Black-shouldered Kite	*Elanus notatus*	O			0							
Brown Goshawk	*Accipiter fasciatus*	F			23	6	4	20	5	5	1	Hunting mainly along edges and nesting in patches of all sizes.
Collared Sparrowhawk	*A. cirrhocephalus*	F			2							Singles observed T11 (6 ha) and F11 (1771 ha).
Wedge-tailed Eagle	*Aquila audax*	F			9	2	0	1	1	2	1	Hunting widely over forest and farmland, nesting in trees T1 (50 ha) and M4 (96 ha).
Marsh Harrier	*Circus aeruginosus*	W			2	1	7	0	0	0	0	Hunting over swamps and nearby forest patches.
Peregrine Falcon	*Falco peregrinus*	F			1							One observed over T31 (9 ha).
Australian Hobby	*F. longipennis*	O			3							Singles hunting over MF (4 ha), in rubbish tip T29 (22 ha) for introduced birds and on edge of M2 (144 ha).
Brown Falcon	*F. berigora*	O			4	1	1	0	1	1	0	Hunting widely over farmland and occasionally on edges of forest patches.
Australian Kestrel	*F. cenchroides*	O	(H)		10	2	1	0	1	2	3	Hunting over farmland and nesting in the larger forest patches.
Stubble Quail	*Coturnix novaezelandiae*	O		S	0							
Painted Button-quail	*Turnix varia*	F			1							One in dry forest T29 (22 ha).
Dusky Moorhen	*Gallinula tenebrosa*	W			0							
Purple Swamphen	*Porphyrio porphyrio*	W			0							
Eurasian Coot	*Fulica atra*	W			0							
Masked Lapwing	*Vanellus miles*	W			5	6	8	0	2	0	0	Mainly in swamps and paddocks or cleared sections of forest patches.
Black-fronted Plover	*Charadrius melanops*	W			0							
Spotted Turtle-Dove	*Streptopelia chinensis*	O		I	4							Scarce in region. Singles at ME (1 ha), MF (4 ha), M14A (40 ha) and recently cleared section of M11 (973 ha).
Common Bronzewing	*Phaps chalcoptera*	F			24	3	9	3	12	5	18	Mainly in dry forest with understorey of Wattles *Acacia* spp.
Brush Bronzewing	*P. elegans*	F			8	0	1	0	3	1	4	Mainly in pine plantations, heathy woodland and some gullies.
Yellow-tailed Black Cockatoo	*Calyptorhynchus funereus*	F	H		10	1	0	10	15	2	3	Mainly in heathy woodland with Banksia, wet forest and pine plantations, moving widely between patches.

Appendix 1. — Continued

Common name	Scientific name	Main habitat	Nesting in tree hollows	Seasonal occurrence	Number of patches species occurred (total 56)	Number of individuals per 100 counts in patches of size: 0-5 ha	6-10 ha	11-22 ha	26-57 ha	94-144 ha	973-1771 ha	Comments
Gang-gang Cockatoo	*Callocephalon fimbriatum*	F	H		14	0	22	6	11	16	12	Feeding in forest canopy, moving widely between patches and often in roadside trees.
Galah	*Cacatua roseicapilla*	O	H		2							Scarce but increasing in region, mainly in low farmland with River Red Gums. A single in T31 (9 ha), and two in T29 (22 ha).
Sulphur-crested Cockatoo	*C. galerita*	O	H		27	31	46	13	17	50	6	Mainly feeding on farmland but sometimes from open forest floor or canopy.
Musk Lorikeet	*Glossopsitta concinna*	F	H		1							A transient flock of 10 feeding in canopy T8 (7 ha).
Crimson Rosella	*Platycercus elegans*	F	H		39	30	38	96	116	97	187	Common in most patches of 3 ha or more, unless heavily grazed.
Eastern Rosella	*P. eximius*	O	H		40	215	188	19	12	22	11	Feeding mainly in paddocks or clearings, nesting and taking refuge in small patches. A few in clearings in larger patches.
Blue-winged Parrot	*Neophema chrysostoma*	F	H	S	2	0	0	0	1	0	5	Breeding pairs in selectively logged forest F11 (1771 ha) and observed in similar habitat T18 (29 ha).
Pallid Cuckoo	*Cuculus pallidus*	F		S	3	3	1	9	2	5	4	Mainly in the more open forest and roadsides, parasitising nests of honeyeaters in canopy.
Brush Cuckoo	*C. variolosus*	F		S	3	0	1	2	2	0	3	Mainly in the wetter patches of 15 ha or more, parasitising nests of other birds.
Fan-tailed Cuckoo	*Cuculus pyrrhophanus*	F		S	17	2	0	3	12	16	14	Common, especially among Wattles, parasitising nests of other birds mainly in the understorey.
Horsfield's Bronze-Cuckoo	*Chrysococcyx basalis*	F		S	11	0	0	2	2	2	4	Mainly in dry or heathy selectively logged forest in patches of 18 ha or more, parasitising nests of other birds in the understorey.
Shining Bronze-Cuckoo	*C. lucidus*	F		S	17	2	1	3	7	5	17	Mainly feeding in canopy, but parasitising nests of other birds in the understorey.
Powerful Owl	*Ninox strenua*	F	H		2							At least one pair resident F11 (1771 ha), feeding on arboreal mammals and birds such as Australian Magpies. Also observed M11 (973 ha).
Southern Boobook	*N. novaeseelandiae*	F	H		14							Widespread in low numbers in patches as small as M10 (1 ha).
Barking Owl	*N. connivens*	F	H		0							Reported in a small patch in nearby pine plantations.
Barn Owl	*Tyto alba*	O	H		0							Scarce resident in farmland only.
Tawny Frogmouth	*Podargus strigoides*	F			18							Widespread in low numbers, mainly on forest edges and roadside trees.
Australian Owlet-nightjar	*Aegotheles cristatus*	F	H		4							Nesting in heathy woodland M14A (40 ha) and present in some larger patches (T1, M1, F11).
White-throated Nightjar	*Caprimulgus mystacalis*	F		S	0							Present in extensive heathy woodland nearby and seen occasionally in roadside strips and pine plantations.
White-throated Needletail	*Hirundapus caudacutus*	O		S	20	4	9	7	26	50	3	Commonly feeding over forest and farmland late summer.
Azure Kingfisher	*Ceyx azurea*	W			1							Observed on stream T2 (15 ha).
Laughing Kookaburra	*Dacelo novaeguineae*	F	H		50	58	79	54	63	69	46	Widespread, often feeding from open ground
Sacred Kingfisher	*Halcyon sancta*	F	H	S	16	1	0	3	7	28	5	Mainly in dry or selectively logged forest in patches 4 ha or more.
Skylark	*Alauda arvensis*	O		I	0							Fairly common in paddocks only.
Welcome Swallow	*Hirundo neoxena*	O			24	39	1	3	3	11	5	Feeding mainly over paddocks and on edges of small grazed patches, nesting mainly in farm buildings.
Tree Martin	*Cecropis nigricans*	F	H	S	27	33	5	2	94	42	4	Feeding mainly over forest or at rubbish tips, nesting in River Red Gums or in small patches or in selectively logged forest.
Richard's Pipit	*Anthus novaeseelandiae*	O			1							Fairly common in paddocks and edges of smallest patches only.
Black-faced Cuckoo-shrike	*Coracina novaehollandiae*	F		S	39	18	35	19	26	30	12	Widespread, feeding mainly in canopy.
White-bellied Cuckoo-shrike	*C. papuensis*	F		S	3	0	0	0	0	1	7	Mainly dark-phase birds in canopy M2 (144 ha), M11 (973 ha), and F11 (1771 ha).

Appendix 1. — Continued

Common name	Scientific name	Main habitat	Nesting in tree hollows	Seasonal occurrence	Number of patches species occurred (total 56)	Number of individuals per 100 counts in patches of size: 0-5 ha	6-10 ha	11-22 ha	26-57 ha	94-144 ha	973-1771 ha	Comments
White's Thrush	*Zoothera dauma*	F		F	4							Few observed mainly in Bracken understorey in autumn.
Blackbird	*Turdus merula*	F		I	16	6	7	19	14	89	16	Mainly in gullies and among blackberries in the wetter patches.
Rose Robin	*Petroica rosea*	F		S	3							Scarce, in Silver Wattle *A. dealbata* in wetter patches T2 (15 ha), F12 (26 ha), and M13A (49 ha).
Flame Robin	*P. phoenicea*	F		P	2							A few in autumn M13A (49 ha) and M23E (134 ha). Breeds in slightly wetter higher forest.
Scarlet Robin	*P. multicolor*	F			37	15	12	34	18	19	24	Mainly in dry forest, feeding from bare ground and canopy.
Eastern Yellow Robin	*Eopsaltria australis*	F			37	25	40	69	106	117	125	Most common in thickets of Burgan and gullies, absent from heavily grazed small patches.
Jacky Winter	*Microeca leucophaea*	F			25	7	13	12	16	30	7	Mainly in dry forest and clearings.
Crested Shrike-tit	*Falcunculus frontatus*	F			11	1	1	3	1	5	1	Feeding from bark, especially among Manna gum/Peppermint in patches of 5 ha or more.
Olive Whistler	*Pachycephala olivacea*	F			7	1	1	1	1	2	3	Mainly in pine plantations and a few gullies and Burgan thickets, or elsewhere on passage, feeding among understorey.
Golden Whistler	*P. pectoralis*	F			29	8	5	19	20	17	30	Feeding in canopy and middle storey mainly in the wetter forest, extending to drier forest and smaller patches in autumn.
Rufous Whister	*P. rufiventris*	F		S	43	33	39	66	112	121	130	Common in canopy, rare in small heavily grazed patches.
Grey Shrike-thrush	*Colluricincla harmonica*	F			42	40	34	83	96	96	106	Common, feeding at all levels, rare in small heavily grazed patches.
Leaden Flycatcher	*Myiagra rubecula*	F		S	15	0	4	4	2	5	6	In canopy in the drier patches only.
Satin Flycatcher	*M. cyanoleuca*	F		S	21	4	1	14	2	40	30	In canopy in the wetter patches and in gullies.
Rufous Fantail	*Rhipidura rufifrons*	F		S	21	3	4	9	1	10	7	In understorey in gullies only.
Grey Fantail	*R. fuliginosa*	F			46	151	193	291	334	308	328	Common in canopy and middle storey except in small heavily grazed patches.
Willie Wagtail	*R. leucophyrys*	O			16	18	13	1	13	9	1	Common in paddocks, nesting mainly in farmhouse gardens but occasionally in small patches or among Red Gums. In larger patches occurs only on edges.
Eastern Whipbird	*Psophodes olivaceus*	F			4	0	3	1	1	0	4	A few in gullies and Burgan thickets.
Spotted Quail-thrush	*Cinclosoma punctatum*	F			4	0	0	3	1	2	0	A few feeding from bare ground in the drier patches only: F24 (11 ha), F7 (19 ha), F19 (44 ha), and M4 (96 ha).
Clamorous Reed Warbler	*Acrocephalus stentoreus*	W			1							One in T2 (15 ha) near stream with willows.
Golden-headed Cisticola	*Cisticola exilis*	O			0							A few in nearby swamps only.
Brown Songlark	*Cinclorhamphus cruralis*	O			0							A few in dry paddocks only.
Superb Fairy-wren	*Malurus cyaneus*	F			41	179	152	232	372	262	168	Common in thickets of Burgan and other Teatree in small and large patches, more widespread in large patches, feeding from bare ground among scattered understorey shrubs, or on edges of paddocks.
White-browed Scrubwren	*Sericornis frontalis*	F			38	50	30	41	109	76	97	Common in dense understorey and absent from heavily grazed patches.
Chestnut-rumped Hylacola	*S. pyrrhopygius*	F			1							Pair in sparsely treed M18 (4 ha) in heath and Bracken understorey.
Calamanthus	*S. fuliginosus*	O			0							A few in swamps and weedy paddocks only.
Brown Thornbill	*Acanthiza pusilla*	F			42	143	103	219	281	246	297	Common in shrubby understorey and absent from small heavily grazed patches.
Buff-rumped Thornbill	*A. reguloides*	F			26	14	26	43	19	30	22	Mainly in drier patches with open understorey or burnt areas but not in small heavily grazed patches.
Yellow-rumped Thornbill	*Acanthiza chrysorrhoa*	O			20	39	19	7	8	27	2	Mainly feeding in paddocks and nesting in farmhouse gardens, but sometimes feeding from ground on edges of forest patches and occasionally feeding in canopy within.

Appendix 1. — Continued

Common name	Scientific name	Main habitat	Nesting in tree hollows	Seasonal occurrence	Number of patches species occurred (total 56)	Number of individuals per 100 counts in patches of size: 0-5 ha	6-10 ha	11-22 ha	26-57 ha	94-144 ha	973-1771 ha	Comments
Yellow Thornbill	*A. nana*	F			1							Irregular visitor to Black Wattles *A. mearnsii* in T29 (22 ha), the closest patch to Traralgon (2 km) where they were resident.
Striated Thornbill	*A. lineata*	F			41	95	137	237	201	219	240	Common in Eucalypt canopy, feeding from foliage, often feeding and nesting in suppressed saplings; absent from small heavily grazed patches.
Varied Sittella	*Daphoenositta chrysoptera*	F			30	24	31	57	45	60	51	Feeding from high branches and wood, flocks moving between small patches, but rarely in small heavily grazed patches, nesting in patches of 12 ha or more.
White-throated Treecreeper	*Climacteris leucophaea*	F	H		43	50	59	160	134	150	188	Common except in small heavily grazed patches, feeding from bark.
Red-browed Treecreeper	*C. erythrops*	F	H		12	0	0	9	18	12	29	Feeding from bark, especially of Narrowleaf Peppermints; bred successfully in F9 (12 ha) and larger patches, sedentary.
Red Wattlebird	*Anthochaera carunculata*	F			39	31	26	37	51	73	23	Feeding from blossom, foliage and bark in canopy and from Banksia flowers.
Little Wattlebird	*A. chrysoptera*	F			2							Irregular visitor to flowering Banksia M4 (96 ha) and M23E (134 ha).
Noisy Friarbird	*Philemon corniculatus*	F		S	2	0	0	0	0	14	0	Regular small breeding population in F1 (108 ha) and occasional visitor to M23E (134 ha), 2 km away.
Noisy Miner	*Manorina melanocephala*	O			34	407	446	20	36	13	0	Common in small patches, especially where heavily grazed, feeding at all levels and from pasture. In larger patches only found on the edge, usually in small peninsulas of forest.
Yellow-faced Honeyeater	*Lichenostomus chrysops*	F		S	41	71	35	106	176	152	146	Widespread, feeding from blossom and foliage mainly in canopy, nesting in understorey.
White-eared Honeyeater	*L. leucotis*	F			34	24	59	22	55	94	90	Mainly in Burgan thickets and heathy patches, feeding from blossom, bark and foliage and nesting in understorey.
Brown-headed Honeyeater	*Melithreptus brevirostris*	F			34	22	36	35	22	43	45	Mainly in canopy feeding from bark and blossom and moving widely between patches.
White-naped Honeyeater	*M. lunatus*	F			30	13	26	85	82	149	196	Common in canopy of Manna Gum/ Peppermint, feeding from foliage, blossom and bark.
Crescent Honeyeater	*Phylidonyris pyrrhoptera*	F			18	6	24	12	18	4	22	Mainly in thickets of Burgan, flowering Banksia or in middle storey in gullies.
New Holland Honeyeater	*P. novaehollandiae*	F			14	10	9	3	22	52	39	Mainly feeding from blossom in heathy thickets, Burgan, Banksia, Manna Gum or Mistletoe, nesting in understorey.
Eastern Spinebill	*Acanthorhynchus tenuirostris*	F			36	34	35	55	84	87	114	Mainly feeding from blossom in understorey.
White-fronted Chat	*Ephthianura albifrons*	O			0							Scarce visitor to wet paddocks only.
Mistletoebird	*Dicaeum hirundinaceum*	F			28	14	5	16	12	17	21	Widespread in Mistletoe, moving widely between patches.
Spotted Pardalote	*Pardalotus punctatus*	F			38	16	50	64	67	36	44	Common in canopy, especially in rough-barked eucalypts, nesting in holes in ground, absent from small heavily grazed patches.
Striated Pardalote	*P. striatus*	F	H		50	67	125	60	56	117	48	Common in canopy, especially in smoothbarked eucalypts, sometimes flying to nest in isolated dead trees in paddocks.
Silvereye	*Zosterops lateralis*	F		S	21	14	13	44	17	11	10	Mainly in wetter patches, feeding in understorey and canopy.
European Goldfinch	*Carduelis carduelis*	O		I	34	21	3	15	32	33	14	Mainly in weedy paddocks and forest clearings, also feeding at blossom of Eucalypts.

Appendix 1. — Continued

Common name	Scientific name	Main habitat	Nesting in tree hollows	Seasonal occurrence	Number of patches species occurred (total 56)	Number of individuals per 100 counts in patches of size: 0-5 ha	6-10 ha	11-22 ha	26-57 ha	94-144 ha	973-1771 ha	Comments
House Sparrow	*Passer domesticus*	O		I	2							Mainly round farm buildings and rubbish tips.
Red-browed Firetail	*Emblema temporalis*	F			16	2	1	24	20	8	21	Mainly in heathy thickets of Burgan and grassy clearings in larger patches and on ungrazed grassy edges of smaller patches.
Beautiful Firetail	*E. bella*	F			2	0	0	0	1	0	6	A few in gullies and heathy thickets F11 (1771 ha) and F12 (26 ha).
Common Starling	*Sturnus vulgaris*	O	H	I	33	165	72	12	7	10	1	Mainly feeding in paddocks and rubbish tips, nesting in small patches and farm buildings.
Common Mynah	*Acridotheres tristis*	O	H	I	4	2	0	0	0	0	2	Mainly feeding along roads, nesting in nearby trees; numerous in rubbish tip T29 (22 ha).
Olive-backed Oriole	*Oriolus sagittatus*	F		S	18	3	8	3	8	5	3	Mainly in canopy, occasional visitor to small patches, only resident in patches of 5 ha or more.
White-winged Chough	*Corcorax melanorhamphus*	F			3	0	15	4	4	0	0	Small colonies in three fairly dry patches with open understorey: MD1 (10 ha), M1 (13 ha) and M13A (48 ha), co-existing with Noisy Miners at MD1.
Australian Magpie-lark	*Grallina cyanoleuca*	O			32	43	55	10	6	3	0	Mainly feeding in paddocks, nesting in small patches.
White-browed Woodswallow	*Artamus superciolosus*	F		S	7							Flocks feeding widely over all areas in summer 1982/83 only (drought).
Dusky Woodswallow	*A. cyanopterus*	F		S	30	20	7	33	60	72	28	Mainly in the drier patches or selectively logged areas, absent from small heavily grazed patches.
Grey Butcherbird	*Cracticus torquatus*	F			37	35	26	17	6	8	4	Present in most small patches particularly where heavily grazed, less common in larger patches.
Australian Magpie	*Gymnorhina tibicen*	O			53	167	215	73	52	81	12	Mainly feeding in paddocks, nesting in small patches, or on edges of larger patches.
Grey Currawong	*Strepera graculina*	F			15	1	1	3	5	2	5	Mainly in heathy woodland, occasionally visiting small patches.
Australian Raven	*Corvus coronoides*	F			31	6	12	7	28	25	15	Mainly in larger patches, but also feeding in paddocks.
Little Raven	*C. mellori*	O			24	12	59	7	16	9	0	Common in paddocks, nesting in small patches and scattered trees.

Noisy Miner, *Manorina melanocephala*

CHAPTER 29

The Bird Community in Eucalypt Woodland and Eucalypt Dieback in the Northern Tablelands of New South Wales

Hugh A. Ford[1]

Woodland in the Northern Tablelands is suffering badly from eucalypt dieback. Although not proven, the most likely cause of dieback is defoliation and other damage by insects. Birds are one group of predators of these insects. The densities, foraging behaviour and diets of birds were recorded in four woodland sites near Armidale. There were about 10 birds per ha in 1981, and about 20 birds per ha in a single site in 1978 and 1979. Birds had perhaps declined during a drought. About half of the birds were insectivorous, and a further 40% were honeyeaters, which also eat many insects. A wide range of insects are eaten by birds, including many that may damage trees.

I estimated that birds consume 13.2-18 6 $\times$ 10^4 Kcals/ha/yr from insects, of which 5.1-6.4 $\times$ 10^4 Kcals/ha/yr is from insects taken from foliage. No estimate of the biomass or productivity of insects in eucalypt woodland has yet been made. I made a crude estimate based on primary productivity and, assuming insects consume 10% of this and convert 10% of what they consume into body tissue, this gives 9 $\times$ 10^4 Kcals/ha/yr of leaf-eating insects. According to these estimates, birds eat of the order of 55-70% of the production of insects. They thus have the potential to influence insect populations, and management to overcome eucalypt dieback should provide habitat for birds. Planting of shrubs, protected from stock, some of which provide nectar or berries, may encourage a wide variety of birds in areas devastated by dieback.

INTRODUCTION

EUCALYPT WOODLAND differs from dry sclerophyll forest in having trees with rounded and deeper crowns and a more open, often broken, canopy. Generally the shrub layer is poorly developed, though herbs and especially grasses are abundant. Woodland occurs in an almost unbroken belt from northwestern Australia across the north, down the east coast and westwards to the Mount Lofty Ranges. There is a further belt in southwestern Australia. It is usually inland from forest, in lower rainfall, though in places woodland reaches the coast. The present paper will concentrate on eucalypt woodland around Armidale in the Northern Tablelands of New South Wales, often called the New England region.

Eucalypt Dieback

Only a small proportion of the eucalypt woodland of New England remains in its original state. European settlers found an open woodland with an abundance of grass and rapidly started using it to rear cattle and sheep. The woodland was partly cleared to encourage grasses, but leaving trees for shelter and shade. This resulted in an attractive parkland, with denser timber on the hilltops. Towards the end of the last century trees started deteriorating and dying. This deterioration and death has worsened in a succession of bouts. By the late 1950's *E. nova-anglica* had been decimated around Armidale. During the late 1970's trees of many species had died over a wide area (Williams and Nadolny 1981). This problem has been termed Eucalypt Dieback or more parochially New England Dieback, to discriminate it from other types of dieback, such as Jarrah Dieback, caused by a fungus. Eucalypt dieback is not confined to the Northern Tablelands of New South Wales; the Southern Tablelands are also badly affected and eucalypt woodlands are suffering from dieback from southeastern Queensland around to South and Western Australia (Old *et al.* 1981; Saunders *et al.* 1982).

[1]Department of Zoology, University of New England, Armidale, New South Wales, Australia 2351.

Pages 333-40 *in* BIRDS OF EUCALYPT FORESTS AND WOODLANDS: ECOLOGY, CONSERVATION, MANAGEMENT ed. by A. Keast, H. F. Recher, H. Ford and D. Saunders. Royal Australasian Ornithologists Union and Surrey Beatty & Sons.

Many possible causes of New England Dieback have been proposed. There could have been subtle changes in climate. Partial clearing of trees and trampling of the soil by stock has altered the microclimate for the trees. Regeneration is prevented by sheep grazing. Chemicals, such as superphosphate, have been added to the soil. Exotic grasses could be competing with the trees. Also the possibility of a fungal disease has not been ruled out. The most conspicuous cause of dieback, though, is defoliation by insects, specially Christmas beetles (Scarabaeidae) (Carne *et al.* 1974; Carter *et al.* 1981; Roberts and Sawtell 1981). However, Old *et al.* (1981) concluded, after a conference on eucalypt dieback, that although insects were often the immediate cause of tree deaths, the underlying cause of mortality was probably environmental stress.

To take the best known example, Christmas beetle larvae live in the soil, feeding on grass roots, whose quality has been improved by clearing trees, adding superphosphate and introducing more nutritious grasses. Larval survival has increased at a time when food for the adults, eucalypt leaves, has declined. It is not surprising that eucalypts are frequently defoliated by Christmas beetles.

Although this explanation is attractive, it is almost certainly too simple. Adult Christmas beetles are only active for a few weeks and are not abundant every year. Eucalypts can regenerate from defoliation. A wide range of other insects, such as leaf-beetles (Chrysomelidae), caterpillars (Lepidoptera) and sawfly larvae (Pergidae) consume eucalypt leaves, and many bugs (lerps, Psyllidae; scale-insects Coccidae) suck sap from leaves and stems. Eucalypts appear to suffer chronically high levels of insect attack (Fox and Morrow 1983). Probably if this overall attack by insects increases, it gradually drains the trees' energy reserves until they can no longer produce fresh shoots.

Even if the cause of dieback is broadened to include the herbivorous insect community, then this does not provide a full answer. It merely prompts the questions: Have insect numbers increased, or if so why, and have the trees' defences against insects weakened, and if so why? One of several reasons why herbivorous insects could have increased is that their natural predators have been reduced in abundance so that they now have less impact on the insect population. Wasps (Scolioidea, Chalcidoidea, Ichneumonoidea) and flies (Tachinidae) parasitize herbivorous insects and ants (Formicidae), robber-flies (Asilidae), assasin bugs (Reduviidae), lacewings (Chrysopidae) and many other predators prey on them (Davidson 1982). Insectivorous birds, reptiles and mammals all eat defoliating and sap-sucking insects. Humphreys *et al.* (1979) summarised what was known about eucalypt dieback in New South Wales. They suggested that the importance of predators, particularly birds, in regulating populations of leaf-eating insects should be studied.

Since 1980 I have been involved in a programme to determine the role of birds in controlling herbivorous insects in eucalypt woodland. This work is not yet complete, though some tentative conclusions can be made. In this chapter I wish to describe the density and composition of the bird community in eucalypt woodland near Armidale and to estimate the biomass of insects consumed annually by birds. I compare this estimate with a crude estimate of the annual production of insects, to obtain a guide to the likely effect of birds on herbivorous insects.

STUDY SITES

Armidale (30°30′S, 151°30′E) has warm summers (mean January maximum — 28°C), cold winters (mean July minimum — 0°C) and receives an average of 790 mm of rain annually, 65% of it from October to March. The late 1970s and early 1980's were dry, with 1980 and 1981 receiving 65% and 82% of average rainfall.

Four sites were chosen within 10 km of Armidale, two in Eastwood State Forest (Eastwood 1, 2) and one each at Hillgrove Creek State Forest and on the campus of the University of New England (UNE). Details of vegetation and relative abundance of tree species are presented in Tables 1 and 2. The first three sites are healthy, but UNE has suffered moderate dieback with many trees dead. The density and composition of trees at UNE differs from the other sites. (Most "shrubs" are sapling eucalypts and cover in the shrub layer (1-5 m) is sparse).

Table 1. Details of vegetation at each site.

	Eastwood 1	Eastwood 2	Hillgrove	UNE
Trees/ha	345	415	365	60
Shrubs/ha	400	240	160	220
Ground cover	41%	52%	22%	83%
1-5 m cover	5%	1%	1%	5%
>5 m cover	16%	32%	24%	7%
Mistletoes/ha	35	75	10	10

Table 2. Numbers of each species of tree at each site, 100 trees counted.

	Eastwood 1	Eastwood 2	Hillgrove	UNE
Eucalyptus caliginosa	44	56	63	0
E. blakelyi	32	20	13	19
E. viminalis	10	7	2	81
E. melliodora	7	11	22	0
E. bridgesiana	0	1	0	0
Acacia filicifolia	6	5	0	0
A. irrorata	1	0	0	0

METHODS

Birds were counted in Eastwood 1 by H. Bell from June to September 1978 and by H. Ford from August to December 1979 (Ford and Bell 1982). The 1978 census was along a variable width transect (Emlen 1971), the 1979 one along a transect 50 m wide. In 1981 birds were counted in grids (50 m squares) in all four sites throughout the year. All covered 9 ha, except for Eastwood 1 which was 8 ha. Grids were covered by the observer (S. Noske, L. Bridges) walking along the midline of each square recording all birds within 25 m either side.

From July 1979 to December 1981 birds were netted and banded in Eastwood 1, and many of the insectivorous species colour-banded. During 1981, specially in the breeding season, the positions of colour-banded birds were plotted on the grid and up to 200 m from it. Home ranges of known birds, and often their unbanded mates or attendants, were plotted. This more accurate estimate of density was compared with that from the transects.

Outside the censuses single foraging observations were made of each bird seen. Foraging height (estimated to nearest metre), substrate, foraging method, plant species, and if possible, food, were recorded. Faecal and stomach samples were collected by flushing captured birds (Ford *et al.* 1982). These and stomach contents of birds accidentally killed near Armidale were identified to order, occasionally to family, by S. Harrington.

The Bird Community of Eucalypt Woodland in the Northern Tablelands

Although eucalypt woodland is probably less diverse structurally and floristically than eucalypt forest, it has more species of birds in northern New South Wales (Table 3). This is perhaps because it harbours species characteristic of both wetter and drier communities. Also the wetter communities have been reduced in area during arid periods in the past, and could have become depauperate.

Table 3. Numbers of breeding bird species in each of five habitats in northern New South Wales (from Kikkawa 1968).

	Wet Forest	Dry Forest	Tall Woodland	Savanna Woodland	Scrub Savanna
Regular	31	18	39	11	33
Occasional	47	59	49	81	61
Total	78	77	88	92	94

Most species that breed in woodland are resident. Of about 80 species breeding around Armidale, only eleven are summer visitors. Several breeding species, mostly honeyeaters and pardalotes, increase in winter. The composition of the community by families is shown in Table 4. Honeyeaters, flycatchers and warblers make up most of the species, as well as most of the individuals (Table 5).

The foraging zones, weights and densities of the more common species are shown in Table 5. Most species are primarily insectivorous, with several

Table 4. Numbers of genera and species in the major families of birds in eucalypt woodland around Armidale.

		Genera	Species
Accipitridae	(hawks)	3	4
Platycercidae	(rosellas, etc)	2	3
Cuculidae	(cuckoos)	2	4
Alcedinidae	(kingfishers)	2	2
Hirundinidae	(swallows)	2	3
Campephagidae	(cuckoo-shrikes)	2	2
Pachycephalidae	(whistlers, flycatchers, robins, etc)	10	16
Maluridae	(fairy-wrens, thornbills, etc)	4	7
Climacteridae	(treecreepers)	1	2
Meliphagidae	(honeyeaters)	8	10
Pardalotidae	(pardalotes)	1	2
Ploceidae	(finches)	2	3
Artamidae	(woodswallows)	1	2
Cracticidae	(butcherbirds, etc)	3	3

Table 5. Niches and densities of some of the commoner species of birds in eucalypt woodland around Armidale.

		Density (birds/ha)		
	Weight (g)	1978	1979	1981
VERTEBRATE-FEEDERS				
Kookaburra *Dacelo novaeguineae*	300	0.01	0.08	0.02
SEED-EATERS				
Ground feeding				
Red-browed Firetail *Emblema temporalis*	11	0.69	0.33	
Diamond Firetail *E. guttata*	18	1.18	0.25	0.03
Tree and ground feeding				
Eastern Rosella *Platycercus eximius*	110	0.17	0.25	0.61
FRUGIVORES				
Mistletoebird *Dicaeum hirundinaceum*	8	0.10	0.21	0.16
VERTEBRATE/INVERTEBRATE FEEDERS				
Ground pouncer				
Sacred Kingfisher *Halcyon sancta*	45		0.17	0.04
Ground/Tree gleaner				
Pied Currawong *Strepera graculina*	300	0.02	0.13	0.10
INVERTEBRATE/HERBIVORES				
Ground gleaner				
Speckled Warbler *Sericornis sagittatus*	13	0.49	0.54	0.26
Ground/Tree gleaner				
Silvereye *Zosterops lateralis*	10	0.36	0.88	0.01
HONEYEATERS				
Nectar and invertebrates				
Eastern Spinebill *Acanthorhynchus tenuirostris*	12	0.11	0.08	0.01
Noisy Friarbird *Philemon corniculatus*	107		0.08	0.11
Red Wattlebird *Anthochaera carunculata*	125	0.05	0.21	0.14

Table 5. Continued.

	Weight (g)	Density (birds/ha) 1978	1979	1981
HONEYEATERS — cont'd				
Invertebrates and nectar				
Brown-headed Honeyeater *Melithreptus brevirostris*	15	0.65	0.38	0.08
White-naped Honeyeater *M. lunatus*	15	4.63	1.00	0.37
Fuscous Honeyeater *Meliphaga fusca*	18	3.07	5.46	3.43
INVERTEBRATE-FEEDERS				
Aerial-feeders				
Tree Martin *Cecropia ariel*	20	—	—	—
Dusky Woodswallow *Artamus cyanoptera*	40	0.01	0.17	0.16
White-browed Woodswallow *A. superciliosus*	40		0.25	
Hawkers				
Grey Fantail *Rhipidura fuliginosa*	8	0.92	0.21	0.09
Hawking/ground				
Willie Wagtail *R. leucophrys*	22	0.02	0.25	0.12
Hawk/drop				
Restless Flycatcher *Myiagra inquieta*	24		0.21	0.04
Ground gleaners				
Australian Magpie *Gymnorhina tibicen*	390	0.03	0.08	0.19
Grey Shrike-thrush *Colluricincla harmonica*	64	0.55	0.50	0.19
Ground pouncer				
Eastern Yellow Robin *Eopsaltria australis*	20	0.68	0.67	0.26
Ground/shrub gleaner				
Superb Fairy-wren *Malurus cyaneus*	9	1.03	1.41	0.72
Shrub-gleaner				
Brown Thornbill *Acanthiza pusilla*	7	0.31		0.01
Ground/bark gleaner				
Brown Treecreeper *Climacteris picumnus*	34	0.01	0.04	0.14
Bark glean/probers				
White-throated Treecreeper *C. leucophaea*	24	1.07	1.13	0.49
Varied Sittella *Daphoenositta chrysoptera*	16	0.20	0.20	0.25
Foliage gleaners				
Cuckoos (3 spp)	23-50	0.04	0.50	
Striated Pardalote *Pardalotus striatus*	12	1.66	0.04	0.53
Spotted Pardalote *P. punctatus*	8	0.33		0.12
Striated Thornbill *Acanthiza lineata*	7	1.58	0.25	0.06
Bark/Foliage tearer				
Crested Shrike-tit *Falcunculus frontatus*	24		0.29	0.20
INVERTEBRATE-FEEDERS — cont'd				
Foliage-snatchers				
Black-faced Cuckoo-shrike *Coracina novaehollandiae*	130	0.09	0.21	0.10
Rufous Whistler *Pachycephala rufiventris*	25	0.15	1.71	0.23
Golden Whistler *P. pectoralis*	24	0.43	0.79	0.05
White-throated Gerygone *Gerygone olivacea*	10	0.12	0.46	0.01
Bark/Foliage/ground gleaner				
Buff-rumped Thornbill *Acanthiza reguloides*	8	1.54	0.92	0.14
All species		23.6	21.5	9.68

facultative nectarivores, a few graminivores and a single frugivore. Several species take vertebrates as well as invertebrates and there is a range of raptors, which take chiefly vertebrates. Raptors are scarce. A few additional parrots, finches and pigeons also occur at low density, somewhat expanding the graminivorous group. In 1981, 47% of the individuals were insectivores, 10% herbivores, 43% honeyeaters and 0.3% vertebrate-feeders.

Most common species differ in either their main food, their feeding site or their feeding method. Where two species occupy the same foraging zone (see Table 5), they usually differ in size, e.g. the firetails, nectarivorous honeyeaters and pardalotes. There are a few cases of two species of similar size foraging in the same manner in the same place. Three such pairs are compared in more detail in Table 6. The whistlers are similar in all respects, but differ in the time of year they are present. The Rufous Whistler is a summer visitor, while the Golden is a winter visitor. The two *Melithreptus* honeyeaters are also similar and both are most common in winter. Although both woodswallows take most of their prey in the air, they differ in that the Dusky Woodswallow does some ground and bark foraging whereas the White-browed Woodswallow takes some nectar.

Table 6. Overlaps in foraging method, substrate, height and tree species for pairs of ecologically similar species (overlap = $\Sigma \min P_{ij}\text{-}P_{ik} \times 100$, where P_{ij} = proportion of observations of species i on niche dimension j etc.).

	Method	Substrate	Height	Tree species
Rufous x Golden Whistler	80%	98%	83%	80%
Brown-headed x White-naped Honeyeater	94%	90%	96%	80%
White-browed x Dusky Woodswallow	73%	73%	57%	*

* Does not apply as mostly aerial foraging.

Estimates of density in 1978 and 1979 were similar but were reduced by half in 1981. The method used by Bell in 1978 is a reasonably accurate one (Bell and Ferrier 1985). In 1981 home ranges of many of the insectivorous species were plotted on the grid for Eastwood 1. These "real" densities are compared with those estimated from the transect method in Table 7. Some species are underestimated; for others the transect method is quite accurate. Thus the decline in density from 1978 to 1981 cannot be entirely attributed to different methods and observers. It probably resulted from two poor breeding seasons due to the drought.

Table 7. Estimates of density for seven species at Eastwood 1 from transects and mapping (birds/ha).

	Transect	Mapping	Transect × 100 / Mapping
Crested Shrike-tit	.12	.15	80%
Speckled Warbler	.51	.77	66%
Rufous Whistler	.54	.88	61%
Superb Fairy Wren	1.34	1.05	128%
Brown Treecreeper	.23	.32	72%
White-throated Treecreeper	.60	.54	111%
Eastern Yellow Robin	.41	.78	53%
ALL	3.75	4.42	85%

Diet

There are biases incurred in estimating diet from either stomach contents or from observation. For this reason, I present data on the diet in the form of an index in Table 8. Most birds eat a wide variety of arthropods; few appear to be specialists, although treecreepers take mostly ants (Noske 1985). Beetles are eaten by most birds and for eleven species are the most frequently identified item in stomach samples. Sacred Kingfishers, Black-faced Cuckoo-shrikes, Noisy Friarbirds and woodswallows all eat Christmas beetles and many birds eat leaf-beetles (Chrysomelids). Perhaps predation by birds on scarab larvae is more important than that on adults (e.g. ibises — Carrick 1959; magpies — Vestjens and Carrick 1982). Larvae are the insects most frequently seen being eaten by birds, perhaps as they require some handling before being swallowed. Hemiptera are less often recorded than beetles or larvae and I have few observations of Psyllidae (lerp-insects) from stomachs. However, psyllids are important foods for honeyeaters (Wykes 1985) and pardalotes (Woinarski 1985). Ants are eaten by a wide range of birds. Thus many of the more common eucalypt defoliators and sap-suckers are eaten by birds.

Energy Demands of the Insectivorous Bird Community

Metabolic rates and daily energy requirements have been calculated for few Australian birds (e.g. Collins and Briffa 1983). However, most studies elsewhere have found a close relationship between metabolic rates, energy requirements and weight

Table 8. Diet of some insectivorous birds in eucalypt woodland in New England. + occasionally eaten; ++ often eaten; +++ major food.

	Bugs	Lerps	Beetles	Scarabs	Leaf-beetles	Larvae	Flies	Ants	Spiders	Others	Lizards
Shining Bronze-cuckoo *Chrysococcyx lucidus*	+					+++	+	+	+		
Sacred Kingfisher			++	++	+	+	+	+	+		++
Scarlet Robin *Petroica multicolor*	+		++		+		+	++	+		
Eastern Yellow Robin	+		++	+	+	++	+	+++	+	+	+
Crested Shrike-tit	+		+++		++	+++		+	++		
Golden/Rufous Whistler	+		+++		+	+++	+	++	+	+	
Superb Fairy-wren	++		++		+		++	+++	+	+	
Speckled Warbler			+++			+	+	++	+	+	
Brown Thornbill	++	+	+++		+	++	+	++	++	+	
Buff-rumped Thornbill	++		+++		+	+	++	+++	+	+	
Striated Thornbill	+	+	+++		+	++	++	+			
Varied Sittella	++		+++		++	+++	++	++	++	++	
Brown/White-throated Treecreeper	+		++		+	+	+	+++	+		+
Fuscous Honeyeater	+	++	++	+	++	+	+	+++	+	+	
White-naped Honeyeater	+	+	++		+	++	+	+	++		
Eastern Spinebill	+		+			+	+++	+	+	+	
Dusky/White-browed Woodswallow	+		+++	+			+	++	+		

(Ricklefs 1974; Aschoff and Pohl 1970). I used the formulae of King (1974) to calculate the daily energy requirements of the birds in my study sites. As King mostly used non-breeding birds at reasonable temperatures, one would expect values to be higher in winter and spring. Collins and Briffa (1983) found that the daily energy requirements of Brown Honeyeaters, *Lichmera indistincta,* in Western Australia were about 50% higher in winter than in summer. Thus I have multiplied my values by 1.5 for winter and spring.

Assimilation rates of birds eating insects are about 60-70% (Ricklefs 1974; Paton 1982), so energy requirements should be again multiplied by 1.5 to estimate energy consumption. My estimates of annual energy consumption for the major groups are given in Table 9. Most insectivores forage on a variety of substrates and I subdivided their energy demand by the same proportions as they subdivided their foraging among substrates. For honeyeaters, the nectarivorous and insectivorous part of their diets was first subdivided, then the latter subdivided among substrates and energy partitioned proportionally. The energy demands of all birds from each substrate are shown in Table 10.

Table 9. Estimated annual energy requirements for the major groups of birds at each site (10^4 Kcals/ha/yr).

	Insectivores	Herbivores	Honeyeaters	Vertebrate-eaters
Eastwood 1	8.45	2.87	3.66	0.40
Eastwood 2	5.77	2.40	4.43	0.33
Hillgrove	6.75	2.36	1.82	0.69
UNE	9.00	7.60	2.79	0.85

Table 10. Energy consumption of insectivorous birds from each substrate (10^4 Kcals/ha/yr).

	Leaves	Bark	Ground	Air	All
Eastwood 1	6.35	3.90	6.35	0.97	17.6
Eastwood 2	6.14	3.12	4.39	0.82	14.5
Hillgrove	5.10	3.18	4.63	0.33	13.2
UNE	5.59	1.87	10.30	0.81	18.6

Estimate of Insect Production

Although work is currently being carried out on the biomass and productivity of insects in eucalypt woodland, no figures are yet available. Therefore I have made a crude estimate from estimates of primary production and the consumption by leaf-eating insects.

Pressland (1982) calculated that, in eucalypt woodland near Armidale, 1480 kg of leaf material falls annually per hectare. This underestimates production by the amount that is consumed on the trees by insects and other herbivores. Fox and Morrow (1983) found that 5-44% of leaf area was lost for a range of *Eucalyptus* species in eastern Australia. Their average value was 15%. Lowman (pers. comm.) estimated that defoliation was about 10% at Eastwood State Forest in 1981. If insects consumed 10% of leaf volume, then 164 kg of the 1640 kg produced (Pressland's figure increased by 1/9) was consumed annually per hectare.

Only about 10% of plant material consumed by insects becomes insect biomass (Gosz *et al.* 1978), thus the production of leaf-eating insects is of the order of 16.4 kg/ha/yr (dry weight). As one gram of insect tissue yields about 5.5 Kcals (Golley 1961), then foliage-eating insects should yield about 9×10^4 Kcals/ha/yr.

Effect of Birds on Herbivorous Insect Populations

I estimate that birds consume 13.2-18.6 $\times 10^4$ Kcals/ha/yr of insects, of which 5.1-6.35 $\times 10^4$ Kcals/ha/yr are taken from leaves (Table 10). I also estimate that the production of leaf-eating insects is about 9×10^4 Kcals/ha/yr. If all insects taken from leaves were herbivorous ones, and leaf-eating insects were taken only from leaves, then birds would take about 55-70% of leaf-eating insects produced annually. This figure could be refined, with better estimates of metabolic rates of birds, and time-budgeting of their activities, and particularly when better data are available on the insects.

Dempster (1983) reviewed studies on the control of Lepidopteran populations. In only three out of 24 cases did he consider that there was any evidence that predators (including birds) regulated populations. Overall though, he questioned the concept of regulation by which he meant the density dependent response of a population around an equilibrium level. Density independent factors, on the other hand, could reduce population levels below carrying capacity. Specifically, he concluded that "the imposition of any large, but constant, mortality will reduce the ability of a pest population from reaching levels where it defoliates its food plant".

The results presented here suggest that birds impose a large, and maybe constant, mortality on herbivorous insect populations in eucalypt woodland. Experiments where birds have been excluded from branches or plants have generally shown that insect populations are higher inside the exclosures than outside (e.g. Solomon *et al.* 1976; Holmes *et al.* 1979). This further indicates birds' ability to depress prey populations. Experiments of this type are being planned in eucalypt woodland near Armidale.

Foliage feeding birds tend to be scarce or absent in areas suffering from eucalypt dieback (Ford and Bell 1982). Thus any effect that birds had on leaf-eating insect populations in healthy woodland would be missing. If insects are convincingly implicated as the major cause of dieback, it seems sensible to consider birds in any programme to regenerate eucalypts. Probably the best way in which this could be done is to provide shrubs. Grazing by stock will have to be removed, at least initially, to allow natural regeneration. In many cases, though, shrubs will have to be planted. At least some of the shrubs should be dense and prickly, and generally they should be planted in clumps. These will provide safer nesting sites and refuges from predators than the bare eucalypts do.

Nectar-bearing shrubs should be included to encourage honeyeaters. Perhaps those that flower in winter would be best as, if nectar is abundant in spring and summer, honeyeaters may feed less on insects at a time when they are increasing. Too many nectar-bearing plants may encourage the aggressive honeyeaters, like wattlebirds, friarbirds and miners, which could exclude smaller insectivorous species. A few shrubs bearing berries, specially in winter, could also be planted, but too many may encourage currawongs, which are nest predators.

Many shrubs will grow well in New England (Hitchcock 1982) and a wide range should be used. Eucalypts may regenerate naturally, or may have to be planted. In time, these should outgrow the shrubs giving a diverse and many-layered woodland. Many of these recommendations, aimed at favouring birds, should also favour other predators and parasites of herbivorous insects. For instance, nectar is important for parasitic flies and wasps (Davidson 1982) and for some mammals, which at times consume large numbers of insects (Smith 1982).

ACKNOWLEDGMENTS

I am grateful to the New England *Eucalyptus* Dieback Research Fund and the University of New England for supporting this research, to Mrs Susan Noske and Ms Lynda Bridges for censusing birds and recording foraging behaviour, to Ms Susan Harrington for identifying insects from birds' stomachs and Dr Stuart Cairns for collecting and identifying insects. The Australian Bird-banding Scheme (CSIRO) provided bands and colour-bands. This chapter was written while on Sabbatical Leave at the Edward Grey Institute of Field Ornithology, Oxford, and the Department of Biological Sciences, University of Stirling. I am grateful to both for their hospitality and to Marianne Burnett for typing the manuscript.

REFERENCES

ASCHOFF, J. AND POHL, H., 1970. Rhythmic variations in energy metabolism. *Fedn. Proc. Fedn. Am. Socs. exp. Biol.* **29**: 1541-552.

BELL, H. L. AND FERRIER, S., 1985. The reliability of estimates of density from transect counts. *Corella.* **9**: 3-13.

CARNE, P. B., GREAVES, R. T. G. AND MCINNES, R. S., 1974. Insect damage to plantation-grown eucalypts in north coastal New South Wales with particular reference to Christmas beetles (Coleoptera, Scarabaeidae). *J. Aust. Ent. Soc.* **13**: 189-206.

CARRICK, R., 1959. The food and feeding habits of the straw-necked ibis, *Threskiornis spinicollis* (Jameson) and the white ibis, *T. molucca* (Curier) in Australia. *C.S.I.R.O. Wildl. Res.* **4**: 69-92.

CARTER, J. J., EDWARDS, D. W. AND HUMPHREYS, F. R., 1981. Eucalypt dieback in New South Wales forests and woodlands. Pp. 27-30 *in* Eucalypt Dieback in Forests and Woodlands ed. by K. M. Old, G. A. Kile and C. P. Ohmart. C.S.I.R.O. Melbourne.

COLLINS, B. G. AND BRIFFA, P., 1983. Seasonal variations in the energetics of an Australian nectarivorous bird, *Lichmera indistincta. Comp. Biochem. Physiol.* **74A**: 731-38.

DAVIDSON, R., 1982. Correcting past mistakes — loss of habitat for predators and parasites of pasture pests. *Proc. 3rd Aust. Conf. Grassl. Invert. Ecol.*: 199-206.

DEMPSTER, J. P., 1983. The natural control of populations of butterflies and moths. *Biol. Rev.* **58**: 461-81.

EMLEN, J. T., 1971. Population densities of birds derived from transect counts. *Auk.* **88**: 323-42.

FORD, H. A. AND BELL, H. L., 1982. Density of birds in eucalypt woodland affected to varying degrees by dieback. *Emu* **81**: 202-8.

FORD, H. A., FORDE, N. AND HARRINGTON, S., 1982. Non-destructive methods to determine the diets of birds. *Corella* **6**: 6-10.

FOX, L. R. AND MORROW, P. A., 1983. Estimates of damage by herbivorous insects on *Eucalyptus* trees. *Aust. J. Ecol.* **8**: 139-47.

GOLLEY, F. B., 1961. Energy values of ecological materials. *Ecology* **42**: 581-84.

GOSZ, J. R., HOLMES, R. T., LIKENS, G. E. AND BORMAN, F. H., 1978. The flow of energy in a forest ecosystem. *Sci. Amer.* **238**: 92-102.

HITCHCOCK, M. (Ed.), 1982. Australian plants for the Northern Tablelands of New South Wales. Armidale: SGAP.

HOLMES, R. T., SHULTZ, J. C. AND NOTHNAGLE, P., 1979. Bird predation on forest insects: an exclosure experiment. *Science* **206**: 462-63.

HUMPHREYS, F. R. AND OTHERS, 1979. Final report to the Minister for Conservation and Water Resources on the New England Dieback. New South Wales Govt. Printer.

KIKKAWA, J., 1968. Ecological association of bird species and habitats in eastern Australia: similarity analysis. *J. Anim. Ecol.* **37**: 143-65.

KING, J R., 1974. Seasonal allocation of time and energy resources in birds. Pp. 4-72 *in* Avian energetics ed. by R. A. Paynter. Publ. Nuttall Orn. Club No. 15.

NOSKE, R. A., 1985. Habitat use by three bark-foragers of eucalypt forest. Pp. 193-204 *in* Birds of Eucalypt Forests and Woodlands: Ecology, Conservation, Management ed. by A. Keast, H. F. Recher, H. Ford and D. Saunders. Royal Australasian Ornithologists Union.

OLD, K. M., KILE, G. A. AND OHMART, C. P., 1981. Eucalypt dieback in forests and woodlands. C.S.I.R.O. Melbourne.

PATON, D. C., 1982. The diet of the New Holland Honeyeater. *Aust. J. Ecol.* 7: 179-98.

PRESSLAND, A. J., 1982. Litter production and decomposition from an overstorey of *Eucalyptus* spp. on two catchments in the New England region of New South Wales. *Aust. J. Ecol.* 7: 171-80.

RICKLEFS, R. E., 1974. Energetics of reproduction in birds. Pp. 152-297 *in* Avian energetics ed. by A. Paynter. Publ. Nuttall Orn. Club No. 15.

ROBERTS, R. J. AND SAWTELL, N. L., 1981. Pests in native and exotic replacement eucalypts in New England. Pp. 87-94 *in* Eucalypt Dieback in Forests and Woodlands. ed. by K. M. Old, G. A. Kile and C. P. Ohmart. C.S.I.R.O., Melbourne.

SAUNDERS, D. A., SMITH, G. T. AND ROWLEY, I., 1982. The availability and dimensions of tree hollows that provide nest-sites for cockatoos (Psittaciformes) in Western Australia. *Aust. Wildl. Res.* **9**: 541-56.

SMITH, A. P., 1982. Diet and feeding strategies of the marsupial sugar glider in temperate Australia. *J. Anim. Ecol.* **51**: 149-66.

SOLOMON, M. E., GLEN, D. M., KENDALL, D. A. AND MILSOM, N. F., 1976. Predation of overwintering larvae of Codling Moth *(Cydia pomonella* (L)) by birds. *J. Appl. Ecol.* **13**: 341-52.

WILLIAMS, J. B. AND NADOLNY, C., 1981. A survey of eucalypt species involved in New England dieback. Pp. 74-86 *in* Eucalypt Dieback in Forests and Woodlands ed. by K. M. Old, G. A. Kile and C. P. Ohmart. C.S.I.R.O., Melbourne.

WOINARSKI, J., 1985. Foliage gleaners of the treetops, the Pardalotes. Pp. 165-75 *in* Birds of Eucalypt Forests and Woodlands: Ecology, Conservation, Management ed. by A. Keast, H. F. Recher, H. Ford and D. Saunders. Royal Australasian Ornithologists Union.

WYKES, B. J., 1985. The Helmeted Honeyeater and related honeyeaters of Victorian woodlands. Pp. 205-17 *in* Birds of Eucalypt Forests and Woodlands: Ecology, Conservation, Management ed. by A. Keast, H. F. Recher, H. Ford and D. Saunders. Royal Australasian Ornithologists Union.

VESTJENS, W. J. M. AND CARRICK, R., 1974. Food of the Black-backed Magpie, *Gymnorhina t. tibicen*, at Canberra. *Aust. Wildl. Res.* **1**: 71-83.

CHAPTER 30

Re-colonization of Restored Bauxite Minelands by Birds in Southwestern Australian Forests

B. G. Collins[1], B. J. Wykes[1] and O. G. Nichols[2]

INTRODUCTION

THE JARRAH, *Eucalyptus marginata,* forest of Western Australia is used for a variety of purposes, such as timber production, water catchment, mining, recreation and the provision of apiary sites. There is pressure from many sources to ensure that these uses are fulfilled, and that conservation of fauna and flora is achieved (Shea and Herbert 1977; Heddle *et al.* 1980; Kabay and Underwood 1981).

Bauxite mining has been carried out in the northern Jarrah forest since 1963, principally by Alcoa of Australia Ltd., whose mining lease covers most of this section of the forest (Fig. 1). Mining by this company is currently carried out at Jarrahdale, Del Park and Huntly, in the western portion of the lease, at a rate of 250-300 ha per year. It is expected that mining will continue in these areas for at least another 30 years, although pockets totalling 7-8% of the entire forest are the only sites that can be mined, and many of these are subject to various environmental constraints.

Alcoa has developed operational and research programmes designed to cover all aspects of rehabilitation following mining. At present, most emphasis is placed on the reconstitution of mined areas so that vegetation is similar to that of the surrounding forest. The aim is to facilitate the return of animal communities that are representative of those found in the unmined forest. Research on a variety of animal groups has already been conducted. This includes studies of collembola (Greenslade and Majer 1980), ants (Majer *et al.* 1983), termites (Nichols and Bunn 1980), reptiles (Nichols and Bamford ms.), mammals (Kabay and Nichols 1980) and birds (Wykes *et al.* 1980; Nichols and Watkins 1984).

In this chapter, we discuss research relating to bird communities within unmined Jarrah forest. This is followed by an outline of bauxite mining and rehabilitation procedures. Re-colonization of rehabilitated sites by birds is then described, avifauna in unmined and restored habitats compared, likely long-term trends discussed, and suggestions made for future research.

THE JARRAH FOREST AND ITS AVIAN COMMUNITY

Those sections of the northern Jarrah forest where studies of avian communities have been conducted comprise laterite ridges, dissected by gullies, some of which penetrate through to the granite bedrock. Vegetation is co-dominated by Jarrah and Marri, *E. calophylla,* throughout, but with various vegetation types represented. These include young, healthy forest, mature forest, riparian thickets and areas opened up by the influence of logging and gravel excavation. Dieback fungus, *Phytophthora cinnamomi*, which is transmitted by the movement of infected soil, has been responsible for the death of large tracts of Jarrah, *Banksia* species and some other understorey vegetation (Shea 1979). Soils within the Jarrah forest are shallow and extremely infertile. The climate is mediterranean, with cool, wet winters and dry summers (Shea 1975). Frequent bush fires are a natural feature of the environment (Churchill 1968).

Shea (1975) has proposed that the presence of a diverse flora (Havel 1975 a, b) and fauna (Kimber 1972) under these conditions makes the Jarrah forest one of the world's most unique ecosystems. This image of a productive and diverse ecosystem, flourishing under harsh conditions, suggests that balance within the system may be fragile, and susceptible to changes that may result from exploitation of the forest. Various aspects of this proposition are examined in relation to the avian community within the Jarrah forest on subsequent pages.

[1]School of Biology, Western Australian Institute of Technology, Kent Street, Bentley, Western Australia 6102.
[2]Environmental Department, Alcoa of Australia Limited, Booragoon, Western Australia 6154.

Pages 341-54 *in* BIRDS OF EUCALYPT FORESTS AND WOODLANDS: ECOLOGY, CONSERVATION, MANAGEMENT ed. by A. Keast, H. F. Recher, H. Ford and D. Saunders. Royal Australasian Ornithologists Union and Surrey Beatty & Sons.

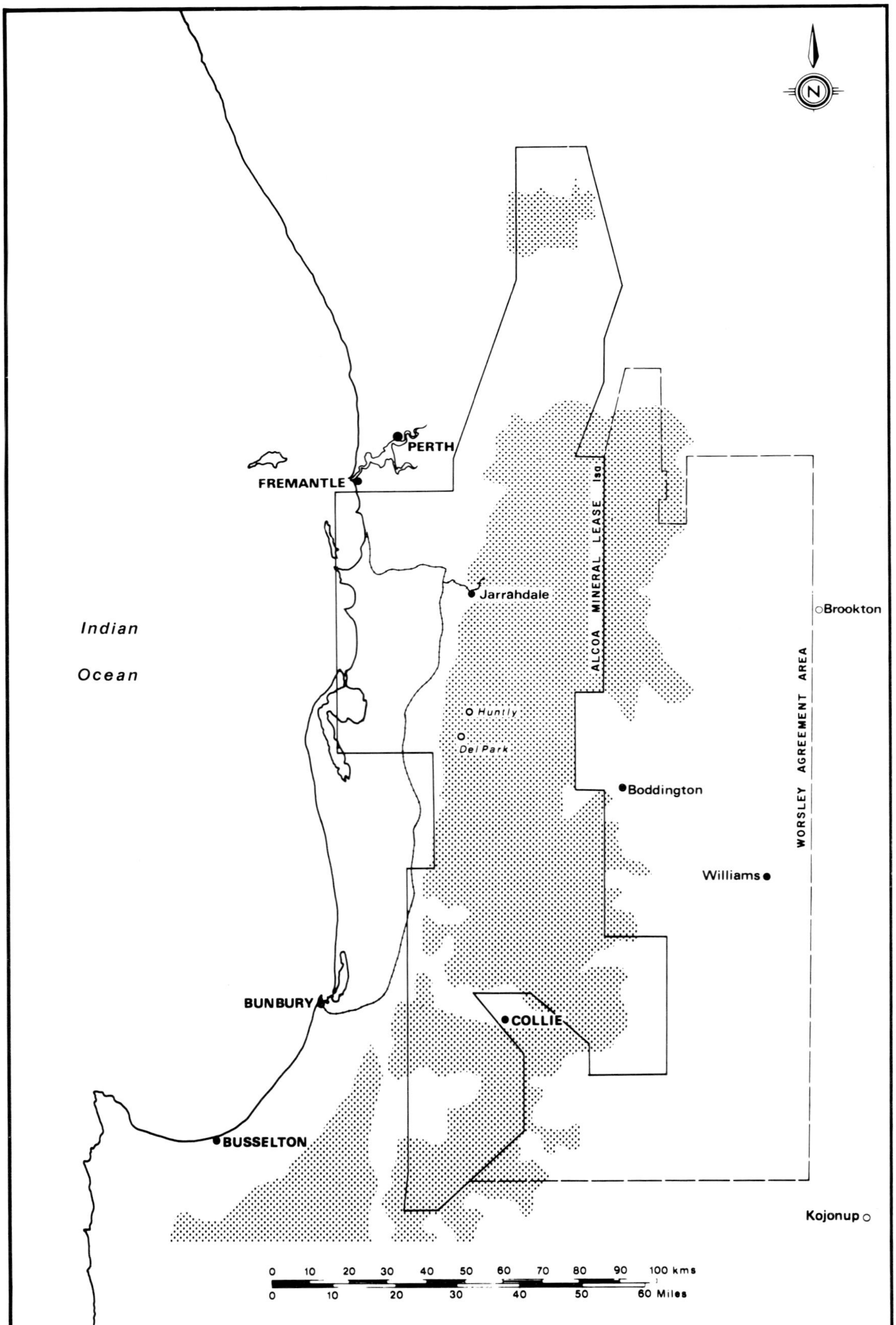

Fig. 1. Extent of bauxite mining leases and locations where mining is occurring in the northern zone of the Jarrah forest. The stippled area approximates the extent of the forest.

Compositon of Avian Community

In a study conducted near Jarrahdale (Fig. 1), from the spring of 1980 to the winter of 1982, the densities and distributions of bird species were estimated using a strip-transect method (14.5 × 0.04 km, covering 53 ha, over three mornings per census; Wykes 1982, 1985). The mean total density of birds during the study period was 52.2/10 ha (Table 1). An average of 27 species per census contributed to this low density, with a total of 42 different species being recorded. Of the 34 species with mean densities equal to or greater than 0.1 birds/10 ha, 71% were primarily insectivores and contributed 75% of the total density. Nectarivores were few in species (9%), but relatively numerous (16%). There were four parrot species (12%), each of which occurred at low densities.

Phenology of Birds and Arthropods

Total avian density fluctuated from a maximum of 66 to a minimum of 38 birds/10 ha over the study period, with an apparent pattern of spring/summer peaks and autumn/winter lows (Fig. 2). This pattern was also shown by species richness, which ranged from 32 to 22 species per census, largely because some species bred in the forest and then moved to alternative habitats.

Table 1. Mean population densities of Jarrah forest avifauna (data based on Wykes 1985). 'Other' = spp. recorded in only one visit and < 1.0/10 ha.

Guild	Species	Bird densities (birds/10ha)	
INSECTIVORE			
a) Ground foraging	Australian Raven*	0.1	
	Grey Currawong*	0.3	
	Scarlet Robin*	0.7	
	Western Yellow Robin*	1.8	
	White-breasted Robin*	0.1	
	Splendid Wren*	3.5	
	Tawny Frogmouth*	0.2	
	Boobook Owl*	0.3	
			7.0
b) Shrub foraging	Western Thornbill*	4.6	
	White-browed Scrubwren*	0.3	
	Red-winged Wren*	1.8	
	Silvereye*	1.5	
			8.2
c) Tree foraging	Broad-tailed Thornbill*	5.2	
	Golden Whistler*	2.3	
	Western Shrike-thrush*	0.7	
	Striated Pardalote*	5.7	
	Spotted Pardalote*	1.2	
	White-naped Honeyeater*	1.8	
	Shining Bronze-cuckoo*	0.2	
			17.1
d) Bark foraging	Rufous Treecreeper*	1.0	
	Varied Sittella*	0.9	
			1.9
e) Air foraging	Tree Martin*	0.7	
	Grey Fantail*	2.6	
	Dusky Woodswallow*	0.1	
	Western Warbler*	1.7	
			5.1
CARNIVORE	Collared Sparrow-hawk*		
	Brown Goshawk*	0.1	
	Laughing Kookaburra*	0.2	
	Little Eagle*	0.1	
			0.4
NECTARIVORE	Western Spinebill*	6.0	
	New Holland Honeyeater*	1.8	
	Brown Honeyeater*	0.3	
	Little Wattlebird*	0.1	
			8.2
GRAMNIVORE	Painted Quail*	0.3	
	Western Rosella*	0.3	
	Red-tailed Black Cockatoo*	0.7	
	Red-capped Parrot*	1.8	
	Port Lincoln Parrot*	1.2	
			4.3
OTHERS	6 species	0.2	
			0.2
TOTAL (mean ± standard deviation)		52.4±10.3	

* See Appendix 1 for Scientific names.

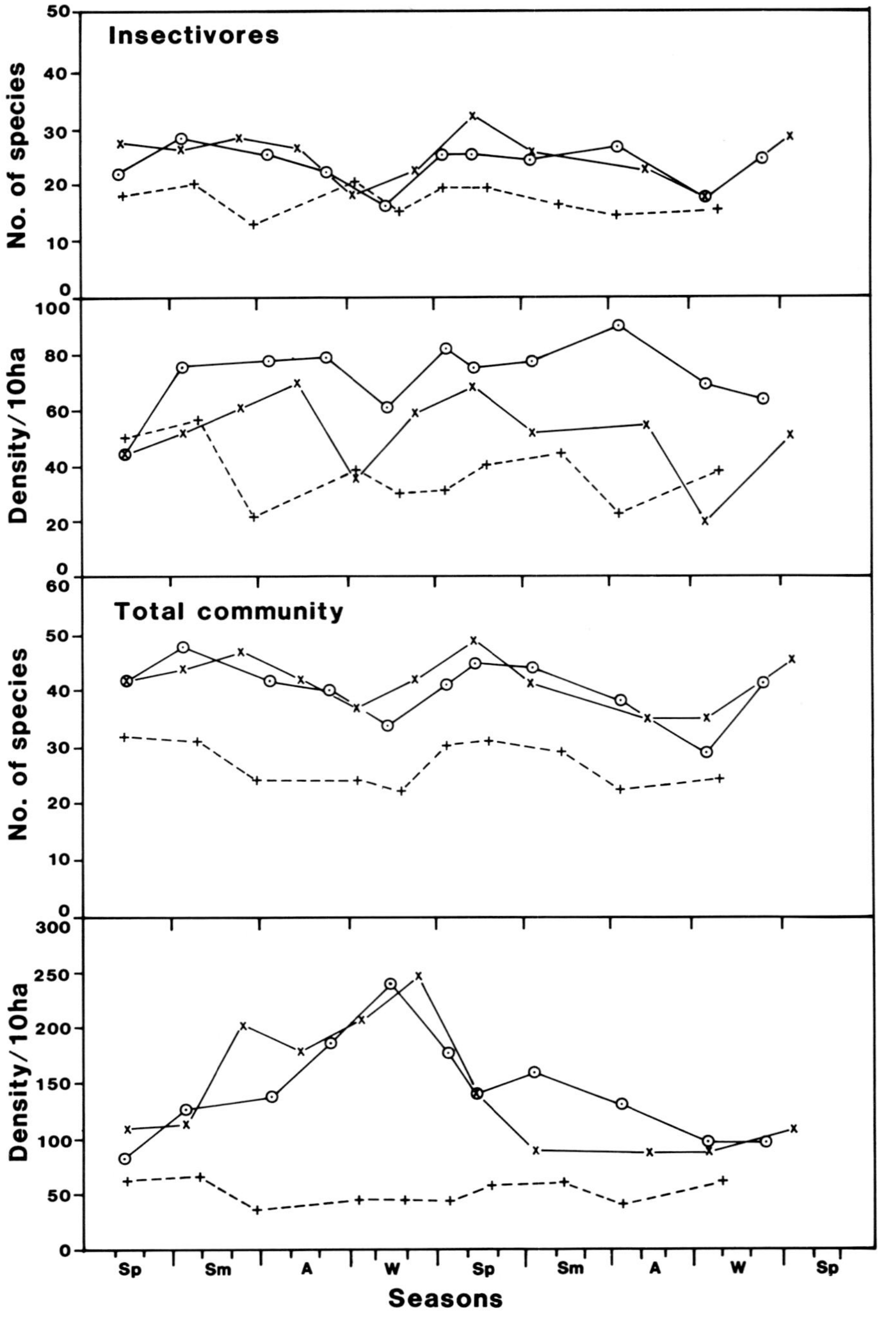

Fig. 2. Relative abundances of insectivores and all birds in Western Australia at Jarrahdale (+), and in Victoria at Brisbane Ranges (⊙) and Tarnagulla (×). Note that in all cases the Jarrahdale values are less than those for the Victorian sites.

Insect abundance and diversity were monitored over the same period, using sticky boards to capture active insects of the eucalypt canopy (10 traps for 72 hrs per sample) and by sweeping through the shrub-layer (20 units of 25 sweeps over 2.4 km per sample, Wykes 1982, 1985). Catches were classified to order level, with species counted and biomass approximated as the product of abundance and size. Biomass estimates indicate that significantly less food was available to insectivores in autumn and winter than in spring and summer (Fig. 3; Wykes 1985). As a consequence, the carrying capacity of the forest for many insectivores was probably lowest in autumn and winter. Not surprisingly, reduced carrying capacity coincided with bird movement out of the Jarrah forest, while breeding occurred in the seasons of peak arthropod abundance. Correlation between changes in density of the birds and in arthropod catches was extremely high for migratory species such as the Western Warbler, *Gerygone fusca*, Tree Martin, *Cecropis nigricans,* and Striated Pardalote, *Pardalotus striatus* (Wykes 1985).

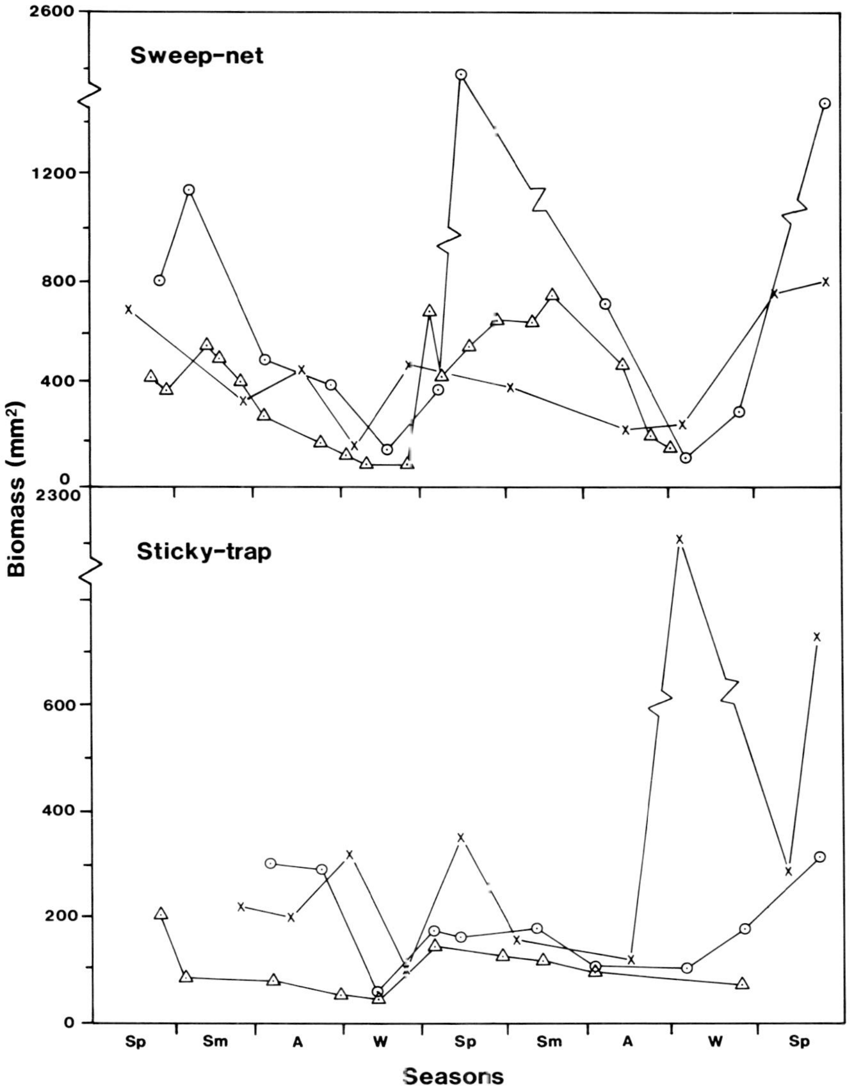

Fig. 3. Phenology of arthropod biomass. Data for Jarrahdale (△), Western Australia, and for Tarnagulla (×) and the Brisbane Range (⊙) in Victoria were gathered between 1980-82, and 1976-78, respectively (Wykes 1982, 1985). Sweepnet biomass was based on the mean catch/50 sweeps on each occasion, sticky trap biomass on the mean catch per trap.

Comparisons With Other Habitats

Density and species richness have been measured for bird communities in some Victorian and New South Wales habitats (Wykes 1975, 1982; Ford and Bell 1982), and provide a basis for further interpretation of information gathered in the jarrah forest (Figs 2, 4, 5). Contrary to expectation, the Jarrah forest avifauna has a relatively low overall density and species diversity, and is composed primarily of insectivores.

Size differences of survey areas in the different habitats may have influenced species richness values. However, Jarrahdale had the smallest mean number of species per visit, despite having the largest strip-transect. Avifaunas with values most similar to those for Jarrahdale occupied coastal heath (Cape Liptrap) and Mountain Ash, *Eucalyptus regnans,* forest (Mt Riddell) habitats, rather than other woodland areas that are superficially similar to the Jarrah forest. Some differences in avian species composition and densities were apparent, e.g., low bark-forager and nectarivore percentages at Jarrahdale and Cape Liptrap. Nevertheless, low absolute values in all categories contributed to the overall depauperacy at Jarrahdale.

Arthropod phenology at Jarrahdale was similar to that for open-forest habitats in the Brisbane Ranges and at Tarnagulla, although very high peaks were sometimes recorded at the Victorian sites (Fig. 3). As expected, insectivore densities and species numbers were lower at Jarrahdale than elsewhere, although similar values were recorded at all sites during winter periods of minimal insect abundance (Fig. 2). Total avian densities show a contrasting trend (Fig. 2), due to the influx of nectarivores at Victorian sites

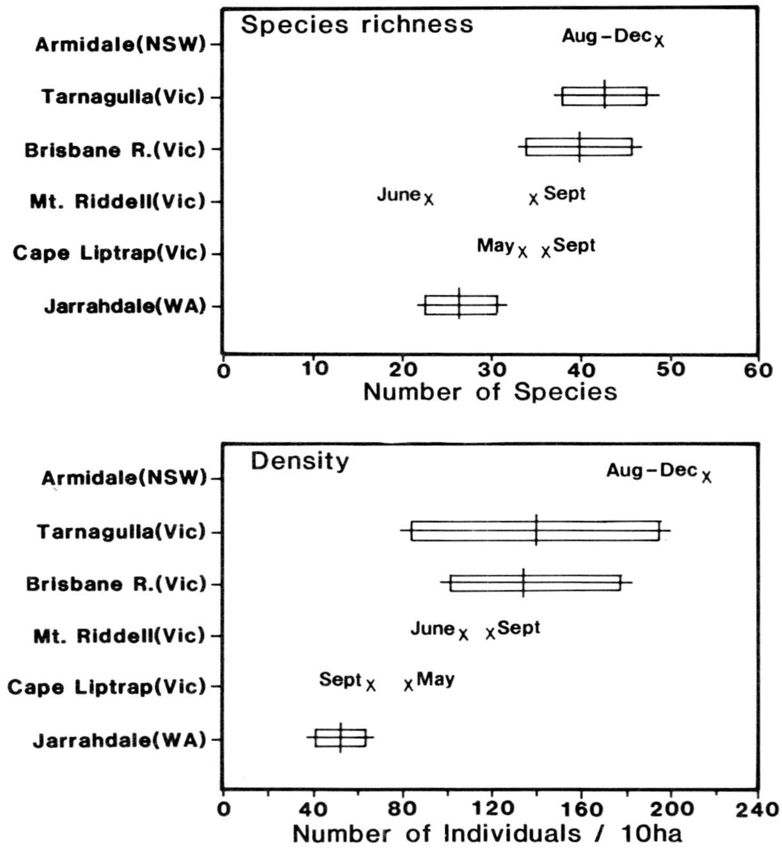

Fig. 4. Species richness and total density values for avian communities in six Australian habitats. All bird counts were obtained using strip transects. Means, standard deviations and ranges for Jarrahdale, Tarnagulla and the Brisbane Ranges are based on data obtained over two years by Wykes (1982, 1985). Data for shorter studies at Armidale (Ford and Bell 1982) and Mt Riddell and Cape Liptrap (Wykes 1975) are also indicated.

during some winters (Wykes 1982). Differences between habitats suggest that there is a smaller non-resident component to the avifauna of the Jarrah forest than elsewhere. This may occur because the southwest of Western Australia offers fewer opportunities for insectivores, nectarivores and partial nectarivores to move between habitats than does eastern Australia, where the Great Dividing Range provides extensive and varied forests.

Historical considerations (Serventy and Whittell 1976) and the above analysis suggest that insectivorous species remaining in the Jarrah forest are those that have survived fluctuating climatic conditions and habitat restrictions during their evolution. It is likely that they are habitat generalists, capable of withstanding seasonal food shortages in a restricted forest region. Analysis of habitat usage and foraging characteristics of insectivores in the study area at Jarrahdale showed most species to be widely distributed among floristic micro-habitats, providing appropriate vegetation strata were available (Wykes 1985). Overall distribution data for these species also indicate their breadth of habitat usage. For instance, 48% of all species found at Jarrahdale are widespread in Western Australia and other parts of the continent; 17% occur as 'identical' forms in eastern and western forests; 36% occur in all forests of the southwest as endemic races or species, while none are confined to the Jarrah forest. Hence, if conservation of races or species is a primary concern of forest management, 36% require particular attention, but none are dependent on Jarrah forest as a habitat.

Discussion to this stage has been concerned primarily with insectivorous birds of the Jarrah forest, although three other groups warrant brief consideration.

Bark-gleaning species

Lack of suitable winter food sources may be a reason for the shortage of bark-foraging *Lichenostomus* honeyeaters in the Jarrah forest, although an absence of eucalypts with decorticating bark sites that harbour insects and their sugar secretions may also be critical (Paton 1980; Wykes 1982). Similarly, the bark-gleaning Western Shrike-tit, *Falcunculus frontatus leucogaster,* does not occur in the Jarrah

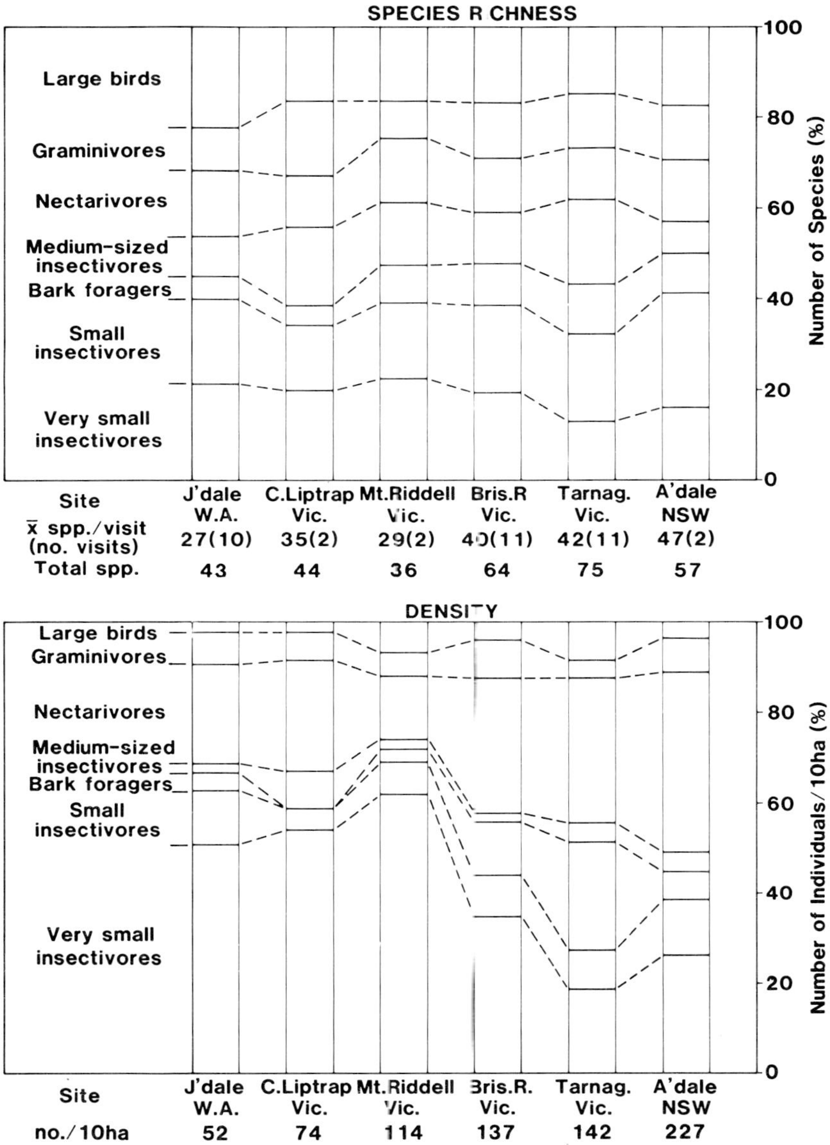

Fig. 5. Relative contributions of various avian guilds to species richness and total density in six Australian habitats. Species classification follows Ford and Bell (1982).

forest, and only one climacterid, the Rufous Tree-creeper, *Climacteris rufa,* is present. The latter species may warrant particular attention by forest managers. It is a log and tree-trunk forager of ants (Wykes 1985), whose density is likely to be regulated by food shortages at particular times of the year, or by the abundance of hollows for nesting. Availability of both resources may be reduced by factors such as spring fuel-reduction burning, changes in vegetation composition following mining, and loss of mature, hollow-providing trees as a result of logging and mining

Nectarivores

In addition to the dominant eucalypts, the most common plant species that are attractive to birds near Jarrahdale are *Adenanthos barbigera* and *Grevillea wilsonii* (year-round flowering), *Banksia grandis* (spring flowering), and *Dryandra sessilis* (autumn to winter flowering) (Collins and Newland ms.). The Western Spinebill, *Acanthorhyncus superciliosis,* is a relatively abundant resident of the forest that uses *A. barbigera* and *G. wilsonii* for much of the year, although other food sources are undoubtedly important at various times (Collins and Briffa 1982; Wykes 1985; Collins and Newland ms.). The other resident nectarivore, the New Holland Honeyeater, *Phylidonyris novaehollandiae*, is a creek-based opportunist that feeds on the nectar of *B. grandis, D. sessilis, Banksia littoralis* and *E. calophylla*, but also obtains manna from *Agonis linearifolia* (Rooke 1979; Wykes 1985; Collins and Newland ms.). A number of other nectarivore species are attracted to the Jarrah forest when nectar is abundant, but none were ever numerous at Jarrahdale (Wykes 1985; Collins and Newland ms.).

Parrots

Little is known about population regulation of the three main species of parrot in the jarrah forest: the Red-tailed Black Cockatoo, *Calyptorhynchus magnificus,* that feeds primarily on the seed of Jarrah and Marri, but seasonally feeds on *Casuarina* seed; the Red-capped Parrot, *Purpureicephalus spurus,* that feeds on seed of *Casuarina* and *Acacia* as well as that of Marri and Jarrah; and the Port Lincoln Parrot, *Barnardius zonarius*, has a highly varied diet of seed, pollen, nectar and lerps (Wykes 1985). All are potentially susceptible to shortages of their specific food requirements at particular times of the year. It is not known whether parrots rely entirely on the Jarrah forest for breeding purposes. If they do, then a reduction in availability of tree hollows as a result of forestry and mining activity could be critical. A particular problem could be that these long-lived, social animals have traditional nest-sites that, when removed, may disrupt populations (Saunders 1977, 1982).

REHABILITATION PROCEDURES

Bauxite mining involves the clearing of forest in pods of 10-20 ha in area. Topsoil and overburden, averaging 50 mm and 0.5 m in depth, respectively, are removed. Caprock is then blasted and the bauxitic layer mined to a depth of 2-5 m. The resultant pit possesses a compacted clay floor, due to the effect of heavy machinery, and vertical pit walls that are adjacent to unmined forest. Methods used to rehabilitate the pits have changed markedly over the years, and are outlined below.

Original Techniques

Early rehabilitation was designed specifically to conserve water supplies and provide a timber resource. Topsoil and overburden were removed together, stockpiled, and replaced after mining. Trees were then planted, but little nutrient applied. The results were unsatisfactory, with steep pit walls remaining and poor tree growth resulting from restricted root penetration through the compacted clay floor. Furthermore, very little understorey vegetation developed because of decreased seed viability and nutrient availability in stockpiled topsoil and overburden. Only a small proportion of mined areas were treated in this way. Most of those that have not been modified since, using modern techniques, still demonstrate poor tree growth, little understorey development and relatively poor re- colonization by fauna (Kabay and Nichols 1980).

Current Techniques

Rehabilitation now involves a variety of earthworks before planting commences. Pit walls are bulldozed to reduce their slope. Where practicable, the top 50 mm of topsoil, containing most nutrients and seed, is removed from another site that is about to be mined and placed immediately on the area that is to be re-vegetated. Soil and clay are then ripped to a depth of 1-2 m, to facilitate tree root penetration. Native seed is spread at a rate of 1 kg per ha, as a supplement to that which is present in the fresh topsoil. Added seed includes that of *Acacia* species, which fix nitrogen and may suppress Dieback fungus (Shea 1979), and other species that do not readily germinate from topsoil. Trees are planted at a rate of 640 per ha. At least half of these are local eucalypt species, the remainder being exotic eucalypts chosen for high growth rate, timber quality and resistance to Dieback fungus. Three or more species of tree are planted in every pit, as a safeguard against the failure of a particular species.

These techniques have resulted in improved soil nutrient stores and better tree development. Understorey plant species richness and diversity are almost as high as in unmined forest habitats (Tacey and Glossop 1980). Re-colonization by fauna in general is superior to that in areas rehabilitated using older techniques (Nichols and Bamford ms.; Nichols and Burrows ms.).

BIRD RE-COLONISATION OF REHABILITATED AREAS

Various reports have emphasised the need to consider fauna in association with land management programmes for the Jarrah forest (e.g. Forests Department 1977; Technical Advisory Group Report 1978; and Alcoa *et al.* 1978). Nevertheless, little was known about the avifauna until Alcoa commenced a research programme in 1978.

Initial investigations were descriptive, indicating which bird species appeared to return successfully to re-vegetated minesites (Kabay and Nichols 1980; Wykes *et al.* 1980). Recent studies, however, have involved the measurement of the density and species richness of birds in rehabilitated minesites and control forest communities. Successful colonisation of rehabilitated areas has been defined in terms of whether values obtained are within the ranges found for the unmined forest. Attention has also been given to the return of particular bird groups or species, such as those that were originally rare, which perform particular roles in the forest ecosystem, or are normally restricted mainly to the Jarrah forest.

Studies of bird colonisation of rehabilitated minesites have encountered several difficulties. The pits have been quite small in area, thus precluding the use of long, straight transect-lines for surveys, and introducing unavoidable edge effects. Visibility has varied greatly between different rehabilitation sites, and between these sites and the forest. The wide range of rehabilitation techniques that have been applied, combined with the youth of plots re-vegetated using more recent techniques, have also limited the number and replicability of suitable sampling sites. Nevertheless, monitoring of avifauna in maturing rehabilitated areas, and consideration of seasonality and breeding success, has permitted prediction of the long-term value of rehabilitation procedures.

Inital Surveys

Initial studies of birds in revegetated areas were designed to identify colonists, and to determine whether these areas were used for feeding, breeding or shelter (Kabay and Nichols 1980). Approximately 80% of all upland Jarrah forest bird species were recorded in re-vegetated areas within 10 years after initial rehabilitation, suggesting that mined pods did not act as barriers to populations in the neighbouring forest. Knowledge of habitat utilization by birds was minimal, although available information was used as a basis for suggesting modifications to rehabilitation techniques so that more species would be attracted. Nevertheless, further research was required in order to justify such manipulative techniques on a large scale.

Quantitative studies that compared bird densities in a variety of rehabilitated and control forest areas were commenced in 1980 (Nichols and Watkins 1984). These demonstrated that early rehabilitation techniques, involving the planting of trees (*Pinus* spp. and eucalypts) without the provision of understorey vegetation, supported very low bird species numbers and densities. In contrast, areas that were seeded and/or given fresh topsoil in order to develop an understorey, supported bird species numbers and total bird densities similar to those at unmined forest control sites within 5-6 years after mining.

Recent Surveys

During the last 2-3 years, additional investigations have been undertaken in recently rehabilitated areas in order to identify those techniques which provide optimal re-colonization by birds of the Jarrah forest. One of these studies, conducted in January 1981, involved a comparison of avian communities at two sites that had been rehabilitated using current techniques with those in an area rehabilitated using an older method, and in four forest control plots (Table 2; Nichols and Watkins 1984). Avian density and species richness values for the recently rehabilitated sites were within the ranges obtained for healthy and dieback-affected Jarrah forest. Values for the site that was rehabilitated using the techniques initially employed compared poorly with those for all other sites.

Bird community composition was similar for the better rehabilitation sites and healthy forest, although there were some differences. For instance, the Red-capped Parrot occurred at all forest sites, but not in rehabilitated areas. The ground-foraging Western Yellow Robin, *Eopsaltria griseogularis,* was uncommon in rehabilitated areas, as were the trunk- and log-foraging Rufous Treecreeper and the bark-gleaning Varied Sittella, *Daphoenositta chrysoptera*. In contrast, various species that are generally confined to dense, riparian understorey vegetation of the Jarrah forest, and not represented in the upland sites surveyed, had colonized the recently rehabilitated site that had very dense understorey vegetation (Site V, Table 2). These included the White-browed Scrubwren, *Sericornis frontalis*, Red-winged Wren, *Malurus elegans*, White-breasted Robin, *E. georgiana* and New Holland Honeyeater. Shrub-layer feeders such as the Western Thornbill, *Acanthiza inornata,* and Silvereye, *Zosterops laterals,* were at least as common in revegetated sites as in the forest. Honeyeaters, whose presence generally depends on the flowering of particular plant species, were detected at only some of the control and rehabilitated sites.

In general, the site rehabilitated using fresh topsoil (Site VII, Table 2) appeared to support a community whose composition was more similar to that of healthy Jarrah forest than did Site V. It might not be desirable, however, to use only one rehabilitation technique, as use of a variety of methods could help to ensure the long-term availability of habitats required by a wide range of bird species.

Other comparisons of Jarrah forest avifauna with that of the more successfully rehabilitated areas were undertaken in 1981/82 (Wykes 1985). Transects were laid out through Marri regrowth (Site VII) and neighbouring healthy (Site VIII) and dieback-affected (Site IX) forest.

Seasonal changes in species richness values for the forest and rehabilitation communities, and variations in density for all three habitats, are shown in Figure 6. Density values obtained over the same period for the Jarrahdale forest community described earlier in this chapter are also indicated. Species richness was similar throughout the year for the Jarrah and rehabilitation habitats, as was avian density for these sites and for the Jarrahdale study area. In contrast, density values for the dieback-affected forest were generally low. Individual species' densities for all sites are included in Table 2.

Bird Breeding

As a first step towards understanding long-term aspects of recolonisation by birds, a preliminary survey of breeding was undertaken in October and November, 1981 (Curry and Nichols ms.). Twenty-one species were found to use revegetated areas wholly or partially during breeding, while 12 species placed nests in the rehabilitation areas surveyed. Among these were the low-nesting Red-winged

Table 2. Avian species density and richness for various Jarrah forest and rehabilitated sites. The dieback-affected sites retained some dieback-resistant Marri. "Tree", "mid-storey" and "understorey" layers were defined as being >10 m, 2-10 m and <2 m high, respectively. No tree species in Sites V, VI and VII had reached a height of 10 m by 1981. Sites VIII and IX were adjacent to Site VII and represented healthy and dieback-affected forest, respectively. The 1981 densities are means based on a single summer census (Nichols and Watkins ms.), while the 1981-82 figures are means for six censuses conducted over a full year (Wykes 1985). Species recorded on only one visit are not included. *See Appendix 1 for Scientific names.

	1981 Summer Survey							1981-82 Year Survey		
Site Number	I	II	III	IV	V	VI	VII	VII	VIII	IX
SITE CHARACTER										
Description	Healthy Forest	Healthy Forest	Severly affected by dieback	Affected by dieback	Rehabilitated 1976 Thick *Acacia* midstorey	Rehabilitated 1970 No understorey added	Rehabilitated 1975 with fresh topsoil (Tacey and Glossop 1980)	Rehabilitated 1975 with fresh topsoil (Tacey and Glossop 1980)	Healthy Forest	Dieback affected forest
Tree species	*E. marginata/ E. calophylla*	*E. marginata/ E. calophylla*	*E. calophylla*	*E. calophylla*	*E. resinifera/ E. calophylla*	*E. microcorys*	*E. calophylla*	*E. calophylla*	*E. marginata/ E. calophylla*	*E. marginata/ E. calophylla*
Tree density (No./ha)	92	88	8	24	0	0	0	–	–	–
Mid-storey density (No./ha)	108	160	32	64	very dense unbroken vegetation	160	200	–	–	–
Understorey (% ground cover	70	40	15	40	10	5	40	–	–	–
Leaf litter (% ground cover)	95	95	10	60	60	20	40	–	–	–
BIRD SPECIES										
Little Eagle*								0.2	0.5	0.3
Painted Quail*	–	–	–	–	–	–	11.7	–	–	–
Common Bronzewing*	–	–	–	–	1.7	1.7	–	1.1	–	–
Western Rosella*	25.0	–	–	1.7	3.4	–	5.0	–	0.4	0.9
Red-capped Parrot*	15.0	5.0	6.7	11.7	–	–	–	1.5	3.8	–
Port Lincoln Parrot*	6.7	–	–	–	3.4	–	–	5.1	1.6	0.8
Laughing Kookaburra*	–	3.4	–	–	–	–	–	–	0.6	–
Sacred Kingfisher*	–	–	3.4	–	–	–	–	–	–	–
Rainbow Bee-eater*	–	–	18.4	31.7	–	–	–	–	–	–
Tree Martin*	–	18.4	–	13.3	–	–	–	0.6	1.0	0.3
White-winged Triller*	–	–	–	8.3	–	–	–	–	–	–
Splendid Wren*	–	11.7	–	–	16.7	–	16.7	6.2	4.7	2.9
Red-winged Wren*	–	–	–	–	5.0	–	–	2.3	1.9	–
Western Warbler*	1.7	1.7	6.7	3.3	–	–	1.7	3.0	2.5	1.8
Broad-tailed Thornbill*	6.7	3.4	–	1.7	8.4	–	11.7	6.1	6.5	3.1
Western Thornbill*	21.7	5.0	6.7	10.0	1.7	–	25.0	5.0	9.7	1.4
White-browed Scrubwren*	–	–	–	–	3.4	–	–	1.2	0.3	–
Weebill*	1.7	3.4	3.4	–	–	–	–	–	–	–
Scarlet Robin*	–	–	3.4	–	–	–	–	0.4	1.9	0.9
Western Yellow Robin*	8.4	3.4	–	–	1.7	–	–	1.0	1.6	–
White-breasted Robin*	–	–	–	–	1.7	–	–	–	–	–
Grey Fantail*	8.4	10.0	–	–	3.4	–	3.4	5.4	6.8	1.2
Golden Whistler*	1.7	–	–	–	5.0	–	–	3.5	3.3	–
Western Shrike-thrush*	5.0	1.7	–	–	3.4	–	3.4	0.6	0.4	–

Table 2. Continued.

	1981 Summer Survey							1981-82 Year Survey		
Site Number	I	II	III	IV	V	VI	VII	VII	VIII	IX
SITE CHARACTER										
Description	Healthy Forest	Healthy Forest	Severely affected by dieback	Affected by dieback	Rehabilitated 1976 Thick *Acacia* midstorey	Rehabilitated 1970 No understorey added	Rehabilitated 1975 with fresh topsoil (Tacey and Glossop 1980)	Rehabilitated 1975 with fresh topsoil (Tacey and Glossop 1980)	Healthy Forest	Dieback affected forest
BIRD SPECIES										
Varied Sittella*	–	15.0	–	8.3	–	–	–	–	1.0	–
Rufous Treecreeper*	6.7	1.7	–	–	–	–	–	–	–	–
Spotted Pardalote*	5.0	–	–	–	–	–	–	0.6	1.0	–
Striated Pardalote*	8.4	21.7	5.0	25.0	–	–	1.7	2.6	5.8	0.8
Silvereye[a]	–	–	–	11.7	1.7	–	15.0	10.6	3.8	2.8
Brown Honeyeater*	–	–	–	–	–	–	18.4	–	–	–
White-naped Honeyeater*	5.0	11.7	–	–	–	–	6.7	0.6	3.6	–
Western Spinebill*	5.0	8.4	–	5.0	1.7	–	–	2.2	2.1	–
New Holland Honeyeater*	–	–	–	–	6.7	–	–	1.5	–	–
Little Wattlebird*	–	–	–	1.7	–	–	–	0.6	–	–
Dusky Woodswallow*	–	–	6.7	–	–	–	–	–	–	–
Black-faced Cuckoo-shrike*	–	–	–	1.7	–	–	–	–	–	–
Total density (birds/10 ha)	132.0	126.0	60.0	135.0	69.0	1.7	120.0	61.9	64.8	17.2
Species richness (No. of species)	16	16	9	14	16	1	12	14	15	8

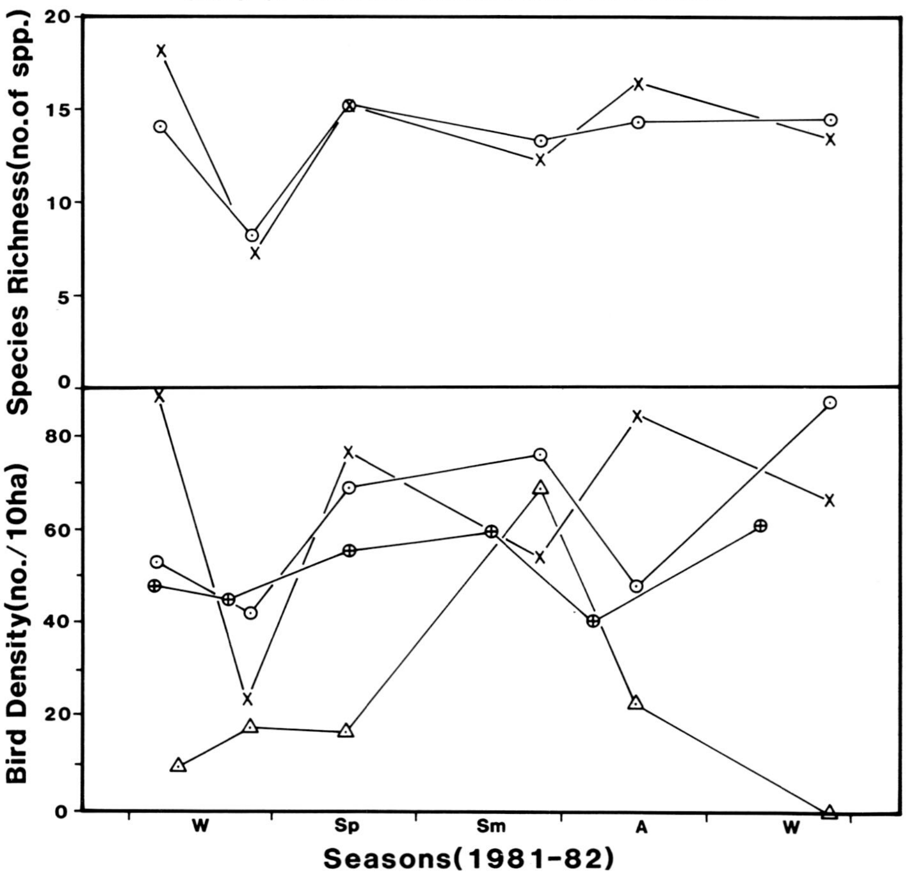

Fig. 6. Phenology of species richness values and densities of avian communities at Jarrahdale (⊕), and at Sites VII (⊙ rehabilitated), VIII (× control forest) and IX (△ dieback-affected forest).

Wren, the shrub-nesting Brown Thornbill, *Acanthiza pusilla*, the mid-storey foliage-nesting Western Warbler, the branch- and twig-nesting Western Spinebill and the fork-nesting Western Yellow Robin.

CONCLUSIONS

Current Re-colonization of Rehabiltated Minesites

Several conclusions can be drawn from the results of research that has been conducted in immature rehabilitated plots. Perhaps the most significant is that there are marked differences between avifaunas at various sites: older plots (those without understorey, or planted with pines) consistently support smaller numbers of species and densities than those in which more advanced rehabilitation techniques have been used.

In addition, it has been demonstrated that bird communities in 6-7 year-old vegetation resulting from advanced rehabilitation techniques have densities and total numbers of species similar to those for healthy forest communities. Species composition varies somewhat between forest and rehabilitated sites, with the latter lacking those species that have specific requirements such as a tree canopy or open litter. Birds that build nests at a variety of sites in the vegetation are capable of breeding in existing rehabilitated areas. Others such as hollow-nesters cannot do so in the short-term unless nest boxes are provided.

Predicted Long-term Trends in Avian Re-colonization

Re-colonization of rehabilitated minesites by fauna clearly depends upon the establishment of suitable habitats. Nevertheless, little is known regarding temporal trends in ecosystem development at such sites, apart from work reported by Majer *et al.* (1983) for ants, and by Nichols and Michaelson (ms.) for plants, in the Jarrah forest and by Fox and Fox (1978) in coastal heath areas.

The most successful bauxite minesites, in terms of colonization by birds, are only 5-6 years old at present. Vegetation at these sites comprises a dense shrub layer, in which short-lived *Acacia* species predominate, together with eucalypt saplings of uniform age. As the vegetation develops beyond this stage, it is predicted that a true tree layer and some open space will arise, and shrub species that are not well-represented currently will become established. Mulch and timber debris on the habitat floor should also form a litter layer that contributes significantly to nutrient cycling.

Responses by birds to successional changes in minesite vegetation are likely to include a reduction in the density and species richness of insectivores that are dependant upon dense shrubbery. Increased abundance of birds such as the Rufous Treecreeper, Varied Sittella, White-naped Honeyeater, *Melithreptus lunatus,* and Striated Pardalote should occur as mature trees develop. Some re-growth of plant species such as *Banksia grandis, Casuarina fraserana* and *Persoonia* sp. is already occurring in rehabilitated areas. If this trend continues, and such species form a middle canopy similar to that which is extensively utilized by birds in the natural forest (Wykes 1985; Collins and Newland ms.), avian diversity in rehabilitated areas should increase.

Despite the optimism expressed above, several factors may prevent rehabilitated areas from acquiring avifaunas similar in composition, density and richness to those for undisturbed Jarrah forest habitats. For instance, it is not known how the eucalypts, many of which are not indigenous to the southwest of Western Australia, will respond in the long-term to climatic and soil conditions in rehabilitated areas. Poor tree-layer development would have marked effects upon the avifauna in these areas.

Species susceptible to Dieback fungus are not planted. However, the disease may attack some self-colonizing shrub species and Jarrah in rehabilitated areas, since it could be introduced by any activity involving soil transfer. This would obviously influence the availability of suitable habitats for certain birds.

Most insectivores of the Jarrah forest are likely to colonize vegetation as it matures, since they tend to be eclectic in habitat requirements. Nevertheless, overall densities may be smaller than in the natural forest if the plant species present support smaller insect populations.

Specific habitat requirements of some bird species may not become available for a considerable time if they are not provided artificially. A deficiency of hollow limb and trunk nesting-sites, and of plant species that provide seed, pollen and/or nectar could adversely affect re-colonization by birds such as parrots and honeyeaters.

Suggested Future Research

There is a continuing need to evaluate rehabilitation procedures, and take appropriate action as sites mature. The most obvious requirement is to promote successional understorey and tree development, and soil nutrient buildup, if ecosystem stability and optimal re-colonization by fauna are to be achieved (Nichols *et al.* 1981).

Continued monitoring of the avian community at rehabilitated sites is necessary if predictions based on present research are to be tested. In addition, further information regarding factors, such as nest-site and food availability, that limit community structure and density is required. Many aspects of the biology of particular avian groups also require additional study. For instance, it is not known to what extent parrots are able to utilize the fruits of introduced eucalypts as distinct from those of Jarrah and Marri.

Trials designed to test whether changes to specific environmental factors will increase the suitability of re-vegetated sites for avifauna have already commenced. Woodchips, vegetation litter, nest boxes or plant species especially suited to the needs of particular birds have been introduced at selected sites. A decision as to whether these techniques should be implemented on a larger scale will depend on future assessment of their influence on avian re-colonization.

An ongoing management plan must be developed as rehabilitated sites become integrated into the surrounding Jarrah forest. Ideally, the long-term objective should be to make these areas self-sustaining and capable of supporting fauna that is no less diverse than that of the unmined Jarrah forest.

REFERENCES

ALCOA OF AUSTRALIA AND DAMES AND MOORE, 1978. Wagerup Alumina Project: Environmental Review and Management Programme.

CHURCHILL, D. M., 1968. The distribution and pre-history of *Eucalyptus diversicolor* F. Muell., *E. marginata* E. Donn. Ex. S.m., and *E. calophylla* R.Br. in relation to rainfall. *Aust. J. Bot.* **16**: 125-51.

COLLINS, B. G. AND BRIFFA, P., 1982. Seasonal variation of abundance and foraging of three species of Australian honeyeaters. *Aust. Wildl. Res.* **9**: 557-69.

COLLINS, B. G. AND NEWLAND, C., 1985. Honeyeater energetics and population changes in the Jarrah forest of Western Australia. Ms.

CURRY, P. J. AND NICHOLS, O. G., 1985. Young rehabilitated bauxite minesites as breeding habitats for birds in the Jarrah forest of southwestern Australia. Ms.

FORD, H. A. AND BELL, H., 1982. Density of birds in eucalypt woodland affected to varying degrees by dieback. *Emu* **81**: 202-08.

FORESTS DEPARTMENT OF WESTERN AUSTRALIA, 1977. General Working Plan, No. 86, Western Australia.

FOX, B. J. AND FOX, M. D., 1978. Recolonisation of coastal heath by *Pseudomys novaehollandiae* (Muridae) following sand mining. *Aust. J. Ecol.* **3**: 447-65.

GREENSLADE, P. AND MAJER, J. D., 1980. Collembola of rehabilitated minesites in Western Australia. Proceedings of the VII International Colloquium of Soil Zoology, Pp. 397-408. U.S. Environmental Protection Agency.

HAVEL, J. J., 1975a. Site-vegetation mapping in the northern Jarrah forest: I. Definition of site-vegetation types. Bulletin of the Forests Department of Western Australia, No. 86.

HAVEL, J. J., 1975b. Site-vegetation mapping in the northern Jarrah forest: II. Location and mapping of site-vegetation types. Bulletin of the Forests Department of Western Australia, No. 87.

HEDDLE, E. M., HAVEL, J. J. AND LONERAGAN, O. W., 1980. Focus on northern Jarrah forest conservation and recreation areas. *Forest Focus* **22**: 3-27.

KABAY, E. D. AND NICHOLS, O. G., 1980. Use of rehabilitated bauxite mined areas in the Jarrah forest by vertebrate fauna. Environmental Research Bulletin No. 8. Alcoa of Australia Limited, Perth.

KABAY, E. D. AND UNDERWOOD, R. J., 1981. Rehabilitation to suit land-use after bauxite mining in the northern Jarrah forest. Australian Mining and Industry Council Workshop, Canberra.

KIMBER, P., 1972. Birds of the Jarrah forest. *Forest Focus* **7**: 5-7.

MAJER, J. D., DAY, J. E., KABAY, E. D. AND PERRIMAN, W. S., 1983. Recolonisation by ants in bauxite mines rehabilitated by a number of different methods. *J. App. Ecol.* **21**: 355-75.

NICHOLS, O. G. AND BAMFORD, M., 1985. Herpetile utilization of rehabilitated bauxite minesites and dieback-affected sites in Western Australian Jarrah, *Eucalyptus marginata,* forest. Ms.

NICHOLS, O. G. AND BUNN, S., 1980. Termite utilisation of rehabilitated bauxite mined areas. Environmental Research Bulletin No. 9. Alcoa of Australia Limited, Perth.

NICHOLS, O. G. AND BURROWS, R., 1985. Re-colonization of revegetated bauxite minesites by predatory invertebrates. *For. Ecol. Man.* (in press).

NICHOLS, O. G., GLOSSOP, B. L. AND SMURTHWAITE, A. J., 1981. A method for assessing the likely effects of bauxite mining on conservation of flora and fauna management priority areas. Environmental Research Bulletin No. 12. Alcoa of Australia Limited, Perth.

NICHOLS, O. G. AND MICHAELSEN, D. M., 1985. Successional processes in bauxite minesites rehabilitated using three topsoil return techniques. Ms.

NICHOLS, O. G. AND WATKINS, D., 1984. Bird utilisation of rehabilitated bauxite minesites in Western Australia. *Biol. Cons.* **30**: 109-31.

PATON, D. C., 1980. The importance of manna, honeydew and lerp in the diets of honeyeaters. *Emu* **80**: 213-26.

ROOKE, I., 1979. The social behaviour of the honeyeater *Phylidonyris novaehollandiae*. Ph.D. Thesis, University of Western Australia, Perth.

SAUNDERS, D. A., 1977. The effect of agricultural clearing on the breeding success of the White-tailed Black Cockatoo. *Emu* **77**: 180-84.

SAUNDERS, D. A., 1982. The breeding behaviour and biology of the short-billed form of the White-tailed Black Cockatoo, *Calyptorhynchus funereus. Ibis* **124**: 422-55.

SERVENTY, D. L. AND WHITTELL, H. M., 1976. Birds of Western Australia. Lamb Publications, Perth.

SHEA, S. R., 1975. Focus on Jarrah Dieback — a threat to W.A.'s unique Jarrah forest. *Forest Focus* **14**: 1-14.

SHEA, S. R., 1979. An ecological approach to the control of Jarrah dieback. *Forest Focus* **21**: 7-18.

SHEA, S. AND HERBERT, E., 1977. Managing Jarrah forest catchments. *Forest Focus* **19**: 1-20.

TACEY, W. H. AND GLOSSOP, B. L., 1980. Assessment of topsoil handling techniques for rehabilitation of sites mined for bauxite within the Jarrah forest of Western Australia. *J. App. Ecol.* **17**: 195-201.

TECHNICAL ADVISORY GROUP REPORT, 1978. Bauxite mining in the Darling Range, Western Australia. A review for the Environmental Protection Authority by the Technical Advisory Group. Department of Conservation and Environment Bulletin No. 44.

WYKES, B. J., 1975. A preliminary comparison of avian social organizations in different habitats. B.Sc. Hons. Thesis, Monash University, Melbourne.

WYKES, B. J., 1982. Resource partitioning and the role of competition in structuring *Lichenostomus* honeyeater (and *Manorina melanophrys*) communities in southern Victoria. Ph.D. Thesis, Monash University, Melbourne.

WYKES, B. J., 1985. The Jarrah forest avifauna and its re-establishment after bauxite mining. Biology Bulletin No. 12, School of Biology, WAIT.

WYKES, B. J., BROWN, L. AND COLLINS, B. G., 1980. Bird recolonisation of rehabilitated bauxite mine pits. Report to Alcoa of Australia Limited.

CHAPTER 31

Human Impact: The Response of Forest and Woodland Bird Communities. Whither the Future? A Synthesis.

D. A. Saunders[1]

THERE IS LITTLE doubt that the landscape of Australia today is vastly different from that when Europeans first settled nearly two centuries ago. The changes brought about in the name of progress have been extensive and dramatic. For example, Peter Smith (1985) points out that the eastern half of New South Wales was covered by eucalypt forest or woodland in 1788, but today only about a third of the original 40 million hectares remains. Many of the changes since European settlement have been to the detriment of the indigenous wildlife, the mammals in particular being very badly affected. Of 245 species of terrestrial indigenous mammals believed to have been present in Australia in 1788, 17(7%) of these have since become extinct or are regarded as so rare they are probably extinct. In certain regions the effect has been very much worse than in others. For example, in the wheatbelt of Western Australia (see Saunders *et al.* 1985, Figure 1e for the extent of this region), an area where a large percentage of the vegetation has been destroyed for agriculture, 46 species of terrestrial indigenous mammals are believed to have occurred before European settlement. Today, a staggering 13 (28%) species are believed to have totally disappeared from this region, nine of them being extinct on the Australian mainland. Less than half (22) of the species are regarded as common in the wheatbelt, although some have very limited ranges. By way of contrast, Tasmania is believed to have had 34 species of terrestrial indigenous mammals at European settlement and 30 of these are still regarded as common with only one, the Thylacine, *Thylacinus cynocephalus*, thought to be extinct. These figures have been collated from Strahan (1983).

Our birds have fared very much better than our mammals. Of 571 species believed to have been breeding species on mainland Australia or Tasmania and their offshore islands at the start of European settlement, only two species (Paradise Parrot, *Psephotus pulcherrimus*, and Night Parrot, *Geopsittacus occidentalis*, are thought to be extinct (figures from Readers' Digest 1976). Occasional reports of the Night Parrot may indicate it is still extant in remote areas. The Kangaroo Island Emu, *Dromaius* sp., if regarded as a separate species has definitely become extinct. Even in the highly disturbed wheatbelt of Western Australia most of the original species still occur. Kitchener *et al.* (1982) surveyed 23 reserves of varying sizes throughout the wheatbelt and found that 133 out of 137 species thought to have occurred before European settlement were still extant. The four species not recorded in the wheatbelt all have viable populations outside the surveyed area. Further work is necessary to evaluate the conservation status of the 133 species still occurring in the wheatbelt.

In addition to the species thought to be extinct, a further five species are regarded as endangered (Orange-bellied Parrot, *Neophema chrysogaster*, Noisy Scrub-bird, *Atrichornis clamosus*, Eyrean Grass Wren, *Amytornis goyderi*, Western Bristlebird, *Dasyornis longirostris*, and the Forty-spotted Pardalote *Pardalotus quadrigintus*). All of these species have been studied over the last decade. The Pardalote is thought to be down to about 850 individuals scattered between eight populations in various parts of southeastern Tasmania (Rounsevell and Woinarski 1983; Woinarski and Rounsevell 1983) while the Grass Wren has an extremely restricted distribution but may be locally abundant within that range (Parker and Reid 1979). The Orange-bellied Parrot has been studied intensively for two years and these studies reveal that the total population breeding in Tasmania and wintering on the extreme southeast of the mainland may be fewer than 200 (Brown and Wilson 1982). Management for this species thus involves conserving both the mainland wintering sites and managing the successional stages of the

[1]C.S.I.R.O. Division of Wildlife and Rangelands Research, Locked Bag No. 4, P.O. Midland, Western Australia 6056.

Pages 355-57 *in* BIRDS OF EUCALYPT FORESTS AND WOODLANDS: ECOLOGY, CONSERVATION, MANAGEMENT ed. by A. Keast, H. F. Recher, H. Ford and D. Saunders. Royal Australasian Ornithologists Union and Surrey Beatty & Sons.

vegetation on the breeding sites in Tasmania as the birds have rarely been recorded feeding in vegetation more than 15 years old. The Western Bristlebird has been studied in association with the intensive study on the ecology of the Noisy Scrub-bird (Smith 1977) and the details of this study and its implications for management are yet to be published. However, the study by Graeme Smith on the Noisy Scrub-bird (1985) highlights the problems associated with dealing with rare fauna and demonstrates that such studies need to be conducted for extended periods and require reliable funding. In this case at least a 10 year study was needed, as it takes that period for vegetation in the drier parts of the bird's habitat to become suitable for the birds after it has been burnt. Smith also shows that it is preferable to learn to manage our rare fauna before numbers fall so low that it is necessary to rely on captive breeding programmes, a point also commented on by Brown and Wilson (*loc. cit.*). In most cases such captive breeding programmes are expensive and it is more economical in the long term to prevent populations from falling to such low numbers that captive breeding programmes are required. For example, Graeme Smith's figures (1985) of nine person years and $40,000 for maintenance and equipment may well have cost the community over $250,000 and he points out that a programme to re-establish birds in the wild would cost many times this figure. There is, of course, little point in organising a captive breeding programme if we do not know how to manage the wild population or are not promised the continuing financial support for such management.

Although we have five species of bird so rare that they are at risk of extinction in the near future and we have studies being conducted on four of these, there is no reason to be complacent about the *status quo*. Many more species are so poorly known that we really do not know their status. As mentioned earlier, changes to the native vegetation are proceeding more rapidly than our capacity to prepare and carry out management plans to lessen the effects of such change. This has been shown by a review of the status of the members of one order, the Psittaciformes. Smith (1979) has shown that of 53 species of parrot in Australia, 38 (72%) species could be regarded as common or abundant, 10 (19%) species are thought to be increasing in numbers and 21 (40%) are believed to have decreasing populations, two of which may already be extinct (Smith *loc. cit.* Table 1). Each of the 10 species with increasing populations have benefitted from changes brought about by agricultural or pastoral activities. Saunders *et al.* (1985) discuss these effects on nine species of two genera (*Cacatua* and *Calyptorhynchus*) and point out that eventually an equilibrium will be reached for all of these species provided that the rapid degradation of our land with its consequent detrimental effect on the vegetation is halted. Unfortunately there are no positive signs that this is happening. In Western Australia the problems of dying woodland and lack of regeneration shown by Saunders *et al.* (1982) are compounded by the increasing salinity of the soils, which has resulted directly from clearing of native vegetation. As recently as March 1984 the Department of Agriculture's Resource Management Division predicted that salt degradation is expected to spread across agricultural lands in Western Australia at a rate of 25,000 ha. per year for most of the 1980's (The West Australian, March 6 1984). Further clearing will only exacerbate this problem. This situation places the future of these and other species in jeopardy.

This trend of deteriorating woodland with little or no regeneration to replace senescent or dying vegetation is a phenomenon which is not restricted to the wheatbelt of western Australia. Hugh Ford raises the problem in his study of the role of birds as predators on insects in eucalypt woodland (Ford 1985), suggesting that birds could take about 55-70% of leaf eating insects annually in healthy woodland. He suggests that if insects are demonstrated to be the major cause of New England Dieback then birds, with their potential to impose large and fairly constant mortality on herbivorous insect populations, should be considered in any management plans to regenerate eucalypts. If this problem of degenerating woodland continues it will be not only the hole-nesting species of bird which will suffer the loss of essential resources. Any species depending on trees for nesting, protection or for feeding will suffer with the decline in woodland. Loss of woodland will affect agricultural land as a result of increased soil erosion by wind or sheet flooding, already serious problems in many parts of Australia. The widespread clearing of native vegetation for agriculture in wetter areas and the extensive pastoral industry in the semi-arid and arid areas poses the major threat facing our endemic fauna and wildlife managers. However, in spite of this the present emphasis of our research does not reflect this fact. Of the nine chapters presented in this section, five (Peter Smith, Rod Kavanagh, Harry Recher, Per Christensen, Brian Collins and their various colleagues) relate to management of birds in forests disturbed by fire, logging or mining. Only three chapters (Saunders *et al.*, Richard Loyn and Hugh Ford) relate to problems associated with agricultural clearing.

With increasing demands being made on our remaining forests it is urgent that suitable management plans are formulated and applied to ensure the continued existence of our diverse plants and animals. Funding for research is coming from some of the industries associated with forest exploitation. With time we should be able to formulate suitable plans. As pointed out by Peter Smith (1985), commercial exploitation of forests and wildlife conservation are not mutually exclusive. It is possible to manage our forests for both exploitation and wildlife but to do so requires compromise. Peter Smith, Kavanagh *et al.* (1985) and Collins *et al.* (1985) all suggest that positive management based on good scientific data is necessary to lessen the adverse effects of development on wildlife. To maintain biotic diversity within intensively exploited forests, adequate areas of all successional stages of forest must be retained as reserves with connecting corridors between them. What exactly are "adequate areas" and "connecting corridors" remains to be determined.

The role of fire in management needs much more research. As pointed out by Christensen *et al.* (1985) "considering the frequency with which fires occur in forests and woodlands and their dramatic impact on the landscape, there is remarkably little information about the effects of fire on fauna or the long term consequences of burning on forest ecosystems". Wildfire has long influenced Australian ecosystems, but the widespread use of frequent control burns to minimise the chance of wildfire needs to be thoroughly investigated. Minor changes in the fire regime may be as damaging as the widespread clearing for agriculture.

Clearing of large tracts of native vegetation has resulted in large numbers of patches of native vegetation of varying size, varying degrees of isolation and varying types of ownership (Crown and private). In most of our agricultural areas the continued existence of many species of birds depends on the maintenance of these patches. This is especially true of birds requiring specific habitats as, for example, Kitchener *et al.*'s (1982) vulnerable species. These are either territorial, insectivorous passerines dependent on native vegetation, or mobile species like the White-tailed Black Cockatoo, *Calyptorhynchus funereus latirostris*, which require large areas of native vegetation surrounding their breeding areas and reconnected by corridors of native vegetation (Saunders 1982; Saunders *ét al.* 1985). The mosaic of native vegetation patches isolated by clearing means that no reserve contains all of the habitats in any region, nor all of the species of birds. This mosaic now comprises our reserve system yet we do not know the efficacy of the system with regard to conserving our biota. There is a real need to address this problem by examining the reserve network and the habitat requirements of our vulnerable species. Failure to carry out this work may well lead to the loss of many species from our woodland ecosystems. Given the rapid degradation of these ecosystems at present there is an urgent need to concentrate our conservation resources on these problems with more studies like those of Richard Loyn (1985), Kitchener *et al.* (*loc. cit.*), Hugh Ford (1985) and Graeme Smith (1985) in addition to the work currently addressing management problems in forest ecosystems. We need to initiate long term studies on breeding biology and population dynamics in relation to the changes Europeans have brought about before it is too late to reverse the current trends.

REFERENCES

Brown, P. B. and Wilson, R. I., 1982. The Orange-bellied Parrot, Pp. 106-115 *in* Species at Risk: Research in Australia ed. by R. H. Groves and W D. L. Ride. Australian Academy of Science: Canberra.

Christensen, P. E. S., Wardell-Johnson, G. and Kimber, P., 1985. Birds and Fire in Southwestern Forests. Pp. 291-99 *in* Birds of Eucalypt Forests and Woodlands: Ecology, Conservation, Management ed. by A. Keast, H. F. Recher, H. Ford and D. Saunders. Royal Australasian Ornithologists Union.

Collins, B. G., Wykes, B. J. and Nicholls, O. G., 1985. Re-colonization of Restored Bauxite Minelands by Birds in Southwestern Australian Forests. Pp. 341-54 *in* Birds of Eucalypt Forests and Woodlands: Ecology, Conservation, Management ed. by A. Keast, H. F. Recher, H. Ford and D. Saunders. Royal Australasian Ornithologists Union.

Ford, H. A., 1985. The Bird Communtity in Eucalypt Woodland and Eucalypt Dieback in the Northern Tablelands of New South Wales. Pp. 333-40 *in* Birds of Eucalypt Forests and Woodlands: Ecology, Conservation, Management ed. by A. Keast, H. F. Recher, H. Ford and D. Saunders. Royal Australasian Ornithologists Union.

Kavanagh, R. P., Shields, J. M., Recher, H. F. and Rohan-Jones, W. G., 1985. Bird Populations of a Logged and Unlogged Forest Mosaic in the Eden Woodchip Area. Pp. 273-81 *in* Birds of Eucalypt Forests and Woodlands: Ecology, Conservation, Management ed. by A. Keast, H. F. Recher, H. Ford and D. Saunders. Royal Australasian Ornithologists Union.

Kitchener, D. J., Dell, J., Muir, B. G. and Palmer, M., 1982. Birds in western Australian wheatbelt reserves — implications for conservation. *Biol. Conserv.* **22**: 127-63.

Loyn, R. H., 1985. Birds in Fragmented Forests in Gippsland, Victoria. Pp. 323-31 *in* Birds of Eucalypt Forests and Woodlands: Ecology, Conservation, Management ed. by A. Keast, H. F. Recher, H. Ford and D. Saunders. Royal Australasian Ornithologists Union.

Parker, S. A. and Reid, N., 1979. Remarks on the status of some Australian passerines. Pp. 109-15 *in* The Status of endangered Australasian Wildlife ed. by M. J. Tyler. Royal Zoological Society of South Australia.

Readers' Digest, 1976. Complete book of Australian birds. Readers' Digest Services Pty. Ltd.: Surry Hills, New South Wales.

Recher, H. F., Allen, D. and Gowing, G., 1985. The Impact of Wildlife on Birds in an Intensively Logged Forest. Pp. 283-90 *in* Birds of Eucalypt Forests and Woodlands: Ecology, Conservation, Management ed. by A. Keast, H. F. Recher, H. Ford and D. Saunders. Royal Australasian Ornithologists Union.

Rounsevell, D. E. and Woinarski, J. C. Z., 1983. Status and distribution of the Forty-spotted Pardalote, *Pardalotus quadragintus*. *Aust. Wild. Res.* **10**: 343-9.

Saunders, D. A., 1982. The breeding behaviour and biology of the short-billed form of the White-tailed Black Cockatoo, *Calyptorhynchus funereus*. *Ibis* **124**: 422-55.

Saunders, D. A., Rowley, I. and Smith, G. T., 1985. The Effects of Clearing for Agriculture on the Distribution of Cockatoos in the Southwest of Western Australia. Pp. 309-21 *in* Birds of Eucalypt Forests and Woodlands: Ecology, Conservation, Management ed. by A. Keast, H. F. Recher, H. Ford and D. Saunders. Royal Australasian Ornithologists Union.

Saunders, D. A., Smith, G. T. and Rowley, I., 1982. The availability and dimensions of tree hollows that provide nest sites for cockatoos (Psittaciformes) in Western Australia. *Aust. Wildl. Res.* **9**: 541-56.

Smith, G. T., 1977. The effect of environmental change on six rare birds. *Emu* **77**: 173-79.

Smith, G. T., 1979. The status of Australian Parrots. Pp. 101-108 *in* The Status of Endangered Australasian Wildlife ed. by M. J. Tyler. Royal Zoological Society of South Australia.

Smith, G., 1985. The Noisy Scrub-bird *Atrichornis clamosus*. Does its Past Suggest a Future? Pp. 301-08 *in* Birds of Eucalypt Forests and Woodlands: Ecology, Conservation, Management ed. by A. Keast, H. F. Recher, H. Ford and D. Saunders. Royal Australasian Ornithologists Union.

Smith, P., 1985. Woodchip Logging and Birds near Bega, New South Wales. Pp. 259-71 *in* Birds of Eucalypt Forests and Woodlands: Ecology, Conservation, Management ed. by A. Keast, H. F. Recher, H. Ford and D. Saunders. Royal Australasian Ornithologists Union.

Strahan, R. (Ed.), 1983. The Complete Book of Australian Mammals. The Australian Museum and Angus and Robertson: Sydney.

Woinarski, J. C. Z. and Rounsevell, D. E., 1983. Comparative ecology of pardalotes, including the Forty-spotted Pardalote, *Pardalotus quadragintus* (Aves: Pardalotidae) in southeastern Tasmania. *Aust. Wildl. Res.* **10**: 351-61.

Appendix 1

Australian Birds —
Common and Scientific Names used in Text

Colour Plates

Australian Forests and Woodlands
Habitat and Birds

Index

to the Volume

APPENDIX 1: Australian Birds — Common and Scientific Names used in Text

Family DROMAIIDAE

Emu	*Dromaius novaehollandiae*

Family PODICIPEDIDAE

Hoary-headed Grebe	*Poliocephalus poliocephalus*
Australasian Grebe	*Tachybaptus novaehollandiae*

Family PHALACROCORACIDAE

Little Pied Cormorant	*Phalacrocorax melanoleucos*

Family ARDEIDAE

Pacific Heron	*Ardea pacifica*
White-faced Heron	*Ardea novaehollandiae*
Great Egret	*Egretta alba*

Family PLATALEIDAE

Sacred Ibis	*Threskiornis aethiopica*
Straw-necked Ibis	*Threskiornis spinicollis*
Yellow-billed Spoonbill	*Platalea flavipes*

Family ANATIDAE

Black Swan	*Cygnus atratus*
Australian Shelduck	*Tadorna tadornoides*
Pacific Black Duck	*Anas superciliosa*
Grey Teal	*Anas gibberifrons*
Maned Duck	*Chenonetta jubata*

Family ACCIPITRIDAE

Black-shouldered Kite	*Elanus notatus*
Whistling Kite	*Haliastur sphenurus*
Brown Goshawk	*Accipiter fasciatus*
Collared Sparrowhawk	*Accipiter cirrhocephalus*
Wedge-tailed Eagle	*Aquila audax*
Marsh Harrier	*Circus aeruginosus*

Family FALCONIDAE

Peregrine Falcon	*Falco peregrinus*
Australian Hobby	*Falco longipennis*
Brown Falcon	*Falco berigora*
Nankeen or Australian Kestrel	*Falco cenchroides*

Family MEGAPODIIDAE

Brush-turkey	*Alectura lathami*

Family PHASIANIDAE

Stubble Quail	*Coturnix novaezelandiae*
Brown Quail	*Coturnix australis*

Family TURNICIDAE

Painted Button-quail	*Turnix varia*

Family RALLIDAE

Dusky Moorhen	*Gallinula tenebrosa*
Purple Swamphen	*Porphyrio porphyrio*
Eurasian Coot	*Fulica atra*

Family CHARADRIIDAE

Masked Lapwing	*Vanellus miles*
Black-fronted Plover	*Charadrius melanops*

Family SCOLOPACIDAE

Latham's Snipe	*Gallinago hardwickii*

Family COLUMBIDAE

Spotted Turtle-Dove	*Streptopelia chinensis*
Brown Pigeon or Cuckoo-Dove	*Macropygia amboinensis*

Family COLUMBIDAE — cont'd

Peaceful Dove	*Geopelia placida*
Diamond Dove	*Geopelia cuneata*
Common Bronzewing	*Phaps chalcoptera*
Brush Bronzewing	*Phaps elegans*
Wonga Pigeon	*Leucosarcia melanoleuca*

Family CACATUIDAE

Red-tailed Black-Cockatoo	*Calyptorhynchus magnificus*
Inland Red-tailed Black-Cockatoo	*Calyptorhynchus magnificus samueli*
Forest Red-tailed Black-Cockatoo	*Calyptorhynchus magnificus naso*
Glossy Black-Cockatoo	*Calyptorhynchus lathami*
Yellow-tailed Black-Cockatoo	*Calyptorhynchus funereus*
Short-billed white-tailed Black-Cockatoo	*Calyptorhynchus funereus latirostris*
Long-billed White-tailed Black-Cockatoo	*Calyptorhynchus baudinii*
Gang-gang Cockatoo	*Callocephalon fimbriatum*
Galah	*Cacatua roseicapilla*
Long-billed Corella	*Cacatua pastinator pastinator*
Little Corella	*Cacatua pastinator gymnotis*
Major Mitchell's Cockatoo	*Cacatua leadbeateri*
Sulphur-crested Cockatoo	*Cacatua galerita*

Family LORIIDAE

Rainbow Lorikeet	*Trichoglossus haematodus*
Red-collared Lorikeet	*Trichoglossus rubritorquis*
Scaly-breasted Lorikeet	*Trichoglossus chlorolepidotus*
Musk Lorikeet	*Glossopsitta concinna*
Purple-crowned Lorikeet	*Glossopsitta porphyrocephala*
Little Lorikeet	*Glossopsitta pusilla*

Family POLYTELITIDAE

Australian King-Parrot	*Alisterus scapularis*
Superb Parrot	*Polytelis swainsonii*
Cockatiel	*Nymphicus hollandicus*

Family PLATYCERCIDAE

Ground Parrot	*Pezoporus wallicus*
Night Parrot	*Geopsittacus occidentalis*
Green Rosella	*Platycercus caledonicus*
Crimson Rosella	*Platycercus elegans*
Eastern Rosella	*Platycercus eximius*
Pale-headed Rosella	*Platycercus adscitus*
Northern Rosella	*Platycercus venustus*
Western Rosella	*Platycercus icterotis*
Yellow Rosella	*Platycercus flaveolus*
Port Lincoln Ringneck	*Barnardius zonarius*
Red-rumped Parrot	*Psephotus haematonotus*
Paradise Parrot	*Psephotus pulcherrimus*
Blue-winged Parrot	*Neophema chrysostoma*
Orange-bellied Parrot	*Neophema chrysogaster*
Turquoise Parrot	*Neophema pulchella*

Family CUCULIDAE

Pallid Cuckoo	*Cuculus pallidus*
Brush Cuckoo	*Cuculus variolosus*
Fan-tailed Cuckoo	*Cuculus pyrrhophanus*
Horsfield's Bronze-Cuckoo	*Chrysococcyx basalis*
Golden or Shining Bronze-Cuckoo	*Chrysococcyx lucidus*

Family STRIGIDAE

Powerful Owl	*Ninox strenua*
Southern Boobook	*Ninox novaeseelandiae*
Barking Owl	*Ninox connivens*

Family TYTONIDAE

Barn Owl	*Tyto alba*
Sooty Owl	*Tyto tenebricosa*

Family PODARGIDAE

Tawny Frogmouth	*Podargus strigoides*

Family AEGOTHELIDAE

Australian Owlet-nightjar	*Aegotheles cristatus*

Family CAPRIMULGIDAE

White-throated Nightjar	*Caprimulgus mystacalis*

Family APODIDAE

White-throated Needletail	*Hirundapus caudacutus*

Family ALCEDINIDAE

Azure Kingfisher	*Ceyx azurea*
Laughing Kookaburra	*Dacelo novaeguineae*
Blue-winged Kookaburra	*Dacelo leachii*
Sacred Kingfisher	*Halcyon sancta*

Family MEROPIDAE

Rainbow Bee-eater	*Merops ornatus*

Family CORACIIDAE

Dollarbird	*Eurystomus orientalis*

Family MENURIDAE

Superb Lyrebird	*Menura novaehollandiae*

Family ATRICHORNITHIDAE

Rufous Scrub-bird	*Atrichornis rufescens*
Noisy Scrub-bird	*Atrichornis clamosus*

Family ALAUDIDAE

Skylark	*Alauda arvensis*

Family HIRUNDINIDAE

Welcome Swallow	*Hirundo neoxena*
Tree Martin	*Cecropis nigricans*

Family MOTACILLIDAE

Richard's Pipit	*Anthus novaeseelandiae*

Family CAMPEPHAGIDAE

Black-faced Cuckoo-shrike	*Coracina novaehollandiae*
White-bellied Cuckoo-shrike	*Coracina papuensis*
Cicadabird	*Coracina tenuirostris*
White-winged Triller	*Lalage sueurii*

Family MUSCICAPIDAE

White's thrush	*Zoothera dauma*
Blackbird	*Turdus merula*
Rose Robin	*Petroica rosea*
Pink Robin	*Petroica rodinogaster*
Flame Robin	*Petroica phoenicea*
Scarlet Robin	*Petroica multicolor*
Red-capped Robin	*Petroica goodenovii*
Hooded Robin	*Melanodryas cucullata*
Mangrove robin	*Eopsaltria pulverulenta*
White-breasted Robin	*Eopsaltria georgiana*
Eastern Yellow Robin	*Eopsaltria australis*
Western Yellow Robin	*Eopsaltria griseogularis*
Lemon-bellied Flycatcher	*Microeca flavigaster*
Jacky Winter	*Microeca leucophaea*
Pale-yellow Robin	*Tregallesia capito*
Crested Shrike-tit	*Falcunculus frontatus*
Olive Whistler	*Pachycephala olivacea*
Golden Whistler	*Pachycephala pectoralis*

Family MUSCICAPIDAE — cont'd

Rufous Whistler	*Pachycephala rufiventris*
Grey Shrike-thrush	*Colluricincla harmonica*
Black-faced Monarch	*Monarcha melanopsis*
Spectacled Monarch	*Monarcha trivirgatus*
Leaden Flycatcher	*Myiagra rubecula*
Satin Flycatcher	*Myiagra cyanoleuca*
Restless Flycatcher	*Myiagra inquieta*
Rufous Fantail	*Rhipidura rufifrons*
Grey Fantail	*Rhipidura fuliginosa*
Northern Fantail	*Rhipidura rufiventris*
Willie Wagtail	*Rhipidura leucophrys*

Family ORTHONYCHIDAE

Logrunner	*Orthonyx temminckii*
Eastern Whipbird	*Psophodes olivaceus*
Spotted Quail-thrush	*Cinclosoma punctatum*

Family TIMALIIDAE

Grey-crowned Babbler	*Pomatostomus temporalis*
White-browed Babbler	*Pomatostomus superciliosus*

Family SYLVIIDAE

Clamorous Reed-Warbler	*Acrocephalus stentoreus*
Golden-headed Cisticola	*Cisticola exilis*
Brown Songlark	*Cinclorhamphus cruralis*

Family MALURIDAE

Superb Fairy-wren	*Malurus cyaneus*
Splendid Fairy-wren	*Malurus splendens*
Variegated Fairy-wren	*Malurus lamberti*
Red-backed Fairy-wren	*Malurus melanocephalus*
Southern Emu-wren	*Stipiturus malachurus*
Eyrean Grasswren	*Amytornis goyderi*

Family ACANTHIZIDAE

Western Bristlebird	*Dasyornis longirostris*
Eastern Bristlebird	*Dasyornis brachypterus*
Pilotbird	*Pycnoptilus floccosus*
Origma	*Origma solitaria*
Large-billed Scrubwren	*Sericornis magnirostris*
Yellow-throated Scrubwren	*Sericornis citreogularis*
White-browed Scrubwren	*Sericornis frontalis*
Chestnut-rumped Heathwren or Hylacola	*Hylacola (=Sericornis) pyrrhopygius*
Fieldwren or Calamanthus	*Calamanthus (=Sericornis) fuliginosus*
Speckled Warbler	*Chthonicola (=Sericornis) sagittatus*
Weebill	*Smicrornis brevirostris*
Brown Warbler	*Gerygone mouki*
Western Warbler	*Gerygone fusca*
White-throated Warbler	*Gerygone olivacea*
Brown Thornbill	*Acanthiza pusilla*
Broad-tailed or Inland Thornbill	*Acanthiza apicalis*
Chestnut-rumped Thornbill	*Acanthiza uropygialis*
Western Thornbill	*Acanthiza inornata*
Buff-rumped Thornbill	*Acanthiza reguloides*
Yellow-rumped Thornbill	*Acanthiza chrysorrhoa*
Yellow Thornbill	*Acanthiza nana*
Striated Thornbill	*Acanthiza lineata*

Family NEOSITTIDAE

Varied Sittella	*Daphoenositta chrysoptera*

Family CLIMACTERIDAE

White-throated Treecreeper	*Climacteris leucophaea*
Red-browed Treecreeper	*Climacteris erythrops*
Brown Treecreeper	*Climacteris picumnus*
Black-tailed Treecreeper	*Climacteris melanura*

Family MELIPHAGIDAE

Red Wattlebird	*Anthochaera carunculata*
Little Wattlebird	*Anthochaera chrysoptera*
Noisy Friarbird	*Philemon corniculatus*
Little Friarbird	*Philemon citreogularis*
Regent Honeyeater	*Xanthomyza phrygia*
Bell Miner	*Manorina melanophrys*
Noisy Miner	*Manorina melanocephala*
Lewin's Honeyeater	*Meliphaga lewinii*
Yellow-spotted Honeyeater	*Meliphaga notata*
Yellow-faced Honeyeater	*Meliphaga (=Lichenostomus) chrysops*
White-gaped Honeyeater	*Meliphaga unicolor*
White-eared Honeyeater	*Meliphaga leucotis*
Yellow-tufted Honeyeater	*Meliphaga melanops*
Helmeted Honeyeater	*Meliphaga melanops cassidix*
Yellow-plumed Honeyeater	*Meliphaga ornata*
Fuscous Honeyeater	*Meliphaga fusca*
White-plumed Honeyeater	*Meliphaga penicillata*
Black-chinned Honeyeater	*Melithreptus gularis*
Brown-headed Honeyeater	*Melithreptus brevirostris*
White-throated Honeyeater	*Melithreptus albogularis*
White-naped Honeyeater	*Melithreptus lunatus*
Brown Honeyeater	*Lichmera indistincta*
Painted Honeyeater	*Grantiella picta*
Crescent Honeyeater	*Phylidonyris pyrrhoptera*
New Holland Honeyeater	*Phylidonyris novaehollandiae*
White-cheeked Honeyeater	*Phylidonyris nigra*
White-fronted Honeyeater	*Phylidonyris albifrons*
Tawny-crowned Honeyeater	*Phylidonyris melanops*
Bar-breasted Honeyeater	*Ramsayornis fasciatus*
Eastern Spinebill	*Acanthorhynchus tenuirostris*
Western Spinebill	*Acanthorhynchus superciliosus*
Dusky Honeyeater	*Myzomela obscura*
Scarlet Honeyeater	*Myzomela sanguinolenta*

Family EPHTHIANURIDAE

White-fronted Chat	*Ephthianura albifrons*

Family DICAEIDAE

Mistletoebird	*Dicaeum hirundinaceum*

Family PARDALOTIDAE

Spotted Pardalote	*Pardalotus punctatus*
Yellow-rumped Pardalote	*Pardalotus xanthopygus*
Forty-spotted Pardalote	*Pardalotus quadragintus*
Red-browed Pardalote	*Pardalotus rubricatus*
Striated Pardalote	*Pardalotus striatus*

Family ZOSTEROPIDAE

Grey-breasted Silvereye	*Zosterops lateralis*

Family FRINGILLIDAE

European Goldfinch	*Carduelis carduelis*

Family PLOCEIDAE

Red-browed Firetail	*Emblema temporalis*
Beautiful Firetail	*Emblema bella*
Diamond Firetail	*Emblema guttata*
Masked Finch	*Poephila personata*
Long-tailed Finch	*Poephila acuticauda*

Family STURNIDAE

Common Starling	*Sturnus vulgaris*
Common Mynah	*Acridotheres tristis*

Family ORIOLIDAE

Olive-backed Oriole	*Oriolus sagittatus*

Family DICRURIDAE

Spangled Drongo	*Dicrurus hottentottus*

Family PTILONORHYNICHIDAE

Satin Bowerbird	*Ptilonorhynchus violaceus*
Regent Bowerbird	*Sericulus chrysocephalus*
Green Catbird	*Ailuroedus crassirostris*

Family CORCORACIDAE

White-winged Chough	*Corcorax melanorhamphos*

Family GRALLINIDAE

Australian Magpie-lark	*Grallina cyanoleuca*

Family ARTAMIDAE

Masked Woodswallow	*Artamus personatus*
White-browed Woodswallow	*Artamus superciliosus*
Dusky Woodswallow	*Artamus cyanopterus*

Family CRACTICIDAE

Grey Butcherbird	*Cracticus torquatus*
Australian Magpie	*Gymnorhina tibicen*
Pied Currawong	*Strepera graculina*
Grey Currawong	*Strepera versicolor*

Family CORVIDAE

Australian Raven	*Corvus coronoides*
Forest Raven	*Corvus tasmanicus*
Little Raven	*Corvus mellori*

FOOTNOTE: At editorial discretion authors followed either the R.A.O.U. Checklist and lists of recommended names or adhered to the 1969 C.S.I.R.O. Checklist. Where the latter was followed this was to maintain continuity of names used in previous publications by the authors.

ANON, 1969. An Index of Australian Bird Names. C.S.I.R.O. Division of Wildlife Reseach Tech. Paper 20.

CONDON, H. T., 1975 Checklist of the Birds of Australia. Part 1 Non-passerines. Royal Australasian Ornithologists Union.

SCHODDE, R., 1975. Interim List of Australian Songbirds. Passerines. Royal Australasian Ornithologists Union.

SCHODDE, R., GLOVER B., KINSKY, F. C., MARCHANT, S., MCGILL, A. R. AND PARKER, S. A. 1978. Recommended English Names for Australian Birds. *Emu* 77: 245-307.

White-rumped Miner, *Manorina flavigula* (A. G. Wells).

△ Mallee woodland, Round Hill Nature Reserve, southwestern New South Wales (H. Recher).

▽ Mallee woodland with spinifex grass, Round Hill Nature Reserve, southwestern New South Wales (H. Cogger).

▽ Snow Gum, *Eucalyptus pauciflora,* woodland in summer in the Snowy Mts, New South Wales (D. Lunney).

△ Gimlet, *Eucalyptus salubris,* woodland, Western Australia (D. Saunders).

▽ Wandoo, *Eucalyptus wandoo,* woodland at Coomallo Creek, Western Australia (D. Saunders).

▽ Snow Gum woodland during winter in the Snowy Mts. (P. O'Connor).

△ Short-billed White-tailed Black Cockatoo, *Calyptorhynchus funereus latirostris* (D. Saunders).

△ Little Falcon, *Falco longipennis* (N. Chaffer).

▽ Hooded Robin, *Melanodryas cucullata* (M. Seyfort).

▽ Spotted Quail-thrush, *Cinclosoma punctatum* (M. Seyfort).

▽ Spotted Pardalote, *Pardalotus punctatus* (M. Seyfort).

▽ Grey-breasted Silvereye, *Zosterops lateralis* (M. Seyfort).

△ 'Auriferous' woodland habitat of Yellow-tufted Honeyeater, *Lichenostomus melanops*, at Tarnagulla, Victoria (B. Wykes).

△ Sub-alpine woodland of Gum-topped Stringybark (Alpine Ash), *Eucalyptus delegantensis*, and Tasmanian Snow Gum, *E. coccifera*, at Pine Tier, Tasmania (D. Milledge).

△ Stringybark, *Eucalyptus caliginosa*, and Red Gum, *E. blakelyi*, open-forest near Armidale in the Eastwood State Forest on the Northern Tablelands of New South Wales (H. Ford).

▽ Pole-stand of Red Ironbark, *Eucalyptus sideroxylon*, in Rushworth State Forest, central Victoria (S. Gilmore).

△ Open-forest of Narrow-leaved Peppermint, *Eucalyptus radiata*, and Mountain Gum, *E. dalrympleana*, near Bombala on the Southern Tablelands of New South Wales (H. Recher).

▽ Open-forest at Wollomombi Falls east of Armidale on the Northern Tablelands of New South Wales. Trees include *Eucalyptus caliginosa*, *E. conica*, and *E. melliodora* (R. Noske).

△ Yellow-tufted and Helmeted Honeyeaters, *Lichenostomus melanops* and *L. melanops cassidix* (C. and D. Frith).

△ Dusky Woodswallow, *Artamus cyanopterus* (M. Seyfort).

△ Brown Thornbill, *Acanthiza pusilla* (M. Seyfort).

△ Red-browed Finch, *Emblema temporalis* (N. Chaffer).

▽ Flame Robin, *Petroica phoenicea* (M. Seyfort).

▽ Grey Fantail, *Rhipidura fuliginosa* (N. Chaffer).

△ Dry, tall open-forest of Blackbutt, *Eucalyptus pilularis,* Pink Bloodwood, *E. intermedia,* and Tallwood, *E. microcorys,* on Middle Brother Mountain near Kendall, New South Wales (D. Milledge).

△ Open-forest of Silvertop Ash, *Eucalyptus sieberi* and White Stringybark, *E. globoidea,* in East State Forest, southeastern New South Wales (H. Recher).

△ Tasmanian Blue Gum, *Eucalyptus globulus,* Manna gum, *E. viminalis,* and Messmate, *E. obliqua,* in dry sclerophyll (open-forest) forest approaching wet sclerophyll near Little Swanport, Tasmania (D. Milledge).

△ Open-forest of Manna Gum, *Eucalyptus viminalis,* and Silver Peppermint, *E. tenuiramis,* at Tinderbox south of Hobart, Tasmania (D. Milledge).

▽ Tree fern understorey in moist tall open-forest in Tennyson Creek Flora Reserve, Bondi State Forest, New South Wales (H. Recher).

▽ Ground vegetation in virgin tall open-forest of Narrow-leaved Peppermint, *Eucalyptus radiata,* on Coast Range Road in East Gippsland, Victoria (H. Recher).

△ Pied Currawong, *Strepera graculina* (M. Seyfort).

△ New Holland Honeyeater, *Phylidonyris novaehollandiae* (N. Chaffer).

△ Tawny Frogmouth, *Podargus strigoides* (M. Seyfort).

△ Striated Thornbill, *Acanthiza lineata* (N. Chaffer).

▽ Laughing Kookaburra, *Dacelo novaeguineae* (M. Seyfort).

▽ Ground Thrush, *Zoothera dauma* (N. Chaffer).

△ Mature Alpine Ash, *Eucalyptus delegatensis,* on Deddick Trail, Victoria (S. Craig).

▽ Virgin Jarrah, *Eucalyptus marginata,* forest on Dixon Tower Road, Western Australia (Forestry Dept. W.A.).

▽ Open-forest of Mountain Gum, *Eucalyptus dalrympleana,* near Campbelltown, Tasmania (H. Recher).

△ Tall open-forest of Narrow-leaved Peppermint, *Eucalyptus radiata,* on Coast Range Road, Victoria (H. Recher).

▽ Open-forest of Broad-leaved Peppermint, *Eucalyptus dives* and Manna Gum, *E. viminalis,* along the Imlay Road, Towamba State Forest, New South Wales (H. Recher).

▽ Open-forest of Silvertop Ash, *Eucalyptus sieberi,* Mt Imlay National Park, southeastern New South Wales (H. Recher).

△ Bell Miner, *Manorina melanophrys* (N. Chaffer).

△ Brown Warbler, *Gerygone mouki* (N. Chaffer).

▽ Rufous Fantail, *Rhipidura rufifrons* (N. Chaffer).

▽ Rose Robin, *Petroica rosea* (N. Chaffer).

▽ Olive-backed Oriole, *Oriolus sagittatus* (N. Chaffer).

▽ Yellow Robin, *Eopsaltria australis* (N. Chaffer).

△ Open-forest on sandstone escarpment, Alligator Rivers Region, Northern Territory (H. Recher).

△ Open-forest on sandstone escarpment, Alligator Rivers Region, Northern Territory (H. Recher).

△ Open-forest on sandstone escarpment, Alligator Rivers Region, Northern Territory (H. Recher).

△ Tropical (monsoonal) woodland on the Magela Creek drainage, Alligator Rivers Region, Northern Territory (H. Recher).

▽ Ironbark and Bloodwood (*Eucalyptus* spp.) woodland at Black Mt near Cooktown, Queensland (D. Milledge).

▽ Open-forest of Spotted Gum, *Eucalyptus maculata,* near Cooloola in southeastern Queensland (H. Recher).

△ Helmeted Friarbird, *Philemon buceroides* (N. Chaffer).

△ Mistletoebird, *Dicaeum hirundinaceum* (N. Chaffer).

△ Brown Weebill, *Smicrornis brevirostris* (N. Chaffer).

△ White-throated Warbler, *Gerygone olivacea* (N. Chaffer).

▽ Forest Kingfisher, *Halcyon macleayii* (N. Chaffer).

▽ Partridge Pigeon, *Petrophassa smithii* (N. Chaffer).

Spiny-cheeked Honeyeater, *Acanthagenys rufogularis* (A. G. Wells).

INDEX: BIRDS OF EUCALYPT FORESTS AND WOODLANDS